THE

DICTIONARY

OF

ECOLOGY

AND

ENVIRONMENTAL

SCIENCE

THE
DICTIONARY
OF
ECOLOGY
AND
ENVIRONMENTAL
SCIENCE

Henry W. Art, General Editor

Foreword by F. Herbert Bormann

Contributing Editors:

Dr. Daniel Botkin • Dr. William C. Grant • Dr. Susan E. Kegley

Dr. Edward A. Keller • Dr. Carl Reidel • Dr. Jean Richardson

• Dr. Jay Shelton

A HENRY HOLT REFERENCE BOOK

HENRY HOLT AND COMPANY

NEW YORK

A Henry Holt Reference Book

Henry Holt and Company, Inc.
Publishers since 1866
115 West 18th Street
New York, New York 10011

Henry Holt® is a registered trademark of Henry Holt and Company, Inc.

Library of Congress Cataloging-in-Publication Data

The Dictionary of ecology and environmental science / Henry W. Art,
    general editor. — 1st ed.
        p. cm.   — (A Henry Holt reference book)
    1. Environmental sciences—Dictionaries. 2. Ecology—
Dictionaries. I. Art, Henry Warren. II. Series.
GE10.D53   1993   92-38526
363.7'003—dc20   CIP

ISBN 0-8050-2079-9

First Edition—1993

Produced by Storey Communications, Inc.
Schoolhouse Road, Pownal, Vermont 05261.
    M. John Storey, President
    Martha M. Storey, Vice President
    Pamela B. Art, Publisher
    Robert D. Spicer, Director of Sales
Editors: Timothy J. Flanagan, Kathleen M. Philbin, Elizabeth P. Stell, Sandra Webb
Editorial support: Andrew Art, Blair Dils, Jessica Storey Dils, Cathy Gee Graney,
    Louise Lloyd, Paul E. McFarland
Book design: Cynthia McFarland
Chart design and production: Nancy Bellnier, Mass Media
Production support: Susan Bernier, Susan Moore
Production management: Caroline Burch

Printed in the United States of America
All first editions are printed on acid-free paper. ∞

10 9 8 7 6 5 4 3 2 1

Contributing Editors

Dr. Henry W. Art: Samuel Fessenden Clarke Professor of Biology and Director of the Science Division, Williams College; B.A. Dartmouth College; M.Phil., Ph.D. Yale University; Visiting Fellow, Botany School, Cambridge University, England; award-winning author of *A Garden of Wildflowers* and five regional editions of *The Wildflower Gardener's Guide* (Garden Way Publishing); Research Ecologist, Department of Interior, National Park Service Atlantic coastal ecosystem research.

Dr. F. Herbert Bormann: Oastler Professor of Forest Ecology, Emeritus, Yale University; B.S. Rutgers; M.A., Ph.D. Duke University; previously served on the faculties at Emory University and Dartmouth College; principal investigator, Hubbard Brook Ecosystem Study; author of *The American Lawn: An Environmental Anachronism* (Yale University Press), *Biogeochemistry of a Forested Ecosystem* (Springer-Verlag Publishers), and *Patterns and Processes in a Forested Ecosystem* (Springer-Verlag Publishers); member, National Academy of Sciences; fellow, American Association for the Advancement of Sciences; former president, Ecological Society of America.

Dr. Daniel Botkin: President, The Center for the Study of the Environment, Santa Barbara, California; Director, The Program on Global Change, George Mason University, Fairfax, Virginia; B.A. University of Rochester; M.A. University of Wisconsin; Ph.D Rutgers; previously served on the faculty of Yale University, School of Forestry and Environmental Studies, and former chair, Environmental Studies Program, University of California,

Santa Barbara; author of *Discordant Harmonies* (Oxford University Press) and *Forest Dynamics: An Ecological Model* (Oxford University Press); first prize, Mitchell International Prize for Sustainable Development, 1991; co-developer, Jabowa Model of Forest Growth.

Dr. William C. Grant: Professor Emeritus, Williams College; B.A. Dartmouth College; Ph.D. Yale University; previously served on the faculties of Gettysburg College, The College of William and Mary, and Dartmouth College; Visiting Scientist, Oxford University, England; Visiting Research Professor, University of Edinburgh, Scotland; National Science Foundation and National Institute of Health grants for studies on the physiological ecology of lower vertebrates at Mt. Desert Island Biological Lab; Fellow, American Association for the Advancement of Sciences.

Dr. Susan E. Kegley: Lecturer and Postdoctoral Research Associate, College of Chemistry, University of California, Berkeley; B.S. University of Richmond; Ph.D. University of North Carolina, Chapel Hill; previously served on the faculty at Williams College; member Sigma Xi, American Chemical Society Environmental and Inorganic Divisions, and Hazardous Materials Control Research Institute; National Science Foundation research on environmental chemistry curriculum development.

Dr. Edward A. Keller: Professor, Environmental Studies and Department of Geological Sciences, University of California, Santa Barbara; B.S., B.A. California State University, Fresno; M.S. University of

California; Ph.D. Purdue University; author of *Environmental Geology* (Charles Merrill Publishing Co.) and, with D.B. Botkin, *Environmental Studies: Earth as a Living Planet* (Charles Merrill Publishing Co.); Hartley Visiting Professor Award, The University of Southampton, England; Sigma Xi National Lecturer.

Dr. Carl Reidel: Director of the Environmental Program and the Daniel Clarke Sanders Professor of Environmental Studies, and Public Administration, University of Vermont; previously served on the faculties of Williams College, Harvard University, and Yale University; B.S. and Ph.D. University of Minnesota; M.P.A. Harvard University; Vice Chairman of the National Wildlife Federation; U.S. Environmental Protection Agency Masters Award on Earth Day 1990; national expert on natural resources policy.

Dr. Jean Richardson: Director of the EPIC Project and Associate Professor of Environmental Studies and Natural Resources, University of Vermont; B.S. University of Newcastle-upon-Tyne, England; M.S. in pollen analysis and Ph.D. in biogeography from the University of Wisconsin, Madison; post-graduate Law Clerkship in Vermont; Fulbright Scholar; Kellogg National Leadership Fellow; elected by the Russians to serve on the Scientific Council of the International Center of Environmental Protection of the Lake Baikal Region of Siberia; Region One Director of the Board of the National Wildlife Federation.

Dr. Jay Shelton: President, Shelton Research, Inc., Sante Fe, New Mexico; Teacher of physics, chemistry, and physical sciences, Sante Fe Preparatory School; B.A. Harvard College; Ph.D. in physics, University of California, Berkeley; previously served on the faculty of Williams College; author of *Solid Fuels Encyclopedia* and *Wood Heat Safety* (both Garden Way Publishing); research on energy efficiencies of wood heating systems and accessories, energy conservation in buildings, and underground heat storage; award, Residential Solid Fuel Industry National Hall of Fame.

FOREWORD

MORE THAN HALF A CENTURY AGO, when I first became interested in ecology, there were no departments of environmental science. In universities the disciplines that today contribute to environmental science were housed in separate departments, often in separate buildings, their studies were published in different journals, scientists attended different professional meetings, and their clients were drawn from different segments of society.

Today, many scientists from the fields of ecology, geology, chemistry, and physics have close connections across fields, attend the same meetings, publish in the same journals, and respond to different aspects of the same environmental problem and consider themselves to be environmental scientists.

What forces gave rise to this amalgamation? During the last half of the twentieth century we began to realize that society's actions may be bringing about the deterioration of the planet. Signs of a steady decline in environmental quality became visible: to name a few, polluted streams and rivers, smog-shrouded cities and urban decay, and hundreds of millions of rural poor in third world countries pushing into forest lands seeking subsistence survival. Scientists began to document more subtle effects: food chains contaminated with biocides and radioactive nuclides, lead poisoning in children, and air masses polluted with a huge variety of wastes. Great debates developed concerning energy and its conservation and the costs and benefits of nuclear power. New technologies allowed us to track more accurately what goes on with the surface of the earth.

What seemed a patchwork of environmental problems in the 1960's and 70's, specific to regions and countries, began to coalesce into global phenomena such as global warming, thinning of the stratospheric ozone layer, and destruction of tropical forests. All this began to reveal the existence of a global environmental deficit, the unanticipated consequence of humanity's alteration of the earth's atmosphere, water, soil, biota, and whole ecological systems. Human activities seemed to be destroying the very life systems on which we were all dependent. It also became clear that the extraordinary growth in human numbers, technology, and in the use of renewable and non-renewable natural resources had, in effect, annihilated space. No longer could we count on limitless space where humans could discharge our wastes knowing that almost infinite dilution would render them harmless or in which we could store them in out of the way places where they could be forever sealed off from human intercourse. We realized that for humanity to survive we had to become a sustainable society.

It was through the examination of phenomena like those discussed and through the development of scientific understanding of man's relationship to nature that environmental science has become a major force in the modern world. A force that can be ignored only at our own peril.

Environmental science evolved first as a documentor of environmental relationships and a predictor of environmental change and more lately as a coequal partner with economics, social science, policy, and management in attempting to design sustainable growth systems in harmony with nature.

This book is a small but important part of environmental science's new role. It will enable anyone to quickly access concepts and definitions central to ecology and environmental sciences.

F.H. Bormann
The Herbarium
East Thornton, New Hampshire

* The author is indebted to Yale Press for permission to use some phrases from the forthcoming book, *The American Lawn: An Environmental Anachronism* by F.H. Bormann, Diana Balmori, and Gordon Geballe.

a

A   See AMPERE.

aa   A Hawaiian term for a blocky, rough-surfaced, slow-moving, viscous basaltic lava flow. Contrast PAHOEHOE.

abaxial   Located away from, or on the opposite side of, the axis of a structure or of an entire organism. In plants, abaxial describes the lower surface of most leaves because they face away from the axis as they develop. Compare ADAXIAL.

abiocoen   All of the geologic, climatic, and other non-living elements of an ecosystem. See HOLOCOEN.

abioseston   A collective term for the density of inorganic particulate matter suspended in water; examples are clays or volcanic ash.

abiotic   Non-living; usually applied to the physical characteristics of biological systems, such as moisture, nutrients, soils, solar radiation, etc. See BIOTIC.

ablation   A general term for the combined processes of glacial melting and evaporation resulting in a net loss of ice and an attendant deposition of particles.

ablation zone   See ZONE OF ABLATION.

Abney level   An instrument for measuring the slopes of surfaces and the angles of objects. An Abney level works like a carpenter's bubble level, but rather than indicating only when a surface is level, the instrument indicates angles of objects above or below the viewer's eye. Abney levels are often used in measuring tree heights and determining slopes in topo-graphical surveys. See CLINOMETER.

above-ground   Referring to components occurring above the soil surface, as opposed to those buried beneath the soil surface. Often the estimates of the biomass of an ecosystem are expressed as above-ground and below-ground components. Compare BELOW-GROUND.

abrasion   1) General: grinding or wearing of rock material by impact of objects in transport. 2) Eolian: wind-accomplished erosion of rock or soil masses by impact of saltating grains. Contrast DEFLATION. 3) Fluvial: mechanical wearing of stream channel by collision of sand and pebbles. 4) Glacial: scouring and grinding of solid rock by action of rock particles embedded in ice.

abscisic acid   A substance found in plants that induces and maintains dormancy. Abscisic acid causes plants to enter dormancy and prevents seeds from breaking dormancy. Formerly called abscisin or dormin, this compound is also known by its abbreviation, ABA.

abscissa   On a standard graph with a horizontal (x) axis and vertical (y) axis, the horizontal distance as measured along the x-axis. Positive abscissa values fall to the right of the y-axis, negative values to the left.

abscission   The separation of a fully developed fruit, leaf, or stem from a twig by the formation of a corky layer of young cells at the base, which protects the twig from decay. Abscission is the mechanism by which deciduous trees drop their leaves in autumn.

absolute age   A determination of geochronology based on a fixed unit of measure, of age in years. Contrast RELATIVE AGE.

absolute date   An age measure in fixed time, such as years, within a probable range of error. Absolute dates are critical for establishing rates of geologic processes, useful in evaluation of geologic hazards. Contrast RELATIVE DATE.

absolute humidity   The total moisture contained in a given amount of air, measured in units of mass per unit volume (such as grams of water vapor per cubic meter of air). Contrast RELATIVE HUMIDITY, HUMIDITY RATIO.

absolute resource scarcity   Lack of sufficient supplies of a particular resource to meet present or future demand. Compare RELATIVE RESOURCE SCARCITY.

absolute time   An interval measured in units of fixed duration, of years.

absolute zero   The lowest temperature possible, 0° on the Kelvin scale or approximately -273°C (-459.67°F).

absorbate   A substance used to soak up another substance.

absorption   The process of one substance being taken up by another, resembling a sponge soaking up water. A solid or liquid may absorb a gas, or a solid may absorb a liquid. Matter also may absorb light. See ADSORPTION.

absorption coefficient   Numerical constant showing the degree to which a material absorbs radiation, either in the form of atomic particles or light rays. It is related to the amount of light (or x or gamma radiation) absorbed by a thin section of material.

absorption field   A trench or pit filled with loose rock fragments that is designed to absorb the effluent from a septic tank.

absorption line   Dark gaps appearing in the continuous spectrum of electromagnetic radiation (such as light) caused by an intervening substance selectively absorbing some of the radiation. Certain wavelengths of energy are taken up, creating dark lines. Because the wavelengths absorbed are specific to each element or molecule, the resulting spectrum is a fingerprint that can be used to determine the molecules present. This enables scientists to determine both the composition of gases in outer space and to use spectroscopes to analyze compounds in the laboratory. See EMISSION LINE, SPECTRUM, SPECTROGRAPHY, SPECTROSCOPE.

absorption spectrum   A plot of wavelength versus intensity of electromagnetic radiation that has been absorbed by a given molecule or atom; it can be produced by examining a substance (or the light from a star) through a spectroscope. Energy is absorbed by molecules at specific wavelengths that are characteristic of each molecule. Energy can be absorbed by a variety of processes, including both electronic and nuclear transitions from a lower energy level to a higher energy level as well as vibrations or rotations of the molecule. The wavelength absorbed is related to the process by which the energy is absorbed. In general, electronic transitions are relatively high-energy (short-wavelength) processes, whereas nuclear transitions, vibrations, and rotations are low-energy (long-wavelength) processes. See SPECTRUM, ABSORPTION LINE, EMISSION LINE.

absorptive capacity   Maximum amount of radiation, water, etc. that can be absorbed.

absorptive or saprobic nutrition   Method of feeding in which an organism absorbs dissolved food molecules across the cell membrane, as is typical of decomposer organisms such as bacteria and fungi. See SAPROPHYTE; compare AUTOTROPH.

abundance The total number of individuals of a species or amount of resources present in a specific area. The term is often used in a qualitative, relative, or subjective manner rather than in reference to exact numbers or amounts. See FREQUENCY, RELATIVE ABUNDANCE.

abyssal A zone of deep oceanic water, generally deeper than 2000 meters, where light does not penetrate.

abyssal ecosystem The collection of organisms and the environmental conditions that exist in the bottom zone of the ocean, at depths of 4000 to 6000 meters. See LITTORAL ZONE, BATHYAL ZONE.

abyssal plain The very flat expanses of the deep ocean floor, usually covered with pelagic mud.

abyssobenthic Having to do with the ocean floor of the abyssal zone.

abyssopelagic Having to do with the open-water region of the abyssal zone.

Ac Chemical symbol for the element actinium. See ACTINIUM.

AC Abbreviation for alternating current (a.c. is also used). See ALTERNATING CURRENT.

acanthite An ore of silver occurring in veins and composed of silver sulfide.

Acanthodii A class of primitive bony fishes that lived from the Silurian through the Permian periods. The Acanthodii had distinct morphological features, including a heterocercal tail and spines in front of the fins.

acaricide A pesticide used to kill spiders, mites, or ticks. Rotenone, when used for control of red spider mites, is an acaricide.

acceleration Rate of change of velocity due to either a change in speed or a change in direction. The units of acceleration are meters per second per second (m/s^2); positive values show increasing velocity and negative values show deceleration.

accelerator Usually refers to high-energy particle accelerators, devices that increase the speed (and therefore the energy) of charged atomic particles to very high levels. Accelerators are used to research the structure of sub-atomic particles, and in medicine to treat some cancers.

accessory mineral A mineral that may be present in small quantities within a rock and that is not included in the formal definition for the given rock type.

accessory species Species that are found in less than half but more than a quarter of the area covered by a plant community.

acclimation To undergo physiological change leading to adaptation to a change in an environmental factor. For example, seedlings are often placed outside for a few hours at a time, or into a cold frame, before being planted in the garden to acclimate them to the outdoor environment and reduce transplanting shock. Compare ACCLIMATIZATION.

acclimatization The physiological and behavioral adaptations of an organism to changes in the total environment. Acclimatization is a short-term adjustment that occurs in individuals, usually taking only days or weeks. Acclimatization refers to an organism's adjustments to changes in several environmental factors, whereas acclimation refers to adjustments in response to a single environmental variable. Compare ADAPTATION, ACCLIMATION.

3

accreted terrane   An area or region that has been added to the edge of a tectonic plate by tectonic processes.

accretion   1) The increase in size of an inorganic body by addition of particles. 2) The enlargement of a tectonic plate by addition of an accreted terrane.

accumulation area   See ZONE OF ACCU- MULATION.

accuracy   Freedom from error.

acetaldehyde CH_3CHO, a colorless, volatile liquid with a strong odor. Acetal- dehyde is produced industrially by oxida- tion of ethanol (ordinary alcohol); it is an important starting material for the produc- tion of many organic compounds.

acetate film   Ordinary photographic film, consisting of an emulsion of photosensitive chemicals deposited on a strip of cellulose triacetate. Also called safety film, because it is so much less flammable than the nitrate film used early in this century.

acetic acid CH_3COOH, a clear liquid with a strong odor and flavor that is the prime ingredient (after water) of vinegar. Acetic acid is a weak acid and an important solvent used to manufacture organic chemicals and plastics.

acetone CH_3COCH_3, a colorless liquid with a strong odor that is one of the most important commercial solvents. Uses of acetone range from nail polish remover and paint remover to the synthesis of organic compounds including plastics, drugs, and disinfectants.

acetylcholine   One of the substances produced by the human body to aid the transmission of impulses from one nerve to another. The body produces acetylcho- line from choline, at the ends of nerve fibers. A lack of acetylcholine is one of the reasons for memory dysfunction in people with Alzheimer's disease. See CHOLINE.

acetyl coenzyme A   (acetyl-CoA) An organic molecule that facilitates the action of enzymes. For both animals and plants, it is an important compound in aerobic respiration and other biochemical reac- tions. See COENZYME, KREBS CYCLE.

acetylene C_2H_2, a poisonous, colorless gas with a strong garliclike odor. It is used as a fuel for soldering and welding in acetylene torches. Many chemicals are manufactured using acetylene as the starting material. Also called ethyne.

ACh   Abbreviation for acetylcholine. See ACETYLCHOLINE.

achene   Any small, dry, hard fruit made up of one seed with a thin outer layer that does not burst open when ripe, such as the sunflower seed. The achene is the simplest of any fruit.

achondrite   A type of meteorite that is stony in nature (as opposed to irons, which consist of iron and nickel) and resembles igneous rocks. Most of the meteorites that land on the earth's surface are stones, but because they soon weather to look like ordinary rocks, meteorites aren't often found unless their fall is directly observed. See CHONDRITE.

acid   1) A substance that forms hydrogen ions upon dissolving in water and neutral- izes bases to form a salt plus water. Acids are characterized by their tendency to lose a proton (or to gain an electron pair) in reactions. Acids turn litmus paper red and can be recognized by their usually sour taste (as in vinegar, acetic acid) and low pH. 2) Acidic, having properties that resemble acids, such as a sour taste. Compare BASE.

acid–base indicator   A chemical substance that changes color depending on the pH (acidity or alkalinity) of a solution. Litmus is a common acid-base indicator: It is blue in basic solutions and red in acidic solutions.

acid deposition   A form of air pollution caused by sulfur dioxide and nitrogen dioxide combining with water vapor in the atmosphere to form sulfuric and nitric acids. This term encompasses acid precipitation (wet deposition) and the fallout of dry particles containing salts of nitrogen and sulfur. See ACID PRECIPITATION.

acidic   Having properties like those of acids, such as a high concentration of hydrogen ions and therefore a low pH value; sour in taste or corrosive in nature. Compare ALKALINE.

acidification   The process of making a substance acidic, lowering its pH or making it "sour."

acid ionization constant, K_a   The equilibrium constant for the dissociation of a particular acid. It expresses the extent to which an acid undergoes ionization. The strongest acids undergo complete ionization, so the higher the value for K_a, the stronger the acid. K_a is calculated by multiplying the concentration of the ions produced by dissociation (H^+ and Cl^- for hydrochloric acid) and dividing the product by the concentration of the acid (HCl) See BASE IONIZATION CONSTANT, EQUILIBRIUM CONSTANT.

acidity   Quality or degree of being acid; "sourness." Also, a technical term for measuring the ability of a base to neutralize an acid, or the intensity of an acid chemical reaction as measured by concentration of hydrogen ions.

acid mine drainage   Surface runoff for groundwater that has become acidic after percolating through tailings from mining activity. Acid mine drainage is a major environmental problem in the coal mining areas of the eastern United States.

acid neutralizing capacity   The ability of a body of water to buffer added acid. The acid-neutralizing capacity of a lake situated over limestone (which is alkaline and can therefore neutralize acids) is much greater than that of a lake over granite. See BUFFERING CAPACITY.

acidophile (acidophil)   1) A cell that stains easily with acid dyes, as compared to basophiles that stain easily with basic dyes. 2) An organism that grows well in an acidic environment.

acidophilous   Describing organisms that prefer and are adapted to an acidic environment; referring to an acidophile.

acid precipitation   Rain, sleet, mist, snow, or other precipitation whose pH is significantly lower than neutral (neutral pH is 7; normal precipitation has a pH of about 5.6 because dissolved carbon dioxide from the atmosphere produces carbonic acid). Normal precipitation becomes acidic (lower in pH) when water vapor interacts with oxides of sulfur and nitrogen in the atmosphere.

acid rain   The common name for acid precipitation, especially in liquid form. See ACID PRECIPITATION.

acid rock   A general term for light-colored or felsic igneous rocks rich in silicate minerals. Granite and rhyolite are typical acid rocks.

acid shock   A term for the biological disruption that occurs when aquatic systems are rapidly acidified. Acid shock may occur from spring runoff produced by a melting snowpack or by a particularly acidic storm event.

acoustic reflex A mechanism in mammals for protection against extremely loud sounds in which muscles in the ear change the position of the auditory ossicles.

acoustics The study of sound waves; also, the physical characteristics of a space that determine the quality of sound production within it. See SOUND.

acquired character A modification of an organ as a result of use or lack of use, or in response to the environment. It is not an inherited characteristic.

Acrasiomycota The cellular slime mold phylum of the kingdom Protista, sometimes referred to as the mycetozoans. During part of the cellular slime molds' life cycle, they exist as amoebalike individuals. Later, these individuals aggregate, forming a macroscopic sluglike mass, which then forms a cellulose-stalked sporophore fruiting body that produces spores. Members of the phylum generally inhabit terrestrial enviroments, where they ingest bacteria and fungi. Familiar members of this phylum are *Dictyostelium* and *Acrasia*. See PROTISTA. Contrast MYXOMCOTA.

acre A unit of English measure of area.

$$1 \text{ acre} = 4840 \text{ sq yd} = 43,560 \text{ sq ft}$$
$$640 \text{ acre} = 1 \text{ sq mi}$$
$$1 \text{ acre} = 0.4047 \text{ hectares (ha)}$$
$$2.4711 \text{ acre} = 1 \text{ ha}$$
$$247.11 \text{ acre} = 1 \text{ km}^2$$

acre-foot The amount of water required to cover an acre of land to the depth of 1 ft. Used in the United States, especially in irrigation engineering.

acrodont Describing teeth that are fixed to the summit of the jaw ridge. Compare BUNODONT, LOPHODONT, SELENODONT.

acrosome A structure at the tip of a sperm cell that produces the enzymes needed to penetrate the egg cell.

acrylic resins A class of polymers used for everything from contact lenses to synthetic rubber. Acrylic resins are durable and transparent; Lucite is an example. They are manufactured from, and get their name from, acrylic acid.

actin A protein capable of linking (polymerizing) into long fibers. It is necessary for muscle contraction but is also found at other sites where cellular movement occurs. See ACTOMYOSIN, MYOSIN.

actinide Any of a group (family) of elements falling between and including actinium (atomic number 89) and lawrencium (atomic number 103) on the periodic table of the elements. These elements are one of two transition series of metals (characterized by partially filled f electron shells); uranium is the most well known. Most actinides do not occur in nature; they are created during nuclear reactions and are among the components of radioactive waste. They are characterized by having properties similar to the element actinium: metallic, heavy, and radioactive with short half-lives.

actinium (Ac) A radioactive, metallic element (atomic number 89; atomic weight 227.0) that is formed in nature by the radioactive decay of uranium. Like uranium, it can be extracted from pitchblende ore; it can also be produced by bombarding radium with neutrons. It is used to produce radiation as it is a good source of alpha particles.

actinomorphic Describing organisms possessing radial symmetry, such as starfish and sea urchins, and the flowers of some plants, such as those in the mustard family. Compare ZYGOMORPHIC.

actinomycin An antibiotic derived from a species of fungus, *Actinomyces antibioticus*. See ANTIBIOTIC.

Actinopoda The actinopod phylum of the kingdom Protista is distinguished by heterotrophic organisms with slender, radially stiffened cytoplasmic projections called axopods. This group of organisms, which include the radiolarians, are found in marine and freshwater plankton, although some are attached. The skeletons of actinopods are composed of silica, chitin, or strontium sulfate. Members of this phylum include the marine *Thassicolla* and the freshwater sun animacules *Actinosphaerium*. See PROTISTA.

actinula A larval form occurring in some jellyfish in which the planula larva develops arms but does not settle before becoming a medusa.

action spectrum A graph showing the range of wavelengths over which a photo-chemical (light-dependent) reaction such as photosynthesis occurs. An action spectrum shows which wavelengths of light are most effective for driving the reaction.

activated alumina Hydrated alumina—a mixture of aluminum hydroxide, aluminum oxide, and water molecules—that has been activated by heating to drive off the water molecules. The resulting mixture has different properties depending on the relative proportions of its two components. The different forms of activated alumina are used as catalysts for reactions and as adsorbents in chromatography.

activated carbon A form of charcoal produced under controlled conditions to maximize its surface area and thus its ability to remove impurities from solids, liquids, and gases through adsorption. It is produced by heating high-carbon materials such as wood to high temperatures in the absence of air. Activated carbon is used in chemical reactions, in water filters and in air filters (including gas masks), and to deodorize and remove dissolved organic matter from public water supplies. Also called activated charcoal. See ADSORP-TION.

activated charcoal The United States Pharmacopeia (USP) term for activated carbon (charcoal) prepared to medicinal standards. Medicinal uses include for absorption of intestinal gases and as a treatment for alkaloid poisoning (the stomach is washed with a slurry of acti-vated charcoal in water to absorb the poisonous compounds that have been swallowed). See ACTIVATED CARBON.

activated manganese oxide process An air pollution control technique, a form of flue gas desulfurization. It uses dry adsorption to remove sulfur dioxide (SO_2) from flue gases and produces ammonium sulfate ($(NH_4)_2SO_4$ as a by-product. Also called an activated manganese oxide scrubber.

activated sludge process A sewage-treatment process that purifies wastewater through digestion of organic matter by microorganisms under aerobic conditions. After an initial screening to remove large particles and a primary settling process to remove smaller particulate matter, the waste is thoroughly aerated (mixed with air to supply abundant oxygen), which enables the bacteria to transform the suspended organic matter into carbon dioxide, water, and biomass. A secondary settling process removes the microorganisms, known as activated sludge. Some of this activated sludge is then injected into the aeration tank to aid in the digestion of the entering waste, while the remainder is digested anaerobically and partially dried for disposal.

activation energy   The minimum amount of energy required to initiate a chemical reaction.

active dispersal   Any mechanism by which plants scatter seeds with force, as opposed to a passive mechanism for distributing seeds, such as wind dispersal. Many species of violets *(Viola)* have fruits that eject their seeds when they are ripe.

active layer   The upper portion of a permafrost soil. The active layer is subject to seasonal freeze-and-thaw cycles.

active lobe (of a delta)   The forefront region of a delta that is prograding into a body of water.

active solar heating system   System for deriving heat from sunlight that incorporates active devices such as blowers and pumps to move the heat from the point of collection to the area of use or of storage. Solar collectors on rooftops, either for hot water or for home heating, are usually active systems. Contrast PASSIVE SOLAR HEATING SYSTEM.

active transport   The process of carrying substances across a cell membrane that requires chemical energy because it opposes the natural flow from high concentration to low concentration (diffusion).

active volcano   Any natural vent or fissure that is issuing lava and volcanic gases, or pyroclastic material.

activity   The activity of a solution is equal to the concentration of the solution in moles per liter multiplied by an activity coefficient, γ. The activity coefficient takes into account the fact that very concentrated solutions, such as seawater, often behave as if they have a different number of dissolved particles in solution than they actually do. The activity is thus a measure of effective concentration of the solution.

activity coefficient   The ratio of the analytic concentration to the effective concentration of a substance, used as a correction for deviations from the behavior of ideal ions and solvents. In many instances, particularly in highly concentrated solutions, such as seawater, the effective concentration of a substance is quite different from the actual concentration due to the interaction of solute particles.

actomyosin   A combination of the proteins actin and myosin, which occurs in muscle cells. Actomyosin shortens when stimulated, causing muscles to contract. See ACTIN, MYOSIN.

acute injury   An injury that is severe, and specific, and has a rapid onset, as opposed to chronic injuries that are of long duration and show only slow progression or little change.

acute toxicity   Any severe poisoning that occurs shortly after exposure to a hazardous substance (within 24 to 96 hours). Acute toxicity causes serious illness, damage, or death.

adaptation   Evolutionary changes (genetic modifications) that allow populations to be better able to exist under prevailing environmental conditions. Adaptations are changes in structure, function, or behavior of populations that occur over a long period — more than one generation. Compare ACCLIMATION.

adaptive radiation   The evolutionary divergence of a species into a variety of different forms, usually as an ancestral form encounters new resources or habitats. For example, see Darwin's finches. See DIVERGENT EVOLUTION.

adaxial   The surface of a leaf, petal, etc., that during early development was situated on the side nearest the axis. Usually, the adaxial side is the upper surface in a fully opened leaf. Compare ABAXIAL.

additive   A chemical substance incorporated into another substance to improve or to preserve its quality. Many additives are used in the food industry, but chemicals incorporated into gasoline and machine lubricants are also additives. See FOOD ADDITIVE, ANTIKNOCK ADDITIVE.

additive mortality   The total mortality caused by different factors affecting a population, such as predation, a fire or other one-time catastrophe, for a given period such as a year.

adelphous   Describing staminate (male) flowers with filaments that are fused together. Compare SYNCARP.

adenine $C_5H_5N_5$, 6-aminopurine. One of the two purine bases that make up the structure of both DNA and RNA; it pairs with thymine in DNA and uracil in RNA. See GUANINE, CYTOSINE, URACIL.

adenosine diphosphate   (ADP) A compound (composed of adenine, the five-carbon sugar ribose, and two phosphate groups) that is involved in the mobilization of energy in cellular metabolism. Energy is stored by adding a phosphate group to ADP to produce ATP, adenosine triphosphate.

adenosine triphosphate   (ATP) A molecule made up of adenine, ribose, and three phosphate groups attached by a high-energy bond and associated with energy transfer in living cells. Energy is released when ATP is converted to ADP and phosphate. See KREBS CYCLE.

adenylic acid $C_{10}H_{14}N_5O_7P$, a compound containing the nucleotide adenosine (adenine plus ribose) and phosphoric acid. Also called adenosine monophosphate (AMP), it is an important intermediate link in biochemical reactions such as energy production (Krebs cycle), cell division, and gene formation.

adhesion   The force of attraction between unlike molecules. Adhesion (in combination with surface tension) is responsible for water moving upward into a capillary tube and similar contacts between other solids and liquids. Contrast COHESION.

adiabatic   Occurring without heat entering or leaving a system.

adiabatic expansion   An increase in the volume of a substance with no exchange of heat between the substance and its surroundings. Adiabatic expansion occurs with steam inside the cylinders of a steam engine, driving the engine. It also occurs with air parcels in the atmosphere.

adiabatic lapse rate   In meteorology, the rate of temperature decrease as a mass of air rises through the atmosphere (calculated as temperature change divided by height change). Adiabatic lapse rate usually refers to dry adiabatic lapse rate. Meteorologists also use a saturated (wet) adiabatic lapse rate. See ADIABATIC.

adipose tissue   A type of connective tissue in which the body stores fat. The cells making up adipose tissue contain globules of fat; the tissue lies under the skin and surrounds some organs.

adit   A horizontal tunnel entering a mine to provide access or drainage.

administrative branch of government An executive unit of government responsible for implementing and enforcing public law, including the development of regulations, and for control, supervision, and implementation of a particular area of the public interest.

adobe A natural mixture of silt and clay, rich in calcium carbonate. Adobe forms in desert basins and is used as a construction material in the southwestern United States.

ADP See ADENOSINE DIPHOSPHATE.

adrenaline (adrenalin) A hormone secreted by the adrenal glands as well as other tissues. Adrenaline speeds up the heartbeat and increases energy and resistance to fatigue; increased production of adrenaline is one of the body's responses to perceived danger. Also called epinephrine.

adsorb To collect and to hold by adsorption. See ADSORPTION.

adsorption The formation of a layer of solid, liquid, or gas on the surface of a solid (or, more rarely, of a liquid). Adsorption occurs in some forms of chromatography (paper, gas–liquid chromatography); it also enables activated charcoal to remove impurities from fluids. Compare ABSORPTION.

adult An organism (or age class in a population) that has reached full sexual maturity and is capable of breeding.

advanced gas-cooled reactor (AGR) A type of nuclear reactor using gas (usually carbon dioxide) to extract heat from the core, and using slightly enriched fuel. The advanced gas-cooled reactor is in use in the United Kingdom; it differs from earlier gas-cooled reactors in that the gas (usually carbon dioxide) is released at much higher temperatures (600°C rather than 350°C). See NUCLEAR REACTOR, NUCLEAR POWER PLANT, HIGH–TEMPERATURE GAS-COOLED REACTOR.

advanced regeneration A term used by foresters and ecologists to refer to seedlings and saplings of tree species already present in a forest at the time of logging or disturbance and, therefore, capable of rapid growth before the germination of seeds of other plants.

advanced sewage treatment Any form of sewage treatment that follows secondary sewage treatment (bacterial decomposition). Such processes are designed to remove inorganic compounds and plant nutrients (especially phosphates and nitrogenous compounds). Also called tertiary sewage (or waste) treatment. Compare ACTIVATED SLUDGE PROCESS.

advection In meteorology, the transfer of heat (or other properties) by the horizontal movement of air masses such as winds. The advection of air masses causes much of the daily weather phenomena. See CONVECTION, ADVECTION FOG.

advection fog Fog caused by warm, moist air moving horizontally across cold ground. Advection fog is a stratus cloud with an extremely low base. See ADVECTION.

adventitious Literally meaning accidental; usually used to describe a structure that develops out of the normal or usual place. Most often refers to plant parts such as adventitious buds or roots that occur at locations or times other than those of the primary growth pattern of the plant.

adventitious roots Roots appearing in places where they usually do not grow, as on stems of plants above the ground. The prop roots of corn and mangroves are examples of adventitious roots.

adventive plant   A species of plant that is not native and has been introduced into the area but has not become permanently established. Compare EXOTIC SPECIES.

AEC   Abbreviation for the Atomic Energy Commission. See ATOMIC ENERGY COMMISSION.

aeolian deposit   See EOLIAN DEPOSIT.

aeon   See EON.

aerate   To introduce bubbles of air into a solution. Aeration is a component of the sewage waste treatment process.

aeration   Process of mixing with air or oxygen. Aeration is the basis for secondary wastewater treatment: Air is bubbled through the water to increase amount of dissolved oxygen. More oxygen promotes the action of aerobic bacteria, which break down organic wastes and thereby purify the water. See SECONDARY SEWAGE TREATMENT.

aerenchyma   Tissue with well-developed air spaces; often found in the stems of aquatic plants. Compare PNEUMATOPHORE.

aerial   Relating to the air or the atmosphere, as in aerial photography or aerial currents. Also, a radio or television antenna.

aeroallergen   An allergen transported by air, such as pollen and mold spores.

aerobe   An organism that can only live and grow in the presence of oxygen, especially a bacterium or similar microorganism requiring atmospheric oxygen. Contrast ANAEROBE.

aerobic   Requiring atmospheric oxygen in order to live and grow. Contrast ANAEROBIC.

aerobic decomposition   Breakdown of organic material by oxygen-consuming organisms; such decomposition can only occur in the presence of atmospheric oxygen. In secondary sewage treatment, aerobic decomposition by microbes breaks down effluent into harmless components. Contrast ANAEROBIC DECOMPOSITION, FERMENTATION.

Aerobic Endospore-Forming Bacteria   A phylum of aerobic bacteria whose members produce highly resistant spores that may endure for years. Some forms produce antibiotics, whereas others are important decomposers capable of breaking down plant material like pectin and cellulose. The important genus *Bacillus* belongs to this group. See MONERA, CELLULOSE, BACILLUS.

aerobic respiration   The cellular process that consumes uncombined oxygen to convert glucose into energy. Aerobic respiration produces large amounts of ATP (adenosine triphosphate), with carbon dioxide and water as by-products. Contrast ANAEROBIC RESPIRATION, FERMENTATION.

aerodynamic drag   The amount of resistance to motion through air. A streamlined automobile or airplane design results in less aerodynamic drag and, therefore, greater fuel efficiency.

aerogenerator   Another (primarily British) term for wind turbine. See WIND TURBINES.

aerosol   A suspension of extremely small particles or tiny droplets of liquids in air or in a gas. Gases under pressure are used to dispense a variety of substances as aerosols, from spray paints and insecticides to asthma medications. The use of chloroflourocarbons (CFCs) as propellants in aerosol cans has been linked to the destruction of the earth's protective ozone layer. See OZONE LAYER, OZONE SHIELD.

aestivation   See ESTIVATION.

afforestation   Reforestation; the process of turning open land into forest.

aflatoxin   Any of a group of poisons produced by species of molds (*Aspergillus*) that grow on peanuts and rice in storage. Aflatoxins are toxic to animals as well as extremely carcinogenic.

afterburner   Auxiliary burner that injects fuel into hot exhaust, resulting in greater combustion efficiency, reduced air pollution, and extra thrust. Afterburners are used in industrial pollution abatement and turbojet engines.

after-ripening   Changes that occur within ripe seeds of some plants after they have left the parent plant and that must be completed before germination can take place. After-ripening prolongs seed dormancy and is not dependent on environmental factors such as moisture or light.

aftershock   Any seismic tremors after a main earthquake event. Aftershocks may be repetitive and are often strong enough to cause additional damage to structures.

Ag   Chemical symbol for the element silver (from its Latin name, *argentum*). See SILVER.

agar   A gelatinous substance derived from some species of seaweed. Agar is a common culture medium for bacteria and fungi in the laboratory, as well as an emulsifier in processed foods and a mild laxative in pharmaceutical products. Sometimes called agar-agar.

agate   A variety of quartz chalcedony with finely crystalline structure. Agate usually has green-to-red banding patterns resulting from iron and manganese impurities.

age distribution   The frequency of individuals of specific age groups within a population of animals or plants. Also called population structure (age structure), it is often presented graphically as an age pyramid.

age (geologic)   1) A subdivision of an epoch in the geologic time scale. The name for a geologic age is taken from the corresponding chronostratigraphic stage. 2) Informally, a reference to a geologic time period known for a particular feature, such as the age of mammals. See STAGE.

Agency for International Development (AID) An agency of the U.S. Department of State responsible for a wide range of international financial and technical aid and development programs.

agency   Governmental body responsible for administering the laws and regulations. See ADMINISTRATIVE BRANCH OF GOVERNMENT.

age ratio   The relative frequencies of various age groups within a population. The age ratio of a population influences its current reproductive status.

age-specific death rate   Mortality rates for specific age groups within a population.

age-specific natality (fecundity)   The birth rates for specific age ranges of a population, often expressed as the number of female offspring per female in the population. Also called specific natality rate.

agglomerate   See VOLCANIC AGGLOMERATE.

aggradation   The process of accumulating

unconsolidated sediment on the land surface or in a river or other body of water.

aggrading   A descriptive term applied to a geomorphic feature that is being built up by aggradation.

aggregate   1) A rock that is composed of a mixture or accumulation of fragments or mineral substances. 2) A group of soil particles that adheres together. 3) The smallest structural unit of a soil. See PED.

aggregation   A dispersion pattern for animals or plants in which individuals occur in groups; individuals are found closer to each other than they would be in a random distribution. Also called clumped or contagious distribution. See DISPERSION, EVEN DISTRIBUTION, RANDOM SPATIAL DISTRIBUTION.

aggregative response   The tendency for a predator to spend more time in areas of a habitat having a high density of prey species; this response causes a correlation of higher predator densities in those areas with higher densities of prey.

aggressive mimicry   Behaving in a manner resembling a harmless species in order to surprise and attack an unsuspecting prey.

Agnatha   Class of vertebrates that lacks true jaws and in which paired appendages are poorly developed or absent. See VERTEBRATA, LAMPREYS.

agonistic behavior   Behavior that occurs between members of the same species in response to a conflict. Agonistic behavior may alternate between aggression (threats or attacks) and fear (appeasement or running away), depending on whether the animal is within or outside its marked territory.

agric horizon   A truncated and homogenous soil layer in which agricultural practices have intermixed original layers within a furrow slice.

agricultural revolution   The shift, roughly 10,000 to 12,000 years ago, of human populations away from dependence primarily on hunting and gathering (and therefore living as relatively small, migrating tribes) toward growing food in settled communities.

agriculture   Work of cultivating the soil, growing crops, and raising livestock; the practice of farming.

agro-, agri-   Prefixes indicating agriculture.

agrochemicals   Substances such as insecticides, fertilizers, and fungicides used in agriculture (especially synthetic compounds).

agroclimatology   A branch of climatology relating to the study of how weather affects crops. Agroclimatologists can recommend which crops, or which varieties, are best suited to a specific region or climate.

agroecosystem   An ecosystem in which some form of agricultural activity is taking place.

agroforestry   A farming method that integrates herbaceous and tree crops.

agronomic   Relating to the management of farmland.

agronomy   The science of managing farmland by studying soil and improving crop production.

Agulhas Current   A warm ocean current off the coast of southern Africa, beginning on the eastern side of the continent as a continuation of the Mozambique Current. It is named for Cape Agulhas.

ahermatypic   Not forming reefs (referring to corals). Contrast HERMATYPIC.

A horizon   A dark-colored and biologically active mineral soil layer nearest the surface characterized by accumulation of humified organic materials and by loss of silicate clay, iron, and aluminum sesquioxides, with attendant development of granular or platy structure. The A horizon is generally known as the zone of leaching. See SOIL PROFILE, also APPENDIX p. 619.

AID   See AGENCY FOR INTERNATIONAL DEVELOPMENT.

AIDS   Acronym for Acquired Immuno-Deficiency Syndrome, a fatal human disease in which the body can no longer defend itself against infections. Victims of AIDS often die from any of a number of opportunistic diseases, including tuberculosis and pneumonia. AIDS is caused by a retrovirus, human immunodeficiency virus (HIV), that is spread by exchange of body fluids. See RETROVIRUS.

air   The mix of gases that makes up earth's atmosphere. See ATMOSPHERE.

air bladder   A sac filled with air that is commonly found in most species of fish and a variety of other animals and plants. The air bladder of a fish, also called the swim bladder, helps fish adjust buoyancy to match the changing pressure of water at varying depths. In some plant species, such as seaweeds, air bladders help to hold the plant erect while submerged in water.

air conditioning   Cooling the air inside a building or vehicle, usually by washing the air, and often reducing humidity as well as temperature. Air conditioning consumes a great deal of energy, and some of the refrigerants used in automobile and other air conditioners are thought to contribute to atmospheric ozone depletion. See REFRIGERANT.

air curtain   A procedure for containing oil spills, or to create a barrier that protects fish from entering polluted water. Air bubbled through a perforated pipe moves water upward; the moving water slows down the spread of oil on the surface.

airflow   The motion of air relative to the surface of a body moving through the air. A streamlined vehicle design creates a smooth airstream as the vehicle moves, reducing drag and increasing fuel efficiency.

air lock   1) A compression chamber, a compartment filled with compressed air that allows access while preventing water or contaminated air from entering the access route. It is used to provide entry to areas that have been contaminated by radioactivity as well as in underwater vessels and spacecraft. 2) A pocket of air that prevents the flow of fluid through a pipeline.

air mass   Large portion of the atmosphere defined by close-to-uniform properties, especially temperature and humidity. Air masses can move great distances across the earth's surface. These parcels of atmosphere are the primary units studied by meteorologists in determining weather patterns. See FRONT, HIGH-PRESSURE AREA.

air monitoring   Repeated sampling of air quality at specified intervals of time to keep track of pollution levels, and to detect failures of pollution-control systems.

air pollution   Contamination of the atmosphere with substances that interfere directly or indirectly with human health and comfort, impair safety by reducing visibility, or impair property by eating away metal or stone. Although air pollution is usually produced by human activity, it also includes natural substances such as pollen, dust, and volcanic emissions.

air pollution episode An increase in air pollution to exceptional levels because of a reduction of atmospheric circulation; an inversion. Air pollution episodes can cause eye and respiratory irritation in healthy individuals and are hazardous to people with respiratory diseases.

air quality The degree of contamination of the ambient atmosphere by airborne pollutants. Air quality standards are the legal limits governing the allowable levels of specific air pollutants over a given period of time for a specified region.

Air Quality Act The 1990 update of the 1970 Clean Air Act (and its 1977 amendment), 42 U.S.C.A. §7521 et seq. This statute goes beyond the earlier legislation in requiring reductions in CFC emissions, requiring industries to use of the best available technology for emissions reduction, requiring cities to meet ozone standards, and allowing companies that reduce their emissions below required limits to sell their "polluting rights" to another company.

Air Quality Control Region (AQCR) Under the 1977 amendments to the Clean Air Act, attainment air quality control regions were established for prevention of significant deterioration (42 U.S.C.A. §§7470–7491. Class I areas include international and national parks, wilderness areas, and national memorial parks that exceed a prescribed acreage. Class II areas are, in essence, all remaining attainment areas. No Class III areas are established by legislation, but the states have the ability to redesignate areas according to a process set out in §7474.

air quality standards See NATIONAL AMBIENT AIR QUALITY STANDARDS.

airshed The atmospheric equivalent of a watershed. The concept was developed for crude calculations of air pollution levels over large areas. See BOX MODEL, WATERSHED.

Aitken counter A device used in meteorology for counting condensation nuclei (called Aitken nuclei) present in a given volume of air. See CONDENSATION NUCLEI.

Al Chemical symbol for the element aluminum. See ALUMINUM.

alabaster 1) A fine-textured white or translucent form of gypsum, suitable for ornamental carvings. 2) A white or translucent calcite or aragonite.

alanine $CH_3CH(NH_2)(COOH)$, an amino acid. Alanine is a common component of animal and plant proteins, although not one of the essential amino acids that must be supplied in the human diet.

Alaska Current An oceanic gyre, or surface current, that circulates along the southern coast of Alaska and the Aleutian Islands.

Alaska Lands Bill, 1980 A federal statute that allocated federal lands in Alaska to state, local, and native governments and designated remaining federal lands as national parks, forests, and refuges.

albatrosses Large oceanic birds with stout bodies, heavy bills and long, narrow wings that are superbly adapted for dynamic soaring close to the sea surface. Albatrosses are distributed in southern seas and the Pacific Ocean. See PROCELLARIIFORMES.

albedo Fraction of light reflected by a surface, such as ice, or by an entire planet. Studying a planet's albedo can help determine the composition of its surface.

albic A soil condition characterized by a pale or white appearance.

albinism A congenital lack of natural pigmentation. The skin and hair of albino animals are very pale or white, and the eyes are pink or red. Albinism is sometimes also used to refer to abnormal plants that lack chlorophyll.

albite A white variety of plagioclase feldspar typically forming in granite rocks. See PLAGIOCLASE, also APPENDIX p. 620.

Alcedinidae The kingfisher family, of the order Coraciiformes, is widely distributed, particularly in the Old World. The birds are stout with relatively large heads and massive, pointed bills. Many kingfishers are brilliantly plumaged with green and blue areas contrasting with white or russet ones. One group of kingfishers feeds principally on fish and on other water life, which they obtain by diving from the air or from perches. Many of the forest kingfisher group, such as the kookaburra of Australia, are seldom found near water and hunt for small birds, reptiles, and mammals in wooded areas or grasslands. Kingfishers of both groups nest in holes excavated in banks or similar locations. See CORACIIFORMES.

Alcidae See ALCIDS.

alcids Common name for the Alcidae family of pelagic birds of the order Charadriiformes with stout bills, large heads, and compact bodies (e.g., guillemots, murres, auks, puffins, dovkies). See AVES, CHARADRIIFORMES, GREAT AUK.

alcohol A class of organic compounds characterized by the presence of a hydroxyl (-OH) group. Methanol and ethanol are common alcohols. See PHENOLS.

alcohol dehydrogenase (ADH) An enzyme that aids in the conversion of alcohols into aldehydes and ketones (by removing hydrogen atoms), and the reverse. Alcohol dehydrogenase helps break down alcohol in the human body; differences in ADH levels contribute to the alcohol intolerance of some individuals.

alcoholic fermentation A form of anaerobic respiration in which sugars are converted to carbon dioxide and alcohol by certain yeasts, molds, and bacteria. Yeasts are used for the alcoholic fermentation essential in yeast-leavened breads, beer brewing, and fermentation of wine. See FERMENTATION.

aldehyde A class of organic compounds produced by the oxidation of alcohols, which results in the formation of a -CHO (aldehyde) functional group. Aldehydes are very reactive. See ACETALDEHYDE, FORMALDEHYDE.

aldrin An insecticide, one of a group known as hard pesticides because of their toxicity (they kill on contact) and persistence in the environment. It is chemically related to chlordane and dieldrin; all are chlorinated hydrocarbons and nerve poisons. Aldrin was formerly used in agriculture to control soil pests such as wireworm and Japanese beetle grubs, among other pests; it was banned for use in the United States in 1975.

aleurone An assortment of minute protein granules found in seeds, especially of cereal grains. In grains, aleurone occurs in the outer layer, known as the aleurone layer.

Aleutian low A semi-permanent region of low pressure over the northern Pacific Ocean, named for the Aleutian Islands off of Alaska. Compare AZORES HIGH.

alfisol any member of a soil order characterized by accumulations of silicate clays in the B horizon, a leached gray-brown upper layer and a red-brown lower layer with aluminum and iron silicate concentrations; typically forms under forest cover in subhumid climates.

algae A general category of photosynthesizing unicellular and multicellular organisms, primarily aquatic. Some are prokaryotes (cyanobacteria, formerly known as blue-green algae) but most belong to the kingdom Protista, eukaryotes that are neither plants, animals, nor fungi. Algae (singular: alga) include many unicellular planktonic organisms, diatoms, and multicellular seaweeds. See CYANOBACTERIA, CHLOROMONADO-PHYTA, CHLOROPHYTA, CHRYSOPHYTA, CRYPTOPHYTA, BACILLARIOPHYTA, DINOFLAGELLATA, EUGLENOPHYTA, GAMOPHYTA, PHAEOPHYTA, RHODO-PHYTA, XANTHOPHYTA.

algal bloom Explosion of algae populations in surface waters usually caused by an increase in nutrients such as nitrates and phosphates. Such high population densities may not last long and may occur periodically throughout the year. Some algal blooms produce toxins that can kill fish and make shellfish toxic to humans. Also called phytoplankton bloom. Compare RED TIDE.

algicide A substance that is toxic to algae, such as chlorine bleach. Algicides are used to prevent algae from growing in swimming pools.

algorithm Mathematical problem-solving procedure using a finite series of steps. In computer programming, the term refers to a plot of a program (a flow-chart) prepared before the actual programming.

alima larva In some types of Stomatopoda (an order of marine crustaceans), a second larval stage that follows the pseudozoeal stage.

alimentary system A system comprised of all the organs associated with digestion, absorption, nutrition, and excretion. It includes the mouth, esophagus, stomach, associated glands, intestines, and rectum in many mammals; ruminany mammals such as cattle and sheep have a more elaborate alimentary system. See RUMINANT.

aliphatic Belonging to a group of hydrocarbon compounds whose molecular structure is linear or branched (as opposed to being linked into closed rings). Paraffin and related substances (simple alkanes), propylene, ethylene, and acetylene are all aliphatic compounds. Compare AROMATIC.

alkali 1) An inorganic compound that dissolves in water to produce hydroxide (OH^-) ions and, thus, an alkaline (or basic) solution, one with a high pH. Alkali soils, usually found in arid regions, are those that are highly alkaline. They contain high concentrations of minerals made of carbonate, bicarbonate, and hydroxide salts; sometimes these concentrations are high enough to inhibit plant growth. Irrigation of alkaline soils dissolves the alkaline mineral salts (although they can be deposited again on the soil surface by evaporation) and can contaminate waters downstream from the irrigated land. 2) An alkali metal. The metals in the first group of the periodic table, including lithium, sodium, and potassium, are known as the alkali metals. They demonstrate typical metallic characteristics (are shiny and malleable, conduct heat and electricity) and are exceedingly reactive with water to produce hydrogen gas and hydroxide (OH^-) ions.

alkaline 1) The opposite of acid, also called basic; having a high pH value and thus a low concentration of hydrogen ions and a high concentration of hydroxide ions. 2)

Containing alkalies, as in alkaline soil or an alkaline chemical reaction. Compare ACIDIC.

alkalinity 1) The degree to which a substance is alkaline or basic; the extent to which its pH value lies above the neutral value of 7. 2) The capacity of a substance to neutralize acid. Contrast ACIDITY.

alkali (sodic) soil A pedocal soil with potentially toxic concentrations of soluble sulfates and chlorides of sodium, calcium, magnesium, and potassium, usually emplaced by capillary transport followed by evaporation of water in arid environments.

alkaloid A diverse group of complex organic compounds containing nitrogen and having alkaline properties, produced by plants and often having commercial uses for humans. Over 1000 different alkaloids are known from 1200 species of plants; some of the more common include nicotine, quinine, caffeine, strychnine, cocaine, morphine, atropine, and mescaline.

alkalosis Condition in which there is a loss of hydronium ion (H_3O^+) concentration in blood and tissue. Alkalosis is an excess of alkalinity in the body; it can be caused by exposure to low-oxygen air found at high altitudes.

alkyl mercury An organometallic compound of the general formula $[R\text{-}Hg]^+$, where R is an aliphatic hydrocarbon group (alkyl group). Methyl mercury, an extremely toxic form of mercury, is an example. See METHYL MERCURY.

alkyl sulfonates Organic compounds of the general formula $R\text{-}SO_3^-M^+$, where R is an aliphatic hydrocarbon group (alkyl group) and M is either a hydrogen atom, an alkali metal cation, or an organic cation. When the alkyl group is longer than eight carbon atoms, these compounds have surfactant properties, where they increase the "wettability" of a surface. Detergents frequently contain alkyl sulfonates. Alkyl sulfonates with linear alkyl groups biodegrade readily, whereas those with branched alkyl groups do not.

allele One of a pair or series of genes that occupies a specific physical position in a specific chromosome; any of the alternative forms of a given gene. Different alleles usually produce different characteristics in an organism, such as brown versus blue eyes. Also called allelomorph.

allelochemical A substance produced by one organism that limits or inhibits the growth of another. Cineole and camphor are toxins produced by two species of shrubs in California chaparral communities; these allelochemicals inhibit the growth of herbaceous plants near the shrubs and thus reduce competition. See ALLELOPATHY.

allelomorph Another term for allele. See ALLELE.

allelopathy The adverse effect on one plant by another plant, which secretes a toxic chemical. See ALLELOCHEMICAL.

Allen's rule A generalization stating that protruding parts of warm-blooded animals (ears, tail, feet) are relatively shorter in the colder ranges of a species than in the warmer ranges. This observation also holds for closely related species and may reflect the need for animals to conserve heat in colder areas. Compare BERGMAN'S RULE.

allergenic Of or having to do with substances that causes an allergic response in particular individuals.

allergy An unusual sensitivity to a particular substance (pollen, food, animal hair, or poison ivy) that causes the immune system to overreact to the substance. Symptoms of allergies may include rashes, hives, itchy eyes, runny nose, and asthma.

allo- Prefix meaning different or other (as in from a different source); from the Greek word for other.

allochthonous material 1) Rock that is not usually found in its current environment, such as a chunk of igneous rock that has fallen off of a cliff into a layer of sedimentary rock. 2) Organic material imported into an ecosystem from outside, in contrast to organic matter produced from photosynthetic activity within the ecosystem. Compare AUTOCHTHONOUS MATERIAL.

allochthonous terranes An older term applied to an accreted terrane. See ACCRETED TERRANE.

allogamy Cross-fertilization between different individuals belonging to the same plant or animal species. Compare HETEROGAMY.

allogenic succession A change in species in a local community (ecological succession) caused by influences from outside the area (such as long-term drought or deposition of riverine sediments) that alter local conditions within the environment. Contrast AUTOGENIC SUCCESSION.

allograft A transfer of tissue from a genetically different donor of the same species as the recipient. Compare HOMOGRAFT.

allometry A condition of different relative growth rates in different parts of an individual. Often used in reference to the study of the distribution of biomass or production of different tissues of organisms.

allomone A chemical compound produced by members of one species that facilitates communication with another species. Compare HORMONE, PHEROMONE.

allopatric Referring to species or populations that do not grow in or inhabit overlapping geographical ranges. Contrast SYMPATRIC.

allopatric speciation The evolution of a new species caused by the accumulation of genetic differences over time in a population that is geographically isolated. When the genetic differences become so great that cross-breeding with other populations is no longer possible, allopatric speciation has occurred. Contrast SYMPATRIC SPECIATION.

allopolyploid Having more than the usual sets of chromosomes (polyploid organisms) in which the different sets originate from two or more different species. Wheat developed into an allopolyploid during its evolution. Compare AUTOPOLYPLOID.

allotetraploid A hybrid organism possessing a diploid set of chromosomes derived from each parent; a double diploid. Also called amphidiploid.

allotropy The ability of an element to exist naturally in different forms in the same solid, liquid, or gaseous state. O_2 is the common form of oxygen gas, but O_3 (ozone) is another, allotropic form of oxygen also found in the earth's atmosphere. Carbon and sulfur are found in two distinct crystalline forms and, thus, exhibit allotropy.

alloy A substance formed by melting together a metal and one or more other elements, usually to change the properties of the metal. Bronze is an alloy of copper, tin, and zinc; steel is an alloy of iron and carbon; dental fillings (amalgam) are an alloy of mercury, silver, and tin.

alluvial   Of or relating to river and to stream deposits. See ALLUVIUM.

alluvial fan   A subaerial landform composed of alluvium deposited by an abrupt decrease in stream velocity because of infiltration of water, increasing sediment concentration, or change in channel geometry. The shape of the landform approximates a segment of a cone. Alluvial fans have important environmental significance because they are sensitive to both climatic and tectonic change and, thus, their study yields information concerning past climatic conditions and tectonic activity.

alluvial soil   A soil formed in material emplaced by the action of running water, such as flood plain, alluvial fan, or deltaic deposits.

alluvium   Unconsolidated clastic material such as sand, silt, and clay deposited on land by geologically recent action of flowing water in rivers and streams. See COLLUVIUM.

alpha diversity   The number of different species in a local area, often called species richness. Contrast BETA DIVERSITY.

alpha particle   A positively charged atomic particle, consisting of two protons and two neutrons (a helium nucleus), given off at a very high speed during the radioactive decay of a larger nucleus such as radium or uranium 238. See ALPHA RADIATION.

alpha radiation   A stream of positively charged particles (alpha particles) released from radioactive isotopes. It is the least penetrating of the three types of nuclear radiation, stopped by paper or clothing. It is, therefore, the least hazardous to humans unless inhaled or ingested. Compare BETA RADIATION, GAMMA RADIATION.

alpine   A structure, process, or environment resembling that of the Alps or other high, rugged mountains.

alpine chain   A descriptive term for a folded mountain-belt region similar to the Alps in structure.

alpine glacier   A glacier that has a source above snowline in the summit area of a mountainous region. Compare MOUNTAIN GLACIER, VALLEY GLACIER. Contrast CONTINENTAL GLACIER.

alpine orogeny   A mountain-building event caused by the collision of continental tectonic plates.

alpine tundra   An ecosystem found on mountains at altitudes above timberline, where climatic extremes are severe and the growing season is very short. Alpine tundra is characterized by mosses, lichens, and low-growing herbaceous plants.

alternate host   One of the two or more unrelated organisms required as hosts in the life cycle of a heteroecious parasite. The white pine blister rust is a fungal disease that kills white pines, but the fungus requires gooseberry or currant *(Ribes)* bushes nearby to complete its life cycle; both the pine and currant are alternate hosts of the fungus.

alternating current   (a.c.) Electricity that flows first in one direction and then reverses, changing direction regularly. In the United States and Canada, power companies usually supply electricity at a frequency of 60 hertz (Hz), or 60 cycles per second. Alternating current changes direction twice in each cycle, so power is usually supplied as current that alternates direction 120 times per second. See DIRECT CURRENT, FREQUENCY, HERTZ.

alternation of generations   A regular alternation of life forms or reproductive methods in consecutive generations of an animal or plant. In some plants and lower invertebrates, successive generations alternate between sexual and asexual propagation or between being in the haploid and diploid states. Also called metagenesis. See GAMETOPHYTE, SPORO-PHYTE.

alternative energy   Energy utilization that does not require the burning of fossil fuels or nuclear fission. See SOLAR ENERGY, GEOTHERMAL ENERGY, HYDROPOWER, WIND POWER.

alternative technology   Decentralized, often nontraditional forms of applied science designed to allow the greatest possible degree of individual control by its users. Alternative technologies often use renewable energy sources and aim to produce minimal pollution or otherwise reduce negative global impacts. Compare APPROPRIATE TECHNOLOGY.

altimeter   A device that measures height above sea level. Most altimeters operate on the principle that atmospheric pressure decreases with altitude. Altimeters are standard features on the control panels of aircraft.

altitude   The vertical distance measured between a point above earth's surface and a fixed datum such as a bench mark or sea level. Altitude also refers to a star or a planet's angular distance above the horizon (its angular height in the sky for an earth-bound observer). Contrast ELEVATION, AZIMUTH.

altocumulus   White or gray layer of clouds occurring between 2900 and 6000 meters and characterized by fleecy clumps or shaded ridges or ripples. See CLOUD, ALTOSTRATUS.

altostratus   Gray-to-blue layer or sheet of clouds occurring between 1900 and 6000 meters, causing a partially overcast sky.

altricial   Referring to those species of birds born blind and often without feathers. Because songbirds are altricial, their young are helpless for some time after hatching and remain in the nest for a relatively long period. Contrast PRECOCIAL.

altruism   Any type of animal behavior in which one animal performs actions that benefit another at the apparent expense of itself.

alumina Al_2O_3, aluminum oxide. Naturally occurring aluminum oxide is also called corundum or emery. Because of its hardness, it is an excellent abrasive; rubies and sapphires are both forms of aluminum oxide containing traces of contamination by other elements (chromium and iron plus titanium, respectively).

aluminosilicate   Any of various compounds containing aluminum, silicon, and various metals or bases. Mica, feldspars, zeolites, and some clays are aluminosilicates; some are used in making glass and porcelain.

aluminous cement   Ordinary (Portland) cement to which up to 50 percent alumina has been added to improve setting ability in low temperatures by causing it to produce more of its own heat as it sets. Aluminous cements harden very quickly and are also more resistant to corrosion by salts and acids.

aluminum   (Al) A very light, metallic element with atomic number 13 and atomic weight 26.98. Aluminum is easily shaped and is an excellent conductor of heat and electricity. Aluminum makes up about 7 percent of the earth's crust but occurs naturally only in combination with other elements. The British spelling for this element is aluminium.

alveoli Tiny sacs at the end of the bronchi-ole tubes in the lungs.

Am Chemical symbol for the element americium. See AMERICIUM.

amalgam An alloy of mercury with one or more other metals. The amalgam used for dental fillings contains mercury, silver, and tin.

Amaryllidaceae The amaryllis family of monocot angiosperms, which some taxonomists lump together with the Lilaceae. It is distinguished from the latter by having flowers arranged in umbels and inferior ovaries. The three petals and three sepals generally look similar and surround the six stamens and ovary of three carpels. Wild hyacinth (*Brodiaea*), daffodils and narcissus (*Narcissus*), and snowdrops (*Galanthus*) are members of the Amaryllideae subfamily, whereas century plants and sisal (*Agave*), dragon tree (*Dracaena*), and others are members of the semi-woody Agaveae subfamily of the Amaryllidaceae. See ANGIOSPERMOPHYTA.

amber A hard and brittle fossilized resin of coniferous trees. Insects or other materials may be trapped and preserved within the resin. Amber is yellow to brown and used as an ornamental jewelery stone.

ambergris A waxy, aromatic substance secreted from the intestines of diseased sperm whales. It is found floating in tropical oceans or along tropical shores, and used as a fixative in perfumes.

ambient Surrounding, or present in the background, as in ambient noise levels, air temperature, or water temperature.

ambient air quality standards Maximum allowable concentrations of specific pollutants in the atmosphere as established by law.

amensalism A relationship in which one individual affects or inhibits another without being affected or inhibited in return. Compare MUTUALISM, SYMBIOSIS, ANTIBIOSIS.

American Fisheries Association A professional society to promote the conservation, development, and wise use of fisheries, both recreational and commercial; organized in 1870; located in Bethesda, MD; membership 8500.

American Forestry Association A citizen organization for the advancement of management and use of forests and related natural resources; organized in 1875; located in Washington, DC.

American Petroleum Institute (API) A private petroleum industry group, established in 1919, which sets standards for, and lobbies on behalf of, the petroleum industry. The group's charter states that API promotes foreign and domestic trade for American petroleum products and the interests of all branches of the industry. Areas of interest include development of equipment, technology, safety, testing and labelling, transportation, and disposal methods.

americium (Am) A radioactive element with atomic number 95 and atomic weight 243. One of the transuranic elements of the actinide series, it exists only when created by bombarding plutonium or uranium with high-energy particles. It is used as a reliable, long-lasting source of alpha particles in scientific research and as an ionizing agent in smoke detectors.

amethyst A violet-purple variety of quartz crystal that is sometimes of gem quality.

ametoecious parasite See AUTOECIOUS PARASITE.

amines   A group of organic compounds based on ammonia (NH_3) but with a hydrocarbon group replacing one or more of the hydrogen atoms. They often have a strong odor and are responsible for the unpleasant smell of rotting fish. Examples include methylamine and dimethylamine. Amines are used in the production of medicinal substances, including local anesthetics and sulfa drugs; the aniline used in fabric dyes is another amine.

amino acid   A group of organic compounds that serves as building blocks for proteins. They contain an amino group ($-NH_2$) plus a carboxyl group ($-COOH$). The basic structure of an amino acid is R-CH $(NH_2)(COOH)$ where R is a variable group, depending on the specific acid. Humans and animals produce many of the amino acids they need to synthesize proteins; those that cannot manufacture must be supplied in their diets and are called essential amino acids. Tryptophan and phenylalanine are examples of amino acids.

ammocoetes   The blind, wormlike larva of a lamprey. See LAMPREYS.

ammonia   A colorless, alkaline gas with a pungent odor; its chemical formula is NH_3. It is formed naturally when bacteria decompose nitrogen-containing compounds, such as proteins. It is used in the manufacture of plastics, explosives, and fertilizers; it is also used directly as a fertilizer, applied as a gas. When liquified by pressure and low temperatures, it is used as a refrigerant; at room temperature, it is used in solution with water as a cleaning agent.

ammonification   The production of ammonia through the decomposition of protein-containing organic matter by soil bacteria. Ammonification is one of the major processes in the conversion of organically bound nitrogen to nitrate. See NITROGEN CYCLE.

ammonifying bacteria   Soil bacteria that carry out decomposition of organic material to release nitrogen in the form of ammonia. Ammonifying bacteria digest amino acids or proteins, producing first urea and then ammonia. Other groups of microorganisms convert ammonia to nitrites and nitrate. See NITRATE-FORMING BACTERIA, NITRITE-FORMING BACTERIA.

ammonite   An extinct cephalopod mollusc of Devonian to Cretaceous age. Both ammanoid and nautiloid molluscs have coiled shells, but the ammonoids are distinguished by a wavy suture pattern joining the chamber partitions.

ammonium   An ion with formula NH_4^+ that does not exist in nature except in combination with other ions (as in ammonium chloride, NH_4Cl). Ammonium behaves like an ion of alkali metal when it combines with other substances; it forms many different ammonium salts, a number of which are used as fertilizers (ammonium sulfate, ammonium nitrate, ammonium phosphates).

ammonium nitrate NH_4NO_3, a salt consisting of the ammonium and nitrate ions. Ammonium nitrate is used extensively as a synthetic source of nitrogen in fertilizers; it is also used in the manufacture of some explosives. Runoff from fields recently fertilized with ammonium nitrate can contaminate surface water and ground water with nitrates. See METHEMOGLOBINAEMIA.

ammonium silicofluoride NH_4SiF_6, one of the compounds used in the fluoridation of municipal water supplies. See FLUORIDATION.

ammonoid   Another term for ammonite (the fossilized, spiral-shaped shell of a now-extinct mollusk). See AMMONITE.

amnion   A membrane surrounding the embryo of some vertebrates. In reptiles, birds, and mammals, the amnion holds fluid that protects the developing embryo from external impacts and from drying out.

amniota   A general term for vertebrates whose embryos possess fetal membranes (amnion, chorion, etc.). Examples are reptiles, birds, and mammals. See CLEIDOIC EGG.

amniote egg   The type of egg laid by birds and reptiles. The hard outer shell protects the embryo. The egg is named for the amnion sac that contains the embryo.

amoeba   Any one-celled organisms of the genus *Amoeba* or related genera of the phylum Rhizopoda that are characterized by dependence on pseudopodia ("false feet," temporary extrusions of the cell) for locomotion. These microscopic creatures are found in fresh and salt water, in soil, and as parasites (including the organism that produces amoebic dysentery in humans). Also spelled ameba; plural is amoebas or amoebae. See PSEUDOPODIUM.

Amoebae   See RHIZOPODA.

amoebiasis, amebiasis   See AMOEBIC DYSENTERY.

amoebic dysentery   A human disease caused by the parasite *Entamoeba histolytica*, an amoeba. The disease comes from drinking contaminated water or eating uncooked food (such as lettuce) that has been washed in such water.

amoebicide   Any substance that kills amoebas, either used medicinally as a treatment for amoebic dysentery or used to sterilize drinking water. Also spelled amebicide and amebacide.

amoebocytes   A cell in a multicellular animal resembling an amoeba in form, or traveling in a manner similar to amoebas. In sponges (*Porifera*), amoebocytes have a variety of different functions but are characterized by their ability to move about within the organism; in starfish and sea urchins (*Echinodermata*), amoebocytes have specialized excretory functions. White blood cells (leucocytes) are also examples of amoebocytes.

amorphous   Having no regular, definite shape or form.

AMP   Short for adenosine monophosphate, a molecule containing adenine, ribose, and one phosphate group that is use in energy-transfer reactions in living cells and in the synthesis of purines and proteins. AMP is often called adenylic acid. See ADENOSINE TRIPHOSPHATE.

ampere   (A) Unit for electric current in the Système International d'Unités (SI). One ampere equals one coulomb per second flowing past a given point. It is one of the seven fundamental SI units, from which all other international scientific units of measurement are derived. See COULOMB.

amphi-Atlantic   A descriptive term applied to populations of organisms found on both sides of the Atlantic Ocean.

Amphibia   A class of carnivorous, ectothermal vertebrates whose living members have a moist, glandular skin that is permeable to water and gases and which, in some species, secretes toxic substances. Most amphibians have a well-defined aquatic, larval stage in their life cycle and then undergo metamorphosis into adults. Depending on the species, adults may occupy aquatic or terrestrial habitats, for example, frogs, toads, and salamanders. See ANURA, URODELA, APODA, ECTOTHERM.

amphibian 1) A member of the vertebrate class *Amphibia*, which includes frogs, toads, newts, and salamanders. 2) Having amphibious characteristics or behavior, such as a plant that can grow either in shallow water or on dry land, or mammals that swim. See AMPHIBIA.

amphiblastula larva A flagellated larval stage that develops from the fertilized egg cell in some species of sponges. It eventually produces bulkier, non-flagellated cells and other cells in the body of the sponge. See PORIFERA.

amphibole A common ferromagnesian, rock-forming silicate mineral characterized by double-chain silica, tetrahedral crystal structure; is usually dark greenish to black. See APPENDIX p. 620.

amphibole group A collective term for the amphibole silicate minerals such as hornblende, tremolite, and actinolite. See APPENDIX p. 620.

amphicoelous Concave at each end, a term used to describe the shape of vertebrae in some reptiles and fishes.

amphimixis Fertilization; the fusion of gametes from two organisms in sexual reproduction. Compare APOMIXIS.

Amphioxus See BRANCHIOSTOMA.

amplitude Half of a total oscillation, the maximum difference from its mean value. For example, an ocean wave is half of the difference between its maximum height above and maximum depth below the water surface. Higher-energy waves have larger amplitudes. Compare FREQUENCY, WAVE PERIOD.

amu See ATOMIC MASS UNIT.

amylases Enzymes that help convert starch into sugar. Amylases are found in saliva and pancreatic juice as well as in plants, especially in sprouting grains and in malt.

anabatic wind Meteorological term for a wind caused when air passing over a sunwarmed slope becomes less dense as it heats up and is displaced upward by cooler, denser air.

anabiont A plant that is perennial, fruiting many times.

anabiosis A temporary state of suspended animation, especially that undergone by certain aquatic invertebrates to survive long periods of drought.

anabolism The stage of metabolism that uses energy to create complex compounds such as proteins from simpler compounds obtained from food. Contrast CATABOLISM.

anadromous Describing a life cycle in which adult individuals travel upriver from the sea to spawn, usually returning to the area where they were born. Salmon and shad are anadromous species. Contrast CATADROMOUS, OCEANADROMOUS FISH.

anaerobe An organism growing in the absence of atmospheric oxygen. Contrast AEROBE.

anaerobic An environment containing no atmospheric (or molecular) oxygen (O_2). Contrast AEROBIC.

anaerobic decomposition Microbial breakdown of organic matter that occurs in the absence of oxygen. Methane is sometimes produced as a by-product of anaerobic decomposition. Contrast AEROBIC DECOMPOSITION.

anaerobic digestion Another term for anaerobic decomposition, used especially to describe processes in which such decomposition is harnessed to produce biogas. See ANAEROBIC DECOMPOSITION.

Anaerobic Photosynthentic Bacteria
A diverse phylum of the Monera that includes green sulfur bacteria, purple sulfur bacteria, and purple non-sulfur bacteria. These bacteria photosynthesize only in the absence of oxygen. They use compounds other than H_2O as hydrogen donors, and liberate compounds other than O_2 in photosynthesis. The sulfur bacteria typically use H_2S as the hydrogen donor and liberate elemental sulfur. See MONERA.

anaerobic respiration A cellular process that converts glucose into energy in the absence of atmospheric oxygen (any oxygen used is scavenged from oxygen-containing compounds). Anaerobic respiration produces much less adenosine triphosphate (ATP), the cellular energy transmitter, than does aerobic respiration. It can produce a number of different substances as by-products, such as lactic acid, when ordinary muscle tissue exhausts its oxygen supply, and ethanol, during fermentation. Contrast AEROBIC RESPIRATION; SEE FERMENTATION.

ana-front Meteorological term for the condition of warm air rising at the edge of a warm or cold front.

anagenesis Evolutionary development within a species that does not result in a separation into distinct groups; linear rather than branching evolution. Contrast CLADOGENESIS.

analog, analogue Operating with signals or numbers representing directly measurable quantities, such as rotations or voltages, as are used in analog computers. Records are examples of analog sound recording because the grooves on a record

vary analogously with the sound signal.

analogous structures Features of organisms that resemble each other in appearance or in function but that have completely different evolutionary origins. Wings on insects and birds are analogous organs. Contrast HOMOLOGOUS STRUCTURES.

analysis of variance (anova) Technique in statistics for separating out observed variations that may be caused by the factors under observation in a system from variations whose causes are statistically independent. It is used in analyzing sources of uncontrolled variations in experimental systems.

anaphase The third stage in cell division (mitosis or meiosis) during which the two sets of daughter chromosomes move to opposite ends of the spindle. Anaphaset occurs after metaphase and before telophase. See MITOSIS, MEIOSIS, METAPHASE, PROPHASE, TELOPHASE.

anatexis The term applied to a high temperature metamorphic process that produces magma.

ANC Abbreviation for acid neutralizing capacity. See ACID NEUTRALIZING CAPACITY.

anchor ice An accumulation of ice formed on a submerged structure or at the bottom of a body of water.

ancient forest 1) Virgin woodland, forest whose trees have never been cut, or old second-growth forest with very large trees. 2) In Britain and Europe, it refers to forests generally in existence before 1700 AD, regardless of their origins. 3) In North America, a synonym for old-growth, primary forest predating European colonialization. See OLD-GROWTH FOREST, SECOND-GROWTH FOREST.

ancient soil   See PALEOSOL.

andalusite   A mineral of aluminum silicate forming orthorhombic crystals by the metamorphic alteration of clay.

Andean floral region   Another name for Pacific South American floral region. See PACIFIC SOUTH AMERICAN FLORAL REGION.

andesite   An intermediate-colored extrusive igneous rock with an aphanitic or porphyritic texture. Andesite has essentially the same mineralogy and composition as diorite.

androecium   The staminate (male) parts of a flower. Compare GYNOECIUM.

androgens   Male sexual hormones; substances that stimulate or enhance masculine characteristics. Compare ESTROGENS.

anechoic   Echo-less or without echoes. Rooms designed for acoustic testing are called anechoic rooms; the walls, floor, and ceiling are constructed to absorb as much sound as possible so that almost none is reflected back into the room to interfere with the testing.

anemia   A medical condition characterized by a lack of hemoglobin or of red blood cells. Anemia has many causes, including disease and poor diet (especially lack of iron or vitamins).

anemochore   Seed or spore dispersed by wind. Contrast HYDROCHORE, ZOOCHORE.

anemometer   An instrument that measures air speed. Wind speed must be measured to determine if a site has potential for powering windmills or other wind generators. See WIND MEASUREMENT, WIND PROFILE.

anemophily   Pollination by wind. Compare ENTOMOPHILY, ORNITHOPHILY.

aneroid barometer   Instrument that measures atmospheric pressure by the movement of a flexible diaphragm at one end of a vacuum chamber. Changes in atmospheric pressure cause the diaphragm to move, which, in turn, moves the needle on a scale. Aneroid means containing no liquid, so aneroid barometers are distinct from mercury barometers. Aneroid barometers are used in altimeters. See BAROMETER, ALTIMETER.

aneuploidy   The condition of having missing or extra chromosomes or parts of chromosomes; having a chromosome number that differs from the standard number for the species but is not a multiple such as a tetraploid. Compare EUPLOIDY.

angiosperm   Any member of the Angiospermophyta (flowering plants), the larger of two major subdivisions of the seed-bearing plants. Grasses, broad-leaved (deciduous) trees, and berry-producing plants are all angiosperms. Compare GYMNOSPERM. See ANGIOSPERMOPHYTA.

Angiospermophyta   A phylum that contains all plants that produce flowers, that is, have a sporophyte generation with seeds developing from ovules that are surrounded by an ovary wall. The gametophyte generation is much reduced, consisting of only a few cells. The angiosperms are the most successful group within the plant kingdom, with greater diversity and dominance than any other plant phylum. Angiosperms are subdivided into two large groups: the monocots, having a single cotyledon, and the dicots, having a pair of cotyledons. Compare GYMNOSPERM.

angle of repose   A measure of the maximum naturally occuring slope of an unconsolidated material. The actual angle of repose is determined by factors such as particle size, degree of rounding, and water pressure in pore spaces.

angstrom unit   (Å) Unit of length equal to 10^{-10} meters that is used in measuring wavelengths of visible light and x-rays. Visible light ranges from 4000 to 7000 Å. Although still used in astronomy, the angstrom unit is no longer the preferred unit in physics; it has been replaced by the nanometer.

angular momentum   Amount of spin of an object around an axis, calculated by multiplying the object's angular velocity times its mass (moment of inertia).

angular unconformity   A type of discordant relationship between geological strata in which the dip of the underlying layers intersects the dip of the overlying layers at a distinct angle. See UNCONFORMITY, DISCONFORMITY, NONCONFORMITY.

angular velocity   ( ) The rate of movement of a body around an axis, expressed as the angular movement over time (e.g., in degrees per hour).

anhydrite $CaSO_4$, a grayish white mineral, the anhydrous sulfate of calcium. In the presence of water, anhydrite readily converts to gypsum. It is made into some forms of plaster (anhydrite plaster). See APPENDIX p. 620.

anhydrous   Not hydrated; refers to salts and oxides containing no water of crystallization or water of combination. Compare HYDRATES.

animal flagellates   See ZOOMASTIGINA.

Animalia   The animal kingdom. To distinguish them from members of other kingdoms, animals can be defined as multicellular, heterotrophic organisms with diploid (paired) chromosomes and a unique type of embryonic development. Most forms have a nervous system and contractile tissues, such as muscle. Originally, unicellular motile hetertrophs, such as amoebas and paramecia, were placed in the animal phylum, Protozoa, but most authorities today classify them with the Protista. See APPENDIX p. 613.

animal unit   (AU) The combined weight of one cow and one calf, set as 454 kg. The measure is used as a standard unit for weighting the grazing pressure from different kinds of grazing animals, assuming that they consume the same kinds of forage. For example, 1 AU = 7.7 white-tailed deer or 5.8 mule deer. See COW-MONTH, SHEEP-MONTH.

anion   An atom or group of atoms with a negative charge. Examples of anions are the hydroxyl ion (OH^-), carboxylic acid functional groups (COO^-), and halide ions (Cl^-, F^-, Br^-, I^-). Contrast CATION.

anisotropic   Having physical properties that differ in different directions. The anisotropy of some crystals allows them to transmit electromagnetic radiation only in certain directions. An anisotropic environment is one that is not homogenous in all directions. See ISOTROPIC.

ankerite   A mineral formed from dolomite in which the magnesium ions have been largely replaced by iron ions. Ankerite is variable in form and color.

annelid   See ANNELIDA.

Annelida   A phylum of animals characterized by elongate bodies that are divided into segments with serially arranged muscles and internal body compartments. Segmentation is reduced in some forms but is always manifest in the nervous system. There is a circulatory system, the gut is complete, and the body cavity is a true coelom. The segment or lobe anterior to the mouth, called a prostomium, encloses the brain. A stout body wall is provided with longitudinal and circular muscles used for crawling, burrowing, and swimming. Externally, the body is covered with a cuticle and bears movable bristles called setae. Three classes are distributed worldwide in soils, seas, and fresh waters. There are approximately 75,000 species (e.g., polychaete worms, earthworms, and leeches). See APPENDIX p. 613.

annual 1) Occuring or happening once every year. 2) A plant that completes its life cycle in only one year or season. Compare BIENNIAL, PERENNIAL.

annual allowable cut (AAC) An established level for permissible harvest.

annual increment A yearly increase in biomass or volume of an organism. The term is frequently used by foresters and ecologists in reference to the amount of annual growth by individuals or by entire communities.

annual rings See GROWTH RINGS.

annulus 1) A ring of tissue on certain plants, such as around the stem of some mushrooms or on spore cases of ferns or mosses. 2) A zoological term for any ringlike structure in an animal. 3) A ring of the sun that is not shadowed during an annular eclipse. Plural is annuli.

anode The positive electrode of an electrolyte cell, the anion-attracting electrode at which oxidation occurs. Also, the negative terminal of a battery or cell. Contrast CATHODE.

anodizing The process that deposits a protective, noncorroding coating of oxide over the surface of a metal, such as aluminum. In anodizing, a metal is made to serve as the anode in an electrolytic cell containing an oxidizing electrolyte such as sulfuric acid.

anomaly The departure of any element or feature from uniformity, from a normal state, or from a long-term average value. Used particularly in meterorology, in connection with temperature; in oceanography; and in connection with gravity. A gravity anomaly is the difference between observed gravity and that computed for an idealized globe. A temperature anomaly is the difference in degrees between the mean temperature (reduced to sealevel) of a station, and the mean temperature for all stations in the latitude. The result is either a positive (higher-than-average) or negative (lower-than-average) anomaly. See BOUGER CORRECTION.

anomaly (magnetic) See MAGNETIC ANOMALY.

Anopheles Genus of mosquitoes, many species of which are vectors of malaria parasites. See DIPTERA.

Anoplura Sucking lice; an order of small, wingless insects, all species of which are mammalian ectoparasites. In these forms, the thorax is fused and the mouthparts are adapted for piercing and sucking. Human body lice, *Pediculus humanus*, and pubic lice, *Pithirus pubis*, are widespread throughout human populations, particularly where unsanitary conditions prevail. Lice may cause intense discomfort and skin hypersensitivity, but more important, they are also vectors of such diseases as typhus and relapsing fever. There are approximately 500 species. See INSECTA.

anorthosite A type of intrusive igneous rock composed of more than 90 percent calcium-rich plagioclase feldspar.

ANOVAR   See ANALYSIS OF VARIANCE.

anoxia   A lack of oxygen, such as in animal tissues (especially blood), or in water bodies. Also called hypoxia.

anoxic   Lacking oxygen, used to refer to water whose supply of oxygen has been depleted. When water bodies are polluted with large amounts of organic matter or nutrients, oxygen consumption greatly exceeds production, leading to an anoxic condition. Anoxic water bodies can no longer support fish and other animals (only anaerobic organisms such as bacteria can survive). See BIOCHEMICAL OXYGEN DEMAND.

Anseriformes   An order of long-necked, robust-bodied, aquatic or semi-aquatic birds that are strong flyers. The order includes approximately 145 species of ducks, geese, and swans, as well as the several species of neotropical screamers, which have shorter bills and longer legs than other members of the order. See AVES, SWANS.

antagonism   A relationship between different organisms such as molds or bacteria in which one inhibits the growth of, or kills, the other, especially by the production of a toxic substance. See ANTIBIOTIC, ALLELOPATHY.

antagonistic effect   Environmental attributes whose interaction produces a combination that provides less resources than when added together separately. For example, with two antagonistic food resources, an increase in the consumption of one food resource by an organism leads to a need for increased consumption of the other food source by the organism. Also called antagonistic resources. Contrast SYNERGY.

Antarctic   Relating to the region near the southern continent of Antarctica.

Antarctic convergence   A zone of oceanic water surrounding Antarctica at about 50° latitude where the cold waters from the south meet and sink below the warmer waters from middle latitudes.

Antarctic treaty   An international treaty that limits national claims to the Antarctic region.

Antarctic divergence   A zone of diverging oceanic currents at about 50° east of the Weddell Sea of Antarctica. The Antarctic divergence forms where upwelling cold water flows both north at the ocean surface and south to descend as Antarctic slope water.

antecedent drainage   A stream or river that cuts across a mountain range or topographic feature. It is inferred that the watercourse is older than the related topographic feature.

antenna   1) Part of a sensory organ consisting of a pair of rodlike appendages on the heads of insects and crustaceans (the latter have two pairs). Antennae are usually sensors of touch, but some antennae can detect other stimuli, such as odor. Also used for similar features on other animals, such as the specialized dorsal fin rays on angler fish. 2) Any device designed to receive (or transmit) electromagnetic waves, such as television or radio signals.

antennal scale   On some species of crustaceans, a short, flattened outer branch of the second antennae. See ANTENNA.

antennule   A small antenna, especially one of the smaller, anterior pair of feelers, one of the two sets on the head of a crustacean.

anterior   Located near the head or forward end on an animal, or toward the front on a human. In plants, anterior refers to the side of a flower or bud facing away from the stem or axis.

anthelion A mirage of the sun caused by ice crystals in the atmosphere refracting sunlight so that a false image of the sun appears opposite the real sun at the same altitude.

anther The pollen-bearing organ of the stamen of a flower. Anthers are usually rounded structures at the top end of a slender stalk (filament) in the center of the flower. See STAMEN.

antheridium The structure producing male gametes (spermatozoids) in lower plants such as mosses, liverworts, and ferns. See ARCHEGONIUM, SPERMATOZOID.

Anthocerotae The class of Bryophyta that contains hornworts. These primitive plants have rounded, leafy gametophyte generations that grow from intercalary meristems at their bases. See BRYOPHYTA.

anthocyanins A large group of flavonoid plant pigments that create red, blue, and purple hues in flowers, fruits, and sometimes leaves.

Anthozoa Class of the phylum Cnidaria containing stout-bodied, marine polyps with a crown of hollow tentacles and gastrovascular cavities divided by mesenteries. There is no medusoid stage. The class contains the sea anemones, which are solitary and usually attached to substrata, and colonial forms such as sea fans and corals. Anthozoans account for more than two-thirds of all species of Cnidaria. See CNIDARIA, CORALS.

anthracene $C_{14}H_{10}$, a colorless, crystalline coal-tar derivative whose structure consists of three connecting carbon rings. Anthracene is fluorescent and carcinogenic. It is used to manufacture alizarin dyes.

anthracite A hard variety of coal with a chonchoidal fracture and metallic luster. Anthracite is formed by the metamorphosis of soft coal. Compare BITUMINOUS COAL, LIGNITE.

anthracnose Fungal diseases of plants that cause circular spots or lesions on leaves or, in extreme cases, defoliation. Anthracnose infects sycamores (*Plantanus*), oaks (*Quercus*), maples (*Acer*), and especially dogwoods (*Cornus*). See DEUTERO-MYCOTA.

anthracosis A lung disease caused by inhalation of coal dust, more commonly known as black lung or coal miner's lung.

anthrax An acute infectious disease caused by a bacterium (*Bacillus anthracis*). Although it usually infects cows, horses, or sheep, humans working with these animals or their hides and waste can become infected (in Britain anthrax is also called woolsorter's disease). In humans it is fatal if not treated with antibiotics; it produces characteristic anthrax boils. Also called splenic fever. Anthrax has been considered for use in biological warfare.

anthropic A term relating to the geologic period in which humans have lived on the earth.

anthropic zone The range in chronostratigraphy in which human remains may be found in the rock record.

anthropocentric Having a human-centered perspective.

anthropochore A plant species whose dissemination depends on human activity. See ZOOCHORE.

anthropogenic Caused by human action, such as changes in vegetation, an ecosystem, or an entire landscape.

anthropology Study of humans, especially of the origins, the distribution, and the relationships between such aspects as physical characteristics, race, and culture.

anthropomorphic Attributing human form or personality to an object.

anti- Prefix meaning opposed to.

antibacterial Destroying, or at least checking the growth of, bacteria. See ANTIBIOTIC.

antibiosis A relationship of mutual antagonism between organisms. Compare SYMBIOSIS, AMENSALISM.

antibiotic Any substance that destroys or inhibits the growth of harmful microorganisms. Antibiotics are often substances produced by bacteria or molds (penicillin), but can also be synthetic; tetracycline and streptomycin are examples. Antibiotic has gradually come to mean only drugs that are used against bacterial infections and that are ineffectual against viral infections.

antibody A protein produced by the body in response to foreign substances (antigens) in order to destroy or deactivate pathogens or poisons. Antibodies are specific to specific antigens; the presence of a specific antibody in human blood indicates prexious exposure to that disease or toxin. See ANTIGENS.

antiboreal convergence See ANTARCTIC CONVERGENCE.

anticaking additive Compounds added to powdered substances such as cake mixes to keep them from solidifying.

anticline An arch in the rock strata in which the layers bend downward and away from the fold axis. The oldest layers are exposed nearest to the fold axis in an eroded anticline. Contrast SYNCLINE.

anticlinorium A composite group of synclines and anticlines forming a large antiformal structure on a regional scale. Contrast SYNCLINORIUM.

anticoagulant Any substance that delays or prevents the normal clotting of blood. Blood-thinning medicines used to prevent strokes are anticoagulants, as are some rodent poisons that work by causing hemorrhaging.

anticyclone A high-pressure air mass. In the Northern Hemisphere, winds circulate clockwise around the high-pressure centers of these air masses; in the Southern Hemisphere the circulation is counterclockwise. Typically, anticyclones travel slowly and are associated with calm weather. Contrast CYCLONE.

anti-degradation clause See PREVENTION OF SIGNIFICANT DETERIORATION.

antiform An anticline-like arching rock structure for which the relative ages of the layers has not been determined. Therefore, an antiform may be an anticline or an inverted syncline. Contrast SYNFORM.

antifouling paint Toxic compounds painted onto ship hulls to discourage barnacles from attaching to the ship. These compounds usually contain tributyl tin, a toxic compound that pollutes sea water as it wears off.

antifungal Destroying or inhibiting the growth of fungi. Antifungal medications are used to treat athlete's foot.

antigens Substances causing the body to produce antibodies. Antigens include pathogenic bacteria, poisons produced by bacteria, and blood cells from an incompatible blood type.

antihistamine Any drug used to minimize or block the actions of histamine and, therefore, to relieve allergies. See HISTAMINE.

antiknock additive   Compound added to gasoline that enhances combustion and prevents engine knocking. Such additives make lower-priced, lower-octane gasoline perform like a higher-octane gasoline. Tetraethyl lead, the best known of these additives, is being phased out because it is toxic (causes lead pollution from automobile exhaust) and because it ruins catalytic converters.

antimalarial   A drug used to prevent or treat malaria.

antimetabolite   A substance whose structure resembles that of an ordinary metabolic compound but which, when taken up by the body in place of its look-alike, is not able to perform the same function. Antimetabolites can, therefore, be targeted to block or to reduce certain metabolic processes. They are used in chemotherapy to treat cancer by interfering with or preventing cell division in tumors.

antimony   (Sb) A silvery, crystalline metallic element with an atomic number 51 and an atomic weight of 121.75. In nature, it usually occurs in combination with other elements, often as stibnite ore (antimony sulfide). The symbol Sb comes from the Latin word for antimony, stibium. Antimony is used in alloys to increase hardness; it is also used in paint, glass manufacture, and medicine. Radioactive isotopes of antimony are used in laboratories as neutron emitters.

antioxidant   Compound added to a substance to delay or to reduce oxidation. Antioxidants added to food delay spoiling; BHA (butylated hydroxy anisole), BHT (butylated hydroxy toluene), and vitamins C and E all delay the oxidation of fats (rancidity). Antioxidants added to paint delay the formation of a skin over the top of the liquid; added to rubbers and plastics, antioxidants help preserve flexibility and delay degradation.

antipodes   Two points exactly opposite each other on the surface of a sphere, for example, the North Pole and the South Pole.

antiserum   A serum containing antibodies for a specific pathogen. Antiserum is developed by injecting a human (or an animal, especially a horse) with the pathogen in order to cause the body to produce antibodies for that pathogen. Serum obtained from that individual will then contain the needed antibodies and be effective in treating another individual lacking the antibodies. The tetanus vaccine is an antiserum.

antistatic agents   Substances used to reduce the build-up of static electrical charges, especially additives and coatings used in plastics.

antivenin   Antiserum used to counteract the poison from the bite of an insect or snake. Also spelled antivenene. See ANTISERUM.

antivenom   Another term for antivenin, but one usually restricted to serum specific for snake venom. See ANTIVENIN.

antler   The bonelike, usually branched, growth on the head of a member of the deer family (moose, elk, deer, etc.). Antlers differ from horns on other animals by being shed annually, with new antlers growing in each year. Contrast HORN.

ants   Belonging to the single family, Formicidae, ants of the order Hymenoptera are among the most numerous and successful of all insects. In many areas, they maintain an extremely high biomass and are essential to the trophic dynamics of ecosystems. Ants form perennial societies that live in galleried nests constructed beneath rocks, in soil, under bark, in wood, etc. They form complex societies of reproductive and sterile female workers. In many species, there is a soldier caste of sterile females as well. Males and virgin queens may have wings but, after fertilization, the wings of queens are shed. Ants may be predators, scavengers, nectar feeders, etc. Some species cultivate fungal gardens. Distribution is worldwide, except for polar and alpine regions. Over 14,000 species exist. See HYMENOPTERA.

Anura The order of the class Amphibia, whose members have elongated, muscular hind limbs and a specialized pelvis capable of producing the thrust necessary for swimming and jumping (e.g., frogs and toads). See AMPHIBIA.

anvil cloud Commonly occurring form of thunderhead (cumulonimbus cloud) whose pointed shape at the top suggests an anvil. See CUMULONIMBUS.

apatetic coloration See CRYPTIC COLORATION.

apatite A common phosphate mineral forming hexagonal crystals. Apatite is typically greenish white and develops as an accessory mineral in igneous rocks.

apetalous Lacking petals, as in poinsettias (*Euphorbia pulcherrima*), willows (*Salix* sp.), and other flowering plants that lack true flower petals although they may have bracts or sepals.

aphagia The inability to swallow and, therefore, to feed.

aphanitic A textural description of igneous rocks in which the crystal size is too small to be seen without magnification.

aphidae See APHIDS.

aphids Any of numerous species of small insects of the family Aphidae, many of which are of economic importance because they infest ornamentals and food crops and transmit some viral plant diseases. Like other members of the insect order Hemiptera, their mandibles and maxillae are modified for piercing and sucking. They may be either winged or wingless. Aphids can produce 20 or more generations in a year. They have a complex life cycle: They can reproduce either sexually or by parthenogenesis, and may either be oviparous or viviparous. Newly matured adults often are spread great distances by winds. See HEMIPTERA.

aphotic zone Any area of the sea or a deep lake that is too deep to receive sunlight, usually below 100 meters. Compare PHOTIC ZONE, EUPHOTIC ZONE.

API See AMERICAN PETROLEUM INSTITUTE.

Apiaceae The parsley or celery family of the angiosperm dicots. The flowers of these hollow-stemmed herbs are arranged in simple or compound umbels, usually with a ring of bracts at the base. The five petals alternate with the five stamens. The inferior ovary is comprised of two carpels that split apart at maturity. Carrot (*Daucus*), celery (*Apium*), parsley (*Petroselinum*), poison hemlock (*Conium*), parsnip (*Pastinaca*), and water-pennywort (*Hydrocotyle*) are members of this family. See ANGIOSPERMOPHYTA.

apical meristem A group of cells at the growing tip of a shoot or root which, through continuous cell division, forms new plant tissue. See CAMBIUM, MERISTEM.

Apicomplexa The sporozoan phylum of

the kingdom Protista. These spore-forming microbes are largely parasites of animals, and include important disease-causing protozoa such as malaria *(Plasmodium)*, coccidia *(Eimeria* and *Isospora)*, and piroplasms of domestic livestock *Babesia* and *Theileria*. They are characterized by an "apical complex" of polar rings, microtubules and other organelles at one end of the solitary cells. Members of this phylum exhibit an alteration of diploid and haploid generations. See PROTISTA, ALTERNATION OF GENERATIONS.

apidae   See BEES.

API scale   An arbitrary ranking system for crude petroleum developed by the American Petrolium Institute and based on specific gravity. Values are expressed in degrees API (or simply degrees). The scale runs from 0 to 100 and low numbers on the scale indicate the heaviest oils; most crude oil falls between 27.0 and 43.0° API.

Apoda   An order of legless, burrowing, or aquatic amphibia of the tropics with reduced eyes and a pair of tentacles located in grooves on either side of the head above the mouth. Both oviparous and viviparous forms exist. The order is often referred to as the gymnophiona. See AMPHIBIA.

Apodiformes   Among the most aerial of all birds, members of this order have small feet, short upper arm bones, and and long distal segments below the elbow. The swifts and crested swifts are insectivorous whereas the hummingbirds may feed on both nectar and insects. There are close to 400 species. See AVES, HUMMINGBIRDS.

apogee   The point on the orbit of the moon or a satellite where it reaches its farthest distance from earth. Contrast PERIGEE.

apomixis   Asexual reproduction, especially the formation of seeds in flowering plants without fusion of gametes. Compare AMPHIMIXIS, PARTHENOGENESIS.

apomorphous   Having distinct structure or form.

aposematic coloration   Markings on an animal that provide protection. Aposematic coloration is a general category that includes markings that protect by signalling caution or warning (sematic coloration) to predators. Compare BATESIAN MIMICRY, PSEUDOAPOSEMATIC COLORATION.

apostatic selection   The action of a predator on the most abundant of several species or several forms in a population. This leads to a stable occurrence of more than one form in a population, or co-occurence of several preferred prey species in a community. Contrast ASPECT DIVERSITY.

apparent polar wandering   The inferred change in position of the geomagnetic pole in relation to the geographic poles as measured over geologic time. Evidence for apparent polar wandering may be due to either a change in the position of the continents or to a true polar wandering. See POLAR WANDERING.

appeal, administrative   To bring a legal case concerning a regulation or a decision of an administrative agency to a higher administrative body.

appeal, judicial   A resort to a higher court for the purpose of obtaining a reversal of a lower court decision.

appetitive behavior   Animal behavior term for initial exploratory actions that precede a goal-oriented sequence of actions. Actions that follow appetitive behavior are called consummatory behavior.

Appleton layer   See F LAYER.

application factor   A quantity used in determining the maximum concentration of a substance that is safe for a particular organism. It is calculated as the ratio between the amount of the substance that produces a response in the organism to the LD_{50} (the dose that is lethal to half of a test group), when given over a specified period of time. See LD_{50}.

applied science   Research applying basic sciences to the solution of practical problems.

appressed   Describing a plant part such as a bud that is flattened up against, but not joined to, another plant part.

appropriate technology   Applied science that is suitable for the level of economic development of a particular group of people. Appropriate technology is decentralized, can be understood and operated by its users (i.e., does not require outside operators), uses fuel and other resources that are either local or easily obtained, and involves machinery that can be maintained and repaired by its users. Often, but not necessarily, it is labor-intensive and involves simple machinery. Compare ALTERNATIVE TECHNOLOGY.

appropriation doctrine, water law   See DOCTRINE OF PRIOR APPROPRIATION.

apterous   Wingless, as in insects such as fleas.

aquaculture   The cultivation of fish or shellfish for human consumption. Aquaculture may take place in a modified natural environment such as stocking ponds in a fish farm, or in an artificial environment such as fish tanks.

aquamarine   A blue, gemstone-quality form of the mineral beryl. Beryl is a very hard beryllium aluminum silicate that forms hexagonal crystals in pegmatite rocks.

aqua regia   A mixture of concentrated hydrochloric acid and concentrated nitric acid in a 3-to-1 ratio by volume of HCl to HNO_3. Aqua regia (meaning royal water) is a very powerful oxidant and is strong enough to dissolve all metals, including gold and platinum.

aquatic   Of or concerning water, especially referring to organisms living in fresh water. Compare TERRESTRIAL.

aquatic ecosystem   Any freshwater-based ecosystem such as a stream, pond, lake, or ocean.

aquatic humic substances   Complex organic molecules found in water as a result of the decomposition of dead plant material. They vary in composition, are usually dark in color, and color the water body or water supplies in which they are found. Aquatic humic substances bind to metals in the environment and can significantly modify the reactivity of these metals, generally making them less toxic. Chlorination of water supplies that contain aquatic humic substances results in the formation of chlorinated organic compounds, some of which have been shown to be mutagenic. See HUMIC ACIDS.

aquatic plant   Plant living in fresh water. See HYDROPHYTE.

aquicide   A substance used to sterilize water, as in swimming pools.

aquiclude   Any type of rock or unconsolidated material that serves as an impermeable barrier to the flow of groundwater.

aquifer   Any hydraulically active body of porous rock or permeable unconsolidated material that is capable of producing water. Aquifers function as natural storage areas for groundwater. See ARTESIAN AQUIFER.

aquifer

aquifer depletion   The reduction in the stored-water volume of an aquifer because of excess pumping or other withdrawal of groundwater. This is sometimes termed mining of ground water and is associated with potential environmental degradation, including desertification and surface subsidence.

aquiherbosa   Herbaceous vegetation that is growing submerged in water. See HERBOSA.

aquitard   Any type of rock or unconsolidated material that restricts the flow of groundwater from an adjacent aquifer. Compare AQUICLUDE.

Ar   Chemical symbol for the element argon. See ARGON.

arable land   Land that is suitable for plowing and growing crops.

Araceae   The arum family of monocot angiosperms. The small, unisexual flowers of this family are arranged on a stalk known as a spadix, subtended by a leafy bract known as a spathe. Fruits are berries. Calla *(Zantedeschia)*, skunk-cabbage *(Symplocaarpus)*, philodendron *(Philodendron)*, water-lettuce *(Pistia)*, Jack-in-the-pulpit *(Arisaema)*, and dumb cane *(Dieffenbachia)* are members of this family. See ANGIOSPERMOPHYTA.

Arachnida   This well-known class of predatory chelicerates includes the spiders. The cephalothorax, which may bear up to eight eyes and has four pairs of walking legs, is joined to the abdomen by a stalk. The chelicerae are jawlike structures provided with fangs attached to venom glands. Gill-like and tracheal respiratory organs occur on the abdomen, as do spinnerets and silk glands. Silk is widely used for a variety purposes including defense, trapping of prey, egg cases, etc. Spiders are distributed worldwide in some 15,000 species. See CHELICERATA, TARANTULAS.

aragonite   A polymorph of the mineral calcite; forms orthorhombic crystals in sedimentary environments. Aragonite is lacking in cleavage and has a higher specific gravity than calcite. Compare CALCITE. See APPENDIX p. 620.

aragonite needles   Clusters of the mineral aragonite occurring in acicular, or needle-shaped, crystals.

arboreal   Referring to, or resembling, trees; also, in the case of animals, living in trees.

arboreal lichen   A lichen that lives on tree trunks or other parts of trees. See MYCOPHYCOPHYTA.

arboretum   A place where many different trees and shrubs are grown for display and scientific study.

arboriculture   Cultivation of trees and shrubs, especially for ornamental use; also, the study of such cultivation methods. Compare HORTICULTURE.

arbovirus   Short for arthropod-borne virus. Any of a large group of RNA-containing viruses that can infect both arthropods and humans and are, therefore, often transmitted to humans by blood-sucking insects (mosquitoes, ticks). Viral and equine encephalitis and yellow fever are caused by arboviruses.

Archaean   The second of three eons in Precambrian geologic time. The Archean eon lasted from approximately 4000 to 2500 million years ago.

Archaebacteria   One of the most unusual phyla in the kingdom Monera, the Archaebacteria include forms which depend on sulfur compounds for energy, methanogens which convert carbon dioxide to hydrogen and methane, and forms which live in hypersaline waters. The composition of their cell walls differ from those of other bacteria and, although they are prokaryotes, elements of their biochemistry and ribonucleic acid (RNA) subunits are more eukaryotic in nature. Because of their peculiarities, some scientists choose to place the archaebacteria in a separate kingdom. See MONERA, PROKARYOTE, EUKARYOTE.

archegonium   The structure producing female gametes in lower plants, including fungi, mosses, liverworts, and ferns; the female organ of most gymnosperms is a modified archegonium. It is analogous to the pistil or gynoecium of flowering plants, and is somewhat similar in shape (i.e., flask-shaped). See ANTHERIDIUM, PISTIL.

archibenthic   Dwelling principally in the deep part of the ocean from about 200 to 1000 m.

Archimedes' principle   A floating object displaces an amount of liquid whose weight is equal to the weight of the object. This relationship is useful for determining an object's specific gravity. It is named for a Greek mathematician and physicist of the third century B.C.

archipelago   A chain of many islands

sharing a common mode of origin.

Archosauria   An important subclass of reptiles including extinct flying reptiles and dinosaurs as well as crocodiles. The skull has two temporal openings behind each eye, the hind legs are larger than the forelegs, and the tail is robust; many forms were bipedal. Some authorities consider the birds to be archosaurs. See DINO-SAURS, CROCODILIA.

arctic   Relating to the region north of the Arctic Circle at 66°32' north latitude, or near the North Pole.

arctic brown earth   A well-weathered and leached inceptisol soil type of periglacial environments.

arctic circle   A circle of latitude paralleling the earth's equator and 23 degrees 28 minutes from the North Pole. It forms the southern boundary of the arctic or frigid zone.

Arctic floral region   One of the phytogeographic regions into which the Holarctic realm is divided according to similarities between plants; it consists of northern (arctic) regions of Siberia, Alaska, Scandinavia, and Labrador. See HOLARCTIC REALM, PHYTOGEOGRAPHY.

arctic sea smoke   Steam fog that forms in arctic regions where the air flowing over ice-covered land masses can be over 20°C colder than the open ocean. As air moves from land to water, intense steaming can cause large expanses of fog. See STEAM FOG.

Arctogea   One of the primary zoogeographical divisions of the earth's surface according to the distribution of animal life (fauna). Arctogea encompasses the Palaearctic, Nearctic, Oriental, and Ethiopian zoogeographic regions. See NEOGEA, NOTOGEA.

arcuate delta   A delta with the frontal lobe recurved into a shape like a bow. See DELTA.

Ardeidae   The family of birds, including herons and bitterns, that belong to the order Ciconiiformes. They usually wade in shallow water while seeking prey. Because of their skeletal structure, the neck is often held in an S-shaped position especially when flying. This characteristic readily distinguishes the herons from the storks and cranes, which fly with their necks extended. See CICONIIFORMES.

Area of Outstanding Natural Beauty (AONB) Land use designation widely used in England to protect a special area.

area source   A small-scale generator of air pollution, defined in the Clean Air Act as, "any stationary source of hazardous air pollutants that is not a major source, i.e., emits less than 25 tons per year of hazardous air pollutants." Compare POINT SOURCES, NONPOINT SOURCES.

area (strip) mining   See STRIP MINING.

Arecaceae   The palm family of the monocot angiosperms, sometimes referred to as the Palmaceae. These semi-woody tropical trees and shrubs have large, fibrous leaves that form a crown at the top of the plant. The inflorescence is large, compound, and subtended by a large bract. Flowers have similar petals and sepals, arranged in two whorls of three. There are usually six stamens and three carpels. Coconut (*Cocos*), date palm (*Phoenix*), palmetto (*Sabal*), saw palmetto (*Serenoa*), California fan plam (*Washingtonia*), and royal palm (*Roystonea*) are members of this family. See ANGIOSPERMOPHYTA.

arenaceous rock   A clastic sedimentary rock that has a particle size ranging between 0.5 and 2.0 mm; examples are sandstone,

arkose, or graywacke.

arenite   A type of sandstone having less than 15% of the rock composed of a mud matrix. Compare GRAYWACKE.

areolation   The subdivision of a biologic structure into distinct compartments or areas (for example, the division of areas between the veins of a leaf).

arete   A narrow, sharp-topped ridge forming as glacial erosion extends the sides of two U-shaped valleys until they meet in the middle. See U-SHAPED VALLEY.

Argentinean floral region   One of the phytogeographic regions into which the Austral realm is divided according to similarities between plants; it consists of Argentina, Paraguay, southern Chili, the Faulklands, and nearby islands. See AUSTRAL REALM, PHYTOGEOGRAPHY.

argentite   See ACANTHITE.

argillaceous   A term applied to rocks or sediments that are composed of more than 50% clay and silt-sized particles.

argillic B soil horizon   An illuvial subsoil horizon characterized by the accumulation of silicate clay as particles or coatings on soil clasts. See ILLUVIATION.

argillite   A compact, hardened, nonfissile argillaceous rock. Compare SHALE.

arginine $H_2N\text{-}C(NH)NH(CH_2)_3\text{-}CH(NH_2)COOH$, an amino acid that makes up plant and animal proteins. It is considered an essential amino acid because humans must get it from their diets as their bodies cannot manufacture it.

argon   (Ar) A colorless, odorless, gaseous element that makes up about 1 percent of earth's atmosphere. Argon is inert (does not combine with other elements); it is the least reactive of its group in the periodic table, the largely inert noble gases. It has an atomic number of 18 and an atomic weight of 39.95. The gas is used in fluorescent light bulbs and was formerly used in radio tubes.

arid Dry, parched, or deficient in moisture. Various definitions include: 1) less than 250 mm (10 in.) of rainfall per annum (USA); 2) insufficient rainfall to support vegetation in any quantity; 3) insufficient rainfall to support agriculture without irrigation; 4) where total evaporation exceeds actual precipitation.

aridisol Any member of a soil order characterized by accumulations of gypsum or lime, with little organic material, and salt layers with concentrations of calcium (Ca), magnesium (Mg), and potassium (K) ions; typically forms in arid climates.

arid zone A dry climatic region in which annual precipitation averages less than 10 inches.

aril An outer covering surrounding the seed coat of some seeds, usually developing from the ovule stalk. It may be fleshy or hairy and usually functions to attract birds or other animals, which act as dispersing agents.

Aristotle's lantern A structure that occurs in the mouths of sea urchins (class Echinoidea). It is composed of hard plates and muscles that support teeth and enclose part of the esophagus.

arithmetic growth A pattern in which increases occur at a constant rate over time. Over a given period, the same increment is added to the previous total (of an entity such as population), giving a pattern such as 1, 2, 3, 4 or 1, 3, 5, 7. Compare GEOMETRIC GROWTH, LOGARITHMIC, J-SHAPED CURVE.

arithmetic mean See MEAN.

arkose A variety of quartz sandstone that contains at least 25% of feldspar minerals in the rock matrix. Arkose usually forms in close proximity to a granitic source area.

armadillos Medium- to small-sized nocturnal mammals of the order Edentata, with plates of bone overlaid by horn covering the upper surfaces of the head, limbs, and body. They range from southern North America throughout the neotropical region. Some species live mainly on insects, whereas others are omnivorous. See EUTHERIA.

aromatic Belonging to a group of hydrocarbon compounds that contain (4n + 2) electons (n is any integer) in a connected system of p-orbitals. Benzene (C_6H_6), toluene (C_7H_8), and naphthalene (C_8H_8) are representative aromatic compounds. They are called aromatic because a number of the earlier ones discovered (such as turpentine, wintergreen oil) have strong odors, although many odorless aromatic compounds are now known.

Arrhenius acid A substance that dissociates in water to produce protons (H^+). Most common acids, such as hydrochloric (HCl) and nitric (HNO_3), are Arrheinus acids.

Arrhenius base A substance that dissociates in water to produce hydroxide ions (OH^-). Most Arrheinus bases are hydroxides of metals, such as sodium hydroxide (NaOH).

arrhenotoky A form of parthenogenesis in which males develop from unfertilized, haploid eggs and therefore have only one set of chromosomes, whereas females are diploid.

arroyo A steep-sided, flat-bottomed gully cut through cohesive sediment deposits in arid regions.

arsenic   (As) A silvery, nonmetallic element that is often found as an impurity in metals, although it also occurs by itself. Atomic number of 33 and atomic weight 74.92. Arsenic is used in making alloys (such as lead shot), solder, semiconductors, dyes, and medicines. Arsenic is very poisonous and is used in rodent poisons and some herbicides.

arsenopyrite   A mineral of iron arsenic sulfide found in hydrothermal veins, often in association with gold and tungsten. The mineral is a principal commercial source of arsenic.

artery   A vessel that conducts blood from the heart to all parts of the body. Compare VEIN.

artesian   The term applied to water that rises under natural pressure above the aquifer that contains it.

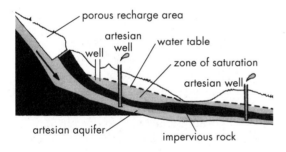

porous recharge area
artesian well   water table
well   artesian well
zone of saturation
artesian well
artesian aquifer   impervious rock

The depth of a well often determines whether it is an artesian well. Note that the wells on the right are deep enough to reach the artesian aquifer, while the left well only reaches the zone of saturation under the water table.

artesian aquifer   An aquifer that overflows onto the land surface or through well holes because of natural pressure built up within the confines of the aquifer.

artesian basin   An artesian aquifer formed within a syncline.

artesian well   A well through which the water flows up, by natural pressure, to a level above that of the aquifer.

Arthropoda   Arguably the most successful and ecologically important animal phylum. The number of arthropod species has been estimated to be between 700,000 and 2 million and they are adapted to every type of habitat and food source. The arthropod body is strongly segmented and is supported by a chitinous exoskeleton that is periodically shed as animals grow and mature. The tubular, jointed legs, which occur in pairs on body segments and give the phylum its name, are the arthropod's most distinguishing feature, not easily mistaken. The number of legs varies with the arthropod group. A well-developed head often bears paired antennae and mouthparts, both of which evolved from legs. Because of the great diversity among arthropods and incompleteness of the fossil record, there is considerable controversy over their classification. Three living subphyla are recognized, Chelicerata (horseshoe-crabs, scorpions, spiders, mites, etc.), Crustacea (varied small groups, barnacles, crabs, shrimp, lobsters, etc.) and Uniramia (myriapods, insects). Some authorities now consider these subphyla to be separate phyla. See CHELICERATA, CRUSTACEA, UNIRAMIA, APPENDIX p. 613.

artificial insemination   (AI) Introduction of semen into the vagina or cervix by a method other than copulation, such as with a syringe. Artificial insemination is used in selective breeding of animals.

artificial recharge   The process of augmenting the natural water supply to an aquifer by pumping water in through boreholes or into catchment basins that drain into the aquifer.

artificial reef   A structure built by humans

that rises above the ocean floor. It creates an artificial habitat to attract shallow marine and benthic organisms.

artificial selection   Using human intervention to steer the processes of evolutionary selection by crossing individuals with specific characteristics in order to produce desired characteristics in the next generation. Artificial selection is used to develop new colors in garden flowers and to breed cattle for increased milk production. Also called selective breeding. See SELECTION, NATURAL SELECTION.

Artiodactyla   The mammalian order of even-toed ungulates. They are characterized by the reduction of all digits except the third and fourth as well as by the teeth and tongue, which are highly modified for cropping and grinding. With the exception of pigs and hippopotamuses, all are ruminant animals. Approximately 170 species exist (e.g., pigs, hippos, camels, llamas, deer, giraffes, and bovids). See RUMINANT, BOVIDAE.

As   Chemical symbol for the element arsenic. See ARSENIC.

asbestos   A fibrous variety of the mineral actinolite, or any fibrous variety of the amphibole and serpentine groups of silicate minerals. In recent years, asbestos has been associated with a variety of environmental problems related to lung disease that may result from contact with the fibrous minerals.

asbestosis   Severe lung damage due to inhalation of asbestos fibers. It often leads to mesothelioma, an otherwise rare form of lung cancer. See MESOTHELIOMA, PNEUMOCONIOSIS.

A-scale sound level   A modification of the decibel scale that approximates the sensitivity of the human ear. Measurements on

this scale are recorded in dBA and are used in identifying noise levels for controlling noise pollution. See DECIBEL, DB.

Ascaris   Genus of nematode worms that is responsible for large, roundworm infection in humans and other animals. *Ascaris lumbricoides* is the major human pathogen, and its incidence of infection in some parts of the world is over 90 percent of the population. The life cycle is direct, requiring no intermediate host. Adult worms lay eggs in the human intestine that are then expelled with the feces and contaminate soils. Where there is little sanitation, these eggs may be ingested by humans. They hatch into larvae that penetrate the intestinal wall into the bloodstream. From there, larvae migrate through the lungs to the esophagous and become mature in the intestine. Fever, pnuemonia, and various intestinal disorders can result, particularly in children. See NEMATODA.

Aschelminthes   A phylum of convenience which is no longer widely in use. It includes a conglomorate of worm-like animal groups whose phylogenetic affinities to each other are unclear. The nematode worms are included in this phylum, but many authorities now prefer to place coherent groups like the nematodes into separate phyla. See NEMATODA.

Asclepiadaceae   The milkweed family of the angiospermophyta. These dicots are usually herbaceous with a milky sap rich in glycosides. The flowers tend to be highly modified and fused with ornate, erect appendages or hoods. Pollen is grouped into sticky blobs (pollinia) that are picked up en masse by pollinating insects. The seeds tend to have a long tuft of hairs at one end. Milkweeds (*Asclepias*), silk-vine (*Periploca*), carnosa wax plant (*Hoya*), and carrion flower (*Stapelia*) are members of this family. See ANGIOSPERMOPHYTA.

ascomycetes Sac fungi, members of a large group of fungi (phylum Ascomycota) that includes yeasts, blue molds, mildews, and truffles. See ASCOMYCOTA.

Ascomycota A phylum of the kingdom Fungi that is characterized by organisms that reproduce by forming a reproductive structure known as an ascus. The ascus is a sac-like structure containing ascospores, haploid cells formed through the conjugation of different mating types producing a diploid nuculeus that then undergoes meiosis. All members of this phylum can also reproduce asexually. The ascomycetes include many cellulose decomposing fungi, mycorrhizal fungi, and economically important fungi such as yeasts (*Saccharomyces*), morels (*Morchella*), truffles (*Tuber*), and the fungus used extensively in genetic research *Neurospera*. See FUNGI.

ascorbic acid Another name for vitamin C (either naturally occurring or synthetic). See VITAMIN C.

aseptic Sterile, free of infection or germs. Home canners sterilize jars before filling them with vegetables, jams, etc. because they want to keep the food they are preserving aseptic. Clothes, rooms, and instruments for surgery are sterilized to create an aseptic environment.

asexual Having no gender; lacking functioning sexual organs.

asexual reproduction Reproduction without sexual fusion, such as by cloning or, in plants, by sending out runners. See APOMIXIS, VEGETATIVE REPRODUCTION.

ash Mineral residue remaining after the complete burning of organic substances; it is usually gray to black in color. See FLY ASH.

ash-fall tuff See TUFF.

ash flow A gas-charged mixture of pyroclastic debris and molten lava ejected in a volcanic eruption. Ash flows are heavier than air and travel downhill at high speeds. From an environmental standpoint, these are extremely dangerous events that have killed many thousands of people, most recently at Mount Unzen, Japan. See NUÉES ARDENTES.

Ashgill epoch A part of the Ordovician subdivision of the Paleozoic Era.

ash, volcanic See VOLCANIC ASH.

asparagine $NH_2\text{-}CO\text{-}CH_2CH(NH_2)$ COOH, a common amino acid that is synthesized by the human body to make proteins. Asparagine also occurs in plants and accumulates in plants infected by some parasites such as rusts.

aspartic acid $HOOC\cdot CH_2\cdot CH(NH_2)\cdot$ COOH. An amino acid. A constituent of proteins, it can be extracted from asparagus, beets, and sugar cane.

aspect diversity Variations in the external form or appearance of species that live in the same habitat and are eaten by hunting predators who choose prey based on visual appearance. Contrast APOSTATIC SELECTION.

asphyxiants Substances such as carbon monoxide that poison by suffocation. Asphyxiants interfere with oxygen uptake and cause asphyxia, a state in which there is too little oxygen in the blood.

assimilation Converting food substances into growth, reproduction, or tissue repair. Assimilation refers both to those metabolic processes, such as photosynthesis, that plants use to convert nonliving substances into protoplasm, and to the conversion of nutrients that have reached an animal's bloodstream after digestion into proto-plasm.

assimilation efficiency 1) The ratio of energy that is absorbed into the bloodstream of an organism to the energy (i.e., food ingested). 2) The ratio of light absorbed by the chloroplasts of a plant to that absorbed at the leaf surface. Compare NET PRODUCTION EFFICIENCY, RESPIRATION EFFICIENCY, ECOLOGICAL GROWTH EFFICIENCY, TISSUE GROWTH EFFICIENCY.

assimilative capacity The ability of a body of water to cleanse itself of organic pollutants, as by uptake into the aquatic biomass.

association 1) A stable grouping of two or more plant species that characterize or dominate a type of biotic community. This term is not used as frequently now in the literature as it was 50 years ago. 2) In the monoclimax theory, a subdivision of a formation, defined by one or several dominant species of climax communities. Compare BIOTIC COMMUNITY, CONSOCIATION. Contrast INDIVIDUALISTIC HYPOTHESIS.

assortative mating Fertilization of plants caused by pollinating insects preferring certain flower types or colors.

astatine (At) A radioactive element that occurs in at least 20 isotopic forms. It is the heaviest in the halogen group, with an atomic weight of 210 and an atomic number of 85. Astatine is produced naturally by the decay of uranium and thorium isotopes.

Asteraceae The aster or sunflower family of the dicot angiosperms. The largest plant family, with over 20,000 species, its representatives have successfully spread to a wide variety of habitats worldwide. The infloresences of members of this family are composed of florets arranged in a dense head with a common receptacle that is surrounded by involucral bracts known as phyllaries. At the center of the head are usually disc or tubular flowers with an inferior ovary and five united petals surrounding the five anthers that form a ring around the two-branched style. The calyx has been reduced to hairs or scaly bracts called pappus. At the periphery of the head are usually ray or ligulate flowers that are similar to the disc flowers except that they are flattened and straplike. The ray flowers are sterile in some members of the family. The family is subdivided into the *Asteroideae*, plants having at least some disc flowers present, and the *Cichoriodeae*, plants having only ray flowers. The *Asteroideae* includes sunflowers (*Helianthus*), tarweeds (*Madia*), ragweeds (*Ambrosia*), sneezeweeds (*Helenium*), asters (*Aster*), ironweed (*Vernonia*), thistles (*Cirsium* and *Carduus*), and burdocks (*Arctium*) among many others. The *Cichoriodeae* is a smaller subfamily but includes familiar plants such as dandelions (*Taraxacum*), chicory (*Cichorium*), hawkweeds (*Heiracium*), and sow thistles (*Sonchus*). See ANGIOSPERMOPHYTA.

asteroid An interplanetary object, larger than a meteoroid but smaller than a planet, in orbit around the sun. Also called a minor planet. Most asteroid orbits lie between Mars and Jupiter. Compare PLANET.

Asteroidea Starfish form a class of free-living, bethnic echinoderms with flattened or domed bodies and five or more arms bearing well-developed ambulacral grooves that radiate from the ventral mouth. Starfish or seastars are not only common on seashores but are also widely distributed elsewhere in the sea, including arctic waters, coral reefs, and oceanic depths. Most are active predators that feed by extruding their stomachs over the surface of their prey. They have commercial impact because they foul nets and prey on shellfish beds. The crown-of-thorns seastar, *Acanthaster planci*, has gained ecological notoriety in recent years for its periodic population eruptions and resulting devastating predation of coral polyps in the tropical Pacific. See ECHINODERMATA.

asthenosphere   The upper zone of the earth's mantle that immediately underlies the lithosphere. The material of the asthenosphere deforms plastically under prolonged strain and permits movement of tectonic plates.

asthma   A chronic condition characterized by episodes during which breathing is impaired by both a narrowing of bronchial tubes and production of thick mucous. It is usually caused by allergies and aggravated by air pollution.

astronomical unit   (A.U.) Unit of measurement equal to the average distance between the earth and the sun (1.496×10^8km or about 93 million miles). It is used to express distances in the solar system.

asymmetrical fold   A bend in layered rock in which the angle formed between one limb and the axis of the fold is different from the angle between the opposite limb and the fold axis. Compare SYMMETRICAL FOLD.

At   Chemical symbol for the element astatine. See ASTATINE.

Athens Treaty on Land-Based Sources of Pollution   An international agreement aiming to reduce pollution of the Mediterranean Sea. It was drafted by 16 countries in 1980 but did not take effect until it was ratified by six of those countries in 1983.

Atlantic North American floral region   One of the phytogeographic regions into which the Holarctic realm is divided according to similarities between plants; it consists of North America below the Great Lakes, east of the Rocky Mountains, to the Gulf Coast. See HOLARCTIC REALM, PHYTOGEOGRAPHY.

atmosphere   The gas layer surrounding a planet. Excluding the water vapor, the earth's atmosphere consists (by volume) of 78 percent nitrogen, 21 percent oxygen, .9 percent argon, .035 percent carbon dioxide (this level is rising), and minute quantities of other gases (including neon, krypton, helium, and air pollutants). Water vapor may constitute up to 3 percent, depending on the relative humidity and temperature of the atmosphere.

atmospheric deposition   The settling out of particulate matter (and also gases) onto the ground surface from the air. See DRY DEPOSITION, WET DEPOSITION. .

atmospheric dispersion   The natural convection and diffusion occurring within the air in which air pollutants and other substances are distributed throughout an air mass.

atmospheric inversion   A reverse of normal atmospheric conditions in which a layer of warm, less-dense air traps a layer of cool, dense air at the earth's surface. Because the layer of warm air prevents normal weather circulation, air pollutants cannot disperse and build up within the cool air layer. Cities located in basins, such as Washington, DC, and Los Angeles, CA, are prone to sustained inversions; when these persist for several days, dangerous levels of air pollutants can accumulate.

atmospheric pressure   The effect exerted by the weight of the air above a point on the earth's surface. The standard value at sea level is 101,325 pascals or 1.01×10^5 newtons per square meter. Variations in atmospheric pressure are measured with a barometer. At sea level, the atmospheric pressure results in a column of mercury that is 760 mm high; as altitude increases, the atmospheric pressure (and the height of a corresponding column of mercury) decreases.

atmospheric tides   Patterns of changing atmospheric pressure resulting from changes in temperature caused by the rotation of the earth. The changing atmospheric pressure causes air masses to move in ways similar to tides.

atoll   A circular reef enclosing a shallow lagoon.

atom   The smallest unit of any chemical element that can take part in a chemical reaction without undergoing a permanent change. The atoms of each element are different by virtue of different numbers of protons in the nucleus, although they share the same basic structure of a nucleus of neutrons and protons surrounded by orbiting electrons. Atoms of elements can be combined to make molecules.

atomic absorption spectrum   A plot of wavelength versus intensity of electromagnetic radiation absorbed by a given atom. Energy is absorbed by atoms at characteristic wavelengths that depend on the electronic structure of the atom. The energy absorbed produces electronic transitions from a lower energy level to a higher energy level within the atom. See ABSORPTION SPECTRUM.

atomic emission spectrum   A plot of wavelength versus intensity of electromagnetic radiation emitted by an atom that has been electronically excited by absorption

of energy. Energy is emitted by atoms at characteristic wavelengths that depend on the electronic structure of the atom. The energy emitted arises from electronic transitions from a higher energy level to a lower energy level within the atom. See ABSORPTION SPECTRUM.

atomic energy   Formerly, another term for nuclear energy. See NUCLEAR ENERGY.

Atomic Energy Commission   (AEC) A federal agency established by the 1946 Atomic Energy Act and responsible for promotion and regulation of nuclear power. In 1974, its duties were divided between the Nuclear Regulatory Commission (NRC) and the Energy Research and Development Administration (ERDA), which since 1977 has been under the auspices of the Department of Energy.

atomic mass unit   (amu) A unit calculated as one-twelfth of the mass of carbon 12 (the most abundant isotope of carbon), equal to 1.660×10^{-27} kg. Also called the dalton, it is used to express atomic mass.

atomic nucleus   The central structure within an atom, containing the positively charged protons and the neutral neutrons. The atomic nucleus contains almost all of the atom's mass; very little is contained in the electrons orbiting the atomic nucleus.

atomic number   The number of protons contained in the nucleus of an atom. This value is used to distinguish elements from each other because each element has a unique number of protons and thus a unique atomic number. Elements are listed by atomic number in the periodic table.

atomic pile   Another term for a nuclear reactor that uses graphite to moderate the speed of atom-splitting nuclei. The name comes from the first experimental self-sustaining chain reaction, which was contained inside a pile of graphite blocks set up underneath the University of Chicago's stadium in 1942. See NUCLEAR REACTOR.

atomic weight   Relative atomic mass; the ratio between the weight of an atom of a chemical element to $\frac{1}{12}$ the weight of the most abundant isotope of carbon (which is set as exactly 12). For elements with more than one isotope, atomic weight is the weighted average of the values for the different isotopes.

atomize   To make into a fine spray. Aerosol cans discharge their contents in atomized form.

ATP   Abbreviation for adenosine triphosphate. See ADENOSINE TRIPHOSPHATE.

ATPase   Enzyme that converts ATP to ADP. See ADENOSINE DIPHOSPHATE.

attenuation   Reduction of the capacity of a pathogen to cause disease. Attenuation is accomplished by heating, treating with chemicals, or growing under unfavorable conditions. Attenuation is used to produce vaccinations that create immunity in recipients without causing disease. Polio, measles, and yellow fever vaccines are prepared through attenuation.

Atterberg limits   The liquid and plastic limits of a soil as defined from standard soil tests.

attractant   A natural or synthetic substance whose scent lures insects or animals. See SEX ATTRACTANT, PHEROMONE.

A.U., AU   1) Abbreviation for animal unit. 2) Abbreviation for astronomical unit. See ANIMAL UNIT, ASTRONOMICAL UNIT.

Au   Chemical symbol for the element gold (from the Latin for gold, *aurum*). See GOLD.

audio frequency   Vibrations falling within the range of human hearing, which encompasses sound waves from 15,000 to 20,000 cycles per second (hertz). See SOUND WAVES.

audiogram   Chart showing hearing as a function of frequency. Audiograms plot the results obtained through use of audiometers.

audiometer   Instrument used in testing hearing ability or loss. It is essentially a controllable tone generator. By generating a series of sounds of known frequency and intensity, audiometers enable thresholds for different frequencies to be measured. See AUDIOGRAM.

aufeis   A sheet of ice that forms on the surface of a river flood plain where drainage is impounded by sand bars or shoals.

aufwuchs   Another term for periphyton, a group of plants and animals attached to or moving about on submerged surfaces. See PERIPHYTON.

augite   A greenish black variety of the mineral pyroxene that typically forms monoclinic crystals in igneous rocks. See APPENDIX p. 620.

aureole   A disk or colored rings around the moon (and, rarely, the sun) created when dust, droplets, or ice crystals in the atmosphere diffract the light as it passes through earth's atmosphere (also called a corona).

auricularia larva   A cylindrical larval form of the marine invertebrates of the classes Holothuroidea and Asteroidea.

auriferous   A substance that contains gold.

aurora   Glowing bands of light visible in the night sky. The changing shapes are often streamers or curtainlike and may be red, green, or white. They are caused by charged particles in the upper atmosphere that stream from the sun and are directed toward the poles by the earth's magnetic field. As a result, the phenomenon is most often observed at high latitudes. Plural is aurorae. See AURORA AUSTRALIS, AURORA BOREALIS.

aurora australis   Southern Hemisphere aurora. See AURORA.

aurora borealis   Northern Hemisphere aurora, the northern lights. See AURORA.

auroral oval   A band around the earth's magnetic poles in which aurorae occur. The region is oval in shape and it changes in size in response to changes in the earth's magnetic field. See AURORA.

Australasian region   One of the primary biogeographical realms into which the earth is divided according to the distribution of animal life (fauna). It includes Australia, New Zealand, Tasmania, New Guinea, and part of the Malay archipelago. Also called Notogea. See ZOOGEOGRAPHY.

Australian floral region   One of the phytogeographic regions into which the Austral realm is divided according to similarities between plants; it covers all of Australia. See AUSTRAL REALM, PHYTOGEOGRAPHY.

Austral realm   One of the phytogeographical realms into which the earth is divided according to the distribution of plant life (flora). The Austral or southern realm is divided into five floral regions: the Argentinean floral region, Australian floral region, New Zealand floral region, South African floral region, and South Oceanic floral region. See FLORAL REALM, PHYTOGEOGRAPHY.

autecology   The ecological study of an individual organism or species and its relationship to its environment. Compare SYNECOLOGY.

authigenic sediment   A sediment having constituent minerals that formed during or soon after deposition.

autochthonous material   A substance that originates in the place where it is found, such as a rock formation or organic matter produced by the photosynthesis of plants within a particular ecosystem under consideration. Contrast ALLOCHTHONOUS MATERIAL.

autocidal control   The release of sterile or genetically altered individuals into wild populations as a means of pest control. Also called sterile male technique.

autoclave   1) Apparatus, usually a strong steel container, using steam under high pressure and temperatures of approximately 250°F (121°C) for sterilization. Medical instruments are sterilized in autoclaves. 2) To sterilize under high-pressure steam, as in an autoclave.

autocopraphagy   An organism's feeding on its own feces. Betsy beetles (*Popilius*) reingest their own fecal pellets after they have been nutritionally enhanced by fungi; rabbits also demonstrate autocopraphagy. Also called refection. Compare COPROPHAGY.

autocorrelation   The expression of the degree to which the value of a signal of a continuous wave form or cyclical phenomenon is determined by the values of the signal at previous times.

autoecious parasite   Parasites, such as some species of fungi, that complete all stages of their life cycles on a single host. Compare HETEROECIOUS PARASITE.

autogamy   Self-fertilization. Autogamy can occur in Paramecium (a protozoa), but more commonly occurs in some plants, in which a flower can be successfully pollinated with its own pollen (self-pollination). See HETEROGAMY.

autogenetic plankton   Species of plankton that naturally occur in an area, living and breeding there, as opposed to allogenetic plankton, those carried into an area by currents but normally existing elsewhere.

autogenic   1) Referring to changes that originate from within a system. 2) Referring to a substance that is self-generated or produced within an organism, such as an autogenic toxin or a vaccine produced from bacteria cultured from the patient who is to receive the vaccine. Also called autogenous. See AUTOGENIC SUCCESSION, ENDOGENOUS.

autogenic succession   A gradual change in a community of plants and of animals that results from environmental changes caused by the members of a biotic community. An example is the decrease in sunlight at ground level in deciduous forests, caused by increased production of tree leaf surfaces as the forests develop in abandoned agricultural fields. Contrast ALLOGENIC SUCCESSION.

autoimmunity   Condition in which an organism produces antibodies that attack its own cells and tissues. Rheumatoid arthritis is a disease caused by autoimmunity.

autoionization   The phenomenon responsible for the normal state of water, in which a very small fraction of the water molecules spontaneously dissociate into hydrogen (H^+) and hydroxyl (OH^-) ions.

autolysis   Internal breakdown of dead tissues of a plant or an animal caused by the action of enzymes present within its own cells; self-digestion. Autolysis is part of normal tissue replacement in organisms. See LYSOSOMES.

autonomic nervous system   (ANS) In vertebrates, the part of the peripheral nervous system controlling smooth muscles and involuntary actions, including the main organs (heart, lungs), digestion, and glandular activity. The ANS is divided into the sympathetic nervous system and the parasympathetic nervous system. See SYMPATHETIC NERVOUS SYSTEM, PARASYMPATHETIC NERVOUS SYSTEM. Compare CENTRAL NERVOUS SYSTEM.

autopelagic plankton   Species of plankton that live at the surface of a body of water.

autopolyploid   An organism containing three or more sets of chromosomes, all of which come from the same species and are, therefore, essentially identical genomes. Many varieties of cultivated flowers are autopolyploids. Autopolyploids with three sets of chromosomes are autotripliods; with four, they are autotetraploids. Compare ALLOPOLYPLOID.

autosome   Any of the paired chromosomes other than the sex-determining (X and Y) chromosomes.

autotroph   An organism that can manufacture its own food from inorganic compounds through photosynthesis or chemosynthesis. Autotrophs include green plants, algae, some Protista, and some bacteria. Also called primary producer. Some autotrophic bacteria use inorganic compounds, such as hydrogen sulfide and ammonium, as their energy source in chemosynthesis. Contrast HETEROTROPH, AUXOTROPH.

autotrophic succession   Ecosystem development in which the total combined rate of photosynthesis (or primary production) is greater than the combined rate of respiration for the members of the ecological community. Autotrophic succession is characterized by the production of autochthonous organic matter (i.e., that formed within the ecosystem).

autumnal equinox   Technically, the point on the sun's path where it crosses the celestial equator when moving from north to south. Also used colloquially for the date the sun reaches this point, on or near September 23 each year, which marks the beginning of Northern Hemisphere autumn. Compare VERNAL EQUINOX.

auxin   Any of a group of plant hormones regulating growth and development. (When plural, it refers to the group; when singular, it may refer only to the most common member of the group, indole-3-acetic acid [IAA].) Natural auxins are synthesized in the young, actively growing leaves and shoots of plants; their effects include cell elongation and initiation of root formation. Some synthetic auxins (2,4-D and 2,4,5-T) are used as herbicides.

auxotroph   A mutant form of an organism that has lost its ability to synthesize one or more substances for its nutrition. Auxotrophic bacteria or fungi can only grow on a nutrient medium containing this substance or substances. See AUTOTROPH.

available concentration   The quantity of a mineral nutrient present in the soil that can be readily taken up and assimilated by growing plants.

available water   See FIELD CAPACITY.

avalanche   A rapid, downhill movement of geologic material, such as snow, ice, or rock, into a precipice. As more and more people venture into Alpine areas during the winter, avalanche occurrence is causing more property damage and loss of human life. See MASS WASTING.

aven   See POTHOLE.

Aves   A class of vertebrates, the birds, which are bipedal, possess feathers, and have forelimbs highly modified to form wings. They are endotherms and their feathers, which provide an excellent cover of insulation by trapping dead air, allow for survival over a wide temperature range. The keel, a prominent elevation of the sternum for the support of flight muscles, is reduced or missing in birds in which flightlessness has evolved. There are approximately 9000 living species. See BIPEDAL, ENDOTHERM, VERTEBRATA.

avicide   Any substance that kills birds.

Avogadro's number 6.023×10^{23}, equal to the number of molecules, ions, or atoms contained in a mole of substance, where a mole is defined as the number of atoms in exactly 12 grams of carbon 12, the most abundant isotope of carbon. Avogadro's number is used as a standard of comparison for measuring the number of particles of a substance rather than the mass of a substance. See ATOMIC MASS UNIT.

awn   A long bristle, especially those surrounding and extending beyond the seeds of grasses such as barley and oats.

axenic   Free of germs or free of other species, as in a pure culture of a single microorganism.

axial plane   An imaginary plane that cuts through an axis (as of a fold in a rock formation), dividing it as symmetrically as possible.

axillary bud   See LATERAL BUD.

axis (of a fold)   An imaginary line through the center of a bend in layered rock. The fold axis connects the points of maximum curvature along the bend.

axon   The single, long projection of a nerve cell, or neuron, that transmits impulses away from the main body of the cell. See NEURON.

axopodium   A pseudopodium of some species of protozoa that contains an axial filament. See ACTINOPODA.

azimuth   The horizontal angle between the sun and due north, measured clockwise.

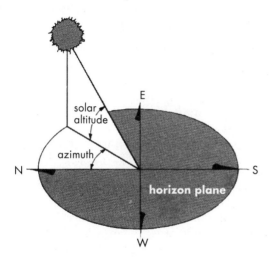

Compare with solar altitude, the vertical angle between the horizon and the sun.

azo dyes   A large group of synthetic dyes produced from coal tar; they are characterized by a double-bonded pair of nitrogen atoms (called the azo group). Many are used as food additives to improve or to change the colors of food.

Azolla   A genus of small, aquatic ferns commonly called water ferns. Water ferns have evolved a symbiotic relationship with nitrogen-fixing cyanobacteria, enabling them to be quite successful, especially in tropical waters.

azonal soil   Any soil lacking well-developed soil horizons. See ENTISOLS.

Azores high   Subtropical high-pressure air mass that forms above the northern Atlantic Ocean, named for the Azores islands found there. See ALEUTIAN LOW.

Azotobacter   Free-living, nitrogen-fixing bacteria; any member of a genus of aerobic bacteria (*Azotobacter*) that can assimilate atmospheric nitrogen and convert it into organic forms for their own use. Compare RHIZOBIA.

b

B   Chemical symbol for the element boron. See BORON.

b.p.   Abbreviation for boiling point. See BOILING POINT.

Ba   Chemical symbol for the element barium. See BARIUM.

baccate   Fleshy; berry-like in appearance.

Bacillariophyta   The diatom phylum of the kingdom Protista. These freshwater and marine algae accumulate silica on their two-part shells, called valves. The upper valve fits over the lower valve like the lid of a box. The phylum is subdivided into the Cetrales class with radially symmetrical members, such as *Thalassiosira* and *Melosira*, and with bilaterally symmetrical members of the class Pennales, such as *Fragilaria* and *Synedra*. Diatoms produce gametes through meiosis that fuse forming a zygote, but can also can reproduce vegetatively by mitotic division. See PROTISTA.

bacillus   1) Any rod-shaped bacterium, such as *Escherichia coli*, found in human and animal intestines, or *Clostridium tetani*, responsible for causing tetanus (lockjaw). 2) When capitalized, a genus of rod-shaped aerobic bacteria belonging to the family Bacillaceae. Most are not pathogenic, although *Bacillus anthracis* causes anthrax. See AEROBIC ENDOSPORE-FORMING BACTERIA.

back-arc basin   A region of thickened crust and extensional tectonics associated with a convergent tectonic boundary between a continental plate and an oceanic plate. The back-arc basin forms between a volcanic island arc and the continent; an example is the Sea of Japan. See ISLAND ARC.

back-arc upwelling   The diapiric upward movement of plutons above the subduction zone of a back-arc basin.

backcross   The mating of an offspring to one of its parents or to an individual of the same strain as one of the parents. Also, the product of such a mating.

backfill   Any deposit of material used to replace a material that has been excavated in a construction or mining operation.

background   Refers to existing environmental conditions or phenomena that potentially confuse attempts to study or quantify other physical phenomena. It is a metaphoric "noise level" existing in any system. Also called ambient or background level. See BACKGROUND CONCENTRATION, BACKGROUND RADIATION.

background concentration   Concentration of a substance usually present in the environment. Background concentration is measured in order to give a more accurate measurement of levels of the substance added by human activity (by subtracting existing levels from the added levels of the substance). See BACKGROUND RADIATION.

background radiation   1) Pre-existing levels of ionizing radiation, which must be taken into account when observing any specific source of radiation. Background radiation comes from a combination of different sources: Some is produced within the earth's crust by naturally occurring radioisotopes; radon releases radiation into the air; cosmic radiation reaches the earth's surface from outer space; our own bodies contain naturally occurring radio-isotopes such as carbon 14; human activities add radiation to the atmosphere.

2) Also called the primordial or cosmic background radiation, the constant low levels of radiation existing throughout the universe that are believed to be remnants from early in the universe's history. See RADON, BIG BANG THEORY.

backing wind   Wind whose direction is changing to counterclockwise (in either hemisphere on earth). Contrast VEERING WIND.

backscatter   Deflection by angles greater than 90°. Specifically, such deflecting of radiation or particles back toward the source.

backshore   The upper portion of a beach that is usually above the high waterline of spring tides. The backshore region is usually dry except when inundated by storm waves.

bacteria   A large group of prokaryotic microorganisms characterized by multiplying via fission or forming spores, and generally lacking chlorophyll or a distinct nucleus surrounded by a membrane. Bacteria are unicellular and often classified by shape and may be spherical, spiral, rodlike, or comma shaped. They may be saprophytic (bacteria are the major decomposers in most ecosystems), parasitic (such as those causing diseases such as pneumonia or typhoid fever), or autotrophic (hydrogen sulfide bacteria). See MONERA.

bacteriochlorophylls   Photosynthetic pigments found in some species of bacteria. They resemble and may be evolutionary precursors of chlorophyll and permit the bacteria to be autotrophs.

bacteriophage   Any of a group of viruses that infects and kills bacteria. These are found in blood, intestines, and other organs as well as in sewage. Often called simply phage.

badlands   A term applied to barren plateau regions that have been intricately dissected by erosion in V-shaped gullies.

bag filter   Another term for baghouse. See BAGHOUSE.

baghouse   Air pollution control device used for large-scale trapping of particulate matter. It consists of a series of up to 300 large bags (in the range of 1.5 x 10 ft) and a suction or blower system to drive the exhaust gases through the bags. Bags of natural fibers (cotton) are used for temperatures up to 90°C; for temperatures up to 200°C, synthetic materials such as nylon must be used. A baghouse can trap over 90 percent of all particulates. Also called a bag filter.

bag limit   The legal number of animals a hunter is allowed to kill in one day. Bag limits may be changed from year to year for some species, depending on habitat and other conditions. See CREEL LIMIT.

bajada   A sloping accumulation of unconsolidated rock debris against the foot of a mountain range. A bajada is typically formed in an arid climate by alluvial fans that enlarge and coalesce.

balanced chemical equation   One in which the numbers of atoms is the same before and after the reaction, though the elements have been rearranged. Aluminum plus hydrochloric acid produces aluminum chloride plus hydrogen gas. The balanced formula for the equation is: $2Al + 6HCl = 2 AlCl_3 + 3H_2$. The formula for a chemical equation must be balanced to follow the law of conservation of matter.

balanced polymorphism   The maintenance of two or more distinct forms (or genes) in a population in the same area over a sustained period of time. Sickle-cell anemia is a form of balanced polymorphism, as it continues to appear in populations where normal genes are also present. See POLYMORPHISM.

baleen   Bonelike plates that grow down from the upper palate of some species of whales. The baleen serves as a strainer for food that the whale takes in.

baling   A way to reduce the volume of solid waste by compacting it and binding it to resemble a huge hay bale.

ballast   The gravel-sized broken stone used to stabilize a roadway or embankment, such as along a railway grade. Compare RIPRAP.

ballistic separator   Device used in composting to remove inorganic matter, which does not decompose, from organic matter. See COMPOST.

ball lightning   Unusual form of lightning which takes the shape of a slowly moving sphere rather than a long jagged streak. Ball lightning has been reported to roll in through a window and along the floor, or to float in the air, before dissipating. See LIGHTNING.

balsatic lava   A molten volcanic material that becomes basalt when cooled. See BASALT.

band   A specific range. Usually a waveband, an energy range for containing electrons or a span of frequencies (wavelengths) of radiation, especially those frequencies used for a specific purpose such as the short-wave radio band.

band application   A method for applying pesticides at the base of plants, just above the soil line.

banded iron formation   A Precambrian age rock sequence showing alternating layers of reddish and then gray-colored hematite bands. The banded iron formations comprise the most commercially significant iron deposits on several continents.

banded ironstone   A rock of the banded iron formation.

banding   An alternating pattern of light and dark or of contrasting colors, as seen in rocks like gneiss or banded ironstone.

banner cloud   An unusual plume-like form of cloud extending from a mountain peak. Banner clouds can form when the air around the peak has a high moisture content and is cooled to the point of condensation by falling air pressure.

bar   An elongated accumulation of sand or gravel in a stream, lake, shallow bay, or surf cone.

barchan dune   A crescent-shaped sand dune having the ends of the arc lying in the downwind direction.

barite   A mineral of barium sulfate having orthorhombic crystals.

barium   (Ba) A soft metallic element with atomic number 56 and atomic weight 137.3. It has chemical properties similar to those of calcium. Barium does not exist by itself but occurs in often-poisonous compounds; it is used as a pesticide in paint compounds. Barium sulfate is used medically in x-ray diagnostics.

bark   1) The protective, often coarse outer layer of trunks, branches, and roots of trees and some other plants. Technically, the bark is all tissue located outside of the cambium. 2) The call of dogs, foxes, and related animals. See CAMBIUM.

barnacles   See CIRREPEDIA.

barn owls   The eleven species of barn owls belong to the family Tytonidae of the order Strigiformes. They differ from the typical owls by having a heart-shaped face, relatively small eyes, and long legs. They frequently nest in abandoned buildings, such as barns, where their screams and chirruping may give such structures a reputation of being haunted. See STRIGIFORMES.

baroclinic Meteorological term for air masses with horizontal temperature gradients, creating surfaces of equal pressure that do not correspond to surfaces of equal density. Contrast BAROTROPIC.

barograph Meteorological device consisting of a recording device attached to a barometer (usually an aneroid barometer). It automatically records changes in atmospheric pressure over a specific time period. See ANEROID BAROMETER, BAROMETER.

barometer Meteorological instrument that measures atmospheric pressure (the pressure exerted by the weight of air above a point on the earth's surface). Barometers are used both for predicting weather changes (by indicating whether higher- or lower-pressure air masses are moving in) and for determining height above sea level. Mercury barometers base their measurements on a column of mercury; they are more accurate but less convenient than aneroid barometers. A standard reading for such a column of mercury at sea level is 760 mm. See ANEROID BAROMETER.

barotropic Meteorological term for the phenomenon of no horizontal temperature gradient occurring at any level, resulting in air masses with equal pressure at their surfaces corresponding to surfaces of equal densities. Contrast BAROCLINIC.

barrens A level area with poor, usually sandy or serpentine soils that are poorly forested or unable to support normal vegetative cover, and that generally have low levels of productivity. Plants growing in barrens are usually much smaller and stunted in comparison to individuals grown on more fertile soils. Frequently, barrens are dominated by specialized groups of endemic plants. Pine barrens, serpentine barrens, and sand barrens are examples.

barrier island An island situated parallel to the coastline with a shallow lagoon separating the island from the mainland.

barrier reef A long ridge built up from the ocean floor by coral. It parallels the mainland but is separated from it by a deep lagoon. For example, the Great Barrier Reef lies off the Queensland Coast of Australia. Compare FRINGING REEF, ATOLL.

bar screen A device consisting of a frame that supports parallel steel bars. A bar screen is used to separate crushed rock into size classes.

basal application The application of pesticides on stems or trunks of plants just above the soil line.

basal area Measure of the area of a forest actually covered by the trunks of trees. It is the sum of the cross-sectional areas of all of the trees within the area measured, expressed in units of square meters per hectare (m^2/ha), or square feet per acre (ft^2/acre).

basal body A structure at the base of some cilia and flagella. The basal body undergoes division along with the nucleus and cell. Also called a blepharoplast.

basal metabolic rate The quantity of energy used and heat produced by an organism at complete rest, measured 12 hours after eating. It is expressed in kilocalories per hour (or milliwatts) per square meter of skin surface. It is usually calculated from the amount of oxygen consumed.

basal metabolism   The amount of energy required for the maintenance of an organism at rest, used as a standard for comparing the effects of different conditions on metabolism. It is measured by the basal metabolic rate.

basal slip (glaciers)   A type of glacial movement in which the entire ice mass slides over the bedrock surface. This motion is usually aided by a thin film of meltwater that develops under pressure from the weight of the glacier. Contrast PLASTIC FLOW.

basalt   A dark gray or black (mafic) extrusive igneous rock having an aphanitic texture. Pyroxene, calcium plagioclase, and olivine are the dominant minerals. Basalt often forms columns in outcrops due to columnar jointing.

base   A substance that forms hydroxyl (OH⁻) ions upon dissolving in water and can neutralize acids to form a salt plus water. Bases are characterized by their tendency to donate an electron pair to a bond (or gain a proton) in reactions. Bases, such as ammonia, can be recognized by slippery feel, usually bitter taste, high pH, and turning of litmus paper to blue. Bases are sometimes called alkalies. Compare ACID.

base ionization constant, K_b   The equilibrium constant for the dissociation of a particular base. It expresses the extent to which a base undergoes ionization in water to form hydroxide ions (OH-). The strongest bases undergo complete ionization, so the higher the value for K_b, the stronger the base. See ACID IONIZATION CONSTANT, EQUILIBRIUM CONSTANT.

base level   In geomorphology, the lowest level of erosion achieved by stream action. Sea level provides a reference point for the base level of streams on a regional scale.

baseline   1) The known line used as a geometrical base in trigonometry. 2) The earliest data base in longitudinal research series. 3) the primary east-west survey line in the principal U.S. land survey system.

basement rock   The metamorphic and igneous rocks underlying the oldest identifiable sedimentary rocks of an area.

base-pairing   The coupling of nucleotides in the formation of DNA and RNA. Adenine pairs with thymine or uracil; guanine pairs with cytosine. The long chain of these stacked base pairs spirals to form the double helix shape of DNA and RNA molecules.

base saturation   In soils, the measure in percent of the chemical exchange sites that are occupied by exchangeable basic cations. See EXCHANGEABLE CATIONS.

basic   Alkaline; demonstrating the qualities of a base. Compare ACIDIC.

basic rocks   A general term for dark-colored or mafic igneous rocks rich in silica. Gabbro and basalt are typical igneous rocks.

basic science   Research that seeks to determine basic principles and theories in scientific disciplines.

basic solution   A water solution in which the concentration of hydroxyl (OH⁻) ions is greater than the concentration of hydrogen (H⁺ ions). As a result, such a solution has a pH value above 7.0.

Basidiomycota   The phylum of the kingdom Fungi that is characterized by organisms that produce a clublike, spore-containing reproductive structure known as a basidium. Within the phylum, the class Heterobasidiomycetae includes jelly fungi, rusts, and smuts, whereas the class Homobasidiomycetae, which bear their basidia on special structures called "hymeia," includes mushrooms (*Fomes, Amanita, Boletus, Agaricus,* and many others), puffballs (*Clavatia*), stinkhorns (*Phallus*), and earthstars (*Geastrum*). See FUNGI.

basin   1) An area of down-warped or bowl-shaped sedimentary strata. 2) The area drained by a stream or river. 3) A down-faulted block of the earth's crust with internal drainage. See GRABEN.

basin-and-range topography   A geomorphological region characterized by block-faulted mountain ranges with intervening alluvial basins.

BAT   See BEST AVAILABLE TECHNOLOGY.

Batesian mimicry   A means of protection in which an edible species mimics or adapts to resemble an inedible or poisonous species in order to be ignored by predators. Viceroy butterflies are avoided by predators solely because they closely resemble Monarch butterflies; the latter are distasteful to predators because they feed on poisonous milkweed. Also called protective mimicry; the actual markings are called pseudoaposematic coloration. Compare APOSEMATIC COLORATION, MULLERIAN MIMICRY.

batholith   A large body of intrusive igneous rock having an areal extent greater than 100 km².

bathyal zone   The deep-water regions of the oceans where very little light penetrates, generally between 300 and 3000 m. Compare ABYSSAL.

bathylimnetic   Relating to the organisms inhabiting the bathyal zone.

bathymetry   The measurement of water-depth relationships in oceans and lakes. Compare TOPOGRAPHY.

bathypelagic   Of or having to do with the bathyal, or deeper, regions of the ocean (from roughly 200 to over 2000 m). Very little light reaches these depths, which are characteristic of the continental slope and continental rise. Bathypelagic organisms live in the water column at great depths.

bats   Belonging to the mammalian order Chiroptera and closely related to the insectivores, the bats are the only flying mammals. Their forearms are highly modified as wings with skin membranes stretching between all digits except the first. At night, bats avoid obstacles and locate prey by emitting high-frequency sounds, the echoes they use for acoustical orientation during flight. See EUTHERIA, INSECTIVORE.

battery   A series of voltaic cells joined together and used to produce an electric current or to store power produced by devices such as windmills and photovoltaic cells.

bauxite   A principal ore of aluminum that is rich in hydrated aluminum oxides, formed in the B horizon of a laterite soil by removal of silica and clays through prolonged weathering. See APPENDIX p. 620.

bay   A part of an ocean or land extending toward the land, generally smaller than a gulf. Compare GULF.

bayhead barrier   A barrier island situated at the mouth of a bay.

baymouth barrier   A barrier beach that partially encloses a bay. A baymouth barrier extends from a landward edge out and across the mouth of the bay.

bbl   An abbreviation for the volumetric measure, barrels.

Be   Chemical symbol for the element beryllium. See BERYLLIUM.

beach   A deposit of unconsolidated sediment, usually sand, at the intertidal zone of a coastline.

beach drift   See DRIFT, LONGSHORE DRIFT.

Beaufort scale   A system for empirical measurement of wind force. Early in the 19th century, naval officer Sir Francis Beaufort correlated a scale of values from 0 to 12 with the effects of different wind speeds on the sails of a sailing ship. Today, other observations such as degree of tree movement are used for a scale ranging from a calm 0 (when smoke rises straight up) to 17 (wind speeds from 12 to 17 are full hurricanes and correspond to over 117 knots). See APPENDIX p. 630.

becquerel   (Bq) International unit in the SI system for measuring the strength of a radioactive source. One becquerel is one disintegration per second. See SYSTÈME INTERNATIONAL D'UNITÉS (SI).

bed   1) A sheet-like layer of rock or sediment. 2) An area at the bottom of a water body, for example a seabed or streambed.

bedding plane   A well-defined flat surface that separates one bed or stratum from another in sedimentary rock or sedimentary deposits.

bedding surface   See BEDDING PLANE.

bed load   The portion of stream-transported material that is moved by dragging, rolling, or saltation along the stream bottom. Compare SUSPENDED LOAD.

bed material   The unconsolidated material forming the bottom of a body of water.

bedrock   The solid rock immediately underlying an unconsolidated material, such as a soil.

bees   Species of Hymenoptera, the most common of which belong to the family Apidae including bumble and honey bees. Most species have highly developed social systems with reproductives of both sexes and sterile, worker females. Nests contain brood cells formed from secreted wax as

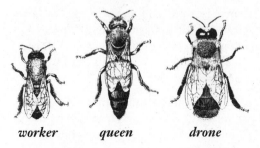

worker   *queen*   *drone*

well as cells for the storage of plant products such as honey. Unique among invertebrates, some bees are endothermic and can raise their body temperature for flight and incubation of broods. Approximately 40,000 species worldwide. See HYMENOPTERA.

beheaded river   A river that has lost part of the headwater area because of piracy by an encroaching watershed, or strike-slip faulting.

bel   Unit of sound intensity. 1 bel equals 10 decibels. See DECIBEL.

belemnite   An order of extinct gastropods of the Mesozoic to early Cenozoic eras. A belemnite shell superficially resembles an uncoiled snail shell.

below-ground   Found underneath the surface of the soil, as opposed to on top of the soil. Ecologists estimating below-ground biomass of ecosystems are attempting to account for the total weight of plant roots, soil-inhabiting organisms, etc. Compare ABOVE-GROUND.

bench mark   A fixed reference point used in land surveys. A bench mark is a point of known elevation.

bends   A condition occurring in divers who return to the water surface too quickly (it also occurs when pilots ascend too rapidly to high altitudes). The rapid reduction in the pressure on their bodies creates bubbles of nitrogen in tissue, causing extreme pain and paralysis. Also known as decompression sickness or caisson disease.

benefit-cost   See COST-BENEFIT ANALYSIS.

benign tumor   An abnormal growth that is not cancerous but consists of cells reproducing at abnormal rates yet remaining within the tissues where they develop. Compare CANCER.

Benioff zone   The region of earthquake hypocenters associated with the inclined plane of convergent motion along a tectonic subduction zone.

benthic   A term applied to any process, organism, or material of the benthos. See BENTHOS.

benthic community   The group of organisms inhabiting the region on the bottom of a body of water, such as a lake or ocean. The benthic community is the living portion of the benthic ecosystem. See BENTHOS.

benthic ecosystem   The environment and organisms that live on the bottom of bodies of saltwater or freshwater.

benthos   The collective term for marine life forms living in or on the deep ocean floor.

bentonite   An absorbent colloidal clay mineral that forms from the decomposition of volcanic ash and tuff. See MONTMORILLONITE.

benzene C_6H_6, a colorless, toxic, and flammable liquid. Benzene is produced from coal tar and is carcinogenic. Its structure—the benzene ring, a six-sided ring of carbon atoms with one hydrogen atom attached to each carbon—names the basic form of aromatic hydrocarbons, most of which are produced from benzene. Benzene is used as a solvent and in manufacturing synthetic rubber, pharmaceuticals, and dyes; it is sometimes called benzol in industrial use.

benzene hexachloride   Another name for gamma-1,2,3,4,5,6-hexachlorocyclo-hexane, more commonly known as Lindane. See LINDANE.

benzopyrene   Another name for benzpyrene. See BENZPYRENE.

benzpyrene   A flammable, potent carcinogen found in polluted air, produced by automobile exhaust, burning tobacco, oil and coal furnaces, and manufacturing of asphalt. Benzpyrene also occurs in coal tar.

Bergeron-Findeisen mechanism   Meteorological theory stating that ice crystals cause precipitation to begin in clouds that are nearly saturated with water droplets.

Bergman's rule   The principle that warm-blooded animals found in cold environments are larger than animals of the same (or similar) species from warmer climates. Compare ALLEN'S RULE.

bergschrund   The gap between the rock head wall of a cirque and the ice of an alpine glacier.

beriberi   A disease caused by a deficiency of vitamin B_1 (thiamine). It causes nerve damage and eventual paralysis or heart failure. Beriberi has become prevalent in the Orient since the last century when white rice replaced unpolished rice as the dietary staple; it was formerly widespread in rice-eating regions of the American South.

beringia   The geographic land bridge area

including the Bering Strait and the adjacent areas of Alaska and Siberia.

berkelium (Bk) A radioactive, metallic, transuranic element with atomic number 97 and atomic weight 247.0. It does not occur naturally but is produced by radioactive bombardment of plutonium, curium, or americium.

berm A nearly horizontal deposit of beach material accumulated by wave action near the water's edge.

Bermuda high An atmospheric high-pressure region air mass over the western Atlantic Ocean; it falls within the subtropic high-pressure belt. During summer months, extensive high-pressure cells form in the region between the Azores and Bermuda. Compare ALEUTIAN LOW, AZORES HIGH.

berry A fleshy fruit having a skin or rind and often containing numerous seeds. Berries are soft throughout, with no stony pit. Contrast DRUPE.

berylliosis Poisoning or permanent lung damage from breathing in beryllium oxide particles. See PNEUMOCONIOSIS, RESPIRATORY FIBROTIC AGENTS.

beryllium (Be) A whitish metallic element with atomic number 4 and atomic weight 9.010. It has chemical properties similar to those of magnesium. Beryllium is used for windows in x-ray tubes, in alloys to make them stronger and lighter, and as a source of neutrons as well as a controller of neutron speed in atomic reactors. Beryllium and its compounds are toxic.

Bessemer process A method for removing impurities in metals. The process involves blowing air through molten metal in a barrel-shaped device called a Bessemer converter. The Bessemer process is used to remove carbon and phosphorus from pig iron in making steel and to remove sulfur and iron from copper.

best available technology (BAT) Term used in enforcement of standards in certain federal and state environmental laws, which require application of most effective non-polluting technology.

best management practice Term used in enforcement of federal and state standards for certain pollutants, such as farmers employing "best management practices" to reduce the amount of non-point source pollutant manure runoff from fields into streams or other bodies of water, to meet state or federal Clean Water Act(s) standards. See BEST AVAILABLE TECHNOLOGY.

best practicable means Similar to Best Available Technolgy (BAT), but modified to meet economic feasibility. See BEST AVAILABLE TECHNOLOGY.

beta diversity An index of the degree of change in species composition of communities along an environmental gradient. For example, although the alpha diversity at the top and bottom of two mountains might be identical, on one the same species might be found at top and bottom (low beta diversity), whereas on the second mountain, completely different species are found at the top and bottom (high beta diversity). Contrast ALPHA DIVERSITY.

beta-naphthylamine A substance derived from coal tar or petroleum and formerly used as an antioxidant in rubber and in the manufacture of dyes. Its use has been discontinued because it is now known to be carcinogenic and is linked with bladder cancer. Also called 2-napthylamine.

beta particle An electron emitted by a radioactive nucleus, or any similarly high-speed electron (roughly between 25 and 99 percent of the speed of light). Beta particles can penetrate ordinary glass, but not a typical sheet of metal, and are harmful to humans, as is all ionizing radiation, especially when inhaled or ingested. See ALPHA PARTICLE, GAMMA RAY.

beta radiation A stream of beta particles; one of the three most common types of radiation from radioactive decay. See ALPHA RADIATION, GAMMA RADIATION.

Betulaceae The birch family of the Angiospermophyta. These monoecious, deciduous dicot trees and shrubs of north temperate environments bear male flowers in complex catkins and female flowers in catkin spikes. The fruits are either a small samara or a nut. Alders (*Alnus*), birches (*Betula*), hazelnuts (*Corylus*), hornbeams (*Carpinus*), and hop-hornbeam (*Ostrya*) are members of this family. See ANGIO-SPERMOPHYTA.

BH Abbreviation for benzene hexachloride, more commonly known as Lindane. See LINDANE.

B horizon A mineral soil layer characterized by the maximum accumulation of any combination of humified organic material, silicate clay, and iron or aluminum sesquioxides, with attendant development of blocky, prismatic, or columnar structure and usually redder or browner than overlying and underlying horizons; generally known as the zone of accumulation. See SOIL PROFILE, also APPENDIX p. 619.

BHT Butylated hydroxytoluene, an antioxidant used by the food industry (often with the related chemical BHA, butylated hydroxy anisole) to retard spoiling of fats.

bias 1) An opinion that strongly favors one side in an argument or one item in a group or series. 2) A statistical error, including a systematic error inherent in a sampling technique.

bicarbonate Any salt containing the hydrogen carbonate ion HCO_3^-. Bicarbonate of soda (sodium bicarbonate, baking soda) is a well-known example. The ion occurs naturally in some bodies of water, where it acts to buffer the acidity of the water.

biennial A plant whose life cycle spans two years, germinating in the first season and maturing to produce fruit in the second season. Parsley is a biennial. Compare ANNUAL, PERENNIAL.

big bang theory The generally accepted model for the formation of the universe. It states that all matter and energy was created from a single, dense mass that exploded 10 to 18 billion years ago, and that the universe has been expanding ever since. The model successfully explains not only the expansion of the universe, but also the abundances of primordial elements (such as helium) and the existence of microwave background radiation throughout the universe.

big cats Common name for cats of the genus *Panthera* whose voice apparatus is so modified that they can truly roar (e.g., lion, tiger, leopard, jaguar, snow leopard). Although most are protected throughout their range, their numbers are decreasing rapidly and some subspecies are highly endangered. The cheetah and clouded leopard are sometimes loosely grouped with the big cats. See CARNIVORE.

bilateral symmetry Divisible into two similar halves (mirror images) by only one imaginary plane running through the center or axis of a structure or organism. Most animals and some flowers (such as peas) show bilateral symmetry, so one side of the body is a mirror image of the other. Compare RADIALLY SYMMETRICAL.

bilayer   Describes the structure of protein-coated lipid membranes in cells according to the fluid-mosaic model. Biological membranes are double layers of phospholipids, with proteins connecting or moving between the two layers.

bilharziasis   Name applied to some forms of schistosomiasis. See SCHISTOSOMIASIS.

biliproteins   See PHYCOBILINS.

billow clouds   A form of altocumulus clouds. Billow clouds are low-level clouds formed by changes in wind speed or direction with elevation; this shearing motion produces long, ropelike clouds that may either be closely packed or widely spaced. Also called billow altocumulus.

bimodal   Having two peaks.

binary fission   A form of cell division in which the nucleus and then the entire cell split into two equal nuclei and cells.

biomass energy   Energy derived from organic matter, usually plants. Biomass energy can come directly from burning, as with wood, or through conversion to gas or liquid fuel (distilled or fermented into alcohol, digested by microbes into methane, or converted to gas or oil through pyrolysis). See BIOMASS FUEL, BIOGAS, PYROLYSIS.

binaural   Of or involving the use of two ears. Binaural hearing allows animals to detect the direction from which a sound comes; stereophonic headphones produce binaural sound.

binocular vision   Sight involving the simultaneous use of both eyes.

binomial classification   Organizational and nomenclature system based on naming

plants and animals using two Latin names. The system, devised by Linnaeus in the 18th century, names a plant or animal by using two Latin words. The first word of a scientific name always starts with a capital letter and denotes the genus; the second is the specific name and is lowercase, as in *Homo sapiens* or *Sequoia sempervirens*. A third name can be added to describe a subspecies. Also called binomial nomenclature. See SPECIES.

binomial nomenclature   See BINOMIAL CLASSIFICATION.

bioaccumulation   The absorption and concentration of toxic chemicals in living organisms. Heavy metals and pesticides, such as DDT, are stored in the fatty tissues of animals and passed along to predators of those animals. The result is higher and higher concentrations of the pesticide in fatty tissue, eventually reaching harmful levels in predators at the top of the food chain, such as eagles. Also called biomagnification.

bioassay   Testing the strength of a drug or other substance by examining its effects on a living organism and comparing it with those of a standard substance.

biocenosis   Another spelling for biocoenosis, or biotic community. See BIOTIC COMMUNITY.

biochemical   Of or having to do with biological chemistry, the branch of science that examines the chemistry of living organisms.

biochemical oxygen demand   (BOD) An indication of the extent to which water is polluted by sewage or other organic waste. It is a measure of the dissolved oxygen consumed by microorganisms as they break down the organic matter in a sample of water; the greater the quantity of organic matter, the greater the oxygen required for its decomposition. It is expressed in parts per million of dissolved oxygen consumed. Also called biological oxygen demand. See ANOXIC, BOD₅.

biocide Any substance that poisons living organisms, especially a chemical that can kill many different kinds of organisms. See ACARICIDE, AVICIDE, FUNGICIDE, HERBICIDE, PESTICIDE, RODENTICIDE.

bioclastic Made up of small fragments of organic material. See CLASTIC.

bioclimatology Study of the effects of climate on living organisms.

biocoen Short for biocoenosis, or biotic community. See ABIOCOEN, BIOTIC COMMUNITY.

biocoenosis A biotic commuity. Also spelled biocenosis. See BIOTIC COMMUNITY.

bioconcentration See BIOACCUMULATION.

biocontrol Short for biological control. See BIOLOGICAL CONTROL.

biodegradable Capable of being decomposed into natural substances (such as carbon dioxide and water) by biological processes, especially bacterial action. Contrast NONBIODEGRADABLE.

biodiversity See SPECIES DIVERSITY.

biodynamic farming A form of organic farming or gardening developed early in the 20th century by Dr. Rudolph Steiner. It is based on a philosophy of interdependence of all living organisms. Practices include avoidance of synthetic pesticides or fertilizers, use of composts made according to specific recipes and often incorporating herbs to guarantee micronutrient and microorganism content, and planting according to the phases of the moon. See ORGANIC AGRICULTURE.

bioengineering 1) Engineering methods using biochemical or genetic processes to produce drugs and food or to recycle waste. 2) Engineering applied in medicine or zoology. See BIOTECHNOLOGY.

bioflavanoids A group of water-soluble micronutrients present in citrus fruits, especially the inner peel. They are necessary for the formation of strong capillaries. Sometimes called vitamin P.

biofuel Gas or liquid fuel made from plants, trees, or organic wastes. Biofuels include ethanol, methanol, biogas, and wood.

biogas Gas produced from the fermentation of organic matter (biomass) that can be used as fuel. Biogas is usually predominanty methane.

biogenesis 1) The principle that living things can only be produced by other living things. 2) Biosynthesis. Compare SPONTANEOUS GENERATION.

biogenically reworked zone An area of sediment that has been disrupted by living organisms. For example, worm burrows and trails may alter the original sediment layering. See TRACE FOSSIL.

biogenic (marine) sediment An ocean floor deposit that is of biologic origin. Limestones are most often formed from biogenic sediment.

biogeocenosis See ECOSYSTEM.

biogeochemical cycles The cycling of chemicals such as carbon, oxygen, phosphorus, nitrogen, and water within (intrasystem nutrient cycles) or between ecosystems (intersystem nutrient cycles) and throughout the biosphere. These compounds are assimilated and broken down over and over again by living organisms. Also called nutrient cycles. See NITROGEN CYCLE, CARBON CYCLE,

PHOSPHOROUS CYCLE, OXYGEN CYCLE, SEDIMENTARY NUTRIENT CYCLES, and HYDROLOGIC CYCLES.

biogeochemical prospecting   Using indicator species to predict where mineral deposits are likely to be found. In some parts of the country, selenium is often associated with uranium ore, so selenium-indicating plants such as *Astragalus* show a good probability of a uranium deposit.

biogeochemistry   The study of the interactions between organic and inorganic compounds and living organisms. See BIOGEOCHEMICAL CYCLES.

biogeography   The study of the distribution of plants and animals over different geographic regions.

biological amplification   Another term for bioaccumulation. Also called bioamplification. See BIOACCUMULATION.

biological bench-marking   The use of biological indicators (indicator species) as a standard of reference (or to establish a baseline) for monitoring the health of an ecosystem or changes within it. See INDICATOR SPECIES.

biological clock   An internal physiological or biochemical mechanism in organisms that controls the timing of activity cycles such as feeding and sleeping and different functions. See PHOTOPERIODISM, CIRCADIAN RHYTHM.

biological community   See BIOTIC COMMUNITY.

biological control   Taking advantage of a pest's natural vulnerability in order to control it; pest control using introduced predators, parasites, disease organisms, or release of sterilized individuals rather than applied pesticides. Also called biocontrol or biological pest control. See INTE-

GRATED PEST MANAGEMENT, STERILE MALE TECHNIQUE, SEX ATTRACTANT.

biological indicator   See BIOMARKER, INDICATOR SPECIES.

biological magnification   See BIOACCUMULATION.

biological monitoring   Observing and measuring changes occurring in an ecosystem.

biological oxygen demand   Another term for biochemical oxygen demand. See BIOCHEMICAL OXYGEN DEMAND.

biological pest control   See BIOLOGICAL CONTROL.

biological resources   Living organisms, as opposed to non-living natural resources such as mineral deposits.

biology   The study of living organisms and life processes, including origins, classification, structure, activities, and distribution. See BOTANY, ECOLOGY, GENETICS, TAXONOMY.

bioluminescence   Emission of light by an organism. Bioluminescence occurs in fireflies and in some species of marine organisms, bacteria, and fungi. See PHOSPHORESCENCE.

biomagnification   See BIOACCUMULATION.

biomarker   A technique of using biological end points in living organisms as indicators of environmental insults. The presence of damaged DNA, stress proteins, altered cell types, or metal binding proteins have all been used as biomarkers. Compare INDICATOR SPECIES.

biomass   1) The total amount of all the biological material, the combined mass of all the animals and plants living in a specific area, or of a given population.

Usually expressed as oven-dry weight per area (grams per square meter, kilograms per hectare, or pounds per acre). 2) Organic matter used as fuel (biomass fuel).

biomass accumulation ratio   The ratio of total biomass to annual net primary productivity in a biotic community. Contrast P/B RATIO.

biomass energy   Energy derived from organic matter, usually plants. Biomass energy can come directly from burning, as with wood, or through conversion to gas or liquid fuel (distilled or fermented into alcohol, digested by microbes into methane, or converted to gas or oil through pyrolysis). See BIOMASS FUEL, BIOGAS, PYROLYSIS.

biomass fuel   Plant and animal matter that can be used directly as a source of heat (e.g., burning wood or dung) or converted to a gaseous or liquid biofuel.

biomass pyramid   A diagram illustrating the relationship between the amount of biomass at different trophic levels. The biomass pyramid shows that the lower levels consisting of producers have the most biomass; above them are herbivores; and carnivores, having the least biomass, are at the top. See BIOMASS, NUMBERS PYRAMID.

biome   Regional land-based ecosystem type such as a tropical rainforest, tiaga, temperature deciduous forest, tundra, grassland, or desert. Biomes are characterized by consistent plant forms and are found over a large climatic area. See BOREAL FOREST, DESERT, GRASSLAND, TEMPERATE DECIDUOUS FOREST, TROPICAL RAINFOREST, TUNDRA.

biomethanation   Method for producing biogas (methane) for use as fuel. See BIOGAS.

bion   An individual living organism within an ecosystem.

bioplex   Short for biological complex, a cyclic system in which waste materials produced by each stage provide the raw materials for the next stage.

bioseston   Particulates suspended in water that are either living organisms, or organic matter that comes from living organisms. See SESTON.

biosphere   The portion of the planet capable of supporting life. It ranges from elevations of approximately 10,000 meters above sea level to the deep ocean, and a few hundred meters below the surface of the soil. The biosphere consists of the hydrosphere, the lower atmosphere (troposphere), and the surface of the lithosphere, which are inhabited by metabolically active organisms.

biosphere reserve   A park set aside to preserve a functioning, large-scale ecosystem for conservation and research in conjunction with the Man and Biosphere Program (an international scientific program of the United Nations) and the independent International Union for Conservation of Nature and Natural Resources (IUCN).

biostratigraphic unit   A body of rock strata that is identifiable by the presence of one or more distinctive fossils without dependence on any lithologic or physical features. The fundamental biostratigraphic unit is a zone established by worldwide correlations; a zone may be divided into subzones and zonules.

biosynthesis   Production of a chemical compounds such as amino acids by a living system.

biota   1) The flora and fauna of a specific

region or period. 2) The total aggregation of organisms in the biosphere.

biotechnology The use of applied science to produce living organisms with particular traits, especially by manipulating different genetic material. Biotechnology ranges from artificial insemination to genetic engineering. See GENETIC ENGINEERING.

biotic Concerning or produced by living organisms, such as environmental factors created by plants or microorganisms.

biotic community All the groups of organisms that share the same habitat or feeding area, usually interacting or depending on each other for existence. Also called biocoenosis, biocoen, or simply, community. Compare ASSOCIATION.

biotic factor The influence of organisms and their activities on the distribution of other organisms.

biotic index (BI) Measure showing the quality of an environment by identifying the numbers of various species present. The biotic index can give an indication of how clean a pond or river is on the basis of the presence of particular indicator species or of groups of species. See INDICATOR SPECIES.

biotic potential (r_{max}) The maximum possible reproductive rate for members of a species under ideal environmental conditions and unlimited resources. Also called innate capacity for increase, it is the maximum value for r, the intrinsic rate of natural increase. Contrast ENVIRONMENTAL RESISTANCE, INTRINSIC RATE OF NATURAL INCREASE.

biotic province Geographical region that covers the distribution of one or more groups of animals (or, less commonly,

plants). Also called biotic region. See BIOME.

biotin $C_{10}H_{16}N_2O_3S$, a crystalline acid of the vitamin B complex found in liver, eggs, and yeast. It promotes growth by acting as a coenzyme for various catalysts. Biotin deficiency is very rare because adequate supplies are usually synthesized by bacteria within animal intestines.

biotite A ferromagnesian sheet silicate mineral of the mica group that; brown-to-black with pearly luster and perfect cleavage in one direction producing thin, elastic, and translucent sheets. Contrast MUSCOVITE. See APPENDIX p. 621.

biotope A small area with uniform environmental conditions (climate, soil) and a characteristic distribution of animal populations.

biotype A group of individuals within a species having the same genotype.

bipedal Having two feet.

bipolar Having two poles with opposite electrical charges, as in some molecules, or having two projections, as in the nerve cells that contain an axon at each end.

biramous Forked; describing an animal structure with two branches, such as the limbs of some crustaceans.

bird's-eye limestone A variety of limestone containing rounded cavities formed by gas bubbles trapped within an intertidal sediment before lithification. The cavities often are filled with calcite, giving an eyespot-like pattern to the rock.

bird's-foot delta A delta in which the deep channels extend seaward in a shape like the outstretched toes of a bird.

Birge's rule The expression measuring the percentage of monochromatic light absorption in a given depth of water; given as:

$$\frac{100\,(I_0 - I_z)}{I_0}$$

where I_0 = irradiance at the lake surface and I_z = irradiance at depth z.

birth control   Any method or device used to reduce the number (or likelihood) of individuals born, especially methods preventing impregnation. Birth control includes celibacy, sterilization, delayed marriage, and use of contraceptives. See CONTRACEPTIVE.

birth defect   An imperfection or weakness that an offspring is born with. See TERATOGEN.

birth rate   The ratio of the number of live births to a total population over a given period of time, usually one year. For humans, the rate is expressed as births per thousand persons per year. See NATALITY, DEATH RATE.

bisexual   See HERMAPHRODITE.

bison   Common and generic name for two species of bovids, the American bison *Bison bison* and the European bison or wisent *Bison bonasus.* Virtually exterminated in the 19th century, the American bison, which ranged from the prairies to northern woodlands, is now preserved in substantial numbers in refuge areas. The wisent became extinct in the wild. It is, however, still maintained in preserves in Poland, the Caucasus, etc. See BOVIDAE.

bituminous coal   A soft variety of coal that has a high proportion of volatile matter and burns with a smoky flame.

Bivalvia   A class of molluscs characterized by the presence of two lateral shells or valves hinged dorsally by a flexible liga-ment. The body is laterally compressed, the head is greatly reduced and the molluscan radula is missing. In many forms, the paired gills or ctenidia are enlarged to form filter feeding structures. Inhalant currents in the mantle cavity draw water over the ctenidia, where particles are filtered out and directed toward the mouth. Bivalves are widely distributed in the world's waters and, being largely enclosed by shell, tend to be sedentary or burrowing. Clams, mussels, scallops, and oysters are essential elements of the shellfish industry. See MOLLUSCA, OYSTERS.

bivoltine   Producing two broods in a year. Compare MULTIVOLTINE, UNIVOLTINE.

black body   A hypothetical object or surface that absorbs all radiant energy falling on it.

black box   An object or entity whose function can be evaluated without knowing or understanding its internal operation or construction.

black lung disease   See ANTHRACOSIS, PNEUMOCONIOSIS.

blanket bog   A form of bog found in cold, wet climates and covering a large area rather than a specific pond. A blanket bog is characterized by a continuous layer of dead but undecomposed organic matter (peat) over waterlogged, acidic, nutrient-poor ground. It derives its mineral nutrients from rainfall. See BOG.

blanket mire   See BLANKET BOG.

blastula   One of the earliest stages of an embryo in most animals, a hollow, spherical layer of cells that forms from the morula, which in turn forms by the division of a fertilized ovum. See GASTRULA, MORULA.

bleicherde   See FULLER'S EARTH.

blepharoplast See BASAL BODY.

blight Common name for any fungal disease that rapidly infects leaves, stems, fruits, or tissues of plants, causing them to wither and die. Fire blight is a serious disease of pears and apples, and the chestnut blight wiped out the tree that once predominated in eastern United States forests. See MILDEW, RUST, SMUT.

block faulting A type of brittle failure of the earth's crust resulting in the movement of large block-shaped masses of rock.

blowdown An extensive toppling of trees by wind within a relatively small area, greatly altering the small-scale climate within the ecosystem.

blowout A section of a sand dune that has been eroded by the wind. Blowouts are often associated with damage to stabilizing vegetation on the dune.

blue asbestos A common name for the amphibole mineral crocidolite. See CROCIDOLITE.

blue-green algae See CYANOBACTERIA.

blue ground A deposit of bluish green kimberlite breccia. Blue ground oxidizes to a yellowish color within surficial deposits. See KIMBERLITE.

blueschists A type of metamorphic schistose rock containing bluish amphibole minerals. Blueschists are formed in the shallow portion of a subduction zone where temperatures and pressures are comparatively low. See SCHIST.

BMR Abbreviation for basal metabolic rate. See BASAL METABOLIC RATE.

BOD Abbreviation for biochemical oxygen demand (also called biological oxygen demand). See BIOCHEMICAL OXYGEN DEMAND.

BOD_5 The biochemical oxygen demand over a 5-day period at a standard temperature of 20°C (68°F) and standard atmospheric pressure. See BIOCHEMICAL OXYGEN DEMAND.

body burden Quantity of radioactive or other toxic material present in the body at a point in time.

body waves In seismology, any earthquake wave that travels through the earth's interior by spreading out in all directions from the central focus point.

bog A poorly drained wetland area with acidic, spongy ground, made up of dead but largely undecayed sphagnum moss (peat) and other vegetable matter. Different kinds of bog include raised bogs, quaking bogs, and blanket bogs. Compare FEN, MOOR, MARSH, SWAMP.

boil To cause a substance to change from its liquid state into its gaseous state through applying heat or reducing pressure so that the liquid's vapor pressure equals the atmospheric pressure. Also, the physical process that occurs when this state-change occurs: Bubbles form in the liquid and the temperature remains constant until all of the liquid has been converted to gas.

boiling point The temperature at which large bubbles form within the liquid state of a substance as it changes to a vapor. Boiling point is a physical constant that helps characterize specific substances. It is usually measured at a standard pressure of one atmosphere; a liquid boils when its vapor pressure is slightly greater that the pressure above it, so when the pressure is lowered, the liquid boils at a lower temperature.

boiling water reactor One type of fission nuclear reactor in which the water used to moderate the fission reaction heats to the boiling point, producing steam that powers turbines to produce energy. It is one of two variations of the light-water reactor design. Compare PRESSURIZED WATER REACTOR.

bole   Trunk of a tree above the root collar and extending along the main axis.

bolide   An unusually luminous meteor, usually one that explodes after entering the earth's atmosphere, producing a number of smaller fragments that reach the earth's surface as meteorites. See METEORITE.

bolson   The basin area of block-faulted mountains composed of the playa, pediment, and bajada. See BASIN-AND-RANGE TOPOGRAPHY.

bomb calorimeter   A device for measuring the calorie content of solids or liquids. It is a strong vessel containing pressurized oxygen into which a sample of the substance to be measured is introduced. When the contents are ignited electrically, the vessel heats up. Measuring the increase in temperature gives the heat of combustion, from which the calorie content can be determined.

bomb, volcanic   See volcanic bomb.

bond (chemical)   The attractive force linking two or more atoms. Molecules consist of two or more atoms bonded together. Bonds can have forces of approximately 1000 kilojoules per mole of atoms and take different forms. Also called chemical bond. See IONIC BOND, COVALENT BOND.

bond dipole   A bond between two atoms of different electronegatives has an uneven distribution of the electrons in the bond. The result is a polarized bond in which the electrons spend more time in the vicinity of the more electronegative atom. These types of bonds are said to have bond dipoles. The O-H bonds in the water have bond dipoles, with the result that the water is very polar. See DIPOLE.

bora   A violent, cold, northerly or northeasterly wind that occurs on and around the Adriatic Sea. See KATABATIC WIND.

Boraginaceae   The borage family of Angiospermophyta. These dicots are largely herbs with actinomorphic, 5-merous flowers arranged in coiled cymes with flowers attached on one side of the floral axis. The two united carpels of the superior ovaries produce four nutlet fruits. Forget-me-not (*Myosotis*), blue-bells (*Mertensia*), fiddlenecks (*Amsinckia*), borage (*Borago*), and heliotrope (*Heliotropium*) are all members of this family. See ANGIOSPERMOPHYTA.

borax   Short for hydrated sodium borate, $Na_2B_4O_7 \cdot 10H_2O$, a white, yellowish, blue-green, or gray alkaline mineral. Borax is used for cleaning and in manufacturing metals, glass, and ceramics.

Bordeaux mixture   One of the earliest fungicides, it is made up of poisonous copper sulfate, lime, and water. It was a standard remedy for many plant diseases until the development of organic (hydrocarbon derivative) fungicides in the 1940s.

borderland   Land and sediments adjacent to continental border.

bore   A tidal floodwave that moves rapidly up a channel in the intertidal zone of a coastline.

boreal   1) Referring to a climatic zone in which winters have snow and summers are short. 2) A climatic period from about 7500 to 5500 BC, generally dry with cold winters and warm summers, indicated by a development of a pine-hazel flora. 3) Short for boreal biogeographic region. See BOREAL FOREST.

boreal forest A vegetation type dominated by coniferous trees (but containing some deciduous broadleaved species such as aspen and birches) stretching across North America, Europe, and northern Asia (regions characterized by short summers and long, cold winters). It is found south of the tundra in the Northern Hemisphere, and often contains peaty or swampy areas. Boreal forests grow in the boreal biogeographic region. Also called northern coniferous forest and taiga.

boric acid H_3BO_4, an acid derived from borax, although also found naturally in volcanic steam vents (fumaroles). Boric acid has many commercial uses: as a mild astringent and antiseptic, in detergents, in welding, and in the manufacture of glass, ceramics, paper, and adhesives.

bornhardt A rounded landform created by surface exfoliation and sheeting in a dome of massive rock and occurring most commonly in tropical regions. The term bornhardt is sometimes compared to the term inselberg, but bornhardt is not technically acceptable as a morphogenetic term. See INSELBERG.

bornite A copper ore mineral that forms from copper iron sulfate in hydrothermal veins. Bornite is characterized by an iridescent purple-and-blue metallic sheen. See APPENDIX p. 621.

boron (B) A nonmetallic, brownish element with atomic number 5 and atomic weight 10.81 that is never found in elemental form in nature. Isotopes of boron are used for control rods and shields in nuclear reactors because of its capacity for absorbing neutrons. Boron is also used in specialized alloys, as in semiconductors.

boss A discordant intrusion of igneous rock, roughly circular in shape, with an outcrop area of less than 25 km². Compare STOCK, BATHOLITH.

botanical 1) Relating to plants or plant life, or botany. 2) When plural, short for botanical pesticides, those extracted from plants such as pyrethrum, ryania, sabadilla, and rotenone.

botany The branch of biology devoted to plants, studying their structure, growth, classification, evolution, ecology, and biochemistry.

botryoidal Describing mineral formations resembling a bunch of grapes.

botryose Branched like a bunch of grapes. Also called racemose.

bottom load See BED LOAD.

bottom sets The lowermost, flat-lying bed deposits in cross-bedded sediment, e.g., at the forefront of a prograding delta or dune. See CROSS-BEDDING.

botulism A dangerous form of food poisoning. It is caused by a toxin (botulin) secreted by a bacterium (*Clostridium botulinum*) sometimes present in food that has not been properly canned or preserved (especially meats and nonacid vegetables). It is frequently fatal.

boudinage A structure of deformed bedrock in which the softer layers are divided by small fractures or distortions into pillow-shaped segments known as boudins.

Bouger correction A gravity measurement that has been corrected for the measuring station's height above sea level and for the gravitational effect of the rock between the station height and datum (reference point).

boulder Any clastic rock fragment that is

greater than 256 mm in average diameter.

boulder clay   A term used in Great Britain to describe glacial till-containing striated rocks embedded in rock flour.

boulder train   A series of glacial erratics from the same source outcrop. Boulder trains usually occur in a straight line or fan shape stretching out from the point of origin.

Bouma sequence   A description of the layers formed in a submarine turbidite deposit. The sequence from the bottom layer up may include: A. graded beds, B. lower parallel laminations, C. current ripple laminations, D. upper parallel laminations, or E. pelitic layers.

boundary layer   In the study of fluid dynamics, the thin layer formed next to a solid body or a surface within which friction causes the fluid to move relatively slowly. There are high-velocity gradients in the boundary layer.

Bovidae   The largest family of ungulates, the bovids, with more than 100 genera, are primarily grazers distinguished by horns with a permanent horny covering over a bony core. Their distribution is worldwide and many forms have been domesticsted by humans (e.g., antelopes, elands, cattle including buffalo, bisons, goats and sheep). See ARTIODACTYLA.

bovids   See BOVIDAE.

box model   The simplest method for estimating the concentration of an air pollutant. It is based on the assumption of homogeneous distribution throughout the box (a specified volume of air) for a given wind speed. Compare AIRSHED.

bp   In geology, the abbreviation for before present.

Br   Chemical symbol for the element bromine. See BROMINE.

brachiolarian larva   The last larval stage in starfish, preceding metamorphosis into the starfish form.

brachypterous   Having rudimentary or very short wings.

Brachyura   Suborder of decapod crustaceans, the true crabs, have broad carapaces often wider than they are long and greatly reduced abdomens folded beneath the thorax region. The walking legs, the first pair of which is provided with pincers, operate to move crabs in a sideways direction. Crabs are widely distributed throughout the marine environment and display a great variety of form and habitat. Hermit crabs, for example, house themselves by insertion of their highly modified, soft abdomens into gastropod shells. The blue crab, *Callinectes sapidus*, of the middle Atlantic coast of the United States is much sought as a luxury food item and a number of other species are also commercially important. See DECAPODA.

brackish   A term describing water with a salinity level between that of fresh water and of sea water.

bract   A modified leaf growing below a flower or cluster of flowers. The showy, white portions of dogwood flowers are actually bracts rather than petals.

bracteole   A small bract.

braided river   A drainage system that follows channels that divide and rejoin in a complex anastomosing network. Braiding forms where a stream or river becomes unable to carry the bed load, or has a steep gradient.

branching decay   A type of radioactivity in which the isotopes involved can disintegrate in more than one way, usually in the context of a decay series. See RADIOACTIVE DECAY.

Branchiostoma A genus, formerly called *Amphioxis*, of the suborder Cephalochordata, which is well known because of its use in biology courses to introduce students to basic chordate characteristics. See CEPHALOCHORDATA, CHORDATA.

Brassicaceae The mustard family of the Angiospermophyta, which is sometimes referred to as the cruciferae because their flowers have four petals in the form of a cross. Members of this family are dicots which have four sepals, four long plus two short stamens, and two carpels that fuse and usually form an elongated fruit. Mustards and cole crops (*Brassica*), radish (*Raphanus*), wallflowers (*Erysimum*), and many herbaceous weeds, ornamentals, and vegetable crops of cool, temperate, Northern Hemisphere regions are in this family. See ANGIOSPERMOPHYTA.

breakwater An artificial offshore structure, such as a wall or a jetty, that serves to protect a shoreline by dissipating the energy of waves.

breccia A collective term applied to rocks composed largely of angular and subangular fragments. Breccias are commonly of sedimentary or pyroclastic origin. Compare FAULT BRECCIA.

breeder reactor A nuclear reactor that produces at least as much fissionable fuel as it consumes. Breeder reactors typically consume a combination of plutonium and uranium and produce plutonium. See NUCLEAR REACTOR, LIQUID METAL FAST BREEDER.

brine Strongly saline water.

British Thermal Unit (Btu or BTU) A measure of heat originally defined as the amount of heat required to raise one pound of water by one Fahrenheit degree. It is now defined as 1055.06 joules. See JOULE.

brittle A physical property of materials that break under strain without significant and prior plastic deformation.

brittle stars Ophiuroidea, a class of echinoderms, the brittle stars have a compact central body and a skeleton of tightly packed plates which are often arranged longitudinally. Unlike many echinoderms which rely on tube feet for movement, locomotion in brittle stars is accomplished by snake-like undulations of the five, flexible arms. In the basket stars the arms are highly branched. Some species are predators, but most scavenge for food. They are easily found by overturning rocks in tide pools or along the shore. In the seas their distribution is worldwide and, like the sea cucumbers, they may be the predominat life form at great depths. See ECHINODERMATA.

broadcast spraying Application of a pesticide over a large area such as a lawn or field, with no attempt to be specific or selective about which plants get doused. Although the term can include crop dusting by aircraft, it usually refers to ground application that covers more than the target species. Compare BASAL APPLICATION.

broadleaf A plant having leaves that are broad and flat rather than needle shaped. Usually used to distinguish evergreen species with large leaves (such as rhododendrons) from conifers such as pines, or to distinguish lawn weeds from grasses. Broadleaf herbicides kill weeds such as plantain and dandelion but do not affect grass species.

broad-spectrum pesticides Chemicals capable of killing a wide range of species, and therefore killing other organisms besides the targeted pest or pests. Also called nonselective pesticides.

Brocken spectre An illusion of a giant in the mist created by the observer's shadow falling onto a cloud of mist, especially from a distance, as by standing on a hill above a bank of clouds.

broken stick model A model describing equitability in which the relative abundance of of species reflects a random division of a line that divides the resources of an environment into segments. Biotic communities conforming to the broken stick model have species that show random subdivisions of the available resources. See EQUITABILITY.

Bromeliaceae The bromeliad family of monocot angiosperms. These frequently epiphytic plants usually form rosettes of fibrous, serrate margined leaves and have an inflorescence with conspicuous bracts. The perianth has a whorl of three sepals and a whorl of three petals surrounding the six stamens and superior ovary. Pineapple (*Ananas*), Spanish moss (*Tillandsia*), and many species of air plants are members of this family. See ANGIOSPERMOPHYTA.

bromide 1) The ionic form of the element bromine, where the bromine atom has gained one electron (Br^-). 2) Any salt containing the Br^- ion, such as silver bromide (AgBr). Many metals readily form bromides. Methyl bromide (CH_3Br) is used in fire extinguishers and as a fumigant against insects, worms, and rodents.

bromine (Br) A red-brown, nonmetallic, halogen element. Extracted from sea water as bromide (Br^-), it is used as an anti-knock fuel additive and in the manufacture of many organic chemicals, including pharmaceuticals and photographic chemicals. Methyl bromide is used in fire extinguishers and as a fumigant against insects, worms, and rodents. See BROMIDE.

bronchial Of or related to the bronchi, the two main branches connecting the windpipe to the lungs, or the bronchioles (smaller subdivisions).

bronchiole Small, branching air passages within the lung that connect the bronchi to the alveoli. Most of the lung is bronchioles. See ALVEOLI.

bronchitis Inflammation of the mucous lining of the bronchi, causing a severe cough. Bronchitis may be caused by viruses, by irritation from allergies, or by inhalation of chemical or physical irritants such as dusts and particulates. See RESPIRATORY FIBROTIC AGENTS.

bronze A yellowish brown metallic alloy of copper and tin that often includes lead, nickel, and zinc.

brood Offspring of a bird or insect that are hatched together and nurtured at the same time.

brood parasitism The act of abandoning offspring to be raised by another species. The cuckoos of Britain and the cowbird in America demonstrate brood parasitism, one form of social parasitism. See SOCIAL PARASITE.

brook A term used for a small stream or watercourse in the upper reaches of a watershed.

brown algae See PHAEOPHYTA.

brown coal See LIGNITE.

Brownian motion Constant small, random, irregular movement of microscopic particles suspended in a liquid or gas, caused by bombardment of molecules of the liquid or gas.

browse   1) Tender parts of woody vegetation that are eaten as food by animals. Twigs are a common winter browse for deer and other animals. 2) To consume browse. Browsing is distinct from grazing because it refers to eating woody material, whereas grazing is usually restricted to nonwoody plants.

browsers   Animals such as deer that consume woody vegetation.

brucellosis   A disease caused by infection from several bacteria of the genus *Brucella*, most common in cattle, pigs, and goats, but also infecting humans through milk. Also called undulant fever when it occurs in humans.

brush discharge   Glowing, threadlike electrical discharge from a conductor that branches out into the atmosphere or surrounding gas, a form of corona. It happens when the electric field near the conductor's surface is high but not sufficiently high to create a spark or arc. St. Elmo's fire, which occurs on airplane extremities and ship masts, is a form of brush discharge. See CONDUCTOR.

Bryophyta   A phylum of the plant kingdom that includes mosses, liverworts, and hornworts. These primitive plants that grow in moist habitats lack a fluid conducting vascular system. The most conspicuous generation of bryophytes is the leafy gametophyte which bear the gametangia. The diploid sporophyte generation, which bears a capsule containing spores, is usually dependent upon the gametophyte for its nutrition. Formerly, bryophytes were included under Pteridophyta.

Bryozoa   A phylum of aquatic, colony-forming invertebrates, most of which live in marine environments. Each individual is only about a millimeter long, but the colonies they form may measure 50 centimeters or more across. The polyplike individuals have cilia-covered circular folds of body wall called lophophores surrounding their mouths to trap the organic particulates on which they feed. Also called Ectoprocta or Polyzoa, they are more commonly known as sea mats or moss animals.

Btu   Abbreviation for British Thermal Unit (also written BTU). See BRITISH THERMAL UNIT.

bubble policy   A complex legal process in the Clean Air Act to permit non-compliance of laws and regulations at some times but only under certain circumstances.

bubonic plague   Disease caused by a bacterium (*Pasteurella pestis*) that is transmitted to humans through the bite of rat fleas, which live on other rodents as well as rats. The usually fatal infection causes great swelling of the lymph glands in the groin and other parts of the body. The Black Death of the Middle Ages was bubonic plague; it still occurs in India and parts of Asia.

bud   1) A small, primordial growth on a plant, consisting of small, dormant cells from which shoots, leafs, flowers, and

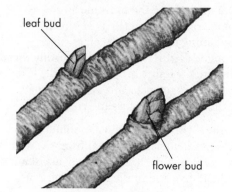

leaf bud

flower bud

other plant organs develop. 2) The swelling on a parent organism that eventually detaches to form a new individual in some species of yeasts and other fungi.

buffer   A substance that prevents rapid change in the acidity or alkalinity (hydrogen ion concentration—pH) of a solution, either when an acid or base is added or when the solution is diluted. Buffers are usually solutions of a weak acid or base with one of its salts. They are used to maintain pH at a constant level, either in chemical reactions or in organisms. Blood is buffered by carbonate salts.

buffering capacity   A measure of the ability of a buffer to neutralize; the amount of acid that a given quantity of buffer can neutralize. Loosely, the ability of a solution to absorb acids (or bases) without changing its pH. Bicarbonate is largely responsible for the buffering capacity of bodies of water.

buffer species   A species used for food that is usually of secondary interest, but which, during times of adverse conditions, becomes a predator's primary food source.

buffer strip   A strip of vegetation left or intentionally planted to screen an adjacent area from an adverse effect. See STRIP-CROPPING.

bugs   A loosely applied and essentially derogatory term for a wide range of organisms, particularly of the "crawly" sort, such as insects and crustacea. However, it is also sometimes used for disease agents like bacteria and viruses and becomes essentially synonomous with germs. True bugs, however, belong only to the insect order Hemiptera. See HEMIPTERA.

bulb   An underground reproductive structure of some plants consisting of a bud surrounded by fleshy scales. Lilies, tulips, and onions all grow from bulbs. Compare CORM, TUBER.

bulk deposition   Large-scale atmospheric deposition. See ATMOSPHERIC DEPOSITION.

bulb

bulk precipitation
A precipitation sample from an open collector; it includes rainwater or snow plus dust and other particulate matter settling out of the atmosphere.

bunodont   Description of mammals' teeth in which the cusps remain separate and rounded. See ACRODONT, LOPHODONT, SELENODONT.

bunt   A disease that infects cereal grains, caused by a fungus that causes the mature grain to become a spore mass. A form of smut.

buoyancy   The upward force a fluid exerts on a body immersed in it. Also refers to an object's tendency to float.

bureacracy   Government by administrative officials.

Bureau of Land Management   (BLM) An agency of the U.S. Department of Interior, with responsibilities for managing public domain lands, and the energy resources of U.S. public lands.

burette   A measuring instrument, a calibrated glass tube that is open at the top and fitted with a small tap at the bottom. It is used to measure small quantities of liquids (usually up to 50 cubic centimeters), and to perform titrations.

burial metamorphism   A type of regional metamorphism that occurs by the deep burial of sediments in the absence of other tectonic compressions or magmatic intrusions.

burn 1) To set on fire; to cause combustion (or rapid oxidation) quickly. 2) Injury to living tissue caused by exposure to high heat, light, drying winds, electricity, radiation, or caustic chemicals. See SLASH-AND-BURN AGRICULTURE. 3) A term used in Scotland and England to refer to a small stream.

burner reactor A nuclear reactor that does not convert material in the reactor's core into fissionable material because there is no fertile material in the core. Compare BREEDER REACTOR.

butadiene CH_2=CH-CH=CH_2, a colorless gas made from butane (from petroleum or natural gas). It is used in synthetic rubber.

butane H_3C-$CH_2CH_2CH_3$, a colorless, gaseous, straight-chain hydrocarbon refined from crude petroleum. It is compressed for use as a fuel, as in lighter fluid.

butte An isolated, flat-topped hill or small plateau, often bounded by steep sides and talus slopes. A butte is commonly taller than it is broad. Compare MESA.

butylated hydroxytolune The full name for the antioxidant BHT. See BHT.

Buys Ballot's law Expresses a meteorological difference between the Northern Hemisphere and Southern Hemisphere. When the wind is coming from behind, the pressure is lower on the observer's left in the Northern Hemisphere. In the Southern Hemisphere, the pressure will be lower to the right of the observer.

BWR Short for boiling water reactor. See BOILING WATER REACTOR.

byssinosis Respiratory disease among cotton, hemp, and flax workers characterized by a recurring tightness in the chest (usually after a break from exposure). It is caused by inhaling fiber dust, bacteria, molds, and fungi. See PNEUMOCONIOSIS, RESPIRATORY FIBROTIC AGENTS.

C

C Chemical symbol for the element carbon. See CARBON.

C_3 plant Those plants processing carbon dioxide directly, through the enzyme ribulose-1,5-diphosphate carboxylase (RuDP) in leaf cells, to produce a three-carbon sugar. Beans and wheat are C_3 plants. Also called a C-3 or Calvin-cycle type plant. Compare C_4 PLANT.

C_4 plant Plants processing carbon dioxide during photosynthesis through the enzyme phosphoenolpyruvate carboxylase. Also called a Hatch and Slack type plant, the process results in a four-carbon sugar and gives such plants much greater productivity from photosynthesis than C_3 plants. Corn is a C_4 (also written C-4) plant. Compare C_3 PLANT.

Ca Chemical symbol for the element calcium. See CALCIUM.

cable logging A timber-harvesting technique of moving felled trees from where they are cut to a landing area by means of a steel cable drawn by a stationary engine. It is usually used in clearcuts or large patch cuts in mountainous terrain. Also called cable yarding.

Cactaceae The cactus family of the Angiospermophyta. See OPUNTIACEAE.

cadmium (Cd) A soft, bluish, toxic, metallic element with atomic number 48 and atomic weight 112.4; it is only found in combination with other elements, especially zinc. Its properties resemble those of zinc. It is used in solders and electroplating, forming a protective layer over steel and steel alloys. It is also used in nicad (nickel-cadmium) batteries. Cadmium is also used in the control rods of nuclear reactors because it is an excellent absorber of neutrons. It is one of the heavy metals that must be removed when treating wastewater, and is found in ash from incinerators. See HEAVY METAL.

caecum Another spelling (British) for cecum. See CECUM.

caenogenetic Having physical features adapted to the needs of immature stages of an animal that disappear when the animal reaches its adult stage. The membranes surrounding vertebrate embryos are caenogenetic.

Caesalpinaceae See FABACEAE.

caffeine A plant alkaloid found in coffee, tea, and chocolate. One cup of strong tea or coffee contains 100 to 150 mg of caffeine. It is a stimulant to the central nervous system. It is used (often in combination with aspirin) in commercial headache remedies and is also a diuretic.

cal or Cal Short for calorie. Lowercase indicates a small calorie; capitalized indicates a large calorie (1000 small calories). See CALORIE.

calc-alkaline A term applied to any rock belonging to a series of igneous rocks, ranging in composition from that of basalt to andesite, dacite, and rhyolite (extrusive forms), or belonging to the range from that of gabbro to diorite, granodiorite, and granite (intrusive forms).

calcareous 1) A soil that contains a high concentration calcium carbonate. 2) A sea floor deposit containing more than 30 percent of calcium carbonate.

calcicole   A plant that grows best on limestone, or soils with a high percentage of free calcium carbonate. Contrast CALCIFUGE.

calcification   The redeposition of calcium carbonate carried by water percolating through the soil and other material. Continuous calcification in soil may lead to formation of caliche. See CALICHE.

calcifuge   A plant that is rarely found on, or unable to grow on, limestone or soils with high percentages of free calcium carbonate. Contrast CALCICOLE.

calcination   1) Heating ores to a high temperature in the presence of oxygen without melting, to drive off water and carbon dioxide or to cause oxidation. Calcination is used in the processing of limestone and for converting metal sulfides to free metals. Also called roasting.

calciphile   Calcium-loving plants; another term for calcicole. See CALCICOLE.

calcite   A calcium carbonate mineral form that is a common constituent of chemical sedimentary rocks. Calcite crystallizes into hexagonal prisms. See APPENDIX p. 621.

calcium   (Ca) A soft, grayish, metallic element with an atomic number 20 and atomic weight 40.08. Calcium occurs commonly in limestone, marble, gypsum, and chalk. Because it is a major component of bones and teeth, it is an essential mineral in the human diet. It is used in plaster and as a reducing (or deoxidizing) agent in chemical reactions.

calcium carbonate $CaCO_3$, a naturally occurring form of calcium that is also called calcite. Chalk and limestone are both calcium carbonate. Dissolved in hydrated form (calcium hydrogen carbonate), it is what makes hard water "hard." It

is used in Portland cement. Calcium carbonate in the form of ground limestone is the most common (and least expensive) form of lime used in agriculture to improve soils. See HARD WATER, DOLOMITE.

calcium chloride $CaCl_2$, a salt of calcium. It is used on walkways to melt ice; it is less toxic to plants than common table salt (sodium chloride). Calcium chloride is a strong deliquescent: It will absorb so much water from the air that the powder turns into a puddle. This property makes it useful as a chemical dehumidifier for damp closets, and effective industrially to remove water from gases or air. Also called muriate of lime.

calcium cycle   The cyclical process in which calcium is taken from the soil by plants, animals eat the plants, and calcium is then passed back into the soil when the animals die and decompose. Calcium is transferred from terrestrial to aquatic and marine ecosystems via river transport. In lakes and oceans, calcium is incorporated in the shells and skeletons of many organisms. At high pH, calcium also readily combines with carbonate ions in the carbonic acid-bicarbonate-carbonate system. The deposition of calcium carbonate shells and skeletal materials is a major process in the formation of limestones on the sea floor. These carbonates may be uplifted to terrestrial landscapes over a geological time scale. Calcium-rich soil particles are often transported between ecosystems through the atmosphere. See BIOGEOCHEMICAL CYCLES, SEDIMENTARY CYCLE.

calcium hydroxide $Ca(OH)_2$, a compound derived from quicklime (CaO, caustic lime) by slaking, mixing with water, which gives off much heat. Calcium hydroxide is used to add calcium to poor soils, as in liming a lawn. It is also used in whitewash, bleaches, wastewater treatment, and glass manufacturing. Also called slaked lime.

calcium oxide   CaO, a compound commonly known as lime, quicklime, or caustic lime. Calcium oxide has been produced for hundreds of years by burning calcium carbonate (in the form of limestone, shells, or bones) to high temperatures in lime kilns. Calcium oxide is used in many industrial reactions, such as in refining metal ores. Different forms of agricultural lime used to neutralize acid soils (which often contain related compounds as well as CaO) are evaluated by comparing their calcium oxide content or calcium oxide equivalent.

calcrete   See CALICHE.

calc-silicate rocks   A term applied to metamorphic rocks that consist of calcium-rich silicate minerals, such as diopside and wollastonite. Calc-silicate rocks are formed by contact metamorphism of dolomite or limestone.

caldera   A large, circular topographic depression, usually measuring a few kilometers across and forming above a volcanic vent. A caldera is formed by the collapse of the roof of a magma chamber or an explosion. Compare CRATER.

caliche   A hardpan soil horizon formed by precipitation of calcium carbonate carried in solution. Caliche typically forms in semi-arid climates.

californium   (Cf) A radioactive, artificially produced, transuranic element with a number of different isotopes. The isotope with the longest half-life (800 years) has an atomic weight of 251.0 and atomic number 98. See CALIFORNIUM.

callus   A hard thickening of skin or plant tissue. In animals, it forms in response to friction or pressure on skin; in plants it forms over wounds and cut stem ends. Callus is sometimes used to refer to any hardened tissues.

calm   Windless; having a wind speed of less than 1 mile per hour or a force of 0 on the Beaufort scale.

calomel Hg_2Cl_2, also known as mercurous bichloride. It is used in very dilute form as a disinfectant and as a medication for intestinal pinworms. It was formerly used as a seed and soil disinfectant and soil fungicide, but this is no longer allowed in the United States.

caloric   Relating to heat (also, relating to the calorie content of foods). In the 18th century, caloric is what scientists called heat.

caloric requirement   Amount of energy as figured in dietary calories that a human or an animal needs each day.

calorie   Quantity of heat required to raise the temperature of 1 gram of water by 1°C. This quantity is also called a small calorie (cal). A large calorie or kilocalorie (Cal) equals 1000 small calories; it is defined as the amount of heat required to raise the temperature of a kilogram of water by 1°C. Calorie refers to the amount of energy (in kilocalories) supplied by food.

calving   The process of iceberg formation in which a glacier flows into a body of water. The buoyancy of ice causes large blocks to break away from the main body of the glacier and drop into the water.

calyptra   See ROOT CAP.

calyx   1) The outer covering of an unopened flower bud, made up of leaflike sepals. 2) A sac that may store eggs in an oviduct.

cambic soil horizon   A weakly developed soil horizon characterized by a slight reddening color change but no movement or other changes.

cambium   Layer of meristematic cells sheathing the main axes of stems or roots that generate secondary tissues in the form of concentric rows of cells. Cambium contains cells capable of dividing in both the radial and tangential planes; it produces vascular tissue in plants (xylem and phloem) as well as new wood and bark on trees. See CORK CAMBIUM, MERISTEM.

Cambrian   The first geologic time period of the Paleozoic era. The Cambrian period spanned from approximately 590 to 505 million years ago. See APPENDIX p. 611.

camouflage   Coloration and patterning (or, rarely, behavior) that provides a disguise from predators.

Campanulaceae   The bellflower family of the Angiospermophyta contains mostly herbaceous (and some woody) perennial dicots. Commoly, the flowers have five sepals, five united petals, five stamens. The flowers may be either bilaterally or radially symmetrical. The inferior ovaries, of two or of five carpels, produce a capsule or berry. Bellflowers and Canterbury-bells *(Campanula)* have actinomorphic flowers characterizing the *(Campanuloideae)* subfamily, whereas cardinal flowers and lobelias *(Lobelia)* have zygomorphic flowers typical of the *Lobelioideae* subfamily. See ANGIOSPERMOPHYTA.

Campbell-Stokes sunshine recorder   A simple meteorological instrument for measuring duration of sunlight. A glass sphere focusses the sun's rays onto a calibrated strip of cardboard, making the heat strong enough to burn through. The track left by the burning rays follows the calibrated hours, which can then be read off directly.

canalization   Evolutionary pathways through which several different genotypes produce the same phenotype.

cancer   Uncontrollable malignant growth of cells in animals or humans producing tumors that invade normal tissues. Cancers can spread by metastasis to other areas of the body. Many cancers are caused or enhanced by environmental factors, from asbestos and tobacco to radiation and chemical pollutants of water or of air. See METASTASIS.

candela   (Cd) The unit of luminous intensity in the Système International d'Unités (SI). It has replaced the old unit, candle. One candela is calculated as 1/60 of the radiating power of 1 square centimeter of a black body at 1772°C.

CANDU   Short for Canadian deuterium-uranium reactor, a nuclear power plant widely used throughout Canada. It uses unenriched uranium as fuel and heavy water (deuterium) as both moderator and coolant. See HEAVY WATER, STEAM-GENERATING HEAVY-WATER REACTOR.

canker   Generally used for any disease in plants that results in decay of the bark and the wood. Technically, any open wound on woody tissues of a plant.

cannel coal   A variety of bituminous coal with a fine-grained, uniform texture. The name "cannel" derives from a local British English variation of "candle" coal. See BITUMINOUS COAL.

canopy   The top layer of a forest or wooded ecosystem consisting of overlapping leaves and branches of trees, shrubs, or both. Compare UNDERSTORY, CROWN.

canopy closure   The stage in plant succession in which trees and shrubs grow large enough to touch or to overlap, resulting in the relatively continuous shading of lower vegetative layers that formerly received sunlight.

canopy cover   Percentage of the amount of ground covered by shadows of tree and shrub leaves; technically, the vertical projection of tree and shrub crowns onto the ground.

cantharophily   Pollination by beetles. See ENTOMOPHILY, MYRMECOPHILY, ORNITHOPHILY.

canyon   A deep and narrow valley.

capacity   A measure of a stream's ability to carry a load of solid sediments.

cap cloud   An unusual stationary cloud formation that forms over a mountain peak. The layerlike cloud is not dispelled by wind blowing through it.

capillary   1) A blood vessel with a very fine, slim opening that joins arterioles (the smallest arteries) with venules (the smallest veins). Also used for similarly small vessels carrying lymph or bile. 2) Adjective describing any slender, hairlike structure in animals or plants. 3) Small pore (micropore) space in soil through which water moves through wicking or capillary action.

capillary action   Movement of a liquid caused by adhesion (molecular attraction between unlike substances, i.e. liquids and solids) and surface tension. Examples include the movement of water through soil, capillary tubes (very fine tubes with small diameters), and blotter paper.

capillary fringe   A zone above the water table where pore spaces of soil or rock are completely filled with water held up either by electrostatic attraction between mineral and water molecules or by osmotic forces.

capillary gas chromatography   A type of gas chromatography used for the separation and quantification of trace levels of volatile organic substances in environmental analyses. A capillary column is a small diameter (0.2 to 0.5 mm) glass column coated on the inside with an adsorbent. A mixture of organic molecules passing through the column during a gas chromatographic run adsorbs onto the walls of the column for different periods of time depending on the nature of the molecule, resulting in the separation of the components of the mixture.

capillary water   Liquid moisture held between soil grains either by electrostatic attraction between mineral and water molecules or by osmotic forces.

Caprifoliaceae   The honeysuckle family of the dicot angiosperms. Mostly woody plants, the members of this family have opposite twig arrangement. The four to five petals are united and have an equal number of attached stamens. The inferior ovary of three to five carpels, each with a single ovule, matures into a berry or drupe. Viburnums (*Viburnum*), elderberries (*Sambucus*), honeysuckles (*Lonicera*), and snowberries (*Symphoricarpos*) are members of this family. See ANGIOSPERMOPHYTA.

Caprimulgidae   A family of birds, commonly known as nightjars, and belonging to the order Caprimulgiformes. They have worldwide distribution and are noted for their persistent, nocturnal calls, which have given some species names such as whip-poor-will and chuck-wills-widow. The small, weak feet are pratically useless for walking. Although most migrate great distances, it has been reported that the poor-will of western North America may hibernate in winter. One group, the nighthawks, have a metropolitan distribution, where they may build nests on the flat roofs of large, city buildings. See CAPRIMULGIFORMES.

Caprimulgiformes An order of birds with swept-back wings and huge mouths adapted for hunting insects at night. See AVES, CAPRIMULGIDAE.

cap rock A layer of impervious rock that obstructs the motion of fluids to or from an underlying layer. Cap rock may restrict the downward motion of groundwater or the upward motion of petroleum.

capsule 1) A dry, budlike casing of seeds that opens when it ripens. Seed capsules form on poppies, iris, and daylilies. 2) A membrane or fibrous covering around an animal organ, also called a theca.

captive breeding Raising animals or plants in a controlled environment, such as a zoo or an arboretum, with the goal of reintroducing the stock into the wild.

capture-recapture See LINCOLN-PETERSON INDEX.

carapace A bony shell shielding the backs of some animals, such as turtles and crabs.

carat 1) A measure of weight, equal to 0.2 grams, applied in assessment of precious stones. 2) In measuring the proportion of gold content in alloys, a carat is equal to 1 part in 24.

carbamate Any of a group of compounds used as herbicides, fungicides, and insecticides. They have the general formula R-O-CO-NH-CH$_3$. Carbamates such as Sevin (carbaryl) have been developed as replacements for chlorinated hydrocarbons (aldrin, DDT, etc.). They are less hazardous to the environment than the chlorinated hydrocarbons; they break down more quickly and do not accumulate in the

food chain. (Sevin, however, is very toxic to bees.) Carbamates include Carbaryl, Barbam, and Asulam, as well as thiocarbamates and dithiocarbamates such as Diallate, Triallate, Zineb, Maneb, and Metam sodium.

carbohydrase Any enzyme catalyzing the hydrolytic breakdown of a carbohydrate. Carbohydrases such as amylase and lactase are essential in converting carbohydrates into usable energy. See HYDROLYSIS.

carbohydrate A class of organic molecules with the general formula $C_x(H_2O)_y$. Plants produce carbohydrates—sugars and starches—through photosynthesis; these substances are eaten by animals as sources of energy. Examples of carbohydrates include glucose, sucrose, starch, and cellulose.

carbon-14 dating (radiocarbon dating) A method for finding the age of geological formations containing organic matter, or archaeological finds. It is based upon the constant levels of a radioactive isotope of carbon ^{14}C present in carbon dioxide in the atmosphere. Plants and animals take up this radioactively tagged carbon dioxide into their tissues. By analyzing how much of this isotope remains in a sample of organic matter, geologists and archaeologists can compare it with the isotope's half-life to determine how much radioactive decay has occurred and to deduce how much time has, therefore, elapsed.

carbon (C) A nonmetallic element with an atomic weight of 12.01 and atomic number 6. It is found in all organic compounds, and is the only element whose compounds are considered an entire branch of chemistry. Pure carbon exists in two different crystalline forms, diamond and graphite, and in numerous noncrystalline forms such as charcoal and carbon black. Its ability to link into complex chains or rings of hydrocarbons and other organic compounds makes it the backbone of biochemistry. See ACTIVATED CARBON, CARBON BLACK.

carbonaceous chondrite A type of stony meteorite that is rich in carbon and lacking in metals. Carbonaceous chondrites are apparently derived from the primitive material of the solar system.

carbonate Any salt or ester of carbonic acid; a substance containing the -CO_3 group (the carbonate ion). Carbonate salts dissolve in water to form basic solutions.

carbonate compensation depth The depth in ocean water at which the supply of solid calcium carbonate from above equals the amount which is being dissolved below. Carbonate rocks generally form above the carbonate compensation depth.

carbonate sediment A clastic or biogenic deposit of material consisting primarily of nonsilicate minerals containing proportions of one carbon atom to three oxygen atoms plus some metallic ions (e.g., calcite and dolomite).

carbon black A pure form of carbon, a fine soot produced by burning natural gas or other hydrocarbon in a chamber with insufficient oxygen to complete the combustion. Carbon black is mixed with rubber to increase its durability (as in tires or rubber soles) and is used in paint and ink as a pigment.

carbon cycle The cyclic pathway through which carbon circulates throughout nature. Plants through photosynthesis fix carbon dioxide (CO_2) from the air, producing carbohydrates and other organic compounds. Animals and plants metabolize these carbohydrates for energy, releasing CO_2 into the air. After the death of organisms, their carbon may be released as CO_2 back into the atmosphere through decomposition. If decomposition is slower than the fixation of CO_2, carbon may accumulate. See BIOGEOCHEMICAL CYCLES.

carbon dioxide CO_2, a colorless gas that is produced by animal respiration, fermentation, and by the burning of hydrocarbons. Carbon dioxide is used in fire extinguishers, and refrigeration (dry ice), and is the gas present in soda water and other carbonated beverages. It is taken up by plants during photosynthesis and expired by plants in the absence of light. Carbon dioxide makes up a very small amount (0.035 percent) of the earth's atmosphere; increasing quantities of CO_2 in the atmosphere may cause a greenhouse effect. See GREENHOUSE EFFECT.

Carboniferous The geologic time period falling between the Devonian and the Permian periods of the Paleozoic era. The Carboniferous period lasted from approximately 360 to 286 million years ago. Carboniferous is the preferred term for international usage because it encompasses both the Mississippian and Pennsylvanian periods commonly used in North America.

carbonization A variety of fossilization in which the tissues of an organism are reduced to a thin film of carbon within a rock.

carbon monoxide CO, an odorless, poisonous gas produced by the incomplete burning of hydrocarbons, as in automobile exhaust and cigarette smoke. A major air pollutant, it is also produced by partially anaerobic decomposition of organic material. It poisons by displacing the oxygen attached to hemoglobin in the blood, preventing cells and tissues from receiving oxygen.

carbon sink   A portion of the biosphere in which carbon dioxide is absorbed faster than it is released, and which tends to keep carbon bound up for relatively long periods of time. Ocean sediments and tropical rainforests are carbon sinks, balancing global carbon dioxide produced by animals and helping to buffer the increasing quantities of carbon dioxide churned out by human activities such as burning fossil fuels. Compare CARBON SOURCE.

carbon source   Source of carbon dioxide that circulates in the carbon cycle, such as human (animal) respiration or combustion. Compare CARBON SINK.

carbon tetrachloride CCl_4, a toxic fluid with dangerous fumes. It was formerly used extensively as a solvent for dry cleaning and in household cleaning fluids. It is still used in some fire extinguishers. Also called tetrachloromethane.

carboxyhemoglobin   A blood compound formed when carbon monoxide is inhaled. Hemoglobin absorbs the carbon monoxide instead of oxygen, and is thus unable to supply oxygen to tissues, causing the asphyxiation characteristic of carbon monoxide poisoning.

carboxyl   The functional group -COOH found in all carboxylic acids. See CARBOXYLIC ACID.

carboxylic acid   Any organic acid containing one or more -COOH (carboxyl) groups. Acetic acid, lactic acid, and citric acid are all carboxylic acids; long-chain carboxylic acids are called fatty acids. See FATTY ACID.

carcinogen   Any substance that produces or promotes cancer.

carcinoma   Any of a group of cancers that occur in epithelial tissue (skin and mucous membranes), capable of spreading to other tissues. Basal-cell carcinoma, usually caused by overexposure to the sun (and in danger of increasing as protective ozone is depleted), is an example. See CANCER, LYMPHOMA, SARCOMA.

cardinal points   The four principal compass points: north, south, east, and west.

Caribbean floral region   Another name for Central American floral region. See CENTRAL AMERICAN FLORAL REGION.

carinate   Having a structure forming a ridge or a keel, as in the breastbone of a bird or the keeled scales of some reptiles.

carnallite   A mineral of potassium magnesium chloride forming orthorhombic crystals in evaporite rocks. Carnallite is a source of potassium and is used in the manufacture of fertilizer.

Carnivora   With over 280 species, carnivores vary in weight from the polar bear (500 kg) to the least weasel (50g) and are usually divided into two suborders, Fissipeda and Pinnipeda. The fissipedes include the families Canidae (dogs, wolves, and foxes), Ursidae (bears, including the giant panda), Procyonidae (raccoons, coatis, etc.), Mustelidae (weasels, badgers, wolverines, martens, skunks, and otters), Viverridae (mongooses, genets, and civets), Hyaenidae (hyenas) and Felidae (big cats, cheetahs, clouded leopards, pumas, and smaller cats such as lynx and servals). With few exceptions, all carnivores have pronounced carnassial dentition in which the first lower molar and fourth upper premolar are modified as shearing surfaces for tearing meat. The pinnipedes, which include the sealions, walruses, and seals, have a highly modified body form for swimming including the paddle-like limbs or flippers. They are sometimes placed in an order of their own because their relationship to other carnivores is controversial. See EUTHERIA, GIANT PANDA, BIG CATS.

carnivore   Organisms that eat live animals, especially a member of the Carnivora order of mammals with characteristic large, sharp canine teeth. Cats, dogs, lions are carnivores. Contrast HERBIVORE, OMNIVORE, FRUGIVORE.

carnivorous plant   A plant that obtains many of its nutrients by trapping and digesting insects, though it also photosynthesizes. Many species, such as the Venus flytrap and pitcher plant, are often found growing on nutrient-poor soils or in wet environments where all of the nutrients present are not readily available. Also called an insectivorous plant.

carnotite   An ore mineral of hydrous vanadate, potassium, and uranium forming monoclinic crystals within sandstones. See APPENDIX p. 621.

Carnot's principle   For any reversible heat engine, efficiency depends only on the temperature range over which it works and is independent of any properties of the working substance. It is derived from Carnot's theorem that heat engines can only be as efficient as a reversible engine working through the same temperature span.

carotene   Any red or yellow pigment of the carotenoid (terpenoid) group found in carrot and other plants and animals. In animals, alpha and beta carotene get converted into vitamin A in the body. See CAROTENOID.

carotenoid   Any of a group of yellow-to-dark red and brown pigments found in many plants and animals. Carotenoids include carotenes and xanthophylls; they provide the characteristic color of vegetables such as tomatoes and some of the colors of autumn leaves (especially yellow). See CAROTENE, PHYCOBILINS, XANTHOPHYLLS, FUCOXANTHIN.

carpals   Bones of the wrist in land-based vertebrates.

carpel   A reproductive unit of the female portion of a flower composed of the ovary, style, and stigma, and often resembling a modified, folded leaf. One or more carpels make up the pistil or gynoecium of a flower. Compare STAMEN.

carr   1) A form of wetland vegetation that includes some trees. 2) A wood composed of alders. See FEN.

carrageenan   1) Irish moss (*Chondrus crispus*), an edible species. 2) The emulsifier extracted from this plant, used by the food processing industry (as in some ice creams).

carrier   1) Any agent that transfers a substance, especially catalytic agents that transfer an element or group from one compound to another and molecules that transport compounds across cell membranes. 2) An individual containing a recessive gene for a genetic disorder (such as hemophilia) in their set of chromosomes, which is transferred to their young. 3) An organism infected with a disease or parasite and showing no symptoms, but still able to transfer the disease or parasite to infect another organism.

carrying capacity   The maximum population of a species that a specific ecosystem can support over long periods of time.

Cartagena Convention   An international agreement between 27 nations calling for an action plan to protect the Carribean. It was signed in 1983 in Cartagenas de Indias, Colombia.

cartilage   A strong, flexible, fibrous material that makes up part of the skeleton of vertebrates. In cartilaginous fish, the entire skeleton is made up of cartilage. In vertebrates, the cartilaginous skeleton of embryos is largely replaced by bones in adults, except in the ear, nose, bone ends, and joints. Sometimes called gristle.

Caryoblastea   Temperate, freshwater organisms resembling giant amoebas comprising a phylum of the Protista. They differ from true amoebas (phylum Rhizopoda) in lacking mitotic spindle microtubules and chromatin granules. *Pelomyxa*, a protist that relies on bacterial endosymbionts that function like mitochondria, is the most well-known member of this phylum. See PROTISTA, RHIZOPODA.

Caryophyllaceae   The pink family of the Angiospermophyta. One of the largest dicot families in North America, it is characterized by herbaceous plants with opposite, simple leaves and swollen nodes. The perianth is made up of five petals, five sepals, five or ten stamens, and usually a single-chambered ovary with free, central placentation. Some of the best-known members of this family are carnations and pinks *(Dianthus)*, chickweeds *(Stellaria* and *Cerastium)*, and campions *(Lychnis)*. See ANGIOSPERMOPHYTA.

caryopsis   A dry, single-seeded fruit grain of grasses and cereals (e.g. wheat).

cascade   1) A small waterfall, especially one composed of a series of small steps. 2) A sequence of devices linked so that the action of one causes the action of the next. The transistors or valves in an amplifier are set up as a cascade. 3) Cascade can also refer to chemical systems for purification in which a substance is run through a sequence of identical steps, each of which results in increased purity of the substance. Such a cascade is used in many industrial processes, including the separation of radioactive isotopes.

casein   Any of a group of phosphorus-containing proteins found in milk; when coagulated by enzymes or acids, they are the main constituent of cheese (which in Latin is *caseus)*. It is used in preparing artists' canvases and in making some forms of paint. Casein is also used in making plastics and adhesives.

case law   See COMMON LAW.

case study   A comprehensive descriptive study of a specific person, place, or issue.

Casparian strip   A band found in cell walls of the endodermis tissue in plant roots, made water-resistant by deposits of lignin and other materials.

Cassiaceae   See FABACEAE.

cassiterite   The chief ore mineral of tin; it forms dipyramidal tetragonal crystals in hydrothermal veins, granites, or pegmatites. See APPENDIX p. 621.

cast   1) Soil and waste excreted by the earthworm. 2) Skin that an insect sheds.

caste   Social position in a hierarchical system. Social insects such as honeybees have different physical forms depending on whether they are of the queen, drone, or worker caste. See BEES.

catabolism   The metabolic process of breaking down complex molecules into simpler ones, releasing energy. Also called destructive metabolism or dissimilation. Compare ANABOLISM.

cataclastic   A textural description applied to

rocks that are deformed by shearing and granulation of minerals during tectonic dislocation. Cataclastic rocks are recognized by an angular fragmental mineral texture. See MYLONITE.

cataclastic metamorphism   An alteration of rock by heat and pressure, which results in the formation of a cataclastic texture.

cataclastic rock   Any rock that has a cataclastic texture.

catadromous   Organisms that live in fresh water but migrate to salt water to lay their eggs. For example, eels are catadromous. Also spelled katadromous. Contrast ANADROMOUS, OCEANADROMOUS FISH.

catalysis   Initiation, speeding up, or (rarely) slowing down a chemical reaction caused by the presence of a catalyst. Digestion depends on catalysis by enzymes. See CATALYST.

catalyst   Any substance that causes or speeds up a chemical reaction without itself undergoing a permanent chemical change. Enzymes are proteins produced by the body that act as catalysts in many physiological processes, such as digestion.

catalytic converter   An air pollution control device. Catalytic converters are attached to the exhaust systems of automobiles and are used in newer models of wood stoves. They contain a catalyst (usually platinum with small amounts of other heavy metals such as rhodium, embedded in aluminum oxide) that promotes conversion of carbon monoxide and unburned hydrocarbons into harmless carbon dioxide and water, and nitrogen oxide into nitrogen gas.

catalytic cracking   The process used to convert crude oil into usable lighter petroleum products. Silica gel or alumina are used, in combination with heat and pressure, to break down the large hydrocarbon molecules in crude oil into smaller molecules. Also called thermal cracking or simply cracking, this process increases the amount of gasoline that can be produced from a given quantity of crude oil. See FRACTION.

catalytic reforming   Any of a group of processes using catalysts to change low-grade hydrocarbons into higher-chain hydrocarbons. Often called simply reforming, this process can also be used to produce benzene from straight-chain hydrocarbons.

catarobic   Describing a body of water containing organic matter that is decomposing slowly enough to avoid depleting the supply of dissolved oxygen.

catastrophe   Any sudden, violent event that causes radical change in an ecosystem, such as a forest fire or massive blowdown. See BLOWDOWN.

catastrophic extinction   Widespread death leading to extinction and caused by a single, cataclysmic event. A number of theorists suggest that catastrophic extinction caused by a huge asteroid colliding with the earth explains the sudden end of the dinosaurs shown by the geologic record.

catastrophic speciation   Rapid evolution into new species resulting from radical changes in an environment caused by a catastrophe.

catchment (area)   See WATERSHED.

catena   A collective term for an area of topographically variable soils that share a common age and history and usually share a common parent material. Catena is usually applicable in non-glaciated topographies and areas of loess deposition.

catenary   A shape naturally assumed by a perfectly flexible chain or cord suspended between two points, or the mathematical curve given by a hyperbolic cosine function.

caterpillars   Larval forms of butterflies and moths of the order Lepidoptera. See LEPIDOPTERA.

cathode   The negative electrode of an electrolytic cell, the cation-attracting electrode at which reduction occurs. Also, the positive terminal of a battery or cell producing electrical current. Contrast ANODE.

cation   1) An atom, or group of atoms, with a positive charge. Examples of cations are the hydrogen ion (H^+) and the ammonium ion (NH_4^+). Contrast ANION.

cation exchange capacity   (cec) The total of all positive, moveable cations that a substance can adsorb. It is usually used in reference to soil and gives an indication of the soil's ability to hold some nutrients long enough to be used by plants. Sandy soils have low cation exchange capacities, but these can be raised by adding organic matter, which has a high cation exchange capacity. Many clays also have high cation exchange capacities. See EXCHANGEABLE CATIONS.

catkin   A flower type consisting of a long spike covered with many stalkless, unisexual flowers (there is usually a large, drooping male catkin and a relatively inconspicuous, erect female catkin). Catkins occur on birch and oak trees and willows. Technically known as an ament.

caudal   Near the tail or rear of an animal, such as a caudal fin on fish or caudal vertebrae.

cauliflory   Formation of flowers directly on trunks and branches rather than the more usual location at the ends of new twigs. Cocoa, figs, and some other tropical trees demonstrate cauliflory.

caustic   Corrosive; causing burning and destruction of living tissue. Caustic substances are usually extremely alkaline (have a very high pH value); lye and Drano (sodium hydroxide) are caustic substances.

caustic scrubbing   An air pollution control process used to remove sulfur dioxide (SO_2) from flue gases. The gases are made to flow through a solution of sodium hydroxide (NaOH) which, with the addition of calcium carbonate ($CaCO_3$), removes the sulfur by precipitating it out in the form of calcium sulfate ($CaSO_4$, gypsum). Remaining in the solution is sodium carbonate (Na_2CO_3), a nontoxic compound; it can be disposed of by diluting and discharging into water bodies. Compare DRY LIMESTONE PROCESS, DRY ALKALI INJECTION.

caustic soda   The common name for sodium hydroxide (NaOH), a very strong and caustic alkali used in the manufacture of bar soap. It is used in many chemical reactions, in papermaking and in the tanning of leather.

cave   A naturally occurring underground chamber that most commonly forms by the dissolution of bedrock material. Caves usually have an opening in the side of a hill or mountain.

cavity   A natural void, pit, or hole in a natural geologic material.

cavity nester   Species of birds that nest in hollowed-out spaces rather than in exposed nests. Woodpeckers are cavity nesters.

CCC   See CIVILIAN CONSERVATION CORPS.

Cd Chemical symbol for the element cadmium. See CADMIUM.

cecum A pouch-like part forming the beginning of the large intestine. The appendix is attached to the cecum.

ceiling Height above the ground of the lowest layer of clouds.

celestial equator The projection of the earth's equator onto the imaginary sphere surrounding the earth on which the stars appear to be located, the celestial sphere.

celestial sphere Imaginary sphere surrounding the earth on which the stars appear to be located.

celestite A principal ore mineral of strontium sulfate that most commonly forms orthorhombic crystals in chemical sedimentary rocks, or is occasionally found in hydrothermal veins with galena and sphalerite.

cell 1) The basic component of animal and plant life, a structure in eukaryotes consisting of a DNA-containing nucleus surrounded by cytoplasm and a cell membrane (plants have a cell wall surrounding the cell membrane). Some organisms (Monera) consist of single cells without discrete nuclei; others consist of colonies of cells or complex arrays of cellular tissues. 2) A portion of a landfill into which solid waste is dumped and, after compaction, covered over with earth. Each cell is separated from others by walls of earth, and every day new solid waste is added and covered with dirt. 3) Meteorological term for centers of pressure (either high or low pressure). 4) A generator that depends on chemical reactions to produce electricity, such as a battery.

cell division The splitting of one cell into two cells; the process requires duplication of the nucleus as well as division of the cytoplasm. Cell division is usually either mitosis or meiosis.

cell membrane See PLASMA MEMBRANE.

cell theory The concept that cells are the basic structures forming all life, and that all cells develop from other cells. The cell theory was developed by Schwann and Schleiden in 1839. Viruses and organelles such as mitochondria and chloroplasts contradict the cell theory because they are not composed of cells but contain genetic material and can reproduce within cells.

cellular slime molds See ACRASIOMYCOTA.

cellulolytic enzymes Enzymes that can hydrolyze cellulose. See ENZYME.

cellulose $(C_6H_{10}O_5)_n$, a carbohydrate polymer that makes up the vast majority of plant matter. Cellulose is the main constituent of most cell walls and, therefore, of woody tissues. It is the raw mate used to make paper and rayon.

cell wall The rigid structure surrounding the cell membrane of plant or algal cells. The cell wall is composed primarily of cellulose (except for fungi and some algae).

Celsius (C) Temperature scale based on the freezing and boiling points of water, and dividing the range between the two into 100 degrees. It is the standard form of measuring temperature in the Système International d'Unités (SI). See CENTIGRADE, KELVIN.

cement 1) Any of various substances used in mortar and concrete to bind other materials (such as gravel) together. Cements are powders that, when mixed with water, set and dry to form a hard, rock-like material. Construction cement usually consists of mineral clays (containing aluminosilicates) mixed with heat-treated limestone. 2) In geology, the substance that glues together loose particles into rock. The cement binding sedimentary rock may be ferruginous, calcareous, or silicaceous. See ALUMINOUS CEMENT.

cementation   The process in which sedimentary particles are fixed together by the chemical precipitation of a binding agent.

cement kiln   A furnace used in the production of Portland cement. Limestone (calcium carbonate) is thoroughly mixed with clay (a source of aluminosilicates) and burned in the kiln to produce silicates and aluminates of calcium. These are ground to produce Portland cement.

cement rock   A form of limestone with a high clay content (over 18 percent).

cenote   A limestone sinkhole having a pool of water at the bottom. Cenotes are common in Yucatan; the name is a Spanish adaptation of a Mayan word. See SINKHOLE.

Cenozoic (era)   The last of three eras of the Phanerozoic Eon in geologic time. The Cenozoic spans from approximately 65 million years ago to the present. The term Cenozoic means recent life. See APPENDIX p. 610.

census   An attempt to count each individual member of a specific population.

center pivot irrigation   Large-scale overhead irrigation of crops using a machine that pivots around a central well. The machine consists of an arm several hundred yards long with one end anchored to a motorized pivot at the well, and the other end supported by wheels. The machine rotates around the central point, spraying water along the length of the arm. This form of irrigation shows from the air as large circles of green. See IRRIGATION.

centi-   A prefix used in the Système International d'Unités (SI) to denote one hundredth. A centimeter is one hundredth (10^{-2}) of a meter.

centigrade   Divided into 100 degrees, especially a thermometer so divided. Centigrade is used colloquially to refer to the Celsius scale, but the name was officially changed to Celsius in 1948. See CELSIUS.

Central American floral region   One of the phytogeographic regions into which the Neotropical realm is divided according to similarities between plants; it consists of the Carribbean islands, Cental America, the southernmost portions of California and Florida, Guyana, Surinam, French Guiana, and most of Colombia and Venezuela. See NEOTROPICAL REALM, PHYTOGEOGRAPHY.

central capsule   The inner region of the protoplast in Radiolaria, containing the nucleus or nuclei and surrounded by a membrane.

central eruption   An early-stage volcanic eruption in which lavas, pyroclastics, or gases are released from a central vent.

central nervous system   (CNS) The portion of an organism's nerve tissues that control all nerve functions. In vertebrates, the CNS is the brain and spinal cord; in invertebrates, it may be simply a few ganglia with their associated nerve cords. Compare AUTONOMIC NERVOUS SYSTEM, PERIPHERAL NERVOUS SYSTEM.

centrifugal   Pulling out from the center. See CENTRIFUGAL FORCE.

centrifugal collector   A pollution control system using "centrifugal force" (or, correctly speaking, the interaction of inertia and centripetal force) to separate and remove dustlike particles from a gas stream such as a smokestack.

centrifugal force   An imaginary force that appears to oppose centripetal force. It may seem as though an actual force is pulling outward on an object moving along a circular path, as when passengers in an automobile rounding a curve at a high speed feel themselves pulled toward the outside of the curve. This sensation is actually caused by inertia, which causes the passengers to tend to continue to move along a straight line as the car in which they sit begins turning. Compare CEN-TRIPETAL, INERTIA.

centrifuge   1) A machine used to separate substances of different densities; the machine whirls the mixture around a central point and gravity pulls the heavier component to the outside of the container. Centrifuges are used to separate cream from milk, and very simple centrifuge extractors are used to remove honey from honeycomb. 2) To process through such a device.

centripetal   Pulling in towards the center and, specifically, the force acting upon an object to change its direction of movement from a straight path to a circular one. In a satellite, gravity provides the centripetal force to keep the satellite orbiting earth instead of continuing on a straight line out into space.

centripetal drainage pattern   A term describing the appearance in map view of stream drainage systems in which all of the streams drain toward a central lowland.

centrosome   A cellular structure located near the nucleus of the cell. During mitosis (cell division), the spindle fibers (microtubules) radiate out from the centrosome.

cephalic index   Measurement of the shape of the skull, calculated by dividing the maximum width by maximum length.

Cephalochordata   A subphylum of small (5 cm), elongated, marine chordates of the littoral zone which lack pigment and have a series of V-shaped, horizontal muscles blocks arranged along the body. The walls of the large pharynx, which has many gill slits, is specialized as a filtering mechanism for small, food particles. Although they normally burrow into sand they can swim rapidly by undulation of body muscles. See CHORDATA.

Cephalopoda   This exclusively marine and carnivorous class of Mollusca contains some of the most interesting, active, and behaviorally advanced of all invertebrates. In this group, the molluscan foot is greatly modified to form a series of tentacles about the mouth as well as a funnel which opens into the mantle cavity. Water in the mantle cavity forcibly expelled through the maneuverable funnel produces a propellent force that can move the body in different directions. Rapid movement in cephalopods is produced by directing the funnel forward and shooting backwards. A stout parrot-like beak and radula are present in the mouth cavity. External shells are present in the single genus *Nautilus*, reduced in cuttlefish, and vestigial or absent in squids and octopuses. Approximately 600 species exist. See MOLLUSCA, NAUTILUS, CUTTLEFISH, SQUIDS, OCTOPODA.

cephalothorax   The head of certain animals, including arthropods such as crabs and spiders, in which the head and thorax are combined.

CEQ   See COUNCIL ON ENVIRONMENTAL QUALITY.

CERCLA   Acronym for the Comprehensive Environmental Response, Compensation, and Liability Act, better known as the

Superfund program. A 1980 statute establishing a fund (financed by federal and state governments, and by taxes on chemical industries) for managing and cleaning up abandoned sites containing significant quantities of hazardous wastes. (42 U.S.C.A. §9601 et seq.).

cere A soft membrane near the upper beak having an opening for the nostrils of some birds, such as parrots.

cerebral hemispheres The two lobes of the cerebrum in vertebrate brains.

cerebroside Any of a group of complex fatty compounds found in nerve and other tissues; a phosphorus-free lipid combined with galactose. Also called galactolipid.

certiorari Legal term meaning to be informed of a means for gaining appellate review. A common law writ, issued from a superior court to one of inferior jurisdiction, commanding the latter to certify and return to the former the record in the particular case.

Cesium 137 A radioactive isotope (half-life, 33 years) of cesium used in medical research and treatment as a source of gamma radiation. (Normal cesium is cesium 133). Cesium 137 is a by-product of nuclear fission and explosions of atomic bombs.

cesium (Cs) A soft, silvery, metallic element with atomic number 55 and atomic weight 132.91. The vibrations of cesium atoms are used to drive the atomic clocks (cesium clocks) used as international standards for time. Cesium is sensitive to light when alloyed with metals such as antimony and gallium, and so is used in photoelectric (solar) cells. Isotopes of cesium are one of the major forms of radioactivity taken up by fish. The British spelling is caesium. See CESIUM 137.

Cestoda Class of the phylum Platyhelminthes whose adults are mainly intestinal parasites of vertebrates. At the anterior end of a tapeworm there is a specialized region, the scolex, which is equipped with suckers and hooks for attachment to the intestinal wall of its host and from which is budded off a long chain of iterated segments called proglottids. Each proglottid develops both female and male reproductive structures as it matures and eventually becomes little more than an egg-filled sac that is sloughed off at the posterior of the worm. There is no digestive tract and all nutrients are absorbed through villi on the tegument. Larval forms develop in secondary hosts during the life cycle. *Hymenolepis nana*, an important human parasite, has a worldwide distibution. See PLATYHELMINTHES.

Cetacea An order of aquatic, carnivorous mammals, the whales are distributed throughout the world's oceans and are among the largest of all mammals, including the gray whale *Eschrichtius robustus* which, at 35 to 40 tons, makes it the largest known animal ever to have existed. The whales, which have a highly modified body form for swimming, are divided into the toothed whales, including dolphins, porpoises, killer and sperm whales, and the baleen whales, such as blue, gray, humpback and right whales. Although international efforts are underway to protect whales, the whaling and fishing industries have dangerously reduced the numbers of many species. See EUTHERIA, BALEEN.

cf 1) Abbreviation for compare. 2) Abbreviation for cubic feet.

Cf Chemical symbol for californium. See CALIFORNIUM.

cfcs Abbreviation for chlorofluorocarbons. See CHLOROFLUOROCARBON.

CFR See CODE OF FEDERAL REGULATIONS.

cfs An abbreviation for a fluid flow measured in cubic feet per second.

CGS system Short for centimeter-gram-second, a version of the metric system in which the centimeter, gram, and second are used as the basic units for length, mass, and time. The CGS system is no longer in common use; it has been replaced by the Système International d'Unités (SI). See SYSTÈME INTERNATIONAL D'UNITÉS (SI).

chaetae Bristles (setae) made of chitin and used for locomotion in some annelid worms, such as earthworms and the class Polychaeta (marine bristleworms). See SETAE, POLYCHAETA.

Chaetognatha A phylum of marine, dueterostomes whose small, elongate bodies are divided into head, trunk, and tail regions, each one of which has one or a pair of coelomic cavities. There is no circulatory, respiratory, or excretory system. A number of spines located anterior to the mouth are used for grasping prey and give the phylum its name. Although seldom recognized, chaetognaths are ecologically important as a major group of planktonic predators. There are approximately 70 species.

chain 1) A common measuring unit used in land surveying. A Gunter's chain, used in the surveys of U.S. public lands, is 66 ft long in 100 links of 7.92 in. 2) A series of landscape features such as lakes, islands, or mountains.

chain reaction A reaction in which the products of one stage facilitate the next stage, making the process self-sustaining e.g., fire (combustion) and nuclear (fission) reactors. See NUCLEAR FISSION.

chalcedony A light-colored fibrous variety of cryptocrystalline quartz. Chalcedony is a major component of nodular cherts. Forms agate. See APPENDIX p. 622.

chalcocite An important ore of copper sulfide in massive form, or found in orthorhombic crystals within veins of hydrothermal rocks. See APPENDIX p. 622.

chalcophile Describing elements that are abundant in sulfide ore deposits. Lead is an example.

chalcopyrite A common ore of copper and iron sulfide forming tetragonal crystals in igneous and hydrothermal rocks. See APPENDIX p. 622.

chalk A fine-grained sedimentary rock composed of calcite crystals biogenically derived from the skeletal fragments of oceanic microorganisms.

chamaephyte A subdivision in the Raunkiaer plant classification system; a plant having overwintering ("perennating") buds located within 25 m above the surface of the soil, including both woody and herbaceous species. See RAUNKIAER'S LIFE FORMS.

chamosite A variety of the chlorite mineral group. Chamosite occurs in sedimentary ironstones and lateritic clays.

change of state A substance's transformation from one phase or state to another, as in water becoming ice or a liquid becoming a gas. Such a change either gives off or absorbs heat. Also called change of phase. See BOIL, FREEZE, MELT, SUBLIMATION.

channel A somewhat linear, concave depression that serves as a drainage route for surface runoff or groundwater flow.

channelization The engineered alteration of a watercourse to improve drainage by excavating to deepen, to widen, or to straighten a natural channel. Channelization may cause environmental damage to the river system affected.

chaos A theoretical mathematical description of bounded systems in which small perturbations are amplified and periodic behavior is often observed. Chaos theory is used to describe the behavior of physical, biological, and social systems as diverse as the circulation of patterns of air currents, fluctuations of populations, and changes in macroeconomic systems.

chaparral Vegetation type dominated by shrubs and small trees, especially evergreen species with thick, small leaves. Chaparral often depends on fires to prevent invasion by trees and is found Mediterranean climates such as south coastal California, parts of Australia, Chile, South Africa, and around the Mediterranean.

char 1) To scorch; to burn incompletely. 2) A solid remaining after a carboniferous substance has been charred, such as charcoal.

character convergence Evolution of a similar form or behavior in unrelated species in order to promote interaction between individuals. Also called social mimicry. Compare CHARACTER DISPLACEMENT, CONVERGENT EVOLUTION.

character displacement Changes in physical appearance between two species as a result of selection pressures on one or both to avoid competition with the other. The changes are most pronounced where their ranges overlap. Compare CHARACTER CONVERGENCE.

character divergence See CHARACTER DISPLACEMENT.

Charadriiformes A widely distributed order of coastal birds, members of which have varied niches and are united by common skeletal and muscular features. Examples are sandpipers, plovers, gulls, terns, skuas, curlews, stilts, woodcock, and the alcids. See ALCIDS, CURLEWS, SNIPES.

charcoal An impure form of carbon produced by burning wood, plant material, or bones in a chamber sealed to limit available oxygen. It retains traces of organic matter from its source. Charcoal is used as a fuel and as a filter, though for specialized filtering purposes purified, activated forms of charcoal are preferred. See ACTIVATED CARBON, ACTIVATED CHARCOAL.

chasmogamous Describing flowers that open to allow pollinization. Contrast CLEISTOGAMOUS.

chela A type of claw that resembles a pincer, as found on arthropods such as lobsters and crabs.

chelate A chemical compound in which multiple chemical bonds secure a metallic cation to another molecule (the chelating agent). Though expensive, chelates are used in horticulture to supply minerals in special situations, such as providing iron to plants growing preferentially in acidic soils. Also used as a verb, to combine or bind to produce a chelate. See SEQUESTERING AGENT.

chelating agent A chemical that forms a chelate with one or more metallic ions. Chelating agents, such as forms of ethylenediaminetetracetic acid (EDTA), are used to treat lead poisoning and other forms of heavy-metal poisoning because they bind to metals and make them unavailable to the body. See LIGAND.

chelation The process of binding a metallic ion to a chelating agent, resulting in the formation of a chelate.

Chelicerata Subphylum of arthropods which lack antennae and have bodies divided into a cephalothorax (head and trunk) and an abdomen of no more than twelve segments. The anterior pair of appendages, the chelicerae, are usually pincer-like, whereas a second pair, the pedipalps, have a variety of forms according to type. Each of the four remaining segments of the cephalothorax has a pair of ambulatory legs. The paired appendages of the abdominal segments, when present, are variously modified but never used for locomotion. There are three living classes, Merostomata (horse-shoe crabs), Arachnida (scorpions, spiders, and mites), and Pyncnogonida (marine, spider-like animals, some groups of which have more than four pairs of legs). See ARACHNIDA, MEROSTOMATA, SCORPIONS.

Chelonia A reptilian order that contains tortises and turtles. These animals have bodies covered by bony plates forming a shell from which the head, tail, and limbs protrude. See REPTILIA.

chemical compound Any substance containing more than one element in definite proportions by weight. These proportions are constant for each compound, regardless of the method of its preparation. Also simply called compound. Contrast ELEMENT.

chemical element Another term for element. See ELEMENT.

chemical energy The potential energy stored in the bonds between atoms of chemical compounds. This energy can be released in chemical reactions, as when the energy in fuel is liberated by combustion, or when food is converted to energy in the body.

chemical equation A representation of the qualitative and quantitative changes in substances that occur during a specified chemical reaction. A chemical equation tells what compounds, and how much of each, are produced by the combination of other compounds or elements. The equation for sulfur trioxide dissolving in water to form sulfuric acid (as in acid rain) is: $SO_3 + H_2O \longrightarrow H_2SO_4$. See BALANCED CHEMICAL EQUATION, CHEMICAL REACTION.

chemical oxygen demand (COD) An indication of the extent of water pollution. It is a measure of the dissolved oxygen required to oxidize compounds (both organic and inorganic) in a given sample of water. The test for chemical oxygen demand usually produces a higher value than the test for biochemical oxygen demand. Compare BIOCHEMICAL OXYGEN DEMAND.

chemical reaction The process by which new substances are formed from compounds and elements that undergo a chemical change. The original substances in a chemical reaction are called the reactants; the new substances are called the products.

chemical remanent magnetism The magnetism of a rock that is acquired as ferromagnetic minerals grow by oxidation of another mineral at low temperatures.

chemical sediment A mineral material that is formed by a chemical process, such as precipitation, rather than being derived from a biogenic or clastic source. Evaporite minerals and some carbonate deposits are examples of chemical sediment.

chemical weathering The chemical decomposition of rock and surface material involving changes in chemical composition occurring at or near the surface of the earth. Contrast MECHANICAL WEATHERING, WEATHERING.

chem(o)- Prefix referring to chemistry or by chemical reaction. Chemotherapy, for example, is medical treatment using chemicals, as in cancer treatments using drugs rather than radiation. Sometimes shortened to chem- when used before words beginning with vowels.

chemoautotroph A microorganism (primarily bacteria, but also some protozoa) that derives its energy from oxidizing nonorganic compounds such as hydrogen sulfide and ammonia. Also spelled chemautotroph; chemolithotrophic and chemosynthetic organisms are additional terms for chemautotrophs. Compare CHEMO-ORGANOTROPHIC, PHOTOTROPHIC.

Chemoautotrophic Bacteria A phylum of the Monera with organisms that gain their energy from oxidizing methane (CH_4) or inorganic nitrogen or sulfur compounds, rather than from sunlight or organic compounds. The bacteria responsible for nitrification, *Nitrosomonas* and *Nitrobacter*, are members of this phylum as are the sulfur-oxidizing bacteria of the genus *Thiobacillus* and the methane-oxidizing *Methylomonas*. See MONERA.

chemolithotrophic See CHEMOTROPHIC.

chemo-organotrophic Describing organisms that derive energy from organic sources. All animals are chemo-organotrophic, as opposed to chemo-autotrophic microorganisms (primarily bacteria) that derive energy by reducing nonorganic compounds such as hydrogen sulfide or ammonia. Compare CHEMO-AUTOTROPH.

chemoreceptor A sensory nerve ending or sense organ that responds to chemical stimulation. The taste buds act as chemoreceptors. See RECEPTOR, EXTEROCEPTOR.

chemosphere The region within the earth's stratosphere and mesosphere where solar radiation causes chemical reactions.

chemostat A system for growing plants, algae, or protists in the laboratory. The system is designed to sustain steady-state growth through harvesting populations and replenishing nutrients.

chemosterilant A pesticide that works by sterilizing its target; it destroys a pest's ability to reproduce.

chemosynthesis Autotrophic process used by some bacteria and other specialized organisms to derive energy from chemical reactions without sunlight or photosynthesis. This energy, produced by oxidizing inorganic compounds such as sulfur or nitrogen, is then used to synthesize carbohydrates. Compare PHOTOSYNTHESIS.

chemosynthetic Of or using chemosynthesis, in which cells produce carbohydrates from carbon dioxide and water with energy obtained from some chemical reaction, rather than from light in photosynthesis. Compare PHOTOSYNTHETIC BACTERIA.

chemotaxis An organism's or cell's movement toward or away from a chemical stimulus. Compare CHEMOTROPISM.

chemotrophic Not phototrophic (photosynthetic); describing organisms that derive energy from consuming phototrophic organisms (or other organic matter, or that produce their own energy from chemical reactions other than photosynthesis. Chemotrophic includes both chemoautotrophic and chemo-organotrophic organisms. See CHEMO-AUTOTROPH, CHEMOORGANOTROPHIC, PHOTOTROPHIC.

chemotropism   An organism's turning or bending toward or away from a chemical stimulus. Compare CHEMOTAXIS.

chenier   A large ridge-shaped deposit of sandy material built up by wave action in a marshy area. Chenier deposits up to 50 m in length are common on the Gulf Coast of North America.

Chenopodiaceae   The goosefoot family of the Angiospermophyta. Members of this dicot family are mostly weeds that inhabit disturbed or alkaline sites and have slightly fleshy leaves that often are scurfy. The small, inconspicuous flowers lack petals and produce an urticle that is often winged. Members of the family include goosefoot or pigweed (*Chenopodium*), Russian thistle and tumbleweed (*Salsola*), saltbushes (*Atriplex*), and agricultural crops such as spinach (*Spinacia*), and beet and chard (*Beta*). See ANGIOSPERMOPHYTA.

Chernobyl   Location in Ukraine (former Union of Soviet Socialist Republics), north of the city of Kiev, where, on April 26, 1986, there were two huge explosions inside one of the four graphite-moderated, water-cooled reactors at the Chernobyl nuclear power station. These blasts blew the 900-metric-ton roof off the reactor building, set the graphite core on fire, and flung radioactive debris several thousands of feet into the air. Winds then carried some of the radioactive material over the Soviet Union and much of Eastern and Western Europe as far as 1,250 miles from the plant.

chernozem   An older term for dark-colored, zonal soils formed in temperate and cool subhumid climates supporting grassland vegetation. See MOLLISOL.

chert   A light-colored, fine-grained rock composed almost entirely of silica. Chert is often found as compact lumps and nodules in sedimentary rock. Compare FLINT.

chestnut soils   An older term for brown-colored zonal soils formed in temperate and cool subhumid to semiarid climates. See MOLLISOL.

Chezy equation   A calculation of river discharge given by the formula $Q = AC(rS)$ where A = cross-sectional area, C = Chezy discharge coefficient, r = hydraulic radius, and S = slope of water surface.

chilling effect   A lowering of global temperature caused by particles in the atmosphere (air pollution or soot from volcanic explosions) shading the earth from the sun's rays. The opposite of the greenhouse effect. See GREENHOUSE EFFECT.

Chilopoda   Centipedes. The class of uniramic arthropods with a segmented, flexible body of many segments that is not divided into thorax and abdomen. The body tends to be flattened and elongate, and each of its segments bears a pair of appendages. The appendages of the first segment are modified into poisonous fangs. There are approximately 2500 species. See UNIRAMIA.

chimera   1) An organism that has two or more tissues of different genetic composition, especially when the different geno-types are detectable in patches. Chimeras are caused by natural or artificial fusing of embryos, or by grafting. 2) A DNA molecule containing nucleotide sequences from more than one different organism. See GRAFT HYBRID.

chimney   1) A vertical structure that serves

as a conduit for a venting geologic material such as gas, mud, or lava. 2) A free-standing column of rock.

chinampas A system of agriculture practiced in Mexico since the days of the Aztecs. It consists of raised beds divided by a grid of canals; the beds are fertilized by periodically dredging the canals and spreading the silt, aquatic weeds, and organic wastes in the water and spreading these on the soil.

China syndrome Colloquial term for a meltdown occurring in a nuclear reactor. The name derives from exaggeration: Molten fuel, so hot that it would melt its way through the containment vessel, is imagined to be able to keep on going—right through the center of the earth to China. See MELTDOWN.

chinook Meteorological name for a warm wind occurring in the northwestern United States, from an Indian word meaning "snoweater." A chinook is a dry wind that travels down the eastern slope of the Rocky Mountains. As the air moves downhill from the high mountains, it is compressed and therefore heated, lowering its relative humidity. While its actual temperature is rarely above 10°C, it often raises winter temperatures by 20°C within a few minutes. As a result it rapidly makes snow cover disappear. See FOEHN.

chiroptera See BATS.

chi-squared test A statistical technique for comparing the frequency distribution of observed data with the expected—theoretically derived—frequencies. The chi-squared test helps determine how well a theory holds up when compared to actual observations from experiments designed to test the theory.

chitin The rigid substance (a polysaccha-ride containing nitrogen) that forms the exoskeleton of arthropods and exists as skeletal material in many other invertebrates. Chitin, which closely resembles cellulose, is also found in some species of fungi.

chloracne A long-lasting rash that results from exposure to large quantities of dioxin, as from an industrial accident. It is very difficult to treat and can last up to 15 years. See DIOXIN.

chlorates Any salts of chloric acid, a very strong but unstable acid, characterized by containing the -ClO_3 group. Chlorates are very strong oxidizing agents and can be very explosive. Compare CHLORIDE.

chlordane An insecticide of the chlorinated hydrocarbon family, containing a six-carbon ring attached to a five-carbon ring with eight chlorine atoms; it is no longer in use in the United States. Also known by various trade names, including Toxichlor, CD-68, Velsicol 1068, Octachlor, and Okta-Klor. See CHLORINATED HYDRO-CARBONS, CYCLODIENE INSECTICIDES.

chloride 1) The ionic form of the element chlorine, where the chlorine atom has gained one electron (Cl^-). 2) Any salt containing the Cl^- anion. Many metals readily form chlorides.

chlorinated Treated (disinfected) with chlorine, or chemically combined with chlorine or a chlorine compound.

chlorinated hydrocarbon insecticides A notorious group of insecticides containing carbon, hydrogen, and chlorine; they act as nerve poisons. Most of these insecticides—DDT, aldrin, dieldrin, lindane, chlordane, heptachlor—are no longer in use in the United States. They have been phased out or banned because they persist in the environment for years and accumulate in the food chain, reaching toxic levels in the fats of high-level predators. About the only one still in use is methoxychlor, a wide-spectrum insecticide of relatively low toxicity that has been used as a replacement for DDT. See CYCLODIENE INSECTICIDES.

chlorinated hydrocarbons Synthetic organic compounds (hydrocarbons) containing chlorine. Usually used to refer to chlorinated hydrocarbon insecticides such as DDT or methoxychlor, but the group also includes PCBs, chloroform, and carbon tetrachloride.

chlorination The use of chlorine as a disinfectant. Chlorination is used to kill germs in food processing and in sewage treatment as well as for sterilizing swimming pools and drinking water. Chlorination is also used in some industrial facilities to oxidize compounds into less harmful forms before discharging them as waste.

chlorinator A device used to add chlorine (in gaseous or liquid form) to drinking water or effluent. See CHLORINATION.

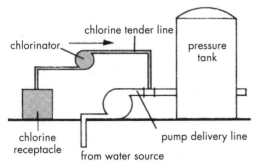

A simple system for chlorinating water.

chlorine (Cl) An element that occurs as a toxic, irritating (and poisonous) greenish gas, with atomic weight 35.45 and atomic number 17. It exists as the diatomic molecule, Cl_2; it is also common as differ-

ent salts in sea water (especially sodium chloride, table salt). A strong oxidizing agent, chlorine (Cl_2) is used in the manufacture of many organic chemicals. It is also widely used for disinfecting and bleaching. It is one of the halogen elements. See HALOGENS.

chlorine-contact chamber The section of a sewage treatment plant where effluent is treated with chlorine to kill pathogens. See CHLORINATION.

chlorine demand Quantity of chlorine required to kill pathogens in a specified quantity of water or sewage.

chlorinity Measurement of the amount of chlorine contained in a given sample of water, such as sea water. Chlorinity is also used to describe the amount of other halogens present.

chlorite Any of a group of greenish, platy, aluminum iron and magnesium silicate minerals formed in metamorphic environments. Chlorite minerals closely resemble mica minerals in composition and structure. See APPENDIX p. 622.

chlor(o)- 1) Containing or referring to the element chlorine. 2) Green in color, as in chlorophyta, the green algae.

chloroethene $CH_2=CHCl$, also known as vinyl chloride. A gaseous compound, chloroethene is used in the manufacture of PVC (polyvinyl chloride). See VINYL CHLORIDE MONOMER, POLYVINYL CHLORIDE.

chlorofluorocarbon (CFC) A class of organic compounds containing carbon, chlorine, and fluorine; Freon is a trade name for a chlorofluorocarbon. These compounds are used as refrigerants (both in refrigerators and air conditioners), as propellants in aerosol cans, as cleaners for computer circuit boards, and in the manufacture of styrofoam. CFCs are not toxic, but are being phased out because they accumulate in the upper atmosphere, where sunlight changes them into chemicals that destroy the earth's protective ozone layer. Also called halocarbons.

chlorofluoromethane   Another name for the chlorofluorocarbon compounds. See CHLOROFLUOROCARBON.

chloroform $CHCl_3$, also called trichloromethane. Chloroform is a clear liquid with a sweet, ether-like odor; it is a carcinogen. It was once widely used as an anesthetic, but has largely been replaced with less risky compounds. Chloroform is a solvent for hydrocarbons such as oils and rubber, and is used as a raw material by the chemical industry. Drinking water often contains trace amounts of chloroform, produced by the reaction of chlorinating agents with organic matter in the water.

Chloromonadophyta   Chloromonads. Unicellular algae without an eye-spot and motile by means of a pair of flagella. See XANTHOPHYTA.

chlorophyll   The green pigment found in leaves and other green parts of most plants. Chlorophyll occurs in several different forms, although one—chlorophyll a—predominates; all are structurally similar to hemoglobin. Chlorophyll molecules absorb light for photosynthesis and are, therefore, essential to photosynthetic plants for the manufacture of carbohydrates from carbon dioxide and water.

Chlorophyta   The green algae phylum of the kingdom Protista. These organisms have cup-shaped, grass green chloroplasts and at least two flagella of equal length at some stage during their life cycles. Green algae reproduce sexually through the production of zoospores. The cell walls in this phylum generally are composed of cellulose and pectin. A wide variety of morphology is exhibited by members of the phylum, which can be found inhabiting marine, freshwater, and even terrestrial habitats: the leafy sea lettuce *Ulva* of marine coasts; the solitary, biflagellated, freshwater *Chlalmydomonas*; the globular colonial *Volvox*, containing hundreds of cells; the macroscopic freshwater lime-encrusting *Chara*; the minute *Chlorella* that is either solitary, colonial, or an endosymbiont; and the common *Chlorococcum*, which grows in soil and on trees, rocks, and other moist surfaces. See PROTISTA.

chloroplast   Organelle containing chlorophyll, found in large numbers in the cells of green plants undergoing photosynthesis. Chloroplasts also occur in all algae except blue-green algae (Cyanobacteria). See PLASTIDS, CHROMATOPHORE, GRANA, STROMA.

chlorosis   A loss of color in green plants that occurs from chlorophyll deficiency. Chlorosis is caused by lack of light or unavailability of iron or magnesium in the soil; it can also be caused by viruses.

cholera   An acute intestinal disease caused by the bacterium *Vibrio comma* (also called *Vibrio cholera)*, transmitted through food or water contaminated by wastes of victims or carriers. Cholera is characterized by intense vomiting and diarrhea, painful muscle cramps, intense thirst, and eventual collapse and death, if untreated. It responds to antibiotics and can be prevented with proper sanitation and vaccination.

cholesterol   A fatty compound produced by animals and found in brain cells, in adrenal glands, and in liver and bile acids. Production of high levels of cholesterol has been implicated in heart disease and in hardening of the arteries. In the body, cholesterol is a precursor in the formation of many steroids, such as sex hormones.

choline $[(H_3C)_3N-CH_2CH_2-OH]^+OH^-$, also known as trimethyl-2-hydroxymethyl ammonium hydroxide. It is a strong base found in many plant and animal tissues combined with lecithin and other phospholipids. It is a member of the B vitamin group, although not considered an essential dietary requirement because it can be synthesized by the body. It acts to regulate the storage of fat in the liver and is a precursor for acetylcholine. See ACETYL-CHOLINE.

cholinesterase Any of a group of enzymes produced by the body to break down and therefore, regulate quantities of choline esters, especially acetylcholine at nerve endings, which control the transmission of information in the form of nerve impulses. See ACETYLCHOLINE.

Chondrichthyes A class of medium- to large-jawed vertebrates with cartilaginous skeletons including both pectoral and pelvic girdles which support the paired fins. The subclass Holocephali (ratfishes) has a single gill opening on each side of the head whereas the Elasmobranchii (sharks, skates, and rays) has multiple gill openings on each side. All 800 or so species are carnivorous and with few exceptions, marine.

chondrite A common variety of meteorite that contains condrules embedded in a stony matrix.

chondrule A spherical, droplet-shaped body of olivine, pyroxene, or plagioclase feldspar that occurs within a chondrite.

Chordata A phylum of organisms characterized by the presence of a stiff, cartilaginous rod located dorsal to the gut and ventral to the nerve cord, the notochord, from which the phylum derives it name. The notochord persists in the adult of many chordates but is lacking in most vertebrate adults, having been eliminated during the formation of the vertebral column. In addition, chordates possess a dorsal, hollow nerve cord and the lateral walls of the embryonic pharynx bear gill slits, which may or may not persist in the adult. The Chordata are divided into several subphyla including the Cephalochordata and Vertebrata. See CEPHALOCHORDATA, VERTEBRATA, also APPENDIX p. 613–615.

C horizon A soil layer composed of oxidized but otherwise unaltered parent material and relatively unaffected by biologic activity and soil-forming processes, which alter the overlying horizon. The C horizon is transitional between unweathered bedrock below and developing soil above. See WEATHERING, SOIL PROFILE, also APPENDIX p. 619.

CHP Short for combined heat and power, a British term for a cogeneration plant, a power plant that uses some of the excess heat (hot water) remaining after generating electrical power for heating living or working space.

chromatid One half of a chromosome during early stages of cell division. The parent chromosomes divide lengthwise to produce two threadlike chromatids, later splitting completely into two daughter chromosomes before the parent cell splits into two daughter cells.

chromatin A substance composed of DNA, proteins, and small quantities of RNA. Chromatin is the basic material making up the nucleus in eukaryote organisms and consolidates to form chromosomes during mitosis. It absorbs stains easily, enabling it to be traced through the stages of mitosis. In its inactive state, it is called heterochromatin; when active, it is sometimes called euchromatin. See HETEROCHROMATIN.

chromatography   A technique for separating, purifying, or analyzing a mixture of chemical substances. The mixture can be gaseous or liquid and is directed across an adsorptive medium to which the different components bind for differing lengths of time, depending on their chemical structure. This process takes many forms, including gas chromatography (GC, also called gas-liquid chromatography or GLC), ion chromatography (IC), high-pressure liquid chromatography (HPLC), and paper chromatography. See GAS CHROMATOGRAPHY, ION CHROMATOGRAPHY, PAPER CHROMATOGRAPHY.

chromatophore   1) Any pigment-containing cell, especially one found in the skin of invertebrates, such as the chameleon, allowing it to change color. 2) Also called chromoplast, a pigment-containing plant organelle. When green, these are called chloroplasts, but they may also contain red, orange, or yellow carotenoid pigments. 3) Pigmented bodies in the cells of certain photosynthetic bacteria. See LEUCOPLASTS, PLASTIDS.

chromic acid H_2CrO_4, an acid containing chromium that does not occur on its own but forms when chromium trioxide (CrO_3, a corrosive solid) dissolves in water. It can also be formed by dissolving sodium dichromate in concentrated sulfuric acid. Chromic acid is a strong oxidizing agent that is used for cleaning glassware.

chromite   The principal ore mineral of chromium occurring in mafic and ultramafic igneous rocks. Chromite is one of the spinel group. See APPENDIX p. 622.

chromium   (Cr) A hard, metallic element with atomic weight 52 and atomic number 24. The metal is electroplated as a shiny outer layer onto other metals, as in bathroom fixtures and automobile parts; it is relatively resistant to rusting and tarnishing. It is also used in steel alloys and photographic chemicals.

chromophores   A group of atoms within molecules of dye compounds, responsible for producing the color of the dyes.

chromosome   Any of the microscopic, threadlike bodies in the nucleus of a cell that become visible during mitosis. Chromosomes carry the genes (DNA strands) that determine hereditary characteristics. They usually occur in pairs in diploid organisms; bacteria and viruses usually contain only one chromosome. See GENE.

chronology   A series of things or events arranged from youngest to oldest; for example, a soil chronology, rock chronology, or flood chronology.

chronosequence   See SOIL CHRONOSEQUENCE.

chrysalis   The casing covering the pupa, a stage in the life cycle of some butterflies and moths. Monarch butterflies form a chrysalis rather than a papery cocoon. Also used for the pupa itself. See PUPA.

Chrysophyta   The golden yellow and yellow-brown algae phylum of the Protista. Their pigments, predominantly yellow and brown, are contained in chromatophores. Members of this phylum, while lacking sexual reproduction, have a single flagellum or pair of undulipodia of unequal length during some phase of their life cycle. Many of the members of this family form colonial masses or branched colonies. Food storage is in the form of oils or leucosin rather than starch. Well-known members of this phylum include the usually braching colonial *Dinobyron*, the globular colonial *Synura*, and the solitary *Mallomonas*. See PROTISTA.

chrysotile A fibrous form of the mineral serpentine. A type of asbestos.

Chytridiomycota The chytrid phylum of the kingdom Protista. Chytrids are parasites and saprobes of aquatic and soil enviroments. They have cell walls made of chitin but a thallus with many nuclei not separated by cell walls. Unlike the Hypochytridiomycotes, they have true sexual repoduction. Brown spot disease of corn is caused by *Physoderma*, a member of the phylum; other members include *Chytria* and *Blastocladiella*. See PROTISTA. Contrast HYPOCHYTRIDIOMYCOTA.

Ci Abbreviation for curie (unit of radioactivity). See CURIE.

cicatrix A scar; the new tissue that is formed in plants or animals when a wound or sore heals. In plants, the cicatrix often marks the location where a structure was attached, such as a leaf scar.

Ciconiidae A family of tall, carnivorous birds, commonly known as storks, of the order Ciconiiformes. They have long legs, large bills, and plumage that is either patterned black-and-white or black. Their distribution is worldwide and many migrate long distances. The white stork of Europe is well known for its habit of nesting on buildings. Almost voiceless, storks make noise by clapping their bills. See CICONIIFORMES.

Ciconiiformes An order of large- to medium-sized, long-billed, wading birds with well-developed powder downs, except in the storks. These downs are never shed and continue to grow as their tips crumble into a fine powder used for dressing feathers. Examples are storks, ibises, spoonbills, herons, flamingos. There are approximately 115 species. See AVES, ARDEIDAE, DOWNS, CICONIIDAE.

cilia Tiny, hair-like projections found on some cells. Cilia line the bronchial tubes; by beating back and forth they can expel dust and mucus from the lungs. Some microorganisms use cilia to move around or to create currents in the surrounding water to sweep food towards them. Cilia is also another name for the eyelashes of mammals.

ciliates See CILIOPHORA.

Ciliophora The ciliate phylum of the kingdom Protista consists of unicellular, heterotrophic protozoa that have both a macronucleus that regulates cellular metabolism and a smaller micronucleus that is involved with sexual reproduction. Typically, ciliates have from one to hundreds of cilia on their surfaces at some stage in their life cycles. Familiar members of the phylum include *Paramecium*, *Stentor*, and *Tetrahymena*. See PROTISTA.

cilium Singular form of cilia. See CILIA.

cinder cone A type of volcanic cone composed of pyroclastic particles and volcanic ash from 0.5 to 1.5 cm in diameter. See PYROCLASTIC CONES.

cingulum A girdle-shaped structure on a animal, such as the clitella of annelid worms or, in mammals, the ridge surrounding the base of a tooth crown.

cinnabar A mineral of mercury and sulfur forming vermillion-red hexagonal crystals in fracture areas associated with volcanic activity. Cinnabar is the principal ore of mercury.

circadian rhythm The approximately 24-

hour cycle of activities of plants and animals. The circadian rhythm is generated internally through the biological clock; it may last slightly longer in the absence of external cues to reset it. See BIOLOGICAL CLOCK, FREE-RUNNING CYCLE.

circum- Prefix meaning around.

circumboreal Surrounding the northern (boreal) regions. The boreal region lies between the arctic and temperate regions. Some species of plants such as aspen and fireweed are circumboreal, found growing in North America, Europe, and Asia.

circum-pacific belt A large area of tectonically active orogenic belts forming the western rim of the Pacific Ocean.

circumpolar Surrounding either the North Pole or the South Pole. The boreal forest is a biome that has a circumpolar distribution in the Northern Hemisphere. Circumpolar stars are always found above the horizon and appear to circle a point above the North Pole or the South Pole. See BOREAL FOREST.

circumscribed halo A variation of a halo, a colored ring around the sun or moon formed by light refracting through ice crystals in the atmosphere. When the ice crystals are a specific shape and orientation, the halo becomes circumscribed, first brightening at the top and bottom and occasionally having larger arcs visible around the smaller halo. See AUREOLE.

cirque A steep-walled, bowl-shaped hollow carved into a mountain by the action of an alpine glacier.

Cirrepedia A class of marine crustacea which are either sessile, filter-feeders, or parasites. In the group of true barnacles, the body is encased in calcereous plates

and there may be a stalk-like structure or peduncle used for attachment to a substratum. They may be free-living or commensal on invertebrates, sharks, whales, etc. and are partuculary prominent in the littoral zone and coastal waters.

cirrocumulus Thin layer of clouds at the highest atmospheric levels, showing small ripples, fine texture, or flocking. A "mackerel sky" consists of cirrocumulus clouds. Contrast ALTOCUMULUS, STRATOCUMULUS, CIRRUS.

cirrostratus Thin, sheet-like cloud layers appearing at the highest atmospheric levels and not showing texture or rippling. Contrast ALTOSTRATUS, STRATUS.

cirrus Ice-crystal clouds, white and often wispy or veil-like, in the highest atmospheric levels. When cirro- is added before other cloud types, it indicates high clouds, found above 6000 meters. Contrast CUMULUS, STRATUS.

CITES See CONVENTION ON INTERNATIONAL TRADE IN ENDANGERED SPECIES.

citizen suit See CLASS ACTION SUIT.

citric acid $HOOC\text{-}CH_2C(OH)(COOH)\text{-}CH_2COOH$, an acid found in lemons and other citrus fruits and contributing to their characteristic tart flavor. It is also formed by cellular respiration. See KREBS CYCLE.

citrine A yellowish variety of crystalline quartz; also known as false topaz. See QUARTZ.

Civilian Conservation Corps (CCC) A federal agency established during the Great Depression to employ young men in a variety of public works and conservation programs.

civil suit Lawsuit in which an individual plaintiff seeks to collect damages for injury to health or for economic loss, to have the court issue a permanent injunction against any further wrongful action, or both. Compare CLASS ACTION SUIT.

Cl Chemical symbol for the element chlorine. See CHLORINE.

cladding A thin layer of metal wrapped around something. Specifically, the material used to "can" or contain the fuel pellets in a nuclear reactor. Cladding prevents the coolant from contacting the fuel and contains most of the fission products.

clade A taxonomic grouping of organisms that share a common ancestor; the basic unit of the classification system known as cladistics. See CLADISTICS.

cladistics A classification system for animals and plants that expresses the branching relationships between species through a phylogenetic tree with ancestral forms at the bottom and recently diverged ones at the top. See PHENETICS.

cladogenesis Evolutionary development within a species that results in a branching out or division into isolated groups. Contrast ANAGENESIS, SPECIATION.

cladogram A branching diagram (or dendrogram) indicating ancestral or evolutionary relationships in cladistics. See CLADISTICS.

clarification The purification of a liquid by removal of sediment or foreign matter by sand filters or centrifugation.

clarifier A device, such as a filter or a centrifuge, that removes sediment or foreign matter from a liquid.

Clark McNary Act A comprehensive federal law establishing state and federal cooperative forest management, fire

protection and extension education programs; also authorized aquisition of private lands for national forests in the eastern United States (1924).

claspers 1) Any organ used by animals to hold on to each other during copulation, such as that found in some fishes and insects. 2) Tendrils; threadlike parts of a climbing plant that grow around something to support the plant. See TENDRIL.

class Name given a group of related orders. For example, all animal orders of mammals—Carivora (carnivores), Cetacea (whales), Rodentia (rodents), etc.,—belong to the class Mammalia; and plant orders of Poales (grasses and sedges), Arales (arums), Lilales (lilies and irises), etc., are members of the class Monocotyledoneae. See ORDER, also APPENDIX p. 613.

class action suit Legal action in which a specific group files a civil lawsuit for a larger group of individuals claiming similar damages such as economic loss or impairment of health; these individuals do not need to be listed individually and each does not need to be represented in court. Compare CIVIL SUIT.

clastic The term describing the texture of rocks composed of preexisting particles or fragments.

clastic rock Any rock that is composed of particles or fragments such as shales, sandstones, and conglomerates.

clastic sediment An unconsolidated deposit of particles or fragments of preexisting rock.

clastic texture See CLASTIC.

clathrate Having a latticelike structure. Clathrate is used both to describe divided foliage of plants and to describe chemical mixtures in which one atom or molecule is enclosed by another molecule without forming a true compound. Clathrate substances are also called cage compounds.

clear air turbulence   Atmospheric turbulence above 600 meters not connected with either cumulus or cumulonimbus clouds. It occurs most often near the jet stream.

Claus kiln   A furnace used to produce the high temperatures required to extract elemental sulfur from sulfur dioxide produced from hydrogen sulfide gas in the Claus process, also called desulfurizing.

clavate   Club shaped; thickened towards the end like a club, such as clavate antennae on insects.

clay   An extremely fine-grained sediment composed of clay minerals or other minerals and having a particle size less than $\frac{1}{256}$ mm in diameter.

clay mineral   Any of a group of hydrous aluminum silicate particles having a sheetlike crystalline structure and platy grain of microscopic size. General chemical formula: $Al_2Si_2O_5(OH)_4$.

claypan   A nearly impermeable rocklike soil layer formed by extremely compacted grains of clay. See HARDPAN.

clay skins   A thin coating of clay-sized particles on the surfaces of a rock particle or ped within a soil.

claystone   A very fine-textured, non-fissile sedimentary rock formed of clay-sized particles. See MUDSTONE, SHALE.

Clean Air Act   The name of the original 1970 federal statute (42 U.S.C.A. §7521 et seq.) establishing federal laws relating to air quality. The Clean Air Act requires the Environmental Protection Agency to set national ambient air quality standards and national emission standards. Amendments made in 1977 required prevention of significant deterioration in areas without much pollution. The act was significantly revised in 1990 as the Air Quality Act (also called the Clean Air Act of 1990). See AIR QUALITY ACT, NATIONAL AMBIENT AIR QUALITY STANDARDS, EMISSION STANDARD, PREVENTION OF SIGNIFICANT DETERIORATION.

cleaning symbiosis   A relationship in which one organism cleans a different organism. Cleaner wrasses are a fish that feed by "cleaning" other fish, ingesting parasites or dead tissues and benefiting both species of fish. See SYMBIOSIS.

Clean Water Act   The name of the original 1972 federal statute (33 U.S.C.A. §1251 et seq.) establishing federal laws relating to water quality.

clear air turbulence   Atmospheric turbulence above 600 meters not connected with either cumulus or cumulonimbus clouds. It occurs most often near the jet stream.

clearcut   To completely harvest a wooded area by cutting down all the trees, regardless of their size or species. In a clearcut, the tops and the branches of trees are usually left on the site, although sometimes disposed of by chipping or burning. See WHOLE-TREE HARVESTING, WHOLE-TREE CLEARCUT.

clearcutting   Method of timber harvest in which all the trees in an area are harvested at one time. Although appropriate for some tree species, clearcutting can result in extensive soil erosion if not done carefully. Compare SELECTIVE HARVEST.

cleavage   The tendency of a mineral to break preferentially along zones of weakness that correspond with planes of weak bonding in the crystal lattice; cleavage is conventionally described as poor, fair, or good, depending on the smoothness of the cleavage surface. Contrast FRACTURE. See APPENDIX p. 620–628.

cleft   See CLEAVAGE.

cleidoic egg   Egg with a protective hard shell that is laid by a land-based animal.

cleistogamous   Describing flowers that do not open and are self-pollinating; the seeds thus produced are genetically identical to the parent. Such flowers are often inconspicuous, as they do not need to attract pollinating insects. Violets have cleistogamous flowers in addition to their regular flowers, which are cross-pollinated. Contrast CHASMOGAMOUS.

cliff   A vertical or nearly vertical ledge or erosional landform in the shape of a precipice.

climate   Aggregate of general weather patterns occurring at a place or in a region over an extended number of years, including average and extreme conditions of temperature, humidity, precipitation, winds, and cloud cover. Climate also takes topography and nearness to oceans or ocean currents into account.

climatic climax   A regional type of plant community that is stable over long periods of time and maintained by climatic factors. The early 20th century plant ecologist F. E. Clements suggested that plant communities, given enough time, would develop into a climax community determined by the regional climate.

climatological station   A weather station, a place where local weather statistics are collected. Climatological stations can be located on ships as well as land.

climatology   Science of climate and its causes. Compare METEOROLOGY.

climax   Final phase of succession in an ecosystem in which populations of animals and plants remain in a relatively self-perpetuating state.

climax community   Biotic community whose different populations have remained relatively stable for many years and will probably continue to remain relatively the same without human interference or climate changes. Different soils, climates, and environmental conditions produce different climax communities.

climograph   Chart displaying one-year distribution of at least two climatic variables occurring at a specific place. See CLIMATE.

cline   A gradual change in one or more individual characteristics (such as size) of a plant or an animal species that corresponds to a gradual environmental change, such as increasing elevation or decreasing shade. See ECOCLINE, GEOCLINE.

clinometer   An instrument used to measure the slope of land or the angle formed between an inclined surface and the horizontal.

clinopyroxene   A general category of rock-forming silicate minerals of the pyroxene group, characterized by monoclinic crystal structures (those with axes intersecting at an oblique angle).

clint   1) A hard, projecting rock ledge. 2) A horizontal surface bounded by vertical joints in a horizontally bedded limestone.

clisere   Succession of one type of vegetation by another due to climatic change. For example, after glacial warming, fir forests were gradually replaced by pine and then by oak forests. See SERE, LITHOSERE, XEROSERE.

clonal dispersal   The separation of parts of a modular organism, or a growth of such parts away from each other without actual detachment. Over time, club mosses may form very large patches through vegetative growth resulting from clonal dispersal. See CLONE.

clone   To reproduce individual organisms asexually, as in propagating a plant through stem or leaf cuttings. Also, an organism or group of organisms (or group of cells) so produced.

closed canopy   Forests in which the crowns of trees have grown to shade over 20 percent of the ground. When a forest reaches the closed-canopy stage, it becomes attractive for commercial harvesting of timber. Compare OPEN CANOPY.

closed system   A system in which there is an exchange of energy with the environment, but no exchange of matter.

cloud   Visible atmospheric condensation, consisting of a high concentration of minute ice crystals or suspended water droplets.

cloudbank   A large mass of clouds.

cloud base   Bottom of a cloud, as in the flattened base of a cumulus cloud.

cloud bow   A very faint rainbow projected onto clouds located below the observer; while difficult to see, this phenomenon is relatively common and accompanies a glory that appears on a cloud below. See GLORY.

cloudburst   Intense, heavy rain that comes on suddenly. Associated with summer and with cumulonimbus clouds.

cloud chamber   Device for showing the paths taken by subatomic particles. When charged particles pass through supersaturated vapor (such as alcohol) in the chamber, the vapor condenses in their wake. They leave a track that can be seen or photographed.

cloud cover   Extent of sky filled with or obscured by clouds.

cloud height   One of two parameters (the other being shape) used to classify clouds. High clouds (cirro-) are those whose bases lie above 6000 meters in the atmosphere. Middle clouds (alto-) lie between 2000 and 6000 meters, and low clouds (stratus) form below 2000 meters. Cloud height is not absolute; in extreme northern areas, or during very cold months, high-type clouds sometimes occur lower in the atmosphere.

cloud seeding   Introducing small particles of dry ice or silver iodide into clouds in order to change their development. Cloud seeding has been used at military air bases and commercial airports to disperse fog or to clear a hole in a thick layer of stratus clouds. It has also been used with less success to attempt to produce precipitation.

cloud streets   Straight or curving parallel lines of clouds, which can extend for many miles. Occurs with forms of cumulus clouds.

Club of Rome   A private international organization of industrialists, which sponsored the landmark 1970 report by the M.I.T. on global environmental projections, *Limits To Growth.*

clumped distribution   The arrangement of organisms where they are closer together than if they has been dispersed at random or spaced equidistant from each other. Also called contagious or aggregated distribution. See AGGREGATION, RANDOM SPATIAL DISTRIBUTION, UNIFORM SPACIAL DISTRIBUTION.

cluster analysis An technique involving classification of data into groups or hierarchies of closely correlated variables and excluding values that are not closely correlated. Cluster analysis is used in numerical taxonomy; it is also used in ecology to uncover patterns and similarities in different sites by analyzing the species lists from the sites.

clutch The group of eggs laid by either a bird or a reptile at one time.

cm Abbreviation for centimeter.

cmd 1) An abbreviation for current meter data in oceanography. 2) An abbreviation for central meridian distance in remote sensing.

cmp An abbreviation for the common midpoint where seismic information from different sources is recorded on a geophone reflector.

cms An abbreviation for a fluid flow measured in cubic meters per second.

Cnidaria Phylum of aquatic, radially symmetrical animals in which the general form of the body is sac-like, with a single opening at the oral end which is surrounded by tentacles and opens into a digestive or gastrovascular cavity. The tentacles bear specialized stinging cells, the cnidocytes which give the phylum its name. Asexual and sexual reproduction occur frequently and may alternate by generations, in which one generation takes a planktonic form, the medusa, and the other, a benthnic or polyp form. There are approximately 10,000 species distributed in three classes. This phylum was formerly referred to as coelenterates. See APPENDIX p. 615.

Cnidosporidia A phylum of heterotrophic microbes, sometimes referred to as

myxozoa, in the kingdom Protista. These microbes are obligate parasites of ectothermic vertebrates (especially fishes) and annelids. Morphologically, the myxozoans form polar capsules with coiled polar filaments. The fish parasite, *Ichthyosporidium;* salmon twist disease, *Myxostoma;* and a disease agent of silk worms, *Nosema,* are all members of this phylum. See PROTISTA.

Co Chemical symbol for the element cobalt. See COBALT.

coadaptation 1) Mutual adaptation or change in two reciprocally dependent organisms, such as predators and prey or flowers and their pollinators. 2) The accumulation of harmoniously interacting genes in a local population's gene pool. See COEVOLUTION.

coal A sedimentary or metamorphic rock consisting primarily of carbon derived from fossil remains. Lignite, bituminous, and anthracite are the principal varieties of coal.

coal-fired power plant Electricity-generating facility that is fueled with coal.

coal gas 1) A naturally flammable gas occurring in coal. 2) A fuel gas obtained by coal gasification.

coal gasification The conversion of coal to flammable methane gas by a chemical reaction with air and steam.

coal liquefaction The conversion of coal to a liquid hydrocarbon by a chemical process of distillation.

coal seam A bed of coal in the natural position parallel to other rock strata.

coal tar A liquid mixture, a crude tar, that can contain hydrocarbon oils, phenols, and bases such as pyridine, and related compounds. Because it contains many aromatic (carbon-ring) compounds, it is a carcinogen. Coal tar is formed by distilling coal; it is a by-product formed in the manufacture of coal gas and in the making of coke for steel. It is used to manufacture many chemicals including dyes and pharmaceuticals.

coal washing   Cleaning process for coal that removes unwanted material, such as shale and pyrite, and reduces fine ash.

coarse-grained   Describing an environment in which resources are plentiful and occur in large groupings, so that organisms can select among the groupings of resources. Compare FINE-GRAINED.

coast   A term describing the boundary area of indefinite width separating the terrestrial and marine environments.

coastal wetland   Land that extends inland from a coastline and is flooded with salt water at least part of the year. Bays, salt marshes, and mangrove swamps are coastal wetlands. Compare INLAND WETLAND, ESTUARY.

coastal zone   1) An area of land and water affected by the biologic and physical processes of both the terrestrial and marine environments. 2) A legal term used in coastal zone management.

Coastal Zone Management Act   (CZMA) Federal laws passed in 1972 (16 U.S.C.A. §1451 et seq.) and 1980 providing federal aid to states and territories bordering the oceans and Great Lakes to assist them in developing protection and management programs for coastlines not under federal protection.

cobalamin $C_{63}H_{90}N_{14}O_{14}PCo$, more commonly known as vitamin B_{12}. It is an essential dietary substance only found in food from animal sources such as milk, meat, and fish. Cobalamin contains a cobalt atom at the center of a ring structure; it is required for cell functions such as the synthesis of RNA and DNA. Vegan diets (those containing no milk, eggs, or other animal products) must be supplemented with synthetic vitamin B_{12} to avoid serious anemia and nerve damage. Pernicious anemia is a disease in which the body is no longer able to absorb B_{12} from ingested food; it is treated with injections of the vitamin. Also called cyanocobalamin.

cobalt 60   A radioactive isotope of cobalt used both as a tracer in x-ray diagnosis and in radiation therapy for cancer.

cobalt   (Co) A hard, gray, metallic element with atomic number 27 and atomic weight 58.93. It is used in many alloys for its strength. Cobalt is used as a blue pigment in glass and pottery; a radioactive isotope (cobalt 60) is used in medicine. See COBALT 60.

cobaltite   A silver-white mineral of cobalt, arsenic, and sulfur forming cubic crystals in hydrothermal veins and contact metamorphic zones. Cobaltite is a principal ore mineral of cobalt.

cobble   Any clastic rock fragment with an average diameter of between 64 and 256 mm.

co-carcinogen   A substance that is not carcinogenic on its own, but which, in combination with another agent or agents, can have a carcinogenic effect. See CARCINOGEN.

coccidiodomycosis   A lung infection caused by inhaling spores of the fungus *Coccidioides immitis*, most commonly found in southwestern United States and Central and South America. Although usually limited to a respiratory infection in humans, in some cases it becomes progressive, spreading to any part of the body and causing death. It can be a chronic infection in animals such as cattle, dogs, or sheep.

coccidiosis   Any of various diseases of domestic animals, birds, and humans caused by the protozoa of the Coccidia order, which usually parasitize intestines and related glands. The diseases rarely occur outside the Far East.

coccolith   A microscopic skeletal disk from a family of unicellular marine planktonic protists, the Coccolithophoridae. Coccoliths may be a major component of pelagic sediment.

Coccolithophorids   See HAPTOPHYTA.

cochlea   The coiled tunnel of the inner ear in mammals, or part of the inner ear of birds and some reptiles. The cochlea contains fine hairs and fluid that translate mechanical vibrations into nerve impulses carried to the auditory nerve.

COD   See CHEMICAL OXYGEN DEMAND.

Code of Federal Regulations   (CFR) Regulations promulgated by the administrative branch of the United States government.

codistillation   A phenomenon in which toxic compounds on land evaporate and are carried by clouds to the ocean, where rain carries them from the atmosphere to the water.

codominant   Two or more species that jointly are the most prevalent or significant species within a plant community. See SUBDOMINANT.

codon   A subunit of a strand of DNA or messenger RNA, consisting of three consecutive nucleotides. The codon specifies the placement of a particular amino acid into a polypeptide during protein synthesis.

coefficient of community   A measurement of the degree to which two plant communities resemble each other, based upon the species composition of each. It is calculated by various ratios of the number of species found in common in two communities to the total number of species that inhabit each. The Jaccard index, Morisita's index of similarity, Sørenson similarity index, Simpson index of floristic resemblance, Gleason's index, and Kulezinski index are all coefficients of community, also called similarity indexes.

coefficient of haze   (COH) Quantification of how much atmospheric visibility is impaired.

Coelenterata   See CNIDARIA.

Coelocanthini   A subclass of fleshy-finned bony fish thought to be extinct until 1938, when a living specimen, *Latimeria chalumnae*, was found off South Africa. Although a number of these living fossils have been collected subsequently, little is known of their physiology and habits. They are predators and give birth to living young. See OSTEICHTHYES.

coenenchyma   The calcium-containing skeleton uniting individual polyps or hydroids in a colony of coral or a similar compound anthozoan. Also called coenosteum. See CORALS, CORAL REEF.

coenospecies   In genecology, the highest order of classification, corresponding to a plant genus. Coenospecies almost never exchange genetic material; if they do, it creates a new species. See GENECOLOGY.

coenzyme   Organic molecules, smaller than proteins, that are loosely associated with enzymes. Coenzymes help regulate the activity of a catalyzed reaction by attaching themselves to inactive proteins to form active enzyme systems. Adenylic acid (AMP) and riboflavin (vitamin B_2) are coenzymes. Also called a cofactor. See COFACTOR.

coevolution   Development of genetically determined traits in two or more species through mutual interactions. The traits in one species have evolved in response to interactions with another species. See COADAPTATION.

coexistence   Two or more species living in the same habitat.

cofactor   Nonprotein compounds that are vital to the normal functioning of enzymes. Cofactors may be organic molecules (coenzymes) or inorganic ions. Some minerals and vitamins act as cofactors. See COENZYME.

cogeneration   Producing usable heat in addition to electricity. Cogeneration plants can either be heating plants designed to drive turbines and produce some electricity on the side, or electrical power plants designed to harness waste heat for heating space or water. (Waste heat is the by-product produced by burning fuel for electricity, ordinarily released to the atmosphere.)

COH   Abbreviation for coefficient of haze, a measurement of the degree to which visibility is impaired in the atmosphere, as by smog and other pollutants.

cohesion   The force of attraction between similar molecules. Cohesion is responsible for a liquid's ability to form drops. Contrast ADHESION.

cohort   Animals that are born in the same year, forming a group of the same age. As a result, they are subject to the same risks and benefits, which may be different from those born earlier or later.

cohort generation time   The average time duration between the birth of a parent organism and the birth of its offspring. It is a rough estimation of generation length that ignores the ability of some offspring to reproduce during the fertile years of the parent.

cohort life table   A life table restricted to illustrating the life span of one group of individuals all born at the same time or within the same year (a cohort), starting from the time of birth to the death of the last survivor of the group. See LIFE TABLE.

coitus interruptus   Sexual intercourse that is interrupted by withdrawing the penis from the vagina before ejaculation. Coitus interruptus is an unreliable method of birth control (although one of the most common worldwide), because sufficient quantities of semen-containing fluid to cause fertilization can be present before ejaculation. See BIRTH CONTROL.

coke   A carbonaceous residue from the incomplete combustion of coal. Coke is used for fuel or as an additive in the metallurgic industry.

col   1) A ridge with a dip in the center formed by the intersecting headwall areas of two cirques. 2) A ridge that connects two summit areas in a mountainous region.

cold front   Forward boundary of a cold air mass that is displacing warmer air as it advances. Dense, cold air is held next to the ground by gravity, so cold fronts push warm air upward along their leading edge. The rising warm, moist air often forms a line of cumulonimbus clouds along the front. See FRONT, WARM FRONT.

cold trough   A southward bend in the jet stream causing an elongated region of low air pressure. The low-pressure trough brings cold air southward. See TROUGH, WARM RIDGE.

Coleoptera (beetles)   The order of beetles and weevils is the largest insect order and may account for more than half of all insect species. As such, beetles are one of the most ecologically important and successful of all animal groups. They have chewing mouthparts and a hard exoskeleton. In most species, the front pair of wings is modified into elytra, hard protective coverings which fold over the membranous, posterior wing pair when at rest. They are distributed worldwide in a wide range of environments including aquatic forms. Most are herbivorous but some are carnivores. Adult beetles and their larvae are among the most destructive of all plant pests, attacking field crops as well as stored grains etc. The cotton-boll weevil, *Anthonomus grandis*, provides a good example of the economically disastrous effects coleopterans can have on farmlands. See INSECTA.

coliform bacteria   Bacteria that live in the intestines of humans and other vertebrate animals, especially *Escherichia coli*. Testing for coliform bacteria is used to determine pollution, because the presence of these harmless bacteria in water (or soil) is evidence of contamination by human or animal feces. Also called fecal coliform bacteria.

coliform index   An estimate for coliform content of a body of water, on the basis of a count of coliform bacteria in water samples, used as an indicator of the water's purity. See COLIFORM BACTERIA.

collagen   The fibrous protein found in the connective tissue (bone, ligaments, carti-

lage, and skin) of vertebrates. Almost a third of the protein in the human body is collagen.

collector-filterer   An aquatic animal that feeds upon small particles of organic matter, filtered from water surrounding and flowing over it. Compare COLLECTOR-GATHERER.

collector-gatherer   An aquatic animal that feeds upon small particles of organic matter contained within sediments. Compare COLLECTOR-FILTERER.

colloblast   One of the cells on the tentacles of comb jellies, jellyfish-like organisms (ctenophores), producing chemicals to help trap prey. Also called a lasso cell.

colloid   An amorphous solid, consisting of minute particles larger than molecules that remain suspended, rather than dissolving or settling. Colloids include solids such as rubber and plastic polymers, gels such as gelatine, and liquid emulsions. Colloidal suspensions frequently form in wastewater treatment plants and are precipitated out by treatment with alum (aluminum sulfate, $Al_2(SO_4)_3$). See MICELLE.

colloidal   Resembling or made up of a colloid.

colloidal electrolyte   A suspension of long-chain hydrocarbons with ionizing groups at the ends of their chains. The ionization of the ends of the molecules makes them behave somewhat like electrolytes. See ELECTROLYTE.

colluvium   A general term for rock debris that has moved downslope to the foot of a hill or mountain by gravity creep.

colonial animal   1) Social organisms that form colonies, interdependent groups of single species. Ants, honey bees, and corals are colonial animals. 2) Groups of animals that are joined together and appear to be solitary organisms, such as Portuguese man-of-war jellyfish (*Physalia*) or corals. See COLONY, CORALS.

colonization   Invasion and establishment of a plant or animal into a area that formerly had none of that species. See INVASION.

colonizer   An animal, plant, or microbe that invades and becomes established in an area, especially one that has adapted to such a role (such as species that routinely sprout first after a major fire). Colonizers (pioneer species) are rugged species that are usually characterized by dispersal methods that cover significant distances; they often don't survive when competition becomes more intense. For example, pioneer tree species may produce so much shade that their seedlings cannot grow fast enough to compete with seedlings of species adapted for shade. Also called early-successional species. See R-SELEC-TION, FUGITIVE SPECIES.

colony   A group of organisms of the same species living in an interdependent group, such as ants. See COLONIAL ANIMAL.

color phase   The color of an animal's fur or plumage at a certain season on animals whose fur or plumage changes with the seasons, such as the ermine.

colostrum   The milk secreted by mammals for two or three days after giving birth; it is more yellow in color than milk secreted later. Colostrum contains serum and white blood cells, helping to confer natural immunity in the infant, as well as strengthening digestion.

Columbiformes   A distinctive order of herbivorous, land birds which feed their young with secretions (milk) regurgitated from their crops (e.g., sandgrouses, doves and pigeons and the extinct dodos). See AVES, CROP MILK, DODOS.

column   1) Any structure resembling a vertical shaft, such as the center portion of an orchid flower or the bundle of nerves running the length of the spinal cord in vertebrae. 2) An upright, cylindrical mass of rock, typical of basalts.

columnar jointing   The distinctive prismlike pattern formed during the cooling of igneous rocks such as basalt. The geometry of interlocking joints and smooth faces give the appearance of vertical columns. See JOINT SET.

combe rock   A term applied to poorly sorted angular rock debris transported on a hillslope by gelifluction.

combined sewer system   An engineered drainage system that receives input from both surface runoff and sewage sources. This older sewage system has many environmental problems, including water pollution during floods.

comb plates   Flagellated platelike structures in comb jellies (Ctenophorea) that assist in locomotion.

combustion   Burning; a rapid form of oxidation, or chemical reaction in which a substance combines with oxygen to produce heat and light, as in the burning of paper.

combustion air   Air that serves as a supply of oxygen for the burning of a fuel, as in a boiler or furnace.

commensalism   A relationship between two organisms in which one lives on or in another species that is not harmed (or benefitted) by its presence. Commensalism includes epiphytic orchids or bromeliads living on the branches of host trees, or small crabs that live inside oysters. See SYMBIOSIS, PARASITISM.

commercial fast breeder reactor A non-research fast breeder reactor, one producing electricity for public use. The United States has no commercial fast-breeder reactor; France has one, called the Superphénix. See FAST BREEDER REACTOR.

comminution Pulverization; the breaking down of large chunks of a substance into small pieces by crushing or grinding. Comminution is used in the processing of mineral ores and in some wastewater treatment plants.

common law System of jurisprudence that originated in England and was later applied in the United States based on judicial precedent rather than statutory laws (legislative enactments).

commons Land set aside for public use, or private land to which the public has access, as for grazing livestock.

Commons, Tragedy of the Depletion or degradation of a resource to which people have free and unmanaged access, as in the depletion of a commercially desirable species of fish in the open sea beyond areas controlled by coastal countries. The concept was popularized by Garrett Hardin in the late 1960s in an essay by the same name.

community All the groups of organisms living together in the same area, usually interacting or depending on each other for existence. Also called biological community. Compare ASSOCIATION.

community ecology The study of groups of populations of different species living in the same area.

compaction 1) The geologic process of reduction in pore space by gravitational compression of rock formation. 2) Artificial densification of soil by mechanical means (an engineering term). See SOIL COMPACTION.

compensation depth The distance below the surface of a body of water where plants' rate of production of organic matter by photosynthesis equals their rate of respiration. At this depth, which depends on how far light can penetrate the water, the quantity of carbon dioxide consumed by plants equals the quantity of oxygen they produce. See COMPENSATION LIGHT INTENSITY, EUPHOTIC ZONE, LITTORAL ZONE, PROFUNDAL ZONE.

compensation light intensity Another term for light compensation point. See LIGHT COMPENSATION POINT.

compensation point Another term for light compensation point (compensation light intensity). See LIGHT COMPENSATION POINT.

compensatory population growth Increased growth of a population that results from a population being reduced below its carrying capacity to a point at which recruitment is at a higher rate. See CARRYING CAPACITY, SIGMOID GROWTH.

competence 1) A measure of the maximum particle size that may be transported in the flow of ice, water, or air. 2) A measure of the capacity of a rock to withstand deformation under pressure.

competition The interaction between two or more organisms, populations, or species that depend on the same limited environmental resource. For example, some plants become tall to compete against shorter plants for light, or two plants (of the same or different species) growing in a desert may be in competition for limited water resources. See DIRECT COMPETITION, INDIRECT COMPETITION, INTERSPECIFIC

COMPETITION, INTRASPECIFIC COMPETITION.

competition coefficient   The depressing effects of one species on the population growth rates of other species with which it competes. See COMPETITION.

competitive exclusion principle   The principle that two or more different species with the same requirements (highly overlapping niches) cannot live indefinitely in the same environment, or depend on the same limited resource. One population will always outcompete the other; the less-competitive one will either dwindle to the point of extinction, or adapt in a way that reduces or eliminates competition. See COMPETITION, COMPETITIVE RELEASE.

competitive release   The increase of a population and its expansion into a broader array of niches than previously occupied, caused by a reduction in inter-specific competition. A species that colonizes an island may prosper if there are no competitors there. Also called ecological release. Contrast COMPETITIVE EXCLUSION PRINCIPLE.

complementary resources   Resources that can replace each other for a given species of consumer; increasing consumption of one reduces the consumption of another.

complete-tree clearcut   A form of clearcut timber harvesting in which more than just the boles of the trees are harvested. All above-ground portions of the trees (branches, bark, twigs, and leaves) are removed from the site. See CLEARCUT, CLEARCUTTING.

complex ion   An ion containing a metal cation in its center that is covalently bound to two or more other ions or molecules. Hemoglobin and chlorophyll are complex ions. See CHELATE.

complex life cycle   Any life cycle in which an organism goes through a series of different larval forms before reaching its adult form. See ALTERNATION OF GENERATIONS, METAMORPHOSIS.

Compositae   See ASTERACEAE.

composite volcano   An explosive type of volcano that is constructed of alternating layers of solidified lava flows and pyroclastic debris. Composite volcanos, such as Mt. St. Helens and Mt. Fujiyama, are also called stratovolcanos. Compare SHIELD VOLCANO, PYROCLASTIC CONES.

compost   1) Organic material that has reached a relatively stable state of decomposition. Compost is often made from a combination of vegetable matter, such as leaves or grass clippings, and animal manures, and makes an excellent soil conditioner and low-level fertilizer. 2) The process of promoting the aerobic decomposition of organic materials, either on a small, backyard scale or as a means of recycling organic components of municipal solid waste.

composting toilet   A toilet which relies on the principles of organic aerobic decomposition to handle human waste, both solid and liquid. Specific examples include Eco-Let and Clivus Multrum (trademark names).

Compound 1080 FCH_2COONa, also known as sodium fluoroacetate. A very toxic substance that acts on the cardiovascular system, lungs, kidneys, and central nervous system. It is used as a rat poison and on rangelands in the U.S. West for killing large predators such as coyotes. Because of its toxicity and persistence in the environment, its use is now restricted.

compound   Short for chemical compound. See CHEMICAL COMPOUND.

compressional wave   See P WAVE.

compressive stress   A one-directional force applied to a body that tends to cause it to strain and to change shape by compressing it along the line of the force. Compare SHEAR STRESS, TENSILE STRESS.

concentration   The amount of one substance dissolved or mixed in a specified quantity of a solution or mixture. Also, a process used to increase the concentration of a substance, such as evaporation of water from a solution.

concentration factor (ore)   A measure of the abundance of a commercially valuable ore in relation to the normal abundance in the earth's crust.

conceptacle   A flask-shaped hollow containing the sex organs in the plant body of some species of seaweed (brown algae), such as *Fucus*. See PHAEOPHYTA.

conceptual model   An abstract, hypothetical representation of a system. Compare DETERMINISTIC MODELS, STOCHASTIC MODELS, PARADIGM.

conchoidal fracture   A breakage in a rock or mineral that leaves a curving fracture like that of chipped glass. Silicate minerals and obsidian break in conchoidal fractures.

concordant intrusion   A type of pluton developing from magma injected parallel to existing layers of the country rock. Contrast DISCORDANT INTRUSION.

concretion   A hard aggregate of particles that are cemented together and form a nodule within a sediment or soil.

condensate   1) A liquid formed from the condensing of a gas onto a cool surface (i.e., water vapor forms as a condensate on a glace of ice water on a humid day). 2) A substance produced by a condensation

reaction. See CONDENSATION.

condensation   1) A change of state from a vapor or gas to a liquid. Condensation in the atmosphere produces clouds. 2) A chemical reaction in which two or more molecules combine to form a larger molecule, with a simpler and smaller molecule (typically water) as a by-product. Contrast EVAPORATION.

condensation level   Height at which condensation begins as rising air undergoes adiabatic cooling and becomes saturated.

condensation nuclei   Minute particles that catalyze the condensation of water vapor into a droplet by providing a surface for deposition. The particles are often salt near a seacoast, microscopic pieces of dust, or smoke particles. See CONDENSATION.

condensation trail   The visible track left by a charged particle or cosmic ray as it passes through a cloud chamber. See CLOUD CHAMBER.

condenser   Device used for condensing gas or vapor into liquid, as in distillation or in a steam engine. In optics, condensers are lenses or mirrors for concentrating light rays into a small area.

condition   1) The state of being of an entity or system. Examples include the condition of one's health or of the environment. 2) To prepare for severe conditions, such as runners having to condition their bodies for a marathon, or the hardening of young seedlings by gradually acclimating them from stable indoor conditions to unstable garden conditions.

conditioned reflex   An automatic, habitual reaction to a stimulus, learned from past experience through repeated exposure to the stimulus or through training. See CONDITIONING.

condition index   A measure of an animal's well-being, usually based on the organism's amount of fat with some form of size adjustment. For example, the fat deposits in bone marrow or around kidneys are a commonly used indication of healthy, well-fed mammals.

conditioning   A repetitive process by which animals learn a specific response to specific stimuli. See CONDITIONED REFLEX.

conductance   (G) A measure of the lack of resistance in a circuit element to electrical current; the opposite of electrical resistance (conductance is equal to the inverse of the electrical resistance). Conductance is also a measure of the lack of resistance to heat flow, and equal to the inverse of the thermal resistance. The SI unit of conductance is the siemens (S).

conduction   Transmission of energy through molecular contact. See ELECTRICAL CONDUCTION, INDUCTION, THERMAL CONDUCTION.

conductivity   A measure of how easily electrical current (or heat) can be made to flow through a material. It is a characteristic of the material itself and not the way the material is shaped or molded. Conductivity is also a measure of how easily heat flows through a material. A material's conductivity is the reciprocal of its resistivity. The conductivity of an aqueous solution containing ionic species is quite high relative to that of pure water. Contrast RESISTIVITY.

conductivity meter   An instrument that measures the ability of a substance (usually a liquid) to conduct electrical current. Conductivity meters are used to determine the concentration of dissolved ionic substances in water. See TOTAL DISSOLVED SOLIDS.

conductor   Material with a relatively good ability to transfer heat (thermal conductor) or electricity (electrical conductor). Materials containing large concentrations of relatively free electrons (metals) are good thermal and electrical conductors. Materials with few free electrons make very poor conductors but excellent insulators. See INSULATE.

cone   1) The reproductive structures on pine, spruce, and other conifers (gymnosperms), especially the hard, scaly female structure that bears the seeds. Technically called a strobilus. 2) In most vertebrates, one of two types of cells of the retina of the eye that respond to light. These cone-shaped cells respond to colors and bright light levels.

cone of depression   The area of a depressed water table forming a funnel-shaped surface surrounding a point of groundwater withdrawal.

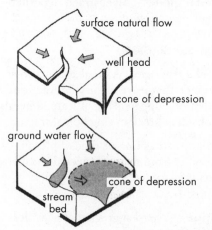

Pumping from a well can affect the surrounding water table, creating a relatively dry cone of depression.

cone sheet   A type of intrusive dike that is shaped like a funnel when viewed in cross section. A cone sheet tapers down and has increasing depth toward the plutonic source rock.

confidence interval An interval formulated to have a specific probability of containing the real value of an unknown parameter. A 95 percent confidence interval has a 95 percent probability of containing the parameter being estimated.

confined aquifer Any aquifer that is both overlain and underlain by impermeable materials.

confining pressure The combined pressure of rock and interstitial pore water at a given depth of bedrock.

confluence A point in a drainage system, or within oceanic or atmospheric circulation systems, where two or more currents flow together.

conformer A species that conforms to external environmental conditions, such as an ectothermic animal whose body temperature reflects its external environment.

conformity A sequence of features indicating a continuous depositional history within a rock.

confusion technique A method of pest control in which sex attractants are introduced into an area to disrupt natural chemical signalling so that males can no longer locate females for reproduction. Only small quantities of the pheromones are needed because the scent can be detected for long distances by the insects.

congelifluction See GELIFLUCTION.

congeners 1) An organism belonging to the same genus as another organism. 2) Chemical elements belonging to the same group in the periodic table of the elements. Compare CONSPECIFIC.

congenital Existing at or before birth, usually a congenital trait or disorder. Congenital defects may be inherited genetically, in which case they are often called genetic defects. They may also be acquired in the uterus or during the birth process. Examples include fetal alcohol syndrome (caused by overconsumption of alcoholic beverages during pregnancy) or exposure to other teratogenic compounds. See TERATOGEN.

conglomerate A sedimentary rock composed of rounded particles in various sizes and including some clasts larger than 2 mm in diameter.

Coniferophyta The phylum of gymnosperms that includes the conifers: cone-bearing, monoecious, woody plants with simple leaves that are often needle-shaped and gathered into bundles. This phylum includes many economically important temperate region evergreen species, including pine, spruce, fir, hemlock, douglas fir, cedar, juniper, cypress, yew, giant sequoia, and coastal redwood. Some species, such as bald cypress and larch, are deciduous. Compare ANGIOSPERMOPHYTA.

coniferous Bearing cones; describing or belonging to the large group of trees and shrubs that bear cones and are typically evergreens, with needles for leaves. Pine, spruce, and hemlock are coniferous; they are sometimes called softwoods because their timber is not as hard as that from most deciduous trees. Larches are also conifers, but are deciduous rather than evergreen.

coning A mechanism for dispersal of atmospheric pollutants in which a plume of emissions spreads out into a conical shape as it flows out of a smokestack. See FANNING, LOOPING.

conjugate acid An acid formed by adding a hydrogen ion (H^+) to a given base. Adding H^+ to ammonia (NH_3) produces ammonia's conjugate acid, the ammonium ion NH_4^+.

conjugate base A base formed by removing a hydrogen ion (H^+) from a given acid. The conjugate base for hydrochloric acid (HCl) is the chloride ion (Cl^-); it is formed by removing the H^+ ion from HCl.

conjugation 1) Union of two prokaryote organisms to transfer nuclear material. Conjugation is the usual reproductive form of ciliated protozoa such as paramecia; it also occurs in some bacteria (such as *Escherichia coli*) and algae (such as *Spirogyra*). 2) The fusion of usually similar male and female reproductive cells to form a zygote, as in isogamy. See ISOGAMY.

connate water Water that has been trapped in the pore spaces of sedimentary rocks since the time the original sediments were deposited.

connectance The interdependence of different organisms within complex food webs.

consequent stream A stream whose course of drainage is determined by the shape of a recent landform.

conservation Management of natural resources to provide maximum benefit over a sustained period of time. Conservation includes preservation and forms of wise use, including reducing waste, balanced multiple use, and recycling. See ENERGY CONSERVATION, SOIL CONSERVATION.

conservation area An area of land set aside for conservation of one or more natural resources.

conservation district An area of land in a town where zoning ordinances resrtict development of specified types. Compare CONSERVATION EASEMENT.

conservation easement A legally binding restriction on allowable uses imposed upon a parcel of land in exchange for a tax break to the landowner. Conservation easements prevent development of a parcel, restricting its use to agriculture, habitat for wildlife, or hiking and other nondestructive forms of recreation. Compare CONSERVATION DISTRICT.

conservationist Person who believes resources should be used, managed, and protected so that they will not be degraded and wasted and will be available to present and future generations.

conservation of energy See LAW OF CONSERVATION OF ENERGY, ENERGY CONSERVATION.

conservation of matter See LAW OF CONSERVATION OF MATTER.

consociation In the monoclimax theory, a unit of vegetation below the association in which the community is dominated by a single species. Compare ASSOCIATION.

consocion A vegetation layer having alternating patches of dominant species.

conspecific Of or relating to the same species. Compare CONGENERS.

constancy 1) Ability of a living system to maintain a particular size. 2) A relative expression of presence or absence of species in different samples (such as individual plots) of a community type. In contrast to frequency—which measures occurence of a species in samples within a specific community—constancy measures presence among different stands. Compare FREQUENCY.

constant 1) A quantity or parameter that retains the same value while other quantities (variables) change. 2) A species always found in a given community, or observed in over half of the sample plots in a given community. Also called a constant species.

constant plankton Species that remain plankton for their entire life cycle (holoplankton), and are present year-round in a given region.

constant-pressure map Chart of the upper levels of the atmosphere showing isobars, the contours of height above sea level at which a given air pressure occurs. See ISOBAR.

constitutional law The original and fundamental principles of law by which a system of law is created and according to which a country is governed. The constitution represents a mandate to the various branches of government directly from the people acting in their sovereign capacity.

constructive plate margin A boundary zone between two diverging tectonic plates where new crust is being formed, usually at a mid-ocean ridge.

consumer An organism that ingests other organisms, either living or dead. All animals and most microorganisms are consumers (herbivores, carnivores, parasites, and detritivores). Fungi and a few plants are consumers, including saprophytes (Indian pipes) and carnivorous plants (pitcher plant, Venus flytrap). Contrast PRODUCER.

consumption efficiency The ratio of energy ingested by a trophic level to the energy in productivity by the previous trophic level. Compare ECOLOGICAL EFFICIENCY, LINDEMAN EFFICIENCY, TROPHIC LEVEL EFFICIENCY, UTILIZATION EFFICIENCY.

contact Any surface where two distinct rock types are directly touching one another.

contact deposits See KAME, CONTACT-HYDROTHERMAL METAMORPHIC ROCK.

contact herbicide Chemical that kills plants by destroying leaves and other tissues that it touches. Paraquat and glyphosate (sold commercially as Roundup) are contact herbicides. Compare SOIL-ACTING HERBICIDES, TRANSLOCATED HERBICIDES.

contact-hydrothermal metamorphic rock A rock produced by contact metamorphism because of hot fluids escaping from a pluton.

contact insecticide Chemical that kills insects on contact. DDT and other chlorinated hydrocarbons are contact insecticides. Compare SYSTEMIC PESTICIDE.

contact metamorphism The alteration or the recrystallization of minerals within a rock by the transfer of thermal energy from a nearby igneous intrusion. The process results from an increase in temperature without any increase in pressure.

contact pesticide Substance that kills pests (plants, animals, fungus diseases) it touches, as opposed to compounds such as stomach poisons causing death by ingestion. Contrast SYSTEMIC PESTICIDE, WARFARIN.

contact potential The voltage (potential difference) that occurs across two different conductors or semiconductors in contact with each other.

contact process A procedure used to create sulfuric acid from sulfur dioxide, the most important of the different methods. Contact process uses high temperatures and a catalyst such as platinum or vanadium to drive the reaction.

contagion   The spreading of viral or bacterial agents by direct or indirect contact with a diseased individual. Also, those viral and bacterial agents causing contagious diseases.

contagious   1) Communicable; referring to diseases capable of being transmitted by direct or indirect contact. 2) Another term for clumped distribution (aggregation). See AGGREGATION.

containment building   Structure surrounding a nuclear power plant that is theoretically designed to prevent release of radioactivity in the event of a major accident. It is constructed of reinforced concrete.

contaminant   An undesirable substance that makes a desired substance impure or unclean; something that causes contamination.

contamination   The process of making a substance unclean, harmful, or impure by the addition of another substance. Drinking water supplies become contaminated if untreated sewage enters them.

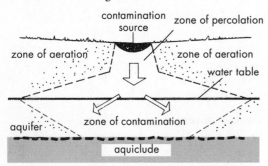

Contamination of groundwater can penetrate upper soil levels and aquifers in this pattern.

contest competition   See INTERFERENCE COMPETITION.

continent   Any large landmass of the earth's crust. The major continents in recent geologic time are Africa, Antarctica, Asia, Australia, Europe, North America, and South America.

continental accretion   See ACCRETED TERRANE.

continental climate   Type of climate characterized by a high annual range of temperature for its latitude. Continental climates have less rainfall, as well as greater temperature extremes, compared with their maritime climate counterparts. See CLIMATE, MARITIME CLIMATE.

continental crust   The portion of the earth's lithosphere that is composed of granitic and dioritic rocks. The thickness of continental crust is usually between 30 to 70 km, with a lower boundary determined by the Mohorovicic discontinuity.

continental divide   A general term for the Rocky Mountain area of North America or for any watershed boundary separating one ocean drainage from that of another ocean drainage.

continental drift   1) A general term for the tectonic movement of the continents. 2) A hypothesis that proposed that continental landmasses moved freely along a plastic zone within the crust of the earth. The continental drift theory was a precursor to plate tectonics.

continental glacier   A glacier maintaining continuous cover over all or most topographic features. The shape and flow characteristics of continental glaciers are generally not constrained by landforms. See ICE SHEET, ICECAP.

continental island   Any island that is geographically related to a continent (for example, the islands of Great Britain are related to Europe).

continentality   Climatic conditions caused by large landmasses. The middle areas of such landmass are removed from the moderating influences of maritime climates and, therefore, experience greater climatic extremes (especially of temperature) than areas at the edges of such landmasses.

continental margin   The zone between the land edge of a continent and the deep sea floor underlain by oceanic crust. The continental margin includes the continental shelf, continental slope, and continental rise.

continental rise   A gently sloping part of the continental margin consisting of an accumulation of sediment between the continental slope and abyssal plain.

continental shelf   An area of the continental margin that slopes gently downward from the land edge of a continent to the top of the continental slope.

continental shield   See CRATON.

continental slope   An area of the continental margin that slopes relatively steeply from the seaward edge of the continental shelf down to the upper edge of the continental rise.

contingency table   A table in which observations or individuals are organized by two variables into two or more rows or columns of data. Data organized into contingency tables can then be analyzed (using the chi-squared test or similar technique) to determine the degree of relationship between the two variables. See CHI-SQUARED TEST.

continuous-reaction series   A sequence of intergrading compositional changes associated with the formation of feldspar minerals in a cooling magma. As the margin cools, placioclase continuously

exchanges calcium ions for aluminum and sodium, leading to a wide range of temperature-related compositions.

continuous spectrum   Spectrum of a source emitting radiation of all wavelengths and showing neither absorption nor emission lines. See SPECTRUM, ABSORPTION LINE, EMISSION LINE.

continuum   An ordering, or gradual change, of populations, species, or communities along an environmental gradient, such as increasing altitude or latitude. See ORDINATION, DISCONTINUOUS DISTRIBUTION, DISJUNCT.

continuum index   Measure of the position of a population or community on a gradient of species composition. See DIVERSITY INDEX.

contour   A line on a map or in the field that connects points of equal elevation above a fixed datum. In more general usage, contour describes the shape of the land.

contour currents   A slow-moving ocean current of the continental rise area of western continental margins. Contour currents are associated with density stratification in cold water near the poles.

contour interval   The change in elevation represented by two adjacent contour lines on a map.

contourites   Any body of sediments composed of sand, silt, or mud that has been deposited by a contour current on a continental rise.

contour line   See CONTOUR.

contour plowing   A soil conservation technique in which cultivation and planting is done in a direction paralleling the contours, rather than up and down the slope. When combined with strip cropping, contour plowing greatly reduces soil erosion and leaching of nutrients. See SOIL CONSERVATION, STRIP-CROPPING.

contour (strip) mining   See STRIP MINING.

contraceptive   Any chemical or device used by humans to prevent unwanted pregnancies occurring from sexual intercourse. Most work by preventing the fertilization of the egg by sperm. See INTRAUTERINE DEVICE.

contract law   Body of law. A contract is a promise or set of promises, for breach of which the law gives a remedy, or the performance of which the law in some ways recognizes as a duty.

contrails   Streaks of cloud formed in the wakes of airplanes (or rockets) flying high in the atmosphere, most often observed in clear skies. The combination of air disturbance and water vapor plus condensation nuclei in the aircraft's exhaust causes the cold, already-humid air to condense in a trail behind the plane.

control   1) General term for manual or automated adjustment of power levels, as in inserting the control rods in a nuclear reactor, or adjusting the throttle of an engine. 2) Something used for comparison for variables tested in an experiment. For example, when testing the toxicity of a chemical in animals, two groups will receive exactly the same treatment except that one group will receive the chemical and the control group will not receive the chemical. See CONTROL GROUP.

control group   A group in an experiment that is subject to all of the same conditions except for the variable being tested. It is used as a standard of comparison for evaluating experimental results. In testing drugs in humans, the control group might be given a fake pill (placebo) while other groups were given pills containing different doses of the drug. Both the control and the experimental groups would be given the same diet, sleep conditions, etc. Also called simply the control.

control rods   Located inside the core of nuclear reactors, control rods are long wands of neutron-absorbing material (such as cadmium or boron) used to regulate the rate of nuclear fission. Removing neutrons slows the chain reaction as there are fewer atomic particles to initiate further fission. Control rods can be removed, allowing the chain reaction to proceed at full speed, gradually lowered to have a gradually dampening effect on the speed of the reaction, or fully inserted to stop fissioning. See NUCLEAR FISSION, NUCLEAR REACTOR.

conurbation   Region where one town merges with the next so that there appears to be one continuous built-up area. Compare MEGALOPOLIS.

convection   Heat transfer by movement of warm gas or liquid through space. Natural convection is caused by a temperature difference in the fluid; forced convection is driven by fans or pumps. Compare RADIATION, THERMAL CONDUCTION.

convection cell   In meteorology, a pattern of air circulation caused by differing temperatures in a gas or liquid. As portions of an air mass or other fluid are heated (e.g., by passing over ground heated by the sun), they expand and rise, causing the cooler, more dense air to sink. See CONVECTION.

convection current   Current created in wind or water by the rising of warmed fluids or the sinking of denser, cooled fluids.

convection streets   Fluid flow patterns that

occur on the leeward side of objects, interrupting the direction of flow.

convective instability   State occurring in a fluid (gas, liquid, plasma) in which a warmer, less-dense layer is created below a colder, denser layer (as from sun-warmed ground). The warm fluid rises and displaces denser, cooler fluid, creating circulation patterns.

conventional-tillage farming   The standard practice of preparing an area for planting by plowing it, breaking up the soil, and then smoothing the surface (harrowing). Soil conservation techniques reduce or modify conventional tillage (conservation-tillage farming); the extreme form of this is no-till farming. See CONTOUR PLOWING, NO-TILL AGRICULTURE.

Convention on International Trade in Endangered Species   (CITES) Acronym for the Washington Convention on International Trade in Endangered Species of Wild Flora and Fauna, an agreement between 103 nations to restrict international commerce involving endangered and threatened species of animals and plants (including tropical birds, rhinoceros horns, orchids, etc.). In 1990, this group banned all international trade in African elephant products such as ivory.

convergence   See CONVERGENT EVOLUTION.

convergent evolution   The development of superficially similar characteristics (form or function) that frequently occurs in unrelated organisms living under the same environmental conditions. Compare DIVERGENT EVOLUTION, PARALLEL EVOLUTION.

convergent margin   See CONVERGENT PLATE BOUNDARY.

convergent plate boundary   A peripheral border between adjacent tectonic plates that are moving toward each other. A convergent plate boundary is typically a region where an orogenic belt, a subduction zone, or both, is formed.

conversion   Process of changing from one form to another. Energy conversion involves changing the form of energy, for example, burning coal to produce electricity or using electricity to run a space heater. Except for some conversions to heat, energy conversion is rarely 100 percent efficient; some energy is always given off to the surrounding environment, usually as heat.

convivium   A differentiated and geographically isolated population within a species; a convivium is usually a subspecies or an ecotype. See ECOTYPE, SUBSPECIES.

coolant   A substance used to draw heat out of a system by convection. Liquid coolant (water) in automobiles absorbs excess heat (which is then radiated—cooled—in the radiator) from the engine to prevent the engine from overheating. Nuclear reactors use water, liquid metals, or gases such as carbon dioxide to extract heat from the fission process.

cool deserts   Deserts characterized by cold winters and usually dominated by sagebrush. They are found in mountain valleys of eastern Washington and eastern Oregon as well as in the Great Basin. See DESERT, HOT DESERT.

cooling pond   Body of water used to dissipate waste heat from industrial processes or electrical energy generation. Mechanisms such as steam condensers use water as a coolant; after circulating through the system, the hot water must be allowed to cool before being reused or released to a stream or river. In a cooling pond, evaporation causes heat to dissipate into the atmosphere, allowing the water to cool.

cooling tower Large structure designed to remove waste heat created by power plants or factories by releasing it into the atmosphere. These towers can either cool steam or air. Huge cement cooling towers are familiar sights next to nuclear and coal-burning electrical power plants.

cool-season plant A plant that does most of its growing during cooler seasons. Compare WARM-SEASON PLANT.

coombe rock A variety of combe rock deposited on a chalky substrate. See COMBE ROCK.

Coordinated Universal Time (UTC) International time standard. It is based upon International Atomic Time (the international scientific standard) but is slower, because leap seconds have been added so that UTC will more closely match the earth's rotation (which is gradually but irregularly slowing down). UTC differs from Greenwich Mean Time because it begins at midnight rather than noon. Often simply called Universal Time. See GREENWICH MEAN TIME.

Copepoda Small crustacea that form an important element of the plankton in the marine environment and in some fresh waters. In free-living forms, head segments are usually fused together and bear a more or less prominent pair of antennules and a pair of biramous antennae. There are five pairs of swimming legs on the thorax and the abdomen terminates in a pair of distinctive, caudal projections or rami. There are also commensal and parasitic forms. Some species of human tapeworms have copepods as intermediate hosts. Size from 0.5mm to about 10mm, although some parasitic forms are considerably larger. See CRUSTACEA.

Cope's rule The principle that states that the body size of animals increases during the evolution of the species.

copolymer A chemical compound formed when two or more unlike monomer units are linked into long chains containing a repeating pattern of the two. A copolymer is a polymer formed of unlike monomers. Compare MONOMER, POLYMER.

copper (Cu) A reddish metallic element with atomic weight 63.54 and atomic number 29. Copper has many uses because it is easily shaped, resistant to corrosion, and is an excellent conductor of electricity and heat. It is an essential mineral in plant and animal metabolisms. See APPENDIX p. 622.

coppice 1) Traditional European form of forest management to produce abundant quantities of small wood (rather than timber). An area of trees is cut and new shoots (stump sprouts) are allowed to grow back. In 10 to 15 years, the area produces wood of small diameter, which is harvested for fuel, fencing, or charcoal. 2) A woodland under such management. Compare POLLARD.

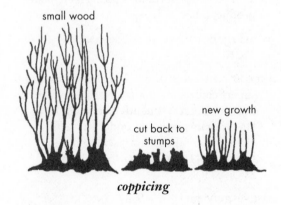

small wood

cut back to stumps

new growth

coppicing

coprolite A fossilized fecal pellet or mass of animal excrement.

coprophagy The act of feeding on dung. Rabbits, some beetles, and many microorganisms derive nutrition by coprophagy.

coquina A form of limestone composed of detrital fragments of the coarse debris of broken calcareous shells.

coracidium A free-swimming larval stage in some species of marine tapeworms in which the egg is surrounded by an embryophore with cilia or flagella.

Coraciiformes This geographically diverse order of birds is characterized by the syndactylous nature of the three front toes, which are fused a good part of their length, and other anatomical characteristics. Includes kingfishers, rollers, todys, motmots, bee-eaters, hoopoes and hornbills. See ALCEDINIDAE, AVES.

coralline algae Species of algae that secrete lime. Also called nullipore.

coral reef A biogenic limestone deposited by living coral and related organisms living in warm and shallow ocean water.

coral reef ecosystem A tropical, shallow-water, marine ecosystem populated by corals whose external skeletons eventually form large reefs. Coral reefs also contain many other species; they are among the most diverse as well as most productive ecosystems. See BARRIER REEF, FRINGING REEF, ATOLL.

corals A general term for members of the cnidarian class Anthozoa, whose polyps deposit skeletal structures of calcium carbonate. The term includes forms with internal skeletons, such as the black coral *Antipathes*, as well as soft corals, which have calcium carbonate spicules in their tissues but no skeletal mass. The stony corals form calcareous exoskeltons called corallums, which are often massive and contribute to the formation of coral reefs in warm, shallow seas. Corals are usually colonial but some species have solitary polyps. All corals are extremely sensitive to pollution and predation; many species are threatened because of their use in the jewelry industry. See ANTHOZOA.

cordate Heart-shaped; rounded and indented on one end and pointed on the other. The common lilac *(Syringa)* has cordate leaves.

cordierite A ferromagnesian aluminum silicate mineral that often forms in metamorphosed argillaceous sediments. Cordierite may form blue orthorhombic crystals of gemstone quality.

cordillera A linear system of mountain ranges formed in the orogenic belt of converging tectonic plates. All of the mountain ranges west of the Great Plains are part of the North American cordillera.

core See CORE SAMPLE.

core area See HOME RANGE.

core-mantle boundary The region of the earth's interior where the composition changes from the iron nickel core to the peridotite of the mantle. See GUTENBERG DISCONTINUITY.

core meltdown See MELTDOWN.

core of the earth The central portion of the earth, composed of nickel and iron. The core is below the mantle, and is divided into a solid inner core with a radius of about 1300 km and a molten outer core with a radius of about 2600 km.

core sample A specimen of rock, soil, or sediment that has been extracted by drilling.

Coriolis effect An appearance created by the earth's rotation that acts like a force causing a curving motion in air and ocean currents (and objects moving over long distances such as missiles). Ocean currents and winds shift clockwise in the Northern Hemisphere; in the Southern Hemisphere, the apparent deflection is reversed. The coriolis effect results in the trade winds. Also called Coriolis force. See TRADE WINDS.

cork   The layer of dead cells surrounding the older stems and roots of many seed plants; cork forms much of the outer bark of trees. Also called phellem.

cork cambium   A layer of cells lying parallel and slightly to the interior of the surface of mature stems and roots. The cork cambium is a site of active cell division; on the exterior side, it produces the protective tissue known as cork (phellum). On its interior side, it produces phelloderm. Also called phellogen. See CAMBIUM, CORK.

corm   A bulblike, swollen underground stem of certain plants, such as the crocus and the gladiolus, that stores food. Corms produce leaves and buds on top and roots grow from the bottom. A corm differs from a bulb because it is harder and is solid, rather than made of layers or overlapping fleshy leaves; it is a modified stem and is replaced by a new corm each year. Compare BULB, TUBER.

dormant corm

active corm

cornucopian premise   A natural resource philosophy that assumes unlimited resources.

corolla   1) The colorful petals of a flower, the outer whorl of the perianth. 2) A collective term for all of the petals of a flower. See PERIANTH.

corona   1) Outermost region of atmosphere surrounding the sun or other star. 2) A coronal discharge, an electrical discharge in the form of a bluish glow surrounding a high-voltage conductor. It is caused by the high voltage ionizing the surrounding air. 3) Another term for aureole. See AUREOLE.

corpus luteum   A small yellow structure that develops in mammals from the follicles remaining after the ova (egg) is released; it secretes progesterone to maintain pregnancy. Corpora lutea are examined to provide fertility information because they show how many ova (eggs) were released by the ovary. See OVA, OVARIAN FOLLICLE.

corrasion   See ABRASION.

correlation   A systematic comparison revealing the degree to which two or more variable quantities are related.

correlation coefficient   (*r*) A measure of the degree to which two variables vary together. A correlation coefficient of -1 indicates a perfect negative correlation (linear relationship), while a value of +1 is a perfect positive correlation.

corridor   1) A strip of land linking one vegetation type to another or providing contiguous wildlife habitat. Corridors may be narrow, but they effectively enlarge animal habitats, even when they connect relatively small parcels. By allowing what would otherwise be separate populations to interbreed, they may keep these populations from dying out. 2) A migratory pathway or channel for animals in annual migrations.

corrosion   Chemical decomposition. Corrosion is used to refer both to the oxidation (rusting) of metals when exposed to moist air or acids and to the chemical erosion of rocks.

corrosive   Causing corrosion; especially, a substance that corrodes. Acid rain is corrosive.

cortex   1) In higher plants, all tissue outside of the vascular tissue but inside the epidermis of roots and stems. Also used for the outer layers of some algae and fungi. 2) The outer layer of organs such as the kidney. 3) The thin, convoluted layer of gray matter surrounding the hemispheres of the brain. Also called cerebral cortex.

cortical shell   The outer part of the skeleton of some species of Radiolaria.

cortisone   A steroid hormone that regulates the metabolism of fats, carbohydrates, electrolytes, and proteins. It is produced by the cortex of the adrenal glands. It is also synthesized for the treatment of inflammatory joint conditions and allergic symptoms.

corundum   A very hard mineral of aluminum and oxygen that usually forms trigonal crystals in metamorphic rocks including marble, schist, or gneiss. Ruby and sapphire are gemstone varieties of corundum. See APPENDIX p. 622.

corvid   See CORVIDAE.

Corvidae   The family of passerine birds (commonly known as corvids), which includes ravens, crows, jackdaws, choughs, magpies and jays. Most are well known in literature and folklore. See PASSERIFORMES.

Corylaceae   A synonym for the birch family. See BETULACEAE.

corymb   A broad, flat-topped cluster of flowers in which the outer flowers blossom before the inner ones. The blossoms of horticultural cherry trees are corymbs. Compare CYME, UMBEL.

cosere   A plant community that overlaps in time and space, rather than replaces, another community. It is an alternative or an equivalent successional stage in succession. See SERE.

cosmetic spraying   Use of pesticides to control insects and diseases that only cause superficial blemishes and do not harm the taste or storing qualities of the fruit.

cosmic   Relating to the entire universe, or cosmos; sometimes used in a more restrictive sense meaning outer space, as in cosmic rays.

cosmic noise   Radio waves produced by astronomical objects other than earth, such as the radio waves associated with sun spots. This "noise" was discovered through research into reducing interference on earth-generated radio transmissions.

cosmic ray   High-energy, penetrating radiation in the form of charged nuclear particles or nuclei traveling through space at high velocities and constantly bombarding the earth's surface.

cosmology   The branch of astronomy studying the structure and origin of the entire universe. Also, a model explaining the universe. See BIG BANG THEORY.

cosmopolitan   A species with a very wide geographic distribution, extending to more than one continent.

cosmos   The universe, from the Greek word for order or world. The term is often used to imply an ordered and systematic universe.

costae Riblike structures. In plants, costae are primary veins, such as the midribs of a leaf. In insects, costae are the primary veins on wings. In vertebrates, a costa is a rib. The singular form is costa.

cost-benefit analysis Estimates and comparisons of short-term and long-term costs (losses) and benefits (gains) from an economic decision. If the estimated benefits exceed the estimated costs, the decision to buy an economic good or provide a public good is considered worthwhile. See RISK-BENEFIT ANALYSIS.

cotyledon A first "seed" leaf, or one of the first pair of leaves in a seedling, growing from a seed; the cotyledon contains food for the growing plant and is thicker than the subsequent true leaves. See MONO-COTYLEDON, DICOTYLEDON.

cotype See SYNTYPE.

coulomb (c) SI unit of electric charge. One coulomb equals the amount of charge transferred by a one-ampere current in one second. See AMPERE.

coulombic forces Electrostatic attraction or repulsion between electrically charged particles.

Coulomb's law States that the electrostatic force of attraction or repulsion between two bodies having an electrical charge is directly proportional to the product of their electrical charges, and inversely proportional to the square of their distance. The proportionality constant for this law is approximately 9.0×10^9 newton meter squared per coulomb squared. See INVERSE SQUARE LAW.

coumarin Any of a group of phenol compounds with a double-ring structure; coumarin ($C_9H_6O_2$) and its relatives. Coumarins are responsible for the scent of new-mown hay and of the herb sweet woodruff; they are used commercially to scent tobacco. Some coumarins are very toxic.

Council on Environmental Quality (CEQ) A group established by public law 91-190, 1970, to assist and to advise the President on national environmental policies. CEQ regulations are found in 40 CFR §1500 et seq.

counteradaptation A character trait or behavior that evolves in response to the adaptations of other newly encountered organisms. Counteradaptation occurs when a new species arrives on an island and, after a period of establishment, is reduced to extinction because the predators, competitors, and parasites on the island have evolved to take advantage of the new arrival. See COEVOLUTION.

countercurrent circulation A heat-exchange mechanism of many boreal and arctic animals in which arteries and veins are so positioned in their extremities that heat flows from warm arterial blood to the cooler, venous blood returning to the core of the animal, therby conserving heat and reducing heat loss.

countershading Type of protective animal coloration in which exposed areas are darker than shaded parts, making the animal's shape difficult to discern. Some caterpillars are protected by countershading. See CRYPTIC COLORATION.

countryside A commonly used British term referring to the combination of rural and village landscapes outside obviously urban, densely populated areas.

Countryside Act A significant 1968 British law to promote comprehensive protection, conservation, and management of rural Britain.

Countryside Commission   The British Quasi-Autonomous Non Governmental Organization (QUANGO) established to administer the Countryside Act.

coupled oscillations   Specific patterns of vibrations produced by two or more systems that influence each other and transfer energy back and forth. Also called coupled modes.

covalent   Referring to the sharing of electrons or an atom that forms covalent bonds.

covalent bond   A link created between two atoms by the sharing of electrons in the region between the atoms. Covalent bonds form between identical atoms and between atoms of different elements. In general, covalent bonds are formed between atoms of similar electronegativity. Covalent bonds may be single, double, or triple, depending on the number of pairs of electrons shared. Compare IONIC BOND.

covalent compound   A chemical substance whose atoms are held together by covalent bonds only. Carbon dioxide (CO_2) and water (H_2O) are both covalent compounds. The electrons in covalent bonds are shared approximately equal between atoms. See COVALENT BOND.

covariance   (covar) A statistical measurement of the degree to which two random variables vary together. It is calculated by multiplying the deviations of the two random variables from their respective means (average values).

covellite   A mineral of copper and sulfur forming in supergene deposits above copper ores. Covellite is also known as indigo copper. See SUPERGENE.

cover   1) The percentage of ground surface covered with plants. 2) A shelter for animals against predators or weather extremes. Brush piles provide cover for many birds and small mammals.

cover crops   Plants that are grown in between commercial crops or during the dormant season to protect the soil against erosion; sometimes they are planted between trees and vines in orchards or vineyards. Cover crops such as winter rye, alfalfa, and clover are plowed under before the next season, improving the soil by adding organic matter. See GREEN MANURE.

cover material   A general term for unconsolidated material above the bedrock.

cow-month   The amount of food or graze needed to maintain a healty, mature cow for 30 days. See ANIMAL UNIT, SHEEP-MONTH.

coxa   In vertebrates, the hip or the hip joint. In arthropods, the short leg segment attaching the leg to the body. Plural is coxae.

Cr   Chemical symbol for the element chromium. See CHROMIUM.

crab   See BRACHYURA, DECAPODA.

cracking   Short for catalytic cracking. See CATALYTIC CRACKING.

crag-and-tail   A topographic landform composed of a glacially quarried outcrop (crag) and a ridge of rock debris (tail) that tapers in the direction of ice travel.

crane   See GRUIDAE.

Crassulaceae   The stonecrop family of the Angiospermophyta. Chiefly herbaceous dicots with succulent stems that lack spines. Flowers are radially symmetrical, commonly with five petals and five carpels. Includes many widely known garden species, including stonecrops (*Sedum*), hen-

and-chicks (*Sempervivum*), and fertility plant (*Kalanchoë*), as well as other succulents such as live-forever (*Dudleya*) and byrophyllum (*Escheveria*). See ANGIOSPERMOPHYTA.

crassulacean acid metabolism (CAM) A specialized form of photosynthesis found in sedums, cacti, and other desert succulents. Opening of stomata and uptake of carbon dioxide (CO_2) occurs at night in these plants; the CO_2 is fixed into malic acid (through nonreductive carboxylation), from which the CO_2 is released the next day. This pathway provides an internal source of CO_2 for photosynthesis so that the plant can keep its stomata closed during the day to reduce evaporation. See PHOTOSYNTHESIS, C_3 PLANT, C_4 PLANT.

crater A roughly circular depression forming at the summit of a composite volcano. A crater has very steep walls contoured by the ejection of volcanic materials. Craters are usually less than 2 km in diameter. Compare CALDERA.

craton A stable interior portion of continental crust that is generally unaffected by tectonic activity. The north central portion of the United States and central Canada form a classic craton example known as the Canadian Shield.

creatine phosphokinase An enzyme that catalyzes the transfer of high-energy phosphate between two molecules, such as adenosine diphosphate (ADP) and adenosine triphosphate (ATP). It is found in the brain and in muscles and is essential to the body's energy metabolism.

creel census A form of information collection for statistics about fish and fishing based on inspecting catches and interviewing anglers. (A creel is the basket an angler uses to carry the catch.) Typically, species and size caught (and the time duration over which they were caught) are recorded and samples of fish scales are collected for analysis of age and structure.

creel limit The legal number of fish that can be caught in a day. It is the equivalent of bag limit in hunting. See BAG LIMIT.

creep 1) A gradual downslope transport of regolith materials moving under the influence of gravity. 2) The deformation or dislocation of minerals during cataclastic metamorphism. 3) Slow displacement without felt earthquakes along faults.

creosote A liquid largely composed of aromatic hydrocarbons. Derived from coal tar, it is used as a wood preservative and as a disinfectant.

crepuscular Becoming active only during periods of twilight (dawn or dusk). Insects that only fly at dusk are crepuscular. Contrast DIURNAL, NOCTURNAL.

crepuscular rays Optical effect in which sunlight streams in spoke-like rays either through gaps in clouds, from below the horizon (as at sunrise or sunset), or through fog.

Cretaceous The last of three geologic time periods of the Mesozoic era. The Cretaceous period lasted from approximately 144 to 65 million years ago. See APPENDIX p. 610.

crevasse A large, open fissure forming in the brittle surface of a glacier as it moves and is deformed. Below a depth of about 40 m, the ice flow is usually plastic enough to prevent crevasse formation.

crevice A narrow void or small aperture within a solid material.

crinoid A member of the Crinoidea, the most primitive living class of the phylum Echinodermata, including the sea lilies and feather stars. Crinoids are important indicator fossils in Paleozoic carbonate rocks.

criteria pollutants   Seven substances that cause most air pollution: carbon monoxide, sulfur dioxide, particulates, hydrocarbons, nitrogen oxides, ozone, and lead. Also called conventional air pollutants, these are listed in the original Clean Air Act.

critical depth   The depth at which the total photosynthesis of all plants at shallower depths equals the total respiration of all of the above plants. See COMPENSATION DEPTH.

critical frequency   Referring to radio waves, a frequency just strong enough to penetrate the ionosphere, the ionized layer of the earth's upper atmosphere that reflects radio waves of very long wavelengths.

critical mass   Minimum quantity (mass) of fissionable nuclear fuel needed to initiate a chain reaction, either in a nuclear reactor or a fission bomb. See CHAIN REACTION.

critical mineral   A mineral that is only stable at certain metamorphic facies and that becomes altered as metamorphic conditions change.

critical minimum area   The minimum plot of land required to maintain a viable population of a plant or animal species. Populations on smaller parcels of land will probably eventually become extinct. Also called critical minimum size.

critical reaction   A self-sustaining nuclear fission reaction. If the number of subsequent fissions caused by one atom fissioning is less than one, then the reaction is subcritical and the chain reaction stops. If the subsequent fissions exceed one, it is supercritical and the chain reaction proceeds explosively. See CHAIN REAC-

TION, CRITICAL MASS, NUCLEAR FISSION.

critical thermal maximum   Temperature at which an animal's thermal tolerance is exceeded; if this temperature continues, it will cause the animal's death. See LOWER CRITICAL AMBIENT TEMPERATURE.

crocidolite   Blue abestos, a basic silicate of the amphibole group.

crocodile   A very large aquatic reptile covered with a thick, scaly hide. Crocodiles were hunted to the point of becoming endangered because their leather-like hides were so valued.

Crocodilia   An order of carnivorous reptiles of the subclass Archosauria adapted to aquatic life and characterized by a long snout with nostrils at the tip, and a secondary palate which separates the mouth from the nasal chamber so animals can breath while submerged with only the tip of the nose exposed. Seriously depleted by humans, many of the 21 species are now protected (e.g., crocodiles, alligators and the gharial or gavial of India). See ARCHOSAURIA.

crop   1) Plants grown for harvest, especially food plants. 2) A sac-like enlargement in the throats of birds and insects where food is prepared for digestion. Also called the craw.

crop milk   Fluid secreted by glands in the crops of females in some species of birds such as pigeons. Crop milk is used to feed the young. See CROP.

crop rotation   Method of agriculture in which fields are sown with different crops from year to year to reduce the depletion of nutrients in the soil, to increase yields, or to reduce disease or pest populations. For example, the first year a field is sown with cotton, tobacco, or corn, all of which take large amounts of nitrogen and other nutrients from the soil. The next year the field is planted with legumes that return nitrogen to the soil, or for two years hay is grown, before the original crops are grown again.

cross 1) The product of cross-breeding, a plant or animal created from breeding two different varieties or breeds. 2) The mating of two selected animals or plants.

cross-bedding A type of overlapping sedimentary structure formed in sand dunes and sand bars. Cross-bedding consists of inclined bedding or layers of sediment at right angles to current direction that caused deposition. See CROSS-STRATIFICATION. Compare CROSS-LAMINATION.

cross-breeding Creating hybrid forms by mating two different breeds, varieties, genotypes, or strains of plants or animals. See CROSS-POLLINATION.

cross-cutting relationships, principle of A relative dating technique stating that a body of rock must be younger than a rock it cuts.

cross-fertilization The mating of a male of one variety with a female of another variety, or between gametes produced by different individuals of the same species. Compare SELF-FERTILIZATION.

crossing Another term for cross-breeding. See CROSS-BREEDING.

crossing-over A mutual exchange of gene segments between homologous chromosomes during early stages of meiosis. Crossing-over alters gene patterns within chromosomes, increasing genetic variability in the resulting gametes. See MEIOSIS.

cross-lamination A type of overlapping sedimentary structure found in small-scaled, bedformed ripple marks. See CROSS-STRATIFICATION. Compare CROSS-BEDDING.

crossover An individual instance of crossing-over. If hyphenated, it is another term for crossing-over. See CROSSING-OVER.

cross-pollination The transfer of pollen from the anther of a flower on one plant to the stigma of a flower on another plant of the same species. See CROSS-BREEDING, CROSS-FERTILIZATION.

cross section A slice of an object made by cutting at right angles to its longest dimension. Thin cross sections of plant tissues are used for studying structure, using a microscope. Also called a transverse section. Compare LONGITUDINAL SECTION, SAGITTAL.

cross-strata Any cross-lamination or cross-stratification that is preserved in a lithified form.

cross-stratification Any type of sedimentary structure formed by progressively overlapping slip faces in bedforms, sand dunes, or sand bars.

crown 1) The leaves and branches above the main trunk or trunks of trees and shrubs. 2) In herbaceous plants, the point at or below the soil surface where the root joins the stem. 3) The portion of a polyp that supports the tentacles and mouth. 4) The chewing surface of a tooth. 5) The outermost portion of a deer's antler. 6) The crest or head of an animal.

crown closure Another term for canopy closure. See CANOPY CLOSURE.

crown fire Severe, intensely hot forest fire, so hot that it spreads from the crown of one tree to the crowns of others. Crown fires destroy much, if not all, vegetation and wildlife and cause an increase in erosion. Compare GROUND FIRE.

crude birth rate Annual number of live births per 1000 persons within a specified population and geographic area at the middle of the year. Compare CRUDE DEATH RATE.

crude death rate Annual number of deaths per 1000 persons within a specified population and geographic area at the middle of the year. Compare CRUDE BIRTH RATE.

crude density The population count (or biomass) divided by the total area of the ecosystem being studied.

crude oil A naturally occurring petroleum composed of fossil remains formed by the anaerobic decay of organic matter. Crude oil in refined form is the major source of industrial petroleum products.

crust The solid outermost layer of the earth above the mantle. The crust is usually categorized as continental or oceanic, depending on composition. Compare LITHOSPHERE. See APPENDIX p. 631.

Crustacea A very diverse subphylum of arthropods about which it is difficult to generalize because of the high degree of structural variability within the group. There is usually a cephalothorax covered by a carapace. There are two pairs of antennae (the second pair called antennules) and three pairs of mouth parts: a pair of mandibles and two pairs of maxillae. Biramous appendages, which vary in number according to type, are born on both thorax and abdomen. Distributed worldwide in almost every conceivable habitat, crustaceans are mostly omnivorous scavengers. They vary in size from a fraction of a millimeter to over 60 cm or more. Although classification of crustacea

is controversial and in flux, eight classes are usually recognized, including Ostracoda, Copepoda (copepods), Cirripedia (barnacles), and Malacostraca (crabs, lobsters, etc.). See ARTHROPODA and separate entries for sub-groups listed above.

cryogenic system Mechanism for producing extremely low temperatures.

cryopedology The study of soils and regolith in periglacial environments.

cryophyte A plant that grows on snow or ice; some algae, mosses, and fungi are cryophytes.

cryoturbation The disruption of soils and regolith in periglacial environments.

crypsis A protective behavior or form that camouflages an organism from its predators.

cryptic coloration Protective coloration or patterning that hides or camouflages an animal. Also called apatetic coloration. See COUNTERSHADING, CAMOUFLAGE.

cryptic species Animals that live in holes in trees or rocks and that are, therefore, usually hidden, or small or inconspicuous plant species that grow hidden from view.

cryptocrystalline Referring to minerals whose crystals are so small that they cannot be seen using ordinary microscopes.

Cryptomonads See CRYPTOPHYTA.

Cryptophyta The cryptomonads, a phylum of kingdom Protista, bear a pair of anterior undulipodia attached to a grooved gullet through which food may be ingested. These organisms may be either pigmented and photosynthetic, storing food in the form of starch, or colorless and autotrophic. Reproduction is by mitosis. Well-known members of this phylum include the freshwater alga *Cryptomonas* and the marine phytoplankton *Pyrenomonas*. See PROTISTA.

cryptophyte An herbaceous plant whose overwintering buds are "hidden," located beneath the soil or under water, as on bulbs or corms. The cryptophyte category is often broken down into geophyte or hydrophyte, depending on whether the buds are under soil or under water (geophyte is often used in place of cryptophyte). The term comes from the Raunkiaer life form classification system. See GEOPHYTE, HYDROPHYTE, HEMICRYPTOPHYTE, RAUNKIAER'S LIFE FORMS.

crystal A stable arrangement of atoms in a solid form.

crystal form The external expression of a mineral form that reflects the orderly internal arrangement of atoms.

crystal fractionation See CRYSTAL SETTLING.

crystal lattice The spatial description, also known as space lattice, of the orderly three-dimensional pattern of atoms or groups of atoms within a crystal.

crystalline Any solids in which the atoms or structural groups of atoms are arranged in stable, regularly repeating geometric patterns.

crystalline rocks A general term for rocks assembled by crystal formation in solid or liquid precursors. Igneous and metamorphic rocks are typically identified as crystalline rocks.

crystallization Any physical and chemical process resulting in the formation of crystals.

crystal settling One process in which early formed minerals in a magma chamber may be removed from the molten portion of the magma. See MAGMATIC DIFFERENTIATION BY FRACTIONAL CRYSTALLIZATION.

crystal structure A description of the configuration of atomic units arranged in a crystal lattice.

Cs Chemical symbol for the element cesium. See CESIUM.

CS gas Abbreviation for orthochloro–benzylidene malononitrile, a form of tear gas. CS gas is very potent, a strong irritant for eyes and mucous membranes; it is used by the military and for riot control.

Cu Chemical symbol for the element copper. See COPPER.

cuckoo See CUCULIDAE.

Cuculidae Birds of the cuckoo family. The cuckoos have worldwide distribution, and many species lay their eggs in the nests of other birds. Their calls tend to be loud and may be repeated monotonously, as in the typical cuckoo call. See CUCULIFORMES.

Cuculiformes An order of birds which, like the parrots, have two toes forward and two toes behind (zygodactylus). The insectivorous cuckoos and the reptile-eating roadrunners of the American southwest belong to this order, as do the fruit-eating, brilliantly plumaged touracos of the tropics. Over 300 species exist. See AVES, CUCULIDAE.

Cucurbitaceae The gourd family of the dicot angiosperms. These are usually herbs, often viny in their growth form, with palmately lobed leaves. Unisexual, five-petaled and five-sepaled flowers, with a large ovary filled with many ovules, mature into a berry-like fruit with a leathery surface and fleshy interior. Gourds, pumpkins, and squashes (*Cucurbita*), watermelon (*Citrullus*), cucumbers and melons (*Cucumis*), and wild cucumbers (*Echinocystis*) are members of this family.

cuesta A ridge-shaped landform having an asymmetrical profile in cross-section, with one side of the hill dipping more steeply than the other. Compare HOGBACK.

Culicidae A family of insects of the order Diptera, commonly known as the mosquitoes, have slender bodies with long legs and mouthparts modified to form a piercing and sucking probosci. Larvae are aquatic. Mosquitoes, as vectors of a wide range of disease organisms, form the singly most important family of insects from a health point of view. Approximately 50 species of the genus *Anopheles* transmit malaria through out wide areas of the world, especially in the wet tropics. Other genera of mosquitoes, such as *Culex* and *Aedes*, transmit diseases such as yellow fever, dengue fever, filariasis, and encephalitis. None, however, transmit malaria. Control of mosquitoes with pesticides and attempts to eliminate breeding areas have only been partially successful in controlling diseases carried by mosquitoes. See DIPTERA, MALARIA.

cull The practice of killing a certain number of individuals as a means of population control (preferably weaker or inferior specimens, but often it is not possible to discriminate). For example, deer culls keep populations from reaching levels that could result in overgrazing and death from starvation. Tree culling in forestry removes commercially less valuable, diseased, or poorly formed trees.

culm The jointed stem, usually hollow, of grasses.

cultivar A variety of a plant produced by selective breeding, a distinct subspecies that does not occur naturally in the wild. Cultivars are indicated by a name following the two-word species name, either preceded by the abbreviation cv. or by placing the name in single quotes.

cultivate 1) To prepare (till) soil for planting by plowing and harrowing or digging; also, to disturb the soil around growing plants in order to uproot weeds. 2) To grow or nurture plants.

cultivation Act of cultivating land; tillage. Land under cultivation is that on which crops are being grown.

cultural eutrophication Overenrichment of aquatic ecosystems caused by human activity, such as industrial pollution, septic tank leachate, or agriculture. Increasing the amounts of nutrients such as nitrates or phosphates causes a rapid increase in plant growth; once-clear ponds fill with algae and the aging (succession) proceeds much more rapidly. See EUTROPHICATION.

culvert A pipe or other artificially enclosed channel that carries a watercourse below ground level (for example, under a road).

cumulate rock A type of igneous rock forming from mineral crystals that accumulates by the gravitational settling in a magma chamber. See CRYSTAL SETTLING.

cumulative reserves The total quantity of proven and probable reserves of an available extractive resource, such as coal or oil.

cumulonimbus A type of cloud that begins as a cumulus cloud but expands vertically into the middle and, eventually, upper-cloud levels until it becomes huge and often produces rain. Cumulonimbus clouds are associated with thunderstorms, hail, cloudbursts, and even tornadoes.

cumulus A type of cloud characterized by a dense, puffy form and often having a flattened base. Cumulus clouds can occur as scattered puffs or in dense, heaped packs. They are associated with fair weather, but are capable of undergoing vast vertical development to become cumulonimbus clouds. Compare CIRRUS, STRATUS.

cuprite A reddish ore of copper, usually forming in the oxidized zone of copper deposits.

curie (Ci) Unit of measurement for radioactivity, originally defined as the rate of decay of 1 gram of radium (3.7×10^{10} disintegrations per second). It is no longer used in the Système International d'Unités (SI), having been replaced by the becquerel (which equals 1 disintegration per second). See BECQUEREL.

Curie temperature Temperature above which a ferromagnetic substance loses its magnetism. Also less commonly used for the lower temperature limit for a substance to retain its magnetism, the lower Curie temperature. Also called the Curie point.

curlews See SCOLOPACIDAE.

current ripples A variety of ripple marks oriented at approximately right angles to water currents. Current ripple marks have a steep slope, or slip-face, on the downstream side. See RIPPLE MARK.

cusp A pointed projection of land or sediment extending seaward from a coastline or beach.

cut bank A steep stream bank maintained by eroding action of streamwater flowing around the outside of a meander bend. Contrast POINT-BAR.

cuticle 1) Outer skin or epidermis of animals. 2) In many plants, the cuticle is a thin layer of waxy material (cutin) covering the outer skin (epidermis). See CUTIN.

cutin A substance forming an outer surface on and impregnating the cell walls of mature cells of the epidermis of plants. It is composed of a mixture of waxy compounds and is related to subarin. Compare SUBERIN.

cutover Clearing a specific area of land of its trees, or an area cleared of trees.

cut terrace 1) An artificially flat land area formed by excavation of a hillside in the engineering practice of cut-and-fill. 2) A stream terrace formed by river planation. See RIVER TERRACE.

cutting cycle The number of years between cutting down large amounts of a forested area, or returning to harvest an area. It is essentially a rotation of harvest.

cutting oils A general term for petroleum-based fluids used to lubricate the cutting tools used in machining metals.

cuttlefish Cephalopods whose reduced shells are imbedded in the mantle and provide a degree of buoyancy. The animals, which rely largely on undulations of the lateral fins for locomotion, are active swimmers. When disturbed, ink sac organs may expel a dark, sepia fluid into surrounding waters. They may be pelagic or benthnic. Some deep-sea forms develop luminescent organs. The shell called cuttlefish"bone" is used commercially by jewelers and pet suppliers. See CEPHALOPODA.

cwm 1) A geologic map symbol abbreviation for a cirque. 2) A steep hollow.

cyanide CN^-, an ion, or any of a number of inorganic salts containing the CN^- anion. Often used to refer to the poison potassium cyanide. Cyanide is used in the manufacture of plastics and in the mining of gold.

Cyanobacteria Formerly called blue-green algae and classified in the plant kingdom, these forms are now classed as a phylum of the kingdom Monera. They photosynthesize like plants but are structurally similar to other photosynthetic bactera. See MONERA.

Cyanophyta See CYANOBACTERIA.

cyanosis A symptomatic blue or grayish discoloring of the skin that shows a lack of oxygen in the blood and may signal the beginning of asphyxiation. A blue baby is one with cyanosis. See ASPHYXIANTS, METHEMOGLOBINAEMIA.

cybernetics The study of communication and control systems in biological nervous systems and machines to further the understanding of both.

Cycadophyta The phylum containing cycads, which are also sometimes misappropriately called sego palms. Cycads bear thick palm-like or fern-like leaves at the top of their scaly stems. A gymnosperm, they produce naked seeds in cones (gynostrobili) rather than being enclosed by ovaries. Cycads are dioecious and are found only in tropical and subtropical regions.

cyclamate Any of a group of compounds derived from cyclohexyl sulfamic acid and formerly used as artificial sweeteners. In the United States, their use has been discontinued because they are suspected carcinogens. They are still used elsewhere because of their sweetening power; they are 30 times as sweet as ordinary sugar.

cycle of erosion An intellectual construct describing landscape development from a condition of recent uplift and rugged topography through reduction to a peneplain of low relief by long-term erosion. The cycle is one period of uplift and erosion of a landscape.

cycle, sedimentary See SEDIMENTARY CYCLE.

cycles per second A unit for frequency of cyclic phenomena such as waves. The modern name for this unit is hertz. See FREQUENCY, HERTZ. .

cyclic replacement An interrupted succession in which a repeated disturbance prevents the ecosystem from evolving into a stable climax. In heaths, wind causes a periodic cycle among bare soil, bearberry (*Arctostaphylos*), heather (*Erica* and *Calluna*), and lichens, without ever "progressing" toward a stable climax. Prairies are maintained by grazing animals or periodic fires. See SUCCESSION.

cyclodiene insecticides A class of chlorinated hydrocarbon pesticides characterized by the presence of a two-ring system containing C-C double bonds and multiple chlorine atoms. Included in this group are Aldrin, Dieldrin, Heptachlor, and Chlordane. These pesticides do not biodegrade readily and persist in the environment for many years. See ALDRIN, CHLORDANE, DIELDRIN, HEPTACHLOR.

cyclogenesis Meteorological term for the creation of a new cyclone (low-pressure region), or the development and intensification of an existing one.

cyclone Low-pressure atmospheric system with winds rotating counterclockwise around its center (in the Northern Hemisphere; in the Southern Hemisphere the wind direction is clockwise). Cyclones are associated with inclement weather. Hurricanes and tornadoes are particularly intense cyclones (tornadoes may be called cyclones, as well as hurricanes occurring over the Indian Ocean), but the term also covers much less intense low-pressure regions. See ANTICYCLONE.

cyclone collector   A device used to remove solids from emissions. The polluted air is spun in a conical device; centrifugal action causes larger particles to settle out of the air stream.

cyclonic vorticity   Propensity for air to move in a revolving, cyclonic direction around a central region of low pressure. Anticyclonic vorticity is the tendency to revolve in the opposite direction around a high-pressure region. See CYCLONE, ANTICYCLONE.

cyclothem   A rock stratigraphic unit composed of a series of deposits laid down by a cyclical transgression and regression of ocean water.

cyme   A flower cluster in which blossoming begins at the center and progresses outward to the periphery; the first flower is often above the others, which form on lateral stems. Cymes are usually broad in shape and flat-topped, such as sweet William (and some other members of the pink family, Caryophyllacea), but may form to one side only (freesias), or two sides (forget-me-not). Also called a cymose inflorescence. Compare CORYMB, RACEME.

Cyperaceae   The sedge family of monocot angiosperms. These grass-like plants have stems that are triangular in cross-section, with 3-ranked leaves that often have sharp edges. A small flower bract subtends the 3 (sometimes 1-6) stamens and single chambered ovary. Papyrus, nut-grass, and umbrella sedge *(Cyperus)*, cotton-grass *(Eriophorum)*, sedges *(Carex)*, and sawgrass

(Cladium) are members of this family. See ANGIOSPERMOPHYTA.

cypsela   A single seed that does not burst open when ripe. It is like an achene but is formed from a double ovary in which only one ovule develops into a seed. See ACHENE.

cyst   1) Any growth enclosed in a pouchlike membrane and filled with fluid or solid material. Cysts are usually abnormal, small structures; they may contain pus or other material produced by inflammation. 2) Any pouchlike structure in animals and plants, such as the dormant spores produced by green algae and other prokaryotes, or the structure surrounding embryonic tapeworms.

cysteine $HS-CH_2CH(NH_2)COOH$, an amino acid found in many plant and animal proteins. It contains sulfur and is an important source of sulfur in metabolism. It is the oxidized form of cystine, another important amino acid.

cystoid   Resembling a bladder.

cytochromes   Pigments found in chloroplasts and mitochondria that transport electrons for cellular respiration. They share structural similarities with hemoglobin.

cytokinesis   The separation of the cytoplasm of a cell into two daughter cells during cell division in animals and plants. Compare KARYOKINESIS.

cytokinins   Any of a group of plant growth substances promoting cell division, bud formation, and enzyme formation, and delaying leaf senescence. They are derivatives of adenine (a component of nucleotides). See KINETIN, ZEATIN.

cytolysis   The destruction of cells, often because of disease. Hemolysis is the cytolysis of red blood cells; bacteriolysis is the cytolysis of bacterial cells.

cytoplasm　The living material (protoplasm) of a cell surrounding the nucleus and inside the plasma membrane (in plants, inside the cell wall). See PLASMA MEMBRANE.

cytosine　One of the pyrimidine bases that make up the structure of both DNA and RNA; it pairs with guanine in both nucleic acids. Cytosine is an organic compound with the formula $C_4H_5N_3O$. See GUANINE, URACIL, ADENINE, DNA, RNA.

cytosome　Another term for cytoplasm. See CYTOPLASM.

cytotoxin　An antibody or toxin that attacks or destroys the cells in some organs or tissues. See NEUROTOXINS.

d

D   Chemical symbol for deuterium, an isotope of hydrogen. See DEUTERIUM.

dacite   A light-to-intermediate–colored extrusive rock with an aphanitic or finely phaneritic texture. Dacite has essentially the same mineralogy and composition as granodiorite.

DALR   See DRY ADIABATIC LAPSE RATE.

dam   A structure that impedes the flow of water by creating an impoundment. A dam may be an engineered structure or may form from natural processes, as occurs with impoundment by a glacial moraine, basalt flow, or landslide.

damping   1) Negative feedback controls that reduce the oscillation of a signal or trend. 2) Decreasing or limiting the amplitude and duration of a vibration or oscillation by means of friction, viscosity, or electrical resistance. 3) Using friction (or electrical resistance) to remove energy from an oscillating particle or system (such as a sound wave) and converting this energy to heat. See NEGATIVE FEEDBACK.

damping off   Wilting, toppling over, and death of young seedlings and cuttings caused by several different soil fungi. Damping off can be controlled by starting seeds and cuttings in a sterile medium such as sterilized soil.

Darcy's law   A description of groundwater flow given by the equation $Q = k \times I \times A$, where Q = volumetric rate of groundwater flow, k = hydraulic conductivity, I = hydraulic gradient, and A = cross-sectional area of flow.

dark minerals   A general term for ferromagnesian minerals. See MAFIC ROCK.

dark reactions   Any of the chemical reactions in photosynthesis that fix carbon dioxide into carbon compounds (sugars). They do not depend on light (can occur either in light or dark), but on the energy produced by the light reactions of photosynthesis. See LIGHT REACTIONS, C_3 PLANT, C_4 PLANT.

Darrieus generator   A wind generator having long, narrow rotor blades that turn on a vertical axis. Darrieus generators are usually used for generating electricity, but they can also be used to perform mechanical work (and so are also called Darrieus rotors). See WIND TURBINES. Compare SAVONIUS ROTOR.

Two styles of Darrieus generators

Darwinian fitness   Reproductive fitness; the success of organisms in reproducing or in transferring their genes to succeeding generations. An individual is more fit if it has a greater number of offspring living to reproduce than other individuals do.

Darwinian theory   The theory of evolution developed by Charles Darwin (and Alfred Russel Wallace). Because natural selection acts upon the variations arising within populations, organisms with traits that are not well suited to the environment tend to die out and those with better-adapted traits tend to survive. Also called Darwinism. See NATURAL SELECTION, NEO-DARWINISM.

Darwin's finches Fourteen species of birds (finches) found only on the Galapagos Islands; they were studied by Charles Darwin and provided evidence needed to support his theory of evolution. Each species had evolved to depend on a different food source, a classic example of adaptive radiation. Also called Galapagos finches.

daughter The nucleus resulting from the radioactive decay of another nucleus. If the daughters are also radioactive, they continue to decay until their daughter products are stable and not radioactive. Also called a daughter isotope or daughter product. See RADIOACTIVE DECAY.

daughter cell Two or more identical cells resulting from mitosis, the primary division of the parent cell.

day-neutral plant A plant that is indifferent to the duration of sunlight and can grow and bloom equally well during short or long days. Newly developed day-neutral varieties of strawberries flower and produce berries over the entire growing season, rather than in just a few of weeks as in traditional short-day varieties. See SHORT-DAY PLANT, LONG-DAY PLANTS, PHOTOPERIODISM.

db (dB), dBA These are both units used for indicating sound intensity, particularly the readings of sound pressure level meters. db is the abbreviation for decibel (the B is often capitalized). dBA refers to the A scale, a curve derived by modifying decibels to approximate the sensitivity of the human ear. Measurements of dBA noise levels are used in identifying and controlling noise pollution. Sounds above 120 to 130 dB cause discomfort; dBA levels for

discomfort are closer to 100. See DECIBEL, A-SCALE SOUND LEVEL.

dde Short for dichlorodiphenyldichloroethylene, one of the toxic compounds formed by the partial breakdown of DDT. Like DDT, it persists in the environment. See DDT.

DDT Short for dichlorodiphenyltrichloroethane, the first and most notorious of the chlorinated hydrocarbon insecticides (also once the most widely used, especially for mosquito control). DDT was one of the first such chemicals to be banned for use in the United States (in 1972) after its hazards to the environment were uncovered. Like its relatives, DDT is dangerous for its persistence as well as its toxicity. It is passed along the food chain and concentrates in the tissues of predators, causing eagles to lay eggs whose shells are too fragile to protect embryos and causing widespread death of songbirds (as recounted in Rachel Carson's *Silent Spring)*.

deactivate To change something so it is no longer active. Catalysts can be deactivated so that they no longer perform their intended function; this happens when leaded gasoline is used in a car equipped with a catalytic converter.

deamination The process of removing amine groups ($-NH_2$) from chemical compounds. Deamination is an essential part of amino acid metabolism, especially the breakdown of proteins and the formation of urine.

death rate The ratio of the number of deaths to the total number of individuals in a specific population during a particular time period, usually one year. For humans, the rate is expressed as deaths per thousand persons per year. See BIRTH RATE.

debris avalanche A type of avalanche in

which the bulk of material in transport is composed of solid rock particles.

debris flow   A type of rapid downslope mass movement of unconsolidated rock particles mixed with mud and other material confined to a channel except where it emerges at a mountain front.

debt-for-nature swap   Agreement in which a certain amount of foreign debt is cancelled in exchange for local currency investments that improve natural resource management or protect certain areas from harmful development in the debtor country.

deca-   A prefix used in the Système International d'Unités (SI) to denote 10. A decagram is 10 grams.

Decapoda   A well-known and varied crustacean order, which ranges from the smallest to the largest crustacea. The carapace, which is fused to and covers the thorax, forms a brachial chamber for the gills. The five pairs of walking legs give the group its name. The first pair of legs may be enlarged and provided with pincers. Distributed worldwide, they are mostly marine, although there are some freshwater forms. Many species are fished commercially (e.g., shrimp, crabs, crayfish, lobsters). Approximately 10,000 species exist. See MALACOSTRACA, LOBSTERS, CRABS.

decay   1) Decomposition of dead organic matter, usually by bacterial or fungal activity. 2) Short for radioactive decay. See DECOMPOSE, RADIOACTIVE DECAY.

decay constant   A number proportional to the probability that, over a short time period, a radioactive nucleus will spontaneously decay. It is inversely proportional to an element's half-life. It is equal to the fraction of a given amount of a radioactive

substance that decays per unit of time, where the time is short compared to the half-life. Also called disintegration constant. See HALF-LIFE.

deci-   A prefix used in the Système International d'Unités (SI) to denote one-tenth. A decimeter is 0.1 meter.

decibel   (db) A unit for relative sound intensity. It is the logarithm of the ratio of the sound's intensity to the intensity of the weakest audible sound. The decibel is dimensionless because it depends on the ratio between two values. Its name indicates that it equals one-tenth of a bel. The b in the symbol is often capitalized because bel was named for Alexander Graham Bell. See DB, A-SCALE SOUND LEVEL.

deciduous   Describing plants that shed their leaves at the end of each growing season. Most broad-leaved trees such as aspens and maples are deciduous; many shrubs and vines are as well. Deciduous can also refer to the shedding of other plant parts such as sepals, stipules, anthers, etc. Contrast EVERGREEN.

declination   One of two values used to pinpoint locations of celestial objects such as planets and stars; it corresponds to latitude on earth. It is the angular distance measured in degrees north or south of the celestial equator. Positive values of declination indicate locations north of the celestial equator; negative values indicate that objects are located below the celestial equator. See CELESTIAL EQUATOR, RIGHT ASCENSION.

declination, magnetic   An angular measure of the variation between the instrument reading of magnetic north and the true north determined by the geographic north pole of the earth.

decline spiral curve   A graphic representation of the collapse of a community or ecosystem as its biological functions are lost at an accelerating rate.

decollement   The surface separating detached, folded, or faulted rocks moving over underlying rocks.

decommissioning   The permanent closing of a nuclear power plant. Nuclear reactors have limited life spans, as do all industrial facilities. Even after the fuel is removed, the reactor vessel and some of the piping and heat exchangers remain radioactive for many years. To protect the public, these components can either be removed or sealed and guarded at the site. See EMBRITTLEMENT, NUCLEAR REACTOR.

decompose   To break down; to rot or decay, especially through microbial action.

decomposer   An organism that derives its nourishment from dead organic matter (animal and plant bodies), breaking down the complex molecules into simpler organic molecules. Decomposers include earthworms, mushrooms and other fungi, and bacteria. Also called microconsumer, detritivore, saprobe, or reducer. Compare CONSUMER, PRODUCER, SAPROBE.

decussate   1) Crossing or interlacing; forming a structure with the shape of an x. 2) Describing plants whose leaves come off the stem in pairs, with each pair at right angles to the next pair. 3) A grain structure of metamorphic rocks in which platy or columnar minerals are arranged in a definite criss-cross pattern.

deductive method   A form of logical reasoning that proceeds step-by-step from known laws or rules to prove that something is true, as in proving that something is a particular case of a general law or scientific principle that is known to be true. Contrast INDUCTIVE METHOD.

deep ecology   A perspective that views human beings as coequals with other species integrated within functioning ecosystems, rather than superior, in a controlling position. The resulting worldview is ecocentric rather than anthropocentric. The term was coined by the Norwegian philosopher Arne Ness in a 1972 article published in *Inquiry*.

deep-focus earthquake   An earthquake that has a focus located deeper than 300 km within the crust of the Earth.

deep-scattering layer   A concentration of microscopic marine organisms or a school of fishes within a horizontal layer of ocean water. A deep-scattering layer may be mistaken for the ocean floor when detected by sonar equipment.

deep sea   A general term for any area of the ocean more than 6000 m deep.

deep-sea fans   The fan-shaped accumulations of sediment forming at the mouth of a submarine canyon or at the base of a continental rise.

deep-sea floor   Any bottom area of an ocean that is deeper than 6000 m.

deep-sea trench   A long, linear depression in the deep-sea floor formed by the high angle subduction of an oceanic tectonic plate.

deep-well disposal   A means for disposing of wastes by injecting them into subterranean fissures. This method is used for low-level radioactive waste, or other hazardous materials, such as oil drilling wastes, municipal waste, or industrial waste. Also called deep-well injection.

deer yard   An area, often sheltered by conifers, where deer in northern climates gather during the winter. Because of the high population density, the area can become severely overbrowsed, especially as the same sites are often used year after year. Also called simply yard.

defaunation   The removal of all animals from an area.

deficiency disease   Plant or animal disease caused by a lack of an essential nutrient in the soil or diet. See BERIBERI, PELLAGRA, SCURVY, RICKETS, CHLOROSIS.

definitive host   The organism on which a parasite lives during the adult stage of its life cycle, when it may reproduce sexually. See INTERMEDIATE HOST.

deflation   The removal or sorting of unconsolidated sand or rock particles by wind erosion.

deflation armor   See DESERT PAVEMENT.

defoliant   Herbicide causing leaves to fall off plants and trees. Agent Orange (2,4,5-T), used by the military to clear jungles of foliage, is a defoliant. See 2,4,5-T.

defoliation   The shedding of leaves in trees or plants, especially when premature or induced (as by gypsy moths or other insects).

deforestation   The practice of permanently removing forest to clear the way for a different land use, such as cultivation or development.

deformation   The process or result of stress acting on a solid material to cause a permanent alteration of shape (strain). Deformation may result in folding or faulting in rocks through application of compression, torsion, or shearing stresses.

degas   To remove gas from a system or environment, as in a test bore in petroleum exploration.

degassing   Gradual release of dissolved or adsorbed gasses. Degassing from molten rock as it cooled contributed to the creation of the earth's atmosphere during our planet's formation. Degassing also refers to the removal of gas from mud when drilling test bores, and from a laboratory vacuum apparatus.

deglaciation   The exposure of the land surface during the ablation or melting of a glacier.

degradable   Substance that can be broken down into smaller, less complex molecules. See BIODEGRADABLE.

degradation, stream   The wearing down of a streambed by erosion processes. Contrast PROGRADE.

degree   1) Unit of temperature. 2) Unit of angular measurement, as for longitude or declination. See CELSIUS, KELVIN, FAHRENHEIT.

dehiscence   The bursting open of a seed pod, anther, or similar organ, usually in a predetermined pattern, to disperse seeds or pollen. An example of dehiscence is jewelweed (*Impatiens capensis*), which gets its other name "touch-me-not" from the way its seed pods quickly rupture and throw out seeds when brushed. Compare INDEHISCENT.

dehydrate   To remove water from a substance. Dehydration can refer to a drying process such as food drying, but is also used to refer to a chemical reaction that removes two atoms of hydrogen for each atom of oxygen removed from a compound.

dehydrogenases   Enzymes that cause removal of hydrogen from other compounds during biological reactions. Dehydrogenases are essential in the reactions that transfer electrons during cellular respiration.

de-inking The process used in recycling paper that removes ink in order to produce a final product that is closer to white rather than dark gray.

Delaney Clause A section of the Food Additives Amendment to the 1938 federal Food, Drug, and Cosmetic Act intended to prevent the addition of carcinogenic substances to foods or cosmetics. It states that "no additive shall be deemed safe if it is found to induce cancer when ingested by man or animal."

delayed density dependent factors Population controls that develop in response to the density of animals in a specific population, but occurring after a time lag, sometimes after several generations. See DENSITY DEPENDENT.

delta A partially subaqueous landform composed of alluvium deposited by abrupt decrease in stream velocity where a stream enters an ocean or lake.

delta front The inclined seaward or offshore edge of a delta. The delta front is the site of active sedimentation in a prograding delta.

deltaic deposits The unconsolidated sediments laid down by a prograding delta, consisting of topset, cross-set, and bottomset beds.

delta plain A nearly horizontal portion of delta that is largely exposed to the atmosphere during low tide or other regression of the water.

deme A local group of organisms that interbreed. See GAMODEME, ECOTYPE.

demersal Living near (or, in the case of fish eggs, sinking to) the bottom of the sea, or very deep water. Contrast PELAGIC.

demetron-s-methyl $C_6H_{15}O_3PS_2$, an organophosphorous insecticide used as a systemic against aphids and spider mites. It is toxic to vertebrates. Also known by the names Demeton and Systox. See ORGANOPHOSPHOROUS PESTICIDES.

demographic process Any function that can alter the size of a population, such as birth, immigration, or death. See DEMOGRAPHY.

demographic transition Gradual pattern of change in human populations of a particular region or country from high birth and death rates to high birth with declining death rates, eventually reaching low birth and death rates. Demographic transition often accompanies overall reduction in population growth with improved living conditions, sometimes reached through economic development.

demography The statistical study of populations (usually human), especially their growth.

denature 1) To change a compound, especially by altering its chemical formula or structure, and thus change its function or properties. Biological molecules such as enzymes and DNA can be denatured by heat, strong acids or bases, and chemicals such as ions of heavy metals, destroying their ability to carry out normal functions. Denaturing of DNA refers to destroying the hydrogen bonds that create its double-helix structure. 2) The addition of another substance that does not change the original substance but prevents improper uses. Denatured alcohol has substances added to it that make it unpleasant to consume or toxic to humans, and uranium 233 can be denatured to make it unsuitable for military use (bombs) by the addition of U238.

dendritic crystals A term describing the

shape of crystals that are precipitated in a branching, treelike shape, as seen in crystals of manganese oxide.

dendritic drainage   A term describing the appearance in map view of stream drainage systems that branch out into the watershed uplands in a treelike shape. Dendritic drainage usually develops in regions of relatively homogenous bedrock, alluvial cover, or glacial deposits.

dendritic pattern   A branching (literally, "treelike") pattern of drainage in which converging tributaries flow into a main river. It is usually found where the underlying rock types do not vary throughout the river basin, or where there are no artificial dams or other structural control.

dendrochronology   The analysis of the annual growth rings to learn the age of the specimen and particular environmental factors occurring during its lifetime. Dendochronology is used by archaeologists to date sites of human habitation, as well as by climatologists as a method of reconstructing past climates. Also called tree-ring dating. See DENDROCLIMA-TOLOGY.

dendroclimatology   The examination of the annual growth rings of trees to gather data about rainfall and other past climatic conditions. See DENDROCHRONOLOGY.

dengue (fever)   Tropical disease characterized by a high fever, pain in the joints, headache, and rash caused by a mosquito-transmitted arbovirus. Rarely fatal. Also called breakbone fever.

denitrification   The reduction of nitrates and nitrites, especially when accomplished by denitrifying bacteria, which require organic matter as an energy source. Denitrification reduces the amount of nitrogen present in the soil; it occurs most

rapidly under warm, anaerobic conditions. See DENITRITFYING BACTERIA.

denitrifying bacteria   Soil organisms that reduce nitrate or nitrite in the soil and convert it to nitrogen gas or gaseous oxides of nitrogen. *Pseudomonas denitrificans* is one species of denitrifying bacteria. The resulting denitrification reduces the amount of nitrogen present in the soil. Contrast NITRATE-FORMING BACTERIA, NITRITE-FORMING BACTERIA, NITROGEN-FIXING BACTERIA.

density   1) The mass per unit of volume of a substance, usually stated in grams per cubic centimeter (or pounds per cubic foot). Density gives an indication of how closely packed the molecules of a substance are; for example, liquid water has a density of 1 gram per cubic centimeter, while liquid mercury has a density of 13.59 grams per cubic centimeter. 2) The size of a population of a given species in a given area.

density current   A flow of water that is produced by differences in water density. Density currents may develop with changes in salinity, temperature, or suspended sediment load. Compare TURBIDITY CURRENT.

density dependent   Describing factors that act in proportion to the density of animals in a specific population. Parasite and predator population levels are often density dependent. As the population of host or prey increases, their populations also increase; likewise, as the density of host or prey populations decrease, the parasite or predator populations decrease. Compare DENSITY INDEPENDENT, NEGATIVE FEEDBACK.

density independent   Referring to a situation in which a population species is not regulated by its size, or factors that are not influenced by population size (such as population reductions caused by weather extremes). Compare DENSITY DEPENDENT.

density overcompensation   The phenomenon in which increases in populations of a species are matched by even greater decreases in the birth rate or increases in the death rate, resulting in an eventual population decrease. Also called overcompensating density dependence. See RESOURCE COMPETITION, UNDERCOMPENSATING DENSITY DEPENDENCE.

dental formula   A method for describing the number and distribution of different kinds of teeth in mammals. In humans, the dental formula is the number of incisors, canines, premolars, and molars for one-half of the upper jaw listed above those for that half of the lower jaw.

den tree   A tree with a hollow that serves as a resting place or a place for mammals to nurture their young. Raccoons often live in a den tree.

denudation   The process of stripping a land surface to bare rock by weathering, erosion, and transportation of particles.

deoxyribonucleic acid   Full name for DNA. See DNA.

Department of Energy   (DOE) A department of the federal government established in 1977, responsible for research and development of energy technology, marketing of federal power, energy conservation programs, nuclear weapons program, and energy regulatory programs.

depauperate   Not reaching normal size or completing normal development. Depauperate may be applied to an individual or a population.

dependent variable   A changing quantity whose variation depends upon that of another quantity, the independent vaiable. Compare INDEPENDENT VARIABLE.

depolarization   Reduction or elimination of polarization, either within ionized solutions, magnets, or sometimes in electrical insulators.

deposit   A general geologic term for a concentration of minerals or aggregate of particles. The term deposit may also be applied to material with potential economic value.

deposit feeder   Aquatic or marine organisms that swallow grains of sediment and assimilate the associated microorganisms.

deposition   The process of laying down sediments after a transportation process. Deposition occurs in areas such as river flood plains, deltas, alluvial fans, or glacial moraines.

depositional remanent magnetism   The fixed component of a rock's magnetism that is independent of externally applied magnetic fields.

deposition velocity   A measure of the water velocity at a point where a stream becomes incompetent to transport a particle of a specified size.

depression   Meteorological term for a low-pressure region; also called cyclone. See CYCLONE.

deranged pattern   A description of the appearance in map view of stream systems that have been altered and disturbed by external influences, such as regional uplift or deposition of glacial materials.

dermal toxicity   Chemicals that poison animals or humans by absorption through the skin.

dermis   The layer of skin below the epidermis.

derris   The extract made from roots of the South American tree *Derris elliptica*, used as a non-synthetic insecticide that kills on contact. The main ingredient of this extract is rotenone. See ROTENONE.

desalinization   The process of removing salt from water or soil.

desert   A biome (vegetation type) that occurs in areas with low precipitation—less than 25 cm (10 inches) per year—usually sandy or rocky, and lacking trees. See BIOME.

desert crust   See DESERT PAVEMENT.

desertification   The process of climate and environmental modification leading to the formation of a more arid-appearing landscape.

desert pavement (deflation armor)   A thin layer of coarse particles left on the surface of unconsolidated sediment after finer particles have been carried away by wind.

desert soils   An older term applied to zonal soils having a light-colored surface horizon underlain by a caliche or hardpan layer. See ARIDISOL.

desert varnish   A smooth, dark, shiny surface of manganese and iron oxides that coats exposed rocks in hot deserts. Desert varnish is formed by biologic and chemical weathering.

desiccant   1) A substance used to control insect pests or mildews by causing them to dehydrate and die. Diatomaceous earth is a desiccant used in agriculture and horticulture as a natural insecticide. 2) A drying agent. See DIATOMACEOUS EARTH, DRYING AGENTS, DESICCANT.

desiccate   To dry out. Some pesticides (desiccants), such as diatomaceous earth, work by desiccating the bodies of insects.

desiccation   The process of drying; desiccants are substances such as silica gel that absorb moisture from the air.

desiccator   1) A laboratory device used for drying substances. It is usually a chamber that is air-tight when closed and uses a chemical drying agent to carry out the process. 2) Chemicals (such as silica gel, calcium chloride, or pesticides such as diatomaceous earth) used for drying are called desiccants or drying agents. See DRYING AGENTS, DESICCANT.

desorption   The opposite of adsorption; the release of materials from being adsorbed onto a surface. Outgassing is a form of desorption. Compare ADSORPTION.

destructive distillation   The process of subjecting complex organic substances to high temperatures in the absence of oxygen that results in their decomposition into a mixture of volatile products. Destructive distillation is used to produce coal tar products, ammonia, and natural gas from solid coal.

destructive plate margin   A contact area where crustal material is destroyed in a subduction zone between converging tectonic plates.

desulfurization   The removal of sulfur from hydrocarbons or iron ore by any of a number of chemical processes. Desulfurization of coal results in much less pollution when it is burned.

detachment fault   A type of low-angle, normal fault in which horizontal displacement is the dominant motion. Detachment faults are often associated with regional uplift.

detachment surface   See DECOLLEMENT.

detection limit The lowest concentration of a pollutant that can be measured by a specific analytical technique. Each method of pollutant analysis has an associated detection limit.

detergent A synthetic substance used to wash away grease and dirt. Like soaps but stronger, detergents work by emulsifying oils and holding dirt in suspension. Detergents high in phosphorous contribute to the eutrophication of water bodies.

deterministic models Mathematical models of systems involving fixed relationships rather than probability. The results of changes introduced given certain initial conditions can be precisely calculated in deterministic models. Contrast STOCHASTIC MODELS.

detoxification Rendering a poisonous substance harmless by removing or denaturing its toxins.

detrital sediments Deposits of clastic particles that were derived from weathering, erosion, and transportation of rock fragments.

detritivore A decomposer or saprobe; an organism that feeds on dead organic matter, including scavengers. Compare HERBIVORE, CARNIVORE, OMNIVORE.

detritus Waste material. Detritus is usually organic material, such as dead or partially decayed plants and animals, or excrement. Detritus can also be small particles of minerals from weathered rock, such as sand and silt.

detritus food chain The transfer of food energy primarily from dead organic matter, which is consumed and broken down into detritus by microorganisms, which in turn are consumed by detritivores and then by predators of detritivores. Contrast GRAZING FOOD CHAIN.

deuterium (D) An isotope of hydrogen that is about twice as heavy as ordinary hydrogen (its atomic weight is 2) due to the presence of a proton in the nucleus. Deuterium is therefore sometimes called heavy hydrogen; water containing deuterium (D_2O) is called heavy water to distinguish it from ordinary water. Less than 0.02 percent of hydrogen atoms in nature are in this isotopic form. See HEAVY WATER, TRITIUM.

Deuteromycota The phylum of the kingdom Fungi that are sometimes called fungi imperfecti, in that they lack clear sexual mating types but reproduce asexually through production of conidia, or oedia, or by budding. This phylum includes the economically important fungi *Penicillium* (including some species used as cheese culture and others from which penicillin is produced), *Aspergillus* (black molds), and the plant pathogens *Verticillium, Gloeosporium, Colletotrichum*. See FUNGI.

deuterostomes Name for a group of related, animal phyla with similar patterns of development including Chaetognatha, Echinodermata, and Chordata. In particular, the mouth in this group is formed from a secondary opening into the gut during development and not from the embryonic blastopore. See CHAETOGNATHA, CHORDATA, ECHINODERMATA.

development rights The legal ability to develop a parcel of land. Transfer of development rights is an important tool in the preservation of prime farmland and other open space in areas where development pressures create high economic incentives for selling off open land. In some areas, the development rights to the land can be sold off, bringing cash or substantial tax breaks to the landowner while allowing them to continue to farm the land or to continue other uses not involving development.

devitrification   The process by which glassy igneous rock is made crystalline. Devitrification converts obsidian into felsites containing very small crystals of quartz and feldspar.

Devonian period   The fourth geologic time period of the Paleozoic era. The Devonian period lasted from approximately 408 to 360 million years ago. See APPENDIX p. 611.

dew   Condensation of water vapor onto objects on the ground. Dew forms when exposed surfaces cool to a temperature below the dew point of ground-level air, as on clear, calm summer nights.

dew point   The temperature at which moist air reaches saturation (the relative humidity reaches 100 percent) and water vapor begins to condense.

dextral fault   The term applied to a right lateral strike-slip fault in which the apparent sense of displacement is to the right. Contrast SINISTRAL FAULT.

dextrose   Another name for glucose (short for dextrorotatory glucose), also known as grape sugar. See GLUCOSE.

diabase   A dark-colored, medium-grained igneous rock commonly forming intrusive sills or dikes. Diabase is intermediate in texture between basalt and gabbro.

diachronous   A geologic term pertaining to a single rock unit that varies in age from one place to another.

diadromous   Organisms that migrate between fresh and salt water, such as eels and carp. See ANADROMOUS, CATADROMOUS.

diagenesis   The alteration of sedimentary rocks and minerals under conditions of near-surface temperatures and pressures. Diagenesis grades into metamorphism at higher temperatures and pressures. Compaction, recrystallization, and cementation are diagenetic processes; sandstone and coal are typical diagenetic rocks.

dialysis   The process of using a membrane to separate colloids or large molecules from small molecules dissolved in a liquid. Dialysis is used in the treatment of some wastes and also medically to duplicate the kidneys' action of removing wastes from the blood for patients whose natural kidneys are no longer fully functioning.

diameter at breast height   (dbh) A standard measurement used to evaluate tree size. It is the diameter of the trunk measured at breast height, which is standardized at 4.5 feet or 1.4 meters above the ground surface. Compare BASAL AREA.

diameter-limit cut   Timber harvest method in which only trees over a certain minimum diameter are cut.

diamictite   A calcareous conglomerate rock composed of unsorted coarse particles interspersed in a matrix of sand and mud.

diamond   A mineral variety of pure carbon forming cubic crystals under extremely high temperature and pressure conditions, as occurs in a kimberlite. Diamond is the hardest naturally occurring mineral, the world's most valuable gem, and an excellent cutting tool. See APPENDIX p. 622.

diapause   A period of dormancy (reduced physiological activity) entered into by some insects to survive adverse environmental conditions. See DORMANCY.

diaspore   1) A naturally occurring mineral containing a mixture of aluminum hydroxide and aluminum oxide, AlO(OH). Diaspore may range from white to violet;

it is found in bauxite. 2) Another term for disseminule or propagule. See BAUXITE, DISSEMINULE.

diastrophism   A general term applied to the deformation of the earth's crust, including processes such as folding, faulting, and mountain building.

diathermal   Describing walls that allow transfer of heat by means of radiation.

diatomaceous earth   A very fine-grained natural material excavated from marine deposits of fossil diatoms. Diatomaceous earth is used in a variety of industrial applications, such as in the manufacture of polishes and filters.

diatoms   See BACILLARIOPHYTA.

diatreme   A vertical pipe or neck of a volcanic fissure that has become filled with angular rock fragments. Diatremes form by explosive upwelling of magma and may contain commercially valuable diamond deposits. See KIMBERLITE.

dibromochloropropane   (DBCP) An organohalogen pesticide whose use in the United States was banned in 1979. DBCP is a suspected carcinogen and may cause sterility. It does not biodegrade readily and has been found in groundwater in agricultural areas.

dichlorodiphenyltrichloroethane   The full name for the insecticide DDT. See DDT.

dichloroethylene $C_2H_2Cl_2$, a group of organic compounds with the same formula. They are used as a source for vinyl chloride, which in turn is polymerized to form polyvinyl chloride (PVC). See VINYL CHLORIDE.

dichogamy   A condition in some plants providing for cross-fertilization by having the pollen-bearing stamens and ovary-containing pistils maturing at different times in the same flower. Only pollen from a different flower will be ripe at the same time and therefore able to pollinate a particular pistil/ovary. Compare AUTO-GAMY, HETEROGAMY. See PROTANDRY, PROTOGYNY.

dichotomous   1) In plant structures such as veins or stems, branching to create two roughly equal parts. 2) Describing a taxonomic key arranged using couplets of questions concerning an organism's characteristics.

dicot   Short for dicotyledon. See DICOTY-LEDON.

dicots   Angiosperm plants that have a pair of cotyledons. Many members of this sudivision often have flowers with parts in multiples of four or five and leaves with a net arrangement of veins. This group includes the buttercups, roses, asters, oaks, snapdragons, cacti, and spurges, among many others.

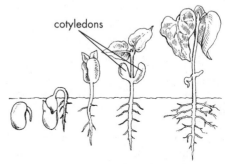

The developing seedling of the bean, a dicotyledonous plant, sends down a root and then raises its cotyledons above ground.

dicotyledon   A plant having two seed leaves (cotyledons) within the seed. Dicotyledons are the larger of the two classes of flowering plants (angiosperms). Most deciduous trees and shrubs are dicotyledons. Dicotyledons are often called dicots. Compare MONOCOTYLEDON.

Dicotyledonae A class of plants characterized by having two cotyledons. See DICOTYLEDON.

dieback 1) A sudden decline in an animal population; a population crash. 2) A condition in plants in which the shoots wither and die, sometimes progressing to other parts of the plant. Dieback may be a symptom of fungal diseases, but may also be caused by unfavorable conditions such as unusual heat, cold, or drought. 3) A gradual and progressive death of trees occurring over a wide area, often progressing from the top and periphery of the crown toward the base of the tree. It is believed to be caused by air pollution, acid rain, or some cause other than a known disease organism. See WALDSTERBEN.

dieldrin One of the better-known chlorinated hydrocarbon pesticides, no longer permitted in the United States. It is a contact poison formerly used for moth-proofing as well as for agricultural pest control. See CHLORINATED HYDROCARBONS, CYCLODIENE INSECTICIDES.

dielectric An electrical insulator; a material through which negligible current can pass. Contrast CONDUCTOR.

dielectric constant A characteristic of the insulating material placed between the two conducting surfaces of a capacitor—specifically, the factor by which the capacitance increases due to the presence of the material (compared to a vacuum). Easily polarized substances such as water have high dielectric constants.

dielectric strength The resistance of an insulating material to electric spark penetration. Specifically, the minimum electric field (voltage per unit thickness) that will cause spark discharges through the material.

diesel A compression-ignition engine, one type of internal-combustion engine, named for its inventor. The heat from compressed air causes the ignition of fuel injected into the cylinder, distinguishing these engines from those requiring spark plugs. The term also refers to the fuel used in such engines. See INTERNAL COMBUSTION ENGINE.

diethylstilbestrol A synthetic estrogen hormone; it is more commonly known by its abbreviation, DES. It is used medically to treat conditions arising from estrogen deficiency. It is no longer used for women during pregnancy to prevent miscarriage, because it was found to increase the cancer rates in the daughters of women who took DES during pregnancy.

differential cooling A reduction of temperature that is not uniform throughout a system, such as occurs in an air mass. See DIFFERENTIAL HEATING.

differential heating Uneven increase in temperature of a system, as when an air mass lies over a rise whose south side is warm but whose north side is cool. See DIFFERENTIAL COOLING.

differential weathering A process by which various mineral components of a rock matrix are weathered at different rates because component minerals differ in resistance to physical and chemical processes. Differential weathering can produce microtopographic forms. See WEATHERING.

differentiation Changes in cells, tissues, and organs during development of an organism that result in modification and specialization into mature tissues.

diffraction   A change in wave direction caused by passing through an opening or around an object. Diffraction allows sound waves to travel around corners.

diffraction analysis   Technique for studying crystal structure based upon the manner in which the crystals diffract (bend or alter) x-rays. Diffraction analysis is used in many fields, including chemistry, biology, physics, and geology.

diffraction grating   Device used for separating light into its component wavelengths. Unlike a prism, diffraction gratings consist of glass or metal with many fine parallel lines or grooves (often in the range of 10,000 per centimeter). These cause incoming light to bend or diffract as it passes through the glass or reflects off the metal. Different colors (wavelengths) are concentrated in different directions. See SPECTROSCOPE.

diffuse   Not concentrated; also, the process of dispersing or allowing to mix together by diffusion.

diffuse competition   See INDIRECT COMPETITION.

diffuse-porous wood structure   Wood characterized by vessels of relatively uniform size distributed evenly throughout each growth ring. Diffuse-porous wood is found in maples and birches. Compare RING-POROUS WOOD STRUCTURE.

diffusion   Transport of matter brought about by the random movement of particles. Diffusion causes the mixing or intermingling of molecules or atoms in solutions of differing concentrations as they move from high concentrations to low concentrations.

diffusion potential   The potential difference (voltage) created when two solutions of differing concentrations (but containing similar substances) are allowed to mix. Differing rates of diffusion of the ions in the two solutions create the potential difference.

digester   1) A device used in secondary wastewater treatment. It consists of a tank in which bacterial action breaks down the organic material into simpler components. 2) A device that uses anaerobic bacteria to produce biogas (methane) from refuse such as manure.

digestible   Describing substances capable of being broken down and converted into elements needed for nutrition.

digestion   The process in which food is broken down into compounds that cells can absorb, store, or oxidize and use as nourishment. 2) The use of anaerobic bacteria to produce biogas (methane) from refuse such as manure. 3) The use of bacteria (usually aerobic) to break down raw sewage into simpler compounds, used in secondary wastewater treatment. See ASSIMILATION, EGESTION, EXCRETION.

digestion efficiency   See ASSIMILATION EFFICIENCY.

digitalis   1) Foxglove; any of a genus of biennial and perennial plants (*Digitalis*) of the figwort family, characterized by long spikes of thimblelike flowers. 2) The dried leaves of *Digitalis purpurea*, or compounds extracted from this or related species (steroid glycosides), used as a heart stimulant in treatment of heart disease.

digitate   Describing compound leaves divided into five leaflets.

digitiform   Having the shape of a finger.

digitigrade   Walking on the toes rather than the entire phalanges or soles of the feet. Terrestrial carnivores such as cats and dogs are digitigrade. Compare PLANTI-GRADE, UNGULIGRADE.

dikaryon   Hyphae or mycelia of fungi containing two nuclei of different genetic type in each cell or segment. Also spelled dicaryon. Contrast MONOKARYON.

dike   A type of tabular pluton that is long and narrow and usually formed in a relatively upright position. A dike is one type of discordant intrusion.

dike swarm   A series of tabular plutons that branch out and cross one another in a complex system.

dilatancy   In geologic science, an increase in the volume of a rock because of grain rotation, fracturing, and increase in pore space. Dilatancy occurs during the deformation of rocks.

diluent   A substance used for diluting or dissolving. Air is a diluent when it is introduced into flues before combustion products are released to the atmosphere.

dilute   Not concentrated; weakened by the addition of a substance that reduces concentration. Vinegar is a dilute acid. Also, to add a substance such as water to thin or weaken a solution. Compare CONCENTRATION.

dilution   The opposite of concentration; the process of adding a substance such as water to thin or weaken a solution. Also, a diluted substance or the volume of solvent in which a given amount of a substance is dissolved (such as a medical preparation specified as a 10-to-1 dilution).

dilution ratio   The ratio of the total volume of a diluted sample to the volume of concentrated sample that was diluted. In studies of water pollution, it is the relationship between volume of water in a stream to the volume of waste introduced

into the stream, expressed as a ratio. The dilution ratio gives an indication of the capacity of a water body to assimilate waste.

diluvium   An accumulation of material deposited by a flood.

dimethoate $(CH_3O)_2$-PS-S-CH_2-CO-NH-CH_3, a moderately toxic organophosphorous compound used as a systemic insecticide for spider mites and aphids. Its trade names include Cygon, Rogor, Fostion, MM, L 395, Roxion, Dimetate, and Pertekthion. See ORGANOPHOS-PHOROUS PESTICIDES.

dimethyl sulfide $(CH_3)_2S$, a gas given off by wastewater that contains large quantities of sewage.

diminishing returns   Orignally an economic term for situations in which each increment of added investment produces a lower increment of production or benefit than the previous investment. Now used generally for situations in which more work, energy, or investment produces increasingly less results.

dimorphism   1) The occurrence of two distinctly different forms within the same animal species. Humans and most species of birds exhibit sexual dimorphism; males and females are two distinct and visually distinguishable forms. 2) Exhibiting two different forms of flowers, leaves, or other structures either on the same plant or within the species. Young eucalyptus trees have round leaves quite unlike the long, pointed leaves of mature specimens. See POLYMORPHISM.

dinitrogen fixation   Another term for nitrogen (N_2) fixation. See NITROGEN FIXATION.

dinitro-o-cresol   Short for dinitroorthocresol, more commonly known as DNOC. It is a dinitro compound used as a fungicide in dormant sprays (oil sprays applied usually early in the spring, while trees are dormant) for fruit trees. It has partially replaced the use of lime sulfur.

dinitro pesticides A group of compounds with a wide range of applications; they are used as powerful fungicides, herbicides, insecticides, and miticides, and to thin blossoms on fruit trees (to produce fewer, larger fruits). Examples include Karathane (dinitrophenolcrotonate), Dinoseb, and DNOC (dinitroorthocresol). See DINITRO-O-CRESOL.

dinoflagelates See DINOFLAGELLATA.

Dinoflagellata A phylum of marine planktonic organisms often associated with red tides, symbiotic relationships with marine coelenterates, and bioluminescence. In some classification systems these orgaisms compose the Pyrrhophyta. Many, but not all, members of this phylum of the Protoctista are autotrophic. The single cells typically have an armored, cellulosic surface with transverse and vertical grooves, as well as two flagella: a trailing, longitudinal undulipodium and a transverse undulipodium that lies in the transvers groove. Familiar members of the phylum include the red tide "algae" *Gonyaulax*, the photosynthetic zooxanthellae of coral reefs (*Symbiodinium* and *Gymnodinium*), and the bioluminescent *Noctiluca*. See PROTOCTISTA, RED TIDE.

dinosaurs A common name for members of two extinct archosaurian lines of evolution, one with lizard-like and the other with bird-like pelvuses. Many were bipedal and it is probable that all, including the quadrupedal forms, evolved from bipedal ancestors. They ranged from chicken-sized species to the gigantic *Seismosaurus* which

was 110 feet long and may have weighed more than 40 tons. Most were herbivorous excepting the theropod or carnosaur order, which includes the sickle-clawed veliceraptors and the mighty tyrannosaurs. Unlike reptiles in general, it is thought that at least some dinosaurs were endothermal. See ARCHOSAURIA.

dioecious Plants whose male and female flowers are borne on separate plants of the same species. Asparagus plants, hollies, and willows are dioecious. Compare MONOECIOUS, HERMAPHRODITE.

diomedeidae See ALBATROSSES.

diorite A general term for an intermediate-colored igneous rock with phaneritic texture. Diorite has less silica than a granite but more silica than a gabbro.

dioxin Any of a large group of poisonous, carcinogenic, and teratogenic chlorinated hydrocarbons not found in nature. They are created as by-products in the manufacture of the herbicide 2,4,5-T, in the improper incineration of plastics, and in paper manufacturing. They are suspected to be among the most toxic substances known and are now widespread pollutants found throughout the environment. See CHLORACNE.

dip 1) The inclination of a planar surface as measured from the horizontal, such as the slope of the bedding surface in an outcrop. In bedrock measurements, the true dip is always measured from a vertical plane, which is perpendicular to the strike of the planar surface feature. 2) A measure of the angular variance of the earth's magnetic field from the horizontal.

dip equator A theoretical line on the earth's surface that connects the points at which the measure of magnetic inclination is zero.

diphyodont   Having two successive sets of teeth. Mammals are diphyodont because their permanent teeth replace an earlier set of milk (or deciduous) teeth. Compare MONOPHYODONT, POLYPHYODONT.

diploblastic   Describing animals whose body walls have only two cell layers, the ectoderm and the endoderm. Organisms in the Cnidaria phylum (jellyfish, coral, sea anemone, and hydra) are diploblastic. Compare TRIPLOBLASTIC.

diploid   Describing a nucleus, cell, or organism that contains two sets of chromosomes, one from each parent. The diploid number is twice the haploid number (n) of chromosomes characteristic of each species, and is therefore indicated as 2n. All normal animal cells except for gametes (reproductive cells) are diploid. See HAPLOID, POLYPLOID, TETRAPLOID.

diplont   An organism in which the diploid phase is dominant during its entire life, and the haploid phase is limited to the production of gametes (germ cells) that again become diploid when they join to form a zygote. Compare HAPLONT.

Diplopoda   A class of uniramic arthropods, commonly known as millipedes, which have most body segments fused in pairs to produce diplosegments each with two pairs of legs. The dorsal portions of the segments are larger than the venters, allowing animals to curl up in flat coils. The exoskeleton is usually hard and partly calcified. Antennae are short and segmented. Compared with centipedes, millipede locomotion is slow. They usually feed on decaying vegetation and are principally distributed in the Southern Hemisphere. Approximately 10,000 species exist. See UNIRAMIA.

dipmeter   An electromagnetic device that measures the occurrence of dipping rock strata or beds exposed within a borehole.

Dipnoi   Subclass of freshwater, fleshy-finned bony fish, commonly know as lungfish, in which swimbladders serve as functional lungs. Although the South American species *Lepidosiren paradoxa* and the four African species of the genus *Protopterus* have poorly developed gills and are dependent on their lungs, the Australian species *Neoceratodus forsteri* respires mainly through its gills. In times of drought, African lungfish survive by aestivating in mud burrows. See OSTEICHTHYES, AESTIVATION.

dipole   Opposite, equal charges separated by a small, specified distance as in an electric dipole or magnetic dipole. Dipole is also used to refer to a molecule with no net charge whose ends have opposite charges. Also called a dipolar molecule.

dipole-dipole interactions   Electrostatic interactions that occur when dipolar molecules are allowed to come in contact with each other. The positively charged end of one molecule aligns with the negatively charged end of another.

dipole field   The electric field pattern around two separated electric charges of equal magnitude but opposite sign. Also, a similar magnetic field pattern.

dipole moment   A value used to describe two separated electrical charges of equal magnitude but opposite sign. It is obtained by multiplying the positive charge and the distance between the charges. Dipole moments are measured in coulomb meters in the Système International d'Unités (SI). Dipole moment can also refer to the degree of ionization of a molecule, or to magnetic fields.

dipping bed   A bedding plane or layer of

rock strata that is inclined from the horizontal.

dip plating   A electroplating process in which a piece of metalwork is immersed in an electrified aqueous solution containing dissolved metallic ions.

dip-slip fault   Any fault in which the sense of displacement is in the same direction as the dip of the fault plane.

dip slope   A geomorphologic term applied to a topographic surface that slopes in the same direction and inclination as the underlying strata. Cuesta topography often provides a good example of dip slopes.

Diptera (true flies)   An important order of insects distinguished by the presence of one pair of functional forewings and a pair of hindwings modified into club-like structures, the halteres, which act as gyroscopic stabilizers in flight. The head is usually large, mobile, and provided with large, compound eyes. Mouthparts are variously modified for feeding on plant and animal secretions, nectar, blood, and decomposition products. Larvae (maggots) may be herbivorous or parasitic or may live on rotting material. Flies are more important medically than any other insect order, particularly as vectors of disease agents such as malaria and yellow fever which are carried by anopheles and culex mosquitoes. Tsetse flies are carriers of trypanosome diseases of both human and domestic animals. Flies are distributed worldwide even in polar regions and occupy a wide variety of habitats. Maggots may be found in environments such as hot springs, saline lakes, and in areas of oil seepage. More than 150,000 species (e.g. flies, horseflies, gnats, midges, mosquitoes, etc.). See CULCIDAE, INSECTA.

direct competition   Exclusion of individu-als from limited resources through aggressive behavior by other individuals or through use of toxins. See INDIRECT COMPETITION, INTERFERENCE COMPETITION.

direct current   (DC or d.c.) Electricity that flows in one direction only. Although it may vary in magnitude, it never reverses direction. Compare ALTERNATING CURRENT.

directional selection   Evolution producing an increase or decrease in the incidence of a particular character trait compared with its present frequency. See SELECTION.

directive evolution   Another term for directional selection. See DIRECTIONAL SELECTION.

direct solar energy   Usable energy obtained directly from sunlight, as through solar cells, solar collectors, and passive solar structures. Contrast INDIRECT SOLAR ENERGY.

disaccharides   A group of carbohydrates characterized by containing two simple sugars (monosaccharides) linked together. Sucrose is a disaccharide; each molecule contains a unit of glucose linked to a unit of fructose. Compare MONOSACCHARIDES.

discharge area   The measured cross-sectional area available for water passage in a stream, channel, or conduit.

discharge, stream   A measure of the total volume of water in a stream passing a given point in a given unit of time. See CFS, CMS, GAGING STATION.

disclimax   Short for disturbance climax, a vegetation type (community) that is not normally the final, stable form that evolves in response to the soils and climate of a particular region, but is artificially stabilized and maintained through human intervention, such as livestock grazing or repeated burning. Sometimes called equilibrium community or plagioclimax.

disconformity   A type of unconformity in which the beds of layered rock above and below lie parallel to the plane of the contact surface. See UNCONFORMITY, NONCONFORMITY.

discontinuity   A boundary separating materials of different densities deep within the earth. A disconformity is detected by measuring a distinct change in the velocity of seismic waves. See GUTENBERG DIS-CONTINUITY.

discontinuous distribution   Disjunct; describing species occurring in two or more areas that are separated (often widely separated) from each other. Compare EVEN DISTRIBUTION, CLUMPED DISTRI-BUTION, RANDOM SPATIAL DISTRIBU-TION.

discordant intrusion   A type of pluton developing from magma injections that cut across the existing layers of the country rock. Contrast CONCORDANT INTRUSION.

discount rate   How much economic value a resource will have in the future compared with its present value.

discrete generations   A series of genera-tions in which each one finishes before the next one begins. Strictly discrete genera-tions are rare; there is usually some overlap between the final stage of one generation and the early stages of the next. See SEMELPARITY, ITEROPARITY.

disease   Illness; pathological condition in which normal physiological function is impaired by improper diet, bacteria, or similar biological agents, or by chemical pollutants.

disharmonic fold   A fold in the bedrock that exhibits geometric peculiarities in wavelength, shape, and symmetry because of the interlayering of competent and incompetent beds.

disinfect   To free an area of germs and bacteria by physical processes (such as boiling) or chemical agents (such as chlorine).

disintegration   A physical process of separation or decomposition into frag-ments, such as radioactive alpha decay. See RADIOACTIVE DECAY.

disjunct   1) Discontinuous, not overlapping, especially in the distribution of popula-tions or species whose geographic ranges are separate. 2) Describing organisms such as ants and wasps that have deep constric-tions between the head, thorax, and abdomen.

disjunction   The separation of pairs of chromosomes, the movement of cen-tromeres to opposite poles of the spindles, that occurs within the nucleus towards the end of meiosis.

dispersal   Active or passive spreading of organisms away from each other, espe-cially away from parent individuals, and into another area. See EMIGRATION, IMMIGRATION, MIGRATION.

dispersal polymorphism   Two or more types of disseminating structures (organs for release of spores, larvae, etc.) found within a species or among the offspring of an individual.

dispersant   Substance used to break up concentrations of organic compounds. It is a chemical tool for cleaning up oil spills; once as much oil as possible has been absorbed and removed, dispersants are added to cause remaining oil on the surface of a water body to scatter.

dispersion   1) Pattern of spatial distribution for animals or plants within a population. 2) Refraction; the separation of electromagnetic radiation such as light into its component wavelengths. See AGGREGATION, RANDOM DISTRIBUTION, UNIFORM SPATIAL DISTRIBUTION, CLUMPED DISTRIBUTION, MORISITA'S INDEX OF DISPERSION, VARIANCE-TO-MEAN RATIO.

dispersion forces   Weak forces between molecules. See VAN DER WAALS FORCES.

displacement activity   Any animal behavior that does not seem to match or relate to a particular situation. Conflict interactions between animals often produce displacement activity. An example of displacement activity is domestic fowl pecking themselves while engaged in fights with others.

display   A pattern of visible or audible behavior, characteristic of a particular species, used to influence the behavior of another organism. Display is commonly used in courtship or intimidation, as in the display of bright plumage by male peacocks before females.

disruptive coloration   Color markings that protect individuals by breaking up solid shapes with stripes or blotches, thereby making the organism more difficult to see from a distance. The markings on zebras, tigers, and giraffes are disruptive coloration. Contrast SEMATIC COLORATION.

disruptive selection   Evolution in which two or more extreme phenotypes (organisms of the same species having recognizably different characteristics) have better success in reproducing than do intermediate phenotypes. The result is an increasingly higher percentage of these phenotypes within the population, and a reduction of the intermediate phenotype, which presumably is less fit.

disseminated mineral deposit   An occurrence of minerals that is highly dispersed throughout an ore rock, rather than concentrated in one place.

disseminule   Detachable reproductive structures of plants, such as seeds, spores, fruits, and fragments of vegetative structures (for asexual reproduction) such as runners. The part of the plant which serves to spur a new individual. Also called propagule or diaspore.

dissolve   To mix a substance with a liquid to form a solution, a homogenous liquid; to cause a substance to become a solute.

dissolved load (stream)   The proportion of the total material in stream transport that is carried in the dissolved state. The principal materials carried in dissolved load are ions of bicarbonate, calcium, chloride, sodium, and sulfate. Compare BED LOAD, SUSPENDED LOAD.

dissolved organic carbon   (DOC) The fraction of carbon bound into organic compounds in wastewater or other water that is made up of the smallest molecules. DOC is one indication of water quality, measured as particles smaller than 0.45 micrometers; it is separated out from total organic carbon by passing it through a filter with pores measuring 0.45 micrometers in diameter. Dissolved organic carbon includes many polar organic molecules such as humic and fulvic acids as well as those synthetic organic compounds that are water soluble. See TOTAL ORGANIC CARBON.

dissolved oxygen   (DO) Amount of oxygen gas contained in water or sewage, usually given in parts per million at a specified temperature and atmospheric pressure. It is a measure of the ability of water to support aquatic organisms. Water with very low dissolved oxygen content (less than 5 ppm), which is usually caused by too much or improperly treated organic wastes, does not support fish and similar organisms. See BIOCHEMICAL OXYGEN DEMAND, CHEMICAL OXYGEN DEMAND.

dissolved solids   Short for total dissolved solids. See TOTAL DISSOLVED SOLIDS.

distal   Spaced far apart; located farthest from the center or away point of attachment to the main structure. Hands are attached to the distal ends of arms. Contrast PROXIMAL.

distill   To boil a liquid causing it to evaporate and then recondensing the vapor. Distilling serves to purify the liquid. See DISTILLATION.

distillation   The process of distilling liquids. Distillation is used to separate liquids into different fractions having different boiling points, as in distilling crude oil into different petroleum products. Distillation is also an energy-intensive means of purifying water. When sea water or other polluted water is boiled, the recondensed steam is pure water, and the minerals (salt) and nonvolatile pollutants are left behind in the boiling vessel.

distilled water   Pure water, prepared using a distillation process to remove all minerals and other contaminants. Using distilled water in household steam irons prevents clogging by mineral build-up; similarly, some industries require distilled water to avoid clogging of pipes or contamination of chemical reactions.

distribution constant, K_D   An equilibrium constant that gives a measure of how a solute distributes itself between two immiscible liquids.

disturbance   An event that changes the local environment by removing organisms or opening up an area, facilitating colonization by new, often different, organisms. When a large tree falls in a mature forest, it opens up a hole in the canopy that lets light fall onto the forest floor; this disturbance allows light-loving plants to grow in a spot that would otherwise be too shaded for them. See CATASTROPHE.

ditocous   Producing two eggs, or bearing two young, at one time. Contrast MONOTOCOUS, POLYTOCOUS.

diurnal   1) Occurring daily; having a circadian rhythm. 2) Active during daylight; describing animals that are not nocturnal or plants whose leaves or flowers open during the day and close at nightfall. Contrast CREPUSCULAR, NOCTURNAL.

divergence   A horizontal spreading of an airstream resulting in a net outflow of air. At low altitudes, diverging winds cause a compensating vertical inward downflow. Divergences at high altitudes contribute to sustaining low-pressure areas and their associated cyclonic air-flow patterns.

divergence zone   See DIVERGENT PLATE BOUNDARY.

divergent evolution   The process of speciation in which a common ancestral population is isolated into separate populations, each of which is subjected to different selective pressures, leading to genetic divergence in their gene pools. See ADAPTIVE RADIATION.

divergent margin   See DIVERGENT PLATE BOUNDARY.

divergent plate boundary   A peripheral border between adjacent tectonic plates that are moving away from each other. A divergent plate boundary is typically a region where a midoceanic ridge is formed.

diversification 1) An increase in variety. 2) In farming, the addition of other crops or products to balance out economic risk and increase income at slack times of the year. Planting Christmas trees on unused land or making maple syrup to supplement income from a dairy farm are examples. 3) A broadening of the economic base of a community by encouraging more, and especially new, forms of business. 4) Applied to niche differentiation processes illustrated by adaptive radiation and the tendency for sympatric species with overlapping niches to diverge through time.

diversity 1) The number of different species, and their relative abundance, in an area. Diversity is a measure of the complexity of an ecosystem, and often an indication of its relative age. Newly established communities are low in diversity; older, more stable communities usually have high diversity. 2) The number of habitats existing in a particular area. See ALPHA DIVERSITY, BETA DIVERSITY, EQUITABILITY, HABITAT DIVERSITY, SPECIES DIVERSITY.

diversity gradient A geographical gradient (gradual change in an attribute over distance) such as latitude, elevation, or moisture along which a change in species diversity is found. A diversity gradient can indicate how far succession has progressed in a particular community because communities generally increase in diversity as they approach climax status.

diversity index A numerical measure of the number of different species, as well as their relative abundance, in a given community or area. Diversity indices can be used to measure the relative health of an ecosystem, or the impact of different human activities. See ALPHA DIVERSITY, SIMPSON DIVERSITY INDEX.

divide A boundary line connecting the high points shared by two watersheds or catchment areas.

division 1) The highest taxonomic grouping within the plant kingdom, equivalent to a phylum of the animal kingdom. Plants of the same division are believed to share a common ancestral form. Divisions are indicated by the suffix -phyta, or -mycota for fungi, as in Spermatophyta and Myxomycota. 2) Horticultural term for propagation by dividing vegetative portions of plants. Compare CLASS, ORDER, PHYLUM.

dizygotic twins Fraternal twins; those that develop from two zygotes and have no more inherited traits in common than siblings born at different times. Also called dizygous. Compare MONOZYGOTIC TWINS.

D-layer The lowest of three subdivisions of the earth's ionosphere. The importance of this region stems from its tendency to absorb significant amounts of radio waves during increases in solar activity, interfering with radio transmissions on earth. See IONOSPHERE, E-LAYER, F-LAYER.

DNA (deoxyribonucleic acid) A complex organic molecule found in all animals, plants, and most viruses that contains the genetic information passed from one generation to the next. DNA is a nucleic acid with a characteristic double helix, a spiral of two parallel strands of the sugar deoxyribose and phosphoric acid. These long chains are bridged by the pyrimidine and purine bases (cytosine, thymine, adenine, and guanine). DNA carries out its functions through messenger RNA (ribonucleic acid). Compare RIBONUCLEIC ACID.

DO Abbreviation for dissolved oxygen content of water. See DISSOLVED OXYGEN.

doctrine   A legal principle or set of principles.

doctrine of prior appropriation   A legal principle in the western United States which holds that entitlement to water rights should be assigned on a "first come, first served" basis as long as the water use is considered to be beneficial.

dodos   Name for a species of extinct, turkey-sized, flightless birds of the order Columbiformes that lived on Mauritius and were eliminated by the introduction of pigs, rats, and monkeys to the island in the late 1680s . Two other species, often called solitaires, which lived on Reunion and Rodriguez islands, became extinct in the 18th century. See COLUMBIFORMES.

doldrums   Region along the equator in which winds are very light or absent. Contrast TRADE WINDS.

dolerite   See DIABASE.

doline   See SINKHOLE.

Dollo's law   The principle that substantial evolutionary changes are irreversible because it is extremely unlikely that genetic changes could occur in exactly the opposite order of those that produced the evolutionary changes.

dolomite   A sedimentary mineral or rock consisting of calcium magnesium carbonate. See LIMESTONE, also APPENDIX p. 623.

dome   A landform or structure in the shape of an inverted bowl. A dome may be formed by upwelling lava, or by the diapiric movement of magma or salt.

dome, structural   A bedrock conformation that has the structure of a dome, with or without the surface expression of the internal form. A structural dome may not appear as a dome-shaped landform.

domesticated   1) Describing formerly wild species of plants or animals that are now raised and bred for human use. 2) Describing an individual wild animal that has been tamed and kept by humans. Compare FERAL.

domestic water   A water source available for household use and usually delivered by a system of pipes.

dominance   1) The condition of a species contributing a major portion of the biomass or numbers within a community, and thereby influencing the abundance of other species in the community. 2) The ability of one gene to prevail over another, different gene in causing a particular trait (phenotype). If T is a dominant trait and t is recessive, the organism with the Tt gene combination will resemble a TT organism rather than a tt organism. Also called the law of dominance. 3) Condition in which behavior establishes more important social positions for some individuals within a population (creating a dominance hierarchy). See MENDEL'S LAWS.

dominance-diversity curve   A graph plotting the importance values, density, abundance, or other measure of importance of the species composing a community (on the y-axis), arranged in descending order on the x-axis. The shape of the resulting curve gives an indication of patterns of equitability in that community. Also called rank-abundance diagrams or Whittaker plots. See EQUITABILITY.

dominance hierarchy   A social ranking among individuals in a group, in which the position of individuals is determined by their degree of submissiveness to the aggressive behavior of others within the group. Also called peck order or pecking order.

Domin scale An abundance rating giving quantitative aspects for a terrestrial plant community. The nonlinear scale is divided into 10 points; it ranges from 1 (1 or 2 individuals), 2 (several individuals), and 3 (covering under 4 percent of total area) to 9 (76 to 90 percent cover) and 10 (91 to 100 percent cover). A + is used to indicate organisms occurring singly.

Donora smog incident One of the worst air pollution episodes, which occurred in 1948 in Donora, PA. The 7-day-long inversion caused 18 deaths and many illnesses.

Doppler effect The apparent change in wave frequency that occurs when either an object emitting radiation or an observer are moving in relation to each other. The Doppler effect is responsible for the rise in pitch of an approaching train whistle, and the lowering of pitch as the train passes and moves away.

dormancy A period of reduced or suspended physiological activity in animals or plants, often to survive a period of harsh environmental conditions (such as winter or drought). Dormancy includes the state of ripe plant seeds before germination as well as animal hibernation. See DIAPAUSE, ESTIVATION, HIBERNATION.

dormant In a condition of suspended activity, such as an inactive volcano, a winter bud on a plant, or an overwintering plant bulb.

dormant volcano A volcano that is inactive. A dormant volcano is not venting lava or pyroclastic material and is not experiencing internal changes because of upwelling magma.

Dorn effect Another term for sedimentation potential, the opposite of electrophoresis. It is the creation of a potential difference when mechanical forces (particularly gravity) cause particles suspended in a liquid to migrate. See ELECTROPHORESIS, SEDIMENTATION POTENTIAL.

dorsal Located on or relating to the back of an animal, plant, or organ, or the side farthest from the ground. Compare VENTRAL.

dorsiventral Possessing discernable upper (dorsal) and lower (ventral) sides.

dose 1) The amount of radiation received or absorbed, or the concentration of a hazardous compound an organism is exposed to, over a given time. 2) The amount of medication to be taken at one time, or the amount of radiation or similar therapeutic agent administered at one time. See MAXIMUM PERMISSIBLE DOSE.

dose rate The amount of nuclear radiation absorbed by an organism, or to which it is exposed, over a period of time.

dose-response curve A graph illustrating the relationship between the concentration of a medicine (or dose of radiation) and the effects on an organism.

dosimeter An instrument for measuring amount of radiation absorbed, or REM, per unit of time. Dosimeters used in hospitals measure total x-rays received. See REM.

double bond A form of covalent bond in which two pairs of electrons are shared between two atoms. Organic molecules containing double bonds are designated as unsaturated. Unsaturated fats contain double bonds. See COVALENT BOND, SINGLE BOND, TRIPLE BOND.

double fertilization A process characteristic of seed plants (angiosperms) in which two male nuclei are required for fertilization. One unites with the egg nucleus; the other usually fuses with two polar nuclei to form a unit that will develop into endosperm, tissue that provides nutrition for developing seedlings.

double layer A region in which electrical charges are divided across a boundary separating two phases (regions that are chemically and physically uniform). An electrode that is positively charged has a double layer consisting of a group of negatively charged ions right next to the positively charged electrode surface. This layer effectively separates the electrode from the rest of the solution until diffusion occurs to remove the charged ions. See GOUY-CHAPMAN MODEL.

double recessive Describing an organism having two recessive genes for one or more traits and therefore showing the characteristics associated with that recessive trait. When only one of the recessive genes is present, the other (dominant) gene masks the recessive characteristic. See DOMINANCE, RECESSIVE.

doubling time 1) The time it takes a specific population to double in size (based on current fertility and mortality rates). 2) The time required in a breeder reactor for fissionable material to produce double the original quantity of fuel. See POSITIVE FEEDBACK.

doubly plunging fold An older term to describe a bedrock fold that exhibits two directions of opposite plunge when viewed from a central point. A doubly plunging fold is more accurately described by detailed use of the terms syncline and anticline.

down draft A downward flow of cool air associated with mature stages of thunderstorms. It is initiated by the motion of falling precipitation and resulting inflow of cool air surrounding the cloud. The cool air can often be felt on the ground before precipitation begins.

downs Grass-covered hills with low bushes and few trees, features characteristic of the limestone regions of southern England. The soil in such areas is often just inches thick, with solid chalk (limestone) underneath.

downstream The direction of a current flow that proceeds from a point of higher potential energy to a point of lower potential energy.

downstream flood An increase in water flow in excess of channel capacity as a result of water accumulation from higher reaches of a catchment area or watershed.

downwash The degree of airflow deflection caused in the wake of an airplane wing or similar aerodynamically designed object.

downwasting A general term for the reduction in the topography of a landscape by weathering, erosion, and transportation of surface materials.

downwelling A point in an ocean current where surface waters move downwards. See ANTARCTIC CONVERGENCE.

dowsing A practice of searching for groundwater by the ritualistic, and presumably paranormal, use of hand-held sticks, wires, pendulums, and other tools.

drag A force of resistance on an object moving through a fluid (liquid or gas). See LIFT.

drainage basin 1) The landform or surface shape of a watershed. 2) The area contributing runoff to a stream system.

drainage density A measurement of the average spatial diversity of a stream system as expressed by the formula $D = L \div A$, where D = drainage density, L = total length of streams, and A = land area.

drainage morphometry   The measurement of the shape of a drainage basin, including features such as area, topographic relief, and drainage density.

drainage net   The system of channels in a drainage basin. See DENDRITIC DRAINAGE, PARALLEL DRAINAGE PATTERN, TRELLIS DRAINAGE.

drawdown management   1) The practice of lowering water levels within an impoundment, such as a pond or lake, in an effort to control the populations of organisms such as aquatic weeds. 2) The regulation of groundwater withdrawal to control the lowering of a water table.

dredge spoils   The discarded sediments or materials that were excavated by dredging.

dredging   The deepening of the bottom area of a water body by the digging, dragging, or hauling out of materials.

drift   1) The general term for any sediment or rock material deposited by a glacier or by water from melting glaciers. 2) The term for sediment transported by near shore currents along a coastal area.

drip irrigation   A watering method used in agriculture and for residences in which a network of small pipes or hoses brings water directly to the soil above the plants' roots, and releases it (through emitter valves) at a slow rate. Drip irrigation is the most efficient form of irrigation because water loss through evaporation is drastically reduced.

dripstone   A deposit formed by precipitation of calcium carbonate as carbon dioxide diffuses into the atmosphere from dripping groundwater, usually within a cave envi-ronment. Dripstones may take the form of stalactites, stalagmites, or curtains.

drive   1) To chase game away from bushes and thickets and into clear, open space. 2) A condition in animals of responsiveness to stimuli and associated activities that usually leads to satisfying some need, as in a hunger drive or sex drive.

droplet   A very small drop of liquid.

drought   Extended period of unusually low precipitation.

dropsonde   A specialized radiosonde that is attached to a parachute rather than a balloon. Its name comes from its being dropped from an airplane above the layer or layers of the atmosphere to be studied. See RADIOSONDE.

dropstone   1) A rock fragment that is released from a melting iceberg and drops through the water to settle in finer-grained bottom sediment. 2) A volcanic bomb that lands in the water and settles in a finer-grained bottom sediment.

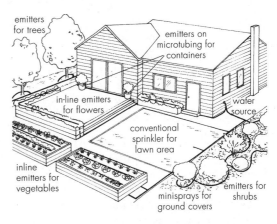

drip irrigation system

By providing water to specified areas, this type of system will save water over more general methods of watering orchards, lawns, and gardens.

drought Extended period of unusually low precipitation.

drought avoidance An evolutionary strategy for survival in periodically dry climates in which an organism enters into a dormant stage or migrates away during the dry season. Desert plants typically bloom and reproduce during brief periods or seasons of rain; when the dry season arrives, many drop leaves, enter dormancy, or die back after seed production to avoid having to sustain normal metabolic levels during the drought. Some African animals migrate away from areas experiencing drought to other regions where water is more readily available. Contrast DROUGHT RESISTANCE, DROUGHT TOLERANT.

drought resistance The evolution of traits that conserve moisture to enable organisms to withstand and remain metabolically active during dry seasons and climates, rather than avoiding them. In plants, adaptations providing drought resistance include narrow leaves (sometimes reduced to only spines), thick waxy cuticles, sunken stomata, coatings of fine hairs, and even modified metabolic processes enabling some plants to keep their stomata (gas exchange pores) closed during the heat of the day (crassulacean acid metabolism). See CRASSULACEAN ACID METABOLISM, XEROPHYTE.

drought tolerant Able to withstand prolonged periods of little precipitation. Xeriscaping, landscape design emphasizing drought-tolerant (xerophytic) plants, is becoming increasingly popular in areas where water use is restricted. See XEROPHYTE, DESERT.

drowned valley A valley that has been inundated by the rising water levels of an ocean or lake. A fiord is an example of a drowned valley.

drug tolerance Capacity for withstanding medicinal substances without adverse affects.

drumlin A streamlined deposit of glacial till shaped like a teardrop in map view. The tapered end of a drumlin points in the direction of glacier movement.

drupe A fleshy fruit whose seed (usually single) is inside a hard pit. Cherries and plums are drupes. See DRUPLET.

druplet A small drupe, especially in an aggregate fruit such as the raspberry. See DRUPE.

dry adiabatic lapse rate (DALR) Rate of adiabatic temperature change in vertically moving, unsaturated air. The rate of cooling or heating for dry air is fairly constant, approximately 1°C for every 100 meters of climb or descent. See ADIABATIC, WET ADIABATIC LAPSE RATE.

dry alkali injection One method for neutralizing sulfuric acid and similarly acidic compounds in flue gases by spraying them with dry sodium bicarbonate. Compare DRY LIMESTONE PROCESS.

dry deposition Fallout of particulate air pollutants from the atmosphere, other than what is carried to the earth's surface by some form of precipitation. See ATMOSPHERIC DEPOSITION, WET DEPOSITION.

dry farming Cultivation techniques for crops that minimize or eliminate the need for irrigation. The Zuñi of the southwestern United States have been successfully dry farming for generations, using windbreaks and embankments to trap meager rains and growing corn in areas where other methods would result in total failure.

drying agents Chemicals (such as silica gel, calcium chloride, or anhydrous sodium sulfate) used for removal of water in chemical or industrial processes. Also called desiccants.

dry limestone process Air pollution control strategy for sulfur oxides (flue gas desulferization). Passing smokestack emissions over limestone neutralizes the sulfur compounds by converting them to less harmful substances. Often called dry scrubbing. See FLUIDIZED BED COMBUSTION, SCRUBBER.

dry weight Measurement of the mass of organic material after water has been removed. Dry weight is more useful than fresh weight in comparing plants because different individual plant specimens or tissues can contain widely varying quantities of water when fresh, especially by the time specimens reach the laboratory.

dry weight rank method (DWR) A technique for estimating the percentage contribution each plant species makes to the total biomass or net primary production of a community.

duct 1) Narrow tube or vessel for carrying liquid, air, or gametes from one part of an organism to another. 2) A similar structure in a building or structure, such as an air conditioning duct.

ductile Refers to a metal's ability to retain its integrity when shaped, especially when drawn out into wire.

ductile deformation The permanent distortion of a material, such as bedrock, by plastic flow without any fracturing or breakage. Ductile deformation of rocks occurs under the conditions of temperature, pressure, and low rates of strain in the deep crustal environment.

duff 1) Decaying organic matter found on the ground in a forest. 2) Fine coal dust, also known as slack.

dune Any landform that is composed of unconsolidated particles transported by the force of wind. Dunes are most commonly formed of sand-sized particles. Barchan, longitudinal, and transverse forms represent the major categories of dunes.

dune stabilization The practice of planting vegetation, such as beach grass, or of constructing drift fences in an effort to control dune migration or erosion.

dunite A dark-colored (mafic) peridotite-like igneous rock in which the proportion of the mineral olivine exceeds 90 percent. See PERIDOTITE.

dust Particulates so fine that they can remain suspended in air.

dust bowl A characterization of an area of the Great Plains in the United States where a combination of drought and inappropriate agricultural practices resulted in severe soil deflation in the 1930s.

dust burden The mass of particulates (dust) suspended within a specified volume of flue gas or other medium.

dust devil Small whirlwind occurring over dry ground that picks up loose dirt from the ground to form a visible pillar.

dust dome Deposition of particulates in a dome-shaped pile. These form in heavily industrialized areas.

dustfall jar A simple instrument for measuring and studying particulate pollutants. It consists of a wide-mouthed, jar-like container left out to collect particulates deposited by the air.

dust storm Strong winds that have lifted large quantities of dust and dirt, caused when large pressure gradients develop across dry ground with little vegetation. Common to parts of the Sahara and, in the 1930s, to the central plains states of the United States.

dust veil A layer of particulates in the earth's atmosphere sufficiently dense to reduce incoming solar radiation and thereby lower temperatures at the earth's surface. Dust veils can be caused by volcanic eruptions, as well as by soot from large-scale fossil fuel combustion.

Dutch elm disease Fungal infection caused by *Ceratocystis ulmi* that can kill elm trees in just a few weeks. It was introduced into North America on elm logs early in this century and was quickly spread by the elm bark beetle. Dutch elm disease has radically changed the planted landscape of many cities and towns, as elms lined many streets at the turn of the century; it has generally spread throughout the natural range of elms as well. The disease is kept in check by burning infected wood and using insecticides to kill the bark beetles but fungicides are very expensive and not always successful.

dynamic equilibrium A balance achieved in a non-static system in which one kind of activity is counteracted by an opposing activity. A chemical equilibrium is a dynamic equilibrium because the forward rate of reaction is equalled by the rate of the reverse reaction, so no measurable change occurs in the system. See EQUILIBRIUM.

dynamic metamorphism The alteration of minerals within a rock along the slip surface of an active fault. The rock is ground to a fine powder and recrystallized by the extreme pressure. Mylonite is a rock formed by dynamic metamorphism.

dynamic pool model An approach to managing harvested animal populations, especially fisheries, that incorporates population structures such as recruitment, growth, natural mortality, and harvesting mortality, all of which may change over time. This approach represents a refinement over surplus yield models. See SURPLUS YIELD MODEL.

dynamic soaring A highly efficient flying technique used by large sea birds, such as albatrosses, to stay aloft for long periods of time. They maintain or even increase their speed by exploiting crosswinds.

dynamic stability Another term for dynamic equilibrium. See DYNAMIC EQUILIBRIUM.

dynamo An electrical generator, usually one that produces direct current. (Direct current is the kind of electricity produced by batteries, as opposed to the alternating current that runs through household wiring.)

dynamometer Device for measuring torque or power. Dynamometers can be designed to measure an engine's power.

dysentery Any of a number of intestinal diseases characterized by inflammation and bleeding of the colon with diarrhea. Now usually restricted to infections caused by the *Shigella* genus of bacteria, the protozoa *Balantidium coli*, or the amoeba *Entamoeba histolytica*, all spread through contaminated food or drinking water.

dysphotic zone A region of an aquatic ecosystem where very little light penetrates, deeper than the euphotic zone but above the dark aphotic zone. Compare EUPHOTIC ZONE, APHOTIC ZONE, PHOTIC ZONE.

dystrophic   Describing a body of water that is brownish from a high content of humus, often acidic, high in organic matter but low in nutrients and thus having a high biological oxygen demand. Bogs and some swamps are characterized by dystrophic water. In dystrophic lakes, which develop from eutrophic lakes and evolve into swamps, organic matter accumulates rather than decomposes because low oxygen levels prevent sufficient populations of decomposing organisms from developing. See EUTROPHICATION.

e

earth The third planet in the solar system, presumably where you are living now. The earth has a mean diameter of 12,742 km and it orbits the sun at a mean distance of 149.6 x 10^6 km. Approximately 70.8 percent of earth's surface is covered by ocean water, and the remainder is occupied by continental landmasses. The entire planet is enveloped in an atmosphere that is less than 200 km thick. The age of the earth is approximately 4.6 billion years. As of this writing, earth is the only place in the universe where life is known to exist. Please handle with care.

earth flow A downslope movement of wet, unconsolidated material not confined to a channel. Earth flow may be initiated by increased pore water pressure between particles.

earthquake A motion of the earth's crust usually associated with brittle failure of rocks that have been subjected to the accumulated strain of tectonic or volcanic forces. Earthquakes are classified according to depth as shallow, intermediate, or deep-focus earthquakes.

earthquake magnitude A measure of the significance of an earthquake. The cultural destructiveness of an earthquake is measured on the Mercalli scale. The Richter scale is used to measure the amplitude of seismic waves. See MERCALLI SCALE, RICHTER MAGNITUDE SCALE.

Earth Resources Technology Satellite (E.R.T.S.) A series of satellites first launched in 1972 that have photographed the entire surface of the earth from a distance of 435 km. In 1975, the series was expanded and renamed Landsat.

Earthwatch Programme An organization established by the United Nations Programme under the terms of its Declaration on the Human Environment. It monitors worldwide environmental trends through the Global Environmental Monitoring System, funds the environmental news and information service Earthscan, and operates the Global Resource Information Database. See GLOBAL ENVIRONMENTAL MONITORING SYSTEM.

earthworms Oligochaete worms that feed by absorbing nutrients from soils which are passed through their intestines by peristaltic contraction of gut muscles. Their feeding and burrowing activities are extremely important to aeration of soils and maintenance of soil structure. Some earthworms of the Southern Hemisphere may attain lengths of several meters. See OLIGOCHAETA.

easement A right created by an express or limited agreement of one owner of land to make lawful and beneficial use of the land of another. Easements allow sewer pipes or gas pipelines to cross private land. See CONSERVATION EASEMENT, UNITED NATIONS DEVELOPMENT PROGRAMME.

East African floral region One of the phytogeographic regions into which the Paleotropical realm is divided according to similarities between plants; it consists of the eastern portions of Africa above Lake Victoria, west to southern Angola and north to southern Mozambique. See PALEOTROPICAL REALM, PHYTOGEOGRAPHY.

easterly A minor element of global wind

patterns. In general, winds moving from high-pressure regions at the poles towards low-pressure regions found roughly above the Arctic and Antarctic circles move from east to west. These prevailing winds are therefore called polar easterlies. See WESTERLY.

ebb tide The time period in which a tidal water level is dropping from high tide to low tide.

ecad Organisms having the same genotype but different forms (height, number of stems, etc.) caused by habitat or environmental differences. Unlike ecotypes, these different forms disappear when different ecads are transplanted to a single site. Also called ecophenes, habitat forms, or environmentally induced variations. Contrast ECOTYPE.

ecdysis Molting; shedding of the outer layers of an animal's skin, common in reptiles and arthropods.

ecdysone A steroid hormone that stimulates molting (ecdysis). It is produced by the prothoracic glands of insects and crustaceans.

ecesis A plant's or animal's establishment in, and adjustment to, a different environment.

Echinodermata An exclusively marine phylum of dueterostome animals with a secondarily derived, five-spoked radial symmetry in adults and without a head or brain. The body may be flattened, globular, elongate, etc., and in some groups may have five or more arms extending from its central mass. A unique, water vascular system draws water through a sieve plate or madreporite into a ring canal encircling the esophagous, from which five radial canals extend setting the basic symmetry. Outgrowths from the canals form tentacles and tube feet, which are located in ambulacral grooves and can be manipulated hydraulically for feeding and locomotion. A spiny skeleton of calcareous ossicles is formed under the epidermis and gives the phylum its name. Included among the phylum's classes are the Asteroidea (starfish or sea stars), Ophiuroidea (brittle stars), Echinoidea (sea urchins and sand dollars) and Holothuroidea (sea cucumbers). Over 6000 species exist. See APPENDIX p. 615.

Echinoidea A class of echinoderms with skeletons roughly formed into the shape of hollow balls. Stiff spines carried on the skeletal plates vary greatly in number, length, and size, according to species. The ventral mouth is provided with a unique, toothed structure that is used for browsing on a variety of foods and surfaces. Although sea urchins have only recently achieved widespread commercial value as gourmet foods, there are already indications that they are being overfished. See ECHINODERMATA.

echo The repetition of sound (or other waves) caused by reflection. Commonly used to refer to a sound reflected back to an observer, but astronomers see and talk about light echoes. See RADAR.

echolocation The technique of sending out high-frequency sounds to discern the distance and direction of objects by how the sounds are reflected by them. Bats and dolphins use echolocation to navigate and locate food. Radar and sonar are forms of echolocation. See RADAR, SONAR.

echo sounder Device used for measuring ocean depths for mapping, or to locate objects on, the ocean floor. The instrument transmits sound waves to the ocean floor. By analyzing the time it takes them to return, the depth (and therefore the rough outlines of large objects) can be determined. Echo sounders can even be used to locate large schools of fish.

eclipse Passage of one astronomical body behind or through the shadow of another. All or part of the planet or star is temporarily hidden from view. In addition to familiar solar and lunar eclipses, stars can eclipse each other.

eclipse plumage Dull, colorless feathers that replace the bright plumage of some birds by the end of the breeding season. Male goldfinches lose their canary yellow color and revert to a dull olive after the end of their breeding season.

ecliptic The sun's apparent yearly path across the celestial sphere. From earth, it appears that the positions of stars remain fixed while the sun moves along this circular orbit. See CELESTIAL SPHERE.

eclogite A coarse-grained mafic igneous rock that has a chemical composition similar to that of basalt. Eclogite is a deep crust rock characterized by the rare minerals pyroxene, omphacite, and almandine-pyrope garnet.

ecocide Deliberate extermination of all or some part of a local or regional ecosystem.

ecocline 1) Change in community composition or ecosystem types along a major environmental gradient (as from polar toward equatorial regions, humid toward arid regions, or elevation gradients in major mountain ranges. 2) A gradient of changes that occurs in different populations of a single species as a result of the different selection pressures of different habitat zones. Ecoclines are a continuous gradation of forms, as distinguished from discontinuous gradations of ecotypes. See GEOCLINE, ECOTYPE.

ecodeme See ECOTYPE.

ecogeographic rules Generalizations describing geographic trends in the distribution of traits in related organisms. See ALLEN'S RULE, BERGMAN'S RULE, HOPKIN'S BIOCLIMATIC LAW, RENSCH'S LAWS.

ecological amplitude See RANGE OF TOLERANCE.

ecological balance Balance of nature; a state in which relative populations of different species remain more or less constant, mediated by the interactions of different species.

ecological efficiency The ratio of the energy ingested by a trophic level to the energy ingested by the previous trophic level in the food web; the percentage of useful energy transferred from one trophic level to the next. See PHOTOSYNTHETIC EFFICIENCY. Compare LINDEMAN EFFICIENCY, TROPHIC LEVEL EFFICIENCY, CONSUMPTION EFFICIENCY, UTILIZATION EFFICIENCY.

ecological energetics Study of the flow, use, and transfer of energy among the components of biotic communities.

ecological equivalents Different species that have similar roles (niches) in similar ecosystems located in different parts of the biosphere. Anteaters in South America, aardvarks in Africa, pangolins in Africa and Asia, and spring anteaters in Australia are all ant-eating animals and therefore ecological equivalents despite their many differences.

ecological gradient See ENVIRONMENTAL GRADIENT.

ecological growth efficiency The ratio of the energy in production to the energy ingested or (absorbed by plant surfaces) by an organism or trophic level. Also called gross growth efficiency. Compare ASSIMI-LATION EFFICIENCY, TISSUE GROWTH EFFICIENCY, RESPIRATION EFFICIENCY.

ecological impact   See ENVIRONMENTAL IMPACT.

ecological models   See APPENDIX p. 603.

ecological release   See COMPETITIVE RELEASE.

ecological time   A time interval measured in roughly 10 generations of an organism's life span, usually several years to several centuries. Compare EVOLUTIONARY TIME, GEOLOGIC TIME.

ecology   The branch of biology that studies the relationships among living organisms and between organisms and their environments. It is derived from the Greek *OIKOS*, meaning house, therefore it is literally the study of the house.

economic efficiency   Benefits of an economic transaction exceed costs.

economic externality   Harmful or beneficial social or environmental effect of producing and using an economic good that is not included in the market price of the good.

economic geology   The branch of geologic science that studies commercially valuable mineral and fuel resources.

economic injury level   The pest population that is cost effective to achieve with a specific pest control strategy. To reduce populations below the economic injury level costs more than the value of the damage done by the pest. (Higher populations justify control on economic grounds because the value of the damage avoided is greater than the cost of controlling the damage.)

ecoparasite   Parasite whose ecological niche is a specific host.

ecosphere   Another term for biosphere. See BIOSPHERE.

ecosystem   A functioning unit of nature that combines biotic communities and the abiotic environments with which they interact. Ecosystems vary greatly in size and characteristics. Also called biogeocoenosis.

ecotone   A transitional area between two (or more) distinct habitats or ecosystems, which may have characteristics of both or its own distinct characteristics. The edge of a woodland, next to a field or lawn, is an ecotone, as are some savanna areas between forests and grasslands. See EDGE EFFECT.

eco-tourism   A term coined relatively recently for tourists who are interested in visiting areas of natural beauty or abundant or exotic wildlife.

ecotype   A locally adapted genetic variant within a species. Different selection pressures of different environments result in the development of different ecotypes within a single species. Unlike ecads, ecotypes retain their physiological and morphological differences when transplanted to a single location. Also called ecological races, physiological races, or ecodemes. Compare ECAD, ECOCLINE. See SUBSPECIES, RACE.

ectocrine   Environmental biochemicals, especially those released by decomposing organisms, affecting the growth of other organisms in the same ecosystem. Penicillin is an inhibiting ectocrine, and some vitamins are stimulating ectocrines; both are often present in the soil. See ALLELOCHEMICAL.

ectoderm   The external cell layer of cells in developing animal embryos. From this layer develops the endodermis in mature animals, including skin, hair, tooth enamel, nails, and parts of the nervous system. See ENDODERM, MESODERM.

ectomycorrhiza See ECTOTROPHIC MYCORRHIZA.

ectoparasite Parasites that dwell on the exterior of their host and feed by piercing the skin. Fleas and lice are examples. Compare ENDOPARASITE.

Ectoprocta See BRYOZOA.

ectotherm Organisms that are thermo-conformers, having low metabolic rates and body temperatures that reflect those of their ambient environments. Also called poikilotherm, and sometimes erroneously referred to as cold-blooded organisms. Contrast ENDOTHERM.

ectotrophic mycorrhiza A symbiotic condition between a fungus and the root of a plant in which the fungus forms a sheath around the root. Some hyphae connecting to this sheath penetrate the host root and spread through the soil surrounding the roots. Also called ectomycorrhiza, these form between many tree species and basidiomycete fungi. Compare ENDO-TROPHIC MYCORRHIZA, VESICULAR-ARBUSCULAR MYCORRHIZA.

edaphic A term relating to the ecologic or biologic influence of the soil, such as the effects of soil moisture and salinity, on a natural community.

eddy A minor reverse flow caused by an obstacle in the primary direction of flow of a fluid such as water or smoke. Also, the small whirlpool or whirlwind often caused by such an interruption.

eddy diffusion A mixing (as of heat) occurring in turbulent regions within the atmosphere. See EDDY.

Edentata Order of mammals including the anteaters, sloths, and armadillos. Anteaters have no teeth and the remaining jaws have relatively uniform (homodont), peg-shaped teeth. With the exception of the nine-banded armadillos, edentate distribution is Neotropical. See EUTHERIA, ARMADILLOS.

edge effect The tendency for an ecotone (transition zone between communities) to contain a greater variety of species and more dense populations of species than either community surrounding the eco-tone, and for ecotones to have different characteristic than either of the systems they are transitional between. See ECO-TONE.

EEC See EUROPEAN ECONOMIC COMMU-NITY.

EEZ See EXCLUSIVE ECONOMIC ZONE.

effective height of emission The distance above the ground at which the plume of emissions from a smokestack levels out to horizontal.

effective radius An alternate measurement of the strength of an atomic bomb that uses area of destruction caused instead of comparing strength to megatons or kilotons of TNT.

efficiency The extent to which a system, or part of a system, is efficient. In general, the ratio of a system's output (or production) to input, as in the useful energy produced by a system compared to the energy put into the system. See ECOLOGICAL EFFI-CIENCY, ENERGY EFFICIENCY.

effluent Waste discharged into the environment by industrial or other human processes. Usually used to refer to a point-source discharge into water, such as sewage and other liquid waste, which may contain suspended solid waste. Compare EMISSION.

effluent charge   A legally imposed fee (tax) for approved amounts of pollution discharge.

effluent standard   Legally allowable amounts of effluent that can be discharged into a body of water. See EFFLUENT.

effluent stream   1) The term applied to waste products entering the environment as the result of an industrial process. 2) In hydrology, a gaining stream fed by seeps or springs from groundwater flow. Effluent streams tend to be perennial.

egestion   The act of discharging undigested or indigestible food or matter. Egestion includes feces and other organic matter that passes through an organism without having been assimilated. Compare EXCRETION.

egg pulling   Gathering eggs of endangered species produced in the wild and transporting them to zoos or research centers for hatching.

E horizon   A light-colored mineral soil layer characterized by maximum loss of silicate, iron, and aluminum sesquioxides in true solution or colloidal suspension. See SOIL PROFILE, also APPENDIX p. 619.

einkorn   A primitive form of modern wheat that was domesticated by humans over 11,000 years ago. It differs from modern durum wheat in that it is diploid rather than tetraploid. See EMMER.

einsteinium   (Es) An artificial, radioactive transuranic element with atomic number 99; the atomic weight of its most stable isotope is 254.0. It is one of the products of hydrogen bomb explosions. The 11 known isotopes of einsteinium are unstable. See ACTINIDE.

EIS   See ENVIRONMENTAL IMPACT STATEMENT.

ejecta   Ash and lava thrown from an erupting volcano or impact crater.

Ekman circulation   A theoretical model to explain the right-spiral circulation pattern of wind-generated ocean currents in the northern hemisphere.

Ekman depth   The lower limit of Ekman circulation at which the current is moving in a direction opposite to that of the wind. The Ekman depth occurs at approximately 100 m, with variation according to latitude.

Ekman flow   See EKMAN CIRCULATION.

Ekman layer   The transitional boundary zone of the atmosphere where shearing stress is constant as wind currents blow over the ocean surface in generating Ekman circulation.

Ekman spiral   The description of water motion in Ekman circulation. In the Ekman spiral, the surface currents flow at approximately 45° to the right of the wind, with increasing right deviation and lower velocities at increasing depths. The net transport of water is approximately 90° to the right of the wind current.

elaioplast   A plastid (plant organelle) whose function relates to the formation of fatty compounds (oils) and, in some cases, starch. Elaioplasts are most common in liverworts and monocotyledons.

elaiosome   A structure on the surface of some seeds containing an oily substance that aids in seed dispersal by attracting organisms such as ants. Bloodroot *(Sanguinaria)* and trout lilies *(Erythronium)* have prominant elaiosomes that enhance their dispersal by ants.

Elasmobranchii   An order of chandrichthyes, commonly known as elasmobranchs, and characterized by multiple gill openings on each side of the head which are not covered by a flap of tissue (operculum). The sharks belong to this group and vary in size from 25 cm in some species to the whale shark, which may achieve a length of 10 m and weigh 10 tons. Skates and rays, which evolved from shark-like ancestors, are generally adapted for bottom dwelling. The body is flattened and locomotion is provided by undulations of the pectoral fins, which are greatly expanded laterally. In many forms, such as the sting ray, the tail is slender and whip-like. See CHANDRICHTHYES.

elastic deformation   A temporary change in shape caused by application of force. The original shape returns once the stress is removed. Contrast PLASTIC DEFORMATION.

elastic limit   The point at which further stress applied to an otherwise elastic material causes measurable, permanent changes in shape. See ELASTIC DEFORMATION.

elastic rebound   The rapid recovery of rocks to a position of lower strain when stress is relieved during the sudden motion of faults.

elastic strain energy   Potential energy contained in an object or material undergoing elastic deformation; this energy is released as the material returns to its original shape. When elastic deformation of the earth's crust occurs, the energy released can cause tremors. See ELASTIC DEFORMATION.

E-layer   The middle of three subdivisions of the earth's ionosphere. It is important because it reflects radio waves of long wavelengths, permitting transmission of certain radio signals over long distances. Sometimes called the Heaviside layer. See IONOSPHERE, D-LAYER, F-LAYER.

electrical conduction   The transfer of net charge through a substance or region. Compare THERMAL CONDUCTION.

electrical double layer   Another term for double layer. See DOUBLE LAYER.

electricity   1) Electric energy or power. Electricity drives chemical changes and creates magnetism as well as powering household lights and appliances. Nonflowing (stationary) electrical charges are called static electricity to distinguish them from "dynamic" electricity. 2) The science of electrical phenomena.

electrochemical   Concerning the interaction between electricity and chemical reactions, especially in liquids. Communication in the human nervous system depends on electrochemical processes.

electrode   A conductor used to bring electric charge to or remove electric charge from a system such as a liquid solution or a circuit. Anodes are positively charged electrodes (as in the positive terminal of a battery) and cathodes are negatively charged electrodes.

electrodialysis   A process for separating ionized compounds from solutions, using electricity. Large-scale desalinization plants (as in Israel) introduce electrodes into salt water to drive ionized molecules across semi-permeable membranes. This separates out the dissolved salt to produce fresh water.

electrojet   Electrical current flowing through the ionosphere, the ionized outer layer of the earth's atmosphere. Usually refers to the current flowing above the earth's equator (equatorial electrojet), but auroral electrojets also occur.

electrolyte   A chemical that is ionized when liquid (either in solution or in molten form). Electrolytes conduct electricity. See IONIZATION.

electromagnet   A magnet constructed of a wire coil carrying electrical current rather than material that is inherently magnetic. Electromagnets are usually made of wire wrapped around an iron core. They are used in electric motors, generators, and audio speakers. See ROTOR.

electromagnetic radiation   Electromagnetic waves of any frequency or wavelength, including radio waves, microwaves, light, infrared radiation, ultraviolet radiation, x-rays, and gamma rays. See ULTRAVIOLET RADIATION, INFRARED RADIATION, RADIOACTIVE EMISSIONS, and VISIBLE RADIATIONS.

electromagnetic spectrum   The complete range of electromagnetic radiation, from wavelengths of less than 1 angstrom (gamma radiation) to many meters (radio waves). See ELECTROMAGNETIC RADIATION.

electromagnetism   Science and engineering focused on properties of, and interactions between, electricity and magnetism.

electromotive force   (EMF) Voltage capable of producing current that can be driven through circuits.

electron   A subatomic particle; one of the parts of the atom. Electrons carry a negative electrical charge and have very little mass. In the atom, electrons orbit the nucleus. They also exist independently of atoms. Compare POSITRON, PROTON, NEUTRON.

electron acceptor   A substance that receives an electron transferred from another substance (an electron donor) and becomes an anion, to form an ionic bond. Nonmetals and oxidizing agents are electron acceptors. See ELECTRON DONOR, IONIC BOND.

electron affinity   The degree to which a gaseous atom or ion attracts electrons, calculated from the energy change when an atom or ion in a gaseous state captures a free electron.

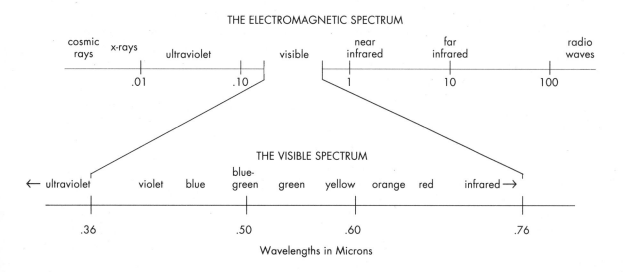

THE ELECTROMAGNETIC SPECTRUM

cosmic rays | x-rays | ultraviolet | visible | near infrared | far infrared | radio waves

.01   .10   1   10   100

THE VISIBLE SPECTRUM

← ultraviolet   violet   blue   blue-green   green   yellow   orange   red   infrared →

.36   .50   .60   .76

Wavelengths in Microns

electron capture detector   A device using spectrophotometric measurement to detect the presence of chemical compounds in the atmosphere. The detector records the quantities of gaseous pollutants (especially halogenated compounds such as chlorinated hydrocarbons) present.

electron configuration   The description of the distribution of the electrons in the different orbitals of an atom, as governed by the different sublevels of energy in the atom. Also called electronic configuration.

electron donor   A substance that gives up an electron to another substance (an electron acceptor) and becomes a cation, to form an ionic bond. Metal elements are electron donors. See ELECTRON ACCEPTOR, IONIC BOND.

electronegativity   The tendency of an atom in a molecule to attract electrons to itself, causing polarization to occur within the molecule. Electronegativity increases across a row and decreases down a column in the periodic table. The most electronegative element is fluorine (F), and the least electronegative element is francium (Fr).

electronic structure   See ELECTRON CONFIGURATION.

electron microscope   (EM) A magnifying instrument whose resolving power is many times greater than that of an ordinary light microscope (achieving resolutions of 0.2 to 0.5 nanometers). Electron microscopes pass beams of electrons through an object, rather than using a beam of light, to create a much larger image of minute objects.

electron transfer   1) The process that occurs during the formation of an ionic bond connecting two atoms. It involves one atom (the electron donor) giving up an electron to another atom (the electron acceptor). 2) Any process by which an electron is transferred from one molecule to another, with or without the formation of a bond. Photosynthesis involves electron transfer between molecules to drive the reaction that builds plant tissues. See IONIC BOND.

electron volt   (eV) A unit of energy equal to the potential energy of an electron at a potential of -1 volt. One eV equals 1.6×10^{-19} joules. See JOULE.

electroosmosis   Process driving a liquid through a permeable membrane by setting up opposite electrical charges on either side of the membrane. Electrodes are used to generate fields, causing the solution to move across the membrane toward the opposite charge.

electrophoresis   A phenomenon in which charged particles move through a solution having an electric field gradient, traveling toward the electrode having the opposite charge (positively charged molecules move toward the cathode and negatively charged particles move toward the anode). For electrophoresis to occur, the particles must be large enough to be associated with a double layer; these particles may be large molecules or colloids. Electrophoresis is used for separating complex molecules such as proteins, especially for the analysis of complicated mixtures. See DOUBLE LAYER.

electropositive   1) Describing substances with a tendency to lose electrons to form positive ions, such as metals. 2) Positive; carrying a positive electrical charge. See ELECTRONEGATIVITY.

electrostatic   Related to static electricity; involving electrical charges that are not moving. See STATIC ELECTRICITY.

electrostatic field   The electrical field produced by stationary electrical charges (as opposed to electrical current, which flows).

electrostatic filter A pollution control system in which a static electrical charge is applied to a filter to improve its efficiency in trapping particulates. Compare ELEC-TROSTATIC PRECIPITATOR.

electrostatic precipitator Device for removing particulates from air emissions. Particulates in the smokestack gases become charged by passing through a region with a high electric field; they are forced by an electric field to the collecting plates of opposite charge. The particulates can then be mechanically removed from the plates.

element Any substance composed of chemically identical particles (atoms) that cannot be subdivided into smaller, simpler particles by normal chemical means. Elements are ordered in the periodic table by their number of nuclear protons (the number of neutrons can vary in different isotopes of the same element). See ISO-TOPE, also APPENDIX p. 608–609.

elephant See PROBOSCIDIA.

elevation A statement of the vertical distance measured between an observation point on the ground surface and a fixed datum such as a bench mark or sea level. Contrast ALTITUDE.

elfinwood Very short, scrubby trees found growing in extreme conditions, common on mountains at the edges of the tree line. Also called elfin forest. See KRUMMHOLTZ.

ellipsoid of rotation A three-dimensional shape (solid) created by spinning an ellipse around its major (longer) axis.

El Niño A current of warm water that periodically flows along the western coast of South America. The name derives from Spanish and means "the child," an association with the Christmas season, which is the usual time that the current forms. See EL NIÑO SOUTHERN OSCILLATION.

El Niño Southern Oscillation (ENSO) A shift in the climate and ocean currents of the Pacific region that occurs about once every 7 years. The equatorial countercurrent of warm water from the the west overrides the cold-water Peru current from the south. The consequent suppression of the nutrient rich upwelling of cold water may result in die off of plankton and fish populations.

Eltonian pyramid See PYRAMID OF NUMBERS.

eluent In liquid chromatography, the eluent is the mobile phase that carries the compounds to be separated through the separation column. See CHROMATOGRAPHY.

elution 1) The separation of substances using a solvent. 2) In chromatography, elution is the use of a liquid to remove the desired ions from the solid resins for retrieval; it also cleans the resins so that they can be used again for another chromatography procedure. See CHROMATOGRAPHY.

elutriation 1) The process of fine particles washing away within a soil, especially as a result of raindrop splash. 2) The process of sorting soil materials by using a current of air or water to separate particles into size classes.

eluviation The removal of material, in true solution or colloidal suspension, from upper layers of soil. Contrast ILLUVIATION. See APPENDIX p. 619.

eluvium A depositional layer of material transported by eluviation.

elytron-to-femur ratio   (e/f) A measurement used to determine the stage of life cycle of an individual locust. It is the ratio of the locust's outer wing length to its hind leg length.

EM   Abbreviation for electron microscope. See ELECTRON MICROSCOPE.

embankment   1) The steep boundary area along the edge of a stream. 2) A ridge or slope separating areas of differing elevations. 3) Artificial emplacement of earth or rock, as in a dike or a levee.

embrittlement   A reduction of strength or elasticity. Embrittlement of metals can be caused by prolonged exposure either to very low temperatures or to high levels of radioactivity. Embrittlement of reactor vessels of nuclear power plants limits their life spans. It was largely responsible for the 1992 shutdown (decommissioning) of Yankee Atomic in Rowe, MA, the oldest commercial nuclear power plant.

emergency core-cooling system   A back-up cooling system for the core of nuclear reactors and power plants. See NUCLEAR REACTOR.

embryo   A plant or animal at the earliest stages of development following fertilization. In plants, the embryo is the structure inside a seed before it sprouts. In lower animals, it refers to the earliest stages of development within the reproductive organs of the mother until hatching or birth. In mammals, the embryo is called a fetus in its later stages. See FETUS.

embryonic   Relating to the embryo or the earliest stages of development.

embryo sac   The female gametophyte in seed plants, a large oval cell in the nucellus in which fertilization occurs. The embryo sac arises from a megaspore (which in turn arose from a megaspore mother cell).

emerald   A green gemstone variety of the mineral beryl.

emergency core-cooling system   A back-up cooling system for the core of nuclear reactors and power plants. See NUCLEAR REACTOR.

emergent plant   A plant whose roots grow in shallow water but whose photosynthesizing structures (stems and leaves) grow mostly above the water surface. Cattails, pickerelweeds, and arrowheads are all emergent plants.

emery   A dark gray variety of the mineral corundum. Crushed emery is used as an industrial abrasive.

emigration   The movement of populations of a species out of a particular area or habitat. Includes the departure of humans from one country in order to relocate in a new country.

eminent domain   The right of the state or sovereign to take private property for public use, with compensation for fair market value.

emission   1) Waste discharged into the environment by industrial or other human processes. Usually used to refer to gaseous discharges, but can also refer to liquid discharge or radioactivity. 2) The giving off of electrons, electromagnetic energy, or radioactivity. Compare EFFLUENT.

emission factor   Average mass of a specific pollutant produced per unit mass of raw material processed. A typical particulate-matter emission factor for fireplaces is about three grams of particulate matter emittted per kilogram of fuel consumed.

emission inventory   Information collected in order to establish emission standards for a district. It catalogues air pollutants by type and source, listing the amounts of each produced daily.

emission line   One of the bright lines appearing in the spectrum given off by a glowing object. Each line represents a given wavelength of radiation that is more intense, and therefore appears brighter, than the background radiation. Emission lines in spectra are used to identify the elements and molecules present in the glowing object. See SPECTRUM, SPECTROGRAPHY, ABSORPTION LINE, ABSORPTION SPECTRUM.

emission standard   Legal acceptable limit of output of specific chemicals. National emssion standards were established by the three Clean Air acts of 1970, 1977, and 1990.

emmer   A wheat species intermediate in evolution between einkorn and modern wheat as we know it *(Triticum durum)*. Emmer differs from modern wheat in that its spike is not a tight cluster but is broken up into segments; also, the chaff doesn't easily separate from the grains, making threshing difficult. Like einkorn, it was domesticated over 11,000 years ago. It is believed to have been produced by a cross between wild einkorn with goat grass (Aegilops) that produced a tetraploid hybrid. See EINKORN.

emphysema   A chronic, irreversible lung disease characterized by shortness of breath and destruction of the alveoli. Emphysema has no one specific cause, but is aggravated by cigarette smoke and air pollution.

empirical   Knowledge based on observation or experiment, not theory.

emulsifier   A substance that causes a mixture of immiscible (not readily mixed) liquids to emulsify, or that stabilizes such a mixture to keep it from separating out. Emulsifiers are used extensively in the food industry, in salad dressing they keep the oil and vinegar from separating and they are added to meats to retain water and thereby increase the weight of the meat.

emulsify   To cause one liquid substance to form a colloidal suspension in another (an emulsion). Detergents cause grease to emulsify in water so that it can be removed from dishes, clothing, etc.

emulsion   A colloidal mixture of two or more immiscible (not readily mixed) liquids; a mixture in which one or more liquids is suspended in another liquid without dissolving. An emulsion such as oil and water usually separates into its separate components unless an emulsifier is added to stabilize the mixture.

endangered species   Any species whose populations have been reduced to the point that it is at risk of becoming extinct over much or all of its range in the near future. Compare THREATENED SPECIES.

endemic   1) Indigenous to, and restricted to, a particular area; also, an endemic plant or animal. 2) Describing a disease regularly found in low levels in a particular area but not epidemic or sporadic, confined to a few regular incidences. Compare EPIDEMIC, EXOTIC.

endergonic   Describes a chemical reaction that requires energy to proceed and therefore does not occur spontaneously. Contrast EXERGONIC.

end moraine   A ridge-shaped accumulation of till forming where the terminus of a glacier is at equilibrium. The rate of ice advance in this condition equals the rate of melting, causing rock debris to pile up at the terminus.

endobiont An organism that lives within cells of a host organism in a relationship that does no apparent harm. Compare ENDOPARASITE, ENDOSYMBIONT.

endocarp The inner layer of the pericarp (fruit wall) surrounding a seed. In drupes such as peaches and plums, the endocarp is a hard stone; in citrus fruits, it is made up of membranes; in many berries it cannot be distinguished from the mesocarp. See EPICARP, MESOCARP, PERICARP.

endocrine Secreting internally; used to describe animal glands (especially those of vertebrates) that secrete hormones directly into the bloodstream (rather than through a duct or directly into an organ). Endocrine hormones are carried in blood to other parts of the body, where they regulate functions such as growth and sexual development.

endoderm The inner layer of the gastrula (an early stage in an animal embryo), from which the alimentary tract and digestive organs develop. See ECTODERM, GASTRULA, MESODERM.

endogamy The union of two similar gametes from closely related parents (animals or plants) or from another flower on the same plant; inbreeding. Compare EXOGAMY.

endogenous Initiated from within an organism, rather than in response to external stimuli; can refer to a substance or organ as well as an internal stimulus. Compare EXOGENOUS.

endogenous plankton Species of plankton found in the area in which they originate. Contrast EXOGENOUS PLANKTON.

endogenous rhythm Internal rhythm; periodic movements or physiological processes not directed by external environ-mental stimuli. Endogenous rhythms, such as the opening and closing of some kinds of flowers, continue even in artificial environments where all fluctuations of temperature and light and dark have been removed. See BIOLOGICAL CLOCK, CIRCADIAN RHYTHM, EXOGENOUS.

endometrium The mucous membrane that forms the inner lining of the uterus.

endoparasite A parasite living in the tissues of its host organism, such as a tapeworm. Compare ECTOPARASITE.

endophagous Feeding on the inside of an animal or plant. Borers are several different insects that tunnel into plant stems and tissues, eating them away from the inside and causing them to collapse.

endopodite In crustaceans, the inner section of a forked limb. See EXOPODITE.

endopsammon Microscopic animals that live in sand and mud.

endoskeleton Internal bony framework, such as that supporting vertebrates or sponges. Contrast EXOSKELETON.

endosperm The triploid food storage tissue in the embryo sac or mature seed of flowering plants supplying nourishment for the embryo. Endosperm is particularly evident in seeds of monocot cereal grains such as corn and wheat.

endostyle A groove in the pharynx of Protochordata (tunicates or sea squirts). It is lined with cilia and helps to trap food particles.

endosymbiont An organism that lives inside of another organism in mutually beneficial coexistence. The bacteria living in human intestines are endosymbionts; they aid in digestion and synthesize essential vitamins. Endosymbionts can also be plants inside of other plants, such as algae growing inside fungi in lichen complexes. See CORALS, ENDOTROPHIC MYCORRHIZA, MYCOPHYCOPHYTA.

endosymbiosis theory   The theory that some organelles such as chloroplasts and mitochondria found in the cells of eukaryotic organisms have evolved from endosymbiotic relationships with unicellular organisms. See ENDOSYMBIONT.

endotherm   A thermoregulating organism (usually an animal) that can control its body heat internally and maintain a constant temperature over a considerable range of environmental temperatures. These organisms generally have high metabolic rates and require large amounts of energy input. Also called homeotherms, and sometimes erroneously referred to as warm-blooded organisms. Contrast ECTOTHERM.

endothermic   Describing a process, especially a chemical change or reaction, in which heat is absorbed from the surroundings. See EXOTHERMIC, THERMONEUTRAL.

endotrophic mycorrhiza   A symbiotic condition between a fungus and the root of the plant in which the fungal hyphae (rootlike structures) grow between and within the cells of the plant roots, benefiting both the fungi and the plants. Many orchids and members of the heath family (Ericacea) cannot survive without endotrophic mycorrhiza. Compare ECTOTROPHIC MYCORRHIZA.

endozoochore   A seed or spore that is dispersed by being swallowed by an animal and later excreted (undigested) after being transported some distance. Wild cherries (*Prunus*) and many similar fruits with hard pits surrounding the seed are endozoochores. Compare ZOOCHORE.

end point   The final goal of titration; the point at which a visible change takes place in the solution to which titrant is added. The end point is usually marked by a sharp change in color or other physical property of the solution. Also called equivalence point. See TITRATION.

endrin   A chlorinated hydrocarbon insecticide that, like others in this group, is extremely poisonous and long-lasting in the environment. Its use is banned or restricted in many countries. See CHLORINATED HYDROCARBONS.

energy   Often defined as the capacity to perform work, but "heat" (internal energy) is all energy even though not all of it can do work. Forms of energy include kinetic, potential, gravitational, electromagnetic, mechanical, thermal, and nuclear (some of these classifications overlap). See KINETIC ENERGY, POTENTIAL ENERGY, HEAT ENERGY, NUCLEAR ENERGY, JOULE.

energy budget   Analysis of the quantity of energy used at different levels in a biotic community or in an industrial process, or of the components of energy flow through an individual organism.

energy cascade   A sequence of biochemical reactions (or sometimes physical reactions) used in respiration and similar metabolic reactions at the cellular level. The series results in a gradual decrease of energy (with corresponding increase in chaos or entropy), following the second law of thermodynamics.

energy conservation   Reduction in consumption of energy accomplished through cutting back on energy use or increasing the efficiency of energy use. Such efficiency can be increasing insulation to minimize heat loss and streamlining car design so less fuel is wasted through air friction.

energy crisis   A significant shortage of energy, causing hardship. When electrical demand outpaced supplies in the mid-1960s, a number of brownouts occurred, which greatly reduced electrical power. In the 1970s, a shortage of oil and gasoline forced people to wait in long lines at gas stations and inspired people to find ways to reduce automobile travel.

energy efficiency   In any biological or industrial process, the percentage of the total energy put into a system that can be converted into useful work and not lost as unproductive heat. Energy efficiency is calculated from the ratio of energy produced to energy consumed. See ECOLOGICAL EFFICIENCY, EFFICIENCY.

energy flow   Transfer of energy from one part of a system to another, especially from one trophic level to another in a food chain or food web. See TRANSFER EFFICIENCY.

energy level   The amount of energy associated with a particular electron surrounding an atom. Energy levels can be thought of as a series of steps; each step corresponds to an orbital in which the electrons possessing that level of energy are predicted to be located. See ORBITAL.

energy pyramid   Diagram representing the loss of available energy at each trophic level in a biotic community. Generally, 90 percent or more of the usable energy in each transfer is lost as waste heat; the rapid narrowing of available energy with increasing trophic level produces a pyramid when graphically portrayed. Also called pyramid of energy flow. See TROPHIC LEVEL.

energy quality   The degree to which a form of energy can be harnessed for useful work. See HIGH-QUALITY ENERGY, LOW-QUALITY ENERGY, ENTROPY.

engineering geology   The application of geologic knowledge to engineering purposes.

enhanced oil recovery   A general term applied to a technology that improves the efficiency of oil extraction. See SECONDARY OIL RECOVERY.

enology   The study of wine and winemaking. Also spelled oenology.

enrich   Increase the quality of; especially, to increase the amount of a desired isotope of an element over what is naturally present.

enrichment   The process of improving the quality of fissile fuel for use in nuclear reactors or weapons. Enrichment of mined uranium by gaseous diffusion increases the concentration of uranium 235 relative to the naturally more abundant isotope, uranium 238. See GASEOUS DIFFUSION.

ensilage   Process of preparing and storing animal fodder under anaerobic conditions. Plant matter, such as grass or corn shoots, is harvested green. It is allowed to ferment in storage (typically in silos, but now often in concrete bunkers, pits, or piles covered with plastic sheeting), producing silage. See SILAGE.

enteric bacteria   Intestinal bacteria, including coliform bacteria such as *Escherichia coli*. Also called enterobacteria.

enthalpy   (H) Heat contained in a given quantity of a substance, measured in joules. It is calculated by measuring the amount of energy used to heat a given mass of liquid at a given temperature to its boiling point under constant pressure. The enthalpy of a chemical reaction is the heat absorbed or released per mole of reactants undergoing the reaction.

entisols   Any member of a soil order

characterized by little or no evidence of soil horizon formation. Entisols are usually recent accumulations such as volcanic ash and river flood plain deposits.

entomology   The branch of zoology devoted to the study of insects.

entomophily   Pollination by insects. See CANTHAROPHILY, MYRMECOPHILY.

Entomostraca   Collective term referring to all the lower (smaller) subclasses of Crustacea. See CRUSTACEA.

entrainment   1) The synchronization of a biological clock to external events such as day length or temperature changes. Some circadian rhythms would run in 25- to 28-hour cycles in artificially constant environments; in nature, these are entrained, reset to day length so that they repeat in 24-hour cycles. 2) Fine droplets placed in a vapor through distillation or evaporation so that the vapor will carry the liquid away. See BIOLOGICAL CLOCK, CIRCADIAN RHYTHM, ZEITGEBER.

entrenched meander   A type of incised meander forming through rapid downcutting, which results in steep to nearly vertical valley sides. Compare INGROWN MEANDER.

entropy   Degree of chaos or disorder in a system. In thermodynamics, entropy is related to the portion of the energy contained in a system that can be converted to usable work.

E number   A classification system for food additives used by the European Economic Community. Food additives are classified as coloring substances, preservatives, antioxidants, emulsifiers and stabilizers, acids and bases, anti-caking additives, and flavor enhancers and sweeteners. For example, food preservatives are given numbers E200-E297.

environment   The whole sum of the surrounding external conditions within which an organism, a community, or an object exists. Environment is not an exclusive term; organisms can be and usually part of another organism's environment.

Environmental Defense Fund   (EDF) A citizen conservation organization whose work spans global issues as wide ranging as ocean pollution, rainforest destruction, and global warming; founded in 1967, EDF's trademark is multidisciplinary teams of scientists, attorneys, and economists to develop economically viable solutions to environemntal problems.

environmental degradation   Depletion or destruction of a potentially renewable resource, such as soil, grassland, forest, or wildlife by using it at a faster rate than that at which it is naturally replenished.

environmental ethics   The application of social ethics to questions of correct behavior toward the environment.

environmental geology   The branch of geological science that deals with the entire spectrum of interactions between humans and the geologic environment (resources, hazards, planning).

environmental gradient   A usually gradual change in environmental conditions between extremes, as the gradation from moist to arid environments or from hot to cold.

environmental impact   Human-induced change in the natural environment.

Environmental Impact Statement   (EIS) An analysis required for all major federal actions by the National Environmental Policy Act of 1968, which evaluates the environmental risks of alternative actions.

environmental law   The body of law pertaining to the environment.

Environmental Law Institute   (ELI) Washington, DC-based research and education organization.

Environmental Protection Agency   (EPA) An independent federal agency established under the National Environmental Policy Act of 1970 (42 U.S.C.A. § 4321 et seq.) responsible for setting and enforcing environmental standards, conducting research on environmental problems, and assisting states and local governments. It manages federal efforts in the United States to control air and water pollution, radiation and pesticide hazards, environmental research, and solid waste disposal, including cleanup of Superfund hazardous waste sites, and regulation of pesticides and toxic substances.

environmental resistance   1) All the factors that inhibit the potential growth (or reduction) of a population such as predators, competition for food or water, weather, disease, and food availability. 2) More specifically, the product of the equations for simple population growth and environmental resistance, expressed mathematically as $(K-N)/K$, where K is the carrying capacity and N the number of individuals present. Contrast BIOTIC POTENTIAL. See LOGISTIC EQUATION.

environmental stress   Any physical environmental factor that has a negative impact on an individual community or ecosystem, such as temperature, salinity, or pollution.

enzootic   Describing animal diseases that are prevalent but not epidemic in a specific climate. Enzootic is the animal equivalent of endemic for humans. Contrast EPIZOOTIC.

enzyme   Biological catalysts; complex proteins produced by plant and animal tissues to initiate or speed up specific reactions between other chemicals without undergoing a permanent structural change. For example, the enzyme pepsin helps digest proteins. Most enzymes can be recognized by the suffix -ase, such as maltase, ribonuclease, and lactase. See CATALYST.

Eocene   The second of five epochs in the Tertiary sub-era of geologic time. The Eocene lasted from approximately 54.9 to 38 million years ago. See APPENDIX p. 610.

eoclimax   The major period of dominance of a group of fossil plants.

eolian deposit   Any unconsolidated accumulation of material which is deposited by wind.

eon   The largest of time units in the geologic time scale. The four eons are called the Priscoan, Archaean, Proterozoic, and Phanerozoic. See APPENDIX p. 610.

EPA   See ENVIRONMENTAL PROTECTION AGENCY.

epeiric sea   A shallow part of an ocean that is occupying the interior regions of a continent. Hudson Bay in North America is an example of an epeiric sea.

epeirogenesis   See EPEIROGENIC.

epeirogenic   The term applied to vertical crustal motions that lead to the transgression or regression of an epeiric sea.

ephemeral   1) Of short duration, as an ephemeral stream that disappears in summer. 2) An organism, such as an insect or herbaceous plant, that completes its entire life cycle, or some aspect of its life cycle, in just a few days or weeks. Many spring wildflowers in the northern United States are ephemerals, blooming early in the season underneath bare trees and entering dormancy soon after the trees leaf out.

ephemeral stream   A stream that forms on a temporary basis following a rainstorm or snowmelt. Ephemeral streams are common in arid lands. An ephemeral stream is above the ground water table.

epibenthic   Concerning plants and animals that live just above the sea floor. Epibenthic organisms (epifauna and epiflora) may live near the bottom of shallow, intertidal zones as well as in deeper waters.

epibiont   An organism that lives on the exterior of another but does not feed parasitically upon its host. Plant epibionts are called ephiphytes; animals (such as limpets living attached to crab shells) are epizoites. See AUFWUCHS, EPIPHYTE.

epibiotic   Living on another organism without causing it harm. Contrast PARASITE.

epicarp   The outer layer of a ripened ovary or fruit, especially when it can be peeled away from the rest of the fruit. An apple skin is an epicarp. Also called exocarp. See ENDOCARP, MESOCARP, PERICARP.

epicenter   The point on the earth's surface where an earthquake is first detected. The epicenter is directly above the focus of an earthquake.

epicontinental sea   See EPEIRIC SEA.

epideictic display   Territorial behavior, intended by an animal to protect its territory.

epidemic   Spreading rapidly and affecting a large portion of a population at one time. Also, an outbreak of disease with such extensive characteristics. Compare ENDEMIC.

epidemic spawning   The phenomenon in which large numbers of individuals release gametes (spawn) at the same time.

epidemiology   The branch of medicine that studies the causes, contributing factors, distribution, and control of infectious diseases.

epidermis   An outer, protective layer forming a skin on animals or plants. In vertebrates, it is the outermost layer of the skin, above the sensitive dermis, and is also called cuticle. In other animals and plants, the epidermis is a layer lying just below the cuticle that generates the cuticle layer.

epifauna   Sessile marine animals, those that live attached to plants or objects found on the bottom of the sea. See INFAUNA, EPIBENTHIC.

epigamic character   A feature on an organism that is attractive to members of the opposite sex.

epigeal   1) Living on or just above the ground (usually insects or plants). 2) Describing plants that germinate so that their cotyledons appear above the ground and function as leaves. Sunflowers are epigeal. Compare HYPOGEAL.

epilimnion   The warm, upper layer of water characteristically found in lakes during the summer in temperate regions. The epilimnion contains more dissolved oxygen than lower layers and is often separated from the colder lower layer (hypolimnion) by a layer with a steep temperature gradient known as the thermocline. See HYPOLIMNION, THERMOCLINE.

epiorganism   A group of individual organisms that functions as an entity, such as a bee colony. Also called a superorganism.

epipedon   The upper portion of a soil profile. The epipedon usually contains the specific diagnostic features used in soil classification.

epipelagic Of or concerning upper levels of ocean waters, the euphotic or well-lighted layer that usually extends to depths of approximately 150 meters. The epipelagic zone is usually on the seaward side of the junction of the continental shelf and continental slope. Epipelagic organisms inhabit the water column rather than the bottom surface. See EUPHOTIC ZONE, PELAGIC. Compare BENTHIC.

epiphyllous Attached to, and often growing upon, a leaf (usually the upper surface).

epiphyte A plant that grows on other plants but is not parasitic, deriving its nutrition from the air and rain instead of from the host that provides structural support. Many lichens, bromeliads, and tropical orchids are epiphytes. Also called air plants or aerophytes. See EPIBIONT.

epiplankton Tiny, floating organisms that dwell in the upper, euphotic zones of water bodies, usually at depths above 200 meters.

epipodite A lateral extension from a basal limb segment on some species of crustaceans.

episode A single event, a specific occurrence of a phenomenon such as an insect population outbreak or high pollution levels.

episome Small pieces of DNA within a bacterium that can reproduce independently of the cell's chromosomes and that are capable of reversibly inserting themselves into the chromosomes. Bacteriophage DNA are episomes. See PLASMID.

epistatic gene A gene that, when present, masks the expression of one or more other genes that are not its allelomorphs (that

are located at different positions on the chromosome). An epistatic gene is analogous to a dominant gene and the recessive gene at the same position (locus). Contrast HYPOSTATIC GENE.

epitheca The upper, larger half of a diatom shell (frustule). See BACILLARIOPHYTA, HYPOTHECA.

epithelium A thin layer of cells surrounding an organ or external structure of an animal or plant, often secreting substances of some form. In vertebrates, the epithelium forms the epidermis in skin and the surface layer of mucous membranes and membranes around organs; it protects, secretes, absorbs, or performs other specialized functions. See MESOTHELIUM.

epithermal A term applied to veins of igneous rock forming under relatively low temperatures in the range of 50° to 200°C. Cinnabar, gold, silver, and stibnite are typical minerals of epithermal veins.

epizootic Describing an outbreak of an animal disease characterized by rapid spread over a wide area; the equivalent of an epidemic in human disease. Contrast ENZOOTIC.

epoch A third-order unit of the geologic time scale. An epoch is shorter than a period and longer than an age. See APPENDIX p. 610.

epontic Aquatic microorganisms that live attached to the surface of objects, a subgroup of periphyton. See PERIPHYTON.

equability See EQUITABILITY.

equation of state An equation describing a gaseous system that gives the relationship between temperature, volume, and pressure. See VAN DER WAALS FORCES.

equator A figurative line on the surface of the earth that divides the planet equally into the Northern and Southern Hemi-spheres. The equator is taken as the starting point, at 0°, in measuring north or south latitude.

equatorial countercurrent   A relatively narrow eastward-flowing surface ocean current located near the equator. The equatorial countercurrent is bounded by the much broader north and south equatorial currents.

equatorial current   The very broad (1000 to 5000 km) westward-flowing surface ocean currents near the equator. The north and south equatorial currents are separated by the relatively narrow equatorial countercurrent.

equatorial submergence   The occurrence of polar shallow-water marine species in the deep, cold waters of an equatorial region. Compare POLAR EMERGENCE.

equilibrium   A state of balance in a system, where opposing factors cancel each other out (or are entirely absent). In thermal equilibrium, heat gain balances any heat loss occurring between an object and its surroundings. A body at rest or moving at constant velocity also demonstrates a balance of opposing forces (or an absence of any forces). See DYNAMIC EQUILIBRIUM.

equilibrium constant, K_{eq}   The equilibrium constant is a measure of the extent of the reaction between the two chemical species. The value characterizes the point of stability of a chemical reaction at a given temperature. At this point of stability or equilibrium, the system has no tendency to change because the forward rate of the reaction is exactly balanced by the reverse rate of reaction.

equilibrium line   An area near the midpoint of a glacier where net gain in ice mass equals net loss. The equilibrium line separates the zone of accumulation from the zone of ablation.

equilibrium theory of island biogeography   A theory of community organization proposed by R. MacArther and E. O. Wilson to explain the numbers of species found on islands of various sizes and distances from continental areas. The number of species present is a function of migration rates and species extinction rates. Large islands have lower extinction rates than small islands, and islands near continents have higher immigration rates than distant islands.

equinox   Either of two points in the earth's path of revolution around the sun at which the ecliptic intersects the celestial equator. The length of night and day are equal at equinox. See AUTUMNAL EQUINOX, VERNAL EQUINOX. Compare SOLSTICE.

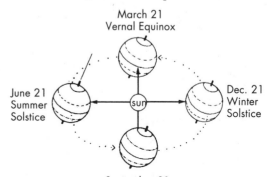

The position of the earth at different times of the year.

Equisetum   The genus containing horsetails and scouring-rushes.

equitability   (H') The degree of evenness in the distribution of patterns of species abundance. The greatest equitability possible is when all species are represented by the same number of individuals. Compare EVENNESS, SHANNON-WIENER DIVERSITY, SIMPSON DIVERSITY INDEX. Contrast SIMPSON DOMINANCE INDEX.

equity   Legal term for the administration of justice according to principles of fairness and conscience, balancing the hardships in those cases where legal remedies and monetary damages would not suffice.

equivalent conductivity   A property of ions that provides a measure of the relative contribution of an ion to the conductance of a solution. For example, Ca^{+2} contributes approximately twice as much to the conductance of a solution as Na^+ because it has an overall +2 charge rather than +1.

era   The first-order unit of the geologic time scale; an era is smaller than an eon and larger than a period. The Paleozoic, Mesozoic, and Cenozoic are the three eras of the Phanerozoic eon. See APPENDIX p. 610.

erathem   A chronostratigraphic unit that is equivalent to the geological time unit "era," for example, the Paleozoic erathem is all the rocks formed during the Cambrian, Ordovician, Silurian, Devonian, Mississippian, Pennsylvanian, plus Permian periods. Erathems rank above "systems" in the hierarchy of chronostratigraphic terminology.

eradication   The total removal or elimination of a species, especially an insect pest or weed. See EXTIRPATION, EXTINCTION.

erg   Unit of energy in the CGS (centimeter-gram-second) system. The erg is no longer in common use. 1 erg equals 10^{-7} joule. See CGS SYSTEM.

ergotism   Poisoning caused by eating bread or other food made of grains that have been contaminated by ergot, a fungus (*Claviceps purpurea*) that grows on wheat and rye. Historically called St. Anthony's fire.

Ericaceae   The heath family, an important group of temperate, boreal, and montane tropical dicots of the Angiospermophyta. The corolla consists of four to five lobed petals that are united at least at the bases and often form an urn or tube. There are twice the number of stamens and carpels as petals. The anthers open by terminal pores. Often the family is subdivided into the wintergreen (*Pyroloideae*), the white-alder (*Clethroideae*), the Indian-pipe (*Monotropoideae*), the heath (*Ericoideae*), and the blueberry (*Vaccinioideae*) subfamilies on the basis of trophy, floral patterns, and plant form. Bearberry and manzanitas (*Arctostaphylos*), madrone (*Arbutus*), rhododendrons and azaleas (*Rhododendron*), heather (*Erica* and *Calluna*), shinleafs (*Pyrola*), Indian-pipes (*Monotropa*), and blueberries and cranberries (*Vaccinium*) are members of this family. See ANGIOSPERMOPHYTA.

ericaceous   Belonging to, or resembling members of, the heath family of plants (Ericaceae). These plants are usually shrubs requiring acid soils and often have leathery leaves. See ERICACEAE.

ericilignosa   Vegetation in which woody shrubs belonging to the genus *Erica* (heaths) dominate. See LIGNOSA.

erosion   The physical removal of rock or soil particles by a transport agent such as running water, wind, glacial ice, and gravity.

erosion unconformity   See DISCONFORMITY.

erratic, glacial   An ice-transported boulder that is not derived from the nearby bedrock.

error   Statistical term for a deviation from the expected value for an observation. See STANDARD ERROR.

ERTS-1   See EARTH RESOURCES TECHNOLOGY SATELLITE.

eruption (volcanic)   A general term for the flow of lava, gas, or pyroclastic material from a volcano.

eruptive   1) A pattern of population growth marked by a sudden increase in numbers, as in a rapid increase in grasshoppers or other insects. 2) Describing rocks formed by volcanic eruptions. 3) Breaking out, as in a skin rash. See IRRUPTION.

erythrism   Unusual redness in the hair of mammals or the feathers of birds. In humans, erythrism (red hair and beard) is usually accompanied by a ruddy complexion.

erythrocytes   Red blood cells; the hemoglobin-containing corpuscles that transport oxygen in the blood (and remove carbon dioxide). The most numerous cell type in the blood, they are important in maintaining the acid-base balance. In mammals, erythrocytes have no nucleus. Compare LEUKOCYTES.

Es   Chemical symbol for the element einsteinium. See EINSTEINIUM.

escarpment   See SCARP.

Escherichia   An important genus of omnibacteria. See OMNIBACTERIA, COLIFORM BACTERIA.

esker   A sinuous ridge of sorted and stratified glacial alluvium. Esker formation is associated with sediment rich meltwater coursing through tunnels within or under glaciers.

essential amino acids   Those amino acids essential to life that cannot be synthesized by a given organism and must be obtained from food. For humans, the nine essential amino acids are histidine, isoleucine, leucine, lysine, methionine, phenylalanine, threonine, tryptophan, and valine. See AMINO ACID.

essential elements   Any of the numerous chemical elements, such as oxygen, nitrogen, phosphorous, and carbon, that are essential to an organism's growth and normal functioning. Although sodium is an essential element for animals, it is not an essential element for plants. See MACRONUTRIENT, MICRONUTRIENT, TRACE ELEMENT.

essential fatty acid   (EFA) Those fatty acids that cannot be synthesized by an organism and need to be supplied in animal diets. For humans, they are linoleic, linolenic, and arachidonic acids, all unsaturated fatty acids. See FATTY ACID, LINOLEIC ACID.

establishment   The stage following immigration or invasion of a new individual or species, in which the newcomer becomes a permanent part of the local community.

esters   Organic compounds produced by the reaction of an acid with an alcohol, creating the -C(O)O-R functional group contained in every ester (where R is a hydrocarbon group). Esters are responsible for the pleasant odors of many fruits and flowers and so are often used in synthetic fragrances and flavors. They are also important industrial solvents and are used in the synthesis of Plexiglas and other plastics. Animal fats and vegetable oils are also esters. Names of organic compounds ending in the suffix -ate usually indicate esters (as in methyl acetate).

estimated reserves   The quantity of an extractive resource, such as coal or oil, that is reasonably expected to be available through the conventional technology and current market state.

estivation   1) The spatial arrangement of

unexpanded leaves, sepals, or petals in plants. 2) A dormant or very torpid state that some animals undergo during summer months. Also spelled aestivation. Compare HIBERNATION.

estrogens   Female sex hormones; any of a group of substances causing the development of secondary sexual characteristics as well as physiological changes associated with sex and reproduction. Estrogens are used medically and in birth control pills. Diethylstilbestrol is a synthetic estrogen. Compare ANDROGENS.

estrous cycle   The periodic bodily changes in sexual and other organs connected with estrus (heat) in nonpregnant females of most mammal species other than primates. Many large mammals have an annual estrous cycle and breed (and later bear young) in a particular season of the year; others are polyestrous and can breed throughout the year. Also spelled estrus. Compare MENSTRUAL CYCLE.

estuary   The lowermost part of a river system where it reaches the intertidal zone of the ocean. Estuarine water is a partly saline mixture of sea water and fresh water.

ethanoic acid   Another name for acetic acid (the acid found in vinegar). See ACETIC ACID.

ethanol CH_3CH_2OH, ethyl alcohol. This is the ordinary alcohol found in whiskey, vodka, etc. It is produced by the fermentation of sugars or starches such as grain or potatoes. Ethanol is also used as a fuel, as in gasohol. See GASOHOL.

ethene   The standard international name for ethylene. See ETHYLENE.

ethics   1) The code of behavior governing the conduct of a group or an individual. 2) A set of moral principles or a system of philosophy that seeks to differentiate

between right and wrong. Contrast MORALS. See ENVIRONMENTAL ETHICS.

Ethiopian region   Biogeographical area, one of the primary regions in which the earth's surface is divided based on distribution of animal life (fauna). It consists of all of Africa south of the Sahara. See ZOOGEOGRAPHY.

ethnobotany   The scientific study of plants that are useful to humans and the interactions between plants and humans. Ethnobotany often bridges the disciplines of anthropology and botany.

ethnozoology   The scientific study of animals that are useful to humans.

ethology   The branch of biology that studies the behavior of living organisms.

ethyl acrylate $CH_3CH_2O-C(O)-CH=CH_2$, a clear liquid used in the manufacture of polymers, especially plastics.

ethyl alcohol   The common name for ethanol. See ETHANOL.

ethylene $CH_2=CH_2$, a flammable gas used as a veterinary anesthetic. A natural plant hormone, ethylene is given off by ripening fruits and has been used commercially to bring fruits in storage to ripeness. It is also used in the manufacture of organic chemicals and plastics. Also called ethene.

ethyne   Another name for acetylene. See ACETYLENE.

etiolation   A condition of plants caused by insufficient light, characterized by unusually long and weak stems, small leaves, and white or pale yellow color occurring from lack of chlorophyll production.

etiology   The study of causes.

et seq   Abbreviation of the Latin *et sequentia* meaning "and the following."

eucaryote Alternate spelling for eukaryote. See EUKARYOTE.

eugenics The study of possible improvements in species (especially humans) through control of hereditary factors in mating.

eugeocline An association of marine sediments formed in shallow waters. Eugeocline sediments are typically composed of volcanic island arc materials, graywackes, and shales. Compare MIOGEOCLINE.

Euglenophyta Members of this kingdom Protista phylum are largely solitary organisms that swim by means of one, two, or rarely, three flagella. While mostly autotrophic, these organisms can also absorb or eat particulate food. Unlike the Chlorophytes, in which some classification schemes have placed them, Euglenophytes lack a rigid, cellulose walls and store food in the form of paramylon rather than starch. Reproduction is by mitosis along a longitudinal plane. Familiar members of this phylum include the spindle-shaped *Euglena* and the flattened, leaf-shaped *Phacus*. See PROTISTA.

euhalabous Describing phytoplankton (or other plankton) that live in salt water, water with a salinity level of 30 to 40 percent. See MESOHALABOUS, OLIGOHALABOUS PLANKTON.

euhaline Describing sea water, or water of equal salinity (approximately 35 percent). See MESOHALINE, OLIGOHALINE.

eukaryote An organism whose cell or cells have a distinct nucleus surrounded by a membrane, as well as a number of distinct organelles within the cytoplasm. All higher unicellular and multicellular organisms are eukaryotes. Also spelled eucaryote. Contrast PROKARYOTE.

eukaryotic cell A cell containing a distinct nucleus surrounded by a membrane; one of the cells of a eukaryotic organism.

Euphausiacea Sometimes called euphausids, this crustacean order contains krill, shrimp-like, planktonic, filter feeders which swim with their abdominal appendages. Found throughout the world's oceans from surface waters to 5000 m, most forms migrate vertically every day, several hundred meters or more are not uncommon. Because they form the diet of many organisms such as baleen whales and many squid and birds species, etc., they play an important role in the trophic dynamics of the sea. Approximately 90 species exist. See MALACOSTRACA.

Euphorbiaceae The spurge family of the Angiospermophyta containing dicot herbs, shrubs, and trees that exude a milky latex or acrid sap when injured. The flowers of members of this family are usually small and exhibit variability among genera in numbers of parts. Often the flowers are surrounded by highly colored bracts. Poinsettia and crown-of-thorns (*Euphorbia*), castor-bean (*Ricinus*), croton (*Croton*), and cassava (*Manihot*) are all members of this family. See ANGIOSPERMOPHYTA.

euphotic zone The layers of a body of water (ocean or lake) that sunlight can penetrate sufficiently for photosynthesis to take place. Also called the photic zone. See COMPENSATION DEPTH, COMPENSATION LIGHT INTENSITY.

euplankton Organisms that spend almost their entire life cycle as plankton. Compare HOLOPLANKTON, MEROPLANKTON.

euploidy Possessing one or more entire chromosome sets, a whole multiple of the haploid number for the species. Euploid organisms can therefore be haploid, diploid, or polyploid. Compare ANEUP-LOIDY.

Euratom   See EUROPEAN ATOMIC ENERGY COMMUNITY.

European Atomic Energy Community (Euratom) A group established by treaty and including all members of the European Economic Community; it works to set basic standards concerning ionizing radiation for protecting workers and the public.

European Economic Community   (EEC) A group of European nations working together to promote free trade, transnational development projects, and other social and political issues. Originally composed of six nations, it now includes Belgium, Denmark, France, Greece, Ireland, Italy, Luxembourg, the United Kingdom, the Netherlands, Portugal, Spain, and Germany.

Euro-Siberian floral region   One of the phytogeographic regions into which the Holarctic realm is divided according to similarities between plants; it consists of portions of Siberia, Russia, and northern Europe south to the Tropic of Cancer. See HOLARCTIC REALM, PHYTOGEOGRAPHY.

eury-   A prefix indicating a wide range of tolerance in an organism to a given environmental factor. Compare STENO-.

euryhaline   Describing organisms that can tolerate a wide range of salt levels or salinity in either water or soils. Contrast STENOHALINE.

eurythermal   Describing organisms that can tolerate a wide temperature range. Contrast STENOTHERMAL.

eurytopic   Able to withstand a wide range of variations in environmental conditions. Contrast STENOTOPIC.

eustatic   The term applied to the vertical changes in sea level that occur worldwide. Eustatic changes occur as a result of tectonic movements or glaciation.

Eustigmatophyta   These photosynthetic, motile protists are similar to xanthophytes, but differ sufficiently in morphology to warrant their designation to a separate phylum of the kingdom Protista. Eustigmatophytes have a single udulipodium with a characteristic swelling at its base, just above an eye spot formed by carotenoids. Members of the phylum include the planktonic algae *Ellipsoidion*, *Polyedriella*, and *Vischeria*. See PROTISTA.

Eutheria   The placental mammals are the most numerous and successful of present-day mammals. The young, which are born in a more advanced state of development than those of marsupials, are weaned in a relatively short time after birth. The orders of mammals include such important groups as the insectivores, bats, rodents, rabbits and conies, even- and odd-toed ungulates, primates, elephants, whales and dolphins, and carnivores. See MAMMALIA.

eutrophication   The process by which a body of water acquires a high concentration of nutrients, especially phosphates and nitrates. These typically promote excessive growths of algae. As the algae die and decompose, high levels of organic matter and the decomposing organisms deplete the water of available oxygen, causing the death of other organisms, such as fish. Eutrophication is a natural, slow-aging process for a water body, but human activity greatly speeds up the process. See CULTURAL EUTROPHICATION. Compare DYSTROPHIC, OLIGOTROPHIC.

eutrophy   To acquire excessively rich nutrient levels; to undergo eutrophication. See EUTROPHICATION.

evaporation   The process of a liquid below its boiling point becoming a gas (like boiling, it is a change of phase from liquid to gas). Compare SUBLIMATION.

evaporation ponds   An industrial containment area designed to allow briny water to evaporate by using solar energy. Commercially valuable evaporites are concentrated in the evaporation ponds.

evaporimeter   A meteorological device for measuring natural rates of evaporation.

evaporite deposits   An accumulation of minerals precipitated by evaporating water, such as occurs in a playa or hot spring. Gypsum, anhydrite, and halite are typical minerals of evaporite deposits.

evapotranspiration   The combined water loss from a biotic community or ecosystem into the atmosphere caused by evaporation of water from the soil plus the transpiration of plants. See EVAPORATION, TRANSPIRATION.

even-aged   Situation where most individuals in a biotic community or forest are about the same age. Often young or recently disturbed biotic communities have an even-aged structure, a condition that may change with time toward an all-aged community.

even-age stand   Forest in which all the trees are about the same age. Usually, these stands contain only one or two types of trees; they are often the result of clearcutting or of a planted stand on a tree farm. Compare UNEVEN-AGE STAND.

even distribution   See UNIFORM SPATIAL DISTRIBUTION.

evenness   (E) The extent to which all species are equally abundant, rather than one or two species greatly exceeding the others in abundance; the opposite of dominance within a community. One measure of evenness is to determine the degree to which equitability (H') of a community approaches the maximum equitability for the number of species present. Compare EQUITABILITY, SIMPSON DIVERSITY INDEX, SHANNON-WEINER INDEX. Contrast SIMPSON DOMINANCE INDEX.

evergreen   Maintaining green leaves or needles throughout the year. On evergreens, the leaves or needles stay on the tree until after new ones form to replace them, a process that may occur gradually throughout the year so the tree is never bare and always has some leaves that are more than one year old. The leaves of many evergreens are adapted to reduce water loss; in conifers such as pine or cedar they are reduced to needles, and in hollies they are leathery with a protective waxy coating. Contrast DECIDUOUS, WINTERGREEN.

evolution   The process by which all existing organisms developed from earlier ones through changes in inherited characteristics over many generations. See DARWINIAN THEORY, NATURAL SELECTION, PUNCTUATED EQUILIBRIUM.

evolutionary time   A period measured in hundreds of successive generations in a population, required for random mutations to show up as evolutionary changes. Evolutionary time is generally hundreds of years to several million years. Compare ECOLOGICAL TIME, GEOLOGIC TIME.

evolutionary tree   Diagram used to show the evolutionary history of relationships between groups of organisms.

exact compensation   The phenomenon in which population increases result in a matched decrease in birth rate or increase in death rate, so that the population is stabilized at its initial density. See DENSITY DEPENDENT, DENSITY OVERCOMPENSATION.

exalbuminous   A seed that contains no endosperm when it reaches maturity. See ENDOSPERM.

excess density compensation   Another term for density overcompensation. See DENSITY OVERCOMPENSATION.

exchangeable cations   The positively charged ions that are adsorbed onto the surface of clay or humus colloids within a soil. Exchangeable cations may replace one another as soil conditions change. Calcium and magnesium are important exchangeable cations in plant nutrition.

excitation   The raising to a higher energy level of an atom or a molecule. Contrast GROUND STATE.

Exclusive Economic Zone (EEZ)   A portion of the continental shelf defined as extending 320 kilometers (200 miles) from the shore of a country. Within this zone, a coastal country or state has jurisdiction of the harvesting of marine resources, including seabed minerals as well as fish and shellfish. EEZs were established by the United Nations Conference on the Law of the Sea. See UNITED NATIONS CONFERENCE ON THE LAW OF THE SEA.

exclusive species   A species whose occurrence is completely, or almost completely, restricted to one biotic community.

excreta   Excretions or excrement; any waste matter released from the body, such as sweat or urine.

excrete   To eliminate toxins or waste matter from the blood or tissues and expel it from the body.

excretion   1) The act of expelling waste matter from the cells and tissues of the body. Unlike egestion, excretion refers to organic matter that has been assimilated by the organism (i.e., has entered the bloodstream). 2) Any waste material so released. Compare SECRETION, EGESTION.

executive branch   One of three branches of the United States government. It includes the President, the Cabinet, federal departments (Department of Health and Human Services, Department of Energy, Department of the Interior, etc.) and their agencies (such as the Forest Service, National Park Service, Soil Conservation Service, Fish and Wildlife Service), federal bureaus (Bureau of Land Management, Bureau of Mines, Bureau of Reclamation, etc.), and federal administrations (Federal Highway Administration, Food and Drug Administration, National Oceanic and Atmospheric Administration, etc.), as well as the military. State governments also have an executive branch. Compare LEGISLATIVE BRANCH, JUDICIAL BRANCH.

exergonic   Describes a chemical reaction that gives off energy and can therefore occur spontaneously. Compare ENDERGONIC.

exfoliation (sheeting)   The spalling off of loose sheet-like rock layers in concentric slabs from physical and chemical weathering.

exhaust gas recirculation   (EGR) An air pollution control method used in boilers and some internal combustion engines to reduce pollutants by preventing the formation of NO_x (nitrogen oxides). Exhaust gases are mixed with incoming combustion air to improve the chemistry of the combustion process and, thus, change the pollutants produced. See COMBUSTION AIR, NO_x.

exogamy The union of two gametes from unrelated (or distantly related) parents; outbreeding. Compare ENDOGAMY.

exogenous 1) Initiated from outside of an organism rather than in response to internal stimuli. 2) Describing a force outside of the defined boundaries of a particular system. In geology, exogenous forces of denudation (weathering, mass-wasting, erosion, transport, deposition) combine with internal (endogenous) forces to produce landforms. Contrast ENDOGENOUS.

exogenous plankton Species of plankton that originate in an area different from that in which they are found; plankton that has been brought into an area by ocean currents, rather than being born there. See ENDOGENOUS PLANKTON.

exopodite In crustaceans, the outer section of a forked appendage. See ENDOPODITE.

exoskeleton Hard supporting or protective external structure. Exoskeletons may be the complete supporting structure for an organism, such as the shells of arthropods (insects, lobsters). Vertebrates supported by an internal skeleton may also have an exoskeleton, as in turtles and armadillos. See ENDOSKELETON.

exosphere The region beyond the ionosphere, the most distant region of the earth's atmosphere. See IONOSPHERE, MESOSPHERE, STRATOSPHERE, TROPOSPHERE.

exothermic Describing a process, especially a chemical change or reaction, in which heat is given off the surroundings. See ENDOTHERMIC, THERMONEUTRAL.

exotic 1) Short for exotic species. 2) Description of a stream that flows across an arid area where it does not receive any additional water. For example, the lower Nile River is an exotic river.

exotic species 1) One that is not native to an area. 2) A plant species that has been introduced to an area or region through human action, but has now naturalized to the point of being self-sustaining. Also called alien species. Compare ENDEMIC, ADVENTIVE PLANT. See IMMIGRATION.

exploitation 1) The use or consumption by an organism of a resource such as food. 2) The use, consumption, or overconsumption of a natural resource by humans, as in the exploitation of coal deposits or food sources. Also called resource exploitation. See EXPLOITATIVE COMPETITION.

exploitative competition Competition between two or more organisms for the same limited resource, especially a food source. Also called exploitation competition.

exponential growth Growth in which some quantity (such as population) increases by a fixed percentage of the whole (rather than by a fixed amount, which is linear growth) in a given period of time. Exponential growth increases its rate greatly over time, unlike linear growth which continues to increase at the same rate over time. Something that increases by only 2 percent a year will show a fourfold increase in 70 years. Also called logarithmic growth and exponential rate of increase. See APPENDIX p. 603.

exposure 1) Being subjected directly to the elements, especially to severe cold or other extremes of weather. 2) The way in which a meteorological instrument is exposed to the elements. In order to compare records from different meteorological stations, this exposure must be standardized.

exsiccate The drying of soil as a result of

groundwater removal. Exsiccation occurs in the draining of a marsh or where deforestation leads to excess evaporation.

extant   Existing; not extinct.

exterior drainage   A term describing any stream system that discharges directly or indirectly into the ocean. Compare INTERIOR DRAINAGE.

externalities   Outside force, such as social benefits and costs not included in the market price of the goods. See ECONOMIC EXTERNALITY.

exteroceptor   A sense organ or nerve ending adapted to receive stimuli from external sources, such as the eye. See CHEMORECEPTOR, RECEPTOR.

extinction   The dying out of a species, or the condition of having no remaining living members; also the process of bringing about such a condition.

extinction coefficient   (ε) A property characteristic of a given substance, related to its ability to absorb light. The amount of light absorbed by a given sample is proportional to not only the extinction coefficient of the substance, but also to how much sample is present (the concentration), and the path length of the light through the sample. This relationship is expressed in Beer's Law: $A = \varepsilon bc$, where A is absorbance, b is the path length of the light through the sample, and c is the concentration of the substance in the sample.

extirpation   Eradication; the loss or removal of a species from one or more specific areas, but not from all areas. Compare EXTINCTION.

extraclinal   Another term for topotype. See TOPOTYPE.

extract   1) To remove a chemical compound or compounds from a solid or liquid substance by dissolving it with a solvent and then evaporating off the solvent, leaving behind the desired compound or compounds. 2) A compound produced through such a process, especially a concentrated form of or the active portion of a drug obtained from plant tissue.

extrafloral nectaries   Nectar-secreting glands found on the leaves and other vegetative parts of plants.

extrapolation   To project or extend current knowledge into a yet-unknown area, as in making a prediction about future economic growth based upon past observations and available data.

extremely low frequency   (ELF) Electromagnetic radiation at the very low-frequency end of the radio spectrum. ELF radio waves have frequencies from 30 to 3000 hertz and, therefore, have very long wavelengths. See EXTREMELY LOW FREQUENCY, HIGH FREQUENCY, LOW FREQUENCY, MEDIUM FREQUENCY, ULTRA HIGH FREQUENCY, VERY HIGH FREQUENCY, VERY LOW FREQUENCY.

extreme ultraviolet   (EUV) Electromagnetic radiation at the shorter-wavelength end of the ultraviolet range. EUV has wavelengths below 200 nm. See ULTRAVIOLET.

extrusive   A term applied to fine-grained igneous rocks that have cooled rapidly at or near the surface of the Earth.

exumbrella   The curved upper surface of a jellyfish (medusa). See SUBUMBRELLA.

eyespot   An area of red or orange pigment, found in protozoa and some unicellular and colony-forming green algae, that is sensitive to light. Also called stigma.

f

F 1) Chemical symbol for the element fluorine. 2) Abbreviation for Fahrenheit temperature. See FLUORINE, FAHRENHEIT.

F₁ generation Symbol for first filial generation. See FIRST FILIAL GENERATION.

F₂ generation Short for second filial generation. See SECOND FILIAL GENERATION.

Fabaceae The bean family, a very large and important group of angiosperm dicots, is characterized by having members that produce leguminous fruit from a single carpel but is often divided into subfamilies based upon floral structures. The *Mimosioideae* have flowers with five sepals and five petals fused into a small tube. Numerous, long stamens project beyond the corolla. For the most part, members of this subfamily are woody, for example, acacias *(Acacia)*, mesquite *(Prosopis)*, and tamarind *(Leucaena)*. The *Caesalpinoideae* also have five sepals and petals, but the upper petal is attached inside the lateral petal and 10 or fewer stamens. Redbud *(Cercis)*, honeylocust *(Gleditsia)*, palo verde *(Cercidium)*, tamarind *(Tamarindus)*, and senna *(Cassia)* are all members of this subfamily. The *Papilionoideae* have bilaterally symmetrical, pea-like flowers with a large upper petal known as a standard, two lateral petals known as wings, and two lower petals fused into a keel that serves as a landing platform for pollinating insects. Lupines *(Lupinus)*, clover *(Trifolium)*, peas *(Lathyrus* and *Pisium)*, beans *(Phaseolus)*, and locoweeds *(Astragalus)* are members of

this subfamily. See ANGIOSPERMOPHYTA.

faciation In the monoclimax theory, a subdivision of the association characterized by recurring groups of dominant species. Compare FORMATION, ASSOCIATION.

facies A set of characteristics that may be collectively used to define the environmental conditions of rock formation. A facies may describe any diagnostic condition of lithology, sedimentation, or faunal composition.

facilitation The process in which species in an early successional phase alters the conditions of a community so that succeeding species will have an easier time becoming established. Contrast INHIBITION, TOLERANCE MODEL.

facultative Capable of existing under different conditions or using different modes for nutrition. Facultative parasites are organisms that can function either as parasites or as saprophytes. Facultative wetland plants can occur in either wetlands or uplands, although they are more abundant in the former.

facultative anaerobes Organisms that are able to live in environments lacking oxygen, but that can also function in environments containing atmospheric oxygen (aerobic environments). Compare OBLIGATE ANAEROBES.

Fagaceae The beech family of the Angiospermophyta. Most of these monoecious woody dicots are trees of the Northern Hemisphere. Small male flowers lacking petals are arranged in catkins or spikes known as aments, whereas the female flowers are surrounded by a cup-like involucre that forms a cap or husk around the developing nut or acorn fruit. Beeches *(Fagus)*, oaks *(Quercus)*, chestnut *(Castanea)*, tan oak *(Lithocarpus)*, and southern beech (*Nothofagus*) are members of this family. See ANGIOSPERMOPHYTA.

Fahrenheit (F) A temperature scale named for its 18th-century physicist inventor. On the Fahrenheit scale, the freezing point of water is 32° and the boiling point (at standard atmospheric pressure) is 212°. Contrast CELSIUS, KELVIN.

Falconiformes The order of raptors or birds of prey which have strong clawed feet, hooked beaks, and a well-developed ability to soar. They are almost exclusively diurnal. Examples include hawks, eagles, kites, old and new world vultures, falcons, ospreys and secretary birds. See AVES, VULTURES.

fallout The deposition of solid particulates formerly suspended in the air; also short for nuclear fallout, the radioactive dust that falls to earth following nuclear explosions or accidents.

fallow Describing land that is not being used for growing crops for an entire growing season or longer. In some crop rotations, land is left fallow after growing crops that greatly deplete the soil (such as corn or cotton) so that the soil can build up nutrients and organic matter again.

fallspeed The velocity of particulate matter or drops of rain descending through the atmosphere.

fallstreak A meteorological phenomenon in which a shaft composed of ice particles falls through a cloud. See VIRGA.

false bedding A bedding layer that is naturally inclined because of the effects of currents and does not indicate a change in depositional history.

family The taxonomic group below an order and above a genus. The names of families can be recognized by their suffixes: the names of animal families end in -idae (as in the genera of bears belonging to the family Ursidae), and names of plant families usually end in -aceae (as in all members of the heath family, Ericaceae). See GENUS, ORDER.

fan A gently sloping cone-shaped deposit of gravel and sand characteristic of the lower end of a canyon or at the foot of some mountains.

fanning An pollution dispersal phenomenon in the atmosphere in which emissions spread out horizontally in all directions from a smokestack. See CONING, LOOPING.

FAO See FOOD AND AGRICULTURAL ORGANIZATION.

farad (F) Unit of electrical capacitance in the Système International d'Unités (SI). The value of the capacitance is equal to the amount of charge (in coulombs) in a capacitor divided by the associated voltage across the capacitor. Because the unit is so large, it is usually used in the form nanofarads (nF) and microfarads (10^F).

Faraday's law A law describing the amount of voltage induced in one or more wire loops when the magnetic flux (related to the magnetic field) passing through the loop(s) is changing. Essentially all our electricity is generated using this law.

fascicle 1) A tight cluster or bundle of flowers, leaves, or roots, such as the clusters of five needles characteristic of white pine. 2) Another term for vascular bundle. See VASCULAR BUNDLE.

fast breeder reactor (FBR) A breeder reactor that is designed to use fast neutrons. See BREEDER REACTOR, FAST REACTOR.

fast reactor A nuclear reactor that uses

predominantly fast neutrons to sustain the nuclear chain reaction. Fast neutrons are neutrons that have not slowed down much from the speeds they have when ejected in the fission process. Ordinary reactors use "slow" neutrons. Both ordinary and breeder reactors can be designed to use fast neutrons. See NUCLEAR REACTOR.

fat   Any of a class of organic chemical compounds, usually mixtures of triglycerides or other lipids, that are solid at room temperature. Fats are common in plants and animals, used to store highly concentrated energy. Fats are usually compound esters of several different acids; they contain carbon, hydrogen, and oxygen, but no nitrogen. See LIPID.

fata morgana   A type of mirage most often seen near the coast, where sharp temperature contrasts help to bend light rays. The result is an image that appears huge or far above its real image, or as multiple images. See SUPERIOR IMAGE.

fatty acid   Any of a large group of organic acids, having the general formula $C_nH_{2n+1}COOH$. Many are essential for metabolism. Long-chain fatty acids are components of lipids and waxes. Acetic acid and oleic acid are both fatty acids.

fault   A fracture in unconsolidated material or, more commonly, rock along which displacement has occurred.

fault block   A unit of rock that is bounded by fault planes on at least two sides.

fault breccia   A rock composed of angular fragments of displaced country rock that was broken during fault movement.

fault drag   The distortion of bedding planes and rock cleavage that occurs by frictional drag along the edges of a fault plane. See TERMINAL CURVATURE.

fault gouge   The finely powdered fragments of country rock, often altered to clay, that was crushed during fault movement.

fauna   All of the animals of a particular region or a particular era. For example, the fauna of New Zealand. Compare FLORA.

FBC   See FLUIDIZED BED COMBUSTION.

FBR   See FAST BREEDER REACTOR.

fe   Chemical symbol for the element iron. See IRON.

featured-species management   A policy geared toward encouraging a single species, even if at the detriment of other species. Featured-species management is often used for endangered species.

fecal   Relating to feces (excrement).

fecal coliform bacteria   See COLIFORM BACTERIA.

feces   Bodily wastes discharged from the intestines; excrement.

fecundity   1) The quantity of gametes (usually eggs) or seeds produced. 2) Fertility; the number of offspring produced by an organism or species per unit of time, sometimes expressed as the number of females produced by each female. 3) More generally, the ability to reproduce prolifically. Compare FERTILITY, NATALITY.

fecundity schedule   A table of data illustrating the patterns of births among individuals of different ages within a population. See NATALITY RATE.

Federal Register   Publication of the U.S. Government that includes a record of all new laws, executive orders, regulations, etc.

feedback   See NEGATIVE FEEDBACK, POSITIVE FEEDBACK.

feeder reservoir   A general term for a

source area for a fluid material. Examples include a magma chamber that feeds an igneous intrusion or a holding tank that feeds a petroleum refinery.

feedlot A small pen where cattle are confined in order to fatten them.

fee-simple ownership A freehold estate of virtually infinite duration and of absolute inheritance free of any condition, limitations, or restriction to particular heirs. Fee simple is the most common way of owning land in the United States.

feldspar A common non-ferromagnesian silicate mineral incorporating sodium, potassium, or calcium as the charge-balancing cations of the crystal structure; forming under a wide range of temperatures and pressures, has two prominent cleavages at approximately 90°, and glassy to pearly luster; hardness = 6.

feldspathoids Minerals that form in the place of feldspars in magmas deficient in silica and rich in alkalis, such as leucite, nephaline, sodalite, and lazurite.

fell A term used in Great Britain to refer to an elongated moorland or an uncultivated open hillside.

felsic rock A general term for light-colored igneous rock rich in silica and aluminum and relatively deficient in ferromagnesian minerals. Contrast MAFIC ROCK.

felsite An inclusive term for varieties of light-colored extrusive igneous rocks composed primarily of quartz and feldspar. Rhyolite is an example of a felsite rock.

female 1) The particular sex in any animal species that is capable of producing ova and bearing offspring. 2) Seed plants or individual flowers having a pistil or pistils but no stamen. Compare MALE, HERMAPHRODITE.

femto- (F) A prefix used in the Système International d'Unités (SI) to denote 1×10^{-15}. A femtogram is 10^{-15} gram.

fen A marshy, low-lying wetland covered by shallow, usually stagnant, and often alkaline water that originates from groundwater sources. Compare BOG, MARSH, MOOR, SWAMP.

fenestra In geology, an irregular cavity formed by gas bubbles or biologic activity within intertidal carbonate sediments. See BIRD'S-EYE LIMESTONE.

feral Describing a domesticated animal that has reverted back to the wild. In some areas, feral dogs traveling in packs are hazardous to deer and herds of sheep, and feral cats are significant predators of passerine birds.

Feret triangle A delta-shaped chart of soil types with each side of the triangle representing one of the three basic soil classifications: sand, silt, and clay. Most soils are a combination of the three types, as represented by the internal sections of the triangle.

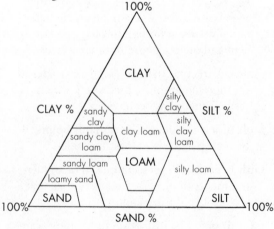

Feret triangle

fermentation The slow chemical decomposition of carbohydrates by yeasts, bacteria, or molds in combination with enzymes. It is an anaerobic process that liberates energy. The fermentation of sugars produces alcohol, a process harnessed for making gasohol; fermentation of milk sugars produces lactic acid.

fermenting bacteria   A phylum of obligate, anaerobic bacteria which produce energy through fermentation of a wide range of organic materials. It includes the lactic-acid bacteria which ferment milk sugars, and others which produce powerful toxins such as *Clostridium botulinum*, the disease agent of botulism. See MONERA.

Fermi-Dirac statistics   A theory of quantum mechanics obeyed by a system in which the Pauli exclusion principle applies; it is essentially the opposite of Bose-Einstein statistics. It states that only one of a particular type of particle may occupy any allowed quantum-mechanical state.

fermium   (Fm) A synthetic, radioactive, transuranic element with atomic number 100; the atomic weight of its main isotope is 257. It is produced from plutonium or uranium and in hydrogen bomb explosions.

fern   See FILICINOPHYTA.

Ferrel's law   A theory explaining global air circulation patterns. Ferrel believed that surface air currents flow towards the east and towards the poles, but at higher levels flows in opposite directions. This 19th-century model does not explain actual circulation patterns. See HADLEY CELL.

ferro-   A prefix referring to iron, usually used to indicate that a substance contains iron (as in ferroconcrete or ferromanganese). The prefix ferri- also indicates iron.

ferromagnesian   An alloy of iron and manganese used to add specified quantities of manganese during the manufacture of very hard forms of steel (as in armor plating) and cast iron.

ferrosilite   An iron-rich variety of the pyroxene mineral group forming orthorhombic crystals. Ferrosilite is composed of iron, silicon, and oxygen.

ferrous   Containing iron, specifying iron in the +2 oxidation state. The suffix -ous indicates a lower valence than the suffix -ic, so ferrous (Fe++) has a lower valence than ferric (Fe+++).

ferruginous   1) A description of the appearance of a rusty surface. 2) Said of a mineral having iron as a component.

fertile   1) Able to reproduce. 2) Describing a soil rich in nutrients and organic matter, capable of producing high yields. Compare INFERTILE.

fertile material (nuclear)   Isotope material which can be converted to fissile material by neutron capture. For example, the fertile material ^{238}U is converted to fissile ^{239}Pu in a nuclear reactor.

fertility   1) The condition of an organism being able to reproduce. 2) Fecundity; the number of viable offspring produced by an individual or group per unit of time. In human population studies, fertility is the average number of children born to women during the span of their normal childbearing years. See GENERAL FERTILITY RATE. Compare FECUNDITY, NATALITY.

fertilization   The fusion of a male and a female reproductive cell (gametes) to form a zygote capable of developing into an individual. Also called syngamy. See ALLOGAMY, AUTOGAMY, SELF-FERTILIZATION, CROSS-FERTILIZATION.

fertilize   To bring about fusion of male and

female reproductive cells (gametes); also, to impregnate or inseminate. 2) To add nutrients to the soil.

fertilizer Any substance that adds nutrients to soil, improving its ability to grow crops and other vegetation. Organic fertilizers are derived from animal or plant sources such as manures, cottonseed meal, bone meal, etc. Inorganic fertilizers can refer either to natural minerals (such as rock phosphate, greensand, and Chilean nitrate) or to synthetic chemical fertilizers.

fetus An animal embryo in the later stages of its development in the womb or in an egg. A developing organism is initially called an embryo and is generally called a fetus once features become recognizable. In humans, the term is used from about three months after conception to birth.

FGD Short for flue gas desulfurization. See FLUE GAS DESULFURIZATION.

fiberglass Glass that has been spun into fine threadlike fibers while in a molten state. It is used as building insulation and, when coated with plastics or resins, as a corrosion-resistant structural material for boats, car bodies, etc., with a high strength-to-weight ratio.

fibrosis Abnormal formation of fibrous tissue in an animal; it may be caused by injury or inflammation or by a restriction of the blood supply to tissues.

fibrous root system A branched system of plant roots comprised o many smaller roots, roughly all the same in size, without any major roots. Grasses have

The fibrous root system of a cardinal flower.

fibrous root systems. Compare TAPROOT SYSTEM.

fidelity 1) The degree to which species are found exclusively in certain types of communities. 2) Description of an animal's long-term preference for or restriction to a particular type of habitat or biotic community. High fidelity indicates a great degree of restriction to, or preference for, a particular community. See RELEVÉ.

field capacity The maximum amount of water a soil in the vadose zone can hold against gravitational forces.

Fijian floral region One of the phytogeographic regions into which the Paleotropical realm is divided according to similarities between plants covering the Fiji islands. See PHYTOGEOGRAPHY.

filament 1) In seed plants, the stalk of a stamen that supports the anther. 2) Any fine, threadlike structure in plants or animals.

Filicinae See FILICINOPHYTA.

Filicinophyta The phylum containing ferns (formerly referred to as part of the Pteridophyta). Ferns, the evolutionary precursors of higher plants, have a more complex vascular system than bryophytes, lycopods, or members of the sphenophyta. As a consequence, they can support larger, more complex leaves, generally called fronds, that arise out of a horizontal stem or rhizome. Some ferns have fronds that are fertile, meaning they bear sporangia, and have other fronds that are sterile, lacking sporangia. The spores give rise to a heart-shaped gametophyte known as the prothallus.

fill terrace 1) A flat-topped, artificial landform created by the engineering practice of cut-and-fill. 2) A natural stream terrace resulting from aggradational processes (valley fill) followed by incision.

film badge A dosimeter made of photographic film, small enough to be worn on a person's wrist or pocket. It is used for estimating radiation exposure for workers in nuclear facilities. See DOSIMETER .

filter feeder Animals that eat small particles of organic matter (or minute organisms) by straining them out of the water.

filtrate Liquid that has been passed through a filter to remove solid particles; the product of filtration.

filtration Causing a liquid to flow through a filter. The filtration process used in wastewater treatment involves passing the effluent through a bed of sand to remove large particles.

fine-grained Describing an environment in which resources are limited to such small patches that the feeding organisms cannot be selective when foraging. Compare COARSE-GRAINED.

fiord (fjord) A long, narrow coastal inlet formed by the post-glacial flooding of a U-shaped valley.

fire Combustion; a visible form of oxidation producing visible flame or glowing.

firedamp A mixture of methane with smaller amounts of similar explosive gases found in coal and therefore in underground mines.

first filial generation (F_1) The first set of offspring in an experimental cross or breeding experiment. F_1 hybrids are the result of crossing genetically different parents. F_1 hybrids are often stronger and more uniform (and, in the case of plants, higher yielding) than either parent variety. See HYBRID, SECOND FILIAL GENERATION.

first law of thermodynamics A particular application of the law of conservation of energy. It states that the change in internal energy of a system equals the amount of heat added minus the amount of work done by the system. See LAW OF CONSERVATION OF ENERGY.

Fish and Wildlife Service The agency of the U.S. Department of the Interior responsible for the conservation of migratory birds, certain mammals, endangered species, and sport fishes. It is also responsible for the management of national wildlife refuges.

fishery 1) A stock of fish, or other aquatic or marine resource such as shrimp, and the economic enterprises that potentially or actually exploit them. Often linked to the area from which the resource is harvested. 2) An area where fish are caught. 3) A country's offshore boundary line marking where fishing is regulated.

fissile Capable of being split by neutrons; fissionable. Fissile is used in reference to heavy isotopes such as U-235 that can be split into lighter elements, releasing large amounts of energy, in the process of nuclear fission. See CHAIN REACTION, NUCLEAR FISSION.

fission The process of splitting; short for nuclear fission. See NUCLEAR FISSION, CHAIN REACTION.

fissionable Capable of being split by neutrons; fissile. Fissionable isotopes are heavy elements such as U-235 that can be split into lighter elements, releasing large amounts of energy, in the process of nuclear fission. See NUCLEAR FISSION, CHAIN REACTION.

fission tracks Trails left by subatomic particles produced during radioactive decay. In geology, the tracks left by naturally decaying U-238 are used in dating rocks and other geological formations. See RADIOACTIVE DECAY.

fissure In volcanology, a long, linear tensional fracture in a shield volcano or cone.

fissure eruption The ejection of lava, pyroclastic material, and volcanic gas through a fissure.

fitness 1) Short for Darwinian fitness. 2) The number of offspring contributed by an individual relative to the numbers contributed by other members of a given population. 3) Extent to which an organism or species is adapted to its habitat and therefore able to survive and reproduce. See DARWINIAN FITNESS.

fixation The process of converting a substance into a more usable form for organisms, usually used to refer to the fixation of carbon, nitrogen, or phosphorous in the soil. See NITROGEN FIXATION.

fjord See FIORD.

flagellum A threadlike extension providing locomotion for a cell.

flame ionization detector (FID) Used to detect substances after they pass through a column. Flame ionization detectors work by burning the compounds, a process that produces ions that are translated into an electrical current which is measured to determine the amount of substance present. A flame ionization detector is used for the analysis of trace levels of hydrocarbon compounds.

flare 1) Energetic eruption on the sun's surface that emits radiation. Flares occur suddenly and last from minutes to hours. Large solar flares can cause power blackouts, short-wave radio blackouts, and auroras on the earth. 2) In industrial processes such as oil refining, to flare means to burn off unusable byproducts such as natural gas.

flash colors Bold markings in bright colors on animals that are usually concealed but are exposed when the animal is disturbed, startling and scaring off predators. Flash colors often resemble vertebrate eyes. Also called startle colors.

flat-plate collector The classic apparatus for active solar energy collection. It consists of a large flat surface oriented to intercept maximum solar radiation. It is black to absorb as much heat as possible. Water or other fluids are run across the black surface to absorb this heat and transfer it to where it is needed or to a storage system such as a water tank. See SOLAR ENERGY.

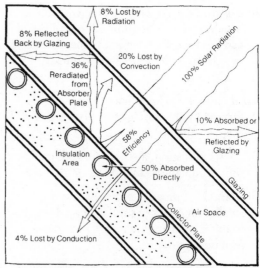

This diagram illustrates why a flat-plate collector is not 100 percent efficient. Only about 50 percent of the heat is transmitted directly to the water in the tubing.

flatworms See PLATYHELMINTHES.

F-layer The outermost subdivision of the

earth's ionosphere; it lies 150 to 1000 km above the earth's surface. Because of its high concentration of free electrons, it reflects high-frequency radio waves and is therefore important for radio transmission. Sometimes called the Appleton layer. Compare E-LAYER, D-LAYER.

F layer (soils)   A term applied to a soil layer in which organic material is decomposing by fermentation.

fleas   See SIPHONOPTERA.

fledgling   An immature bird, one that has just become capable of flight.

flint   A dark-colored, fine-grained rock composed almost entirely of silica. Flint is often found as compact lumps and nodules in sedimentary rock. Compare CHERT.

floc   A lump of solids formed from a liquid suspension, especially from biological or chemical treatment of sewage. An individual product of flocculation.

flocculation   The process of causing colloidal solids that are suspended in a liquid to clump together into larger particles, usually in order to make them precipitate out so the solids can be removed from the liquid. Flocculation is used in mineral extraction and in sewage treatment. See COLLOID.

flock   A grouping of birds (or herbivorous animals) kept together by social interactions. See HERD.

floe   A mass of floating sea ice. A floe is smaller than an ice field.

flood   A flow of surface water in excess of channel capacity. A flood overflows the bank level and inundates the surrounding topography (usually the floodplain).

flood basalt   A vast accumulation of basaltic lavas on the earth's surface by an outpour-ing of lava from large lateral fissures.

flood-frequency curve   A graphical analysis that plots the predicted size of a flood against the probable frequency of occurrence. A flood-frequency curve is calculated independently for each point in a drainage system.

flood lava   See FLOOD BASALT.

floodplain   The relatively flat land adjacent to a river channel that is constructed of unconsolidated sediment deposited by periodic flooding and lateral migration of the river channel.

floodplain zoning ordinances   Land-use constraints on development in areas likely to flood. The National Flood Insurance Program requires local governments to have a comprehensive plan for floodplain regulation, including zoning ordinances, in order to qualify for federal flood insurance.

flood tide   The period during which a tidal water level is rising from low tide to high tide.

floodway district   The designated area of a floodplain which could theoretically contain the discharge of a 100-year flood with only a 1-foot rise in water elevation above the height of an unrestricted flood. The hydrologic calculation of the floodway district illustrates the consequences of filling and building within the floodway fringe district. Federal development loans are not available for building within the floodway district. See FLOODWAY FRINGE DISTRICT.

floodway fringe district   The designated area of a floodplain that would be inundated by the 100-year flood. The floodway fringe district flanks the floodway district. Insurance regulations and federal development loans require new buildings in the floodway fringe district to be flood-proofed. See FLOODWAY DISTRICT.

flora   All of the plants of a particular region or a particular era. For example, the flora of the Florida Keys. Compare FAUNA.

floral formula   A sequence of numbers used to indicate the structure of a flower, giving the numbers of parts contained in the perianth or calyx and corolla, androecium (male reproductive structures), and gynoecium (female reproductive structures); it also indicates the position of the ovary and the extent to which parts are fused together or separate.

floral realm   The highest level of relationship recognized by plant geographers, who have divided the world into four biogeographical areas according to the plant life. The floral realms are the Holarctic realm, Neotropical realm, Paleotropical realm and Austral realm; they are in turn divided into 30 floral regions, which are divided into provinces (domains), and finally into districts. Floral realms are also called floral kingdoms. See BOREAL, PALEOTROPIC REALM.

floret   An individual flower in a large flower head (inflorescence), as in grasses or one of the small flowers within the flower head of a composite plant such as the daisy.

flow banding   A textural feature of metamorphic and igneous rocks characterized by alternating layers of differing appearance and composition. The presence of flow banding indicates the direction of forces acting upon minerals during rock formation. See GNEISS.

flower   The often colorful reproductive structure of an angiosperm, the part of a plant that produces the seed. A flower usually consists of one or more pistils and stamens, corolla (collectively all the

petals), and calyx (collectively all the sepals), although any of these may be absent in a particular kind of flower. See PERIANTH.

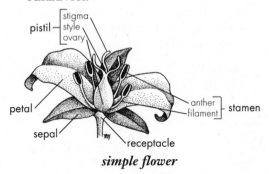

simple flower

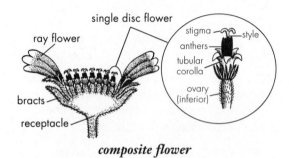

composite flower

flowmeter   Any device used to measure the rate of flow of a liquid or gas through a pipe.

flow stone   See DRIPSTONE.

flow structure   1) A feature indicating a submarine slump or flow during the formation of a sedimentary rock. 2) A fluid structure in igneous rocks.

flue gas   Emissions from industrial processes that are discharged through a chimney; combustion exhaust from boiler furnaces and similar devices. Flue gas is usually a mixture of carbon dioxide, carbon monoxide, oxygen gas, nitrogen gas, water vapor, nitrogen oxides, and sulfur oxides plus particulates. Analyzing flue gas shows how efficiently a furnace is running.

flue gas desulfurization   (FGD) Air pollution control technique, a process of removing oxides of sulfur from exhausts created during the combustion of hydrocarbons, especially coal. Forms of flue gas desulfurization include caustic scrubbing, dry alkali injection, and dry limestone process. See CAUSTIC SCRUBBING, DRY ALKALI INJECTION, DRY LIMESTONE PROCESS, SCRUBBER.

flue gas scrubber   Another name for scrubber. See SCRUBBER.

fluid injection   A technique of forcing pressurized fluid into an oil reservoir to mobilize the flow of oil into a well.

fluidized bed combustion   (FBC) A technique in which solid fuels are made to behave as liquids. A hot air stream passing up through powdered coal suspends the fuel. This provides for efficient combustion. It also provides excellent mixing of two solids, as when mixing powdered limestone and coal. This process removes almost all sulfur dioxide produced by burning coal, as well as improving combustion. See DRY LIMESTONE PROCESS.

fluid-mosaic model   A model for the structure of cell membranes in which two layers of lipids (phospholipids) contain proteins either floating within them or connecting across them. The proteins control the movement of substances in and out of the cell (giving the membrane its semipermeable qualities) by their interaction with other proteins and compounds. See BILAYER, PHOSPHOLIPID, SEMIPERMEABLE MEMBRANE, UNIT MEMBRANES.

fluke   1) Any of several parasitic flatworms that often have complex life cycles (alternation of generations), with one stage living in a mollusc such as a snail and another parasitizing animals. Flukes parasitic in humans may infest the liver, blood, lungs, or intestines, depending on the species. 2) One of the flatted tail divisions of whales. 3) A flatfish of the genus *Paralichthys*. See SCHISTOSOMIASIS.

flukes   See TREMATODA.

flume   An artificial stream channel constructed to allow the measurement of hydrologic flow characteristics. See GAGING STATION.

fluorescence   The transformation of radiation such as ultraviolet light by natural materials into lower-frequency radiation such as visible light. See PHOSPHORESCENCE.

fluoridation   The addition of fluoride in low concentrations to drinking water in order to reduce tooth decay. Sodium fluoride, NaF, is often used, but other fluoride sources used include sodium silicofluoride ($NaSiF_6$), hydrofluosilicic acid (H_2SiF_6), and ammonium silicofluoride (NH_4SiF_6). See FLUORIDE.

fluoride   1) The ionic form of the element fluorine, where the fluorine atom has gained one electron (F^-). 2) A solid compound containing ionic fluorine as F^-. Many metals form fluorides; natural fluorides often contain sodium, tin, or potassium. As air pollutants, fluorides are destructive to vegetation and therefore detrimental to agriculture. Fluoride is often added to toothpaste and to municipal drinking water supplies at low concentrations (1 ppm) because it has been shown to reduce tooth decay. See FLUORITE, FLUORIDATION.

fluorine   (F) A gaseous, poisonous, halogen element existing as the diatomic molecule F_2; it has an atomic number of 9 and atomic weight 19.00. It is the most chemically reactive of all the elements; as a result, it is never found alone in nature and is very dangerous to work with (explosive and extremely corrosive) in pure form. It is used to form uranium hexafluoride (UF_6) in the separation of uranium isotopes for nuclear fuel and in the manufacture of the many organic compounds containing fluorine (including CFCs).

fluorite   A variably colored mineral of calcium and fluorine that forms cubic crystals in veins of igneous rock. Fluorite is a fluorescent mineral that glows in ultraviolet light. See APPENDIX p. 623.

fluorocarbon   Any chemical compound containing carbon and fluorine. Fluorocarbons are relatively stable under high temperatures and relatively inert, making them excellent lubricants. They are also used as solvents, in polymers, and as artificial blood. Teflon (tetrafluoroethylene) is a fluorocarbon. See CHLOROFLUOROCARBON.

fluoroscopy   The examination of objects using x-rays and a fluorescent screen; the objects appear as shadows on the screen. The technique is used to examine airline luggage and to inspect structures and welding for mechanical faults.

fluorosis   Condition of low-grade fluorine poisoning, appearing as mottling of the teeth and causing reduced milk yields from dairy cattle. It is caused by excessive levels fluoride in drinking water or by consuming plants grown on high-fluoride soils.

fluorspar   A crystal of fluorite.

fluvial   1) A term that indicates a relation to a stream process. 2) Comprehensive term for river processes.

flux   1) A continuous flow, as in a stream, or succession of changes, as in a state of flux. 2) Rate of flow, as of a liquid or other fluid

or radiation, across a given area.

fly ash   Solid particles, created during or remaining from combustion processes, that are fine enough to be swept along in exhaust fumes. Fly ash includes soot, dust, and partially combusted substances (paper, coal dust, etc.).

flysch   A sedimentary facies indicating deep-sea deposition of clastic material, as in a turbidite. See FACIES.

flyway   A route repeatedly traveled by migrating birds; major flyways exist along the eastern and western coasts of the United States.

Fm   Chemical symbol for the element fermium. See FERMIUM.

focus (earthquake)   The subsurface point of brittle rock failure during an earthquake. Also called hypocenter. Compare EPICENTER.

fodder   Plants grown as food for animals, such as grass or clover. Fodder can be fed as dry hay, or fermented into silage to provide green plant food to animals throughout the winter. See SILAGE.

FOE   See FRIENDS OF THE EARTH.

foehn   A relatively dry and warm wind blowing down the leeward side of mountain ranges. High-elevation air is warmed by adiabatic heating (compression) as it descends down the slope of the mountains. The Santa Ana winds of southern California are foehn winds. See ADIABATIC, CHINOOK.

fog   A cloud or thick mist situated at or very close to the ground. A fog is distinguished from a mist by containing slightly larger droplets of water and by causing greater impairment of visibility.

fogbow   An optical phenomenon, similar to

a faint rainbow, sometimes occurring in a fog in the opposite direction from the sun.

fold   A curve introduced in any planar feature of a rock by lateral compression, differential shearing, differential vertical movement, or thrusting. Synclines and anticlines are typical examples of folds.

fold-and-thrust belt   An area of bedrock that is regionally deformed by the tectonic action of folding and thrust faulting. A fold-and-thrust belt is usually found in the foreland of an orogenic belt.

foliage   All the leaves on a plant or in a plant community.

foliage height diversity   A mathematical formula for representing degree of vertical stratification or layering of plants in a community.

foliated rock   Any rock that has a continuous fabric of platy or tabular minerals aligned in a preferred subplanar orientation. Foliation is a common definitive feature of metamorphic rocks.

folic acid   A water-soluble vitamin in the B complex, also called folacin. Essential for human metabolism, it is needed to synthesize DNA and RNA, and to manufacture and break down amino acids. The best source of folic acid is dark green, leafy vegetables, although much is destroyed by cooking or long storage. It is also synthesized by intestinal bacteria. Folic acid deficiency results in lack of growth and anemia, because the body cannot produce healthy blood cells.

follicle   1) Any tiny cavity in an animal, such as the saclike structure surrounding hair roots. 2) The Graafian follicle, a small structure formed in the ovaries of mammals that contains the developing ovum (egg). The rupture of the follicle to release

the egg starts the process of ovulation. 3) In plants, a dry fruit formed from one carpel that splits open along one side when ripe to reveal many seeds. Follicles form in delphiniums and milkweeds.

follicle-stimulating hormone   A human hormone produced in the pituitary gland to stimulate production of egg cells in the ovaries or the production of sperm cells in the testes. Abbreviated FSH.

food   Any material that can be ingested and assimilated by an organism to supply nutrients and energy for physiological processes such as growth and repair of tissues. Plant and animal tissues provide carbohydrates, proteins, fats, vitamins, and minerals to organisms consuming them. See MACRONUTRIENT, MICRONUTRIENT.

food additive   Any substance added to foods to improve or to preserve their quality, including colorings, flavorings, and preservatives. Mono- and diglycerides are examples of food additives; they are naturally occurring compounds added to breads and other baked goods to prevent them from stiffening in the process of becoming stale.

Food and Agricultural Organization (FAO) An agency of the United Nations responsible for a wide range of programs in food, agriculture, forestry, and related matters.

food chain   The system of feeding (trophic) levels found in any biotic community. Members of one level feed upon members of the level below and are in turn eaten by organisms in the next level above. The lowest trophic level in any food chain ultimately contains plants, which fix inorganic compounds into plant tissues that can be digested by animals. Food chains may also be fueled by dead organic matter originating from photosynthesis outside the community; organisms dying within the community serve as a source of energy for decomposing or reducing organisms. See DETRITUS FOOD CHAIN, GRAZING FOOD CHAIN, TROPHIC LEVEL.

food chain efficiency Another term for ecological efficiency. See ECOLOGICAL EFFICIENCY.

food web A complex feeding system comprised of linked food chains in a particular ecosystem. See FOOD CHAIN.

fool's gold A common name for the mineral pyrite. A foolish prospector might mistake pyrite for gold.

footwall The planar rock surface of a fault block that lies below the fault plane. Compare HANGING WALL.

forage Food for animals, especially that obtained by grazing or browsing. Also, to look for food.

foraging strategy Method by which animals select their food and allocate energy to seek out, capture, and eat their food. See OPTIMAL FORAGING THEORY.

Foraminifera The foram phylum of the Protista kingdom consists of marine organisms with perforated, chambered, calcium carbonate shells. The deposition of these shells is an important contribution to marine sediments, the white cliffs of Dover and the limestones of the pyramids of Egypt being composed primarily of forams tests. The typical life cycle of forams includes both sexual and asexual phases, somewhat similar to members of the kingdom Plantae. Familiar members of the phylum include the planktonic *Globigerina*, the warm water *Discorbis*, and many others. See PROTISTA.

forb Any herbaceous (non-woody) plant having broad leaves, and therefore excluding grasses and grasslike plants. Forb is used especially to distinguish non-grass species when discussing grasslands and prairies.

force A push or a pull. The metric unit for force is the newton. The net force on an object equals the product of the mass of the object and the acceleration produced on that body by the force (Newton's second law). See NEWTON.

forced convection Heat transfer by the bulk movement of a liquid or gas driven by a pump, fan, or blower. An example is forced hot air heating in homes. This process uses blowers to circulate heat from a central furnace through ducts to heat the living spaces. Contrast NATURAL CONVECTION.

Forchhammer principle An oceanographic principle stating that the relative proportions of the principal dissolved ions in sea water remain constant even during variation in the total salinity.

foredeep A downwarped tectonic basin filled with clastic sediment derived from a nearby orogenic uplift.

foreset bed 1) An individual tilted layer of a cross set bed. 2) A sloping deposit at the forefront of a delta.

foreshock Seismic activity that occurs as a precursor to a larger earthquake or volcanic eruption.

foreshore The part of a shoreline that lies within the normal intertidal range. The foreshore is typically a gently sloping surface of a beach.

forest A large group of trees, especially (but not necessarily) those growing close enough that the tops of most touch or overlap, shading the ground below. Forests may or may not have extensive undergrowth. Compare WOODLAND.

forest decline   Poor health and increased death in woodlands that cannot be attributed solely to disease or insect infestations. Because the incidence of forest decline has increased since the 1970s, it is believed to be caused or promoted by air pollution (of various forms). In Europe it is called *Waldsterben* because it has been extensively documented in German forests. Also called dieback. See WALDSTERBEN, DIEBACK.

forest dieback   Another term for forest decline or *Waldsterben*, dieback occurring in wooded areas. See DIEBACK, FOREST DECLINE.

forest floor   1) The ground level in a forest community. 2) The surface of the soil, including its organic matter, in a forest ecosystem.

forest management   A system of planned, scientific use of forest resources for sustainable harvest (long-term productivity), multiple use, regeneration, and maintenance of a healthy biological community.

forestry   The practice of planting, managing, and caring for forests or heavily wooded areas for human use, usually for timber or firewood. See FOREST MANAGEMENT.

Forest Service   The agency in the U.S. Department of Agriculture that administers, and is responsible for resource management of, national forests and national grasslands. Established by Congress in 1907, the Forest Service also conducts research in forestry and wildlife (game) management.

form   The lowest taxonomic rank, often below subspecies, based on differences in a particular trait, such as different flower colors that are possessed by varieties of a subspecies. An example is the white and purple flower-color forms of garden phlox. See VARIETY, RACE.

formaldehyde CH_2O, a strong-smelling gas that irritates the eyes and the upper respiratory tract and is a probable carcinogen. Formaldehyde is used as a disinfectant and preservative and in the formation of many plastics and adhesives. It is used extensively in plywood, fiber board, and particle board. As buildings have become more air tight, formaldehyde has become recognized as an indoor pollutant. In addition to being given off by construction materials, it is produced by incomplete combustion, and so is found in cigarette smoke, automobile exhaust, wood smoke, and emissions from power plants and incinerators. Also known as methyl aldehyde and methanal (not to be confused with methanol, the alcohol). See UREA FORMALDEHYDE.

formation   1) An obsolete term roughly equivalent to biome. 2) In the monoclimax theory, the major unit of vegetation. See BIOME.

formation, geologic   The basic unit of lithostratigraphy. Formations are subdivided into members or combined into groups.

formicidae   See ANTS.

fosse   An elongated waterway such as a trench, ditch, or canal.

fossil   Any remains or evidence of life existing in geologic time. See GEOLOGIC TIME.

fossil assemblage   A group of fossil organisms that lived in a particular environment at a particular time.

fossil fuel   Any deposit of fossil organic material that is combustible enough to be used as fuel. Coal, oil, and natural gas are fossil fuels.

fossilization   Any process of fossil formation. Common varieties of fossilization are carbonization, permineralization, and recrystallization.

fossorial   Describing animals or their structures that are adapted for digging or burrowing. A mole is a fossorial animal.

Foucault pendulum   A device for demonstrating the rotation of the earth on its axis. It incorporates a heavy mass hanging on a long wire, which is free to rotate at its top. The orientation of its back-and-forth swing slowly changes; this apparent rotation of the pendulum is really a reflection of the earth's rotation. A large Foucault pendulum is on permanent display at the Smithsonian Institution's National Museum of American History in Washington, DC.

founder principle   Another term for founder's effect. See FOUNDER'S EFFECT.

founder's effect   The principle that organisms starting a new colony have only a fraction of the total genetic variation of the population from which they come; this fraction determines the genetic attributes of the new population.

Fr   Chemical symbol for the element francium. See FRANCIUM.

fraction   1) A petroleum product produced by catalytic cracking of crude oil. Different fractions are characterized by the lengths of the carbon chains characterizing their primary component: Natural gas molecules contain chains of one to four carbon atoms, gasoline has four to twelve carbons in the chain, and paraffin molecules have chains containing 23 to 29 carbon atoms. 2) A portion of a mixture collected from a distillation or a chromatographic run. See CATALYTIC CRACKING.

fractionation (magmatic)   See MAGMATIC DIFFERENTIATION BY FRACTIONAL CRYSTALLIZATION.

fracture   The irregular breakage of a mineral crystal, showing no correspondence with planes of weakness in the crystal structure; fracture is conventionally described as smooth, splintery, fibrous, or conchoidal depending upon the appearance of the broken surface. Any break in rocks. Contrast CLEAVAGE.

fracture zone   1) A region of the deep-sea floor containing a series of related fractures that cross segments of ocean crust that differ in age and depth. 2) A zone, usually linear, containing a number of fractures.

fragipan   A dense rocklike subsurface soil layer, nearly impermeable to fluids. See CLAYPAN, HARDPAN.

fragmentation   A form of asexual reproduction (propagation) in which an organism divides into two or more pieces that develop into individuals. Fragmentation is common in filamentous algae, some coelenterates, and annelid worms.

francium   (Fr) A radioactive element, the heaviest of the alkali metal group. It has an atomic number of 87. None of its isotopes are stable; the atomic weight of its most stable isotope (with a half-life of 22 minutes) is 223.

fraternal twins   Twins of the same or opposite sex that develop from two separately fertilized egg cells rather than from one egg cell. Also called dizygotic twins.

Fraunhofer lines   Distinct, narrow dark lines appearing in the continuous spectrum of the sun, stars, or other glowing bodies. These absorption lines identify the presence of specific elements or molecules surrounding the body emitting the spectrum. See ABSORPTION LINE, SPECTRUM.

free acceleration test   A method for determining the emissions produced by automobiles and similar vehicles, often part of statewide automobile inspections. While the vehicle is standing still, the engine is accelerated; the exhaust gases produced during acceleration are sampled and their emissions content analyzed.

free energy   1) The energy in a system that is available to perform useful work. 2) In chemistry, the energy given off or absorbed in a reversible chemical reaction (G), also called Gibbs free energy. It is derived by subtracting the product of a system' entropy and (constant) temperature from the enthalpy of a system. Changes in free energy (G) are used to determine if a chemical reaction will proceed spontaneously (without being driven externally). Compare FREE ENERGY OF MIXING.

free energy of mixing   The change in the free energy of a system (a measure of the system's ability to do work) when one substance is mixed with another substance. Free energy of mixing is calculated from the enthalpy (heat of mixing) and the entropy of the combined system at a constant temperature. See FREE ENERGY, HEAT OF MIXING.

free field   Electromagnetic radiation (especially sound waves) flowing directly from a source without being deflected by an object or electric gradient.

freemartin   A female calf with birth defects, usually sterile and possessing some male characteristics, born as a twin to a male.

free oscillation   Mechanical or electronic vibrations occurring without an oscillatory driving force, that is, determined by the inherent properties of the oscillating material rather than external forces. See OSCILLATION.

free radical   An atom, molecule, or functional group of atoms having an unpaired electron that makes it very reactive. As a result, free radicals rarely exist on their own. The hydrogen atom ($H\cdot$) and the methyl group ($\cdot CH_3$) are free radicals.

free-running cycle   The time period over which a circadian rhythm repeats without the aid of external time cues, such as day and night, to reset it. See BIOLOGICAL CLOCK, CIRCADIAN RHYTHM, DIURNAL. Compare ENTRAINMENT.

freestone   Any type of sandstone or limestone that may be worked easily in any direction. Freestone is used as a building material.

freeze   To change from a liquid phase to a solid phase. The freezing point of a substance is the temperature at which it solidifies from a liquid.

freeze drying   A process for preserving food without requiring refrigeration. As the name implies, it is a process of dehydration in which material is first frozen, and then dried in a vacuum to remove ice. It produces a higher-quality product than simple food dehydration.

freezing nuclei   Solid particles in the atmosphere with a shape resembling ice crystals. They provide a core around which supercooled water droplets crystalize into ice.

freezing point   The temperature at which a substance undergoes a change of state from liquid to solid. This temperature is identical to a substance's melting point. See FREEZE, MELTING POINT.

freon   Commercial name for dichlorodifluoromethane, CCl_2F_2. It is one of the most well-known chlorofluorocarbons, used in refrigeration systems including automobile air conditioners. See CHLOROFLUOROCARBON.

frequency   1) Number of wave cycles or repetitions passing a given point per unit of time. In the Système International d'Unités (SI), frequency is measured in hertz (cycles per second). Frequency is the reciprocal of period. Compare AMPLITUDE. 2) The likelihood of finding individuals of a particular species in a given area, often expressed as the number or percentage of sample plots or points in which a species occurs. If the percent occurrence is used, the term is called relative frequency. See ABUNDANCE, IMPORTANCE VALUE. Compare CONSTANCY.

frequency dependence   1) Describing a characteristic that varies in response to how often it occurs in a population. In frequency-dependent selection, reproduction rates for a given type change in response to increasing numbers of that type appearing in a given population, because the relative fitness of that type compared to the rest of the population changes as that type becomes more abundant. 2) Describing the tendency of a predator to prey disproportionately upon the most abundant species.

freshwater   Of, relating to, or living in water that contains very little salt (less than 0.05 percent, as compared with brackish water that has 0.05 to 3.0 percent), such as that in rivers, ponds, and lakes.

friability   A term describing the physical consistency of a soil or the degree to which a soil crumbles when handled.

friction layer   The layer of the atmosphere lying next to the earth and extending for up to one kilometer. The friction between air masses and the earth's surface exerts a considerable influence over wind speed and direction (and therefore weather patterns) within this layer.

Friends of the Earth   (FOE) An independent global advocacy group based in Washington, DC. It was originally formed in 1970 by David Brower, formerly of the Sierra Club. In 1990 it merged with the Environmental Policy Institute and the Oceanic Society.

fringing reef   A coral reef that forms near the shoreline. See BARRIER REEF, ATOLL.

frond   1) The large, compound leaf of a fern, cycad, or palm. 2) A leaflike part in liverworts, lichens, or some algae.

front   The sloping boundary surface or zone between two distinct air masses differing sharply in temperature and therefore density. The warmer air mass usually has a higher moisture content than the cooler of the two air masses. Fronts are often associated with significant changes in weather, temperature, clouds, and wind direction. See COLD FRONT, WARM FRONT.

frontal slope   In meteorology, the angle of inclination of an advancing front, caused by the drag of friction from the ground. Warm fronts have a gradual slope (about 1:200); the slope of cold fronts is usually twice as steep (about 1:100). See FRONT.

frontal wave   A wavelike airflow pattern that occurs at the leading edge of a warm front.

frontal zone   In meteorology, the band of transition between two air masses. These zones can range from 15 to 200 kilometers wide, but the fronts extend laterally for great distances. See FRONT.

frost   A freezing at ground level, especially of dew, of water vapor, or of plant tissues, that can occur when air temperatures fall to or below 0°C (32°F). Frost can be a major problem for fruit growers in warm climates. Frost also refers to the layer of ice crystals (often called white frost, or in British usage, hoarfrost), that often forms with such a temperature drop.

frostbite   Freezing of living tissue caused by exposure to extremely cold temperatures or severe wind chill. It commonly occurs at the extremities (hands, feet, and nose).

frost heave   A layer of loose rock, soil, or pavement that is lifted by the expansion of freezing water.

frost pocket   A small, low region particularly prone to frost damage. Cold, dense air collects here from surrounding higher topography; as radiational cooling lowers air temperatures, these areas freeze first.

frost point   The temperature to which air has to be cooled for its water vapor to be in equilibrium with ice crystals. See SATURATION POINT.

fructose   Fruit sugar; the simple, extremely sweet sugar found in honey and many fruits. It has the same formula as glucose—$C_6H_{12}O_6$—but the atoms have a different, three-dimensional arrangement in space (fructose is an isomer of glucose). Compare GLUCOSE.

frugivore   An animal that feeds on fruit, especially those primates that cannot digest the cellulose in leaves and therefore restrict their plant diet to fruit. Contrast HERBIVORE, OMNIVORE, CARNIVORE.

fruit   The ripe ovary of a seed plant, including the seed or seeds and connected tissues, and their coverings. Sometimes used for the seeds of certain gymnosperms, especially if fleshy, as in juniper and yew. Different types of fruit include achene, caryopsis, drupe, drupelet, hesperidium, pome, samara (key), schizocarp, silqua, silicula, and nut.

frustule   The hard, silica-containing cell wall of a diatom that has two parts resembling the top and bottom of a tight-fitting box. The two parts are called the epitheca and hypotheca. See BACILLARIOPHYTA.

frutescent   Another term for fruticose. See FRUTICOSE.

fruticose   Shrubby; having a shrublike form.

fry   A young fish or a small adult fish in a large school of fish.

FSH   See FOLLICLE-STIMULATING HORMONE.

fucoxanthin   The major brown carotenoid pigment in brown algae, giving them their characteristic color. See CAROTENOID.

fuel   Any material that can produce usable energy, either through burning (gas, wood) or through atomic reactions (fission, fusion). See FOSSIL FUEL, FISSION, FUSION.

fuel assembly   A grouping of fuel rods, used to produce a chain reaction during nuclear fission. Fuel assemblies, together with control rods, make up the core of a nuclear reactor. See NUCLEAR REACTOR, CONTROL RODS.

fuel cell   Device producing electricity directly from chemical reactions in a galvanic cell wherein the reactants are replenished. A fuel cell is one type of galvanic cell. See GALVANIC CELL.

fuel efficiency A measure of the degree to which a combustion engine produces useful energy (instead of waste heat or incomplete combustion), thereby limiting pollution and making the most economical use of fuel. Specifically, it is the desired energy output as a percentage of the fuel energy input. Fuel efficiency is used in rating automobiles and oil furnaces.

fuel rod A unit of fissionable material, such as uranium, used in nuclear reactors. Fuel rods consist of slender canisters up to twelve feet long filled with small (one-inch) pellets of enriched isotopes. Fuel rods are grouped together into fuel assemblies. Also called a fuel element. See FUEL ASSEMBLY, NUCLEAR REACTOR.

fuelwood Wood raised for or prepared for burning in a furnace or wood stove.

fugacious Any flower, fruit, or leaf that falls or fades early in the growing season. See DECIDUOUS.

fugitive species Species that inhabit an area for a short period of time only. Ragweed, for example, grows best in disturbed habitats and usually disappears quickly as succession progresses.

fulgurite A glassy, rootlike mass of fused sand particles formed by a lightning strike.

Fuller's earth A natural, fine-grained clay material used as an industrial absorbent. Fuller's earth is typically composed of the clay mineral montmorillonite.

fulvic acids A mixture of complex organic molecules that, together with humic acids, comprise humic substances. Both fulvic and humic acids are decompositioin products of plant material and are found in soils and in water. Fulvic and humic acids are separated by an acid/base extraction.

fumarole A volcanic vent associated with escaping gas or steam.

fumigant A substance used as a gas to poison insects or other pests, as in flea bombs used to rid houses of flea infestations. Organophosphorous insecticides and formaldehyde are used as fumigants in greenhouses for soil sterilization. Hydrogen cyanide (hydrocyanic acid), methyl bromide, and ethylene oxide are also used as fumigants.

function 1) Statistical (and mathematical) term for a quantity whose value and variation is controlled by the value and variation in another related quantity or quantities. 2) The physiological action or activity of an organism, or a part of an organism. 3) Rate of flow through an ecosystem, such as the rate of energy flow or nutrient cycling.

functional response The reaction of a predator to changes in the availability of prey; the increase or decrease in the number of prey consumed in a given period of time compared to the increase or the decrease in the number of prey available. Compare NUMERICAL RESPONSE.

fundamental frequency The lowest frequency of vibration or oscillation of a mechanical or an electronic system. Also, the lowest frequency component of a periodic wave, such as a sound wave. See HARMONIC.

fundamental niche The largest theoretical niche that an organism or a species could occupy in the absence of competition from other species. See NICHE; Contrast REALIZED NICHE.

fundamental particles The basic units comprising all matter. As science progresses, the theory of which particles are fundamental changes. Fundamental particles are subdivisions of the proton, neutron, electron, photon, neutrino, and positron; these were called elementary particles (and sometimes still are) before it was discovered that they, too, were made up of smaller units. See SUBATOMIC PARTICLES.

Fungi The kingdom of organisms consisting of eucaryotes that absorb their food, form spores that develop directly into slender tubes called hyphae or single vegetative cells, have cell walls composed of chitin, and lack flagella at all stages of their life cycles. Until recently, fungi were often included in the plant kingdom, but are now considered to occupy their own kingdom based on their distinctive life cycles, mode of nutrition, and pattern of development. The fungi kingdom has five phyla: Zygomycota (bread molds and fly fungi), Ascomycota (yeasts and sac fungi), Basidiomycota (mushrooms, smuts, rusts, puffballs, and stinkhorns), Deuteromycota (imperfect fungi), and Mycophycophyta (lichens). The taxa composing slime nets, slime molds, chytrids, and oomycetes have been included the fungi in some past classification systems, but are now more frequently included in the kingdom Protista. See ZYGOMYCOTA, ASCO-MYCOTA, BASIDIOMYCOTA, DEUTERO–MYCOTA, MYCOPHYCOPHYTA, PROTISTA. Also see APPENDIX p. 616.

fungicide Any substance that kills fungi. See BORDEAUX MIXTURE, SULFUR.

funnel cloud The narrow, columnar vortex cloud extending downward from a massive, dark layer of cumulonimbus clouds in a tornado. See VORTEX.

furan C_4H_4O, a liquid organic compound with a ring structure. It differs from aromatic compounds because it is a five-atom ring with one oxygen and four carbon atoms in the ring. It is used in making copolymers and is part of the structure of some simple sugars. See GALACTOSE.

fuse 1) To join separate entities together to form a single whole. 2) A safety device that prevents overloading of an electrical device or circuit. It consists of a section of wire made from an alloy having a low melting point. If excess current flows through the circuit, the wire heats and melts, breaking the circuit. 3) A detonating device for an explosive charge or bomb.

fusiform Describing a spindle-shaped animal or plant structure, one characterized by tapering from a broad middle to narrow ends.

fusion 1) The process in which two light nuclei are joined to form one nucleus of greater atomic mass, releasing vast quantities of energy as a result. This process powers the sun and the hydrogen (thermonuclear) bomb, and may some day be used to generate electric power. 2) A melting or melting together. See HEAT OF FUSION, NUCLEAR FISSION.

fusion reactor A reactor in which the controlled nuclear fusion of deuterium nuclei into helium takes place. Although no commercially viable fusion reactors yet exist, they are the focus of much research since the potential energy resource is much larger than for fission reactors and they produce less radioactive waste. Also called a thermonuclear reactor. See DEUTERIUM.

g

g   1) Abbreviation for gram. 2) When capitalized (G), short for the prefix giga-. See GIGA-, GRAM.

gabbro   A general term for dark-colored (mafic) igneous rocks with phaneritic texture. Olivine, pyroxene, and calcium plagioclase are the predominant minerals. Gabbro is used as an ornamental, dark building stone.

gaging station   A place where streamflow is measured. A gaging station consists of devices to record water depth or stage of flow. At specific times, velocity of flow and cross-sectional area of flow are measured to calculate discharge.

Gaia hypothesis   The proposition that the biosphere functions as a single system that maintains homeostasis in the same way as a single organism, rather than as an assembly of discrete systems. According to the Gaia hypothesis, the earth's flora and fauna, climate, and biogeochemical cycles are interconnected so that changes in one part of the system affect the biosphere as a whole. Proposed in 1979 by J. E. Lovelock, the theory was named for Gaia, the Greek earth-mother goddess.

galactose   A simple sugar (monosaccharide) with the same formula as glucose— $C_6H_{12}O_6$—but with a different structural arrangement of these atoms (it is therefore an isomer of glucose). Solutions of galactose often contain a furanose form of the molecule, a form containing the furan ring.

galaxy   Immense cluster of stars, dust, and gases held together by gravity. Often used to refer specifically to the Milky Way Galaxy, the one containing our solar system. There are billions of other galaxies in the universe.

gale   A strong wind, one measuring 7 to 9 on the Beaufort Scale (32 to 54 mph, moderate to strong gales). Gales are strong enough to cause difficulty for people walking into the wind; they also break twigs off trees and cause slight structural damage. See BEAUFORT SCALE, also APPENDIX p. 630.

galena   A very dense, lead sulfide mineral forming cubic crystals with a brilliant metallic luster in hydrothermal veins. Galena also occurs as a replacement mineral in limestone and dolomite. See APPENDIX p. 623.

gall   An abnormal growth or swelling on a plant caused by insects, parasitic bacteria, or fungi. Different organisms create galls of a characteristic shape or location on leaves, stems, or roots of a particular species of plant. See WITCHES' BROOM.

Galliformes   The order of fowls or fowl-like birds whose members are largely gound dwellers with sturdy bodies and legs. Able to fly well for only short distances, they are usually non-migratory (e.g., currasaws, grouse, quail and pheasants, peacocks, guinea fowl, turkeys, and jungle fowls from which the common chicken is descended). See AVES, GAME BIRDS.

gallinaceous bird   Any bird belonging to the order Galliformes, characterized by having feet adapted for running, like those of a chicken. Pheasants, grouse, and turkeys are gallinaceous birds. See GALLIFORMES.

galvanic cell   A device for producing electricity from chemical reactions. Such cells usually consist of two electrodes made of different metals immersed in an electrolyte. Also called a voltaic cell. See ELECTROLYTE, FUEL CELL.

galvanize 1) To subject to an electrical current. 2) To coat (especially iron or steel) with zinc. Galvanized nails are resistant to rusting.

game birds Birds belonging to several families of the Galliformes including grouse, pheasants, and turkeys. See GALLIFORMES.

gametangium Any cell or organ where gametes (reproductive cells) are produced. See ANTHERIDIUM, ARCHEGONIUM, OOGONIUM.

gamete A mature reproductive cell (haploid) that can unite with another haploid cell of the opposite sex to form a fertilized cell, which can develop into a new organism. Egg cells and sperm are gametes. Also called germ cell. See OVUM, SPERM, SPORE.

game theory A method of applying mathematical analysis to decision making and strategy, using the probabilities for different outcomes from different choices. Although it originated for selecting the best strategy to minimize one's maximum losses or to maximize one's minimum winnings in a competition, it is now applied much more broadly.

gametocide Any drug or other agent that destroys gamete-producing cells (gametocytes). Gametocides are used in treating malaria.

gametophyte The individual plant that produces gametes, or (where alternation of generations occurs), the generation that undergoes sexual reproduction and bears gamete-producing organs. The gameto-

phyte is haploid. In plants such as mosses, it is the dominant phase of the life cycle, but in seed plants, it is reduced to the pollen grain (male) and embryo sac (female). Compare SPOROPHYTE.

gamma radiation The most penetrating of the three basic types of ionizing radiation. Gamma rays can pass right through a human body and require more than a few centimeters of lead to stop them. See ALPHA RADIATION, BETA RADIATION.

gamma ray The highest-energy form of radioactivity. Unlike alpha or beta particles, gamma rays have no electrical charge or mass; they resemble x-rays but have shorter wavelengths and higher energies. See ALPHA PARTICLE, BETA PARTICLE.

gamodeme A group of organisms that are able to interbreed.

Gamophyta The phylum of conjugating green algae of the kingdom Protista. This phylum consists of freshwater, filamentous, or solitary green algae that reproduce by producing haploid, non-flagellated, amoeboid gametes that fuse after conjugation (two cells aligning side-by-side and connecting to each other) to produce a zygote. Two classes make up this phylum, the filamentous Euconjugatae, such as *Spirogyra* and *Mougeotia;* and the solitary Desmidioideae, with halves of the cells forming mirror images, such as *Micrasterias* and *Cosmarium.* See PROTISTA.

ganglion A mass of nerve cells that forms a nerve center, usually located outside of the brain or spinal cord in vertebrates. In invertebrates, ganglia constitute much of the central nervous system.

ganglioside Any of a class of organic compounds (glycolipids) derived from cerebrosides and containing neuraminic acid. Gangliosides are found in the nerve cells and in the spleen.

gangue   The portion in a geologic ore deposit that is not of economic value or that does not contain the ore mineral to be mined.

gannister (ganister)   A fine-grained quartz sandstone often occurring beneath coal deposits. Gannister is used to produce silica brick for use in furnace linings.

gap phase replacement   Successional development in small areas that are disturbed (by the falling of a tree, for example) within a stable plant community; filling in of a space left by a disturbance, not necessarily by the species eliminated by the disturbance.

garden city   A term used in land use planning to denote a planned new town designed to include green space. The term that originated in the late 19th century in England with the work of Ebeneezer Howard.

garnet   A metamorphic ferromagnesian silicate mineral similar to olivine in structure; variable color is often brown to deep red; crystals may be of gemstone quality, with conchoidal fracture and glassy luster. See APPENDIX p. 623.

garnierite   A variety of the mineral serpentine forming pale green monoclinic crystals that may be of gemstone quality.

garrigue   A form of scrub vegetation common in Mediterranean regions such as Provence, characterized by open areas of bunchgrasses and areas of low shrubs. See MAQUIS.

gas   1) The state of matter between liquid and plasma as temperature increases. Molecules in a gaseous state are relatively far apart, resulting in gases having relatively low density. Gases expand to fill any container holding them. 2) A light (low molecular weight) form of oil, such as that used to fuel automobiles (the British term for such gas is petrol). 3) A derivative of natural gas. Compare LIQUID, SOLID, PLASMA.

gas chromatography   A chromatographic process in which a mixture of volatile (e.g., boiling points less than 400°C) components are passed through a column of adsorbent (called the *stationary phase*), using a carrier gas to sweep the molecules through the column. The gas is known as the *mobile phase*. The instrument is constructed such that the column is placed in a heated oven with adjustable temperature. Molecules that have different physical properties (structure, polarity, or boiling point) stick to the column for different periods of time, resulting in a separation of the components. See CHROMATOGRAPHY, CAPILLARY GAS CHROMATOGRAPHY, GAS-LIQUID CHROMATOGRAPHY, GAS-SOLID CHROMATOGRAPHY.

gas chromatography-mass spectrometry (GC-MS) An analytical technique used for environmental analysis of organic compounds. A mixture of compounds is passed through a gas chromatograph, where separation into individual components occurs. The separated components are then passed one at a time directly into a mass spectrometer, which produces a mass spectrum that is characteristic of each molecule in the mixture. Computer matching of mass spectra can be done in order to identify the components of the original mixture. See GAS-LIQUID CHROMATOGRAPHY, GAS-SOLID CHROMATOGRAPHY, MASS SPECTROMETER.

gas-cooled reactor   (GCR) A nuclear

reactor using a gas (such as carbon dioxide or helium) instead of a liquid as a coolant, and using slightly enriched fuel. See ADVANCED GAS-COOLED REACTOR, HIGH-TEMPERATURE GAS-COOLED REACTOR, MAGNOX REACTOR.

gaseous diffusion One method for enrichment of uranium (increasing the proportion of the most desirable isotope U-235). Uranium is converted to gaseous uranium hexafluoride. The lighter U-235–based gas travels slightly faster than the U-238–based gas and therefore diffuses faster through a permeable barrier, allowing it to be separated out. See ENRICHMENT.

gas field A large area of land underlain by a reservoir of commercially valuable gaseous fossil hydrocarbons.

gasification The conversion of a solid hydrocarbon such as coal or oil shale into a liquid or gas. The resulting synfuel can be transported through a pipe instead of by truck or train. See SYNFUELS.

gasifier An instrument for producing synfuels out of solids such as coal or oil shale. See GASIFICATION, SYNFUELS.

gas-liquid chromatography One form of gas chromatography in which the volatile compounds are separated out by being passed through a non-volatile liquid. See CHROMATOGRAPHY, GAS CHROMATOGRAPHY.

gasohol A mixture of gasoline and 10 to 20 percent ethanol (or sometimes methanol), used as a substitute for pure gasoline. Gasohol burns more cleanly than ordinary automobile fuel.

gas/oil ratio An estimate of oil reservoir composition expressed as cubic feet of gas per barrel of oil at 14.7 psi at 60°F.

gasoline A blend of different liquid fractions of petroleum used to fuel cars, trucks, and other internal combustion engines. See OCTANE NUMBER.

gas-solid chromatography One form of gas chromatography in which the volatile compounds are separated out by being passed through a long column filled with an absorbent solid such as charcoal. See CHROMATOGRAPHY, GAS CHROMATOGRAPHY.

gastric Of or relating to the stomach, as in gastric juice.

Gastropoda Class of molluscs with a body divided into a foot and visceral mass and a head which usually bears eyes and tentacles. In one major group of gastropods, snails and whelks, the visceral mass is turned approximately 180 degrees during development, producing a basic body assymetry. A single, assymetrically coiled shell is common to most species, but may be tubular or cup-shaped. In another gastropod group of slugs and their relatives, the body form is symmetrical and shells are either greatly reduced or missing. Gastropods have a worldwide distribution on land and in water. Many species are important crop pests and some function as intermediate hosts of human disease agents such as liver flukes. Approximately 35,000 species exist. See MOLLUSCA, NUDIBRANCHIA.

gastrula One of the earliest stages of an animal embryo, a structure that deveops from the blastula and shows the beginning of differentiation into layers. See ECTODERM, ENDODERM, MESODERM.

GATT Short for the international convention, General Agreement on Tariffs and Trade. It attempts to remove import restrictions and tariffs in order to facilitate international trade.

Gause's principle   The rule that two competing species will not exist in exactly the same niche in a stable environment; either one species will eliminate the other, or they will undergo some kind of differentiation of their realized niches to limit their similarity. See COMPETITIVE EXCLUSION PRINCIPLE.

gauss   A unit in the CGS system for magnetic field strength, equal to 10^{-4} tesla. The tesla is now the preferred unit. See CGS SYSTEM, TESLA.

Gaussian distribution   See NORMAL DISTRIBUTION.

Gaviiformes   The loons or divers of holarctic waters, which only come on land to nest and possess many primitive characteristics. They dive deep in pursuit of fish, propelled by muscular, rear-positioned feet and by expelling air. They can ride so low in the water that only their heads protrude. Four species exist. See AVES.

GC-MS   See GAS CHROMATOGRAPHY-MASS SPECTROMETRY.

gedanken experiment   A thought experiment, one that can only be carried out in the mind (as in some applications of the theory of relativity). *Gedanken* means thought in German.

gegenions   Literally, counter-ions (from the German). Ions produced by dissociation of a colloidal electrolyte. They have an opposite charge from the charge of the original colloidal ion.

Geiger counter   An instrument that measures ionizing radiation. The level of audible clicks increases as a Geiger counter moves into a region of higher radiation. (Geiger counters also detect cosmic rays, which are not related to radioactivity.)

gel   A solid material formed from a colloidal solution, a solution in which small particles remain suspended rather than actually dissolving. A gel has the consistency of jelly. Although gels may contain mostly water, they behave more like a solid than a liquid. Gels are used in electrophoresis. See ELECTROPHORESIS.

gelifluction   The downslope movement of saturated rock debris in the active layer of permafrost soils. Gelifluction may occur even on very gentle slopes. Compare SOLIFLUCTION.

gemma, gemmae   Budlike reproductive structures in some plants (and animals that reproduce by budding) that, when detached from the parent plant, can develop into new individuals. Liverworts, mosses, and some clubmosses and algae can reproduce asexually by means of gemmae.

GEMS   See GLOBAL ENVIRONMENTAL MONITORING SYSTEM.

gene   The biochemical structure that transfers inherited characteristics in an organism. A gene is a unit of DNA that occupies a specific place (locus) on an individual chromosome. Each gene controls a particular characteristic; it works by controlling the synthesis of one protein (often, an enzyme). The particular set of genes inherited by an organism determine what characteristics develop from a fertilized egg cell. Compare ALLELE, CHROMOSOME.

genealogy   1) The study of the descent or development of plants and animals from progenitors or older forms. 2) The study of family lineage in human populations.

gene bank   A collection of tissues (germ plasm) designed for long-term storage of genetic material. It may be a collection of cell cultures, frozen pollen, frozen sperm and eggs, or a collection of living plants.

Gene banks are tools for preservation of gene pools and resources for animal and crop breeding. See GENE POOL, GERM PLASM.

genecology   A branch of ecology that studies population genetics in relation to habitat, especially to find explanations for patterns of distribution over time or over particular regions. Genecology is especially important in plant ecology.

gene flow   The movement of genes within a population and from one population to another.

gene frequency   The relative commonness of a particular allele in a particular population; the number of times the form of the gene is found in the gene pool divided by the theoretical maximal number of times it could occur.

gene locus   The position on a segment of a chromosome where a particular allele (gene) is found.

gene pool   The total collection of genes within a population; each gene represents a number of different alleles. See GENE BANK.

genera   Plural of genus. See GENUS.

general fertility rate   An indication of the age structure of a population and the resulting fertility. For humans, it is calculated from the crude birth rate multiplied by the percentage of women between the ages of 15 and 45 years. See FERTILITY.

generalist   A species that has broad food or habitat preferences, or both, and that, as a result, can live in many different environments.

generation   1) The average length of time between the birth of a parent and the birth of its offspring. 2) A group of organisms having the same parents. 3) A single step or level in a direct line of descent, occupied by individuals within a species that share a common ancestor and are all the same number of broods away from that ancestor.

generation time   The time between the birth of a new individual to the birth of its first offspring, usually estimated at about 20 years for humans.

generic   1) Characterizing a genus; sometimes used for a broader group of organisms, such as a class. 2) Not protected by a registered trademark, as in generic drugs.

genet   A genetically unique plant, the individual that develops from a particular zygote plus any clones produced from that organism. Genet is used to describe genetically distinct individuals in clonal or modular organisms. Compare RAMET, CLONE.

genetic control   Pest control achieved either through breeding resistant plants or animals, or by introducing individuals carrying harmful genes (such as sterility) into the pest population.

genetic damage   Deleterious effects caused by the disruption of the chromosomes, or negative mutations. Genetic damage can be caused by exposure to radiation or hazardous chemicals, or from spontaneous mutations. See MUTAGEN, TERATOGEN.

genetic diversity   Variability in the genetic makeup among a group of individuals in a population. Also called genetic variability.

genetic drift   Random fluctuations in gene frequency occurring in an isolated population from generation to generation. Genetic drift is the result of chance combinations of different characteristics and is independent of the forces of selection; it shows up most clearly in small populations.

genetic engineering   Altering genes or genetic material to produce desirable new traits in organisms or to eliminate undesirable ones. It is accomplished primarily through gene-splicing, artificially transferring genes from one organism to a similar or entirely different organism. (It is sometimes also used to refer to controlled breeding.) Also called genetic manipulation or recombinant DNA technology. See RECOMBINATION, TRANSGENIC, VECTOR.

genetic feedback   A process of natural selection in which one species acts as the primary selection force on a second species and then the second species subsequently acts as a selection force on the first. Genetic feedback is responsible for the coevolution that occurs between predators and their prey as well as between hosts and parasites; it can sometimes lead to mutual accommodation between two such organisms, as in the myxoma virus and in rabbit populations. See COEVOLUTION, MYXOMATOSIS.

genetic load   The proportion of deleterious genes present in the gene pool of a population.

genetic polymorphism   A variety of characteristics within a population or species (such as different colors of eyes or fur) caused by different alleles for the same gene. See POLYMORPHISM.

genetic resistance   An organism's ability to withstand insect pests or diseases to a greater extent than others of the same species because of an inherent genetic strength. Through careful selective breeding, varieties of plants and animals have been developed that have much greater genetic resistance than their parents. New apple varieties such as Liberty, bred to produce sound fruit with little or no spraying, are outstanding examples.

genetics   The branch of biology that studies heredity, how variations are transferred to successive generations and how information contained within genetic material is expressed in individuals.

genic selection   Selection occurring at the individual gene level so that an allele is transmitted to more (or less) than 50 percent of the gametes of a heterozygous individual. Sometimes called genetic selection. See DARWINIAN THEORY, NATURAL SELECTION.

geniculate   Describing an animal or plant structure that is bent like a knee; geniculate antennae are those that bend sharply at an acute angle.

genital   Referring to reproductive organs or associated accessory structures such as ducts.

genital segment   The segment (somite) or fused segment pair in copepods where the the genital openings are located. Also galled genital somite.

genome   1) All of the genetic material in one organism. 2) The sum of all the DNA sequences in a single (haploid) set of chromosomes, and thus the set of genetic material inherited from either parent.

genotype   The genetic composition of an organism, the particular combination of alleles found in a given individual. Contrast PHENOTYPE.

gentrification   A process whereby previously poor urban areas with dilapidated houses are slowly improved through urban renovation, and the urban poor are replaced by residents who are wealthier.

genus   A group of similar or closely related species; a taxonomic grouping for organisms that ranks below family and above a species. The genus name is the first name in the scientific (Latin) name for a particular species. Genus names always start with a capital letter. Plural is genera. See SPECIES, FAMILY, CLASS, ORDER.

geo   A narrow inlet on a cliff-edged coastline.

geo-   A prefix indicating a relationship to the earth or the earth's processes.

geochemical cycle   A description of the continuous stages of change in which compositional elements are exchanged between the lithosphere, hydrosphere, biosphere, and atmosphere. Examples: carbon cycle, silica cycle.

geochemical dispersion, primary   The distributive movement of chemical elements below the earth's surface related to the formation of igneous and metamorphic rocks. Primary geochemical dispersion occurs by magmatic, metamorphic, and hydrothermal processes.

geochemical dispersion, secondary   The distributive movement of chemical elements at or near the earth's surface related to the formation of sedimentary rocks. Secondary geochemical dispersion occurs by processes of weathering, erosion, transport, and deposition.

geochemistry   The study of the patterns of abundance and distribution of chemical elements or their isotopes within the earth.

geochronology   The determination of time intervals relating to events in the earth's history. Geochronological determinations are made by either absolute or relative dating techniques.

geocline   Gradual changes that occur in a species over a geographical gradient, a series of gradual changes in the geographical environment such as increasing altitude. See ECOCLINE.

geode   A roughly spherical sedimentary rock containing a cavity lined with crystals of quartz or calcite. Geodes are valued for the beauty and the perfection of the crystals that grow within the geode cavity.

geodesy   The sequence of measuring the size and the shape of the earth's surface or its gravitational field.

geographic information system   (GIS) A computer mapping software system that links geographically referenced data with graphic map features. These systems allow for manipulation, analysis, and display of spatial information in either raster-matrix or vectorline formats.

geographic isolation   The separation of a population from other populations of a species and thus the inability of the population to migrate or interbreed with other populations, caused by barriers of geography such as rivers, mountain ranges, etc. The unique flora and fauna of the individual Galapagos Islands developed because of their geographic isolation.

geographic speciation   The evolution of a new species from a geographically isolated population. The variations that develop within the isolated population gradually make it increasingly different from interbreeding populations of the same species; eventually so many variations accumulate that the population is no longer considered part of the same species. Also called allopatric speciation.

geoid   A figurative description of the shape of the earth, as if the entire surface existed at sea level with no mountains, basins, or topographic features.

geologic age   See AGE, GEOLOGIC.

geologic column   A description of the array of rocks in vertical sequence at a particular location. The geologic column description typically includes age, composition, and thickness of the rocks in each unit of a corresponding map.

geologic cross section   A representation of the succession of rock units found in a particular area as seen in vertical profile. The end points of a geologic cross section usually correspond to the end points of a straight line drawn on a geologic map.

geologic cycle   Any repetitive, large-scale earth process. See GEOCHEMICAL CYCLE, CYCLE OF EROSION, WILSON CYCLE.

geologic map   A two-dimensional representation of the areal extent of geologic map units. Each map unit may represent characteristics such as age, composition, structure, or history of a rock type.

geologic record   The collective history of the earth as interpreted through the study of rocks.

geologic time   The intervals of the earth's history from the formation of the earth through the beginning of written history.

geology   The science of the structure, process, composition, character, and history of the earth.

geomagnetic   A term referring to the magnetism of the earth, or the expression of earth's magnetism as preserved in rocks.

geomagnetic field   The description of the intensity and distribution pattern of the earth's magnetic forces.

geomagnetic storm   A global disturbance in the earth's magnetic field that is more

dramatic than daily patterns of change.

geomagnetism   The magnetic qualities of the earth and its atmosphere. Also, the science devoted to studying earth's magnetism.

geometric growth   Growth (or rate of increase) with a constant ratio between successive quantities.

geometric rate of increase   See GEOMETRIC GROWTH.

geomorphology   The study of the forms, characteristics, and processes related to the landforms on the earth.

geophyte   An herbaceous plant whose overwintering buds are located in the soil, as on bulbs or corms. The term comes from the Raunkiaer life form classification system. See CRYPTOPHYTE, HYDROPHYTE, HELIOPHYTE, RAUNKIAER'S LIFE FORMS.

geopressurized natural gas   Naturally occurring, gaseous fossil hydrocarbons that are confined in bedrock reservoirs under pressures significantly higher than those of the atmospheric norm.

geostrophic flow   A fluid movement of the oceans or atmosphere in which the Coriolis force is exactly balanced by a horizontal pressure force so that the current flows parallel to the pressure gradient. The Gulf Stream is an example of a geostrophic flow.

geostrophic wind   An air current produced by a geostrophic flow.

geosynchronous   Moving with earth. A satellite above the earth's equator with a geosynchronous orbit remains above the same spot on earth at all times because its orbital period is 24 hours.

geosyncline   Any part of the earth's crust that is downwarped over time and contains a thick sequence of sediments. The term has been replaced in modern usage by the terms eugeocline and miogeocline.

geotaxis   The traveling of an organism or a cell toward or away from the center of the earth; the tendency to move in response to the force of gravity. Compare GEOTROPISM.

geothermal   A term relating to the internal heat of the earth.

geothermal energy   A utilitarian energy source generated by the heat of the earth.

geothermal gradient   The increase in heat potential corresponding to increased depth within the earth.

geotropism   The movement of a plant (or an attached animal, one not capable of traveling) in response to gravity. The tendency for plants to grow upwards, opposite the force of gravity, is one form of geotropism. See CHEMOTROPISM, PHOTOTROPISM, TROPISM.

Geraniaceae   The geranium family of the Angiospermophyta, containing dicot herbs that are usually 5-merous. Flowers usually have five petals, five sepals, five, 10, or 15 stamens. The five carpels are joined around an elongated axis, giving rise to the common family name, cranesbill. Geraniums (*Geranium*, *Pelargonium*) and filaree (*Erodium*) are members of this family. See ANGIOSPERMOPHYTA.

germ   1) A microscopic organism, especially a bacterium that causes disease. 2) Any structure that will develop into a new individual, such as a fertilized egg or spore. 3) The embryo of a cereal grain, as in wheat germ.

germ cell   Another term for gamete. See GAMETE.

germicide   Any substance that destroys microscopic organisms. Disinfectants and fungicides are germicides. See DISINFECT, FUMIGANT, FUNGICIDE.

germination   The process of beginning to grow or develop, especially after a dormant period, as in a spore or sprouting seed.

germ plasm   Protoplasm containing genetic material (seeds, plant parts, animal embryos, eggs, or sperm), preserved for possible future use, as for plant or animal breeding, genetic engineering, treatment of human or preservation of endangered species. Seed banks, sperm banks, and gene banks are all repositories for such tissues.

gestation   Pregnancy in mammals; the development of offspring in the uterus from conception to birth.

geyser   A forceful eruption of steam and water powered by the geothermal heating of a natural aquifer.

Ghyben-Herzberg principle   A quantification of the relationship between a freshwater lens and the underlying sea water within oceanic islands. The Ghyben-Herzberg principle is used to calculate the thickness of the freshwater lens, d, given by the equation $d = P_w \div (P_m - P_w)$ where P_w = groundwater density and P_m = sea water density.

giant panda   A large, bearlike mammal from eastern China with distinctive black-and-white markings. It lives in forests of bamboo, which provide the main source of its diet. A wrist bone of the forepaw is enlarged into a pseudo-thumb for grasping food. Once thought to have affinities with the raccoon family, pandas are now usually considered members of the Ursidae, or bear, family. Unfortunately, this endangered species probably numbers less than a thousand individuals in the wild and does not breed well in captivity. See CARNI-VORE.

gibberellins A group of plant hormones that are usually synthesized in leaves. They promote growth through stem elongation; they also promote seed germination and initiate growth in dormant buds. Compare AUXIN, CYTOKININS, KINETIN, ZEATIN.

Gibbs free energy See FREE ENERGY. Compare FREE ENERGY OF MIXING.

Gibbs-Helmholtz equation An equation used in thermodynamics to determine the useful work available in a reversible process occurring at a constant temperature. It is used in conjunction with voltaic cells.

gibbsite A mineral of aluminum hydroxide formed by the alteration of laterite or bauxite soils.

giga- (G) Prefix in the Système International d'Unités (SI) indicating one billion, 1×10^9. One gigawatt equals one billion (10^{-9}) watts.

gigawatt (gw) A unit for measuring large quantities of electrical power. One gigawatt equals one billion (10^{-9}) watts.

gill 1) The structure in fish, aquatic invertebrates, and amphibian larvae that enables them to breathe in water. It consists of membranes across which gas exchange occurs, allowing uptake of oxygen and release of carbon dioxide. 2) One of the thin, radiating sheets holding spores underneath the cap of mushrooms.

Ginkgophyta A phylum of gymnosperms that contains a single species *Ginkgo biloba*. The ginkgo, or maidenhair tree, is a deciduous, broadleaved, dioecious species, the females bearing rancid-smelling seeds

that are covered with a fleshy covering that arises from the outer layer of the ovule rather than from a ripened ovary wall. During the Mesozoic Era members of this phylum were far more numerous than they are at present.

girdle The removal of bark and the underlying living tissues from a ring around the stem or trunk of a tree or shrub. Girdling kills trees by destroying the tissue that transports photosynthesized food compounds down to the roots. It is a technique used frequently by foresters to kill unwanted trees in culling. Also called ring barking. See CULL.

GIS See GEOGRAPHIC INFORMATION SYSTEM.

gizzard 1) The second stomach of a bird, receiving food from the crop and grinding it up by mixing it with bits of sand or gravel; The gizzard is technically called the proventriculus. 2) A similar organ located below the crop in earthworms and insects.

glabrous Smooth-skinned; describing animals lacking hair or down, or smooth plant surfaces lacking hairs or bristles.

glacial Relating or pertaining to the features, time, processes, or effects of a glacier.

glacial abrasion The mechanical erosion of bedrock through the action of rock tools embedded in the ice of a glacier.

glacial drift A general term for any kind of rock material deposited by the action of a glacier. Particles of glacial drift may range in size from fine clay and silt to large boulders.

glacial marine sediment The clay-rich sediment of glacial origin deposited on continental shelves or the deep ocean floor. This fine-crushed rock sediment is usually associated with meltwater but it may also contain rock fragments dropped from melting icebergs.

glacial striations   The distinct grooves or scratches carved in bedrock as glacial movement drags rock fragments over the land. Glacial striations may serve as indicators of the direction of glacial movement. Compare GLACIAL ABRASION.

glacial surge   A rapid and temporary increase in the velocity of glacial ice that is often due to increased basal slip through fluid pressure with freezing and thawing of meltwater. See BASAL SLIP (GLACIERS).

glacial till   The unstratified and unsorted glacial debris deposited directly by a glacier.

glaciation   1) The formation of a glaciers with attendant effects on the landscape. 2) The onset of an ice age.

glacier   A land-bound mass of moving glacial ice formed by the accumulation, compaction, and recrystallization of snow. Glaciers move across the landscape under their own weight by plastic flow and sliding. See CONTINENTAL GLACIER, ALPINE GLACIER.

gland   Any of various structures at or near the surface of plants, or one-cellular to multicellular organs in animals, that secretes a specific substance. Examples include endocrine glands, exocrine glands, and secretory cells in animals, as well as water glands and nectaries in plants.

glassy texture   A description of a rock or mineral that has a smooth and shiny appearance like that of glass.

glauconite   A dark green clay mineral formed on continental shelf environments where sedimentation rates are low.

Gleason's index   An index of similarity

(coefficient of community) that is a modification of the Jaccard index. It uses some measure of dominance or importance of species common to both communities and of unique species rather than just the number of species. See JACCARD INDEX.

gleization   The process that creates a gley condition within a soil.

gley   A mottled gray-and-brown condition forming in saturated soils by anaerobic reduction of iron-bearing minerals.

gley soil   A soil that contains mottles formed by gleization.

glide (transitional slide)   See TRANSLATION (SLAB) LANDSLIDE.

Global 2000 Report to the President   Report requested by President Carter that assessed future world conditions by the year 2000.

Global Environmental Monitoring System (GEMS) An organization established to collect global data on trends or changes in climate, renewable resources, and oceans, and to monitor long-distance movement of pollutants and human health. It was created as part of Earthwatch by the United Nations Environment Programme. See EARTHWATCH.

global positioning system   (GPS) A network of satellites developed by the U.S. Department of Defense, and now available to the general public, as surveying and navigational aids. By using a receiver that simultaneously tracks the signals emitted by several of the satellites, the position of a location on the earth's surface can be determined to within several meters or even centimeters.

globigerina ooze   A widespread type of calcareous marine sediment containing a composition of at least 30 percent plank-tonic Foraminifera remains. The primary Foraminifera constituent in cooler waters is *Globigerina pachyderma.*

Gloger's rule   The general principle that as climate gradually becomes cooler from the equator toward polar regions, endothermic animals become paler due to reduced pigmentation.

glory   Series of rainbow-colored rings seen on the surface of clouds. It most often appears around the shadow of an airplane on underlying clouds, but can appear as a halo surrounding an observer's head if the shadow of the head falls on a fog bank.

glowing avalanche (ash flow)   See NUÉES ARDENTES.

glucose   The most commonly occurring simple sugar in plant and animal tissues. It has the same formula as fructose—$C_6H_{12}O_6$—but its atoms have a different spatial arrangement (it is an isomer of fructose). It is also called dextrose. Starch and cellulose are polymers of glucose.

glume   One or two stiff, modified leaves found at the base of the whole spikelet of grasses.

glutamic acid $HOOC-CH_2CH_2-CH(NH_2)COOH$, an amino acid found in many plant and animal proteins. It is the only amino acid metabolized by the brain. Monosodium glutamate is manufactured from this amino acid.

glutamine $O=C(NH_2)CH_2CH_2-C(NH_2)H-COOH$, an amino acid found in the juices of many plants and necessary for breaking down proteins in animals.

glycerides   Esters containing glycerol and fatty acids, also known as glycerin esters or glycerol esters. These are fats and oils and include monoglycerides, diglycerides, and

triglycerides. Triglycerides are blood fats that, like cholesterol, are implicated in heart disease.

glycine $H_2C(NH_2)COOH$, the simplest amino acid. It is used by the body to form purine, which in turn is used to form the basic components of DNA. It is also present in many plant proteins.

glycogen $(C_6H_{10}O_5)_n$, a polysaccharide resembling starch. It is a polymer of glucose, used by the body to store energy; the body stores it in the liver and muscle tissue metabolizes it into glucose when energy is needed. It is found in animals, bacteria, and blue-green algae but not in plants.

glycolysis   The biochemical process by which glucose from carbohydrates is converted into pyruvate, releasing energy in the form of ATP. Glycolysis precedes the tricarboxylic acid (Krebs) cycle in aerobic respiration.

glycophyte   Plant that only grows on soils containing little salt (sodium chloride, or other sodium salt). Most plants are glycophytes because few have adapted to tolerate salt. Contrast HALOPHYTE.

glycoside   Any of a large group of organic compounds that contain simple sugars (monosaccharides). They are present in (and derived from) plants; when broken down they yield a simple sugar plus another compound.

glyphosate $HO-CO-CH_2NH-CH_2-P(OH)_2O$, a common broad-action, contact herbicide, approved for use by home gardeners. It is more commonly known by its trade names Roundup and Kleenup.

GMT   See GREENWICH MEAN TIME.

gneiss   A general term for metamorphic

rocks that have a coarse banding pattern resulting from the separation of light and dark minerals during regional metamorphism. Varieties of gneiss are named for constituent minerals; an example is hornblende gneiss.

Gnetophyta A gymnosperm phylum that consists of diverse cone-bearing plants, mostly growing in desert environments. The male, staminate cones typically have anthers, whereas the female, ovulate cones are complex with bracts that cover the ovules. This phylum lacks the integrity of other phyla, and appears to have characteristics closest to angiosperms. The phylum includes *Ephedra*, Mormon tea; *Gnetum*, a tropical genus; and *Welwitschia*, a desert plant of southern Africa.

gnotobiotic Describing a controlled laboratory environment for raising organisms. A gnotobiotic environment is initially free of germs and any microorganisms needed for study are then introduced, avoiding side effects from unknown contaminants. Gnotobiotic conditions are necessary for studying some plant and animal diseases.

GNP See GROSS NATIONAL PRODUCT.

goethite A massive, earthy-textured form of hydrous iron oxide usually occurring in association with limonite or hematite.

gold (Au) A shiny, soft, heavy metallic element with atomic number 79 and atomic weight 197.0. It is a precious metal, valued because it is so easily shaped and so resistant to corrosion or other alteration. It is used in electronics and dentistry as well as for jewelry. See APPENDIX p. 623.

gonad Type of organ in the male and female where reproductive cells develop; a sex gland such as an ovary or testicle.

gonadotropic hormones Animal growth substances in mammals that stimulate the reproductive activity of the gonads (sex glands). In mammals, these compounds are produced in the pituitary glands and include the female follicle-stimulating hormone (FSH) and the male luteotrophic hormone (LTH). Also called gonadotrophins. See FOLLICLE-STIMULATING HORMONE, PROLACTIN.

gonangium A modified polyp in hydroid coelenterates from which medusae buds develop. See HYDROZOA.

Gondwanaland A presumed supercontinent of the Southern Hemisphere during the Mesozoic era. Gondwanaland formed from the southern half of Pangaea during the late Paleozoic era. Gondwanaland broke up during the Mesozoic era and forms landmasses of Africa, Antarctica, Australia, India, Madagascar, New Zealand, and Sri Lanka.

gorge A deep narrow ravine or canyon with steep or vertical sides.

gossan A deposit of hydrated iron oxide formed by the leaching of sulfur and metals from the upper layers of a sulfide deposit.

Gouy-Chapman model A model for electrical conduction in an ionized medium that describes how electrical charge diminishes with distance from the electrode, depending on the charge of individual ions and the total ionic charge of the solution. This model ignores forces other than electrostatic interactions that influence molecular adsorption, ignores the limits of an ion's size, and ignores the changes in dielectric constant created by the higher ion concentrations near the electrode.

GPS See GLOBAL POSITIONING SYSTEM.

graben   A fault block that has subsided vertically between two high-angle normal faults. A graben may be located between two horsts. See HORST.

grab sampling   1) A technique of using a remote-controlled device to collect geologic material from a hazardous environment. For example, a research submarine may rely on grab sampling to collect material from the deep-sea floor. 2) The collection of a wastewater treatment sample that represents conditions at a certain place and time in the treatment process.

graded aggregates   An accumulation of clastic particles that has the characteristics of graded bedding.

graded bedding   A sequence of sedimentary layers in which the average particle size is finest at the top and coarsest at the bottom.

graded layer   A well-sorted sedimentary layer that has the finest particles at the top and the coarsest particles at the bottom.

graded slope   A theoretical description of a hillside in a condition of dynamic stability; the addition of material to the slope is balanced by loss of material from the slope.

graded stream   A theoretical description of a stream in which there is a rough balance between the load imposed and the work done. A condition of dynamic equilibrium with little or no net aggradation or degredation of the channel.

grade (metamorphic)   An analysis of the intensity of metamorphic alteration of a rock based on the constituent minerals. For example, a rock that has been exposed to temperatures and pressures sufficiently high for garnet formation is said to have reached the garnet grade.

gradient   1) A gradual change in any quantity or system through space. A temperature gradient is a change in degrees over a physical distance; a moisture gradient is a progressively increasing or decreasing difference in the water content of a system such as the soil. 2) The slope of a hill; the degree of vertical rise or fall over a horizontal distance.

gradient analysis   The analysis of species composition of a plant community along a changing array of environmental conditions. In gradient analysis, the different species are plotted on a graph, with an environmental gradient (such as increasing altitude) along one axis.

gradient, geothermal   See GEOTHERMAL GRADIENT.

gradient wind   A curving pattern of air movement caused by the interaction between the Coriolis "force," pressure gradient force (air movement caused by differing air pressures), and centrifugal "force." The gradient wind causes a cyclonic curvature to air flowing past a low-pressure center (counterclockwise in the Northern Hemisphere); around a high-pressure center it causes an anticyclonic flow (clockwise in the Northern Hemisphere). See CORIOLIS EFFECT, CENTRIFUGAL FORCE, CYCLONE, GEOSTROPHIC FLOW.

graft   To transfer tissue from one organism to another, or from one part of an individual to another. Skin grafts are used on humans to treat severe burns. Buds or shoots of plants (scions) are often grafted into a slit of another kind of plant (rootstock), eventually growing as a single plant but having greater vigor or different qualities from either parent. Most dwarf fruit trees are produced by grafting a scion of the desired fruit variety onto a genetically short rootstock.

graft hybrid   A type of plant chimera that sometimes forms when one plant is grafted onto a plant of another species. In a graft hybrid, an intermediate plant with characteristics of both parents results, rather than maintaining the form of either plant. See CHIMERA.

grain   1) Common term for a caryopsis, the fruit of grasses. 2) The characteristic small-scale pattern seen in finished wood (such as furniture) from different kinds of trees. Oak has a much coarser grain than maple. 3) A unit of mass equal to 0.065 grams, the smallest unit in the avoirdupois, troy, and apothecaries' systems of weights. Originally based on the weight of a grain of wheat, it is now rarely used except by pharmacists. See CARYOPSIS. 4) A small fragment or particle in a sediment or a rock. 5) A characterization of the crystalline fabric of a rock.

gram   Unit for mass in the centigrade-gram-second (CGS) system, a variant of the metric system. One gram is one-thousandth of a standard kilogram, or approximately the mass of one cubic centimeter of water at 4°C. See CGS SYSTEM, MASS, KILOGRAM, METRIC SYSTEM.

Gramineae   The grass family. See POACEAE.

gram-negative   See GRAM'S REACTION.

gram-positive   See GRAM'S REACTION.

Gram's reaction   A method for classifying bacteria based on whether or not they retain a violet color when stained with gentian violet (or crystal violet) and Gram's iodine solution. Gram-positive bacteria are those that retain the violet color, and include the disease organisms causing diphtheria, scarlet fever, botulism, and tetanus. Gram-negative bacteria do not hold color after the treatment, and include those organisms causing cholera, gonorrhea, typhoid, and whooping cough.

grana   Structures within the chloroplasts of plant cells consisting of stacks of thin disks (thylakoids) containing chlorophyll and the enzymes needed for photosynthesis. Forty to sixty grana may be present in a single chloroplast. Singular is granum. See STROMA.

granite   A general term for light-colored (felsic) igneous rocks with phaneritic texture. Quartz, feldspar, and biotite are the predominant minerals. This common rock type is often used in construction.

granite-greenstone terrane   An accreted terrane consisting of a greenstone belt.

granitization   The process by which crustal rocks are converted to granite by the introduction or the removal of minerals in a fluid solution. See METASOMA.

granivore   An animal whose diet consists mainly of grain or seeds. Compare HERBIVORE.

granodiorite   An igneous rock with phaneritic texture and a composition intermediate between granite and diorite.

granofels   A nonfoliated metamorphic rock containing uniformly sized quartz and feldspar crystals and only small amounts of mafic minerals such as pyroxene and garnet. Compare HORNFELS.

granophyre   A granitic rock with a porphyritic texture. The groundmass of granophyre is composed of fine-textured, interlocking quartz and feldspar crystals.

granular   A textural description applied to rocks with visible mineral grains.

granulation   The reduction in crystal grain size that occurs during rock deformation.

granule   Any rock particle that is between 2 mm and 4 mm in size. Granules are larger than sand particles and smaller than pebbles.

granulite   A general term for coarse-grained metamorphic rocks containing equiangular crystals of quartz and feldspar and variable amounts of minerals such as pyroxene and garnet. Granulite may be categorized as a fine-grained gneiss or as a gneissic granite.

graphite   A very soft gray or black mineral of pure carbon. Graphite occurs within metamorphosed carbon-rich sediments or in veins of intrusive rocks. Graphite is used as a lubricant. See APPENDIX p. 623.

GRAS   Acronym for Generally Regarded As Safe. The GRAS list of food additives includes those that have been found by repeated testing to be safe for human consumption.

grassland   A biome comprised mainly of indigenous grasses or grasslike plants, the predominant vegetation type in semi-arid temperate and tropical latitudes. Grassland is used to refer to semi-natural meadows and cultivated farm meadows as well as to natural prairies. See PRAIRIE, SAVANNA.

graupel   Meteorological term for pelletized snow. Some rain is produced by graupel melting on its way down through the atmosphere.

gravel   A clastic rock particle between 2 mm and 60 mm in average diameter.

gravid   Pregnant; carrying eggs or a fetus.

gravimeter   An instrument used either to measure variations in a gravitational field or to measure the specific gravity for a substance. The former consists of a variation of a balance scale.

gravimetric   1) Describing a procedure for chemical analysis that not only separates out the different constituents of a substance, but also measures the mass of each constituent of the substance. Gravimetric analysis provides information on the relative quantities of the different compounds of a substance as well as identifying what the compounds are. 2) Describing the variations in gravitational fields measured with a gravimeter. See GRAVIMETER.

gravitation   Another term for gravitational force. See GRAVITATIONAL FORCE.

gravitational acceleration   The rate of change of velocity experienced by an object in free fall (subjected to no other force other than gravity). Near earth's surface, this acceleration is approximately 9.8 meters per second squared.

gravitational force   One of the fundamental forces of the universe, along with the electromagnetic and weak forces (often unified into the electroweak force), and the strong force. It is the attractive force between all masses due to their mass. Whereas on a molecular level, the gravitational force is the weakest of the four, on the scale of the universe as a whole, it is the dominant force.

gravitational water   Water that is moving through the soil under the influence of gravity. A soil cannot become saturated until gravitational water has settled. See FIELD CAPACITY.

gravity   The effect of the gravitational force at the surface of a planet, especially the earth. It takes into account the rotation of the planet as well as the gravitational force. See GRAVITATIONAL FORCE.

gravity anomaly   Any variation in gravitational acceleration that is not accounted for in a predicted density model for a homogeneous earth.

gray (Gy) Unit for dose of ionizing radiation absorbed in the Système International d'Unités (SI). The gray is gradually replacing the older unit, the rad. 1 Gy = 1 joule/kilogram = 100 rad. Compare RAD, REM, REP.

gray-brown podzolic soils An older term describing eluviated soils that have a distinctive, clay-enriched B horizon. See ALFISOL.

graywacke A sandstone that contains more than 15% of clay minerals. Graywackes are usually a partly sorted mixture of quartz and feldspar grains, pebbles, clay, and carbonate muds and are collectively deposited by turbidity currents.

gray water A general term for domestic wastewater that does not contain sewage or fecal contamination. Effluent from washing clothes is an example of gray water.

grazer An herbivore that eats non-woody plant matter.

grazer-scraper Water-based animals that eat the organic layer of algae, microorganism, and dead organic matter found on stones and other objects on the floor of a body of water.

grazing The act of animals feeding on fresh grass and herbaceous plants. Contrast BROWSE.

grazing food chain Transfer of food energy in which live vegetation or algae is eaten by animals (herbivores) that in turn are eaten by carnivorous animals. The grazing food chain is one of two basic types of food chain, the other being the detritus food chain. See DETRITUS FOOD CHAIN.

great auk A 2-foot tall, flightless alcid *Pinguinus impennis* of the North Atlantic whose habits closely paralleled those of penguins of the Southern Hemisphere. Newfoundland populations may have migrated by sea as far south as Florida during the winter. Helpless on land, they were slaughtered by seamen and the last pair was killed in 1844. See ALCIDS.

green algae See CHLOROPHYTA, GAMOPHYTA.

green belt An area around a city designated for minimal development as part of a comprehensive land use planning scheme. Used especially in England.

greenfield site British term for agricultural land chosen for development. Because they are level, have good drainage, and do not need to be cleared of trees, agricultural fields are often attractive to developers of industrial or housing developments.

greenhouse A building constructed of clear material, formerly glass but now usually forms of plastic or fiberglass, for climate-controlled growing of plants.

greenhouse effect Heating of the earth's atmosphere that is loosely analogous to the glass of a greenhouse letting light in but not letting heat out. (Greenhouses are actually heated more by the reduction of convention than by this process.) Radiation from the sun easily enters the atmosphere as light waves. It heats the earth's surface, causing it—like any warm surface—to emit infrared radiation. Gases such as carbon dioxide absorb infrared radiation, preventing its energy from leaving the earth. Greenhouse effect is used to describe the theoretical rise in global temperatures that may be occurring from the great increases in global carbon dioxide caused by human activities such as combustion.

greenhouse gases Those gases in the

earth's atmosphere that contribute to the greenhouse effect. Carbon dioxide and water vapor are the primary greenhouse gases; methane and chlorofluorocarbons also contribute.

green manure   An agricultural technique for increasing the content of organic matter in soil. Rapid-growing plants, such as buckwheat, alfalfa, and winter rye are planted, allowed to grow, and plowed under while still green. This adds large quantities of organic matter to the soil, which decomposes, improving soil structure, encouraging beneficial microorganisms, and increasing fertilizer-holding capacity in much the same way that adding well-rotted animal manures does.

Greenpeace   Greenpeace U.S.A. Inc., a citizen organization dedicated to environmental protection with emphasis on the protection of sea mammals and other endangered species, and nuclear and toxic pollution.

green revolution   Development of new varieties of domestic animals and crop plants such as wheat and rice, combined with intensive use of fertilizers, mechanization, and management techniques to greatly increase harvests and food production, especially in tropical climates. Originally referred to breeding high-yielding plants, but now used more generally to refer to all aspects of agriculture, including animals. Compare AGRICULTURAL REVOLUTION.

Greens   A relatively new term used for political parties with strong environmental platforms.

greenschists   Greenish, chlorite-bearing schistose rocks formed by the regional metamorphism of mafic igneous rocks. Greenschists are associated with orogenic belts of converging tectonic plates.

greenstone belts   A region of mostly Archean age rocks that represent ancient back-arc basins that have been intruded by granitic rocks and regionally metamorphosed in a tectonic convergence.

Greenwich mean time   (GMT) Local time along the meridian that passes through Greenwich, England (the "prime meridian"). It is now derived from International Atomic Time. It is often called Coordinated Universal Time, a name chosen in 1928 to distinguish it from the earlier form of Greenwich mean time, whose day began at noon rather than midnight. See COORDINATED UNIVERSAL TIME.

grid   1) A network of evenly spaced lines or squares, such as those used on maps to locate positions. 2) The network of electrical lines that delivers electricity from a power plant to users such as individual homes and businesses. 3) When capitalized, the acronym for the Global Resource Information Database of the Earthwatch Programme. See EARTHWATCH PROGRAMME.

Grignard reagents   A group of compounds containing an organic (carbon) group and a halogen plus magnesium. They are essential in a large number of organic synthesis reactions, including the production of secondary and tertiary alcohols and alkyl and aryl derivatives of halogen compounds.

grike   A tapering vertical rock joint that has been enlarged by solution weathering in limestone. Compare CLINT.

groin   A breakwater structure that extends seaward at a right angle to the shoreline. Groins are designed to inhibit the longshore drift of sediments.

gross   1) An overall total that includes no deductions. Gross income is the total amount of money coming in; when expenses are subtracted from this value, it is called net income. Gross primary production is the total photosynthesis of a biotic community; subtracting the energy consumed by respiration changes the gross to net primary production. 2) Twelve dozen (144) items.

gross national product   (GNP) The total market value in current dollars of all goods and services produced by one nation's economy for final use during a year.

gross primary production   (GPP) 1) The total photosynthesis of a biotic community. 2) The amount of energy stored through photosynthesis plus that consumed through animal and plant respiration in a specific community or ecological system. Generally expressed as kilocalories per square meter per unit of time, or dry organic matter in grams per square meter per unit of time. Compare NET PRIMARY PRODUCTION.

gross production   See GROSS PRIMARY PRODUCTION.

ground cover   Any grass, legume, or other plant that is grown in order to keep soil from being blown or washed away. Ornamental ground covers such as pachysandra and vinca are often used to replace lawn grass, either to reduce mowing or to cover ground in areas where grasses do not grow well.

ground fire   A fire in which organic matter underneath the surface of the soil (such as peat) burns. Compare CROWN FIRE, SURFACE FIRE.

ground ice   Ice that forms in the interstitial spaces of soils. Ground ice is responsible for frost heaving.

ground layer   The lower levels of vegetation in a forest community; the low-growing herbaceous plants, shrubs, and woody vines growing beneath the trees (in the understory). Tree seedlings and saplings are not included as part of the ground layer. Compare CANOPY, UNDERSTORY. See HERB LAYER.

ground-level event   (GLE) Abrupt, dramatic increase in levels of cosmic rays reaching the earth's surface. When these relatively rare events occur, they follow flare eruptions. See COSMIC RAY.

ground moraine   An extensive, blanketlike layer of till deposited across the landscape by a melting ice sheet.

ground state   The state of lowest energy of a system, usually used in reference to an individual atom, molecule, or ion. Also called ground level. Contrast EXCITATION.

groundwater   All water that is contained in the pore spaces of rock and soil below the elevation of the water table.

groundwater storage   A volumetric measurement of groundwater in an aquifer.

group selection   A controversial concept that natural selection may operate at the level of an entire group as well as for individuals, as in social insects that have elaborate caste systems or some bird species that appear to reproduce at less than maximum rates (benefitting the population as a whole over individual birds). See NATURAL SELECTION, SOCIOBIOLOGY, SPECIES SELECTION.

growth   An irreversible change in an organism, an increase in volume and weight, caused by cell division, cell enlargement, or cell differentiation and utilization of nutrients. Growth can also refer to an increase in the size (population) of a colony of microorganisms, especially when cultured.

growth rate   1) The speed at which a plant, animal, or colony of microorganisms increases in size. 2) The increase (or decrease) of a population over a period of time. In humans, the rate of change is calculated from the number of births minus the number of deaths per thousand in a population, with net migration added; it is expressed as percentage.

growth rings   1) Concentric growth layers seen in a cross-section of the stems and trunk of a tree or shrub. These reveal the tree's age, and sometimes the environmental conditions during the tree's life span, because a each ring represents one year's growth. Also called annual rings, but that term is misleading because sometimes trees show missing rings or extra rings depending on environmental conditions. 2) Similar structures (annuli) on some species of animals that repeatedly form ringlike growths. See DENDROCHRONOLOGY.

grubs   Common name for insect larvae generally belonging to the order of beetles, Coleoptera. See COLEOPTERA.

Gruidae   A family of Gruiformes, cranes inhabitat or marshland and open grasslands. They have long bills, necks, and legs and display bare patches or plumes on the head. The windpipe, which is elongated and convoluted, gives the cranes their loud, trumpetlike voices. They tend to do elaborate "dances" during courtship. The North American whooping crane *Grus americana* has been the object of a highly publicized campaign of species preservation, which is beginning to show modest signs of success. See GRUIFORMES.

Gruiformes   A diverse and ancient order of birds, many of which inhabit wetlands.

They tend toward compact bodies, long necks and oblate wings (e.g., cranes, rails, coots, sun bitterns, bustards). See AVES, GRUIDAE.

grumosols   An older soil classification term for vertisols.

guanine $C_5H_5N_5O$, 2-amino-6-oxypurine, one of the primary components of DNA and RNA; it pairs with cytosine. Guanine belongs to a group of compounds called purine bases. See DNA, RNA, PURINE BASE.

guano   Bat or bird fees, sometimes collected and used as organic fertilizer because of its high nitrogen and phosphate content.

guard cell   Kidney-shaped cells found in pairs surrounding the stomata, pores in the outer layer of cells on plants. The guard cells open and close the stomata, regulating gas exchange; they swell open when water enters them and close when they become flaccid due to water loss. See STOMATA.

guild   Group of the same species of plants or animals living in the same type of environment and sharing a similar form or appearance. Compare SYNUSIA.

gulf   1) A large portion of an ocean that is mostly enclosed by landmasses, such as the Gulf of Mexico. 2) A deep stream valley with steep, rugged sides.

Gulf Stream   The major warm ocean current of the western Atlantic Ocean. The Gulf Stream flows north from the eastern coast of Florida to the western coast of Europe. The Gulf Stream is part of the large gyre that spirals in a clockwise flow around the Atlantic Ocean.

gully   A preferred drainage channel eroded so deeply in the soil that cultivation does not restore the original surface contours.

gully erosion   The enlargement of rills to

form deeper channels in the soil as surface water carries away soil particles.

gunpowder A mixture of potassium nitrate, charcoal, and sulfur used as an explosive to propel ammunition in guns and artillery, also for blasting.

gust Sudden burst of wind, of shorter duration than a squall.

Gutenberg discontinuity A boundary surface between the core of the earth and its mantle and located at a depth of approximately 2600 km. Seismic waves traveling through the Gutenberg discontinuity experience a rapid change in velocity that is attributed to the differing densities of the core and mantle.

guttation The extrusion of water droplets from healthy plants, usually from the leaf. Guttation occurs most often when nighttime humidity is high and may be restricted to structures at the edges called hydathodes. See HYDATHODE.

guyot A submarine seamount or volcano at a depth of 1000 to 2000 m that has a distinctive shape like a plateau. The flat-topped form is attributed to erosion that occurred when the guyot stood at sea level.

gw Short for gigawatt, one billion (10^{-9}) watts.

Gy See GRAY.

gymnophiona See AMPHIBIA.

gymnosperm Plants that bear "naked seeds," that is, the seeds that contain the embryos that give rise to the sporophyte generation are produced by the pollination of ovules that are not enclosed in a ripened ovary wall (fruit), but may be associated with cones or fleshy arils. This group includes four phyla of the plant kingdom. See CONIFEROPHYTA, CYCADOPHYTA, GINKGOPHYTA, GNETOPHYTA.

gynandromorphism An abnormal condition in which an animal (such as an insect) exhibits both female and male characteristics; it may or may not possess functional sexual organs of both sexes. See HERMAPHRODITE.

gynoecium The carpel or cluster of carpels (female parts) of a flower, also called pistil. Compare ANDROECIUM.

gypsum A rock or mineral composed of calcium sulfate and water. Gypsum rock originates either from the evaporation of water from a saline lake or playa or from the hydration of anhydrite. See APPENDIX p. 624.

gyre A large-scale circular ocean current that is generated principally by the force of surface winds. A gyre exists and is centered at approximately 30° latitude in each major ocean basin of the Northern Hemisphere and Southern Hemisphere.

h

H' The symbol for equitability. See EQUITABILITY.

H Chemical symbol for the element hydrogen. See HYDROGEN.

H Abbreviation for henry, a unit of inductance. See HENRY.

ha See HECTARE.

Haber process A procedure discovered in 1908 for synthesizing ammonia. Under high pressures and temperatures, iron oxides catalyze a reaction combining nitrogen with hydrogen to produce ammonia. This process was used by Germany to produce explosives during World War I; it is still the most important method of nitrogen synthesis. Also called the Haber-Bosch process.

habit The usual mode of growth for a plant species, whether climbing, erect, prostate, etc.

habitat The place where an animal or plant normally lives or grows, usually characterized either by physical features or by dominant plants. Deserts, lakes, and forests are all habitats. Compare NICHE.

habitat breadth Distribution of a species among different types of environments, and thus the range of habitats that it can inhabit.

habitat diversity The range of habitats within a region.

habitat island An isolated type of habitat surrounded by a large area of differing habitats. An alpine tundra on top of a high, isolated mountain represents a habitat island.

habitat loss The destruction of living places for animal and plants, especially by human activity such as development. Filling wetlands and replacing a forest with a large parking lot for a shopping mall result in habitat loss.

habitat selection The capacity of an organism to choose a certain habitat in which to live from the range of habitats within a region.

habitat suitability index (HSI) A mathematical procedure for evaluating a particular habitat. A value (ranging from 0.0 to 1.0) is assigned on the basis of available food and cover requirements of a species. See HABITAT UNIT.

habitat unit (HU) A mathematical procedure for evaluating for habitats. It is derived by multiplying the habitat suitability index (HSI) by the number of acres available with that HSI. Habitat units are used for assigning the relative value of habitat to different species. See HABITAT SUITABILITY INDEX.

habituation A behavioral change in which animals can become accustomed to unnatural components in their environments. Deer feeding alongside highways may become habituated to traffic, or at least to traffic noise.

hackle 1) Any of the long, fine feathers on the neck of male domestic birds such as roosters or peacocks. 2) The hairs on the neck and back of an animal such as a dog that bristle when the animal is ready to fight.

hackly A description that characterizes the roughened or jagged surface of minerals which break with a brittle fracture. Broken cast iron has a hackly texture.

hadal zone The part of an ocean floor that lies in a deep trench below the general level of the abyssal zone.

hade The angular measure between a vertical axis and the trend of a mineral vein or a rock feature.

Hadley cell A pattern of global atmospheric circulation driven by convection. In each hemisphere's cell, air flows along the earth's surface from just beyond the tropics toward the equator. Near the equator the air rises and heads back toward the poles. At about 30° latitude, the air descends again.

hail Precipitation that falls as pellets of ice, usually somewhat rounded. Hail is associated with thunderstorms.

halarch succession The development of an ecological environment that originates in a saline environment. Also called a halosere. See HYDRARCH SUCCESSION, XERARCH SUCCESSION.

half-graben A fault block that has subsided vertically along a high-angle normal fault on just one side. The formation of a half-graben results in tilting of the fault block.

half-hardy Describing ornamental species of plants that can withstand some cold but not severe winters. See HARDINESS.

half-life A property used to characterize radioactive substances. The half-life is the period of time required for half of a quantity of a radioactive nuclide to undergo decay. Half-life is independent of conditions such as temperature and chemical form, and specific to each nuclide. See RADIOACTIVE DECAY, NUCLIDE.

half-value thickness Thickness of a given material that will reduce the intensity of a beam of radiation by one-half.

halide Compounds containing halogens (chlorine, bromine, fluorine, iodine) in ionic form (as Cl^-, Br^-, F^-, and I^-). See CHLORIDE, FLUORIDE.

haliplankton Plankton dwelling in the open ocean. See EPIPLANKTON, PLANKTON.

halite A mineral composed of sodium chloride and that forms perfect cubic crystals in evaporite environments. Halite is known as rock salt. See APPENDIX p. 624.

halo- A prefix indicating salinity.

halocline A zone of ocean water in which salinity increases with depth. A halocline is usually well developed in coastal waters that receive a significant influx from freshwater river systems.

halogenated hydrocarbon An organic compound containing a halogen (chlorine, bromine, fluorine, or iodine). Chlorinated hydrocarbon insecticides and CFCs are halogenated hydrocarbons. See CHLORINATED HYDROCARBONS, CHLOROFLUOROCARBON.

halogenation The process of adding a halogen to an organic compound. Halogenation is used to produce a wide variety of organic compounds, including chlorinated hydrocarbons such as DDT and PCBs.

halogens The group of nonmetallic elements consisting of chlorine, bromine, fluorine, iodine, and astatine (a radioactive element). All exist in elemental form as compounds containing two atoms: C_2 and F_2 are gases, Br_2 is a liquid, and I_2 and At_2 are solids. Their structure is one electron short of being stable; they therefore combine easily with most metals, abstracting an electron and forming salts.

halokinesis A general term for the formation and movement of salt domes.

halomorphic soil A term applied to soils that are strongly influenced by neutral or alkaline salts.

halophyte A plant that naturally grows in salty soils (as near the ocean) or in the mud of saltwater environments; one that has developed adaptations allowing it to tolerate high concentrations of sodium in soil and in ocean spray, such as sea oats or saltmarsh hay. Some halophytes require high salt concentrations around their roots whereas others are facultative. Contrast GLYCOPHYTE.

h-alpha The spectral line occurring at 6563 angstroms. It is part of the pattern of lines (called the Balmer series) produced when hydrogen is excited from a lower energy state to a higher energy state. H-alpha filters are used for viewing solar activity such as prominences. See ABSORPTION LINE.

hamada A desert region characterized by a surface topography of bedrock and boulders and few or no other surface deposits.

handling time The amount of time spent by a predator in hunting, capturing, and eating its prey and the time it takes to prepare for hunting additional prey.

hanging valley 1) A U-shaped tributary to the main valley of an alpine glacier. 2) A valley which is significantly higher than the valley it joins at the point of intersection.

hanging wall The planar rock surface of a fault block that lies above the fault plane Compare FOOTWALL.

haploid Having the same number of sets of chromosomes as a gamete (germ cell), half as many as in ordinary adults of the species (which have the diploid number of chromosomes). The haploid number is represented by n (as opposed to 2n for diploid). Contrast DIPLOID.

haplont A plant that has the haploid number of chromosomes (half the diploid number, usually one unpaired set) throughout its life and is only diploid as a zygote. See DIPLONT.

Haptophyta Members of this kingdom Protista phylum all have an unique filiform organelle, called a haptoneme, inserted between a pair of undulipodia. In each cell there are usually two saucer-shaped chloroplasts rich in fucoxanthin. These organisms have a resting coccolithophorid stage during which calcium carbonate crystalizes on their surfaces. The sedimentation of coccoliths has played an important role in the formation of many marine chalk deposits. Well-known members of this phylum include *Coccolithus* and *Emiliania*, both abundant in marine chalk deposits. See PROTISTA.

haptotropism See THIGMOTROPISM.

hardening 1) The process of cold acclimation in plants. 2) To intentionally acclimate plants raised indoors to outdoor temperature fluctuations and winds by gradually increasing their time outdoors.

hardiness The range in minimum winter temperatures that a plant tolerates. The United States Department of Agriculture publishes a hardiness zone map showing average temperature ranges for different regions of the country, to assist gardeners and farmers in choosing plants of the appropriate hardiness for their location. See APPENDIX p. 612.

hardness   1) A physical property of a mineral that is controlled by strength of bonding between atoms; describes the ability to resist scratching or abrasion. See APPENDIX p. 620–628. 2) A measurement of the content of dissolved calcium and magnesium in water. See HARD WATER.

hardpan   A nearly impermeable rocklike soil layer formed by cementation of soil grains with calcium carbonate or iron oxides; is difficult to dig or drill. See FRAGIPAN.

hard water   Groundwater containing dissolved mineral salts, usually calcium carbonate (or a combination of calcium and magnesium). Hard water does not lather well with soap, forms deposits (scale, $CaCO_3$ or $MgCO_3$) in teakettles and hot water tanks, and in more extreme cases may block plumbing by forming calcium deposits in water pipes. Treating with water softeners removes most of the calcium and magnesium by means of ion exchange. Typically, hard water is mixed with sodium chloride (common table salt); the sodium "softens" the water by replacing much of the calcium during the ion-exchange process. See ION EXCHANGE.

hardwood   A deciduous, broad-leaved tree (an angiosperm) such as oak or walnut characterized by dense, closely grained wood, as opposed to the coniferous softwoods such as pine. Also, the wood from such a tree (although the wood of some "hardwoods" such as sumac and balsa is considerably softer than that of some "softwoods" such as southern yellow pine). Contrast SOFTWOOD.

Hardy-Weinberg law   The principle stating that the relative frequency (ratios) of different genes and genotypes in a large population remains the same as long as mating occurs randomly, the environment is stable from one generation to the next, and no mutation, migration, or natural selection occurs. See HARDY-WEINBERG RATIO.

Hardy-Weinberg ratio   The ratio of different genotypes produced by random mating or random union of gametes. The resulting ratios for two alleles (alternative forms of genes at one location) are p^2:$2pq$:q^2, where p is the frequency of one allele and q is the frequency of the other. See HARDY-WEINBERG LAW.

harem   Several females associated with one male in a polygynous mating system. The harem is a social unit maintained by a male fending off other males and preventing them from gaining access to members of the harem.

harmattan   A dry, winter wind that blows from the Sahara across West Africa. It often carries dust and sand.

harmonic   A repeating vibration or wave whose frequency is a whole multiple of the fundamental frequency. Sound harmonics are also called overtones. According to Fourier's law, any wave or sound can be broken down into a series of harmonics and a fundamental frequency. See FUNDAMENTAL FREQUENCY.

harmonic tremor   An earthquake wave with a vibration frequency that is an integral multiple of the fundamental wave frequency.

harvest   1) A technique for measuring the productivity of an ecosystem by weighing the growth produced in a single season. 2) The time when crops are gathered, or the crops that are gathered.

haustorium   A specialized root or mycelium of a parasitic plant or fungus, which penetrates and absorbs nutrients from host tissue. The parasitic dodder plants have haustoria that enter the vascular systems of their host plants.

Hawaiian floral region   One of the phytogeographic regions into which the Paleotropical realm is divided according to similarities between plants, consisting solely of the Hawaiian islands. See PALEOTROPICAL REALM, PHYTOGEOGRAPHY.

Hawaiian-type volcano   See SHIELD VOLCANO.

hazard (natural hazard)   A natural phenomenon or process that poses a danger to human welfare.

hazardous chemicals   Any chemicals that are potentially dangerous to humans or to the environment. Hazardous chemicals include many substances that are not harmful if used correctly, such as paint thinners and many pesticides.

hazardous waste   Any by-product or refuse that is harmful to humans or the environment when improperly handled. Hazardous wastes are often products of industrial processes such as nuclear power production or chemical synthesis, but ordinary fertilizers can also be hazardous wastes because they can pollute groundwater sources with nitrates.

haze   A lack of clarity in the atmosphere caused by airborne particulates (pollution, dust) or a fine mist.

Hb   Abbreviation for hemoglobin. See HEMOGLOBIN.

HCH   See LINDANE.

He   Chemical symbol for the element helium. See HELIUM.

head (groundwater pressure)   A measure of the gravitational potential energy of water in an aquifer or impoundment. Head is usually stated as a linear unit of elevation, such as feet, that is equivalent to the height of a free-standing column of water, or the sum of pressure head and elevation head.

headland   1) A high part of a landscape overlooking a body of water. 2) Part of the landscape that extends out into a body of water such as a lake or ocean.

headstream   A stream that is in the uppermost regions of a watershed or catchment area.

headwater   The highest reaches of a stream in a drainage basin.

heart rot   Decay of the dense wood in the center of the trunk of a tree, caused by a fungus.

heartwood   The non-living, oldest woody tissues in the center of the trunk of a tree. Heartwood is denser and often darker than the newer wood that surrounds it because of the accumulation of resins and oils near the center of the trunk. Also called duramen. Compare SAPWOOD.

heat   The energy of random molecule motion and internal potential energy that moves between objects or substances due to a difference in temperature. Heat can be transmitted by conduction, convection, or radiation. See CONDUCTION, CONVECTION, RADIATION.

heat capacity   Ability to absorb heat, expressed as a ratio between heat absorbed and resulting temperature change. The units for heat capacity are usually joules or calories per degree C. See SPECIFIC HEAT.

heat conduction   Transmission of heat energy through molecular contact. Also called thermal conduction. Compare ELECTRICAL CONDUCTION.

heat energy   Another term for heat. See HEAT.

heat engine   A device converting heat into mechanical work. Heat engines include internal combustion engines and steam turbines.

heat exchanger   A device transferring heat from one fluid to another without mixing the two fluids. A heat exchanger can be used to remove excess heat from a system (as in a car radiator), or to transfer heat already in a system.

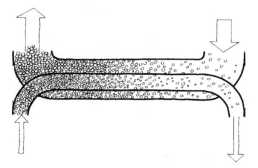

In a counterflow heat exchanger, two liquids move past each other in opposite directions. Heat is transferred from one tube to the other across the barrier.

heath   A vegetation type found on level areas with dry, sandy, or well-drained acid soil, characterized by low shrubs and no trees. In European countries, heaths are usually covered by heather, heath, or gorse; elsewhere other narrow-leaved species predominate. Heaths are members of the *Erica* species, often found in such areas and giving their name to the vegetation type. Compare MOOR.

heat island   A phenomenon in which the temperatures in an urbanized region are consistently higher than those of surround-ing areas. Urban haze can trap heat from pavement and tall buildings, in turn intensifying the haze and therefore pushing temperatures even higher. High pollution levels can build up unless the heat island is disrupted by high winds. Also called urban heat island.

heat of combustion   Amount of heat released when a specified amount of substance is completely burned. Heats of combustion can be expressed in joules/kg, Btu/lb, or cal/gm. See LATENT HEAT.

heat of formation   Amount of heat ab-sorbed or released by a chemical reaction that produces one mole of a specified compound, where the reactants at the beginning and the products at the end are at standard temperature and pressure.

heat of fusion   Quantity of heat required to bring about the change of phase from solid to liquid (fusion) of one gram of a sub-stance at its melting point, without raising the temperature of the substance.

heat of mixing   The enthalpy, heat ab-sorbed or released per mole of substance, when two substances are mixed but do not undergo a reaction to form a third com-pound. It is used to calculate the (Gibbs) free energy of mixing. See FREE ENERGY OF MIXING, HEAT OF SOLUTION.

heat of solution   Amount of heat absorbed or released when one mole of a substance is completely dissolved in a large volume of solvent.

heat of vaporization   Quantity of heat required to bring about the change of phase from liquid to gas (evaporation) of a given quantity of a substance at its boiling point, without raising the temperature of the substance. See HEAT OF FUSION.

heat pump   A device that, by compression or by absorption, raises low-temperature air or water to a higher temperature. The principle is that of a refrigeration machine used in reverse.

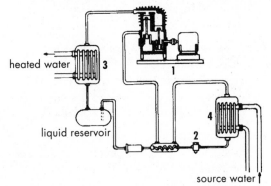

heated water   **3**

1

liquid reservoir

4

2

source water

In this heat pump, a refrigerant is circulated in a closed loop by the compressor (1). A pressure release valve (2) keeps it under pressure at the hot end of the condensor (3) and under reduced pressure at the cold end of the evaporator (4). When compressed, the temperature of the refrigerant is increased and it will liquify, giving off heat at the condensor (heat out). Passing through the pressure release valve, it rapidly evaporates and drops in temperature taking up heat from its environment (heat in).

(Courtesy Don Watson, *Designing and Building a Solar House*)

heat recovery   The capture and use of heat that would otherwise be lost as waste heat. For example, heat produced as a by-product of electrical power generation could be used to warm living space instead of being released to the atmosphere.

heat transfer   The conveyance of heat from one system or substance to another as a result of a difference in temperatures. See HEAT CONDUCTION, CONVECTION, RADIATION.

heavy liquids   Liquid organic compounds or solutions of salts with high densities. Heavy liquids are used in the extraction of minerals from ores; they are added to suspensions of minerals to adjust the specific gravity and separate out the suspension into light fractions that float and heavy fractions that sink. Carbon tetrachloride (CCl_4) and saturated solutions of calcium chloride ($CaCl_2$) are used as heavy liquids.

heavy metal   Metals with moderate to high atomic numbers such as lead, cadmium, silver, arsenic, chromium, and mercury and that are toxic in relatively low concentrations. They persist in the environment and can accumulate to levels that stunt plant growth and interfere with animal life. Mining and industrial wastes and sewage sludge are all sources of potentially harmful concentrations of heavy metals.

heavy oil   A viscous crude oil that has a relatively high boiling point.

heavy water   Water containing a large proportion of molecules having the deuterium isotope of hydrogen instead of ordinary hydrogen (written as D_2O or HDO). Such molecules are found in very small quantities in ordinary water. Heavy water, also called deuterium oxide, is used as a moderator in some nuclear reactors. See HEAVY-WATER REACTOR.

heavy-water reactor   Fission nuclear reactor using as its moderator water in which ordinary hydrogen atoms have been replaced by the heavier isotope deuterium (heavy water). Heavy-water reactors include the Canadian deuterium-uranium reactor (CANDU) and the steam-generating heavy-water reactor. See CANDU, STEAM-GENERATING HEAVY-WATER REACTOR.

hectare   (ha) Unit of measurement of area.
    1 ha = 2.47 acres = 10,000 m^2
    100 ha = km^2

hecto- (h) Prefix in the Système International d'Unités (SI) indicating one hundred, 1×10^2. One hectogram equals one hundred (10^2) grams.

hedenbergite A black variety of the pyroxene mineral group that forms short prismatic crystals in contact metamorphic zones of limestones or granitic rocks.

hedgerow A row of shrubs or trees forming a hedge along the edge of a field. In Europe, hedgerows are living fences constructed to keep livestock in pastures, whereas in North America the term usually refers to a line of trees growing alongside a road or separating fields or pastures.

hekistoplankton Plankton that live in waters of very low temperatures, as in the Arctic Ocean.

hekistotherm Areas with very low temperatures, as in the Arctic and Antarctic.

helio- Prefix referring to the sun.

heliophobe An organism that cannot tolerate bright light and therefore is found in a dark environment.

heliophyll A sun leaf; the form of leaf a plant develops when growing in bright light as opposed to shade (usually smaller but thicker and often having a tougher outer layer). Compare SCIOPHYLL.

heliophyte A sun plant, one that grows best in full sunlight.

heliosis Sunstroke; overexposure to the sun or extreme heat.

heliotaxis An animal's movement in response to sunlight.

heliothermism A method for regulating body temperature in ectothermic organisms, whose body temperature is roughly the same of that in their immediate environment. Heliothermism is changing the body's orientation throughout the day to change the amount of solar radiation received, thereby adjusting temperature. See ECTOTHERM.

heliotropism The turning or bending of a plant towards the sun, especially to follow the sun's path. It is a form of phototropism. See PHOTOTROPISM.

helium (He) An inert, odorless gaseous element with atomic number 2 and atomic weight 4.003. Helium is the second lightest of all elements and the second most prevalent throughout the universe. It does not readily react with any other elements. In addition to its use in filling helium balloons, it is used to create artificial, unreactive atmospheres for welding and manufacturing semiconductors, as a coolant in some nuclear reactors, and as a refrigerant at temperatures approaching absolute zero (as required for superconductors).

helix A spiral with a diminishing circumference, as in a corkscrew. The structure of DNA is a double helix.

helm wind A head wind; wind blowing against the direction in which an airplane, sailboat, or other vessel is moving.

helophyte An herbaceous marsh plant whose overwintering (perennating) bud lies buried in mud. The term comes from the Raunkiaer life form classification system. See RAUNKIAER'S LIFE FORMS.

hematite A common reddish brown ore of iron oxides that forms in a wide variety of environments including hydrothermal veins and sedimentary concretions, and also forms as a replacement mineral. Hematite is a major source of commercial iron ore. See APPENDIX p. 624.

hemicryptophyte   An herbaceous plant whose overwintering (perennating) bud is at soil level. The term comes from the Raunkiaer life form classification system. See CRYPTOPHYTE, RAUNKIAER'S LIFE FORMS.

hemimetabolous   Of or relating to hemimetabolism, a partial or incomplete metamorphosis in insects in which the pupal stage does not occur. Compare HOLOMETABOLOUS.

hemiparasite   A green plant that is partially parasitic (such as mistletoe, *Phoradendron*, or cow-wheat (Melanpyrum), or that may become parasitic under some circumstances (a facultative parasite). Also called a partial parasite.

hemiplankton   Another term for meroplankton. See MEROPLANKTON.

Hemiptera   Insect order which includes the true bugs. In this group mandibles and maxillae are modified into a piercing, sucking beak which can be directed posteriorly to lie on the ventral surface of the thorax when not in use. The body is usually flattened and bears two pairs of wings which can be laid back over the abdomen. In most species the basal portion of the forewings is hard and leathery. Bugs are largely terrestrial but there are also truly aquatic forms as well surface species such as the familiar water-striders. Ecologically, bugs are important herbivores but they also are highly destructive crop pests. Some forms such as bedbugs and kissing bugs are not only a nuisance but also carriers of human disease. Approximately 35,000 species. See INSECTA, BUGS.

hemoglobin   The red pigment found in red blood cells of vertebrates that transports oxygen from the lungs to the tissues and transports carbon dioxide from the tissues to the lungs. Hemoglobin is made up of compounds whose structure resembles that of chlorophyll. See ERYTHROCYTES, OXYHEMOGLOBIN.

henry   (H) Unit for inductance in the Système International d'Unités (SI) and metric (MKS) system. One henry equals the inductance in a circuit when the application of one ampere per second results in an electromotive force of one volt. See INDUCTION.

Henry's law   Henry's law states that at a constant temperature, the amount of a gas that will dissolve in a specified amount of liquid depends on the pressure of the gas above the liquid. This law assumes the gas does not react with its liquid solvent.

Hepaticae   The class of Bryophyta that contains liverworts. These primitive plants have flattened leafy gametophyte generations with single-celled rhizoids. See BRYOPHYTA.

heptachlor   A chlorinated hydrocarbon insecticide related to aldrin and dieldrin in structure. Like most such persistent, toxic chemicals, its use is no longer allowed in the United States. See CHLORINATED HYDROCARBONS, CYCLODIENE INSECTICIDES.

herb   1) A flowering plant whose stem is usually not woody; most therefore do not grow as large as most shrubs. Herbs may be annual or perennial. 2) A plant (usually of temperate climates) whose leaves, flowers, or roots are or were historically used for seasoning, medicine, or fragrance.

herbaceous   Resembling an herb, a green, leafy plant that does not produce persistent woody tissue. Herbaceous plants form the lowest layer of vegetation in most plant communities.

herbarium   1) A collection of pressed plant specimens. 2) A place where such a collection is stored or displayed.

herbicide   Any substance used to kill unwanted plants. Herbicides work in different ways; some sterilize the soil (these are used to keep railroad rights-of-way clear of all growth), others prevent seeds from sprouting, others kill plants once they have sprouted. Some are plant hormones that disrupt the growth control mechanisms of the plant. See SOIL-ACTING HERBICIDES.

herbivore   Any of a large group of animals whose diet consists of plants. Granivores and frugivores are also herbivorous animals. Contrast CARNIVORE, OMNIVORE, XYLOPHAGOUS.

herb layer   A low (generally less than 1 m) stratum of a plant community, usually dominated by herbaceous or low-growing woody plants. Also called herbaceous layer. See GROUND LAYER. Compare SHRUB LAYER, UNDERSTORY, CANOPY.

herbosa   Vegetation or plant communities made up of non-woody (i.e., herbaceous) plants. See AQUIHERBOSA, SPHAGNIHERBOSA, TERRIHERBOSA. Contrast LIGNOSA.

herd   A group of animals of the same kind, especially large herbivores, that live and travel together.

heredity   The transference of physical characteristics and traits from parent to offspring through chromosomes.

heritable character   A physical trait such as a disease that is transmitted from parent to offspring through a particular gene.

hermaphrodite   1) A plant that produces both male and female gametes, especially a flowering plant whose individual flowers contain both male and female organs. 2) An animal with both male and female reproductive organs, exhibiting gynandromorphism. See GYNANDROMORPHISM, MONOECIOUS. Contrast DIOECIOUS.

hermatypic   Building reefs (referring to corals). Contrast AHERMATYPIC.

herpetology   The study of reptiles and amphibians, a branch of zoology.

hertz   (Hz) The unit for frequency in the Système International d'Unités (SI); 1 hertz = 1 cycle per second.

hesperidium   A type of fleshy fruit (berry) having a leathery rind on the outside and divided into segments on the inside. The fruits of citrus trees such as oranges and grapefruit are hesperidia.

Hess's law   A principle governing heat change in chemical reactions. The heat given off or absorbed in a chemical reaction depends only on the initial reactants and final products; it is independent of any equivalent series of steps taken to convert the same reactants into the same products (i.e., whether the products are produced in a single step or series of partial steps). Also called the law of constant heat summation.

heterochromatin   The condensed form of chromatin found in the nuclei of cells that are not undergoing division or similar metabolic activity. Its DNA molecules are tightly coiled and have temporarily stopped transcription of the genetic code. See CHROMATIN.

heterochrosis   An abnormal coloration of skin, fur, or plumage.

heterocyclic compound   Any organic compound with a ring structure that contains other atoms such as oxygen, nitrogen, or sulfur in addition to the carbon atoms. Examples include purine, pyrimidine, furan, thymine, and a number of plant alkaloids. Contrast HOMOCYCLIC COMPOUNDS.

heterodont Possessing different forms of teeth. Mammals are heterodonts because they have canines, incisors, and molars for teeth. Contrast HOMODONT.

heteroecious parasite An organism that lives on or in not one organism, but organisms of two unrelated species during its life cycle. White pine blister rust is a fungal disease that kills white pines, but the fungus requires gooseberry or currant *(Ribes)* bushes nearby to complete its life cycle. As a result, white pines can be protected from the disease by removing all relatives of currants from the area. Also called metoecious parasite. See PARASITE, AUTOECIOUS PARASITE.

heterogametic sex The gender of an organism that contains both (is heterozygous for the) sex-determining chromosomes. In mammals, the male has the X and Y chromosomes, and therefore determines the gender of offspring by contributing the determining sex chromosome (females contribute one of their X chromosomes). In birds, the female is the heterogametic sex. Compare HOMOGAMETIC SEX.

heterogamy Reproduction by two unlike gametes; describing species that produce dissimilar male and female gametes such as human egg and sperm, or plants that produce distinct male flowers and female flowers. Also called anisogamy. Contrast ISOGAMY.

heterogeneous Made up of a number of elements different from each other, a mixture of dissimilar ingredients. Contrast HOMOGENEOUS.

heterogeneous accretion A hypothetical model for the relatively slow accretionary formation of planets in the solar system. Planets formed by heterogeneous accretion are presumed to have accreted in successive layers, each composed of different materials, during the cooling of a solar nebula Compare HOMOGENEOUS ACCRETION.

heterogeneous phase reaction A chemical reaction in which all of the chemical components are not in the same physical phase, that is, a combination of liquids and solids or solids and gases. See PHASE, PHASE TRANSITION.

heterokaryon An individual cell containing more than one nucleus, each from a different species and therefore containing different genetic information. Heterokaryons are hybrids (somatic cell hybrids) produced by fusing cell tissue from different species. Also called a heterocaryote. Compare MONOKARYON.

Heterokonta A group of algae consisting of the phyla Xanthophyta, Chrysophyta, and Phaeophyta. See XANTHOPHYTA, CHRYSOPHYTA, PHAEOPHYTA.

heterolytic bond cleavage The breaking of a chemical bond between two atoms where the two electrons in the bond both go with the atom, making it a negatively charged species, and leaving the other atom positively charged. The transfer of a proton in acid/base reactions is an example of a heterolytic bond cleavage: H-$Cl \longrightarrow H^+ + Cl^-$. Compare HOMOLYTIC BOND CLEAVAGE.

heterophyte A plant which obtains its food from other animals or plants, living or dead; a parasitic or saprophytic plant.

heterosis See HYBRID VIGOR.

heterospory The production of multiple

kinds of asexual spores, such as megaspores and microspores. See MEGASPORE, MICROSPORE, HOMOSPORY.

heterostyly   Describing plants that produce styles (structures attached to the ovary) of two or more different lengths on different flowers. This prevents the styles and anthers (structures bearing male pollen) from being the same height in any one flower, discouraging self-fertilization and promoting pollination from other flowers from different plants.

heterothallism   Describing fungi and lower plants with two (or more) haploid mating types that act as male and female; sexual reproduction can only occur between different types. In heterothallism, each type is self-incompatible. Compare HOMOTHALLISM.

heterotherm   See ECTOTHERM.

heterotroph   An organism that cannot produce its own food, and derives its nutrition by consuming the complex organic molecules produced by plants or present in other living or decaying organisms. Most organisms other than green plants—animals, fungi, most bacteria, some algae, and parasitic plants—are heterotrophic. Contrast AUTOTROPH.

heterozygous   Possessing different alleles in a pair of genes, such as the allele for tallness and shortness (rather than two identical— homozygous—alleles for tallness or two identical alleles for shortness). Heterozygous organisms can pass on either of the heterozygous genes, so they do not always breed true to type. Contrast HOMOZYGOUS.

hexoses   Simple sugars containing six carbon atoms; these are common in plants and animals and serve as the basic unit for synthesizing carbohydrates. Glucose,

fructose, and galactose are all hexoses. See PENTOSES.

HF   See HIGH FREQUENCY.

Hg   Chemical symbol for the element mercury. See MERCURY.

hibernaculum   1) A plant structure that protects the embryo or a growing tip during its dormant season, such as a bulb or bud. 2) A shelter for an overwintering animal, such as an insect pupa or den for hibernating snakes.

hibernation   A somewhat sleeplike condition of partial or total inactivity, lowered body temperature (capable of falling to 1° or so above the ambient temperature, even down to freezing), and greatly reduced metabolism that some animals enter into for the winter. Chipmunks, marmots, and hedgehogs hibernate. Compare ESTIVATION, TORPOR.

hiemilignosa   Deciduous woody vegetation, both trees and shrubs, found in monsoon climates and characterized by small leaves that are shed during the hot, dry season. See LIGNOSA.

hierarchy   A system of ranking by grade, class, or position, especially a social ranking (pecking order) in a group of animals. See DOMINANCE HIERARCHY.

high density polyethylene   (HDPE) A specialized plastic, a translucent form of polyethylene that can be melted and reshaped several times, and therefore is easily recycled. It is used in blow-molded products such as beverage bottles, in injection-molded products such as cassette tape holders, and in containers for oils and gasoline. It withstands higher temperatures better than low-density and medium-density forms of polyethylene. See POLYETHYLENE.

high frequency   (HF) A subdivision of the radio spectrum, including radio waves from 3 to 30 megahertz. Compare LOW FREQUENCY, EXTREMELY LOW FREQUENCY.

high-grade metamorphic rock   Any metamorphic rock that forms under conditions of high temperatures and high pressures. The upper limit of high-grade metamorphism occurs when minerals melt and the rock becomes igneous.

high-grade ore   A general term for ore bodies having an economically significant concentration factor.

high-grade terranes   A term applied to a belt or a tectonic terrane that has been subjected to high-grade regional metamorphism.

highgrading   1) A method of forest harvesting in which only the most commercially valuable trees are cut, usually without attention to maintaining sustainable stocks of desirable species for future harvests. Also called highgrade. 2) A similar mineral extraction technique, in which only the highest-quality ores are mined.

high latitude   A general geographic reference to polar and subpolar regions.

high-level inversion   A thermal or temperature inversion occurring at high atmospheric levels (above roughly 6000 m). Subsidence inversions are often high-level inversions. See THERMAL INVERSION, SUBSIDENCE INVERSION.

high-level radioactive waste   Waste from nuclear reactors and similar facilities that contains high levels of radioactivity. High-level radioactive wastes are specifically spent fuel rods or wastes from reprocessing spent fuel rods. They require constant cooling to remove heat from their radioactive decay and provide challenging storage problems. Contrast LOW-LEVEL RADIOACTIVE WASTE, INTERMEDIATE-LEVEL WASTES.

highly stratified estuary   A type of estuary community that may occur where large rivers meet the ocean. If the river currents are strong enough to counteract the tides, fresh water flows over the dense salt water for some distance, creating two horizontal layers with a sharp change in the salt content where they meet. The mouth of the Mississippi River contains a highly stratified estuary. Also called a salt-wedge estuary. See ESTUARY.

high-pressure area   An atmospheric region with high barometric pressure; an anticyclone. High-pressure areas are often associated with clear weather. Contrast LOW-PRESSURE AREA.

high-quality energy   Concentrated energy that can easily be harnessed to perform useful work. Electricity and hydrocarbon fuels provide high-quality energy. Contrast LOW-QUALITY ENERGY.

high-sulfur crude   A general reference to crude oil having a relatively high proportion of the element sulfur, usually in the range of 2 percent to 5 percent. Compare LOW-SULFUR CRUDE.

high-temperature gas-cooled reactor (HTGR) A fission nuclear reactor that operates at core temperatures approaching 1000°C and uses a gas such as helium as the coolant. Such temperatures are much higher than those of advanced gas-cooled models. Operating HTGR reactors are located at Fort St. Vrain, Colorado, and Winfrith, Dorset (United Kingdom). See ADVANCED GAS-COOLED REACTOR, NUCLEAR REACTOR.

high tide   The maximum water level in a

cycle of tidal fluctuation. High tide marks the point in the tidal cycle when flood tide ceases and ebb tide begins.

Highway Trust Fund   Federal trust account funded by fuel taxes, used to finance transportation system construction, principally highways.

high-yielding variety   (HYV) Plants, especially hybrids, developed to produce larger crops than the original varieties. Many F_1 hybrids are high-yielding varieties.

hinge fault   A type of rotational fault that moves in an apparent scissorlike motion around an axis that is perpendicular to the fault plane.

hinge joint   An animal joint such as the elbow that permits motion around a central axis, usually having the convex end of one bone fitting into the concave end of another.

hingeline   A figurative line that connects all the points of maximum curvature on the surface of a fold.

Hirudinoidea   A class of annelids, the leeches have suckers at the head and tail end and a fixed number of 33 body segments,. These, however, are obscured by secondary annulations on the body surface and loss of segmentation in most body regions except the nervous system. Like the earthworms, they have a clitellum which secretes a cocoon for the eggs. Species may be terrestrial or aquatic and are either usually parasites or predators of vertebrates. Leeches are sometimes included with the oligochaetes in the class Clitellata. Approximately 500 species. See ANNELIDA.

histamine   A compound that animals produce from the amino acid histidine in response to injury and during allergic reactions. It is responsible for the sneezing and itchy eyes of hay fever. Histamine also dilates blood vessels, stimulates gastric and other secretions, and contracts smooth muscle tissue (producing anaphylactic shock in extreme allergic reactions). See ANTIHISTAMINE.

histic   A term applied to the upper layers of a soil that are high in organic carbon and saturated with water for at least part of the year.

histidine $HOOC\text{-}CH(NH_2)CH_2\text{-}C_3H_3N_2$, one of the essential amino acids that must be supplied in the human diet. Present in many proteins, it is used for tissue repair and growth.

histo-   Prefix meaning tissue.

histogens   Regions at the tips of plant roots and growing shoots (meristems) where tissue differentiation occurs.

histogram   A vertical bar graph representing a frequency distribution of a variable, such as seasonal frequencies of rainfall. Amounts are plotted on the x-axis, with the scale of frequencies as the y-axis.

histolysis   The destruction and disintegration of tissues, a part of normal, ongoing tissue growth and renewal in organisms.

historic preservation   A process of identifying and establishing mechanisms for the long-term or permanent protection of buildings and districts of historical significance.

historic preservation law   The statutory laws established to protect historic buildings and historic districts.

histosol   Any member or a soil order characterized by large concentrations of organic matter, usually from accumulated plant remains in swamps and peat bogs.

HIV Abbreviation for human immunodeficiency virus, the retrovirus that causes AIDS. See AIDS.

hoarfrost Another term for frost, a layer of frozen dew. (Hoarfrost is the preferred term in British usage.)

hogback A ridge-shaped landform having a symmetrical profile in cross section, with each side of the hill dipping at approximately equal angles. Compare CUESTA.

Holarctic realm One of the phytogeographical realms into which the earth is divided according to the distribution of plant life (flora). The Holarctic realm is all land north of the Tropic of Cancer. It is divided into eight floral regions: Arctic floral region, Atlantic-North American floral region, Euro-Siberian floral region, Hudsonian floral region, Irano-Turanian floral region, Pacific-North American floral region, Mediterranean floral region, and Sino-Japanese floral region. See FLORAL REALM, PHYTOGEOGRAPHY.

Holarctic region One of the biogeographical realms into which the earth is divided according to the distribution of animal life (fauna). The Holarctic region includes North America (stopping at the northern edge of the Mexican plateau), the Arctic islands, Europe, most of Asia (not Afghanistan, Iran, southern India, or the Malay peninsula), and Africa north of the Sahara desert. The Holarctic region is sometimes subdivided into Palaearctic and Nearctic regions. See ZOOGEOGRAPHY.

holdfast A rootlike structure, not a true root, that is used by some species of algae to cling to something for support.

holistic 1) Relating to, or examining, entities as wholes rather than breaking them down into component parts. Environmental studies and some branches of geography have a holistic approach in viewing phenomena not as individual entities but as interrelated complexes. 2) Referring to the view that in nature functional organisms are produced from individual structures that act as complete wholes. Contrast REDUCTIONISM. See GAIA HYPOTHESIS.

hollow A small valley or lowland incised in a mountainous region.

Holocene The most recent epoch of the geologic time scale. The Holocene epoch spans from 10,000 years ago to the present.

holocoen A living community considered as a whole, emphasizing the interdependence of all parts and without reference to any causal factors. Also spelled holocenosis, this consists of the biocoen plus the abiocoen; it is the British equivalent of ecosystem. See BIOCOEN, ABIOCOEN.

holometabolous Describing insects that go through the complete set of instars, stages of metamorphosis (egg, larva, pupa, and adult). Compare HEMIMETABOLOUS.

holoparasite Parasitic plants that do not contain chlorophyll and are dependent on their host plants to supply water, minerals, and photosynthesized carbon compounds. Dodder (*Cuscuta*) is a holoparasite.

holopelagic Describing organisms that live in the open waters of the ocean for their entire life cycle.

holophytic Producing food through photosynthesis; plants that are not parasitic or saprophytic are holophytic. Compare HOLOZOIC.

holoplankton Organisms that spend their

entire lives as plankton. Contrast MEROPLANKTON, EUPLANKTON.

Holothuroidea A highly specialized class of soft or leathery bodied echinoderms with reduced skeletons and a horizontal axis of symmetry with a mouth at one end. Water vascular canals are located in five radii which extend along the sides from mouth to anus in most species. Tube feet may occur in radii or be scattered over the body and modified feet form tentacles about the mouth. Commonly known as sea cucumbers, they are largely burrowing or sedentary, feed on small organisms or food particles. When disturbed some species eject viscera which are later regenerated. They are widely distributed throughout the seas, including deep trenches where they may be the predominant life form. See ECHINODERMATA.

holotype See TYPE SPECIMEN.

holozoic Feeding by ingesting and digesting complex organic substances, as most animals do. Compare HOLOPHYTIC.

homeostasis An organism's or a cell's ability to maintain a constant internal environment, a balance of conditions such as internal temperature or fluid content, through regulation of physiological processes (negative feedback) and adjustments to changes in the external environment. See NEGATIVE FEEDBACK.

homeotherm See ENDOTHERM.

home range Area that an animal or family group occupies on a daily basis. The home range does not necessarily correspond to territory, which is the area defended by an animal or a group. See TERRITORY.

Homestead Act U.S. federal law of 1864 providing for the transfer of U.S. government lands to private ownership in small tracts for cash payment or work on the claimed lands in lieu of payment.

homing Navigational ability permitting an accurate return to the place of origin, as in the homing pigeon.

Hominidae A family of primates, Hominidae, whose single living species *Homo sapiens,* the humans, differs from its closest relatives, the pongids, in having smaller canine teeth, shorter arms, a greatly enlarged brain, and fully developed bipedal locomotion. The lower jaw is bow shaped and the entire face is flattened except for the prominent nose and chin. See PRIMATES, PONGIDAE.

hominids See HOMINIDAE.

homocaryon See HOMOKARYON.

homocyclic compounds Organic compounds with ring structures that contain only one kind of atom, usually carbon. Examples include benzene and cyclohexane. Compare HETEROCYCLIC COMPOUND.

homodont Having teeth that are all the same type. Contrast HETERODONT.

homogametic sex The gender of an organism that contains two identical (homozygous) sex-determining chromosomes. In mammals, females are the homogametic sex because they have two X chromosomes, as opposed to the X and Y chromosomes present in males. In birds, males are the homogametic sex. Compare HETEROGAMETIC SEX.

homogeneous Formed of elements that are all of the same kind or size, a mixture of similar ingredients. Contrast HETEROGENEOUS.

homogeneous accretion A hypothetical model for the relatively rapid accretionary formation of planets in the solar system. Planets formed by homogeneous accretion are presumed to have accreted uniformly from material that was in equilibrium with the environmental conditions of a cooling solar nebula.

homograft   Taking tissue from an individual of one species and grafting it with a genetically identical individual (as from the same inbred genetic line) of the same species. Also called an isograft. Compare ALLOGRAFT.

homoiohydric   Describing plants that are able to regulate their rate of water loss (as through stomata and guard cells) in order to survive periods of little precipitation. Most land-based vascular plants are homoiohydric. Compare POIKILOHYDRIC.

homoiosmotic   Describing aquatic organisms that are able to maintain a constant concentration in their body fluids, regardless of how salty or dilute the water is in their immediate environment. Most fish and marine mammals are homoiosmotic. Compare POIKILOOSMOTIC.

homoiotherm   See ENDOTHERM.

homokaryon   An individual cell containing more than one nucleus, each from a the same species. Homokaryons are hybrids (somatic cell hybrids) produced by fusing cell tissue from different individuals of the same species. Compare HETEROKARYON.

homologous chromosome   One of a pair of similar chromosomes that carry genes for the same traits. Homologous chromosomes join into pairs during the early stages of meiosis; they separate by the end of meiosis so that each germ cell carries one of the pair. Diploid organisms contain pairs of homologous chromosomes, one derived from the female parent and the other from the male.

homologous structures   Organs or parts of organisms that share common ancestry and develop in similar ways, even if they have different appearances or functions. Bat wings, dolphin flippers, and human hands are homologous structures that originated in paired pectoral fins of an aquatic ancestor.

homolytic bond cleavage   The breaking of a chemical bond between two atoms where the two atoms each end up with one of the electrons that used to constitute the bond, leaving two neutral radical species. The decomposition of CFCs proceeds via a homolytic bond cleavage reaction that produces the chlorine radical, a reactive species indicated in the destruction of ozone in the stratosphere: $CF_2Cl_2 \longrightarrow \cdot CF_2Cl + Cl\cdot$. Compare HETEROLYTIC BOND CLEAVAGE.

homopolymer   A polymer made up of identical subunits (monomers). Polyethylene is a homopolymer; the term is also used in biochemistry to refer to strands of DNA or RNA made up of nucleotides that are all the same.

Homoptera   An herbivorous insect order with close affinities to the true bugs, Hemiptera. They usually have well-developed eyes and a beak which emerges from the back of the head and is held beneath the body. The membranous wings, when present, have reduced venation. The order includes many important crop pests and vectors of plant diseases. One group, the cicadas, have unusually long life cycles which, as in the case of the 17-year "locusts" of North America may extend over many years (i.e., cicadas, leaf hoppers, whiteflies, aphids, scale insects, mealy bugs, etc.). Approximately 45,000 species. See INSECT, APHID.

homospory   The formation of only one

kind of asexual spores. See HETEROSPORY.

homothallism   Describing plants with only one haploid mating type, so that all gametes from the same organism are self-fertile (can mate with each other). See HETEROTHALLISM.

homozygous   Possessing two identical alleles for a particular genetic characteristic. Female mammals are homozygous for the sex-determining chromosome; they possess two X chromosomes (as opposed to males, which have an X and a Y chromosome and are thus heterozygous). Contrast HETEROZYGOUS.

honeydew   Sweet secretions from aphids and leaf hoppers (Hemiptera insects), found on the leaves of some plants. Honeydew produced by aphids is fed upon by some species of ants.

Hooke's law   A physical principle stating that the strain on an elastic material is proportional to the stress applied. Hooke's law only applies below the elastic limit of a solid. See ELASTIC LIMIT.

hook order   A social heirarchy of horned ungulates. See DOMINANCE HIERARCHY.

hookworms   Agents of hookworm infection and disease, hookworms of several genera are among the most widespread and important nematode parasites of humans. They are particularly prevalent in moist, tropical areas. Mature worms live in the small intestine and their eggs are deposited in stools. Larval forms develop in soils and penetrate the host skin causing acute dermatitis in many cases. Eventually, worms reach the intestine where they attach themselves to its wall with hooklike mouthparts and suck blood. Fatigue, digestive disorders, and severe anemia may result. See NEMATODA.

Hopkin's bioclimatic law   The generalization that in North America, spring phenological (seasonally dependent) events take place four days later per each degree of latitude as one progresses northward, per 5° longitude eastward, or per 400 feet in elevation. Autumn events progress four days earlier in the same corresponding directions. See ECOGEOGRAPHIC RULES.

horizon   1) An identifiable layer of a soil. 2) A body of rock strata with identifiable upper and lower boundaries. 3) The apparent intersection between the land and the sky as seen in a horizontal view of the topography. See APPENDIX p. 619.

hormone   1) Any of a variety of compounds secreted in very small quantities by endocrine glands (or other cells) into the bloodstream to control the activity of distant organs, tissues, or metabolic processes. Insulin and adrenalin are hormones. 2) A plant growth substance such as gibberellic acid.

hormone weedkillers   Herbicides containing high concentrations of synthetic auxins, plant growth substances. Examples include 2,4-D (commonly used against broad-leaved weeds in lawns) and 2,4,5-T. Most act on contact (translocated herbicides). See HERBICIDE.

horn   1) Keratin; a hard, strong, slightly translucent substance found on animals in hooves, feathers, claws, and horns. 2) A pointed, sometimes curved or branched projection from the head of some mammals; also, a similar-looking structure on other animals. See KERATIN.

hornblende   The most common member of the amphibole silicate minerals; a dark green-to-black mineral with prominent cleavage in two directions at approximately 60° and 120°; hardness = 6; specific gravity = 3 to 3.5. See APPENDIX p. 624.

hornfels   A very dense, dark rock with dull luster and chonchoidal fracture; formed by contact metamorphism of argillaceous sedimentary rock such as siltstone, shale, or slate.

horn, glacial   A steep, pyramid-shaped peak forming between three or more adjacent glacial cirques. The Matterhorn in the Swiss Alps is one of the most famous examples.

horse latitudes   The two belts of high-pressure regions on either side of the equator characterized by calm or light and variable winds. They occur between 20 and 35 degrees latitude.

horsepower   (HP) A traditional unit for measuring the ability of an engine to do work in the foot-pound-second system, now usually replaced by the watt. One horsepower equals 550 foot-pounds per second and is equivalent to 745.7 watts. See WATT.

horse-shoe crabs   See MEROSTOMATA.

horst   An uplifted fault block that is bounded on two or more sides by high-angle normal faults. A graben is often formed between two horsts.

horticulture   Cultivation of flowers, fruits, vegetables, or ornamental plants; also, the study of such cultivation methods. Compare ARBORICULTURE.

host   1) An organism that supports a parasite, often to its own detriment. 2) An organism that receives a grafting of another organism, usually used for embryos. See PARASITE, DEFINITIVE HOST, INTERMEDIATE HOST.

host-parasite relationship   An association between organisms in which one (the parasite) lives on or within another (the host); the parasite derives its nutrition from the host, sometimes causing little damage but often impairing the health of the host. See PARASITISM.

hot brine   A body of spring water that is heated geothermally and thus rich in dissolved minerals.

hot desert   Type of desert found in the American southwest and Mexico, which is characterized by intense heat during the summers and often having shrubs and cacti predominant forms of vegetation.

hot dry rock   A granitic rock with a high rate of internal heat production from the decay of radioactive elements. Hot dry rock may be used as a source of geothermal energy.

hot spot   A general term for an area of high volcanic activity or a stable area of rapid heat transfer from the mantle to the crust of the earth.

hot spring   A surface discharge of geothermally heated groundwater. Evaporite mineral deposits are often associated with a hot spring.

HPLC   High-performance (or high-pressure) liquid chromatography. HPLC is a chromatographic technique used in the analysis of pesticides and pesticide residues. A solvent is used to elute a mixture of compounds through a column, where the liquid is forced through the column with a high-pressure pump. Separation of the compounds in the mixture occurs in the column and the amount of each component is quantified using a detector. See ELUENT, CHROMATOGRAPHY, GAS CHROMATOGRAPHY.

Hudsonian floral region   One of the phytogeographic regions into which the Holarctic realm is divided according to similarities between plants; it consists of parts of Alaska and Labroador, reaching south to the Great Lakes. See HOLARCTIC REALM, PHYTOGEOGRAPHY.

Humboldt Current   See PERU CURRENT.

humic acids   A mixture of complex organic molecules that, together with fulvic acids, comprise humic substances. Both fulvic and humic acids are decomposition products of plant material and are found in soils and in water. Fulvic and humic acids are separated by an acid/base extraction. See AQUATIC HUMIC SUBSTANCES.

humic coal   A variety of lignite coal composed of woody fragments from trees and shrubs.

humidity   General term for the amount of moisture (water vapor) in the atmosphere. See ABSOLUTE HUMIDITY, RELATIVE HUMIDITY.

humidity ratio   The ratio of the mass of moisture to the mass of dry air. Also called mixing ratio and moisture content. Contrast RELATIVE HUMIDITY, ABSOLUTE HUMIDITY.

humification   The breakdown (rotting) of organic matter from plants and animals into soil; the formation of humus.

hummingbirds   The hummingbirds of the order Apodiformes are noted for their hovering flight and brilliant plumage. The over-300 species are restricted to the Western Hemisphere with the majority occurring in the neotropics. In proportion to their size and weight, the flight muscles are the largest found in any birds. They are capable of producing wing beats which may range from over 50 per second in hovering flight up to 200 during courtship displays. Because of their small size and high metabolism, species in cooler areas may go into a state of torpor at night to conserve energy. They feed on nectar, sap,

and insects. See APODIFORMES.

humus   An organic soil material so thoroughly decayed that the identity of the biologic source cannot be recognized.

hundred-year flood   A flood of a specified dimension that has a statistical probability of occurring once in every 100 years.

hunter-gatherer   Group of people dependent primarily on gathering wild food and hunting, as opposed to cultivating crops and domesticated animals.

hurricane   A severe tropical cyclonic storm characterized by wind speeds registering 12 and above on the Beaufort scale (exceeding 73 mph, 115 km/hour) and heavy rains. In some areas, hurricanes are called typhoons or cyclones. See BEAUFORT SCALE, also APPENDIX p. 630.

hyaline   1) Translucent, not fibrous or granular, and lacking in color. Hyaline cartilage is the material covering the ends of bones in joints. 2) A description of the translucent or glassy appearance of a mineral.

hybrid   The offspring of two genetically distinct organisms, either different species, varieties, or breeds; a crossbreed. The mule is a hybrid between a female horse and a male donkey.

hybridization   Crossbreeding; the production of a hybrid, either by mating (crossing) two organisms, or through new technologies such as protoplast fusion.

hybrid swarm   A continuous series of morphologically distinct hybrids resulting from interspecific crosses followed by back crosses in subsequent generations.

hybrid vigor   An increase in growth, disease resistance, fertility, or size seen in organisms produced by crossbreeding genetically different parents. Also called heterosis.

hydathode   A specialized plant structure, a pore near the edge of some leaves that allows plants to secrete excess water when atmospheric conditions do not favor transpiration. See GUTTATION.

hydrarch succession   The development of an ecological community that starts in fresh water or a wet area, such as a pond. See HALARCH SUCCESSION, HYDROSERE, XERARCH SUCCESSION.

hydrates   Refers to salts and oxides containing water (water of crystallization) bound within the crystal lattice of the solid. If this water is removed, the substance becomes anhydrous. Crystallized copper sulfate is a hydrate. Compare ANHYDROUS.

hydration   The process of adding water to anhydrous compounds or to an organic compound.

hydraulic   Relating to the flow of fluids, or driven by fluid under pressure. A hydraulic brake, such as that sometimes used on automobiles, operates using brake fluid under pressure

hydraulic geometry   A description of the changes in width, depth, flow velocity, gradient, friction, and sediment load of a stream in relation to discharge.

hydraulic gradient   A measure of the slope of a water surface between two points along a stream channel.

hydrazine H_2N-NH_2, a poisonous, very strong liquid base. It is a strong reducing agent, making it very useful in organic synthesis. It and its derivatives are used as rocket propellants.

hydrocarbon   Any compound containing only carbon and hydrogen. Hydrocarbons form a large group of organic chemicals that includes most petroleum and coal tar products such as benzene, propane, and propylene. See CHLORINATED HYDRO-CARBONS.

hydrocarbon cracking   Another term for catalytic cracking. See CATALYTIC CRACK-ING.

hydrocarbon emissions   Air pollutants made up of compounds containing carbon and hydrogen. They comprise a large proportion of automobile and other combustion emissions; gasoline is made up of hydrocarbons, and unburned gasoline causes the emission problems. In the presence of sunlight hydrocarbons form toxic, irritating secondary pollutants. They contribute to the formation of ozone and of smog. See SECONDARY POLLUTANTS.

hydrochloric acid   HCl, a solution of hydrogen chloride in water. It is a strong acid with an acrid odor and is highly corrosive. It dissolves many metals to form chloride salts, giving off hydrogen gas. It is secreted by the body for digestion in the stomach; it also has a great many industrial uses. Also known as muriatic acid.

hydrochore   A seed or spore dispersed by means of water. Contrast ANEMOCHORE, ZOOCHORE.

hydrocompaction   Settling and compaction that occurs when sediments become saturated with water.

Hydrocorallinae   A classification of marine invertebrates that includes the coelenterate groups Milleporina and Stylasteria.

hydroelectric   Relating to electricity derived from turbine generators powered by falling water, either from a dam or a waterfall.

hydroelectric power   Electricity produced by turbine generators driven by falling water. Hydroelectric power plants often use a dam to build up water pressure, but originally were driven by waterfalls.

hydrofluoric acid   HF, a solution of hydrogen fluoride in water. Like hydrochloric acid, it is highly corrosive and dissolves many metals, giving off hydrogen gas. It is used in cleaning compounds and for etching glass.

hydrofluosilicic acid H_2SIF_6, one of the compounds used for the fluoridation of drinking water. See FLUORIDATION.

hydroforming   A petroleum refining technique that forms high-octane motor fuel additives by a catalytic reaction with hydrogen.

hydrofracturing   The injection of a pressurized gas or liquid to cause fracturing of joints and bedding planes in bedrock. Hydrofracturing is used to increase bedrock permeability and improve the yield of oil and gas wells.

hydrogen   (H) The lightest of all the elements, hydrogen is gaseous and has an atomic number of 1 and an atomic weight of 1.008. Its elemental form is a diatomic molecule, H_2. Hydrogen is much more abundant throughout the universe than any other element. It is found in water and in most organic compounds. Deuterium and tritium are two isotopes of hydrogen.

hydrogen bond   A weak bonding interaction between a hydrogen atom bound to an electronegative atom (fluorine, oxygen, chlorine, or nitrogen) and another electronegative atom in a different molecule. Hydrogen bonding is about one-tenth as strong as covalent bonding (the force holding together the atoms within a single water molecule). Hydrogen bonding affects physical properties such as boiling points, is responsible for the secondary structure of proteins (and thus how they interact with other proteins), and holds together the base pairs in strands of DNA.

hydrogen cyanide   HCN, a very poisonous liquid or gas with an odor like almonds. Hydrocyanic (prussic) acid is a solution of this liquid in water. Uses include fumigation, electroplating, the extraction of gold and silver from ores, and in the manufacture of synthetic fibers, pigments, and plastics such as acrylonitrile.

hydrogen ion   Hydrogen ions are responsible for the strength of acids, and the properties of solutions associated with acids; the greater the concentration of hydrogen ions, the stronger the acid. See ACID, PH.

hydrogen ion concentration   The concentration of hydrogen ions (H^+) in a solution. This concentration determines a solution's acidity; it is the entity measured with pH values. The pH of a solution is defined as the negative logarithm to the base ten of the hydrogen ion concentration expressed in moles per liter of aqueous solution. See PH.

hydrogen peroxide H_2O_2, a colorless liquid rarely found in concentrations greater than 90 percent because it is so explosive. It is a strong oxidizing and bleaching agent, usually used in very dilute form. A three percent solution is used medically as a disinfectant.

hydrogen sulfide H_2S, a flammable, poisonous gas with a characteristic smell of rotten eggs. Its presence in some drinking water supplies makes them unpalatable. It has many uses in the chemical industry. Oil-burning power plants as well as a number of industries including refineries, produce it as a toxic by-product and air pollutant. It is also a component of volcanic gases.

hydrography   The collection and analysis of hydrologic data. Hydrograpy is used in the preparation of a streamflow hydrograph.

hydroid   In Cnidaria that show alternation of generations, an organism in the asexual, benthic stage. See CNIDARIA.

hydrologic cycle   A natural solar-driven cycle of evapotranspiration, condensation, precipitation, and runoff. The hydrologic cycle controls the water movement between the atmosphere, oceans, aquatic, and terrestrial environments.

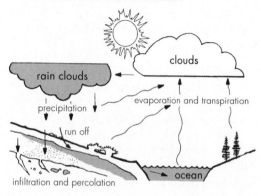

A schematic representation of the hydrologic cycle.

hydrologic gradient   1) The measure of the slope of the surface of water that is flowing in an open channel or through a porous substrate. 2) The change in hydraulic head per unit distance of groundwater flow.

hydrology   The science of water in the hydrologic cycle.

hydrolysis   1) Chemical decomposition in which organic compounds react with water molecules to form other compounds. Many metabolic reactions involve hydrolysis and are catalyzed by a number of different enzymes. 2) A chemical reaction of a compound with water, such as the reaction of some salts with water to form a solution of an acid plus a base.

hydrometer   A device for determining the specific gravity of liquids. Hydrometers often consist of a floating bulb attached to a stem marked with gradations. The hydrometer is allowed to float in the liquid, and the value read off the exposed stem is used to calculate specific gravity. See SPECIFIC GRAVITY.

hydromorphic soil   A term applied to soils predominantly influenced by poor drainage and lengthy periods of saturation.

hydronium ion   The hydrated form of the hydrogen ion, H_3O^+, which is the actual form in which H^+ most often occurs when acids ionize in water. The hydronium ion behaves as H^+. See HYDROGEN ION.

hydrophilic   Literally means water-loving. A molecule or part of a molecule that is readily solvated by water molecules. Polar substances are generally hydrophilic. The polar head groups of a micelle are hydrophilic. Contrast HYDROPHOBIC. See MICELLE.

hydrophobic   Literally means water-fearing. A molecule or part of a molecule that is insoluble in water. Non-polar substances such as gasoline are hydrophobic. The non-polar tails in a micelle are also hydrophobic. Contrast HYDROPHILIC. See MICELLE.

hydrophyte   1) A plant that grows partly or completely underwater, or in very wet soils. 2) In the Raunkiaer life form classification system, an aquatic herbaceous plant whose overwintering (perennating) buds lie at the bottom of a body of water. Compare MESOPHYTE, XEROPHYTE, CRYPTOPHYTE. See RAUNKIAER'S LIFE FORMS.

hydroponics   Agricultural (or laboratory) practice of growing plants in water or in an inert, soil-less medium saturated with water (such as coarse sand, vermiculite, Perlite, etc.). The water is mixed with a controlled nutrient medium (fertilizer). Also called soil-less gardening.

hydropower See HYDROELECTRIC POWER.

hydrosere The succession of plant communities that begins in water or other wet conditions. A hydrosere is any of the phases or stages in hydrarch succession.

hydrosphere The portion of the biosphere consisting of the liquid water environments of oceans and freshwater bodies. See BIOSPHERE.

hydrostatic Referring to the behavior of fluids at rest and under pressure, and forces generated by such fluids.

hydrotaxis An organism's movement (traveling) in response to moisture.

hydrothermal Relating to heated water. Hydrothermal energy is the most common form of geothermal energy. See GEOTHERMAL ENERGY.

hydrothermal metamorphic rock Any rock in which the principal agent of alteration is a hot-water solution rather than the heat and pressure associated with typical metamorphic rocks. See METAMORPHISM.

hydrothermal mineral deposit A concentration of minerals deposited in the bedrock after transport by hot-water solutions. These deposits form veins within the faults and fractures of the bedrock.

hydrothermal ore A concentration of metallic elements precipitated after transport by hot-water solutions in the bedrock. Many commercially significant deposits of gold, silver, copper, lead, and tungsten are hydrothermal ores.

hydrothermal solution A hot liquid rich in dissolved salts and minerals. Hydrothermal solutions are heated by magma and migrate through cracks or porous rock until changes in temperature and pressure cause the precipitation of dissolved minerals.

hydrothermal vents A fissure or opening through which hydrothermal fluids escape from the earth's crust.

hydrotropism The response of a plant (or non-motile organism) to moisture by turning or bending toward or away from it. See TROPISM.

hydroxide 1) OH^-, the anion resulting from loss of a proton (H^+) from a water molecule or from addition of a proton to the oxide anion (O^{-2}). The hydroxide ion is responsible for the properties of solutions associated with bases. It is also called the hydroxyl ion. 2) A class of compounds containing the hydroxide anion. Examples are calcium hydroxide—$Ca(OH)_2$, lime—and sodium hydroxide—$NaOH$, lye. Hydroxides dissolve in water to give hydroxide ions (OH^-), which produce a basic solution. See BASE.

Hydrozoa Class of of the phylum Cnidaria with a range of life cycles and body forms. Many species show alternation of a sexually reproducing, medusa generation with an asexual, polyp generation in which medusae are usually small and inconspicuous. In some species, medusae remain attached to the polyp forms, whereas in others the medusoid stage is missing altogether. Many species are colonial and may deposite chitenous or calcareous exoskeletons. The common freshwater genus, *Hydra*, has solitary polyps and no medusoid stage. In contrast, the Portuguese man-of-war, *Physalia*, is a colonial form showing extreme polymorphism. The gas float is a specialized medusa, whereas the tentacles and reproductive gonozooids are modified polyps. Approximately 2700 species. See CNIDARIA.

hygrometer Device for measuring moisture content of air (relative humidity). Hygrometers can use a bundle of fibers, such as human hair, that changes length as it absorbs or releases humidity (a hair hygrometer), or a combination of wet and dry thermometers (a psychrometer), or other principles. See PSYCHROMETER, RELATIVE HUMIDITY.

hygropetric Describing organisms that live on wet rocks.

hygrophilous Of or relating to plants and animals that inhabit wet areas.

hygroscopic Absorbing water. Hygroscopic chemicals and compounds such as silica gel absorb water from the air and can therefore be used as desiccants. Small particles of such chemicals in the atmosphere are sometimes called hygroscopic nuclei because their affinity for water makes them excellent condensation nuclei. See CONDENSATION NUCLEI.

Hymenoptera A very important order of insects including many species which display moderate to highly evolved social behavior. Two pairs of membraneous wings are present, although these may be absent at some point in the adult stage of a number of species. Mouthparts may be for chewing plant material or elongated for sucking nectar. There are two suborders, the Symphyta (saw flies and their allies), which are thick-waisted and in which females have elaborate ovipositors, and the Apocrita (wasps, bees and ants), which display the typically constricted "wasp waist" and whose females often have ovipositors adapted as stinging organs. Except for extremely cold environments, hymenopterans have a worldwide distribu-

tion. Over 130,000 species. See INSECTA, WASPS, BEES, ANTS.

hypabyssal A general term for medium-grained intrusive igneous rocks that have crystallized at relatively shallow depths

hyper- Prefix indicating above, beyond, or excessive.

hyperkinesis Abnormally high muscular movement, as in the hyperactivity of some children. Also called hyperkinesia.

hyperosmotic Having a higher salt concentration than the surrounding medium. Water will pass from hyperosmotic cells to surrounding tissues because of the concentration gradient.

hyperparasite An organism parasitizing (living on) another parasite.

hyperphagia An abnormal desire to eat; an eating disorder in which animals consume vast quantities of foods as well as nonfood substances.

hyperplasia An abnormal increase in the number of cells present in a tissue or an organ of a plant or animal, usually from disease.

hypersaline A term applied to water that is less saline than a predicted or comparative value.

hypersensitivity An organism's unusually strong response to a stimulus, such as some food, or sunlight. In humans, it usually refers to an unusually strong allergic response to a substance (antigen).

hyperthermia An abnormal, unusually high body temperature in an organism; a very high fever. Contrast HYPOTHERMIA.

hyperthyroidism Abnormal condition of high thyroid hormone levels that increases the metabolic rate, which may result in loss of weight, tiredness, or heat intolerance. Hyperthyroidism is sometimes associated with a form of goiter. Contrast HYPOTHYROIDISM.

hypertonic solution   A liquid that has a greater osmotic pressure than another solution (especially a standard solution, or normal level as in blood) to which it is compared. Contrast HYPOTONIC SOLUTION.

hypertrophy   An increase in the size of an organ or tissue in animals or plants due to an increase in the size of its cellular components. Hypertrophy is usually caused by disease, but does not involve the formation of a tumor. See HYPERPLASIA.

hypha   One of the long, thin filaments that forms the structural basis of fungi, either in form of a loose network of rootlike mycelia or packed together to form the fruiting body of a mushroom. See MYCELIUM.

hyphalomyraplankton   Species of plankton found in brackish water.

hypo-   Prefix indicating below, beneath, or at a reduced level from what is normal.

hypocenter   1) The point from which an earthquake develops, also called the focus. See FOCUS (EARTHQUAKE). 2) The point directly underneath the center of the explosion of a nuclear bomb.

Hypochytridiomycota   A phylum of the kingdom Protista that includes funguslike aquatic microbes that are either parasitic or saprobic, invading their food source by means of filaments. They differ from fungi in that they produce flagellated zoospores, although they are not clearly sexual. *Phizidomyces*, a parasite of water molds, and *Hypochytrium*, which grows on conifer pollen, are members of this phylum. See PROTISTA. Contrast CHYTRIDIOMYCOTA.

hypocotyl   In a seed, the lower portion of an embryo from which the primordial root arises.

hypodermis   The layer of cells below the epidermis in plants and animal (in arthropods, the hypodermis is the layer beneath the cuticle). See EPIDERMIS.

hypogeal   Growing or maturing beneath the surface, especially having cotyledons that never emerge above the surface of the soil. Also, having fruiting bodies that mature within the soil, as in corn. Also spelled hypogeous. Compare EPIGEAL.

hypogene   Rock that was formed below the earth's surface, such as granite, or a process at work below the earth's surface. Compare SUPERGENE.

hypolimnion   The lower layer of cool water in a thermally stratified lake. The hypolimnion undergoes oxygen depletion during the summer.

hypoosmotic   Having a concentration of salt less than that of the surrounding medium. Water will pass into hypoosmotic cells from surrounding tissues because of the concentration gradient. Compare HYPEROSMOTIC.

hypoplankton   Demersal species of plankton, those found near the bottom of a body of water. See DEMERSAL.

hypostatic gene   A gene that is masked by one or more genes that are not its allelomorphs (that are located at different positions on the chromosome). The relationship between a hypostatic gene and the epistatic gene that hides it is analogous to that between a recessive gene and the dominant gene at the same position (locus). Contrast EPISTATIC GENE.

hypothalamus   The region of the brain that secretes substances which control body temperature, water balance, sugar and fat metabolism (and hunger), and produces the hormones controlling the endocrine glands. It is located beneath the thalamus.

hypotheca   The lower (smaller) half of a diatom shell (frustule). See BACILLARIO-PHYTA, EPITHECA.

hypothermia   A significant reduction of body temperature (below 35°C). Hypoth-ermia can result from prolonged exposure to cold (especially cold, wet weather); it is also induced for medical reasons such as to reduce oxygen consumption during surgery. Compare HYPERTHERMIA.

hypothesis   A conceptualization to explain a given phenomenon, based upon limited observations of the phenomenon. Hypoth-eses are used to design experiments to prove or further understand the behavior of a system or phenomenon. If a hypoth-esis withstands experimental tests, it may be elevated to a theory. Compare THEORY, LAW.

hypothyroidism   Low thyroid hormone levels that result in listlessness and other symptoms of reduced metabolic rate. Contrast HYPERTHYROIDISM.

hypotonic solution   A liquid that has a lower osmotic pressure than another solution (especially a standard solution, or normal level as in blood) to which it is compared. Compare HYPERTONIC SOLU-TION.

hypoxia   See ANOXIA.

hypsithermal interval   An interglacial period of the earth's history.

hypsodont   Describing teeth with high crowns and deep sockets, occurring in mammals. See ACRODONT, BUNODONT, LOPHODONT, SELENODONT.

hypsometric curve   A graph showing the relationship between the temperature at which water boils and height above sea level. The curve plots the results of a hypsometer, an instrument that measures water's boiling point. It can be used to calculate altitude or (at a known altitude and therefore known atmospheric pres-sure) to calibrate thermometers.

Hz   Abbreviation for the unit hertz. See HERTZ.

i

I Chemical symbol for the element iodine. See IODINE.

IAA Short for indole-3-acetic acid, C_8H_6N-CH_2-COOH, a natural plant hormone. Also called auxin, it is the best known of the auxin group of hormones, which regulate plant growth and development. The synthetic auxins 2,4-D and 2,4,5-T used as herbicides are structurally related to IAA.

IAEA See INTERNATIONAL ATOMIC ENERGY AUTHORITY.

IBP See INTERNATIONAL BIOLOGICAL PROGRAMME.

ice accretion The build-up of ice on objects surrounded by air with supercooled water droplets. Ice accretion causes hail to increase in size; it also occurs on airplane wings, telephone and electrical lines, and ship rigging.

ice age A time in geologic history when large areas of landmasses were occupied by continental glaciers and lowland valleys were occupied by alpine glaciers. See INTERGLACIAL.

iceberg A floating block of ice broken off by calving where a glacier terminates in a body of water. See CALVING.

icecap A relatively small continental glacier occupying a land area less than 50,000 km².

ice-contact stratified drift A deposit of glacial drift modified by meltwater action during or after deposition in contact with glacial ice. The latter is recognized by chaotic slump structures and inclusions of glacial till. Compare GLACIAL DRIFT, STRATIFIED DRIFT.

ice dome The central mass of an ice sheet or ice cap. An ice dome is thickest at the center, often measures more than 3000 m, and has a convex surface form.

ice field 1) A large floating sheet of ice, typically occurring in polar oceans. 2) A large sheet of ice resting on the land, typically occurring in mountainous regions.

ice sheet A relatively large continental glacier occupying a land area greater than 50,000 km².

ice shelf A floating sheet of ice still connected to the glacial source, as can be seen in Antarctica where the continental glacier flows into ocean water.

ice-wedge polygon A regolith feature in periglacial climates characterized by interlocking ice fractures that form a geometric pattern in permafrost soil. A network of sediment-filled cracks is formed by contracting ground, frost heaving, and ice-sorting of particles. See PATTERNED GROUND.

ichthyofauna General term for all kinds of fish.

ideal gas A hypothetical gas that responds to forces exactly as the gas laws predict. True gases behave more like ideal gas at high temperatures and low pressures.

igneous The term for processes related to the solidification of rock from a molten state.

igneous rock An aggregrate or mass of minerals that have crystallized while cooling from a molten state. Compare METAMORPHIC ROCK, SEDIMENTARY ROCK.

ignimbrite An accumulation of pyroclastic

debris deposited in a flow of volcanic pumice. Ignimbrites have complex flow structures composed of trapped gas bubbles and welded pumice shards. See ASHFLOW.

illuviation   The vertical or lateral movement and deposition of soil material in a soil profile, for example, downward movement and deposition of clay in the B horizon. Contrast ELUVIATION. See APPENDIX p. 619.

ilmenite   A black ore mineral of iron and titanium oxide that forms trigonal crystals in igneous rocks. Ilmenite is a weather-resistant mineral that may be a clastic component in sedimentary rocks. Principal ore of titanium.

imago   An adult insect, one that has completed all stages of metamorphosis, often having wings. See INSTAR.

imbibition   The absorption of water by colloidal and capillary action (rather than by osmotic action) in plants, which results in swelling of the tissue. Imbibition is the first stage in seed germination.

imbricate   1) Showing an overlapping pattern as in scales on the buds of many trees, or the scales of a pine cone. 2) The geological phenomenon in which wedges of rock pile up on one another as a result of an extreme form of overthrust. Rocks are thrust over in slices without folding, each separated by a thrust plane, with which each slice makes an oblique angle. 3) A pattern found in sedimentary deposits such as pebble beds and conglomerates, where a strong current in the river responsible for deposition caused the long axes of the pebbles to flow downstream and become cemented in that position.

Imhoff tank   A sewage treatment tank that is constructed with separate compartments for a digestion chamber and a settling chamber.

immature   Still in its youth or early stages of development, not ripe or mature. See MATURATION.

immigration   The movement of an organism, population, or species into a new area. In population dynamics models, immigration is a factor contributing to population growth. Contrast EMIGRATION, MIGRATION.

immiscible   Not capable of being mixed to form a homogenous liquid or a single phase. Oil and water is the classic example of two immiscible liquids.

immune response   The body's reaction to foreign bodies such as disease organisms, viruses, transplanted tissue, allergens, or toxins: the production of specialized white blood cells (antibodies) and macrophages. See ANTIBODY, IMMUNITY.

immunity   The body's ability to fight against disease germs or their toxins and other destructive substances by the production of specific antibodies that combine with the foreign substances (antigens) to render them harmless. Immunity may be inherited or acquired through an earlier infection or contact with the antigen.

immunization   Protecting the body from disease by artificially inducing immunity, usually by injecting weakened organisms or their toxins (active immunization), or by injecting serum containing pre-formed antibodies from another organism (passive immunization).

impact cratering   The disruption of a planetary surface by the collision impact of falling meteors. Roughly circular craters with sloping aprons of fractured debris are typical landforms generated by impact cratering.

impedance (Z) A measurement of apparent resistance of a circuit to the passage of alternating current. Impedance, like resistance, is measured in ohms.

imperfectly drained A description of a soil or sediment that is frequently saturated with water.

impermeable Not allowing the passage of fluids such as water or gas. Landfills are now required to have impermeable liners underneath them to prevent toxic substances from leaching out and possibly contaminating groundwater.

implantation 1) The process in which an embryo becomes embedded in the uterine wall in higher mammals, or attaches to the yolk in other vertebrates, in order to obtain nourishment. 2) The introduction of tissue or other living or artificial matter into an organism, such as a tissue graft or a fertilized egg.

importance value An evaluation of the role of a species within a community, calculated by adding the relative density, relative dominance (either relative cover or relative basal area), and relative frequency at which the species is found in the community.

impoundment A natural or artificial body of water that is held back by a dam.

impoverishment Depletion from overexploitation; the act of reducing strength or fertility, of making poor. Bad agricultural practices cause impoverishment of the soil.

impregnate 1) To fertilize by introducing sperm. 2) To cause something to become saturated, permeated, or filled. Pressure-treated wood is wood that has been impregnated (under pressure) with com-

pounds that are toxic to decay organisms.

imprinting A learning process occurring shortly after birth that enables newborn animals to recognize and to form attachments to another animal, which it sees as its parent.

impulse turbine A form of steam turbine in which steam is directed through shaped tubes at rotor blades. It results in minimal loss of steam pressure and therefore greater efficiency. See PELTON WHEEL.

impurities Foreign matter present in small quantities in another substance. Impurities can be undesirable contaminants, as in drinking water. But the impurities present in some crystals and metals are sometimes beneficial; they are intentionally added to semiconductors to achieve specific properties of conductivity.

inbreeding Mating of closely related organisms. Inbreeding is done intentionally to reduce variation within a desirable species, but it eventually leads to a higher incidence of inherited defects. Contrast CROSS-BREEDING, OUTBREEDING.

inbreeding depression A decline in desirable characteristics such as fertility, general vigor, or yield produced by repeatedly crossing related organisms (inbreeding). Inbreeding depression can be seen in some specimens of purebred pets. Contrast HYBRID VIGOR.

inceptisols Any member of a soil order characterized by weakly developed or poorly expressed diagnostic horizons, usually denoting the early stages of soil development.

incidence 1) Occurrence of an event. 2) The rate at which something occurs or affects people or things.

incinerate To burn waste materials. Air-pollution regulation has reduced the use of incineration for municipal waste disposal. However, a specialized form of incineration (euphemistically called "resource recovery") is now used to burn municipal garbage and "recover" the resulting heat for energy generation or space heating.

incised meander   A stream meander that has formed a sinuous steep-sided valley or ravine shape by downcutting through alluvium or bedrock. An incised meander is an inherited pattern that forms when a meandering river experiences regional uplift or base level lowering.

inclined fold   A bedrock fold in which the dip of the axial plane is between 10° and 80°.

included niche   The situation in which the niche of a particular species is found completely within the niche space of another species.

inclusive fitness   The sum of an organism's Darwinian fitness plus kin selection relative to the fitness of its relatives. See KIN SELECTION.

incompatibility   1) The phenomenon in which the body rejects transplanted tissue, transfusions of a different blood type, or a similar mismatch causing an immune, physical, or chemical reaction. 2) A similar phenomenon in plants in which an attempted graft is rejected. 3) The inability for fertilization to take place between one flower (or algae or fungus) and other flowers on the same or genetically similar plants. 4) Mutually antagonistic state between two substances, such as medications. See SELF-INCOMPATIBLE.

incompetent bed   A layer of rock that deforms under pressure more easily than do adjacent layers.

incubation period   The time interval between exposure to a disease organism and the manifestation of the symptoms of the infection.

indehiscent   Describing a seed or other plant structure that does not split open when ripe. Acorns are indehiscent. Compare DEHISCENCE.

independent assortment   The principle of Mendelian genetics stating that if allele pairs for two different traits are located on different chromosomes, they will separate randomly during the formation of reproductive cells (gametes). The random separation enables the traits to be passed along to the next generation independently of each other. Also called the law of independent assortment. See MENDEL'S LAWS.

independent variable   A quantity that changes without being directly influenced by other quantities or changes in those quantities. Compare DEPENDENT VARIABLE.

index fossil   A fossil that is found within a known age range. Called a "guide fossil" as it helps establish stratigraphic position and chronology of the rocks it is found in.

index mineral   A mineral that serves as an indicator of metamorphic grade within regionally metamorphosed rocks. Index minerals form within a definite range of temperatures and pressures and thus serve as a diagnostic indicator of metamorphic environments.

index species   See INDICATOR SPECIES.

Indian floral region   One of the phytogeographic regions into which the Paleotropical realm is divided according to similarities between plants; it covers most of the Indian subcontinent. See PALEO-

TROPICAL REALM, PHYTOGEOGRAPHY.

indicator species An organism whose presence (or state of health) is used to identify a specific type of biotic community, or as a measure of ecological conditions or changes occurring in the environment. A number of different plants have been used in the western United States to appraise water and soil conditions. Also called a biological indicator indicator or index species. See BIOGEOCHEMICAL PROSPECTING, BIOLOGICAL BENCH-MARKING, BIOTIC INDEX. Compare BIOMARKER.

indifferent species Species that can live in many different communities or ecosystems. Contrast FIDELITY.

indigenous Native; originating or growing naturally in a specific region. Compare EXOTIC SPECIES.

indirect competition Limitations imposed on the population size or fertility rates of two or more organisms, populations, or species because of a shared dependence on the same limited environmental resource. See COMPETITION, DIRECT COMPETITION.

indirect solar energy Energy derived from sources that are at least one step removed from the energy of sunlight; this includes most traditional energy sources. Wood is a form of indirect solar energy; trees capture the sun's energy through photosynthesis, producing carbon compounds that release energy when burned. These carbon compounds can, over millions of years, become coal or oil. Hydroelectricity, wind, all biomass energy, and clean thermal energy conversion are examples of indirect solar energy. Contrast DIRECT SOLAR ENERGY.

individual distance The maintenance of a specific separation between animals in a social group (flock or herd) through a balance between aggression and forces of social cohesion.

individualistic hypothesis The theory that a community is a collection of species that live together simply because of similarities in their requirements rather than as a result of a long history of co-evolution. Each species has its own unique niche characteristics and geographic distribution, therefore no two biotic communities are identical. Contrast ASSOCIATION.

indole-3-acetic acid The most commonly occurring type of a group of plant growth substances called auxins. Indole-3-acetic acid is a natural compound that promotes the formation of roots, elongation of cells, and other phenomena depending on its concentration. Also called auxin and IAA. See AUXIN.

induced dipole The creation of a dipole (a molecule with no net charge whose ends have opposite electrical charges) in atoms or molecules by the application of an electrical or magnetic field to a solution, or by the presence of an adjacent molecule that can act to polarize the molecule.

inducer A substance that, when it is present, initiates or increases production of an enzyme or other protein.

induction 1) Changing the action of an enzyme, also initiating or increasing its production. 2) Bringing on labor through artificial stimulation (usually, drugs) of the mother. 3) The initiation of structural differentiation in an embryo by the movement of a chemical signal to a specific location. 4) The process by which a conductor (an object capable of conducting electricity) becomes electrified by proximity to—rather than direct contact with—another charged object. 5) The analogous magnetization of an object by proximity to a magnetic field or magnetic flux. 6) The production of an electromotive force within a circuit by changing its associated magnetic field. Compare CONDUCTION.

inductive method   A form of logical reasoning that generalizes from a specific case or cases to produce a universal rule or scientific principle. Contrast DEDUCTIVE METHOD.

induration   1) The formation of brittle hardpan layers within a soil. 2) The hardening of a geologic material by heating, compaction, or cementation.

industrial melanism   The development of (and increase in) populations of a darker variety of a species as an evolutionary adaptation to a darkening of the local environment caused by soot and by air pollution. Sometimes used to refer to the incidence that led to the coining of the term, the appearance of dark forms of a formerly light-colored tree-trunk moth in heavily industrialized areas of England where tree trunks were darker than elsewhere because smog had killed the lichens normally found on the tree trunks. See MELANISM.

inert   Unreactive; not readily changed chemically and therefore unlikely to form compounds with other elements. Argon is inert because it forms no known compounds. See NOBLE GASES.

inert gas   1) Another name for the noble gases. 2) Used generally to refer to gases that are relatively unreactive; nitrogen gas (N_2) behaves as an inert gas in many situations. See NOBLE GASES.

inertia   The tendency for matter to resist any change in motion. Inertia is propor-

tional to an object's mass, so it takes a greater force to put a heavy boulder into motion than a small one.

inertial confinement   A method for short-term containment of the hot plasma in a nuclear fusion reactor. Pelletized fuel is bombarded by lasers (or electron beams) from all directions, causing the solids to vaporize rapidly and then form a plasma. Compression of the plasma produces the high densities and temperatures required to initiate fusion. Compare MAGNETIC CONFINEMENT.

infant mortality rate   Death rate for human babies, calculated as the number of deaths of infants less than one year old per thousand live births.

infauna   The animals that live in the sediments on the floor of the ocean. See EPIBENTHIC, EPIFAUNA.

infection   Condition of disease in animals, humans, or plants from invasion by pathogenic organisms such as bacteria, protozoa, viruses, fungi, or animal parasites. Infection can spread by contact with diseased organisms, airborne microbes, and animal carriers such as mosquitoes.

infertile   1) Unable to produce viable young. 2) Describing a soil that lacks sufficient quantities of nutrients and organic matter, producing stunted crops with poor yields. Contrast FERTILE.

infiltration   The seepage of meteoric water or a fluid into the ground.

infiltration-runoff ratio   An expression of the amount of meteoric water that infiltrates the ground in comparison with the amount that runs off the surface.

inflammatory response   A complex reaction to injury or infection of animal tissues involving increased flow of blood to the area, the production of lymph swelling, plus localized reddening and increased temperature in the affected tissues. Also called inflammation.

inflorescence   A stem on a flowering plant supporting no leaves but including the flower or flower clusters, bracts, and any minor flower branches. Also, the arrangement of flowers on such as stem; different types of inflorescence include the corymb, cyme, raceme, umbel, spike, spikelet, catkin, and panicle.

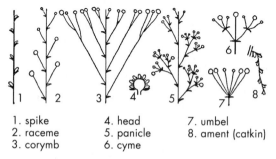

1. spike   4. head   7. umbel
2. raceme   5. panicle   8. ament (catkin)
3. corymb   6. cyme

Types of inflorescence

inflow   A measure of the amount of water entering a steam or hydraulic system.

influent stream   A stream which is topographically above the groundwater table and loses water through infiltration into the streambed.

information   1) Facts, data, or other inputs to be stored in memory. 2) A measure of the complexity and uncertainty in a system, sometimes expressed as the logarithm to the base 2 of the possible combinations.

information theory   The detailed study of transmission and reception of electrical communications. Also used metaphorically for any networks of information flow.

infralittoral   The region of the ocean below the littoral zone. See LITTORAL ZONE.

infrared radiation   (IR) Electromagnetic radiation with wavelengths from about 7000 angstroms to 1 millimeter. In the electromagnetic spectrum, infrared starts just beyond (has longer wavelengths than) visible red light. Water vapor in the earth's atmosphere absorbs much infrared radiation.

infrasonic   Having sound frequencies below what is audible to most human ears (less than 20 hertz).

infrastructure   1) Foundation. 2) All of the permanent, engineered and constructed portions of a community, such as the highways, bridges, and storm sewers.

ingrown meander   A type of incised meander that forms from relatively gradual downcutting and is accompanied by sideward cutting of the stream channel. Ingrown meanders have asymmetric V-shaped valley sides Compare EN-TRENCHED MEANDER.

inheritance   The reception of characteristics transmitted from one generation to the next through genetic material; also, the total group of genetic material and associated characteristics inherited.

inhibit   To limit the growth of cells, tissues, organisms, the activity of an enzyme, or other metabolic process such as the function of a gland or organ.

inhibition   1) The tendency of earlier successional species to resist the invasion of later species. 2) A model of succession based on the premise that biotic communities tend to be dominated by whatever species arrive at a site first and inhibit other species attempting to invade, rather than an orderly, predictable process. This model has also been called the initial floristic composition, a term coined by F. E. Egler in 1954. Contrast FACILITATION, TOLERANCE MODEL.

initial floristic composition   See INHIBI-
TION.

injunction   An order or command to stop or
to start something; used in in civil law
under State and Federal Rules of Civil
Procedure.

inland wetland   Land that is covered all or
part of the year with fresh water, or that
has a water table just below the soil surface.
Examples include swamps, marshes, fens,
and bogs. Compare COASTAL WETLAND.

inlet   A geographic term for a narrow area
of water extending in a landward direction.

inlier   An area of bedrock that is completely
surrounded by areas of newer rock.

innate capacity for increase   Another
expression for biotic potential (r_{max}). See
BIOTIC POTENTIAL.

inner core   See CORE OF THE EARTH.

inoculation   1) The act or process of
introducing microorganisms, cells, or
spores into surroundings that are appropriate for their growth, such as a culture
medium. In humans, this usually refers to
vaccination with weakened disease organisms or antibody serums. 2) To coat seeds
of legume species with nitrogen-fixing
bacteria *(Rhizobia)* before planting in order
to assure nitrogen fixation by the mature
plants. See RHIZOBIA.

inorganic   1) Not coming from animal or
vegetable sources, as in inorganic fertilizers. 2) Describing compounds that either
contain no carbon or contain only carbon
bound to elements other than hydrogen.
Compare ORGANIC MATTER, ORGANIC
FERTILIZER.

inositol $C_6H_6(OH)_6$, an alcohol with a six-
carbon ring structure, also called
cyclohexanehexol. It is contained in some

phospholipids that help to control such
cellular phenomena as the secretion of
adrenaline and the contraction of smooth
muscle tissue. It is sometimes considered a
member of the B-complex vitamin group,
but is not an essential nutrient because it is
synthesized by most animals.

input   Information, energy, or materials
flowing into or placed into a system.
Examples include nutrient deposition into
an ecosystem, data entered into a computer, and power fed into an electronic
device. Contrast OUTPUT.

inquiline   An animal that takes up residence
in the nest or home of another animal.

insect   See INSECTA.

Insecta   A class of uniramous arthropods
with a body distinctly divided into head,
thorax, and abdomen. The thorax bears
three pairs of legs and in most species, two
pairs of wings. Structural and habitat
variability among insects is enormous,
although none is truly marine. Their
ecological importance is manifest; insects
are not only pollinators of more than half
of flowering plant species but also major
crop pests as well as vectors of numerous
plant and animal diseases. The recognized
number of insect species is close to a
million but as many more may be unknown. Among the 27 insect orders are the
Odonata (dragonflies, etc.), Orthoptera
(roaches, grasshoppers, crickets, etc.),
Isoptera (termites), Anoplura (sucking lice),
Hemiptera (true bugs), Homoptera
(aphids, cicadas), Coleoptera (beetles),
Lepidoptera (moths and butterflies),
Hymenoptera (bees, wasps and ants),
Siphonoptera (fleas), and Diptera (true
flies).

insecticide   Any natural or synthetic
compound (pesticide) used to kill insects.

See CHLORINATED HYDROCARBONS, ORGANOPHOSPHORUS PESTICIDES, PYRETHRUM, ROTENONE.

insectavora See INSECTIVORE.

insectivore Any animal or plant that consumes insects as its main food source. The insectivores belong to the most primitive order of placental mammals, the Insectivora. Members of this order are usually small and nocturnal and have craniums with many primitive features including dentition (e.g., shrews, moles and hedgehogs). See EUTHERIA.

inselberg An isolated hill of barren rock standing above a nearby plain in arid and semiarid regions. The term inselberg in technical use is applied to landforms created by a specific morphogenetic process in savanna climates.

in situ In its original, natural, or present position, from the Latin.

insolation Incoming solar radiation. Also, a measurement of this solar energy, the

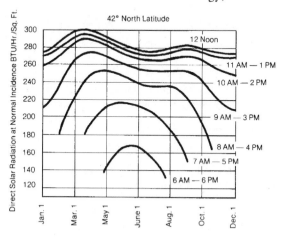

This graph shows the energy per hour that falls on a square foot of collector surface perpendicular to the sun's rays at 42 degrees N. during the year.

amount falling on a surface perpendicular to the sun's rays, of a specified size and over a specified period of time.

instar The stage between moltings for an insect or other arthropod in the larval phase of its development (before formation of a pupa and metamorphosis).

instinct Inborn tendency to respond to stimuli; natural rather than learned behavior.

insular Of or relating to an island or an islandlike entity such as a detached population.

insulate To protect from undesired loss of heat, electricity, or sound, usually by surrounding with a nonconducting material (insulation). Electrical wires are insulated to prevent electricity from leaving the circuit and to reduce the hazards of fire and of electrocution. Homes are insulated both to keep heat in during cold seasons and to keep heat out in hot weather.

integrated pest management (IPM) A strategy for controlling pests that incorporates cultivation methods such as timing of planting or crop rotation and biological controls, such as natural predators, with a greatly reduced use of carefully targeted chemical pesticides. The goal of IPM is not only to reduce costs and pollution associated with chemical pesticides, but also to lower pest levels sufficiently to avoid economic loss while maintaining small populations sufficient to sustain populations of beneficial insects. See BIOLOGICAL CONTROL.

intensity scale See MERCALLI SCALE.

intensive forest management The form of timber cultivation using the maximum amount of human intervention. It consists of first clearing all vegetation from an area, planting a single tree species on it, and using fertilizers and pesticides to raise the resulting even-aged stand of trees to maturity. At maturity, all of the timber is harvested and the cycle is started again.

inter- Prefix meaning between.

inter-arc basin A type of back-arc basin that is formed in oceanic crust and has an abundance of volcanic and turbidite sediments See BACK-ARC BASIN, CONTRAST RETRO-ARC BASIN.

interbedded Material found within or between parallel sedimentary beds. Example: coal interbedded with shales

interbreeding Crossbreeding; the mating of two different varieties or species of animals or plants.

intercalary meristem A meristem (area of concentrated cell division) located somewhere between the base and top of a plant; intercalary meristems are responsible for the growth of blades of grass. Contrast APICAL MERISTEM.

interception 1) The reception of rainfall by leaves, branches, etc., thereby preventing it from falling directly to the ground. 2) The absorption of solar radiation by leaf surfaces.

intercropping Growing two or more different crops at the same time on a piece of land, either to maximize yields or to protect the soil. Intercropping is also practiced in home gardens, where quick-growing vegetables are planted between slower-growing vegetables to make use of the space before it is filled by the slower-growing varieties. Also called interplanting.

interdependent Describing the condition in which different elements in any system interact and depend on each other, as opposed to being completely separate (independent) or having a one-way (dependent) relationship.

interference The superposition of more than one wave, which can result in a combined wave of either greater or smaller amplitude than any of the original waves. Interference is applicable to light, sound, and water waves. It is also used loosely to refer to energy that interrupts signal reception, or the signal "noise" resulting from such interruption. Sixty-cycle current in household wiring can cause a "hum" of interference in audio equipment.

interference competition Competition between two organisms in which one physically prevents the other from occupying a portion of habitat and hence from having access to the resources there. Also called contest competition or direct competition. Some consider allelopathy and other direct forms of inhibition to be forms of interference competition. Compare RESOURCE COMPETITION.

interferon Any of a group of proteins produced by animal cells in response to infection by a virus. Interferon disrupts replication of the virus; it is also able to confer resistance to the virus in other cells that have not been exposed to the virus.

interfluve Land area between two streams flowing in the same direction.

interglacial 1) A term referring to events within an interglacial period. 2) The time between glacial events.

interior drainage A term describing any stream system that discharges directly or indirectly into an enclosed basin and does not ultimately reach the ocean. Compare EXTERIOR DRAINAGE.

intermediate-focus earthquake Any earthquake in which the focus, or point of brittle failure of rock, is located at a depth of between 70 and 300 km in the crust of the earth.

intermediate host An organism on which immature forms of a parasite live, or on which adult parasites live during resting stages between times on a primary (definitive) host. The parasite may reproduce asexually on an intermediate host, but sexual reproduction occurs on the definitive host. See DEFINITIVE HOST. Compare ALTERNATE HOST.

intermediate-level wastes Wastes from nuclear reactors whose radioactivity is too high to be considered low-level waste, but is not as dangerous as high-level waste. Intermediate-level wastes include materials used to clean reactor effluent before discharging and materials from storage areas. See HIGH-LEVEL RADIOACTIVE WASTE, LOW-LEVEL RADIOACTIVE WASTE.

intermediate rock A general term for igneous rocks that are intermediate in composition between mafic rocks and felsic rocks. Andesite and diorite are intermediate rocks.

intermedin A hormone secreted by the pituitary gland to control production of melanin; it is therefore capable of causing changes in skin color. Also called melanocyte-stimulating hormone. See MELANOCYTES.

intermittent stream A stream that flows only part of the year after precipitation events and receives some water during that time from groundwater sources.

internal combustion engine A form of engine using cylinders to contain the combustion of the fuel (as opposed to the furnace of a steam engine). The combustion produces pressure to drive a piston and produce mechanical work. Internal combustion engines are typically two-stroke (as in some motorcycles) or four-stroke (automobiles). See DIESEL.

International Atomic Energy Agency (IAEA) A group formed in 1957 as a private agency for exchange of information about working with radioactive materials; it was later adopted by the United Nations. Headquartered in Vienna, Austria, it now deals with atomic energy and all commercial and scientific uses of radioisotopes, including safety and nuclear proliferation issues.

International Biological Programme (IBP) A worldwide study of biological productivity of terrestrial, freshwater and marine communities, conservation, and human adaptability carried out from 1964 to 1974 by the nongovernmental International Council of Scientific Unions, with participation by over 40 countries.

International Monetary Fund (IMF) An international organization whose purpose is the stabilization of national currencies. It grew out of the 1945 United Nations Monetary and Financial Conference at Bretton Woods, New Hampshire. Compare WORLD BANK.

International Whaling Commission (IWC) A 36-nation association to provide for the conservation of whale populations through regulation and orderly development of the whaling industry. It was established under the 1946 International Convention for the Regulation of Whaling. The IWC has set annual quotas since 1949 and in 1986 established a five-year moratorium on commercial whaling, but it has no authority to enforce these restrictions.

interphase A resting period for a cell between the end of one cell division and the beginning of another. See MITOSIS.

interplanetary magnetic field (IMF) A magnetic field carried by the stream of charged particles known as the solar wind. The IMF can cause magnetic storms on earth.

interplanting Another term for intercropping. See INTERCROPPING.

interradial Located between radiating structures, such as the arms of a starfish.

intersex An organism having sexual characteristics that are intermediate between those of the average female and male organisms of the species. An intersex is often the product of abnormal chromosomes or abnormal hormone production. See FREEMARTIN, GYNANDROMORPHISM.

interspecific Involving or occurring between two separate species, as in an interspecific cross or interspecific competition. Contrast INTRASPECIFIC.

interspecific competition Competition for resources between members of different species. Contrast INTRASPECIFIC COMPETITION.

interstadial 1) A relatively brief interval of warmer climate within a glacial period. 2) The deposits and the time during an interglacial period.

interstate carrier water supply A federally regulated water supply that may be used for human consumption aboard common carriers, such as aircraft, trains, and buses, while traveling between states.

interstate waters A term applied to rivers and watersheds or catchment basins that reside within two or more state political boundaries.

interstitial 1) A term applied to water existing in pore spaces within a soil. 2) A term applied to minerals filling spaces between the earlier formed minerals in a rock.

intersystem nutrient cycle The circulation of nutrients between ecosystems characterized by inputs and outputs. Intersystem transfers may be broken down into meteorological, geological, and biological components. Compare INTRASYSTEM NUTRIENT CYCLE.

intertidal An area of the coastal environment that is between the mean high-tide and mean low-tide water levels.

intolerant Not capable of enduring a specific condition. Calcifuge plants are intolerant of lime and will not survive if grown in alkaline soils. Foresters and forest ecologists use the term to refer to plants requiring relatively high light intensities for their survival, species that are intolerant of shade. Contrast TOLERANT.

intra- Prefix meaning within.

intraneustonic Species of neuston that live suspended from the underside of the surface film of a body of water.

intraspecific Within a single species; involving members of the same community of species, as in intraspecific competition. Contrast INTERSPECIFIC.

intraspecific competition Members of the same species competing for the same resources. Intraspecific competition is usually more intense than interspecific competition because individuals of the same species have nearly identical niches. Contrast INTERSPECIFIC COMPETITION.

intrasystem nutrient cycle The transfers of nutrients within a particular ecosystem. Intrasystem transfers occur among the various living and abiotic components of an ecosystem. Compare INTERSYSTEM NUTRIENT CYCLE.

intrauterine device   (IUD) Birth control device that is inserted into the uterus. Intrauterine devices are very effective in preventing conception; they may be inert, or coated with copper to enhance their contraceptive effect.

intrazonal soil   An older soil classification term for soil orders that are strongly characterized by parent material or age and less influenced by climate, topography, or biologic activity.

intrinsic rate of natural increase   (r) The reproductive rate of increase of a population that has reached a stable age structure without restraints such as competiton. See BIOTIC POTENTIAL.

introduction   Bringing a species of plant or animal into an area where it was not previously found.

introgression   Short for introgressive hybridization, the introducing of one or more genes of one species into the gene pool of another species through hybridization (natural or artificial) between the species. Introgression also occurs between distinct populations of the same species, through repeated backcrossings.

intrusion   A body of magma that penetrates older rock and cools deep below the earth's surface. Compare EXTRUSIVE.

intrusive contact   The surface of contact between an intrusive igneous rock and the surrounding country rock.

intrusive igneous rock   Igneous rocks that cool and crystallize from magma beneath the surface of the earth. Granite is an intrusive igneous rock. Compare EXTRUSIVE.

intrusive relationships, principle of   A relative dating technique stating that an intruding body of magma must be younger than the rock it is disrupting. See CROSS-CUTTING RELATIONSHIPS, PRINCIPLE OF.

intrusive rock   See INTRUSIVE IGNEOUS ROCK.

invasion   1) The introduction and rapid population increase of an organism into a region it had previously not inhabited, often with negative effects. 2) The entry of a disease organism into the human body, and its spread throughout tissues in the early stages of a disease.

inverse density dependence   The phenomenon in which population levels are inversely proportionate to specific environmental factors, that is, the population levels increase as the factors decrease (or the reverse). See DENSITY DEPENDENT.

inverse square law   Any law in which the magnitude of a quantity is proportional to the reciprocal of the square of the distance from the source of the quantity. It has applications in gravity (Newton's law of gravitation), electrostatics, and radio and light waves. See COULOMB'S LAW.

inversion   1) Short for temperature inversion (thermal inversion). 2) The reversal of a portion of a chromosome, so that the DNA sequence for that particular segment is reversed from its previous order. See THERMAL INVERSION.

inversion layer   A horizontal region of the atmosphere that is warmer than the region below it. Inversion layers are responsible for thermal (or temperature) inversions. See THERMAL INVERSION.

invertebrate   An animal without a backbone or internal bony skeleton. All animals except for the phylum Chordata (vertebrates) fall into this category, including insects, crustaceans, worms, corals, and molluscs. Also used as an adjective to describe such organisms. Contrast VERTEBRATA.

inverted relief   A description of a topographic condition in that the surface of the land develops a form which is the reverse of the underlying bedrock structure. A hill developing over a syncline or a valley developing over an anticline is a case of inverted relief.

involucre   1) A protective structure in members of the Asteraceae such as daisies and dandelions, consisting of a circle of bracts (modified leaves called phyllaries) immediately below a flower cluster or compound flower. 2) A sheath in bryophytes (mosses and liverworts) as structures surrounding a cluster of archegonia or antheridia.

involution   1) A regressionary (as opposed to evolutionary) change, such as a decrease in the size of an animal organ, or a degeneration from the usual morphological type in bacteria grown under unfavorable conditions. 2) Describing a rolling inward from the edges of a plant structure, usually a leaf; also a similar rolling inward of developing cells in animal embryos. 3) An irregular, contorted, or distorted surface of unconsolidated regolith or soil. Involution typically occurs as a consequence of frost wedging in periglacial soils.

iodide   1) The ionic form of the element iodine, where the iodine atom has gained one electron (I-). 2) Any salt containing the I-ion, such as potassium iodide (KI). Many metals readily form iodides.

iodine   (I) A blackish-purple halogen element that exists in elemental form as the diatomic molecule I_2. It has atomic number 53 and atomic weight 126.9. It is less reactive than the other halogens fluorine, chlorine, and bromine and is found naturally in combination with other elements. Iodine, dissolved in alcohol, is used as an antiseptic. It is also used extensively in manufacturing organic chemicals. Iodine is an essential nutrient in trace amounts, required for healthy thyroid metabolism.

ion   An atom or molecule that has become either positively or negatively charged by losing or gaining an electron. A solution of ions will conduct electricity. See ANION, CATION.

ion chromatography   A technique for separating a mixture of ions for analysis, and for measuring or estimating the quantities of the different ions present. Typically, an aqueous sample containing ionic substances is passed over an adsorptive medium (such as an ion exchange resin) that selectively binds the component ions for differing lengths of time. Each component ion has a characteristic retention time. See ION-EXCHANGE RESIN, RETENTION TIME.

ion-dipole interactions   Phenomena caused by the electrical attraction of dipoles for ions of the opposite charge. While weaker than the bonding that occurs between ions, ion-dipole interactions are important in the creation of active sites on complex metabolic molecules such as enzymes.

ion exchange   A trading of ions between a solution and a solid (such as an ion exchange resin) that comes in contact with the solution. Ion exchange is used to soften hard water by exchanging sodium for calcium. Ion exhange also occurs in soils when fertilizer is applied and percolates through the soil, exchanging ions between a water solution of the fertilizer ions and the surface of the soil particles. See HARD WATER.

ion-exchange capacity The potential for substitution of ions in the crystal lattice of a mineral without any disruption of the lattice. See EXCHANGEABLE CATIONS.

ion-exchange resin Any of several synthetic polymers used (in ion chromatography and in water softeners) for their ability to trade ions with a solution that is poured over them. Such resins are usually organic (carbon-based). The ion-exchange resin is the medium in which the process of ion exchange takes place.

ionic bond A link between two atoms created by the transfer of an electron from one to the other, resulting in electrostatic attraction between the cation (positively charged ion) and the anion (negatively charged ion). Ionic compounds bound by such attraction usually dissolve readily in water, have high melting points and boiling points, and form solutions that conduct electricity. See ELECTRON DONOR, ELECTRON ACCEPTOR. Compare COVALENT BOND.

ionic radius The size (radius) of an ion contained in a solid crystal. The value for this radius is calculated from x-ray measurements of distances between the nuclei of atoms in crystals, using the assumption that ions are spherical and have a definite size. The largest atomic radii are about one nanometer (10^{-9}m).

ionic strength A value for the intensity of the electrical field in a solution of electrolytes. Ionic strength is calculated by multiplying the concentration per unit of solute of each ion by the square of the charge of that ion.

ionization Formation of atoms, molecules, or parts of molecules carrying a net electrical charge. Ionization occurs when ordinary atoms gain or lose an electron and thus become ions.

ionization chamber An instrument for measuring ionizing radiation. Somewhat like a Geiger counter, it consists of a sealed vessel containing a gas and electrodes. As radiation is passed through the vessel, it ionizes the gas. The number of ions generated is measured by collecting the ions' charge on electrodes.

ionization energy The amount of energy needed to remove an electron from a neutral, isolated atom in its ground state. See ENERGY LEVEL, GROUND STATE.

ionizing radiation Forms of radiation with enough energy per interaction to ionize matter. Ionizing radiation includes streams of subatomic particles such as electrons, as well as short-wave radiation such as x-rays and gamma rays. Although, ionizing radiation can cause mutations, it also has medical uses such as cancer treatment as well as x-ray diagnosis.

ionosphere The outer layer of the earth's atmosphere, in which atoms and molecules have been ionized by solar radiation. It is sometimes called the thermosphere because the absorption of solar radiation causes high temperatures here; the temperature increases with altitude, in contrast to the lowest layer of earth's atmosphere. See D-LAYER, E-LAYER, F-LAYER.

ionospheric storm Turbulence in the outer layer of the ionosphere. Ionospheric storms are associated with disruptions in the earth's magnetic field and disrupt short-wave radio transmission.

ion pair A positive ion and negative ion produced from the same starting molecule. An example is the formation of the ion pair $CH_3^+I^-$ from the molecule CH_3I. Ion pairs are typically short-lived, either reacting quickly with other molecules or recombining to form the starting material.

IPM See INTEGRATED PEST MANAGEMENT.

Irano-Turanian floral region One of the phytogeographic regions into which the Holarctic realm is divided according to similarities between plants; it consists of central Asia north of the Himalayan mountains, east to the Black Sea, and west to the middle of China. See HOLARCTIC REALM, PHYTOGEOGRAPHY.

Iridaceae The iris family of monocot angiosperms. These herbaceous plants have straplike leaves with bases that overlap. The three sepals are often petaloid and frequently have a stamen attached to each. The three styles are also sometimes petaloid. The three-chambered inferior ovary produces a capsule fruit at maturity. *Crocus, Iris, Fresia, Gladiolus*, and blue-eyed grasses *(Sisyrinchium)* are members of the iris family. See ANGIOSPERMOPHYTA.

iron (Fe) A magnetic, metallic element with atomic weight 55.85 and atomic number 26. It is the most common and most used of all the metallic elements, incorporated into many commercial metals. Iron is found at the center of the hemoglobin molecule that transports oxygen throughout the human body.

iron meteorite A meteorite primarily composed of iron and nickel, as well as smaller amounts of silicate minerals. Iron meteorites are far less abundant than stony meteorites.

iron ore Any deposit containing commercially valuable iron minerals. See HEMATITE, LIMONITE, BANDED IRON FORMATION.

iron pan A soil hardpan layer that is cemented by iron oxides. See HARDPAN.

iron pyrites See PYRITE.

irradiate To expose to radiant energy such as ultraviolet rays or ionizing radiation such as gamma rays or x-rays.

irrigate To deliver water artificially to cropland. Irrigation water may be sprayed directly on plants, dripped on soil, or supplied as runoff from pipes and canals.

irrigation A system for, or the process of, bringing water to plants to help them grow and to supplement scant natural rainfall. See CENTER PIVOT IRRIGATION, DRIP IRRIGATION.

irritability The ability of a living plant or animal to register, interpret, and respond to stimuli. Also called sensitivity.

irruption An unusually sudden, rapid surge in population growth of an animal species.

isallobaric winds Winds showing equal atmospheric pressure over a period of time.

island arc A curving sequence of volcanoes that forms on the upper crustal plate of a tectonic subduction zone. The volcanic rocks are usually of intermediate composition and are derived from the thermal activity of the Benioff zone or rising diapirs.

island biogeography The study of distribution of species and community composition on islands. See EQUILIBRIUM THEORY OF ISLAND BIOGEOGRAPHY.

iso- From the Greek word meaning equal, used as a prefix to lines on maps linking points with similar values or quantities.

isobar Line on weather maps connecting points with the same barometric pressure (corrected for height above sea level).

Groups of isobars show distribution of high- and low-pressure regions and are therefore useful in weather prediction. Compare ISOTHERM.

isocline A geology term for tightly packed parallel overfolds, all limbs dipping at approximately the same angle, and in the same direction. If the folding is still more intense, the folds rupture and become imbricated. See IMBRICATE.

isocryme A figurative map line connecting points that share the same mean temperature at the coldest time of the year.

isogamy The union of two similar gametes; describing species whose male and female gametes are similar. Compare HETEROGAMY, OOGAMY.

isogenic Having identical genes, used to describe a strain of genetically identical, although not necessarily homozygous, individuals. Also called syngenic or isogeneic.

isograd A figurative map line connecting points that share the same temperature conditions. Isograds are commonly used to illustrate metamorphic grade on a map of regionally metamorphosed rock.

isohaline Having equal amounts of salinity.

isohel Line drawn on a map joining points of equal amounts (average number of hours) of sunshine.

isohyet Line drawn on a map joining points receiving equal average amounts of rainfall.

isolateral Possessing mirror-image symmetry in two different directions; bilaterally symmetrical in two different planes. Used to describe leaves that are not dorsoventral but have palisade tissue (long, narrow cells) underlying both the upper and lower surfaces. Compare DORSIVENTRAL.

isolating mechanism Any factor that restricts or prevents exchange of genetic material between different populations of a species, including geographic isolation and genetic differences that interfere with interbreeding. See GEOGRAPHIC ISOLATION.

isolation 1) Lack of competition between two species because of differences in food, habitat, activity period, or geographical range. 2) Lack of interbreeding between different populations of a species. Isolation may be caused by geographic factors, or differences in physiology or behavior.

isoleucine $CH_3CH_2CH(CH_3)CH(NH_2)COOH$, an essential amino acid found in body tissues and in plant proteins. It must be supplied through diet.

isoline Line drawn on a map joining points of equal value. Isolines include isobars, isohels, and isohyets. Also called an isogram or isopleth.

isolume Line drawn on a map joining points of equal light intensity. Also called isolux.

isomer Chemical compounds with the same molecular formula but a different physical arrangement of atoms. Many organic compounds exist as different isomers, such as the hydrocarbons butane and isobutane.

isometric contraction A form of muscle contraction in which the muscle becomes taut but is prevented from shortening in length. In an isometric contraction, all of the energy is converted into heat and no mechanical work is performed.

isopach A type of contour line used to illustrate the thickness of strata on a geologic map.

Isopoda A crustacean order with members having dorsoventrally flattened bodies without a carapace and with the first 1 or 2 thoracic segments fused to the head. The order includes marine, freshwater, and terrestrial forms. Some, such as pill bugs, are commonly found in leaf litter, rotting wood, etc. They are also called wood lice or sow bugs. Their distinctly segmented body can be rolled into a ball. There are seven pairs of walking legs. See MALACOSTRACA.

Isoptera Termites, sometimes incorrectly called "white ants", form this order of insects. Termites have evolved an elaborate social structure which includes soldier, worker, and reproductive castes. All species live in some kind of nest and are capable of digesting cellulose either on their own or, in more primitive groups, through the aid of cellulose-digesting symbionts which live in their gut. Some species build huge nests of mud reinforced by saliva, which may be in or above ground or in trees. They are an important link in tropical ecosystems because of their ability to break down plant products, but this very ability also makes them economically disastrous in the destruction of wood products, houses, fences etc. Approximately 2000 species. See INSECTA.

isoseismic line A figurative line drawn on an earthquake map to show points of equal earthquake intensity or points of equal earthquake frequency.

isostasy A theoretical model to account for the thickness and elevation of the earth's crust. In isostasy, the weight of the upper crust is compensated by the buoyancy of a mass of deeper crustal material.

isostatic adjustment The flexural compensation of the earth's lithosphere in response to changes in the gravitational load of the crust. The gravitational load is effected by glaciation, erosion, or sedimentation.

isotach Line drawn on a map joining points of equal wind speed.

isotherm Line drawn on a map joining points of equal temperature.

isotonic Describing solutions having the same osmotic pressure.

isotope Atoms of the same element that have different numbers of neutrons in the nucleus, while retaining the same number of protons and electrons. The different number of neutrons gives each isotope a different atomic weight and slightly different properties (such as radioactive half-life). Most elements exist naturally as different isotopes; hydrogen (0 neutrons) also exists as deuterium (1 neutron) and tritium (2 neutrons). Compare NUCLIDE.

isotropic Uniform in all directions.

isthmus A narrow strip of land bounded by water on two sides that connects two larger landmasses. The Isthmus of Panama, which connects the continents of North America and South America, is a classic example.

itai-itai disease Literally, pain-pain in Japanese; a bone disease characterized by great pain and caused by prolonged, low-level cadmium poisoning. Compare MAD HATTER'S DISEASE, LEAD POISONING.

iterative evolution The repeated, independent evolution of the same adaptive type found more than once from a single ancestral group.

iteroparity The production of offspring in a series of separate events, occurring two or more times during the life span of an organism. Contrast SEMELPARITY.

IUCN Abbreviation for International Union for the Conservation of Nature.

Izaak Walton League of America A membership organization promoting public education in conservation, mainte-nance, restoration, and protection of U.S. natural resources, including soil, forest, air, and water. The group promotes outdoor recreation and enjoyment of these natural resources.

j

J Abbreviation for the unit joule. See JOULE.

Jaccard index An index of similarity (coefficient of community), a measurement of the degree to which two plant communities resemble each other based upon their species composition. It is calculated two different ways. It can be calculated as a percentage derived from dividing the number of species common to two communities by the sum of the number of species unique to one community plus the number unique to the other plus the number of species they have in common. It can also be calculated by dividing the number of species in common by the sum of the total number of species in each community minus the number of species in common. Compare SØRENSEN SIMILARITY INDEX, SIMPSON'S INDEX OF FLORISTIC RESEMBLANCE, GLEASON'S INDEX, MORISITA'S SIMILARITY INDEX, KULEZINSKI INDEX.

Japan Current See KUROSHIO CURRENT.

jasper A characteristically red quartz mineral that is used as an ornamental gemstone. The color of jasper is due to iron impurities in the cryptocrystalline quartz structure.

jet 1) A stream of fluid moving at high speed and issuing from a small opening. 2) JET is short for Joint European Torus. See JOINT EUROPEAN TORUS.

jet stream Fast-moving, relatively narrow flow of air in the upper troposphere (10 to 15 miles above the earth's surface). These wandering air currents usually travel from west to east at speeds over 100 mph and affect weather patterns. Sometimes called the polar jet stream. Compare SUBTROPICAL JET.

jetty A structure to stabilize the location of a channel, shield vessels from waves, and affect sand movement at an inlet or the mouth of a river. See GROIN.

Joint European Torus (JET) An experimental nuclear fusion facility at Culham in the United Kingdom, sponsored by the European Economic Community. It uses a tokamak (a torus, or doughnut, shape) for magnetic confinement of the plasma for fusion. See TOKAMAK.

joint set Repeating pairs of brittle fractures arranged at fixed angles to one another in a body of rock. Intersecting joint sets divide the rock into regular block or prismatic shapes.

Jordan's laws 1) The most closely related species are located nearest to each other, but separated by a geographical barrier. 2) Within a given species of fish, individuals inhabiting cold environments tend to develop more vertebrae then those in warm environments.

joule (J) The unit for measuring work, energy, and heat in the Système International d'Unités (SI). One joule equals 10^7 ergs. A 1-watt electric clock uses 1 joule of electrical energy each second.

Jovian planets The giant planets, those planets resembling Jupiter in size and gaseous structure: Uranus, Neptune, Saturn, and Jupiter. Contrast TERRESTRIAL PLANETS.

J-shaped curve A curve that illustrates exponential (or geometric) growth, especially in a population of organisms, in which a small number quickly accelerates to a much larger population. When plotted, such growth produces a shape resembling the letter J. Compare S-SHAPED CURVE.

judicial branch   One of the three branches of government in the United States, it deals with lawsuits and appeals and includes both federal and state systems. Independent of the legislative and executive branches, it is part of the checks and balances built into the structure of the U.S. government. The federal and state systems exist side by side, exercising exclusive jurisdiction in some areas, and overlapping or concurrent in others. The federal system, which includes the Supreme Court, is less complicated than the various state judicial systems. No two states have exactly the same court system. Compare EXECUTIVE BRANCH, LEGISLATIVE BRANCH.

judicial review   The review by a court of law of some act or failure to act by a government official or entity, or by some other legally appointed person or organized body; the review of the decision of a trial by an appellate court.

Juglandaceae   The walnut family of the Angiospermophyta. These monoecious, deciduous dicot trees bear male flowers in catkins whereas female flowers are in clusters. The fruits are a large, single-seeded bony nut covered by a husk. Walnuts and butternuts (*Juglans*) and hickories and pecans (*Carya*) are members of this family. See ANGIOSPERMOPHYTA.

Juncaceae   The rush family of monocot angiosperms. These grasslike plants have flattened or rounded stems that are solid, rather than hollow as in members of the Poaceae. The small flowers have two whorls of three bracts that surround the three-to-six stamens and single pistil of three carpels. Rushes (*Juncus*) and wood rushes (*Luzula*) are members of this family. See ANGIOSPERMOPHYTA.

jungle   A dense tropical forest characterized by much undergrowth, with many shrubs and vines in addition to trees. The term is associated with the early phasesin recovery of a tropical forest following disturbance such as clearing, agriculture, or forestry. Compare RAINFOREST.

Jurassic   The second of three geologic time periods of the Mesozoic era. The Jurassic period lasted from approximately 200 to 144 million years ago. See APPENDIX p. 610.

jurisdiction   The authority to hear and determine a case.

jury   A group of people summoned and sworn to decide on the facts in issue at a trial. Composed of the peers of the parties involved, or a cross-section of the community.

juvenile   Not adult; physiologically undeveloped or immature. Compare ADULT.

juvenile hormone   A hormone produced by insects (secreted by the corpus allata) that suppresses development of adult features in all larval stages of development, those before metamorphosis into adulthood. Juvenile hormones are used in biological pest control; when synthesized and sprayed on crops, they can prevent all of the larvae of a pest species in the area from maturing into breeding adults, greatly reducing the size of the subsequent generation of the pest. Also called neotenin. Compare ECDYSONE.

juvenile water   Water that is chemically derived from magma during the process of mineral formation. Juvenile water has never been circulated in the water cycle.

k

K 1) Symbol for prefix kilo. See KILO. 2) Symbol for kelvin. See KELVIN. 3) Chemical symbol for the element potassium. See POTASSIUM. 4) (K) Symbol for ecological carrying capacity, the equilibrium population size (the upper asymptote of an S-shaped growth curve). See CARRYING CAPACITY, S-SHAPED CURVE.

kala-azar Hindi for black fever, an infectious tropical disease caused by a protozoan, *Leishmania donovani*, that is transmitted to humans by the bites of the intermediate host, sandflies *(Phlebotomus)*. Kala-azar is characterized by liver and spleen enlargement, fever, and general deterioration; it is often fatal. Also called LEISHMANIASIS.

kame A deposit of stratified glacial drift in isolated mounds or steep-sided hills. Kames usually form in openings between blocks of ice.

kame terrace A deposit of stratified glacial drift formed in an opening between a block of ice and the wall of a valley.

kaolin A naturally weathering product of feldspar that forms a variety of clay minerals in old or acidic soils. Kaolin is used in ceramics, to make glossy papers, and in a stomach medicine Kaopectate.

kaolinite The principal clay mineral of kaolin. Kaolinite forms monoclinic crystals of aluminum, silicon, and oxygen linked in a sheetlike crystal lattice. See APPENDIX p. 624.

kaolinization The process of kaolin formation in a soil or a rock.

Karst A region of Yugoslavia that has a characteristic topography dominated by the formation of caves and sinkholes in limestone.

Karst topography A description of landscapes characterized by the formation and collapse of caves and sinkholes because of carbonation weathering of a limestone bedrock.

karyogamy The union of cell nuclei.

karyokinesis The even splitting of the cell nucleus, as occurs in mitosis. Compare CYTOKINESIS.

karyotype The chromosome array characterizing a particular individual or species; the arrangement, number, size, and shape of the chromosomes making up the set found in each cell within the organism. The human karyotype consists of 22 pairs of chromosomes common to men and women, plus one pair designated XX in women and one pair designated XY in men.

katabatic wind A local wind consisting of cold, dense air flowing down the side of a mountain or plateau because of gravity. Katabatic winds can be extremely strong at the edges of the polar ice caps. Sometimes called a fall or gravity wind. Katabatic winds include the mistral of France, the levanter that blows through the Strait of Gibralter, and the pampero of Argentina. See MISTRAL.

katadromous See CATADROMOUS.

kelvin (K) Unit for temperature in the Système International d'Unités (SI). One degree kelvin equals one degree Celsius, but is measured from absolute zero rather than the freezing point of water so that 0° C = 273.15K. See CELSIUS, FAHRENHEIT.

keratin   A tough, complex protein that is the primary constituent of horn, nails, hair, and feathers. Keratin is unaffected by most enzymes that break down protein; it is also insoluble in water and weak acids or bases.

keratinized   Toughened or hardened; describing soft tissue that has hardened to a consistency more like horn.

kerogen   The component of bituminous fossil hydrocarbon found in oil shales. Kerogen can be extracted and refined for oil production.

kerosene   A very light liquid fraction distilled from petroleum and used as a fuel and a solvent. Also called paraffin oil, it is a mixture of hydrocarbons having 11 or 12 carbon atoms.

ketones   A group of very reactive organic compounds containing a carbonyl (-CO) group between two hydrocarbon groups. They can be recognized by the suffix -one in their name; acetone (H_3C-C(O)-CH_3) is a familiar ketone.

ketose   Any member of the group of sugars containing a carbonyl (-C=O-) radical connecting two smaller carbon chains.

kettle   See KETTLE HOLE.

kettle-and-kame topography   An irregular rolling topography characterized by glacial kame deposits pitted with kettle holes. See KAME, KETTLE HOLE.

kettle hole   A depression in an outwash plain or body of glacial drift. Kettle holes form when a block of ice is isolated from a receding glacier, becomes buried in outwash or till, leaving a void as the ice melts.

kettle lake   A water-filled kettle hole.

key bed   See MARKER BED.

key fruit   See SAMARA.

keystone species   Organisms that play dominant roles in an ecosystem and affect many other organisms. The removal of a keystone predator from an ecosystem causes a reduction of the species diversity among its former prey.

kg   See KILOGRAM.

khamsin   A very hot wind that periodically blows northward from the Sahara over Egypt and the Red Sea. It is most common in the spring and early summer months.

k horizon   See K SOIL HORIZON.

kidney iron ore   A form of hematite occurring in kidney-shaped concretions.

killing power   An expression of the intensity of age-specific mortality rates in a life table. It is calculated by subtracting the log of survivors in the next age class from the log of survivors in the previous age class. See K-VALUE.

kilo   (K) 1) Prefix used in the Système International d'Unités (SI) to denote one thousand, 1 x 10^3. A kilowatt is 10^3 watts. 2) Short for kilogram. See KILOGRAM.

kilocalorie   (kcal) One thousand calories; the amount of heat required to raise a kilogram of water by one degree Celsius. Kilocalories are also called Calories, especially when describing the energy content of foods. See CALORIE.

kilogram   (kg) One thousand grams; the basic unit of mass in the MKS (a variant of the metric) system. It is the only basic unit still defined in terms of an object, a cylinder of platinum-iridium used as the international standard. See GRAM.

kilometer   (km) Unit of distance in the Système International d'Unités (SI), equal to 1,000 meters.
   1 km = 1094 yards = 0.621 miles.

kilowatt (kw) One thousand watts. The kilowatt is a more convenient unit than the watt when discussing large quantities of energy such as the output of electrical power plants. One kilowatt is approximately 1.34 horsepower. See WATT.

kilowatt-hour (kwh) A unit for measuring electrical energy. One kilowatt-hour is the energy delivered by 1000 watts during one hour, or 3.6 million joules. See WATT.

kimberlite A brecciated ultramafic igneous rock containing minerals formed at high pressures in the mantle of the earth. Diamond is one of the valuable crystals found in kimberlite.

kinase An organic substance that can activate a proenzyme, causing it to become an active enzyme. See ZYMOGENOUS.

kinesis An animal's involuntary movement or reaction to an environmental stimulus. The extent of kinesis is proportional to the strength of the initial stimulus, but (unlike a taxis) the movement is not oriented toward or away from the stimulus. Compare TAXIS.

kinesthetic sense The perception of body position, muscle effort, and movement (extent, direction, etc.) within an organism. The sensors of kinesthesia are located at the ends of muscles and tendons and in joints.

kinetic energy The energy by virtue of mass in motion. The kinetic energy of a non-relativistic moving object equals half the product of the object's mass and the square of its velocity. Contrast POTENTIAL ENERGY.

kinetin A plant growth substance (a synthetic cytokinin) that causes cell division in plants. Its technical name is 6-furfurylaminopurine. See CYTOKININS.

kingdom The highest taxonomic division of living organisms. Historically, organisms have been classified into the plant kingdom and the animal kingdom, the former made up of a number of divisions, the latter made up of a number of phyla. Today, organisms are typically grouped into five kingdoms: prokaryotes (Monera), one-celled and simple eukaryotes (Protista), Fungi, plants (Plantae), and animals (Animalia).

kingfishers See ALCEDINIDAE.

kin selection A type of natural selection favoring a particular altruistic trait in an organism because the trait enhances the survival of individuals (kin), and these carry the same trait (alleles) by common descent. See NATURAL SELECTION, SOCIOBIOLOGY.

klendusity Avoiding disease by growth habit, as by maturing before a fungus produces spores.

klepto-parasite Oraganisms that steal the food of another organism.

klippe A geologic outlier that was emplaced by tectonic action or faulting and which was later isolated by erosion.

klippen See KLIPPE.

km See KILOMETER.

knephoplankton Species of plankton that are found at depths from 30 to 500 meters.

knot A unit of speed equal to one nautical mile per hour (1.15 mph). It is used to describe the speed of winds or ships.

kopje A small steep-sided hill, often forming as an erosional remnant of a granitic bornhardt.

Kr Chemical symbol for the element krypton. See KRYPTON.

Krebs cycle   The repeating series of chemical reactions within cells that converts food into energy. It is the primary reaction of aerobic respiration in plants and animals and produces ATP (adenosine triphosphate). Also called the citric acid cycle or tricarboxylic acid cycle. See ADENOSINE TRIPHOSPHATE.

krill   Small, shrimplike animals (Order Euphausiacea) that live in cold regions of the oceans and comprise the basic diet of many marine animals such as whales. See EUPHAUSIACEA.

krummholtz   Stunted form of trees found growing above the timberline of the subalpine coniferous forest, in the ecotone between the alpine boreal forest and the alpine tundra. The term is German for crooked wood.

krypton   (Kr) A relatively inert gaseous element with atomic weight 83.80 and atomic number 36; it is one of the noble gases. Krypton is used in lasers and bulbs for fluorescent lamps, and is found in minute quantities in the earth's atmosphere.

K selection   Factors (selection pressure) favoring species having lower rates of population growth but an increased ability to compete against other species in more densely populated communities characteristic of later stages of ecological succession. The factors contributing to keep a population at close to its carrying capacity (K) include large size, delayed reproduction, iteroparity, a limited number of offspring, and a high degree of parental attention. Compare R SELECTION.

K soil horizon   A term applied to a soil horizon in which the grains are cemented with $CaCO_3$ to form a hardpan layer.

K-strategists   Species adapted to the conditions of K selection, having lower rates of potential population growth but higher competitive abilities. K-strategists tend to be large organisms that live a long time, have few offspring and care for them, and usually have populations that stabilize at the carrying capacity. Also called K-selected species. Compare R-STRATEGISTS.

Kulezinski index   An index of similarity (coefficient of community), a measurement of the degree to which two plant communities resemble each other based upon their species composition. It is calculated by dividing the number of species common to two communities by the sum of the total number of species in each community. Compare SØRENSEN SIMILARITY INDEX, SIMPSON'S INDEX OF FLORISTIC RESEMBLANCE, GLEASON'S INDEX, MORISITA'S SIMILARITY INDEX, JACCARD INDEX.

Kuroshio Current   A western boundary surface current of the Pacific Ocean. The warm Kuroshio Current flows from the Phillipines northward along the coast of Japan, and out into the main gyre of the Pacific Ocean.

k-value   An equation describing the effects of intraspecific competition on mortality, fecundity, and growth. K-value is calculated by subtracting that of the final density from that of the initial density. Also called killing power.

kw   Symbol for kilowatt. See KILOWATT.

kwh   See KILOWATT-HOUR.

kyanite   An aluminum silicate mineral that forms bluish triclinic crystals in relatively high-pressure, low-temperature, metamorphic environments. Kyanite is chemically equivalent to andalusite and sillimanite. See APPENDIX p. 624.

l

l Abbreviation for liter in the metric system. See LITER.

L$_x$ noise levels A level of noise that exceeds acceptable (or ordinary) levels for a specified percentage of time; L$_{25}$ noise levels are those that are excessive for 25 percent of the time.

Labrador Current An ocean current that carries arctic water southward along the western coast of Greenland. The Labrador current converges with the Gulf Stream along the coast of Newfoundland.

Labyrinthulamycota The slime net phylum of the kingdom Protista. These colonial organisms of spindle-shaped cells move on slimeways they secrete from specialized pits. The labyrinthulids are primarily parasites of marine algae and higher plants, although some inhabit aquatic environments and soils. *Labyrinthula*, a parasite of the marine angiosperm eel-grass *(Zostera)*, is perhaps the best known member of the phylum. Die-off of eel-grass has produced subsequent declines in waterfowl and mollusk populations. See PROTISTA.

laccolith A large concordant pluton having a mushroomlike shape. The intrusion of a laccolith causes a dome-shaped bulge to form in the overlying country rock.

Lacertilia The reptile suborder commonly known as lizards. They are adapted to a wide range of habitats, including extremely arid regions. Most species are quadrupedal, but some can run on their hindlegs, and a number such as the slow worms are legless. Most small lizards are insectivorous whereas larger ones may be either herbivorous or carnivorous. The Komodo monitor is the largest living lizard and is capable of preying on deer and pigs. Over 3300 species include geckos, iguanas, agamas, skinks, beaded lizards, monitors, etc. See SQUAMATA.

lactate dehydrogenase (LDH) An enzyme that catalyzes the oxidation of lactate (a salt or ester of lactic acid) into pyruvic acid, a process that occurs during glycolysis. Also called lactic dehydrogenase. See GLYCOLYSIS.

lactic acid $CH_3CHOHCOOH$, an organic acid with two isomers. It is produced by bacterial action in souring milk and as a by-product of glucose in muscle exertion.

lactoflavin See RIBOFLAVIN.

lactogen See PROLACTIN.

lactose $C_{12}H_{22}O_{11} + H_2O$, also known as milk sugar. It is a disaccharide consisting of glucose linked to galactose and found only in milk. Lactose is converted into lactic acid by bacteria in the souring of milk and in the formation of cheese.

lacustrine A term describing a relationship to a lake or lake process. For example, lacustrine silts are those deposited on the bottom of a lake.

lag 1) A period of time occurring between a stimulus and the reaction to the stimulus. 2) The latent period, or period of slow growth, following the introduction of bacteria into a culture medium. 3) A similar interval of inaction or slow growth in any system.

lag deposit A residual deposit of unconsolidated particles that remains in place after a current has carried away the finer particles from the original mass.

lagomorph Any member of an order of mammals (Lagomorpha) having two pairs of upper incisor teeth, one pair behind the other. These gnawing animals include rabbits, picas, and hares. See RABBITS.

lagoon A body of shallow ocean water that is connected to an ocean by a narrow channel. A lagoon is usually formed by growth of an enclosing reef.

lahar An energetic and, therefore, potentially destructive mud flow on the slopes of a volcano. Lahar material may be a combination of pyroclastic debris, water, and sometimes soil, glacial ice, or snow-melt.

LAI See LEAF AREA INDEX.

laissez-faire A doctrine that an economic system or plan functions best when there is little or no interference by government. From the French for leave alone.

lake A body of fresh water that is deep enough so that rooted plants do not grow all the way across the bottom.

Lamarckism A theory proposed by the French biologist Jean de Lamarck (1744–1829) to explain evolution, stating that habits or traits acquired in a single generation can be incorporated into the genetic material and transferred to later generations. Lamarckism explains the long neck of a giraffe as resulting from the stretching (and the resulting lengthening) of an individual giraffe's neck being passed onto its offspring, which undergo the same process until a very long neck develops. Contrast NATURAL SELECTION, LYSENKOISM.

lambert A unit for measuring brightness. One lambert equals the brightness at a uniformly diffusing surface that emits or reflects one lumen per square centimeter. It is often more convenient to use millilamberts. See LUMEN.

Lamiaceae The mint family of dicot angiosperms, sometimes referred to as the Labiatae, because of their two-lipped flowers. The stems in this family are usually square in cross-section and the usually aromatic leaves are attached in an opposite arrangement. The flowers are mostly highly zygomorphic and usually two-lipped, the upper having two lobes and the lower having three lobes. There are two or four stamens that are attached to the upper lobes, and the five sepals are united. The superior ovary froms one-to-four nutlets at maturity. Sages *(Salvia)*, mints *(Mentha)*, thyme *(Thymus)*, self-heal *(Prunella)*, rosemary *(Rosmarinus)*, and dead-nettle *(Lamium)* are members of this family. See ANGIOSPERMOPHYTA.

lamina A very small-scale bed form that is less than 1-cm thick within a layered deposit or rock.

laminar flow Smooth, nonturbulent flow of fluids in which adjacent layers undergo very little mixing. Compare TURBULENCE.

lamination A sedimentary deposit structure composed of lamina.

lampreys A group of jawless vertebrates of the class Agnatha, which lack paired fins and have naked, elongated bodies bearing well-developed eyes and seven pairs of gill openings. Adult lampreys have funnel-shaped, fleshy discs surrounding their mouths, which are studded as are their tongues, with numbers of horny teeth. Predatory species feed on other vertebrates. Its oral disc attaches a lamprey to its prey and, together with the tongue, is used to erode the prey's integument so that its blood and other fluids can be rendered. The marine lamprey *Petromyzon marinus*, which was introduced to the Great Lakes in the 19th century, caused a serious decline in lake trout and continues to affect the lake fishing industry. All lampreys must migrate up freshwater tributaries to breed. See AGNATHA.

land breeze   A local wind occurring at the coast in good, otherwise calm weather. As the land cools at night, the cool air begins to blow toward the warmer, less-dense air above the sea. During the afternoon the process reverses: As the land warms, cool air moves in from the water to create a sea breeze.

land bridge   A terrestrial connection between adjacent landmasses. See BERINGIA.

land ethic   A conservation philosophy developed by Aldo Leopold (author of *Sand County Almanac*) promoting a respectful attitude toward not only soils but also water resources, plants and animals: "We abuse land because we regard it as a commodity belonging to us. When we see land as a community to which we belong, we may begin to use it with love and respect." See ENVIRONMENTAL ETHICS.

landfill   See SANITARY LANDFILL.

landform   A general description for any feature of topography that has a characteristic shape.

landmass   Any area of land that is geographically large or distinct.

land race   1) Native species of plant or animal which has not been cultivated. 2) An ancient form (cultivar) of a modern crop plant. Einkorn and emmer are land races of wheat.

Landsat   A remote-sensing project consisting of a series of satellites carrying multi-spectral scanners that are used primarily in vegetation surveys and thematic mapping of the earth.

landscape ecology   A branch of ecology that studies the effects of topography and soils on biological communities and succession, as well as the regional distributions of biotic communities.

landslide   A general term for mass wasting characterized by a downslope movement of earth materials. See MASS WASTING.

land subsidence   The term applied to a localized sinking of the ground surface. Land subsidence may be due to natural phenomena, such as karst activity, or to an artificial cause such as groundwater withdrawal.

land use planning   Process for determining the preferred present and future use of each parcel of land in an area. Land use planning may be carried out by local, regional, or state bodies.

land use planning law   Law establishing a process for land use planning and/or for implementation of a land use plan.

langley   A unit for solar radiation. One langley equals one calorie of solar radiation falling on one centimeter.

lapilli   Volcanic pyroclastic particles the size of gravel. Lapilli may also be composed of sticky accretions of finer ash. Singular is lapillus.

lapse rate   Meteorological term for amount of decrease of temperature with increase in altitude. The average lapse rate is approximately 6.5°C for every kilometer of height in the lowest layer of the earth's atmosphere, but actual lapse rates can vary widely. See ADIABATIC LAPSE RATE, WET ADIABATIC LAPSE RATE.

larva, larvae   Any of the immature (premetamorphosis) forms of organisms that undergo metamorphosis. Tadpoles, grubs, and caterpillars are all larvae (larval forms), radically different from the adult frogs (or toads), beetles, and butterflies that they will become after metamorphosis.

laser   Acronym for Light Amplification by Stimulated Emission of Radiation. A laser is a device emitting radiation (usually light) that is very uniform in wavelength and phase, usually in a narrow beam. Lasers are powerful tools for cutting, heating (as in nuclear fusion), and transmission of communications signals.

latent heat   Amount of heat required to change a substance from a solid state to a liquid or gaseous state, or to change a liquid to a gas, without causing a change in temperature. See HEAT OF FUSION, HEAT OF VAPORIZATION.

latent heat of fusion   See HEAT OF FUSION.

lateral bud   A bud growing in an axil, the crotch formed between the base of a leaf and the stem to which it is attached. Also called axillary bud.

lateral diffusion   Diffusion occurring in a horizontal direction, as within a layer of the atmosphere. See DIFFUSION.

lateral fault   See STRIKE-SLIP FAULT.

lateral moraine   A ridge-shaped accumulation of till along the side of a glacier. The orientation of a lateral moraine is generally parallel to the direction of ice movement.

lateral stem bud   Any meristematic embryonic shoot that forms on the side of a plant stem.

laterite   A type of tropical soil rich in hydrated aluminum and iron oxides; may be regarded as an extreme form of chemi-

cal weathering; name is derived from Latin *latere* for brick, because it is typically red. Can form ores of iron, aluminum, or nickel. See PEDALFER.

lateritic soil   See LATERITE.

laterization   The process of forming a laterite soil. See LATERITE.

latex   1) A complex, milky fluid secreted by some plants such as poppies and rubber trees. The latter is processed into natural rubber. 2) A stage in the manufacture of synthetic rubber. 3) A form of house paint in which a latexlike fluid (a water emulsion of a synthetic rubber or plastic polymer) replaces the oil used in oil-based paints. These paints can be cleaned up with soap and water instead of the solvents required for oil-based paints.

latitude   The angular measure of distance from the equator of the earth, where the equator is taken as 0° and the poles are taken as 90°.

latosol   See LATERITE.

lattice   The repeating three-dimensional arrangement of atoms, molecules, or ions constituting a crystal.

Laurasia   A postulated continent of the early Mesozoic era. Laurasia formed from the northern part of Pangaea and later broke up to form the landmasses of Asia, Europe, Greenland, and North America.

Laurasian distribution   Describing organisms with a widespread distribution throughout the Northern Hemisphere, as a result of sharing ancestors that existed in Laurasia, the northern component of the single landmass (supercontinent) Pangaea that, through the process of continental drift, evolved into our current continents. See LAURASIA.

laurilignosa   Woody vegetation dominated

by laurel and similar species. See LIGNOSA.

lava A general term for molten rock material which is extruded on the surface of the earth. Compare MAGMA.

lava dome A type of shield volcano in the shape of an inverted bowl. Lava domes are built up by sequential eruptions of lava from a central vent. See SHIELD VOLCANO.

law A basic scientific principle explaining how a phenomenon (or phenomena) works, one that has been repeatedly tested and found to be consistently reliable in making predictions concerning the phenomenon.

law of conservation of energy A fundamental principle of physics stating that the amount of energy within any isolated system, including the universe, is always constant. Energy can be neither created nor destroyed but only changed into a new form (which could be matter). See FIRST LAW OF THERMODYNAMICS. Compare LAW OF CONSERVATION OF MATTER.

law of conservation of matter A fundamental principle of physics stating that mass, the amount of matter within any isolated system (such as a chemical reaction in a closed vessel), is always constant. In a closed system, matter can be neither created nor destroyed but only changed into a new form (which could be energy). Compare LAW OF CONSERVATION OF ENERGY.

law of cross-cutting relationships See CROSS-CUTTING RELATIONSHIPS, PRINCIPLE OF.

law of dominance See MENDEL'S LAWS.

law of faunal succession A biologically based principle of chronostratigraphy used in dating rocks that states that fossil-bearing rocks may be correlated over long distances when each of the rocks contains particular faunal assemblages that are known to be part of a chronostratigraphic sequence.

law of independent assortment See MENDEL'S LAWS.

law of limiting factors The principle that each physical variable in an ecosystem has maximum and minimum levels (tolerance limits) outside of which a particular species cannot survive. Within these tolerance levels, and under steady-state conditions, the variable in shortest supply will effectively limit the growth of the population. This principle is an expansion of Liebig's law of the minimum. See LIEBIG'S LAW OF THE MINIMUM, LIMITING FACTOR, TOLERANCE LIMITS.

law of original horizontality A stratigraphic principle stating that sedimentary rocks are usually deposited in a continuous sequence of horizontal strata.

law of parsimony See OCCAM'S RAZOR.

law of superposition A stratigraphic principle stating that in an undisturbed sequence of sedimentary rocks, each layer of rock is progressively younger than the layer beneath it.

law of the minimum Although sometimes equated with the law of limiting factors or law of tolerance and used for any environmental variable, this is usually used to refer to Leibig's law of the minimum.

law of tolerance The existence, abundance, and distribution of a species under steady-state conditions are dependent upon whether the levels of one or more physical or chemical factors fall above or below the levels tolerated by the species. See LIMITING FACTOR, TOLERANCE LIMITS.

lawrencium (Lr) A radioactive, transuranic, metallic element with atomic weight 257.0 and atomic number 103. A member of the actinide group, lawrencium has a half-life of 8 seconds, too short to be utilized; several isotopes with similarly short half-lives have been synthesized.

LD₅₀ A measure of toxicity. The dosage (lethal dose) of a substance that will kill half of a sizeable population of a species. Also called mean lethal dose and median lethal concentration and LD_{50}. See APPLICATION FACTOR.

LDH See LACTATE DEHYDROGENASE.

leach To remove material in chemical solution by the action of percolating liquid.

leachate The chemical solution obtained by process of leaching.

leach field A system of drainpipes used to treat wastewater by the process of percolating effluent through a soil.

leaching 1) An intrasystem nutrient transfer in terrestrial ecosystems that occurs when precipitation is channeled down plant surfaces and stems (stem flow) as well as washing through the canopy directly to the soil surface (throughfall). 2) The movement of dissolved nutrients from surface soil horizons to deep soil horizons by groundwater infiltration.

lead (Pb) A soft, heavy, metallic element with atomic weight 207.2 and atomic number 82. It is found in several different ores (primarily galena, but also anglesite and cerussite) and used in a wide variety of alloys. Lead makes an excellent shield for radioactive materials. Lead compounds are toxic and will accumulate in the body and cause poisoning if levels as low as 0.5 mg are absorbed daily. Low-level or chronic lead poisoning is common; symptoms include muscle cramps, abdominal pain ("lead colic"), and eventual nervous system and brain damage. Lead was formerly widely used in house paint (especially white paint) and as plumbing solder; both uses are now banned because they resulted in dust, paint chips, and drinking water with significant lead levels.

lead poisoning Common, chronic poisoning caused by inhalation, absorption through the skin, or swallowing of substances containing lead. Sources of lead poisoning include dust and chips from lead paint, industrial exposure, acidic tap water traveling through lead pipes, lead glazes on ceramics, and exposure to automobile fumes. Lead poisoning causes abdominal pain, constipation, damage of the nervous system, and brain damage in developing children.

lead ratio A comparison of the proportion of radiogenically produced lead isotopes to their radionuclide precursors. The lead ratio is used in radiometric dating to estimate the age of rocks and minerals. See RADIOMETRIC DATING.

leaf area index (LAI) A measure of potential photosynthesis in a community, expressed as the ratio of the total leaf surface of individual plants, crops, or communities to the total ground surface. The LAI has no dimensions because the units (square feet/square feet or square meters/square meters) cancel each other.

leaf forms See APPENDIX p. 628–629.

leaf structure Viewed in cross-section, leaves have an upper cuticle and epidermis composed of tightly packed, hexagonal-shaped cells. Light passes through these layers and is absorbed by chlorophyll contained in the palisade cells. Beneath the palisade layer are spongy parenchyma cells interspersed with intercellular space that contains water from the root system and air entering from the stomata, the tiny openings on the bottom layer of the leaf.

Cross-section of a leaf

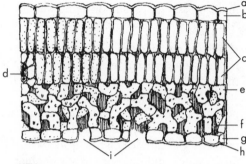

KEY
a. cuticle
b. upper epidermis
c. palisade parenchyma
d. bundle sheath
e. spongy parenchyma
f. intercellular space
g. lower epidermis
h. cuticle
i. stomata

lean-burn engine   A gasoline engine that can achieve near-complete combustion by including a higher proportion of air in the mixture of air and gasoline, mixtures too "lean" or low in fuel for ordinary spark plugs. These engines are used in some automobiles to increase fuel efficiency and to reduce polluting exhaust emissions (NO_x and unburnt hydrocarbons).

Le Chatelier principle   The tendency for any system at equilibrium to change, when a stress is applied, in a way that balances out or reduces the effects of the stress.

lecithin   Any of a group of phospholipids, fatlike substances containing phosphorous found in animals and higher plants. It is present in cellular membranes, blood, bile, and many tissues, including the brain. Also called phosphatidyl choline, lecithin is widely used in the food industry as an emulsifier. See EMULSIFIER.

lecithotrophic larva   Aquatic larvae that subsist on nutrients provided by the yolk of their eggs, rather than feeding on plankton.

lectotype   A replacement for a missing holotype (type specimen used in defining a species), chosen from the same original material as the holotype. See TYPE SPECIMEN.

lee   The side away from the direction in which the wind is blowing; also, a sheltered spot. See LEEWARD, WINDWARD.

leeward   Away from the direction in which the wind is blowing; downwind. Contrast WINDWARD.

lee waves   A disturbance of waveflow caused by the presence of an obstacle in the path of the flow. The disturbance takes the form of a standing wave on the leeward side of the object. Lee waves can form in air currents flowing over mountains and cause difficulties for airplanes. See WAVE CLOUD.

left-lateral fault   See SINISTRAL FAULT.

legislative branch   One of three branches of the United States government, the portion comprising elected representatives of the people with responsibility to make laws (legislate). Also called the legislature, at the federal level it includes both houses of Congress. State governments also have a legislative branch. Compare EXECUTIVE BRANCH, JUDICIAL BRANCH.

legume   1) Any plant of the pea family (Fabaceae). Many are important crops for green manure (clover, alfalfa, soybeans) because of their ability to fix atmospheric nitrogen through bacteria in their root nodules. 2) A peapod, a fruit consisting of a dry, flat pod that opens on two sides. 3) The edible members of the pea family, such as peas, beans, lentils, etc., many of which are very high in protein. See FABACEAE, GREEN MANURE, NITROGEN FIXATION.

Leguminosae   See FABACEAE.

lek   A common area used for display and courtship by some animals (usually birds) during their mating season. Males congregate in this small area and display themselves to attract mates. Also called booming or dancing ground, or arena.

lemurs   Restricted to the island of Madagascar, the lemurs are primitive, nocturnal, arboreal primates with relatively long, bushy tails, pronounced snouts, forward-directed eyes, prehensile fingers and toes, and comparatively small cerebrums. All are threatened with habitat destruction.

lentic   Relating to or living in still, slow-moving, or stagnant water, including ponds and swamps. Contrast LOTIC.

lenticel   A raised pore located on stems and in the bark of some plants. When present, it is used for gas exchange, much as the stomata in leaves. The name comes from its usually flattened, lenslike shape; it is especially noticeable on the bark of cherry trees. See STOMATA.

Lepidoptera   An order of insects including the butterflies and moths. They are soft bodied and usually have prominent eyes. Most adults have mouthparts formed into a coiled proboscis used for extracting nectar from plants. The body and wings are covered with scales which are often highly pigmented. The larvae or caterpillars have functional mandibles and are, for the most paret, herbivorous. Most butter-

flies and moths belong to the suborder Ditrysia. In general, diurnal species are called butterflies and nocturnal ones, moths. Mimicry, in which edible species display color patterns and behaviors similar to those of non-edible forms, is prevalent in many butterfly species. Many caterpillars are major plant pests such as the gypsy moth *Lymantria dispar* whereas others like the silk "worm" *Bombyx mori* have great economoc value. Many tropical butterflies are now being raised in captivity for display in butterfly gardens. See INSECTA, MIMICRY.

Leslie matrix   A model proposed by P. H. Leslie in 1945 for predicting the future age structure of a population based upon the known age structure at a given time and survival and birth rates for different age groups. See AGE DISTRIBUTION.

lethal gene   Any gene whose presence causes premature death of an organism. Conditional lethal genes are those that only cause death in a specific environment, such as a particular cultural medium for bacteria.

leucine $(CH_3)_2CH\text{-}CH_2CH(NH_2)COOH$, an amino acid that is a constituent of many plant and animal proteins. Leucine must be supplied in the diet and is therefore called an essential amino acid.

leucocratic   A term applied to light-colored igneous rocks having a proportion of visible dark minerals that is less than 50 percent.

leucoplasts   One of the colorless plastids in the cytoplasm of plant cells that often contain reserves of starch, protein, or oil. See PLASTIDS.

leukemia   A type of cancer characterized by uncontrolled overproduction of nonfunctional and undeveloped white blood cells and malfunction of the blood-forming bone marrow. Once generally fatal, many forms of leukemia can now be treated, resulting in a high rate of remission. See CANCER, CARCINOMA, LYMPHOMA, SARCOMA.

leukocytes   White blood cells; any colorless blood corpuscle that contains a nucleus. Leukocytes are capable of amoebalike movement. They act as scavengers and ingest bacteria and other particles in the blood. They are specialized into several different forms. Also spelled leucocytes. See ERYTHROCYTES, LYMPHOCYTES, MONOCYTES, PHAGOCYTE.

levee   A river bank structure composed of flood plain sediments that slope gradually away from the river channel. A levee may be constructed by humans as a flood protection strategy.

Lewis acid   A molecule, ion, or atom that can accept a pair of electrons. This is the most complete definition for an acid; Lewis acids include some compounds that do not qualify as Arrhenius acids. It is based on a chemical's behavior during a specific reaction rather than physical properties, such as sour taste. Aluminum trichloride, $AlCl_3$, is a Lewis acid that is not an Arrhenius acid. See ARRHENIUS ACID, LEWIS BASE.

Lewis base   A molecule, ion, or atom that can donate a pair of electrons. This is the most complete definition for a base; Lewis bases include some compounds that do not qualify as Arrhenius bases. It is based on a chemical's behavior during a specific reaction rather than physical properties, such as high pH. Ammonia is a Lewis base, but not an Arrhenius base. See ARRHENIUS BASE, LEWIS ACID.

Lewis dot structure   A method for representing elements, ions, or molecules using one or more dots (or, less often, dashes) to represent the valence shell electrons of the element. The electrons are usually grouped in pairs; argon, for example, is represented as Ar surrounded by two dots above, below, and on each side for a total of eight electrons in its outer shell. See OCTET RULE.

LH   See LUTENIZING HORMONE.

Li   Chemical symbol for the element lithium. See LITHIUM.

liana   General term for vines and climbing plants, especially tropical woody vines that grow around tree trunks in order to get greater access to light. In temperate climates, lianas include grapes, kudzu, and poison ivy.

lichens   See MYCOPHYCOPHYTA.

lichen dating   The estimation of relative age by lichenometry.

lichenometry   The measurement of the size and growth of lichens. Lichenometry is used as a dating technique in subarctic and alpine regions.

lidar   Acronym for light detection and ranging. A laser instrument used by meteorologists to study clouds. A laser beam is directed at the cloud formation, and the degree of scattering or reflection of the beam gives information about the structure of the clouds. Lidar is also used to track smokestack plumes, as it can detect them long after they are invisible to the eye.

Liebig's law of the minimum   The theory that states that the growth or the survival of a population is dependent on the chemical requirement (nutrient, oxygen, etc.) that is in the least or most limiting supply. As originally proposed by Liebig (a 19th-century soil scientist), it is that the "growth of a plant is dependent on the amount of foodstuff which is presented to it in minimum quantity." See LAW OF LIMITING FACTORS.

life 1) The property that distinguishes organisms from inorganic objects and organic dead bodies; animate existence characterized by active metabolism, growth, reproduction, and response to stimuli. 2) General term for living organisms, as in the animal life in a certain region.

life cycle The complete series of changes or developments that an organism progresses through, from one stage in a generation (usually the adult) through a reproductive stage to the same stage in the subsequent generation. In higher organisms, it is often considered to start with the fusion of gametes, progressing to the production of gametes by adult organisms. See ALTERNATION OF GENERATIONS.

life-cycle cost The total cost of an item amortized (spread out) over its entire life. Also called lifetime cost.

life expectancy The probable number of years (or other time period) that members of a particular age class of a population are expected to live, based on statistical studies of similar populations in similar environments.

life form 1) The overall body type (morphology) of a plant or other organism. A given life form usually characterizes a particular biome; for example, grasslands are dominated by grasslike plants rather than a particular species, and tropical forests may be dominated by trees and lianas. 2) Part of a specific plant classification system developed by Raunkiaer in 1934. See RAUNKIAER'S LIFE FORMS.

life span An individual lifetime; also, the longest period of time that an organism can be expected to live.

life table A chart listing mortality and survivorship data for different age groups in a population. Often, age-specific fecundity is also included in a life table. See MORTALITY RATE, SURVIVORSHIP.

life zone A region characterized by a uniform type of plant or animal life, and generally one type of climate. Originally based on temperature zones, life zones are now based on distribution of organisms and are equivalent to divisions or subdivisions of biomes. See BIOME.

lift An aerodynamic term for the major force that supports a flying airplane. Lift is the force perpendicular to existing airflow that is created by relative airflow over a structure like a wing. Often called aerodynamic lift. See DRAG.

ligand An atom, anion, or molecule bound to the central metal atom of a complex ion (an ion containing a metal cation at its center, bonded to two or more other ions or molecules). Hydroxide can act as a ligand to iron to form $Fe(OH)_3$. Chelating agents are ligands that can form more than one bond with the central metal. See CHELATING AGENT.

light Visible electromagnetic radiation; radiation having wavelengths between about 4000 and 7000 angstroms (400 to just over 700 nm), between infrared and ultraviolet radiation. See ELECTROMAGNETIC RADIATION, WHITE LIGHT.

light compensation point The light level at which rates of photosynthesis and respiration are equal. At the light compensation point, uptake of carbon dioxide by plants during photosynthesis equals the release of carbon dioxide by respiration, so there is no net increase or decrease. Many tree seedlings reach a light compensation point at approximately 30 percent of full sunlight; the term also applies to regions below the surface of aquatic ecosystems. See COMPENSATION DEPTH, EUPHOTIC ZONE.

light detection and ranging   Full name for lidar, a cloud-observing technology used in meteorology, usually referred to by its acronym. See LIDAR.

light minerals   See FELSIC ROCK.

lightning   The series of electrical discharges between or within clouds or between the earth and clouds that we see as a flash or bolt.

light reactions   Any of the chemical reactions in photosynthesis that require light, such as the photolysis of water molecules that produces oxygen (O_2). Light is absorbed by chlorophyll (or similar pigments) and provides the energy for generating ATP and reduced NADP. The products of these reactions are used to fix carbon dioxide during the dark reactions of photosynthesis. See DARK REACTIONS.

light saturation   1) The optimum light level (light intensity) in an ecosystem. The light saturation point for shade-adapted plants is lower than that for sun-loving plants; some plants and animals show inhibited growth when light levels are much higher than the saturation level. 2) The amount of light that can be used by a species photosynthesizing at its maximum rate. See LIGHT COMPENSATION POINT.

light water reactor   (LWR) A fission nuclear reactor using ordinary water as coolant and moderator instead of deuterium (heavy water). Boiling water reactors and pressurized water reactors are both light water reactors.

lignin   A complex organic polymer that increases the rigidity of plant cell walls. Cellulose and lignin, often combined, make up a large percentage of woody tissue, and therefore of harvested wood.

lignite   A very soft type of coal, midway between peat and bituminous coal in its physical and chemical properties.

lignosa   Vegetation or plant communities in which woody plants dominate. See HIEMILIGNOSA, LAURILIGNOSA, PLEURILIGNOSA. Contrast HERBOSA.

Lilaceae   The lily family of monocot anigosperms, which some taxonomists lump together with the Amaryllidaceae. The three petals and three sepals generally look similar, and surround the six stamens and superior ovary of three carpels. The fruit is a capsule or berry. Lilies (*Lilium*), onions (*Allium*), yucca (*Yucca*), day lilies (*Hemerocallis*), trout lilies (*Erythronium*), fritillaries (*Fritillaria*), aloes (*Aloe*), and many other genera are included in this family. See ANGIOSPERMOPHYTA.

lime   A general term for material that is mostly composed of calcium carbonate.

limestone   Any chemical or organic sedimentary rock that is primarily composed of calcite minerals.

limit cycle   A regular cyclic pattern oscillating between upper and lower limits. Limit cycles show long-term stability, such as those in predator-prey systems when destabilizing tendencies are balanced by stabilizing ones.

limiting factor   Environmental variable, or combination of variables, that inhibits the rate of growth of an organism or an entire population. Also, a variable such as temperature or quantity of enzyme present, that limits the rate of a physiological process. See LAW OF LIMITING FACTORS.

limnetic zone   Layer of water in a lake, away from the shore, in which there is sufficient light for photosynthesis. It reaches from the surface of the open water down to the compensation depth, which is typically the depth at which light intensity is about 1 percent of full sunlight. The limnetic zone is inhabited by animals and plants. See COMPENSATION DEPTH, LITTORAL ZONE, PROFUNDAL ZONE.

limnology   The study of inland bodies of fresh water, especially lakes. Limnology includes the meteorological as well as physical, chemical, and biological aspects of lakes.

limonite   A yellowish mineral of iron oxide and water. Limonite occurs as a secondary mineral in the weathering products of iron-rich rocks. See APPENDIX p. 625.

Lincoln-Peterson index   Animal census formula used to estimate population size. A sample of the animal population is captured, marked, and released. After the released animals are allowed to mingle with the population, another sample is taken. The Lincoln-Peterson index uses the ratio of marked to unmarked animals to derive a figure for the total population. Also called the Lincoln index, or the capture-recapture method.

lindane $C_6H_6Cl_6$, a chlorinated hydrocarbon insecticide once used against cockroaches, as a pre-treatment on seeds to prevent infestation before seed sprouting, and on plants against sucking insects (especially aphids) and leaf miners. Its use is no longer allowed in the United States. It was known by many different trade names, depending on the company that

marketed it, including benzene hexachloride (BHC), Ben-Hex, Gamma-BHC, HCH, Gexane, Jucutin, Streunex, Tri-6, Aparasin, Aphtiria, Lorexane, and HGI. It is still allowed for medical treatment of infestations of head lice and scabies; the medical preparation goes by the name Kwell. See CHLORINATED HYDROCARBON INSECTICIDE.

Lindeman efficiency   The ratio of energy assimilated by one trophic level to the energy assimilated by the previous trophic level. Compare ECOLOGICAL EFFICIENCY, CONSUMPTION EFFICIENCY, TROPHIC LEVEL EFFICIENCY, UTILIZATION EFFICIENCY.

lineament   1) A landscape characterized by lines of gently sloping depressions. 2) Large-scale tectonic patterns observable from space. 3) Linear features identified from aerial photography.

linear   1) Describing any motion or process in which an effect is directly proportional to a cause (and thus if graphed would produce a straight line), as in linear growth. 2) A term used in statistics to indicate the degree of correlation between two or more variables. A perfect linear relationship is one with the maximum value for correlation coefficient (+1). 3) The description of an equation or function in which the variables are of the first degree. See LINEAR GROWTH, CORRELATION COEFFICIENT.

linear energy transfer   The energy transferred per unit length of travel by ionizing radiation as it passes through matter such as living tissue. Alpha radiation has high linear energy transfer; gamma radiation has low linear energy transfer. For the same amount of energy deposited in tissue, high linear energy transfer radiation tends to be more harmful.

linear erosion   A downward stream incision resulting in the deepening of an existing valley.

linear growth   Growth or rate of increase in which a quantity increases by some fixed amount during each unit of time.

linear ocean-floor magnetic anomalies   A pattern of magnetic stripes, as determined by instrument measurements, lying parallel to mid-oceanic ridge spreading centers. These linear patterns correlate with a magnetostratigraphic time scale to serve as indicators of sea-floor spreading history. See MAGNETIC ANOMALY.

lineation   Any feature of rock that appears in parallel alignments. Lineations may be expressions of parallel minerals, surface scratches, mineral cleavages, or crenulations.

line squall   A severe, localized storm or system of storms such as a thunderstorm occurring along a line that can extend hundreds of miles. Line squalls can form along a cold front, but in the central and southwestern United States they often occur along a squall a hundred miles ahead of a cold front. They often produce heavy rainfall or hail.

linguoid ripple marks   Asymmetric ripple marks that have a sinous crest. A linguoid ripple mark is typically shaped like a tongue.

linkage   The tendency for two or more characteristics to be inherited together because the genes that control them are located on the same chromosome.

linkage group   A group of genes that are known to be transmitted together because of the pattern of inheritance of their associated traits.

Linnaean system   A system for classifying and naming organisms devised by the Swedish naturalist, Carolus Linnaeus (Carl von Linné, 1707–1778). His system of assigning two-word Latin names, the first for the genus and the second for the species, is still the basis for modern binomial nomenclature. See BINOMIAL CLASSIFICATION.

linoleic acid $H_3C-(CH_2CH=CH)_3-(CH_2)_7-COOH$, a polyunsaturated fatty acid. Linoleic acid is an essential nutrient supplied by many plant oils. It occurs in linseed oil and helps give it the drying properties that make linseed oil useful in oil paints.

lipase   Any of a group of enzymes produced by the body to digest fats (lipids) by changing them into fatty acids and glycerin.

lipid   Any of a large, diverse group of fatty organic compounds containing long hydrocarbon chains as part of their structure: oils, fats, waxes, steroids, and related compounds. Their functions include making up cell membranes, storing energy for the body, and in hormones helping to control the body.

liposomes   1) Fatty or lipid droplets in suspension within the cytoplasm of a cell. 2) Synthetic capsules made by mixing an aqueous solution into a phospholipid gel, used for administering vaccines and other drugs.

liquefaction   1) The conversion of a gas into a liquid by lowering its temperature or increasing pressure. Liquefaction is also used to refer to the conversion of a solid to a liquid. 2) The rapid, temporary loss of shear strength in a soil. Liquefaction results in a change from solid engineering properties to liquid characteristics. Liquefaction may be caused by mechanical shock.

liquefied natural gas   A gaseous fossil fuel consisting primarily of methane. Liquified natural gas is cooled to the liquid phase (approximately -100°F), for transport and storage.

liquefy   To convert into a liquid state, especially from a gas. LNG or liquified natural gas is cooled to very low temperatures in order to change it from gas to liquid for easier transport.

liquid   The state of matter between solid and gas. Liquids take on the shape of the vessel containing them. Unlike gases, they are also characterized by not changing much in volume when pressure is applied. The term liquid differs from fluid because the latter can include gases. Compare SOLID, GAS.

liquid crystal   A flowing, liquid substance that has certain physical properties, especially optical, that are characteristic of crystals and not characteristic of ordinary liquids. Some liquid crystals respond to electrical voltages like a crystal. In liquid crystal displays, such as those on digital watches or calculators, a very thin layer of liquid crystal is sandwiched between two layers of electrical conductors. Using very little energy, a change in applied voltage changes the polarity of transmitted light; this phenomenon can be used to make digital numbers appear or disappear in the display.

liquid metal fast breeder   (LMFB) A nuclear reactor producing more fissionable fuel than it consumes and using liquid sodium as a coolant. The "fast" refers to the speed of the neutrons causing the fission; these reactors use no moderators. Liquid metal fast breeders use plutonium, or a uranium-plutonium oxide mixture, as fuel. Also called a fast breeder reactor.

liter   (l) Unit of volume in the MKS (metric) system. (The preferred SI unit is the cubic meter.)
  1 liter = 1.507 quarts (liquid measure)
1 U.S. gallon = 3.785 liter

lithification   The process of sedimentary rock formation by the cementation of unconsolidated materials

lithium   (Li) A very light, soft, metallic element with atomic weight 6.941 and atomic number 3. Lithium is incorporated into many alloys and lubricants. It is used in high-energy fuels and many chemical processes such as tritium production. Lithium salts are used as drugs to combat depression.

lith-, litho-   Prefix referring to rock or stone, from the Greek *lithos*, rock.

lithology   A description of the visible texture and composition of a rock.

lithophile   Describing chemical elements with an affinity for silicates, and therefore found in mineral silicates. Also called oxyphile.

lithosere   Succession of plant communities on rock surfaces. See SERE, CLISERE, XEROSERE.

lithosol   A shallow soil composed of rock fragments and lacking a well-developed horizon.

lithosphere   1) The portion of the biosphere consisting of the upper layer of rocks interacting with the hydrosphere and the atmosphere. 2) The crust of the earth, specifically the sial and crustal sima layers above the Mohorovicic discontinuity. The lithosphere is the zone of earthquakes, because it is characterized by a resistance to shear waves. See BIOSPHERE.

lithostatic pressure   The gravitational

confining pressure that is generated by the weight of bedrock at a specified depth.

litmus   A substance derived from lichens and used as an acid-base (pH) indicator. It turns red in strong acids and turns blue in bases. See PH.

lit-par-lit   A term describing a rock that has been disrupted along the bedding planes or foliation by narrow intrusions of granitic rock.

litter   1) Dead and partially decomposed leaves and other recognizable plant residues on the soil surface of the forest floor, also called the litter layer. See APPENDIX p. 619. 2) A group of mammalian offspring born at one time to one mother, as in a litter of puppies. 3) Trash discarded by people. See HUMUS.

litterfall   The amount of dead plant matter added to the forest floor in a given community over a given time period, usually a year. Annual litterfall, when coupled with measures of the humus layers, can be used to estimate rates of soil metabolism (respiration) and activity of detritivores.

Little Ice Age   A period of unusually cold weather in Europe from roughly 1645 to 1715. It corresponds to the Maunder minimum, when there was an almost total lack of sunspots during that period. A drought in the southwestern United States also occurred over the same period. This period is also referred to as the Fernau glaciation. See MAUNDER MINIMUM.

littoral drift   The clastic debris transported by longshore currents or beach processes.

littoral zone   1) The zone in a body of fresh water where light penetration is sufficient for the growth of plants. 2) The zone of a marine environment inhabited by intertidal organisms.

living fossil   A plant or animal that is an example of a group of organisms or species descended from a now-extinct species. The coelacanth is a fish that was known only from the fossil record until it was discovered alive in 1938.

lizards   See LACERTILIA.

L layer   The layer of organic litter on the surface of a soil.

lm   Symbol/abbreviation for lumen. See LUMEN.

LNG   See LIQUEFIED NATURAL GAS.

load   1) The amount of a substance carried by a system, as in the sediment load of a stream. 2) Total demand for electric power from a given supply such as an electrical generator. 3) Mechanical force applied to an object.

load-on-top   A general term for a pattern of mine excavation in which the ore body is approached from below by a series of adits, cross-cuts, and raises.

loam   A general term for a soil mixture containing sand, silt, and clay in nearly equal parts.

lobsters   A general name for a group of robust, decapod crustaceans which have prominent abdomens bearing full sets of appendages. In the American lobster and its European counterpart, the first pair of legs bear large, heavy pincers. Lobsters walk with their legs but may also propel themselves rapidly backward by forceful flexion of their abdomens. Both species are heavily trapped commercially throughout most of their range as specialty foods. The rock or spiny lobsters and the Spanish lobster lack pincers. They are also edible and their abdomens are widely marketed as frozen lobster tails. Crayfish are freshwater forms, closely related to the lobsters. See DECAPODA.

local stability   The tendency of a community to return to its original state after experiencing small disturbances. See HOMEOSTASIS.

loch   1) The Scottish word for lake. 2) The British term for a cavity within a vein of ore (for example, in a lead mine).

locus   The physical position of a gene (and its alleles) on a chromosome. On different chromosomes in a homologous pair, the genes at identical locations control identical traits.

locusts   Common name for species of large grasshoppers belonging to the insect order Orthoptera, particularly of the genus *Schistocerca*. Solitary when young, they may aggregate at maturity to form huge, migratory swarms of millions which fly over the countryside causing widespread damage to crops wherever they land. See INSECTA, ORTHOPTERA.

lode   A general reference to a commercially valuable vein of mineral ore.

loess   An unconsolidated sediment deposited by wind. Loess is usually composed of unstratified fine sand or silt.

logarithmic   1) Relating to or involving logarithms, the exponential power to which a base number (usually 10) must be raised to give a specific number. 2) Describing measurements in which an increase in one unit represents a power (exponential) increase in the quantity involved. Logarithmic phenomena are often graphed using a logarithmic scale on the vertical axis and time intervals as the horizontal scale. Such a graph converts the characteristic J-shaped curve of a logarithmic increase into a straight line. 3) Exponential; logarithmic growth is another

term for exponential growth. See EXPONENTIAL GROWTH.

logarithmic reproductive curve   See J-SHAPED CURVE.

logarithmic wind profile   The logarithmic relationship between increase in wind speed and increase in altitude under neutral (nonconvective) conditions. See WIND PROFILE.

logistic curve   See S-SHAPED CURVE.

logistic equation   A theoretical mathematical formula that describes the idealized growth of a population subject to a density-dependent limiting factor. The population growth begins at a slow rate, but soon increases to logarithmic rates. As the population reaches its carrying capacity, determined by the limiting factor, population growth levels off (at what is called an asymptotic value) and approaches zero. The logistic equation is the product of the equation for simple population growth and environmental resistance; it produces an S-shaped curve when graphed. In reality, the logistic equation may describe the early phases of population growth, but it doesn't accurately describe the fluctuations that characterize the population dynamics of most species. See ENVIRONMENTAL RESISTANCE, S-SHAPED CURVE.

logistic population growth   The idealized rates of increase in a population subject to a density-dependent limiting factor, such as a finite food supply. Logistic population growth reaches logarithmic rates at early stages in the development of the community, but rates of increase level off as the community reaches a steady-state level. See LOGISTIC EQUATION.

log normal   A graph of frequency distribution of species abundance with abundance shown on the horizontal axis and expressed on a logarithmic scale.

log-normal distribution   Short for logarithmic normal distribution, a statistical distribution in which the logarithm of a variable or measured quantity displays a normal (bell-shaped) distribution. Compare RANDOM DISTRIBUTION, NORMAL DISTRIBUTION.

long-day organism   Organism that needs long periods of light in order to mature and undergo sexual reproduction. See PHOTOPERIODISM.

long-day plants   A plant that flowers only when it receives daily exposure to light for a relatively long period of time, followed by a relatively short period of darkness, an example of photoperiodism. It is actually the short night period rather than the long days that controls flowering. Many grasses, henbane, and petunias are long-day plants; spinach bolts in midsummer as much because it is a long-day plant as from the heat. Contrast SHORT-DAY ORGANISM, DAY-NEUTRAL PLANT.

long-distance dispersal   The spread of reproductive structures or young over a wide area. Mechanisms for long-distance dispersal include lightweight seeds that can be carried long distances by the wind, seeds that must pass through animal intestines in order to sprout and so may be carried long distances by birds and other animals, and seeds carried by ocean currents. See ANEMOCHORE, ENDOZOOCHORE, ZOOCHORE.

long-distance transport   Pollution dispersion over long distances.

longevity   Life span; the length of life for an individual, or an average for a species or population.

longitude   The distance around the earth measured in degrees (angular distance) around the equator, or parallel to the equator, from the reference line running from the poles through Greenwich, U.K. (the prime meridian, zero degrees longitude). Also refers to celestial longitude or right ascension. Compare RIGHT ASCENSION, LATITUDE.

longitudinal dune   Any linear or elongated curving dune that is oriented parallel to the prevailing wind. Longitudinal dunes are also known as seifs. See DUNE.

longitudinal profile   1) A study of a population or other phenomenon through time. 2) A stream cross-section diagram, or profile that follows the valley gradient or thalweg of the stream.

longitudinal section   A slice of an object made by cutting parallel to its longest dimension. Compare CROSS SECTION, SAGITTAL.

long-range forecast   Any prediction that covers a long period of time. Usually used in reference to weather predictions for periods beyond three to five days. Long-range forecasts usually include temperatures as well as precipitation, and may range for a week or a month.

longshore   A term describing a structure, feature, or condition that is in an orientation parallel to a coastline.

longshore current   A water flow generated along a coastline by waves approaching at an oblique angle to the shore. Material is transported in the longshore direction by both the breaking and backwash action of waves.

longshore drift   The movement of unconsolidated materials and debris by longshore currents.

longwall method A mining technique in which the excavation of walls is angled to undercut the mine roof. The roof is allowed to collapse and the fallen ore is removed.

loons See GAVIIFORMES.

looping A form of dispersion of atmospheric pollutants in which the plume of emissions from a smokestack is carried by air currents down to the ground level before rising again. See FANNING, CONING.

lophodont Having side teeth that are ridged on the chewing surface, as in some mammals. See ACRODONT, BUNODONT, HYPSODONT, SELENODONT.

lophophore A tentacle covered with cilia, or an organ made up of such tentacles, located near the mouth and used to set up currents that carry tiny organisms and particles of food into the mouth. Lophophores are found on brachiopods, bryozoans, and some other animals.

lopolith A large tabular pluton in which the floor of the magma chamber sags so as to give the overall shape of a bowl. A lopolith is one type of concordant intrusion.

lorica 1) A hard, thickened body wall, as found on some microscopic aquatic invertebrates (Rotifera). 2) A protective case or sheath, as on some unicellular animals (protozoa). Also called a theca.

lotic Relating to or living in moving water, such as a river or stream. Contrast LENTIC.

Love Canal A suburb of Niagara Falls, New York, where industrial wastes were buried from the 1940s into the 1950s. It was declared a disaster area in 1978 because some of the metal drums had corroded, releasing toxic chemicals. Over 200 families were evacuated, although in 1982 parts of the area were declared again fit for habitation. The Environmental Protection Agency has renamed the area Black Creek village.

love wave Geological term for seismic waves traveling across the earth's surface, within the top 20 miles of crust. Love waves travel at speeds of 4 to 4 ½ km/sec. See PRIMARY WAVE, SECONDARY WAVE, RAYLEIGH WAVE.

lower critical ambient temperature Surrounding temperature below which endothermic animals must generate heat in order to maintain their body temperature. Compare CRITICAL THERMAL MAXIMUM.

lower crust The portion of the earth's crust adjacent to the Mohorovicic discontinuity and composed of mafic minerals.

low frequency (LF) General term for the portion of the radio spectrum having long wavelengths (in the range of 1 to 10 km); LF radio frequencies fall in the range of 30 to 300 kilohertz. See EXTREMELY LOW FREQUENCY, HIGH FREQUENCY, LOW FREQUENCY, MEDIUM FREQUENCY, ULTRA HIGH FREQENCY, VERY HIGH FREQUENCY.

low-grade metamorphic rock A rock formed under low-grade metamorphic conditions. Slate is an example of low-grade metamorphic rock.

low-grade ore A mineral ore that has a marginal economic value because of a low mineral concentration ratio.

low-head hydropower Generation of electricity using small headwater dams (which form a body of water that has a low head, or height, above the turbines). The small headwater dams used in low-head hydropower cause less disruption of the aquatic environment than large-scale dams.

low-level radioactive waste   Radioactive waste that is not classified as high-level waste, transuranic waste, spent nuclear fuel, or by-product material (mill tailings). Such waste tends to contain radioactivity of low intensity or with a short half-life. It requires minimal special handling but is still unsafe (too radioactive) for ordinary methods of waste disposal. Low-level radioactive wastes include items such as protective clothing worn by workers in nuclear facilities, and materials from hospitals. Contrast HIGH-LEVEL RADIOAC-TIVE WASTE.

low-pressure area   An atmospheric region with low barometric pressure; a cyclone. Low-pressure areas are often associated with inclement weather. Contrast HIGH-PRESSURE AREA.

low-quality energy   Energy in a form that is dispersed or dilute and therefore not easily harnessed to perform useful work. Compare HIGH-QUALITY ENERGY.

low-sulfur crude   A general reference to crude oil with a relatively low proportion of the element sulfur, usually less than 2 percent. Compare HIGH-SULFUR CRUDE.

low-velocity zone   A region of partial melting in the upper mantle of the earth that varies between 40 and 160 km deep and where seismic waves are slowed or partly absorbed by the presence of fluid phase materials.

Lr   Chemical symbol for the element lawrencium. See LAWRENCIUM.

LTH   Abbreviation for luteotropic hormone, another term for prolactin. See PROLACTIN.

lumen   1) The central cavity within the cell wall of a plant, especially if empty because the protoplast is no longer present. 2) The space within tubelike structures in animals, including arteries, veins, and intestines. 3) (lm) Unit of measurement for light (luminous flux) in the Système International d'Unités (SI), equal to the amount of light given off in a unit solid angle by a uniform point source with intensity equal to one international candle.

luminous flux   Total visible light energy emitted by a source per unit of time. See LUMEN, LUX.

lungfish   See DIPNOI.

lunule   A marking shaped like a new moon, such as the white, crescent-shaped region near the base of a fingernail. Also spelled lunula.

luster   A description of the quality of reflected light or surface sheen of a mineral, such as metallic, vitreous, pearly, resinous, earthy, or adamantine.

lutenizing hormone   (LH) A hormone that stimulates the development of the corpus luteum in females and stimulates the testosterone-producing cells in male testes. It is secreted by the anterior lobe of the pituitary gland. Also called the interstitial cell-stimulating hormone (ICSH). See CORPUS LUTEUM.

luteotropic hormone   See PROLACTIN.

lux   (lx) The unit for measuring illumination (light intensity, or density of luminous flux) in the Système International d'Unités (SI). One lux equals one lumen falling uniformly on an area of one square meter.

L wave (long wave)   A type of surface seismic wave characterized by horizontal particle motion in the same direction of wave propagation See SURFACE WAVE. Contrast RAYLEIGH WAVE.

LWR See LIGHT WATER REACTOR.

lx Symbol for the measurement unit lux. See LUX.

lycopod See LYCOPODOPHYTA.

Lycopodium The genus that contains club-mosses. The leaves of these lycopods are arranged in tight whorls on the aerial branches. Sporangia of members of this genus may arise either from the axils of the leaves or on separate clublike clusters of fertile leaves known as strobili. See LYCOPODOPHYTA.

Lycopodophyta A phylum of the plant kingdom that contains club-mosses, spikemoss, and quillworts. Lycopods contain a true, but poorly developed vascular system, and reproduce by spores. The gametophyte generation is reduced to a small, usually subterranean, individual. The conspicuous above-ground leafy sporophyte, with reduced microphyllus leaves, generally spreads by branching horizontal rhizoids.

lymphocytes Specialized white blood cells (leukocytes) produced by the lymph nodes to help fight infections. They contain large nuclei, leaving very little room for cytoplasm in each cell. Lymphocytes usually make up from a quarter to half of all white blood cells. The colorless cells of the blood and lymphatic system, which defend the body against infection, are produced by lymph glands. Compare MONOCYTES.

lymphoma Any of a group of cancers or growths that originate in lymph tissue and are capable of spreading to other tissues.

Hodgkin's disease is a lymphoma. See CANCER, CARCINOMA, SARCOMA.

Lysenkoism An obsolete theory proposed by Soviet agronomist T. D. Lysenko (1898–1976) that contradicted Darwinian theories of evolution. Lysenkoism holds that genes do not exist, and explains all variable characteristics of organisms as caused by environmental changes that can be transmitted to subsequent generations. Lysenkoism was the official genetic theory of the former Soviet Union for two decades. Compare LAMARCKISM, NATURAL SELECTION.

lysimeter 1) A device for measuring evapotranspiration in vegetated soils. 2) A device for collection of soil water in the vadose zone.

lysine $H_2N\text{-}(CH_2)_4\text{-}CH(NH_2)COOH$, an essential amino acid. Lysine cannot be synthesized by the body and is needed for growth and repair of body tissues. It is also present in plant proteins.

lysogenic bacterium A bacterium that contains a prophage, a form of bacteriophage virus. The bacteriophage can destroy other bacterium of the same type that do not contain the virus, but somehow protects its host from further infection by the same phage, but not other phages.

lysosomes Organelles found in the cytoplasm of most cells that contain hydrolytic enzymes bound within membranes. They act as a digestive system within the cell, breaking down larger components that enter the cell (such as fats, proteins, or nucleic acids) into compounds that can be used in cellular metabolism.

m

m   See METER.

MA   An abbreviation for a million years.

MAB   See MAN AND THE BIOSPHERE PROGRAMME.

MAC   Short for maximum allowable concentration.

Macaronesian floral region   One of the phytogeographic regions into which the Paleotropical realm is divided according to similarities between plants; it consists of the islands off of West Africa. See PALEOTROPICAL REALM, PHYTOGEOGRAPHY.

macaws   The largest and most brilliantly colored parrots of the order Psittaciformes, which are found from Mexico south through the neotropics. Their desirability as cage birds has seriously depleted wild populations. See PSITTACIFORMES.

Mach number   The ratio of an object's speed to the speed of sound in the medium in which the object moves. Mach number can also refer to the speed of flow of a fluid. An airplane traveling at Mach 1 is traveling at the speed of sound; at Mach 2 the speed is twice that of sound (supersonic). See SUBSONIC, SUPERSONIC TRANSPORT.

macro-   Prefix meaning large or on a large scale, as in macroeconomics. Contrast MICRO-.

macroalgae   Species of algae large enough to be visible with the unaided eye, such as the kelps, rockweed (*Fucus*), and sea lettuce (*Ulva*). See MACROPLANKTON, ALGAE.

macrobenthos   Larger (macroscopic) organisms that live attached to objects or in the sediment at the bottom of bodies of water, such as clams and snails. See BENTHOS, MACROPLANKTON, MEIOBENTHOS, MESOBENTHOS, MICROBENTHOS.

macrobiotics   A specialized vegetarian diet that avoids consumption of animal products, consisting primarily of whole grains, nuts and seeds, and fruits and vegetables in season. See VEGETARIAN.

macroclimate   A large-scale climate system, one covering an entire region. Contrast MICROCLIMATE.

macroevolution   Large-scale evolution, above the level of individual populations, such as the development of new species or larger groupings. Macroevolution occurs over geologic time, such as the evolution of humans from primate progenitors.

macrofauna   The macroscopic animals of a particular region or time.

macrogamete   The larger of two fusing gametes of an organism that reproduces by means of unlike gametes (heterogamy); also called megagamete. See MICROGAMETE.

macromolecule   A large, complex molecule containing many atoms. Some examples of macromolecules include hemoglobin, containing about 10,000 atoms, as well as natural and synthetic polymers such as RNA, DNA, or plastics with high molecular weight.

macronucleus   The larger of the two nuclei present in some protozoa (*Ciliophora*); it controls metabolic functions within the unicellular organism. Compare MICRONUCLEUS.

macronutrient   A nutrient required in relatively large quantities for healthy growth, such as calcium, phosphorous, nitrogen, and potassium. Sometimes used exclusively to refer to plant nutrients. See ESSENTIAL ELEMENTS, MICRONUTRIENT.

macroparasite   A parasite that grows in its host, but often in body cavities or between cells rather than inside of host cells. Macroparasites breed by producing disseminating stages that are released from the host and infect new hosts. Compare MICROPARASITE.

macrophage   Any of a group of motile, large phagocytes that contain lysosomes and are capable of ingesting and digesting particles ranging from some microbes to worn-out cells. They are also capable of storing colloidal material. See PHAGO-CYTE, MONOCYTES.

macrophagous   Feeding on relatively large food pieces. Compare MICROPHAGOUS.

macrophyte   A plant, usually an aquatic species such as *Potomageton* (pond weeds), that is macroscopic in size.

macroplankton   Plankton large enough to be seen by the naked eye or caught in a plankton net, generally those with a diameter between 20 millimeters and 20 centimeters. Compare MICROPLANKTON, MESOPLANKTON, MEGAPLANKTON.

macropterous   Characterized by very large fins or wings.

macrozooplankton   Species of animal plankton large enough to be seen by the naked eye or caught in a plankton net, generally those with a diameter greater than 1 mm. Compare MICROPLANKTON.

Madagascan floral region   One of the phytogeographic regions into which the Paleotropical realm is divided according to similarities between plants; it consists of Madagascar plus adjacent offshore islands. See PALEOTROPICAL REALM, PHYTO-GEOGRAPHY.

Mad Hatter's disease   Insanity produced by long-term exposure to low levels of mercury. Hatmakers originally used mercury for making felt; the low levels gradually accumulated in their bodies until symptoms unrecognized as mercury poisoning appeared. Compare ITAI-ITAI DISEASE, LEAD POISONING.

mafic rock   A general term for a dark-colored igneous rock rich in ferromagnesian minerals; name derives from Latin for magnesium and iron *(ferrum)*. Contrast FELSIC ROCK.

maggots   Common name for insect larvae belonging to the order of true flies, Diptera. See DIPTERA.

magma   A general term for the molten material from which rock forms while contained within the earth. Magma is usually composed of molten silica with dissolved gases and crystallized minerals. Compare LAVA.

magma chamber   A reservoir containing magma surrounded by solid rock.

magma genesis   The process of magma formation.

magma tap   A conduit or fissure that intrudes through a country rock from a magma chamber.

magmatic deposits (ore)   Concentrations of economically significant minerals and metallic elements by one or more pro-cesses of differentiation. See MAGMATIC DIFFERENTIATION BY FRACTIONAL CRYSTALLIZATION, MAGMATIC DIFFER-ENTIATION BY PARTIAL MELTING.

magmatic differentiation by fractional crystallization The diversification of igneous rocks forming within a magma chamber by the sequential removal of early-formed minerals. A basaltic magma may become more rhyolitic through this process.

magmatic differentiation by partial melting The diversification of igneous rocks by the removal of minerals through incomplete melting of existing materials. The early molten fraction is separated from the still-solid portion by fluid motion.

magnesite A whitish mineral of magnesium carbonate that forms by chemical precipitation or by alteration of limestone, dolomite, or serpentine.

magnesium (Mg) White, metallic element with atomic weight 24.31 and atomic number 12, found in nature only in compounds with other elements. It is used in many alloys and in magnesium flares and fireworks. It is an essential nutrient for both plants and animals and is an essential part of the chlorophyll molecule.

magnet An object such as a piece of metal or a stone that has the property of attracting iron or some forms of steel. Objects can be natural magnets, or magnetism can be induced as in an electromagnet. See ELECTROMAGNET.

magnetic Possessing the properties of a magnet, capable of being magnetized, or relating to a magnet. See MAGNET.

magnetic anomaly Any unexplained magnetic field remaining after adjustment has been made for predicted magnetic patterns. See LINEAR OCEAN-FLOOR MAGNETIC ANOMALIES.

magnetic confinement Containment of a plasma by a magnetic field, one of two primary methods for containing hot plasma in order to control nuclear fusion. One form of magnetic confinement is a tokamak, which uses magnetic fields to confine a plama in a doughnut shape in fusion research. See TOKAMAK, INERTIAL CONFINEMENT.

magnetic declination The deviation between true north and magnetic north (as read on a compass) at a specified location. Also called magnetic deviation. Although this value is not caused by local interference, it varies during the year and in different locations.

magnetic epochs The fundamental geochronologic unit of the magnetostratigraphic time scale. Each magnetic epoch, also known as a polarity chron, marks an interval of the earth's history when the geomagnetic pole was relatively constant.

magnetic events Any short interval of alternate geomagnetic polarity occurring within the time interval of a magnetic epoch. A magnetic event is also known as a polarity subchron in the magnetostratigraphic time scale.

magnetic field The region surrounding a magnet (or moving electrical charge, such as current flowing through a wire) in which the magnet exerts force. A magnetic field is said to exist in a region of space if there would be a magnetic force on a moving, charged particle in that region of space. The strength of the field is measured in gauss (the term magnetic field is also sometimes used to refer to this field strength).

magnetic polarity time sequence A sequence of rocks that documents the magnetic pole reversals of the earth's magnetic field through natural remanent magnetism.

magnetic resonance imaging (MRI) A diagnostic technique for providing information about the medical condition of internal organs and tissues in humans without cutting into the body or the use of x-rays. It uses nuclear magnetic resonance to map sections of the body. One of its many uses is imaging the brain to detect the presence of tumors. See NUCLEAR MAGNETIC RESONANCE.

magnetic reversal A 180° change in the polarity of the earth's magnetic field. In a magnetic reversal, the field polarity at the North pole is exchanged with the polarity at the South pole.

magnetic storm A dramatic fluctuation or disturbance of the earth's magnetic field. Magnetic storms occur with solar flares and cause corresponding disturbances in the ionosphere.

magnetic stripes See LINEAR OCEAN-FLOOR MAGNETIC ANOMALIES.

magnetite A black iron oxide mineral forming cubic crystals in igneous rocks. Magnetite, also known as lodestone, is a magnetic iron ore. See APPENDIX p. 625.

magnetohydrodynamic generator (MHD generator) A device harnessing the flow of a plasma through a transverse magnetic field to produce electricity. Also called a magnetoplasmadynamic generator (MPD generator).

magnetohydrodynamics (MHD) The study or application of how electrically conducting fluids such as a plasma are affected by magnetic fields. Magneto-hydrodynamics help in understanding the structure of the sun, but also has possible applications for generating electrical energy using plasmas.

magnetometer Instrument for measuring the magnitude and direction of a magnetic field or one of its components.

magnetopause The boundary of the earth's magnetosphere and the solar wind. See MAGNETOSPHERE.

magnetosphere Magnetic region surrounding a planet, especially earth. Within this region, the earth's magnetic field dominates over the sun's magnetic field. The magnetosphere deflects the solar wind around the planet. It is much smaller on the sunward side where it encounters the solar wind, and expands greatly on the side of the planet away from the sun.

magnification 1) The process of making larger, increasing the magnitude of an entity (such as a problem). 2) To make an image appear larger, as by looking through a microscope. See BIOACCUMULATION.

magnitude (earthquake) A measure of the energy released in an earthquake as determined by the amplitude of seismic waves. Earthquake magnitude is usually measured on the Richter scale.

Magnoliaceae An angiosperm family, considered by many to contain the most primitive living flowering plants, that contains magnolias (*Magnolia*) and the tulip-tree (*Liriodendron*). These woody dicots have alternate, simple leaves and numerous, spirally arranged stamens and carpels. See ANGIOSPERMOPHYTA.

magnox reactor Nuclear reactor in which magnox is used for the canisters surrounding the fuel. Magnox is any one of several alloys of magnesium oxide. Magnox reactors use natural uranium as fuel, with graphite as the moderator and carbon dioxide as the coolant. The first commercial nuclear reactor, built in Calder Hall, Cumbria, UK, in 1956, used this design. See NUCLEAR REACTOR.

make up water   Water required to replace that used in or lost from a system. Make up water includes that used to replace water leaking out of an irrigation system and, more commonly, that lost in a cooling tower used in power generation.

malachite   A bright green mineral of copper carbonate and water that forms botryoidal masses in the oxidized zone of copper ore deposits. Malachite is used as an ornamental jewelry stone.

malacology   The study of molluscs. See MOLLUSCA.

Malacostraca   A large and important class of crustacea whose members generally have a head of five, a thorax of eight, and an abdomen of six segments. Antennae are prominent and the thorax is covered with a carapace in most groups. Eyes are frequently stalked and the well-developed appendages are usually distinctly biramous. The abdomen often terminates in a flattened plate, the telson. They are widely distributed throughout seas and freshwaters and some forms are terrestrial. Familiar types include pill bugs (wood lice), krill and Decapoda (shrimps, crabs, and lobsters). Approximately 20,000 species exist. See ISOPODA, KRILL, DECAPODA, LOBSTERS.

malaria   Common tropical disease caused by four different species of a parasite, *Plasmodium*, that lives in human red blood cells. It enters the blood through the bite of certain mosquitoes (of the genus *Anopheles)* that serve as its alternate host. It is characterized by recurring fever. Also called paludism or swamp fever.

malathion $(CH_3O)_2$-(PS)-S-$CH(CH_2COOCH_2CH_3)$-$CO_2CH_2CH_3$, an organic phosphate insecticide. Malathion controls many pests, yet has low toxicity for humans and when applied correctly is relatively safe for the environment because it breaks down quickly.

Malaysian-Papuan floral region   One of the phytogeographic regions into which the Paleotropical realm is divided according to similarities between plants; it consists of the southern Malaysian peninsula, Indonesia, and Papua New Guinea. See PALEOTROPICAL REALM, PHYTOGEOGRAPHY.

male   1) An animal that produces sperm or a corresponding form of gamete that fertilizes ova or their analogues. 2) A plant or flower that produces microspores or pollen, or has functioning stamens without having functioning carpels. Compare FEMALE.

maleic hydrazide   A plant growth retardant. Its uses include stunting the height of grains to discourage lodging (falling over, making them impossible to harvest by machine) and preventing root crops such as potatoes from sprouting during storage.

malignant tumor   Cancerous growth in which cells multiply uncontrollably and invade surrounding healthy tissue (metastasis); a tumor capable of causing eventual death. Contrast BENIGN TUMOR.

mallee scrub   Scrub vegetation in which shrub forms of eucalyptus dominate, found in dry, subtropical regions of southern Australia. See MAQUIS, CHAPARRAL.

malnutrition   Lack of sufficient food (starvation) or lack of specific nutrients caused by poor diet or a malfunction preventing the body from absorbing nutrients. See DEFICIENCY DISEASE.

Malpighian tubules   Tube-shaped glands in insects, some arachnids, and centipedes and millipedes. They open into the alimentary canal and help to excrete wastes. Also called Malpighian tube or vessel.

Malthusian growth A population explosion followed by a population crash, because of exhaustion of food resources; also called irruptive growth.

Malthusian theory of population The theory proposed by English clergyman and economist Thomas R. Malthus (1766–1834) that the world's population tends to increase at a geometric rate, thus outpacing the food supply, which increases at an arithmetic rate. Humans are thus destined to suffer hunger unless they limit population growth, or unless growth is limited for them by famine, epidemics, wars, etc. See NEO-MALTHUSIAN.

maltose $C_{12}H_{22}O_{11} \cdot H_2O$, also called malt sugar, a disaccharide consisting of two linked glucose molecules. It is produced by germinating grains, converting stored starch into glucose to provide energy for the emerging seedling. Maltose is also formed by the body when it digests starch.

Malvaceae The mallow family of the Angiospermophyta with plants often having palmately veined leaves with stellate tufts of hairs. The five sepals and five petals of these dicots surround the numerous stamens that are united into a tube that encases the elongated style. Hollyhock (*Althaea*), mallows (*Malva*), hibiscus (*Hibiscus*), and cotton (*Gossypium*) are members of this family. See ANGIOSPERMOPHYTA.

mamma Mammary (milk-producing) gland in mammals.

Mammalia A class of endothermic vertebrates whose integument provides an insulating cover of hair as well as claws, nails, horns, and glandular elements for the postnatal feeding of young, the mammary glands, which give the class its name. The mammalian kidney is uniquely adapted to maintain the constancy of body fluids, and the possesion of a diaphragm greatly increases the efficiency of breathing. See VERTEBRATA, ENDOTHERM.

Man and the Biosphere Programme (MAB) A scientific program of the United Nations Educational, Scientific, and Cultural Organization (UNESCO) for protection of significant biological reserve areas. It is attempting to create a worldwide network of preserves, one for each of the world's 194 biogeographical provinces, for conservation and scientific research.

manganese (Mn) A brittle, metallic element with atomic weight 54.94 and atomic number 25. It is used in the manufacture of steel and industrial chemicals. It is also an essential trace nutrient for animal and plant life and so is included in some fertilizers.

manganese oxide nodule A submarine concretion typically composed of manganese, iron, nickel, and cobalt. Manganese oxide nodules are widely distributed in the abyssal plains of oceans and other submarine environments with little sedimentation.

mangrove swamp ecosystem Subtropical (or tropical), coastal marine community dominated by mangrove trees (*Rhizopora* and *Avicennia*), which can withstand high salt concentrations. Their extensive root systems have "knees" (pneumatophores), large bends of roots that stick up above the water, to allow gas exchange. See ADVENTITIOUS ROOTS, PNEUMATOPHORE.

mantle The internal layer of the earth that is between the core and the crust. The mantle is approximately 2300 km thick and composed of rocks such as peridotite.

mantle cavity   An internal space within the body of some animals. Found in sea squirts (Tunicata or Urochordata), where it is called an atrial cavity, as well as in molluscs, brachiopods, and barnacles.

manubrium   1) Any animal structure that is shaped like a handle. 2) In mammals, the upper bone of the sternum of mammals; also a small, handle-shaped project of the malleus in the ear.

maquis   A vegetation type characterized by dense, impenetrable scrubby shrubs with few, if any, trees. It is found in Mediterranean regions on poor soils that receive little rainfall (except in winter), as on the island of Corsica. See CHAPARRAL.

marasmus   Emaciation; extreme malnutrition of children, generally under two years of age, in which the abdomen becomes extended but the rest of the child is reduced to skin and bones. Marasmus may be caused by lack of food, or an inability to absorb food following acute diseases such as extreme forms of diarrhea.

marble   A non-foliated rock that is formed of metamorphosed limestone or dolomite. Marble is hard enough to be cut and polished and is commonly used in building and sculpture.

marginal value theorem   A proposed principle regarding a predator hunting groups of prey that it depletes, which states that the predator should kill roughly the same number of individuals from each group, hunting at the maximum average overall rate for that environment as a whole.

mariculture   Saltwater fish farming; the cultivation of any saltwater organisms (fish,

shellfish, etc.) in ocean water.

marine   A term relating to the environment of the ocean.

marine humus   The decomposing organic material of the ocean floor.

marine regression   A lowering of ocean level on a regional or global scale.

marine terrace   A terrace-shaped landform that was formed by wave erosion in a marine environment. See TERRACE.

marine transgression   A raising of ocean level on a regional or global scale.

maritime climate   A climate in which the ocean plays a major role, moderating other effects to produce a climate with mild winters and relatively cool summers. Also called a marine climate. Contrast CONTINENTAL CLIMATE.

marker bed   A bed that may serve as a chronostratigraphic datum in the determination of relative geologic age. Marker beds have unique or distinctive characteristica that may be traced over a wide geographic area.

Markovian model   See STOCHASTIC MODELS.

marl   An unconsolidated sediment composed mostly of calcium carbonate and clay.

marsh   Lowland occasionally covered by water. A marsh differs from a swamp in that it is dominated by rushes, reeds, cattails, and sedges with few, if any, woody plants. It differs from a bog in having soil rather than peat as its base.

marshland   An extensive marsh. See MARSH.

marsupials   See METATHERIA.

marsupium   1) A pouch or skin fold on the abdomen in which female metatherian (marsupial) mammals carry their young. 2) A similar structure found in some fish and crustaceans that is used for carrying eggs or young. See METATHERIA.

mascon   An abbreviation of a mass concentration of dense material below a planetary surface.

masking   A reduction in the ear's sensitivity to specific sounds when other, louder sounds are present. Masking can also refer to other situations in which one sound (or even another phenomenon such as smell) overpowers another.

mass   The measurable quantity of matter in an object. Mass is usually determined by comparing an object to one of known mass. It can also be derived either from the object's inertia or, more commonly, its weight. Unlike weight, mass is a constant value; an astronaut floating between galaxies may be weightless, yet he or she will still have the same mass as on earth. Compare WEIGHT.

mass balance (of a glacier)   A calculated relation between the net ice accumulation and the net ice loss within a glacier.

mass burn   Solid-waste incineration, with minimal sorting or pretreatment of the garbage.

mass extinction   Large-scale extinction in which large numbers of species disappear in a few million years or less. Examples include the mass extinction of the dinosaurs (possibly from an asteroid collision) and that of many North American large mammals coincident with the rise of human populations on that continent.

massif   A large, rigid structure or mass of the earth's crust. The crystalline rock components of orogenic belts are often described as massifs.

mass mortality   Large-scale die-off of most of a population or of many species within a community. Mass mortality occurs in rivers if industrial accidents spill toxic chemicals into the river.

mass movement   See MASS WASTING.

mass number   The total number of protons and neutrons in the nucleus of an atom. This value identifies different isotopes of an element. Mass number follows the name of the element when describing a particular isotope, as in carbon 14. It is also often written as a superscript before the element symbol (^{238}U) or hyphenated after the element symbol (U-235).

mass spectrometer   An instrument for identifying isotopes and chemical compounds. A combination of electric and magnetic fields are used to separate a stream of ions by charge and mass. This charge-to-mass relationship specifies not only which elements are present but what the relative proportions of the different isotopes of the elements are, and also the masses of larger chunks of molecular fragments.

mass transit   Public bus and train systems for moving large numbers of people between places on a regular schedule.

mass wasting   A general term for downslope movement of earth material. Mass wasting is usually categorized as a flow, slide, fall, or creep of material, depending on the speed and conditions of movement. See LANDSLIDE.

mast   1) The collective name for the nuts of forest trees, such as beechnuts and acorns, that often become food for wildlife (or livestock such as pigs) when they fall to the ground. 2) The phenomenon in which oaks and beeches (and sometimes other members of the Fagaceae) produce a superabundance of fruits once every several years (during a mast year) with intervening years of low fruit production. It is believed to be an adaptation to periodically ensure reproductive success by producing more seeds than could possibly be consumed by seed predators.

mast cells   Cells resembling granular leukocytes but found in connective tissue. When stimulated, mast cells release the histamine that causes allergy symptoms; they can also produce the anticoagulant heparin.

mating system   Pattern within a population of individuals pairing off to mate and to produce offspring. The mating system includes how mates find each other, number of simultaneous mates, degree of bonding with each mate, and permanence of pair bonding. See MONOGAMY, POLYGAMY, POLYANDRY.

matrix   1) A web or system of interconnecting or interdependent parts or functions. 2) In geology, the material (rock) in which a fossil, pebble, crystal, or mineral is embedded. 3) A mold or structure within which something is cast or shaped. 4) Data arrayed in rows and columns. 5) A set of elements in a rectangular arrangement.

matter   General term for anything having mass and occupying space. On earth, matter is recognized by having weight, as well as mass and volume. See MASS.

mattoral   Spanish term for a Mediterranean scrub woodland found in semi-arid climates. See CHAPARRAL, MAQUIS, MALLEE SCRUB.

maturation   1) The development in germ cells in preparation for fertilization, from meiosis through final stages. 2) The completion of development, reaching adult (mature) form.

Maunder minimum   The period of time from 1645–1715 when almost no sunspots were visible and the sun showed an abnormal lack of activity of any kind. See LITTLE ICE AGE.

maxillae   1) The jawbones of mammals and most other vertebrates. Inferior maxilla is the lower jawbone, also called mandible; superior maxilla is the upper jawbone. 2) A pair of appendages behind the mandibles of insects, crustaceans, centipedes, and millipedes; they assist with feeding. Also called maxillipeds.

maxillipeds   Modified anterior limbs of crustaceans and other arthropods that assist with feeding. Also called maxillae.

maximum contaminant level   (MCL) A federal standard for the highest amount of an impurity allowed in public supplies of drinking water. States also set their own standards, which may not be identified by this term.

maximum permissible level   Standards for radiation limits set by the Nuclear Regulatory Commission and the Environmental Protection Agency to protect workers at nuclear facilities and the public. Limits are set for the absorbed dose, total amount of various nuclides in a person's body, and concentrations in water and air.

maximum sustainable yield   (MSY) The highest rate of harvest of a renewable resource, such as lumber, that can be taken each year over a long period of time. If harvests above this level are taken, the system will not be able to regenerate in time and yields will eventually decline. Compare CARRYING CAPACITY. See SURPLUS YIELD MODEL.

mb Abbreviation for the meteorological term millibar.

MCL See MAXIMUM CONTAMINANT LEVEL.

meadow An area of land covered primarily with grass, either through cultivation or naturally occurring grassland.

mean 1) The average; statistical term for the average of a series of values, calculated by adding together all of the values and dividing by the number of different values. Also called arithmetic mean. 2) A point or quantity equally far from two extremes. Compare MEDIAN.

meander Tendency for a river to have a sinuous channel pattern characterized by a series of bends. Meanders usually form on flood plains composed of unconsolidated alluvium.

meander belt The region of a floodplain along which meanders develop.

meander scroll The curved ridges of coarse material deposited parallel to the bends of a meandering stream.

mean free path 1) The average (mean) distance a gas molecule can travel freely, that is, without hitting another molecule. 2) A similar measurement for the movement of sound waves in an enclosed space, used to calculate reverberations.

mean intensity of infection The average number of parasites per host in a community.

mechanical turbulence Movement of air characterized by constant small-scale changes in direction and speed, and caused by physical obstructions such as buildings. See TURBULENCE.

mechanical weathering The physical disintegration of rock and surface material by reduction in particle size without significant chemical change. Contrast CHEMICAL WEATHERING, WEATHERING.

medfly Short for Mediterranean fruit fly (*Ceratitis capitata*), an alien species that feeds on wild and cultivated fruit. It is an economic pest for many fruit growers in regions without frost, especially in citrus groves in California.

medial moraine A ridge-shaped accumulation of till carried down the center of a glacier where two tributary alpine glaciers converge.

median 1) Statistical term for the middle point in a series of values; above and below this point are an equal number of values. 2) A line or plane that divides a bilaterally symmetrical organism equally into right and left halves. Compare MEAN.

median valley The fault-bounded central valley located in the rift axis of a mid-ocean ridge. Median valleys are as deep as 3 km and usually form at relatively slow-moving, mid–ocean spreading centers.

mediation A method of settling disputes outside of a court setting; the imposition of a third party to act as a link between the parties.

Mediterranean floral region One of the phytogeographic regions into which the Holarctic realm is divided according to similarities between plants; it consists of southern Europe, North Africa, and Mediterranean seashores. See HOLARCTIC REALM, PHYTOGEOGRAPHY.

mediterranean-type climate A colloquial term for a dry-summer subtropical climate. This is a transition between the marine west coast climate (or maritime) and tropical steppes. It differs from other climates in getting most of its precipitation during winter months; in coastal areas it is associated with cool summers but further inland summers can be warm.

medium frequency (MF) Refers to the portion of the radio spectrum between 0.3 and 3 megahertz. See EXTREMELY LOW FREQUENCY, HIGH FREQUENCY, LOW FREQUENCY, ULTRA HIGH FREQUENCY, VERY HIGH FREQUENCY, VERY LOW FREQUENCY.

medulla The central section of an organ or tissue.

medullary rays Radiating vertical plates within stems and roots of plants that store and transport food. They run through the vascular bundles and extend from the pith to the cortex or bark.

Medusae Planktonic forms of the phylum Cnidaria, commonly called jellyfish. See CNIDARIA, SCYPHOZOA.

mega- 1) Prefix meaning large. 2) Prefix in the Système International d'Unités (SI) indicating one million, 1×10^6. One megahertz equals one million hertz (cycles per second), or 1×10^6 Hz.

megafauna The largest size category of animals in a community.

megagamete Another term for macrogamete. See MACROGAMETE.

megalecithal Describing eggs that contain large amounts of yolk. Contrast MICROLECITHAL.

megalopolis A thickly populated region in which several cities have expanded to overtake, merge with, or butt up against other cities. This creates a huge urbanized area; the corridor stretching between Washington, DC, and Boston, MA, has become a megalopolis (sometimes called the megalopolitan corridor). The North-

West European megalopolis extends from Germany's Ruhr west to Paris, France, and north to Manchester and Liverpool in the United Kingdom.

megaplankton Any plankton that is large enough to be seen with the unaided eye, generally used for those organisms larger than 20 centimeters in diameter or length and therefore larger than the macroplankton. It includes seaweeds (such as *Sargassum)* and large jellyfish. Compare MACROPLANKTON, MESOPLANKTON, MICROPLANKTON.

megaspore In bryophytes and vascular plants producing two different types of spores, the larger haploid spore, which is capable of developing into a female gametophyte. Compare MICROSPORE.

megasporophyll In flowering plants and pteridophytes (ferns, club-mosses, horse-tails), the leaflike structures (sporophylls) supporting megaspores (the larger of the two different kinds of spores. Compare MICROSPOROPHYLL.

megatherm Describing climates having very high temperatures. See MICROTHERM, MESOTHERM.

megawatt (mw) One million watts. This unit is more convenient for describing electrical generating capacity than watts or kilowatts. See WATT.

meiobenthos Organisms (meiofauna or meioflora) living in or on bottom sediments and ranging in size from 0.1 to 0.5 millimeters, including many kinds of small invertebrates and larger protozoans. Compare MICROBENTHOS, MESOBENTHOS, BENTHOS, MACROBENTHOS.

meiosis The form of cell division in which the number of chromosomes is reduced from the diploid number to the haploid number, occurring in reproductive cells of sexually reproducing organisms. Meiosis involves two consecutive cell divisions with only one replication of chromosomes, thereby producing four haploid gametes or spores. Contrast MITOSIS.

melange A group of varied rock types in a chaotic association of tectonic origin, often subduction zones.

melanin A dark brown or black pigment present in the skin, hair, and eyes of humans and many animals. It provides some protection from bright sunlight, and is produced in cancers such as melanoma.

melanism Abnormal darkness of color caused by an unusual increase of melanin in the skin, hair, or plumage of animals or humans, resulting in a portion of a normally light-colored population being black. See ALBINISM, INDUSTRIAL MELANISM.

melanocratic A term applied to dark-colored igneous rocks having a proportion of visible dark minerals that is greater than 50 percent.

melanocytes Cells found in the skin, eye, and hair or coat that manufacture melanin. See MELANIN.

melanoma Malignant tumor originating in melanin-producing cells in the skin and eye. Melanomas are usually molelike growths on the skin that increase rapidly in size, and are among the most lethal types of tumor. Melanomas occur most commonly in fair-skinned people living in sunny regions and exposed to large amounts of sunlight.

melanophores Cells (chromatophores) containing black or dark pigment. Some lower vertebrates have melanophores in their skin that can expand in cold temperatures to increase absorption of sunlight.

melilite An intermediate member of a group of sorosilicate minerals containing magnesium or aluminum and calcium or sodium in the crystal lattice. Melilite occurs in basalts and metamorphosed limestones.

melt To cause a change in phase from solid to liquid using heat. Melting is sometimes called fusion, as in heat of fusion.

meltdown An accident in a nuclear reactor in which the core overheats because of a failure in the cooling system. Eventually, so much heat can build up that the fuel rods melt. A partial meltdown occurred at Pennsylvania's Three Mile Island nuclear plant in 1979. See CHERNOBYL.

melting point The temperature at which a substance begins to undergo a phase change from solid to liquid usually at the standard pressure of one atmosphere.

meltwater Surface runoff derived from a melting glacier.

member A stratigraphic term for a unique subdivision of a geologic formation. See FORMATION.

membrane A thin, pliable sheet tissue in animals or plants that lines, covers, connects, or separates organelles, cells, organs, or other structures. See PLASMA MEMBRANE, SEMIPERMEABLE MEMBRANE

menarche The onset of menstruation, indicating sexual maturation in a female.

mendelevium (Md) An artificial, radioactive element with atomic number 101; the atomic weight of its main isotope is 258. It is a transuranic element in the actinide group.

Mendelian population Reproductive group that shares a common gene pool.

Mendel's laws   The principles governing inheritance of characteristics developed by Austrian botanist and monk Gregor Johann Mendel (1822–1884) through experiments with peas. His law of segregation states that characteristics are inherited independently of each other. The law of independent assortment states that traits combine randomly in offspring, causing successive generations of crossbreeds to exhibit different combinations of inherited characteristics, each combination in a specific proportion of individuals. The law of dominance states that in characteristics showing dominant and recessive forms, recessive traits will only appear if the gene pair contains two recessive genes. See DOMINANCE, INDEPENDENT ASSORTMENT, SEGREGATION, AND LINKAGE.

menstrual cycle   The repeating series of changes in hormones and sex organs that occur in females between puberty and menopause. The cycle has four stages that repeat over intervals of roughly one month: maturation of the egg and the beginning of estrogen-induced thickening of the lining of the uterus, ovulation (the rupture of the follicle to release the egg), formation of the corpus luteum and associated production of progesterone, and menstruation. See MENSTRUATION.

menstruation   The roughly monthly discharge of blood from the uterus that results when no fertilization has occurred and the thickened uterine lining is shed. Menstruation is the most obvious phase of the menstrual cycle. See MENSTRUAL CYCLE.

Mercalli scale   A measure of the intensity of seismic waves generated by an earthquake. The Mercalli scale is a context-sensitive measure of the destructiveness of earthquake events at a given location. The scale ranges from level I, at which an earthquake is not felt, to level XII, a total devastation of human structures. Compare RICHTER MAGNITUDE SCALE.

mercaptans   A group of common organic compounds containing an -SH (thiol) group attached to a hydrocarbon group. They are structurally like alcohols with the sulfur atom replacing the oxygen atom of the alcohol. Mercaptans have characteristic unpleasant odors: ethyl mercaptan is responsible for the smell of garlic and another mercaptan is responsible for essence of skunk. Also known as thiols.

merchantable timber   Trees that are of a suitable species and size for harvest and sale as lumber.

mercury   (Hg) A heavy, poisonous liquid metallic element with atomic weight 200.59 and atomic number 80. Mercury is a solvent for most metals, producing amalgams. It is used in thermometers, barometers, light switches, paints, and batteries. Once in the environment, mercury persists and concentrates as it moves up the food chain, reaching particularly high levels in fish and shellfish. Prolonged exposure to mercury, either through inhaling or swallowing, can damage the central nervous system. See AMALGAM.

mercury fungicides   Extremely toxic organic compounds containing mercury that have been used as soil disinfectants, to treat seeds and bulbs, and for lawn diseases. The use of calomel (mercurous bichloride) and similar compounds is no longer allowed in the United States.

meridian   One of the imaginary lines connecting the earth's North and South poles; a line of longitude. Contrast PARALLEL OF LATITUDE.

meristem   The undifferentiated, growing

parts of plants, consisting of groups of cells capable of actively dividing. See APICAL MERISTEM, CAMBIUM, INTERCALARY MERISTEM.

merogony The fertilization and growth of an embryo from a portion of an egg or an egg lacking a nucleus.

meromictic A term describing lakes that are permanently stratified into epilimnion and hypolimnion layers.

meromixis The process leading to the meromitic condition in a lake.

meroplankton Plankton that spend only a part of their lives as plankton, such as the juveniles of many nekton. Contrast HOLOPLANKTON, EUPLANKTON.

Merostomata A class of the phylum chelicerata, with a single, living order, *Xiphosura*. These marine animals, commonly known as horseshoe crabs, have bodies divided into cephalothorax, abdomen and tail spine. The cephalothorax, which is covered by a dish-shaped shell or carapace through which two, lateral eyes protrude, is hinged to the abdomen. Five of the eleven abdominal segments bear paired appendages modified as gills. One common species, *Limulus polyphemus*, is found in coastal waters from the Canadian maritimes to Yucatan, while the remaining three species are in found in the western Pacific, Bay of Bengal and southeast Asia. See CHELICERATA.

-merous Having a specified number of parts. For example, members of the Caryophyllacea (pink family) are 5-merous: They have five sepals, five petals, and stamens in multiples of five.

mesa An isolated flat-topped small plateau that is often bounded by steep sides. A mesa is usually broader than it is tall.

mesic Describing an environment having moderate rainfall and moderately moist, well-draining soils. Mesic plants are those that require moisture. Contrast XERIC.

mesobenthos A term applied to the ocean bottom at a depth of 180 to 900 meters.

mesocarp A central layer of fruit wall (pericarp) present in some fruits. The fleshy part of a plum, the fibrous layer in a coconut, and the hard woody tissue surrounding an almond are all examples of mesocarp. See PERICARP, EPICARP, ENDOCARP.

mesoclimate A local variation in climate only found in an area restricted to a few miles in diameter. Compare MACRO-CLIMATE, MICROCLIMATE.

mesoderm The central layer of cells formed during the development of the embryos of animals, lying between the endoderm and ectoderm. From this layer the muscular, skeletal, circulatory, lymphatic, and connective tissues develop. Also called mesoblast.

mesohalabous Describing plankton, especially phytoplankton, that live in brackish water, that having a salinity range of 5 to 20 percent. See EUHALABOUS, OLIGOHALABOUS PLANKTON.

mesohaline A term describing water that has a chlorinity value between 0.3 percent and 1.0 percent.

mesolithic A transitional period between the Paleolithic and Neolithic cultures of the Stone Age.

meso-, mes- Prefix meaning middle or intermediate.

mesopause The boundary between the earth's mesosphere and the ionosphere. See MESOSPHERE.

mesopelagic   Of or pertaining to the area of the sea that is between 200 and 1000 meters, a region characterized by some light, but insufficient levels for much photosynthesis to occur. Also called dysphotic.

mesophilic   Describing organisms that need moderate temperatures (2 to 40°C) for growth. Compare PSYCHROPHILIC, THERMOPHILIC.

mesophyll   The inner tissue of a leaf, containing chlorophyll and surrounded on top and bottom by layers of epidermis. See PALISADE MESOPHYLL, SPONGY MESOPHYLL.

mesophyte   A plant that can tolerate average moisture levels and only occasional dryness. Compare HYDROPHYTE, XEROPHYTE.

mesoplankton   Plankton that are intermediate between macroplankton and microplankton in size, ranging from 0.2 to 20 millimeters in diameter. See MACROPLANKTON, MICROPLANKTON.

mesosphere   The region of the earth's atmosphere between the stratosphere and the ionosphere, extending from 64 to 80 km above the earth's surface. In this layer, temperatures drop rapidly as altitude increases.

mesothelioma   A form of cancer caused by the inhalation of asbestos fibers. Even minute quantities of asbestos lodging in the lungs can lead to this rapid, lethal form of cancer. See ASBESTOSIS.

mesothelium   Cells that develop from the mesoderm to line the primitive body cavity in vertebrate embryos. The mesothelium continues to differentiate, becoming the epithelium in adults. See EPITHELIUM.

mesotherm   Describing climates characterized by warm, rather than hot or cold, temperatures. See MEGATHERM, MICROTHERM.

mesotrophic lake   Lake that contains moderate amounts of nutrients (normal levels for a healthy lake). Mesotrophic lakes have evolved beyond the nutrient-starved oligotrophic stage but have not yet reached the overenriched eutrophic state. See EUTROPHICATION, OLIGOTROPHIC LAKE.

Mesozoic era   The second of three eras of the Phanerozoic eon in geologic time. The Mesozoic era lasted from approximately 248 to 65 million years ago. The term Mesozoic means middle life. See APPENDIX p. 610.

messenger RNA   A form of RNA (ribonucleic acid) that carries genetic information transcribed from the DNA in the cell nucleus to special regions in the a cell (ribosomes), where it directs the manufacture of a particular protein. See RNA, TRANSFER RNA.

Messinian crisis   A term applied to the events of the Miocene epoch that led to the closing of the Strait of Gibraltar and the isolation of the Mediterranean Sea.

meta-   Prefix meaning after, between, among, or changeable.

metabiosis   See COMMENSALISM.

metabolic   Produced by or relating to metabolism.

metabolic rate   The level of energy expenditure of an organism.

metabolic reserve   Lower half of rangeland grass plants that can grow back if they are not consumed by herbivores.

metabolism   All of the chemical and physical activities that sustain an organism.

Metabolism involves the breakdown of organic compounds to create energy, which animals use to grow, repair tissues, provide heat, and engage in physical activity. Plant metabolism involves the creation as well as utilization of organic compounds. In some organisms, metabolism includes metamorphosis.

metabolite Any material produced by metabolism.

metagenesis Alternation of generations, especially reproduction involving a regular alternation of sexual and asexual generations. See ALTERNATION OF GENERATIONS.

metal A class of elements characterized by being good conductors of heat and electricity, typically existing as lustrous solids, and easily shaped by casting, hammering, or drawing out into wires. Metallic elements readily give up electrons to form positive ions.

metaldehyde $C_4O_4H_4(CH_3)_4$, an organic compound of low toxicity used in baits and sprays to kill slugs and snails. Metaldehyde is also used as a fire starter and as a fuel for portable stoves.

metalimnion See THERMOCLINE.

metallic Having the characteristics of a metal, or containing metals. See METAL, NONMETALLIC.

metallic bond The form of covalent bonding characteristic of metals and their alloys. In this type of bonding, the electrons are not confined to discrete regions but are able to move freely between a number of different atoms. The electrons' freedom of movement gives metals their good electrical and thermal conductivity.

metallic luster A term applied to a mineral with a surface that reflects light like the sheen of a metal.

metallic mineral Any mineral composed primarily of elements that form a metallic bond.

metalliferous A term applied to rocks that have an extractable mineral component.

metallogenic provinces A region that is characterized by earth processes that have resulted in ore deposition.

metalloid An element having both metallic and nonmetallic properties, such as antimony or arsenic. See METALLIC, NONMETALLIC.

metamere A body segment, one of a series of similar parts or segments along the long axis of the body of certain animals, such as earthworms. Also called a merome or somite.

metameric segmentation Another term for metamerism. See METAMERISM.

metamerism The division of the body of animals such as earthworms into a series of similar segments (metameres or somites) along its long axis. Each segment contains the same organs.

metamorphic aureole A zone of metamorphosed country rock surrounding the area of contact with a pluton.

metamorphic belt An area of regionally metamorphosed rocks sharing a common tectonic history.

metamorphic facies A set of characteristics that may be collectively used to define the conditions of a metamorphic environment. Rocks that have formed along a metamorphic facies have been exposed to the same metamorphic grade.

metamorphic rock Any rock that has been recrystallized by the action of heat and pressure. These altered rocks are categorized as regional, contact, cataclastic, or burial metamorphic types, according to the context and the environmental conditions of formation.

metamorphic zone An area enclosed by two isograds within a metamorphic region. See ISOGRAD.

metamorphism The process of rock recrystallization by conditions of heat and pressure. See METAMORPHIC ROCK.

metamorphosis A significant change in the shape and structure, and usually the habits, of an animal during a fairly short period, between the embryonic and adult stages. Tadpoles metamorphose into frogs and caterpillars metamorphose into butterflies.

metam sodium H_3C-NH-CS-S-Na, a pesticide used as a fumigant for soils. It is also called N869, Vapam, VPM, SMDC, Metam, or Trimaton.

metaphase The second stage in cell division (mitosis or meiosis), in which the nuclear membrane disappears and the chromosomes align along the center of the cell and attach to the spindle fibers. It occurs after prophase and before anaphase. See ANAPHASE, MEIOSIS, MITOSIS, PROPHASE, TELOPHASE.

metaphysics A branch of philosophy that deals with the nature of existence and of truth and knowledge.

metaplasm Lifeless matter derived from protoplasm and found in a cell, especially stored food reserves such as carbohydrates. See PROTOPLASM.

metaquartzite A hard and massive rock formed by the metamorphic recrystallization of quartz sandstone.

metasediment A general term applied to a rock formed of metamorphosed sediment.

metasomatism, metasoma The alteration of rocks by removal or addition of minerals and volatiles in hot solution. The rocks are altered without passing through phases of melting and recrystallization.

metastasis The spreading of cancerous body cells, bacteria, or a disease from one part of the body to other organs or tissues. Metastasis usually occurs through the blood vessels or the lymphatic system. See MALIGNANT TUMOR.

Metatheria A subclass of mammals, commonly called marsupials. They have a number of distinct anatomical features; their most striking characteristic is that the young, which are not fully developed at birth, complete development in the mother's marsupium, a pouch which covers the mammary glands. Once widespread, the marsupials are now mainly restricted to the Australian region although some forms occur in North and South America (e.g., kangaroos, wallabies, bandicoots, wombats, koalas, opposums, etc.). Approximately 250 species exist. See MAMMALIA.

metaxenia Any effect caused by male pollen contacting tissues of female plant organs. Compare XENIA.

meteoric water Water recently derived from atmospheric processes (e.g., rain, snow, sleet).

meteorite A body of extraterrestrial solid material that has fallen through the atmosphere to the earth's surface. The principal compositional varieties of meteorites are classified as stony irons, iron nickel, chondrites, and achondrites.

meteorology The study of the atmosphere, weather, and climate.

meter (m) Unit of length in the Système International d'Unités (SI) and metric system (MKS). The standard meter (also metre) is the distance between two lines on a platinum-iridium alloy bar at 0°C kept in Paris, France. It is now defined precisely in terms of the wavelength of the orange light emitted by krypton[86]. 1 meter = 39.37 inches.

methane CH_4, the simplest hydrocarbon. It is sometimes called marsh gas because it is produced by decomposing organic wastes. Natural gas is nearly pure methane. See FIREDAMP.

methanol CH_3OH, wood alcohol or methyl alcohol. Its structure is like methane but with an -OH group substituted for one of the -H in methane. As the name wood alcohol implies, it was formerly made from distilling wood; now it is usually synthesized from coal, natural gas, or hydrogen gas and carbon monoxide. Methanol is important to the chemical industry as a solvent and in the manufacture of many compounds such as urea formaldehyde resins.

methemoglobinaemia A medical condition in which abnormal quantities of the blood hemoglobin has been converted into methemoglobin through oxidation of the iron ion at the center of the hemoglobin. Methemoglobin is usually caused by poisoning, especially from nitrate-contaminated drinking water (contamination from chemical fertilizers or animal manures), or other toxins such as potassium chlorate and aniline dyes. Because methemoglobin cannot carry oxygen, methemoglobinemia eventually causes cyanosis and asphyxiation. See CYANOSIS, NITRATE.

methionine $CH_3S(CH_2)_2CH(NH_2)$-COOH, an essential amino acid used in protein synthesis. Present in egg white, cheese, yeasts, and some plant proteins, it supplies necessary sulfur in animal diets.

methoxychlor A chlorinated hydrocarbon insecticide, about the only one of these compounds whose use is still allowed. It is a wide-spectrum insecticide of relatively low toxicity that has been used as a replacement for DDT. See CHLORINATED HYDROCARBON INSECTICIDES.

methyl alcohol Another name for methanol. See METHANOL.

methylated spirits Denatured alcohol; ethanol (grain alcohol) to which methanol or similar poisonous or nauseating chemicals and coloring have been added to make it unfit for drinking.

methyl chloride Another name for chloromethane. See CHLOROMETHANE.

methylene blue $C_{16}H_{18}ClN_3S \cdot 3H_2O$, a compound used as a textile dye, as a biological stain, and as a test for sewage effluent to determine whether it has become fully oxidized. It is also an antidote for carbon monoxide and cyanide poisoning. Also known as methylthionine chloride. See METHYLENE BLUE STABILITY TEST.

methylene blue stability test A chemical test run on sewage effluent to determine if it has reached a fully oxidized, stable state. If the effluent has stabilized, the dye remains blue for the duration of the testing period.

methyl isocyanate (MIC) A very toxic gas used to manufacture insecticides. Methyl isocyanate was released at the Union Carbide plant in the Bhopal, India, disaster of 1984.

methyl mercury $[CH_3HG]^+$, an extremely toxic form of mercury for humans formed in natural systems from biological action on the elemental form of mercury. A variety of microorganisms that live in the bottom sediments of surface waters are capable of transforming elemental mercury into methyl mercury. Methyl mercury belongs to the category of compounds known as alkyl mercuries.

metoxenous parasite   See HETEROECIOUS PARASITE.

metric system   1) Common name of the standard international system of measurement, formally known as the Système International d'Unités or SI. 2) The original French decimal system of measurement upon which the SI is based. It is more correctly called the MKS or MKSA (meter, kilogram, second, ampere) system because it is based on the units meter, kilogram, second, and ampere as basic units. See SYSTÈME INTERNATIONAL D'UNITÉS, APPENDIX P. 605.

metropolis   The chief city of a country or region.

mev   Abbreviation for the unit megaelectron volt, one million electron volts. Such large quantities of electron volts are useful in quantifying nuclear reactions. See ELECTRON VOLT.

mg   Abbreviation for the metric unit milligram, one thousandth of a gram. See GRAM.

Mg   Chemical symbol for the element magnesium. See MAGNESIUM.

mgd   An abbreviation of million gallons per day.

MHD generator   See MAGNETOHYDRODYNAMIC GENERATOR.

mHz   Abbreviation for the metric unit megahertz, one million hertz (cycles per second). See HERTZ.

mic   Short for methyl isocyanate. See METHYL ISOCYANATE.

mica   Any of a group of sheet silicate minerals having a characteristic platy cleavage. Biotite and muscovite are typical forms of mica, and many intermediate varieties are formed by substitution of potassium (K), sodium (Na), aluminum (Al), magnesium (Mg), and iron (Fe) ions in the crystal lattice.

micelle   A colloidal-sized, polar aggregate of molecules (especially polymers such as starch). Surfactants form micelles, as when detergent is added to water or bile acids prepare fatty compounds for digestion. See COLLOID.

micro-   1) Prefix meaning small. 2) Prefix in the Système International d'Unités (SI) indicating one millionth, 1×10^{-6}. One microgram equals one millionth of a gram, or $10^{-6}g$.

microbe   A microorganism, an organism that can only be seen with the aid of a microscope, especially a bacterium producing disease or fermentation.

microbenthos   Very small (microscopic) and often single-celled plants, animals, and bacteria that live on or in sediments at the bottom of a body of water. See BENTHOS, MACROBENTHOS, MEIOBENTHOS, MESOBENTHOS.

microbial   Produced by or relating to microbes.

microbial metallurgy   The study of the deposition and alteration of metallic elements by the activity of microorganisms. The "desert varnish" found on rocks in arid climates is produced in part by the microbial alteration of metallic elements.

microbivore   An organism that feeds on microorganisms.

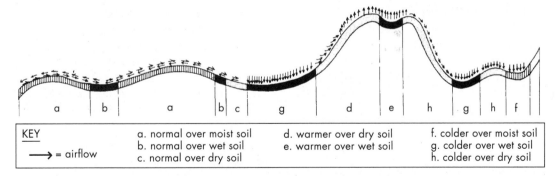

KEY			
→ = airflow	a. normal over moist soil b. normal over wet soil c. normal over dry soil	d. warmer over dry soil e. warmer over wet soil	f. colder over moist soil g. colder over wet soil h. colder over dry soil

The effect of local topography and vegetation on microclimate.
(Adapted from, *Designing and Building a Solar House* by Donald Watson. Copyright © 1977.
Reprinted with permission from Storey Communications, Inc.)

microclimate   A very small-scale variation from the overall climate pattern, usually caused by local physical conditions such as topography. A frost pocket is an example of a microclimate, as is the shade under a tree, or a small pond. Microclimates have a great influence on ecological niches. Contrast MACROCLIMATE, MESOCLIMATE.

microconsumer   Another term for decomposer. See DECOMPOSER.

microcosm   Literally, a little world. A small-scale ecosystem, especially one created in a container, that is used as a model in the study of large ecosystems. Microcosms may range in size from a testtube to an artificial river channel.

microcrystalline   A term applied to a crystalline rock texture that is observable only under microscopic examination.

microenvironment   The sum of all conditions within a very small area (often as little as several square meters), such as the environment surrounding individual organisms. A tree creates a shady microenvironment underneath it, usually quite different from that above the tree.

microfauna   The microscopic animals of a particular region or time. Contrast MACROFAUNA.

microfiltration   Another term for reverse osmosis. See REVERSE OSMOSIS.

microflora   The microscopic plants or plantlike organisms of a particular region or time.

microfossil   A fossil that is small enough to require a microscope for study. Microfossils in pelagic sediment, such as the Foraminifera, have become important in exploration for fossil fuels.

microgabbro   See DIABASE.

microgamete   The smaller, usually the male, of two fusing gametes of an organism that reproduces by means of unlike gametes (heterogamy). See MACROGAMETE.

microgranite   A medium-grained igneous rock of granitic composition having an average crystal diameter between 1 and 5 mm.

microhabitat   The portion of the more general habitat actually frequented by a specific organism. See HABITAT.

micro-hydro generators   Small-scale hydroelectric generators capable of producing electrical power in relatively shallow rivers. Such generators typically produce enough power for four to six dwelling units.

microlecithal   Describing eggs that contain very small amounts of yolk. Contrast MEGALECITHAL.

micrometer   An instrument for measuring small objects or anglular distances.

micron   (μ) Unit of length equal to one millionth of a meter (10^{-6}m). It has been replaced in the SI by the unit, micrometer (μm).

micronucleus   The smaller of the two nuclei present in some protozoa (*Ciliophora*); it contains genetic material and controls sexual reproduction within the unicellular organism. Compare MACRONUCLEUS.

micronutrient   A mineral (usually a trace element) or vitamin used only in small amounts by organisms but necessary for their well being. Micronutriens often are toxic in high concentrations. See ESSENTIAL ELEMENTS, TRACE ELEMENT, MACRONUTRIENT.

microorganism   A microscopic or submicroscopic organism, one that is too small to be seen by the naked eye. Bacteria, viruses, protozoans, and some fungi and algae are microorgansims.

micropaleontology   The scientific of study of microfossils.

microparasite   A microorganism that lives as a parasite on or in another, larger organism. Compare MACROPARASITE.

microphagous   Feeding on relatively small food pieces. Compare MACROPHAGOUS.

microphyllous   Of or relating to a microphyll, a small leaf with a single, unbranching vein down the center found in some primitive plants such as clubmosses and selaginella. Sometimes used to describe any plant having leaves only a few millimeters long; many desert plants have evolved microphyllous leaves as an adaptation to low moisture levels.

microplankton   Plankton made up of microscopic organisms, too small to be seen with the naked eye and smaller than 0.5 millimeters in diameter. Compare MACROPLANKTON, MESOPLANKTON.

micropopulation   The population of microorganisms living in a given community.

microrelief   The small-scale topographic variations in a land surface. Mounds, swales, and boulders are examples of microrelief features.

microsere   Ecological succession occurring on a very small scale, such as in a small depression in a rock outcrop. See SERE, CLISERE.

microspecies   One or more slight morphological variants of a plant species that breed true to type within a larger, inbreeding group of the species. Also called a Jordanon (as opposed to Linnean) species.

Microspora   See CNIDOSPORIDIA.

microspore   In bryophytes and vascular plants producing two different types of spores, the smaller haploid spore, which is capable of developing into a male gametophyte. In flowering plants this is the pollen grain. See MEGASPORE.

microsporophyll   A leaflike plant part bearing microspores. In seed plants, the microsporophyll is the anther. Compare MEGASPOROPHYLL.

microtherm Describing climates characterized by cold temperatures. See MEGATHERM, MESOTHERM.

microtopography The investigation or characterization of topography on a very small scale. Microtopography is measured in units of meters and centimeters.

microwave burst Increased radiation from the sun in the radio spectrum, usually at wavelengths around 10 cm (2700 megahertz) but often spanning from much smaller to much larger wavelengths. Microwave bursts are associated with solar flares, including x-ray flares. See FLARE.

mictium A mixture of species that is not a standard community, such as that which occurs when there is a transition between two habitats. See ECOTONE.

mid-Atlantic ridge The mid-oceanic ridge located centrally within the Atlantic Ocean.

middle latitudes Latitudes between 20° and 50°, falling between equatorial and high latitudes. (The United States is located in the middle latitudes.) The term is used in describing large-scale climate patterns and activity in the earth's magnetic field. See LATITUDE.

middle shore A general term for the area of a beach that extends from the low-tide water level to the upper reach of storm wave action.

mid-oceanic ridge A topographic feature of a tectonic spreading center between diverging oceanic plates. New crustal material is formed by upwelling magma as the plates diverge; therefore, the relative age of the sea floor is progressively older with increasing distance from the mid-oceanic ridge. A median valley is often formed at the center of the ridge.

migmatite A coarse-grained metamorphic rock with a combination of foliated gneissic and non-foliated granitic textures. Migmatite forms by the partial melting of rocks during high-grade metamorphism.

migmatization The process of partial melting that produces a migmatite rock.

migration 1) The seasonal movement of animal populations from one region to another and back again, often covering tremendous distances. 2) The daily vertical movements of some freshwater plankton several meters downward during the day and back to the surface at night. 3) Movement of individuals into or out of local populations (includes emigration and immigration).

migrational homing The navigational ability allowing animals to return to the exact areas year after year during their seasonal migrations.

migration rate A measurement of how immigration and emigration affect the size of a human population. It is the difference between how many people enter a particular country and how many people leave per thousand in the population per year. See ISLAND BIOGEOGRAPHY.

migratory Characterized by migrating; moving from one place to another, such as migratory species of birds.

migrule Another term for disseminule and propagule. See DISSEMINULE.

Milankovitch cycles A theory that climate is influenced by variations in the earth's orbit, proposed by the Yugoslavian astronomer Milutin Milankovitch. He proposed that the amount of solar radiation that reaches the earth is affected by variations in the shape of the earth's orbit, the obliquity of tilt of the earth's axis of rotation in relation to its plane of orbit, and the precession of the equinoxes. Analysis of deep-sea sediments supports the claim that these variations influenced the onset and retreat of the ice ages. See PRECESSION OF THE EQUINOXES.

mildew Fungus, especially a plant disease such as powdery mildew, in which the mycelium is visible on the surface, often looking like a thin powdery dusting. See OOMYCOTA.

milk sugar Another name for lactose. See LACTOSE.

milk teeth The first set of teeth, those that are eventually replaced by permanent teeth, in many mammals. Also called primary or deciduous teeth. See DIPHYO-DONT.

milli- Prefix in the Système International d'Unités (SI) indicating one thousandth, 1×10^{-3}. One millimeter equals one thousandth of a meter, or 10^{-3}m.

millimicron A unit of measurement (one thousandth of a micron) that has now been replaced by the nanometer (10^{-9}m).

milling A preliminary treatment of ore to separate the economic mineral from the gangue.

millions of tons of coal equivalent (MTCE) A measurement for energy available from a source other than coal (for example, solar energy). MTCE is used to indicate how much coal that source could replace.

millipedes See DIPLOPODA.

millisievert Unit of radiation dose, equal to 0.001 sievert. See SIEVERT.

mimic 1) To copy or closely resemble something. 2) An animal that imitates another organism, such as birds that repeat the calls of other birds.

mimicry The external resemblance of an animal or plant to its surroundings or to some other plant or animal. One species may copy habits, coloration, calls, or shape of another species to hide from predators or otherwise derive protection. See BATESIAN MIMICRY, MÜLLERIAN MIMICRY.

Mimosaceae See FABACEAE.

Minamata disease Mercury poisoning from eating contaminated fish. The name comes from a village in Japan where people exhibited an epidemic of unusual symptoms believed at first to be a new disease. Symptoms of Minamata "disease" include mental disturbances, tremors and spastic movements, coma and eventual death, plus babies born with severe birth defects.

mineral A naturally occurring inorganic substance with relatively consistent chemical formula and physical properties, including a characteristic atomic structure that is usually crystalline. The term is also used more loosely for ores, rocks, and similar substances extracted by mining or quarrying. See APPENDIX p. 620–628.

mineral assemblage A group of minerals that are commonly found together in an igneous or a metamorphic rock.

mineral deposit See DEPOSIT.

mineralization The breakdown of organic matter into its inorganic chemical components (minerals), such as nitrogen, phosphorous, potassium, and sulfur. Mineralization is a stage in the biogeochemical cycles of most minerals during which they are released back into the environment.

mineralogy The science of physical, chemical, economic, and crystallographic properties of minerals.

mineral soil A soil that is composed primarily of mineral material. The characteristics of a mineral soil are determined more by weathering processes than by biologic processes.

mine spoil Any heap of solid waste material from the excavation of a mining operation.

minimal area The smallest area in a community that can be used for sampling vegetation in order to find individuals of all species present. See SPECIES-AREA CURVE.

minimum factor Another term for limiting factor.

minimum-tillage farming Cultivating crops with less disturbance to the soil than traditional methods involving extensive plowing and harrowing. Crop residues and litter are left on top of the soil (or mixed just into the surface, as with a chisel plow) rather than being turned under, providing some protection against erosion for the soil underneath. Minimum-tillage farming reduces soil erosion, and because tractors drive over the fields fewer times, it reduces labor and energy use. The most extreme form of minimum-tillage farming (also called conservation-tillage) is no-till farming. See NO-TILL AGRICULTURE.

minimum viable population The smallest-size interbreeding group of a particular species that can sustain itself over time; if the group becomes any smaller, it will fail to replace itself successfully and slowly die out.

mining water 1) The aqueous brine or slurry containing the dissolved or solid mineral materials that are extracted during a fluid mining process. 2) Overdraft of groundwater resources.

Miocene The fourth of five epochs in the Tertiary sub-era of geologic time. The Miocene lasted from approximately 24 to 5.1 million years ago. See APPENDIX p. 610.

miogeocline An association of marine sediments formed within a nearshore portion of a tectonic basin floored by continental crust. Miogeocline sediments are typically composed of carbonates, sandstones, and shales. Miogeocline sediments are lacking in the volcanic material that is typical in eugeocline sediments. Compare EUGEOCLINE.

mire Another (chiefly British) term for bog, although sometimes used loosely to also include marshes. See BOG.

Mirror Fusion Test Facility (MFTF) A facility experimenting with use of magnetic "mirrors" (strong, very specially shaped magnetic fields) to contain the plasma within a straight vessel, as opposed to using a shaped tokamak. See FUSION, TOKAMAK.

Mississipian The fifth geologic time period of the Paleozoic Era. The Mississippian Period lasted from approximately 360 to 320 million years ago. See CARBONIFEROUS, also APPENDIX p. 611.

mist A cloud, usually just above the ground, made up of fine, suspended droplets of a liquid (water vapor). A mist is distinguished from a fog by having smaller droplets of water vapor and by not impairing visibility as much as a fog does.

mistral A cold, katabatic wind blowing northwest over France from the high elevations of the Alps toward the western Mediterranean Sea. See KATABATIC WIND.

mites Mites and ticks form a loose and extremely diverse subclass, Acarina, of the class Arachnida. So varied are they that it is not easy to define common characteristics. Chelicerae are present and there is a unique structure, the subcapitulum, formed from pedipalpal and cheliceral parts. Cephalothorax and abdomen are usually fused and, except for appendages, body segmentation is not apparent. Mites are small: Many are microscopic and few exceed a length of 1 mm. Their distribution is worldwide. Many are parasitic (scabies, chiggers, etc.) and others are vectors of human and animal diseases (spotted fever, relapsing fever, Lyme disease, etc.). There are at least 30,000 species, but this is probably a highly conservative estimate. See ARACHNIDA, CHELICERATA.

mite

mitochondria   Specialized, capsule-shaped or threadlike organelles in the cytoplasm of cells that contain genetic material and many enzymes. Mitochondria are the site of the Krebs cycle, which produces ATP to provide energy for cell metabolism. They are also involved in the synthesis of proteins and in lipid metabolism. See KREBS CYCLE.

mitosis   The form of cell division that produces two nuclei containing the same number of chromosomes as the original nucleus. Mitosis is the means by which all nonreproductive cells in multicellular organisms multiply. The four stages of mitosis are prophase, metaphase, anaphase, and telophase. See PROPHASE, METAPHASE, ANAPHASE, TELOPHASE, MEIOSIS.

mixed cropping   See POLYCULTURE.

mixed forest   Woodland containing conifers as well as deciduous trees, such as is sometimes found in a northern hardwood forest.

mixed-hardwood forest   Woodland in which a mixture of hardwood deciduous species dominates, especially the mixed-mesophytic plant community of the Appalachian Mountains, which may have as many as 20 co-dominant species, including maples, basswood, beech, ironwood, musclewood, and dogwood.

mixed liquor (waste water)   An aqueous mixture of activated sludge, primary effluent, and return sludge within an aeration tank of a sewage treatment process.

mixed perennial polyculture   A form of sustainable agriculture in which a mixture of different perennial plants are grown together, consciously designed to imitate the diversity of a natural system and to reduce energy use. Permaculture, a name given to mixed perennial polyculture by some of its adherents, involves planting perennial shrubs mixed in with tree crops (for example, pomegranate and fig underneath palm trees), or a mixture of herbaceous perennials such as grains and legumes. Compare MONOCULTURE.

mixing depth   1) A limnologic or oceanographic measure of the depth at which water circulation occurs between stratified layers. 2) The height above the earth's surface in which convection is effective in circulating localized air. A large mixing depth causes dilution of pollutants and improves local air quality.

mixotrophic   Using two or more basic forms of nutrition to survive, such as combining photosynthesis and saprotrophy, or partial parasitism.

ml   Abbreviation for milliliter, the metric unit of volume equal to one-thousandth (10^{-3}) of a liter.

mm   Abbreviation for millimeter, the metric

unit of length equal to one-thousandth (10^{-3}) of a meter.

Mn   Chemical symbol for the element manganese. See MANGANESE.

Mo   Chemical symbol for the element molybdenum. See MOLYBDENUM.

mobile belt   See OROGENIC BELT.

mobile source   Moving source of pollution, such as an automobile.

mode   1) Statistical term for the value of a variable that occurs most frequently in a set of values (frequency distribution). 2) The pattern of vibration of electromagnetic waves, or one of several wave frequencies generated by a source. 3) A way of doing something, as in a mode of operation. Compare MEAN, MEDIAN.

model   A simplified representation of a system or a structure, usually on a smaller scale than that of the original. A theoretical model is a mental construct that may be formalized into mathematical equations or verbal descriptions. If accurate, it may be used to make predictions about the original system. Models can also be physical; a flowchart is a two-dimensional model of a system, and three-dimensional models or prototypes are often made of airplanes and other vehicles in the process of development. See APPENDIX p. 603.

moderately stratified estuary   A term applied to estuaries that develop a stratification pattern that is subject to partial mixing by waves and by wind.

modified Mercalli scale   A version of the Mercalli scale of earthquake intensity that has been adapted to North American conditions of structural design.

modular organism   An organism that develops by repeatedly producing similarly shaped and usually branching parts, as in the polyps of a coral or plants that form clones.

modulation   1) The process of adjusting, altering, or regulating. 2) Causing a change in either the frequency or the amplitude of a telecommunication or an audio signal of a given frequency. The AM in AM radio stands for amplitude modulated signal; FM stands for frequency modulated signal.

Moho   See MOHOROVICIC DISCONTINUITY.

Mohorovičić discontinuity   (Moho) The boundary between the earth's crust and mantle. The depth of this boundary is 10 to 12 km under ocean basins and 40 to 50 km under continents. Earthquake waves (P waves and S waves) increase sharply at the Moho due to the composition differences between the crust and mantle. Named after the Yugoslavian geophysist who discovered it in 1909.

Mohs hardness scale   A numerical array of minerals arranged by a relative and nonlinear measure of hardness, ranging from 1 for the softest mineral, talc, to 10 for the hardest mineral, diamond.

moisture gradient   The change in water content of soil (and thus availability of water to plants) with vertical or horizontal distance in the soil.

molality   The concentration of a solution expressed as the number of moles of a substance dissolved in 1 kg of solvent. See MOLE, MOLARITY.

molarity   The concentration of a solution expressed as the number of moles of a substance dissolved per liter of a solution. See MOLE, MOLALITY.

molasse   An older term describing sediments deposited in a shallow marine basin after an orogenic uplift of the land. These sediments are more likely to be classified as syntectonic than post-tectonic in practical application, thus the term is no longer widely used.

mold   1) A fungal growth on a surface, often velvety or powdery and usually growing on damp and decaying organic matter or living organisms. 2) Soil very rich in humus, as in leaf mold. See DEUTER-OMYCOTA, OOMYCOTA, ZYGOMYCOTA.

mole   (mol) The unit for the amount of substance in the Système International d'Unités (SI). A mole of any substance contains the same number of elemental units (atoms, molecules, ions, etc.) as there are atoms in 12 g of ^{12}C (which is 6.02 x 10^{23}, known as Avogadro's number). For any chemical compound, one mole corresponds to the mass equal to its molecular mass in grams. See AVOGADRO'S NUMBER.

molecular weight   The sum of the atomic masses of the different atoms present in a molecule. Proteins and nucleic acids have vastly greater molecular weights than simple, small molecules such as water (H_2O) or methane (CH_4). See ATOMIC MASS UNIT.

molecule   An atom or a group of atoms that constitutes the basic unit of a substance. Molecules are capable of existing independently as well as of forming compounds with other molecules (the basic unit of the new combined substance is also called a molecule). Some examples are H_2, CH_4, hemoglobin, and DNA.

mole fraction   A ratio expressing the relationship between the number of moles of one substance in a mixture to the total number of moles of all of the different substances contained in the mixture. The mole fraction can be used to express the relative amounts of gases in a sample of air in terms of the actual number of molecules present, rather than the relative weights of the different gases. See MOLE.

mollic   A term applied to a relatively deep soil horizon that is dark due to the presence of finely distributed organic matter. A mollic soil horizon is the diagnostic feature of a mollisol.

mollisol   Any member of a soil order characterized by well-developed, dark-surface horizons with high concentrations of lime and moderate-to-high amounts of organic material, typically forming in grasslands.

Mollusca   With approximately 100,000 species, this phylum is one of the largest in the animal kingdom and is structurally and ecologically one of the most diverse. Members of the phylum have a mantle (a dorsal, body fold that secretes calcium carbonate) and a radula (a rasping, tongue-like organ). Because of its many modifications, molluscan struture can only be described in very general terms. The soft, unsegmented body is usually humped or dorsoventrally flattened and has a fleshy, muscular structure, the foot, on the ventral surface. A distinctive head with sense organs and tentacles may be present. The gut is complete and gills are usually present in the cavity under the mantle. The dorsal surface has a proteinaceous covering under which the mantle may secrete one or more calcareous shells. Species are distributed on land, and in fresh waters and seas. A few are parasitic. There are a number of relatively obscure classes and four major ones: Polyplacophora (chitons), Gastropoda (snails and whelks), Bivalvia (clams, mussels, etc.) and Cephelapoda (nautiloids, squids, and octupuses). See APPENDIX p. 615.

molluscicide A substance used to kill molluscs. Molluscicides are used in tropical countries to control the snails that serve as intermediate hosts for human blood flukes (causing bilharziasis or schistosomiasis).

molt 1) To shed the outer covering before new growth on an animal begins. Reptiles and arthropods shed their entire skin when they molt; birds merely lose one set of feathers. 2) The process of shedding, or an individual occurrence of shedding. Also called ecdysis.

molybdenite MoS_2, the most common ore of molybdenum; molybdenum disulfide. Small quantities occur in granites and associated rocks.

molybdenum (Mo) A heavy metallic element with atomic weight 95.94 and atomic number 42. It is unreactive and, unlike most metals, is not dissolved by most acids. It resists heat but conducts electricity well. Molybdenum is added to steel to strengthen it. In small quantities, it is an essential nutrient for plants and is required for nitrogen fixation.

momentum The amount of motion of a body. Momentum is calculated as the product of an object's mass and its velocity. Momentum can be linear or angular, depending on whether the object is traveling a linear or a circular path. See ANGULAR MOMENTUM.

monadnock A geomorphology term applied to an isolated mountain that stands above a peneplain. Mount Monadnock in Jaffrey, NH, is the type locality for the term.

monazite A mineral of thorium, rare earths, and phosphate that form monoclinic crystals as an accessory mineral in granites and pegmatites. Monazite is a source of rare earth elements such as cerium (Ce), lanthanum (La), thorium (Th), and yttrium (Y).

Monera The kingdom, Monera, includes bacteria and other similar prokaryotes, but there is much dispute over the precise classification of many of its subgroups. Bacteria, although morphologically simple and microscopic, are metabolically the most diverse of all life forms and their activities as nitrogen fixers, decomposers, and pathogens, etc. make them essential players in the ecology of the biosphere. Phyla include fermenting bacteria, spirochaetes, archaebacteria, cyanobacteria, aerobic endospore-forming bacteria, omnibacteria, and nitrogen-fixing bacteria, etc. See PROKARYOTE, also APPENDIX p. 616.

monocarpic Describing plants that bear fruit only once and then die, as in annual and biennial plants. Century plants (*Agave*) live for many years as a large rosette, then they produce a large, spectacular flower stalk, and after fruiting they die.

monoclimax theory The idea that all successional sequences lead to one characteristic climax vegetation in a given region, produced by a particular combination of soil and climate. It was popularized by F. E. Clements early in the 20th century but is not widely embraced at present. The monoclimax theory categorizes vegetation into formation (equivalent to biome), association, consociation, faciation, and society. Contrast POLYCLIMAX THEORY.

monocline A flexure in stratified rock. A monocline has only one limb of displacement, which is characterized by a uniform low-angle dip.

monocots Angiosperm plants that have a single cotyledon. Many members of this sudivision often have flowers with parts in multiples of three and leaves with parallel veins. This group includes grasses, sedges, orchids, lilies, irises, palms, and bromeliads, among others. Compare DICOTYLEDON. See MONOCOTYLEDONAE.

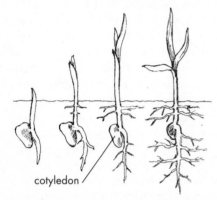

cotyledon

The single cotyledon of corn (a monocot plant) remains below ground as the seed germinates and the seedling grows.

Monocotyledonae   A class of plants characterized by having a single cotyledon.

monocropping   Another term for monoculture. See MONOCULTURE.

monoculture   A method of farming in which only one type of crop is grown on a large area over a number of years, or a plantation devoted to one species of trees. In areas of the U.S. Midwest, for example, thousands of acres are devoted exclusively to corn *(Zea mays)*. Monoculture results in a reduction in the diversity of associated animal species, including beneficial insect predators; it increases pest and disease infestation. Compare MIXED PERENNIAL POLYCULTURE, POLYCULTURE.

monocytes   The largest white blood cells (leukocytes) found in vertebrates. Produced in the bone marrow, they are carried by the blood to the tissues, where they become macrophages and ingest cell debris and bacteria. See PHAGOCYTE.

monoecious   1) Describing plants with distinct male and female flowers appearing on the same plant, not having functional stamens and functional pistils within the same flower. 2) Describing mosses and algae that produce both male and female gametes in one organism. Compare HERMAPHRODITE, DIOECIOUS.

monogamous   Describing species in which individuals mate with only one other individual during their life; of or relating to the practice of mating with a single individual.

monogamy   Practice of mating with only one other individual for at least one reproductive cycle. Monogamy usually is associated with parental care being provided to the offspring.

monogenetic   1) Autoecious; describing parasites that have only one host during their life cycle. 2) Reproducing asexually.

monohybrid inheritance   The pattern of inheritance that typically results from crossing parents that have identical genes except for one pair, which in one parent is two dominant genes for the trait and in the other parent is two recessive genes. The first generation produced by such a cross produces only the dominant phenotype, although each individual carries one dominant and one recessive gene for the trait. The subsequent generation produces the characteristic monohybrid ratio. See MONOHYBRID RATIO.

monohybrid ratio   The three-to-one ratio of dominant-to-recessive phenotypes for one trait found in the second generation of offspring of monohybrid individuals (or in the first generation of individuals that each contain one dominant and one recessive gene for a particular trait). Monohybrids are organisms that have identical chromosomes except for one pair, which in one parent is two dominant genes for the trait and in the other parent is two recessives. Contrast HARDY-WEINBERG RATIO.

monokaryon   Hyphae or mycleia of fungi containing one haploid nucleus in each cell or segment (all the nuclei are identical). Also spelled monocaryon. Contrast DIKARYON.

monomer   A molecule that can combine with other similar or dissimilar molecules to form polymers. See POLYMER.

monomolecular layer   A film of a substance that is only one molecule thick.

monomorphic   Describing organisms with only one form throughout their life, or only one genotype.

monomyarian   Describing bivalve species of molluscs that have only one adductor muscle.

monophagous   Eating only one type of food. Also called monotrophic. Compare POLYPHAGOUS.

monophyletic   Describing organisms that are presumed to have descended from a single, common ancestral form, usually resembling the existing group. Contrast POLYPHYLETIC.

monophyodont   Describing organisms having only one set of teeth during their lifetime. Compare DIPHYODONT, POLYPHYODONT.

monosaccharides   Simple sugars, the basic units that make up carbohydrates. Unlike more complex sugars (polysaccharides), monosaccharides cannot be hydrolyzed (broken down) into simpler sugars. Examples include glucose, fructose, and galactose. See DISACCHARIDES, POLYSACCHARIDE.

monosodium glutamate   (MSG) HOOC-CH(NH$_2$)CH$_2$CH$_2$COONa, a compound derived from the amino acid, glutamic acid. MSG is a food additive used to enhance flavor and is used in almost all commercial chicken soup and some Chinese foods. Some people develop reactions after consuming it. Like table salt, it is a source of sodium in the diet if added to food.

monosome   A single, unpaired, sex chromosome, especially the one that is not the sex-determining chromosome (i.e., the X chromosome in humans).

monospecific   Having only one species, as in a genus that contains but a single species. The genus *Ginkgo* and the order Ginkgophyta both contain only one species, *Ginkgo biloba*.

monothetic   A method of defining taxonomic groups based on one single key characteristic. Compare POLYTHETIC.

monotocous   Bearing one offspring at each birth, as is usually the case in humans. Contrast DITOCOUS, POLYTOCOUS.

monoton plankton   Describing a group of plankton in which one species strongly predominates.

monotypic   Having only one representative in a taxonomic group; especially a genus containing a single species (monospecific genus).

monovular twins   Another term for monozygotic twins. See MONOZYGOTIC TWINS.

monoxenous   Describing parasites that have only one species as their host. Compare POLYXENOUS.

monoxide An oxide whose molecules contain a single oxygen atom, as in carbon monoxide.

monozygotic twins Identical twins, those produced by the splitting in two of a single fertilized egg. Also called monovular or uniovular twins. Compare DIZYGOTIC TWINS.

monsoon A seasonal reversal of prevailing wind direction associated with large continents. It is usually used to refer to southern Asia and to the Indian Ocean. The wind blowing from sea to land (bringing heavy rains) from April to October reverses to blow from land to sea in winter.

monsoon forest Rainforest found in tropical regions that experience a monsoon season of rains. Monsoon forests are distinguished from other forms of tropical rainforests by their distinct wet and dry seasons. See MONSOON.

montane A cool subalpine upland region located just below timberline. Montane regions are characterized by coniferous vegetation.

montane coniferous forest A type of woodland vegetation in which conifer species predominate, found on mountains, as in the Pacific Northwest and in parts of the Appalachians. It contains zones of different forest communities at different altitudes. Compare BOREAL FOREST.

montmorillonite A hydrous aluminum silicate clay mineral. Montmorillonite is produced by the decomposition of volcanic ash in a marine environment or by the weathering of intermediate and mafic rocks. The mineral absorbs water into the crystal structure and is a common component of soils that swell when wet.

moor A site with vegetation made up of ericaceous plants (heaths, heathers, and similar plants) and sedges with wet, acid subsoil and lateral water movement. Moors are found throughout Great Britain and much of northern Europe. Sometimes called moorland. Compare HEATH.

mor A surface soil horizon of humus formed above acid soils. Mor is typically lacking in microbial decomposition, and organic material is not mixed with the underlying mineral horizons. Mor is also known as ectohumus. Compare MULL.

Moraceae The mulberry family of the Angiospermophyta. A dicot family with clusters of tiny unisexual flowers borne in catkins or inside elaborate structures as in figs (*Ficus*). The fruits of members of this family vary from achenes to nuts to drupes to aggregates. Breadfruit (*Atrocarpus*), Osage orange (*Maclura*), and mulberry (*Morus*) are all members of this family. See ANGIOSPERMOPHYTA.

moraine A mass of till either carried by an active glacier or deposited on the land after a glacier recedes.

morbidity rate The number of cases of a disease (morbidity) per 100,000 people in a region. Compare MORTALITY RATE.

Morisita's index of dispersion (I) An index of dispersion that is independent of the sample mean. See DISPERSION.

Morisita's similarity index (C_γ) An index of similarity (coefficient of community) that is weighted by the population sizes of species present in the two communities. Compare JACCARD INDEX, SØRENSEN SIMILARITY INDEX, SIMPSON'S INDEX OF FLORISTIC RESEMBLANCE, GLEASON'S INDEX, KULEZINSKI INDEX.

morph 1) A particular form of a species,

especially a variant of the typical form of an organism. Ecotypes are different morphs of one species, as are the variations found on an ecocline or a geocline. 2) A distinct flower color form of a species.

morphactins   Synthetic plant growth substances derived from fluorene rings containing carboxylic acid. Most disrupt and inhibit plant growth, apparently by interfering with natural growth substances; they also inhibit germination in seeds. Their inhibition of the elongation of stem internodes can be used to cause dwarfing or bushiness in young, developing plants. Morphactins are selective, do not persist in the soil, and are not particularly toxic to animal life.

morphallaxis   The growth of one part into another during regeneration. In some crustaceans, an antenna will grow as the replacement for an eye.

morphogenesis   The origin and evolution of the overall form and structure of an organism; development through growth and differentiation of cells and tissues into the characteristic mature form of an organism.

morph ratio cline   A gradual change in the frequency of morphs over the geographical range of a population, often associated with gradual changes in ecological conditions. See MORPH, GEOCLINE, ECOCLINE.

mortality factors   Environmental variables that affect (increase or decrease) the death rate in a population, such as predation.

mortality rate   The number of deaths per 100,000 people in a population over a year. Compare MORBIDITY RATE, NATALITY RATE.

morula   A solid, sphere-shaped mass of cells produced by repeated division (segmenta-tion) of the fertilized egg. The morula is the earliest stage in the development of an animal embryo. See BLASTULA.

mosaic   A pattern or picture made up of many different interspersed elements.

mosquitoes   See CULICIDAE.

mother cell   Any cell that undergoes division, producing two daughter cells.

mother-of-pearl clouds   Common name for nacreous clouds. See NACREOUS CLOUD.

motor end plate   The flat region at the end of a motor nerve fiber. The motor end plate connects a nerve to a muscle fiber. See NEURON.

mountain chain   A group (chain) of mountains sharing physiographic trends with similar geologic features.

mountain glacier   An alpine glacier that has a shape and a direction of flow controlled by the topography of the flanks or the summit area of a mountain. Compare VALLEY GLACIER.

mountain range   A group of mountain peaks and ridges that are closely related in geographic position.

mountain waves   Lee waves formed adjacent to mountains. See LEE WAVES.

mp   See MELTING POINT.

MRI   See MAGNETIC RESONANCE IMAGING.

MSG   See MONOSODIUM GLUTAMATE.

MTCE   Abbreviation for millions of tons of coal equivalent. See MILLIONS OF TONS OF COAL EQUIVALENT.

m 3000   Radio waves at that frequency within the high-frequency (3 to 30 MHz) range that make a single bounce off of the earth's ionosphere. This "single-hop transmission" is very useful in radio communication because the waves can travel over a range of 3000 km. See HIGH FREQUENCY.

muck soils   See HISTOSOL.

mucoid feeding   A technique used by some molluscs to trap food by pushing mucus out of the mouth, then retrieving and reingesting it after food particles have adhered to it.

mud   An unconsolidated mixture of clay and silt particles with water.

mud crack   A vertical fissure in the surface of a desiccated mud. Mud cracks often form patterns of polygons as the mud shrinks while drying out.

mudflats   A broad, level area of a coastline or shoreline where silt and clay sediment are deposited. Mudflats often form within sheltered areas of an intertidal zone.

mudflow   A rapid type of mass wasting, usually confined to a channel, characterized by heterogenous debris with a mud matrix. Can transport very large boulders. Can be a very hazardous event to people and property.

mudstone   A nonfoliated, clay-rich sedimentary rock formed of lithified mud. Contrast SHALE.

mulch   A natural or synthetic covering that is placed around the base of plants to conserve moisture in the soil, to reduce weeds, or to moderate temperature changes. Natural mulches such as leaves, straw, or bark gradually decay and enrich the soil. Synthetic mulches such as special plastics are increasingly used in agriculture.

mull   A surface soil horizon that is well aerated and favorable to organic decomposition. Mull horizons are composed of organic material that is well mixed with mineral material. Compare MOR.

Müllerian mimicry   Resemblance among two or more species, when all are inedible or unpalatable, that benefits each of the species. Müllerian mimicry helps predators learn what color pattern to avoid. See MIMICRY. Compare BATESIAN MIMICRY.

multiple factors   Theoretical interactions between two or more pairs of genes that result in a particular trait or variation in a trait, such as skin or fur coloration or size. See POLYGENES.

multiple gene inheritance   The control of a quantitatively varying trait by the combined effects of more than one interacting allele. Multiple gene inheritance controls traits showing gradual variation, such as height. Also called multifactorial inheritance. See POLYGENES, QUANTITATIVE GENETICS.

multiple resistance to pesticides   The development in an organism of a tolerance to more than one (usually synthetic) pesticide, requiring a change in pest-control technique.

multiple use   Principle of managing public land such as a national forest so that it is used simultaneously for a variety of purposes such as timbering, mining, recreation, grazing, wildlife preservation, and soil and water conservation.

multiple working hypotheses   A situation in which more than one hypothesis explains a system or a phenomenon. Also refers to more than one hypothesis being used in an experiment. See HYPOTHESIS.

multituberculate   Describing teeth having many small points. Compare HYPSODONT.

multivariate   1) Incorporating or involving

two or more variable quantities. 2) Describing a statistical analysis of several measurements made on more than one attribute of each entity under observation.

multivoltine   Describing animals that produce more than one brood in one year. Compare BIVOLTINE, UNIVOLTINE.

Musci   The class of Bryophyta that contains the true mosses. These primitive plants have leafy gametophyte generations with multicellular rhizoids, but lack true roots. See BRYOPHYTA.

muscovite (white mica)   A non-ferromagnesian sheet silicate mineral of the mica group. Light in color and with pearly luster, it has perfect cleavage in one direction producing thin, elastic, transparent sheets. See APPENDIX p. 625. Contrast BIOTITE.

mushroom rocks   A pedestal-shaped rock feature. The mushroomlike top of the pedestal is undercut by chemical and wind erosion near the base.

muskeg   A boreal landscape characterized by peat bogs and tussock meadows.

mustard gas $(CH_2Cl\text{-}CH_2)_2S$, a very irritating, poisonous gas also known as dichlorodiethyl sulfide. It is used as a military weapon. Mustard gas is a vesicant, or substance that causes blisters; tannic acid is an antidote.

mutagen   Any substance capable of producing genetic changes (mutations), or increasing the rate of mutation above the naturally occurring (spontaneous) level. Ionizing radiation, ultraviolet light, and many chemicals are mutagens. Also called mutagenic agent.

mutagenic   Causing (or capable of causing) a mutation. See MUTATION.

mutant   1) A genetic variant of a plant or animal produced by mutation, and therefore passing on the genes for the variant trait or traits to subsequent generations. 2) A gene in which a mutation has occurred. See MUTATION.

mutation   An alteration within a gene or chromosome in animals or in plants that results in an inheritable variation in one or more characteristics of the organism. Spontaneous mutation occurs throughout nature, but mutations can also be induced by some toxic chemicals or by exposure to radiation (mutagens).

mutation rate   The frequency at which an allele is changed to a different form during a particular time period, such as a generation. Mutation rate is usually expressed as the number of mutants per gamete per generation.

mutual antagonism   A relationship between two or more organisms in which each has a negative effect on the other. Compare SYNERGY, SYMBIOSIS.

mutualism   A mutually beneficial relationship between two organisms, especially one in which neither organism can survive without the other. See OBLIGATE MUTUALISM, PROTOCOOPERATION, SYMBIOSIS.

mya   An abbreviation for a million years ago.

mycelium   The vegetative structure of a fungus, comprised of a network of fine filaments called hyphae. Mycelia can either appear as a branching series of fine white hairs, or be densely massed into the body (thallus) of a mushroom. See HYPHA.

mycetocytes   Specialized cells found in a mass of tissue (the pseudovitellus) in the abdomen of some insects, containing symbiotic microbes. Protozoans in the guts of termites and of cockroaches enable them to digest cellulose from wood or paper; the protozoans are not capable of existing outside of their insect hosts, and the insects cannot digest cellulose without them.

mycetophagous   Feeding upon fungi.

Mycetozoans   See ACRASIOMYCOTA.

mycobacterium   Any member of a genus (*Mycobacterium*) of nonmotile, aerobic, rod-shaped bacteria. The group includes many saprophytes as well as the organisms responsible for leprosy and tuberculosis.

mycology   1) The branch of biology that studies fungi. 2) The fungal life of a certain region or country.

Mycophycophyta   The phylum of the kingdom Fungi that includes the lichens. Lichens are symbiotic organisms formed through the association of a fungus (usually an ascomycete) with a green alga or a cyanobacterium. Lichens are found in a diversity of environments ranging from deserts to boreal forests and from polar plains to tropics. They often grow in harsh conditions that are too exposed for other plants. Because of their sensitivity to air pollution, many are useful as indicator species. Lichens are generally classified on the basis of the growth habit of their thallus (vegetative fungal body) as crustose (a tight, low crust), foliose (a leafy structure), or fruticose (a bushy mass). Reproduction of lichens is through spores clustered together in fruiting bodies. Common members of this phylum include the fruticose lichens reindeer-moss and British soldiers *Cladonia*, the crustose lichens *Parmelia*, and the foliose lichens. See FUNGI.

mycoplasmas   A varied group of primitive bacteria usually lacking cell walls; they are the smallest independent organisms known. Formerly called pleuropneumonia-like organisms (PPLO), they cause viral pneumonia (now technically primary atypical pneumonia) in humans.

mycorrhizae   The symbiotic relationship between the mycelia of some species of fungi and the roots or other structures of some flowering plants. The fungal mycelia help the plant absorb minerals and in return absorb energy compounds produced by the plant. Many tree species such as beech cannot grow without their associated mycorrhizae; others such as pines can grow on otherwise barren soil with the assistance of mycorrhizae. See ECTOTROPHIC MYCORRHIZAE, ENDOTROPHIC MYCORRHIZAE, VESICULAR-ARBUSCULAR MYCORRHIZAE.

mycorrhizal fungus   Any species of fungus that participates in a symbiotic relationship with a flowering plant (which is thereby known as a mycotrophic plant), forming mycorrhizae. See MYCORRHIZAE, MYCOTROPHY.

mycotrophy   Deriving nutrition through mycorrhizae; the symbiotic relationship between a mycorrhizal fungus and its associated mycotrophic plant.

myiasis   Infestation of animal tissues by maggots (fly larvae).

mylonite   Any type of rock produced by cataclasis in tectonic fault zones. Mylonite has a characteristic texture of crushed and sheared mineral grains. See CATACLASTIC ROCK.

myofibril   The individual contracting element (filament) connecting cells in a muscle fiber. The myofibril contains actin, myosin, and other associated proteins. Also called sarcostyle. See SARCOLEMMA.

myoglobin   A protein related to hemoglobin that is present in large quantities in mammalian muscle cells. Myoglobin absorbs and stores oxygen; it enables marine mammals to swim underwater for relatively long periods without replenishing their oxygen supply. Also called myohemoglobin.

myosin   One of two proteins in muscle cells that aids in the elasticity and the contraction of muscles. Myosin makes up over half of the protein in muscles; it is also found in some other cells. See ACTIN, ACTOMYOSIN.

myrmecochore   A seed or a spore that depends on ants for its dispersal. See ELAIOSOME, ZOOCHORE.

myrmecophily   A beneficial relationship between an organism and ants, including pollination by ants and more involved symbiosis. See MYRMECOPHYTE.

myrmecophyte   A plant that has a symbiotic relationship with ants, such as some species of acacia. The acacias provide a nesting location, extrafloral nectaries, or starchy food (in food bodies on the leaves) for the ants; the ants in turn attack herbivores that might feed on the acacias.

Myrtaceae   The myrtle family, an important angiosperm family with dicot trees and shrubs, mostly of tropical distribution. It has flowers with four to five sepals and four to five petals, numerous stamens, and three to four carpels forming a single-chambered ovary. Stoppers *(Eugenia)*, eucalyptus *(Eucalyptus)*, guava *(Psidium)*, and bottlebrush *(Callistemon)* are members. See ANGIOSPERMOPHYTA.

myxobacteria   A group of saprophytic bacteria characterized by motility, lack of rigid cell walls, and formation of a slime in which their colonies grow. Myxobacteria, also called gliding bacteria, grow in soil and animal feces.

Myxogastria   See MYXOMYCOTA.

myxomatosis   Viral disease of rabbits that is highly contagious and characterized by growth of soft, tumorlike tissue underneath the skin. Though once generally fatal, genetic feedback between myxoma virus and rabbit populations has resulted in greater rabbit resistance and less virulence in the virus. As the virus selects for resistant rabbits and rabbits select for less virulent strains of the virus, there is accommodation by both species over time. See GENETIC FEEDBACK.

Myxomycota   The true or plasmodial slime mold phylum of the kingdom Protista. Like the cellular slime molds, these organisms have amoeba like cells that aggregate periodically into colonial masses. However, members of the Myxomycota have an alternation of generations and form plasmodia, multinucleate masses of protoplasm that develop into the sporophore fruiting body. Familiar members of this phylum, such as *Stemonitis* and *Echinostelium*, inhabit dead wood and humus. See PROTISTA.

myxotrophic   A term applied to organisms that combine two or more modes of nutrition; for example, some protozoan Mastigophora are partial parasites.

Myxozoa   See CNIDOSPORIDIA.

n

n- A prefix used in organic chemistry to indicate a straight-chain hydrocarbon group in a compound.

N Chemical symbol for the element nitrogen. See NITROGEN.

Na Chemical symbol for the element sodium. See SODIUM.

nacreous cloud Iridescent ice-crystal clouds resembling cirrus. They occur in northern latitudes at altitudes of 25 to 30 km (15 to 20 mi). Also called mother-of-pearl clouds.

naked dinoflagellates Species of Dinoflagellata that lack protective plates on the cell surface. See DINOFLAGELLATA.

nanism Dwarfism; unusually small size in animals, plants, or humans.

nano- Prefix in the Système International d'Unités (SI) indicating one billionth, 1×10^{-9}. One nanosecond equals one billionth of a second, or 10^{-9} sec.

nanometer (nm) A metric unit of length equal to one billionth of a meter. Nanometer has replaced the term millimicron, formerly used for the same quantity.

nanoplankton Very small microscopic plankton, ranging from 0.2 to 20 micrometers.

nanotesla (nT) Unit equal to one billionth (10^{-9}) of a tesla, the unit for magnetic field in the Système International d'Unités (SI). One nanotesla also equals 10^{-5} gauss. See TESLA.

nappe A massive thrust-faulted and folded sheet of bedrock in which the fold axes and limbs are approximately horizontal.

narcotic Any substance that causes dullness, stupor, or sleep; most are habit-forming. In small doses, narcotic drugs such as morphine and codeine reduce pain by dulling the nervous system; large doses cause unconsciousness or death.

NASA See NATIONAL AERONAUTICS AND SPACE ADMINISTRATION.

nastic response Reversible or permanent movement of a plant in reaction to a stimulus, but not simply toward or away from the stimulus. The action of a Venus flytrap, and the opening and closing of flowers in response to temperature (thermonasty) are nastic responses. See NYCTINASTY, PHOTONASTY, SEISMONASTY, THERMONASTY, TROPIC RESPONSE.

natality 1) Birthrate; the production of new individuals per unit time in a population, also called fertility or fecundity (although those terms also have different definitions). 2) The inherent ability of a population to increase through sexual or asexual reproduction.

natality rate Number of births per thousand in a specific human population. Compare MORTALITY RATE.

National Aeronautics and Space Administration (NASA) The agency of the U.S. government that is responsible for space exploration and research and development supporting nonmilitary space technology. It oversees satellites such as Landsat.

National Ambient Air Quality Standards (NAAQS) Maximum allowable level, averaged over a specific time period, for a certain pollutant in outdoor (ambient) air.

The levels are set by the Environmental Protection Agency, as required by the Clean Air Acts of 1970, 1977, and 1990. See PRIMARY AMBIENT AIR QUALITY STANDARDS, SECONDARY AMBIENT AIR QUALITY STANDARDS.

National Audubon Society   A citizen organization founded in 1905 to protect bird populations and habitat and now active in a wide range of environmental issues; its membership is over 500,000. It has several chapters in different states and Latin America, although some states have only independent organizations with similar names (such as the Massachusetts Audubon Society). It was named for the American naturalist and artist John James Audubon (1785–1851).

National Environmental Policy Act of 1969   (NEPA) A comprehensive federal environmental law (42 U.S.C.A. §4321 et seq.) declaring that the federal government has responsibility for restoring and maintaining environmental quality and establishing the Environmental Protection Agency (EPA). NEPA requires all federal agencies to file an environmental impact statement for any project (including proposed legislation) having the potential for a significant effect on environmental quality.

national forest   A term for the public land areas set aside as forest or grassland reserves for management under the principle of multiple use as a renewable resource. Besides timber harvesting, other activities permitted on these lands include grazing, agriculture, mineral extraction, drilling for oil and natural gas, recreation, and sport hunting. The almost 300,000 acres administered by the U.S. Forest Service include 156 national forests and 19 national grasslands. See MULTIPLE USE.

national park   A term for public land often, but not always, in areas removed from human development, and set aside for the protection of wildlife and the natural landscape. In the United States, the 125,000 acres of national parks are administered by the National Park Service.

National Park Service   An agency of the U.S. Department of the Interior that administers national parks, monuments, memorials, battlefields, parkways, trails, rivers, seashores, lakeshores, and historic areas. These areas are set aside to preserve scenic or unique natural areas, to protect wildlife habitat, to preserve and to interpret historic and cultural sites, and to allow some forms of recreation.

National Parks and Access to the Countryside Act   An act of the British government in 1949 that established their national parks and laid the foundation of the Countryside Planning Acts.

National Wildlife Federation   (NWF) A not-for-profit citizen conservation organization with 5 million members and supporters, and affiliate organizations in the United States and territories, involved in all aspects of environmental advocacy and environmental education. It was founded in 1936.

national wildlife refuge   An area of public land set aside to protect habitats for waterfowl and sport game (and occasionally to protect specific endangered species from extinction). In the United States, the 452 refuges are administered by the Fish and Wildlife Service.

native element   A simple natural substance composed of a single chemical constituent in a free, uncombined state; examples are gold, arsenic, sulfur, and carbon.

native species   Endemic species; those

indigenous to a specific area. Compare EXOTIC SPECIES.

natric A term applied to an argillaceous soil horizon with a blocky texture and an exchangeable sodium ion saturation of greater than 15 percent.

natural chemical control The use of naturally occurring chemicals, especially pheromones or hormones, to control pests. See JUVENILE HORMONE, PHEROMONE.

natural convection Heat transfer by the bulk movement of a liquid or a gas driven by natural forces such as buoyancy rather than by pumps or blowers. Natural convection is the principle behind passive solar heating systems, as opposed to the forced convection of active solar heating systems. Contrast FORCED CONVECTION, RADIATION.

natural gas A naturally occurring gaseous hydrocarbon that is principally composed of methane. Natural gas is valued as an efficient and clean-burning fuel.

natural history The study of nature and natural objects (animals, plants, rocks); usually refers to popular study and amateur field study rather than laboratory investigation or thorough controlled scientific experiments.

natural increase Growth of a particular population when the number of births exceeds the number of deaths over a given time period.

natural ionizing radiation See BACKGROUND RADIATION.

naturalize The establishment of a species in a region where it had not previously lived so that it becomes an established part of the community, successfully reproducing and competing with native species.

natural radioactivity Radioactivity from naturally occurring isotopes as opposed to artificially made isotopes. See RADIOACTIVITY.

natural resource Any material supplied by the environment that is utilized by humans, such as fuels (wood, coal, etc.), mineral resources, or timber.

Natural Resources Defense Council (NRDC) A nonprofit membership organization founded in 1970 to combine legal action (litigation and lobbying for legislation) and citizen education to protect natural resources and to improve the quality of the human environment.

natural selection The mechanism proposed in Darwinian theory to explain evolution. New traits or variations in traits arise from spontaneous genetic changes within populations. Those traits that persist are those best adapted to the current environment, because organisms with those traits are more apt to survive and to reproduce than organisms without them. See DARWINIAN THEORY, EVOLUTION.

nature 1) General term for organisms and the environment they live in; the natural world. 2) The essence or characteristic qualities of a thing or a person.

Nature Conservancy, The A national, nonprofit membership organization that works to preserve biological diversity through protecting areas of land that are particularly rich in biological diversity. It works with individual states, as well as other conservation agencies, through national heritage programs to identify natural areas with ecological and biological significance and manages over 1000 nature sanctuaries throughout the United States.

nature-nurture General term for the

debate as to whether animal behavior traits are inborn instincts (nature) or learned behavior (nurture).

nature preserve   A specific area set aside for the protection of wildlife and ecosystems. Commercial uses within the reserve, when allowed, are controlled to protect the animals and the plants (limited tourism in African game parks is an example). Also called nature reserve.

nauplius   The primary stage of development of most crustaceans; an unsegmented, egg-shaped larva that hatches from a crustacean egg.

nautical mile   A unit of length used in navigation equal to 6,075 feet or 1.852 kilometers. It was originally determined as one-sixtieth of a degree of latitude.

Nautilus   The only living genus of the nautiloid group of molluscs. The external shell is coiled in one plane and divided into a number of linear chambers. Only the outermost of these is occupied by the nautilis, the remainder being gas filled and providing buoyancy. The glossy shells with their bright, pearly interior are sought for their decorative value. See CEPHALOPODA.

n-dimensional hypervolume   A theoretical model proposed by G. E. Hutchinson and used to describe the ecological niche of a species. Different variables are assigned axes, becoming vectors or dimensions interacting in a multidimensional (theoretical) space. The interaction of these vectors or dimensions describes the niche.

neap tide   A time of periodically low tidal ranges that coincide with the first and last quarter phases of the moon. The tidal range at neap tide is lower than the range of other phases because, during a neap tide, the gravitational pull reaches a minimum as the sun and moon align at a right angle to the earth. Compare SPRING TIDE.

Nearctic region   One of the biogeographical realms into which the earth is divided according to the distribution of animal life (fauna). The Nearctic region comprises North America (to the northern edge of the Mexican plateau) and Greenland. It is sometimes considered a subdivision of the Holarctic region. See HOLARCTIC REGION, NEOTROPICAL REGION, PALAEARCTIC REGION.

nebula   An interstellar cloud of dust and gas (it sometimes includes stars) that may be either brighter or darker than the space around it. Plural is nebulae.

necromass   The total weight of dead organisms, commonly for a given unit of land or volume of water. Sometimes used to refer to dead portions of living organisms, such as the bark of trees. Compare BIOMASS.

necroparasite   A parasite that kills its host and continues to live off of the dead organism.

necrosis   Death of localized cells within a living organism; the appearance of small areas of dead tissue surrounded by living tissue. Necrosis in plants can be a symptom of deficiencies of any of a number of different minerals, or of viral diseases.

nectarivore   A nectar-eating organism. Also spelled nectivore.

nectophore   A swimming bell in jellyfish. Necto- is from the Greek *nektos*, meaning swimming. Also called a nectocalyx. See HYDROZOA.

needle   1) A spire-shaped mass of rock. 2) An acicular mineral crystal. 3) A snow crystal that is at least five times longer than it is wide. 4) A device for setting a fuse in blasting powder within a borehole. 5) A slender, pointed leaf of an evergreen plant.

NEF See NOISE EXPOSURE FORECAST.

negative feedback 1) Describing any system in which a response to a stimulus or a cause exerts an effect that counteracts the stimulus or cause. Negative feedback is at work when a physiological reaction is slowed down by the increased concentration of its product, or sped up by a decreased concentration of its product. 2) In population dynamics, a factor that works to stabilize population growth. Contrast POSITIVE FEEDBACK. See HOMEOSTASIS.

negligence Legal term for failure to exercise that degree of care which an ordinary person of ordinary prudence would exercise under the same circumstances. The term refers to conduct which falls below the standard established by law for the protection of others against unreasonable risk of harm.

nekton Marine animals that swim, moving under their own power, in contrast to plankton that only drift or float. Fish, marine mammals, and jellyfish are nekton. Sometimes spelled necton. See NEUSTON, PLANKTON.

nematicide A substance that kills nematodes, especially soil-dwelling species. See NEMATODA.

nematoblast A cell that contains or will develop a nematocyst. Also called a thread cell or cnidoblast. See NEMATOCYST.

nematocyst A stinging device consisting of a sac filled with irritating fluid and a long threadlike structure. Groups of many nematocysts are responsible for the sting of hydra and jellyfish; they are used for defense and to capture prey.

Nematoda Phylum of the nematode or round worms whose species are distributed widely as free-living forms in waters and soils throughout the world or as animal and plant parasites. Although the number of nematode species has been estimated above 500,000, their structural diversity is limited to a rather basic plan featuring a bilaterally symmetrical, cylindrical, unsegmented body covered with a layered cuticle. Nematodes are psuedocoelomate and lack vascular and respiratory systems. They range in size from microscopic to several meters in length and display a wide variety of feeding habits. Many groups, such as the hookworm genus *Ankylostoma*, are serious pathogens of crops, domestic animals, and humans. See HOOKWORMS, ASCARIS.

neo- Prefix meaning new, recent, or a new form of.

neo-Darwinism A modification of the Darwinian theory of evolution that incorporates modern genetic discoveries. Neo-Darwinism adds the laws discovered by Mendel as well as population genetics to explain how the mutations and variations develop that lead to natural selection. See DARWINIAN THEORY, MENDEL'S LAWS, WEISMANNISM.

Neogea Another term for Neotropical region. See NEOTROPICAL REGION.

neoglaciation The readvance of mountain glaciers during the little ice age. See LITTLE ICE AGE.

neo-Malthusian Describing modern supporters of, or the modern version of, the Malthusian theory: Human populations will continue to grow until they reach the maximum carrying capacity of the environment, or until conditions of famine, poverty, and associated political upheaval become intolerable. See MALTHUSIAN GROWTH.

neoplasia   The development of new tissue; also, the formation of neoplasm. See NEOPLASM.

neoplasm   New, abnormal development of tissue; the formation of a growth or a tumor. Neoplasms do not contribute to the organism; they are detrimental to varying degrees depending on whether they are benign or malignant.

neossoptiles   The covering of newly hatched birds, consisting of fine feathers that resemble down.

neoteny   The retention of immature features or an entire juvenile body form in the adult, sexually mature stage of an organism. For example, some amphibian species resemble tadpoles in their adult form.

Neotropical realm   One of the phytogeographical realms into which the earth is divided according to the distribution of plant life (flora). The Neotropical realm is divided into three floral regions: the Pacific-South American (Andean) floral region, Central American floral region, and Parano-Amazonian floral region. See FLORAL REALM, PHYTOGEOGRAPHY. Contrast NEOTROPICAL REGION.

Neotropical region   One of the biogeographical realms into which the earth is divided according to the distribution of animal life (fauna). The Neotropic region includes South America, the West Indies, and Central America from the Mexican plateau south. Sometimes called Neogea. See FAUNA, ZOOGEOGRAPHY. Contrast NEOTROPICAL REALM.

NEPA   See NATIONAL ENVIRONMENTAL POLICY ACT.

nepheline   A variably colored feldspathoid mineral that forms hexagonal crystals in association with silica-poor igneous rocks. Nephaline is used in ceramic industries.

nephelinite   A fine-grained or porphyritic extrusive mafic rock containing nephaline, pyroxene, and titanomagnetite.

nephelometer   An instrument that uses light absorption to measure the concentration of substances suspended in water, such as total suspended solids present in samples of drinking water.

nephelometric turbidity units   (NTU) The units used to express the measurements obtained with a nephelometer. See NEPHELOMETER.

nephridium   A primitive, segmented organ that functions like a kidney to remove wastes in some invertebrates such as molluscs and lower vertebrates. It functions in a similar manner as the kidneys of higher animals, and in some cases also serves in reproduction.

neptunium   (Np) A radioactive, transuranic, metallic element with atomic number 93; the atomic weight of its most stable isotope is 237.0. It occurs in very small amounts in uranium ore and is one of the by-products of nuclear reactors.

neritic zone   A zone of shallow marine water less than 200 m deep. The neritic zone is usually highly populated by marine organisms.

nerve impulse   Information in the form of electrical signals transmitted along the axon of a nerve fiber (neuron). See NEURON.

ness   A British term for a headland.

nest parasitism   A form of parasitism in which one species lays eggs in the nest made and used by another (host) species. The host species then rears the young of the parasite as one of its own. The cowbird is a nest parasite that lays eggs in the nests of redwing blackbirds.

net above-ground production   The net primary production of plant parts growing above the soil surface only. See NET PRIMARY PRODUCTION.

net below-ground production   A measurement of net primary production of only those parts of plants growing below the soil surface. See NET PRIMARY PRODUCTION.

net energy   Total useful energy generated from an energy resource or energy system minus the energy that was used, lost, or wasted in finding, processing, concentrating, and transporting it to a user. Also called net useful energy.

net energy yield   Amount of usable energy remaining after all the energy spent on prospecting, extracting, refining, and transporting (to the power plant and from the power plant to the consumer) are subtracted from the total energy. Also called net useful energy.

net migration rate   The overall increase or decrease in population for a particular region caused by movement in or out of the region. It is calculated as the difference between the numbers of people immigrating and emigrating during a year (or other period of time) per thousand persons in the population.

net primary production   (NPP) The total amount of energy produced and stored through photosynthesis in a specific community or ecological system (gross primary production), minus the quantity consumed during respiration by the photosynthetic organisms. NPP is expressed as oven dry weight (grams) of tissues/area or energy content of tissues

(Kcal)/area. Also called apparent photosynthesis, or net assimilation, and often used interchangeably with net primary productivity (although that is technically the productivity per year). Compare GROSS PRIMARY PRODUCTION, NET PRIMARY PRODUCTIVITY.

net primary productivity   (NPP) The rate per year of net primary production, which is the rate of useful chemical energy production by plants in an ecosystem. It equals the rate of total energy production by the plants in the ecosystem minus the rate at which they use some of this energy through cellular respiration. NPP is expressed as oven dry weight (grams) of tissues/area/time or energy content of tissues (Kcal)/area/time. Often used interchangeably with net primary production, although technically net primary productivity is for a time period of one year. Compare NET PRIMARY PRODUCTION.

net production   See NET PRIMARY PRODUCTION.

net production efficiency   Percentage of total energy consumed or produced through photosynthesis that is incorporated into growth and reproduction. Also called tissue growth efficiency. Compare ASSIMILATION EFFICIENCY.

net reproductive rate   (R_o) The number of offspring (usually, the female offspring) a typical female produces in a lifetime. If immigration equals emigration and R_o is 1.0, the population is stable; if R_o is less than 1.0, the population is declining and if more than 1.0, it is increasing.

neuromuscular junction   The point at which a nerve joins a muscle through several motor end plates. Also called myoneural junction. See MOTOR END PLATE.

neuron A nerve cell; one of the impulse-conducting functional units of the brain, the spinal cord, and the rest of the nervous system. Each neuron has a cell body, which contains the nucleus, usually several branched structures (dendrites) that carry impulses to the cell body, and a long nerve fiber (axon) that projects beyond the cell body and transmits impulses away from the cell.

neurotoxins Substances that can damage or destroy nerve tissue. Lead and mercury are both neurotoxins, as are some snake venoms.

neurula The stage of development in vertebrate and amphibian embryos in which organs begin to form, with rapid differentiation. The neurula follows the stage known as the gastrula. See GAS-TRULA.

neuston, neustron Organisms that live on the surface of a body of water; they may be either floating or swimming organisms. Duckweed (*Lemna minor*, a small plant) and water-striders (*Gerris*, an insect) are examples of neuston. Compare NEKTON, PLANKTON.

neutral 1) Having no net positive or negative electrical charge. The charges of cations and anions cancel each other out when they combine to form neutral salts. 2) Neither acid nor alkaline; having a pH value of 7. Lime is added to acid soils to raise their pH to neutral, which improves plant growth. See PH.

neutral alleles Alleles that do not contribute significantly to the fitness of an organism, compared with similar alleles.

neutral fat A true fat, an ester of a fatty acid combined with glycerol. Neutral fats in animals include tristearin, triolein, and tripalmitin; olive oil is a neutral fat made up of glycerol and oleic acid.

neutralism The lack of any effect of one organism on another coexisting organism. In an association between populations characterized by neutralism, neither population affects the other.

neutralization The reaction that occurs when strong acids are mixed with strong bases, which combine to form a salt plus water. The hydrogen ions (H^+) responsible for acidity combine with the hydroxyl ions (OH^-) responsible for alkalinity; the product is neutral water (H_2O).

neutron A subatomic particle with no electrical charge that resides in the atomic nucleus (hydrogen is the only element with no neutrons, except in two of its isotopes). The neutron's mass is roughly the same as the mass of a proton. Neutrons can be separated from atomic nuclei (they are released in nuclear fission), but they are no longer stable once free. See PROTON, ELECTRON.

neutrophilus Describing cells that stain readily with neutral dyes, such as some forms of white blood cells (leukocytes). See POLYMORPHONUCLEAR LEUKOCYTES.

neve A term applied to an area of perennial snow in the accumulation zone of a glacier.

New Caledonian floral region One of the phytogeographic regions into which the Paleotropical realm is divided according to similarities between plants; it consists of the island of New Caledonia (Vanuatu). See PALEOTROPICAL REALM, PHYTO-GEOGRAPHY.

new global tectonics A term pertaining to plate tectonic interpretations on a global scale.

new red sandstone A reference to the red-bedded sandstone deposits of the Permian and Triassic ages exposed in northwest Great Britain.

newton   The unit for force in the Système International d'Unités (SI). One newton equals the amount of force required to produce an acceleration of one meter per second per second in a mass of one kilogram. See FORCE.

Newtonian mechanics   The study or application of mechanics in which Isaac Newton's three laws of motion are assumed to be valid. Also called classical mechanics, it is used explain many large-scale interactions of forces on earth. Newton's laws are not adequate to explain circumstances involving speeds approaching that of light, particles the size of individual small molecules, and regions in which gravity is especially strong. Contrast QUANTUM MECHANICS.

New Zealand floral region   One of the phytogeographic regions into which the Austral realm is divided according to similarities between plants; it consists of New Zealand plus offshore islands. See AUSTRAL REALM, PHYTOGEOGRAPHY.

NGO   See NONGOVERNMENTAL ORGANIZATION.

Ni   Chemical symbol for the element nickel. See NICKEL.

niacin   Nicotinic acid, a water-soluble vitamin sometimes called vitamin B_3. Niacin is required for all of the energy reactions in the body. Severe deficiencies of niacin cause pellagra. See PELLAGRA.

niche   An organism's physical location and function within an ecosystem. Also called ecological niche. See FUNDAMENTAL NICHE, REALIZED NICHE, N-DIMENSIONAL HYPERVOLUME. Contrast HABITAT.

niche breadth   See NICHE WIDTH.

niche complementarity   The tendency for coexisting species that live in similar areas to share one or more aspects or dimensions of a niche (such as altitude) but to differ in another (such as diet). Compare NICHE DIFFERENTIATION.

niche differentiation   The tendency of species sharing similar habitats to differ in at least some aspects (niche requirements) of their ecological roles. See COMPETITIVE EXCLUSION PRINCIPLE. Compare NICHE COMPLEMENTARITY.

niche diversification   See NICHE DIFFERENTIATION.

niche expansion   An increase in one or more dimensions of an organism's ecological role (niche). Niche expansion and niche differentiation are different adaptive strategies for dealing with increased competition between species.

niche overlap   The sharing of one or more dimensions of the ecological role (niche) of different species; lack of total competitive exclusion. An example is many plant species coexisting within the same habitat or space. Compare COMPETITIVE EXCLUSION PRINCIPLE.

niche packing   The tendency for coexisting species to expand their niches to the available volume along important niche dimensions.

niche width   The expression of the degree to which an organism can occupy a segment of a particular niche dimension. Species with exact requirements and very little flexibility are said to occupy narrow niches; generalist species with wide tolerances are said to occupy wide niches. Also called niche breadth. See NICHE, N-DIMENSIONAL HYPERVOLUME.

nickel (Ni) A hard, metallic element with atomic weight 58.71 and atomic number 28. Nickel is easily shaped and used in numerous alloys (including some coins) and for electroplating. It is also used as a catalyst by the chemical industry, as in the production of adipic acid, a starting material for the production of nylon.

nickel carbonyl $Ni(CO)_4$, a volatile compound formed when carbon monoxide comes in contact with hot nickel. Further heating reverses the reaction, recreating its original components. Nickel carbonyl is used in extracting nickel from natural ores and as a catalyst in the preparation of acrylic esters, used in plastics. It is extremely toxic. It decomposes in air to form carbon monoxide and very fine nickel dust.

nick point A point or zone in a stream channel where a concave upward-stream section abruptly changes gradient as a rapid or a waterfall.

nicotine $C_5H_4N\text{-}C_4H_7NCH_3$, a poisonous plant alkaloid found in tobacco leaves. It has long been used as a contact insecticide and fumigant, especially as nicotine sulfate (trade name, Black Leaf 40). Because it is so toxic to humans, it has been replaced with less hazardous insecticides. It is also believed to be a cocarcinogen.

nicotinic acid The chemical name for the vitamin niacin. See NIACIN.

nidicolous Describing bird species that reside in nests for a relatively long period after hatching because they remain highly dependent upon the parent birds for some time (are altricial). Contrast NIDIFUGOUS.

nidifugous Describing bird species that leave the nest in a relatively short period after hatching. Contrast NIDICOLOUS.

night-jars See CAPRIMULGIDAE.

nimbostratus Low layer of shapeless, dark gray clouds, usually associated with heavy precipitation.

nimby Acronym for "not in my backyard," a term of recent origin, now commonly used to refer to people who refuse to have something possibly or potentially environmentally harmful close to their home.

nitrate 1) The ion NO_3^-. 2) A salt or ester of nitric acid, which contains the nitrate ion. Nitrates are important as concentrated sources of nitrogen in fertilizers (potassium nitrate and sodium nitrate); they are also used in explosives. Nitrates are very water soluble and easily leached from soils, resulting in contamination of drinking water supplies. An excess of nitrate in drinking water (more than 10 milligrams per liter) can cause a condition known as methemoglobinemia in infants. See NITRITE, METHEMOGLOBINAEMIA.

nitrate-forming bacteria Any of several bacteria (genus *Nitrobacter)* that oxidize nitrites (produced by nitrite-forming bacteria) to nitrates in the soil. The process is one stage in nitrification, the conversion of nitrogen present in organic matter into inorganic forms that can be utilized by plants. See NITROGEN CYCLE, NITRIFICATION. Compare DENITRIFYING BACTERIA, NITRITE-FORMING BACTERIA, NITROGEN-FIXING BACTERIA.

nitric acid HNO_3, a very corrosive strong acid. It is an important chemical in the manufacture of organic compounds, sulfuric acid, fertilizers, and explosives. The old name for nitric acid was *aqua fortis*, Latin for strong water. See NITRATE.

nitric oxide NO, a gas formed in internal combustion engines and similar forms of combustion under heat and pressure. Although not problematic itself, when released into the atmosphere, nitric oxide reacts with oxygen (O_2) to form the hazardous pollutant, nitrogen dioxide (NO_2). Nitric oxide also destroys ozone and is thus harmful to the earth's ozone shield. Nitric oxide is also known as nitrogen monoxide. See NITROGEN DIOXIDE.

nitrification The processing of converting ammonia (the form of nitrogen produced by decaying organic matter) into nitrites or nitrates, inorganic forms of nitrogen that can be assimilated by plants. Nitrification is a form of oxidation; the various stages are carried out by a group of organisms collectively called nitrobacteria or nitrifying bacteria.

nitrify 1) To convert soil ammonia into nitrites or nitrates by bacterial action. 2) To add nitrates, as by applying fertilizer.

nitrite 1) The ion NO_2^-. 2) A salt (or ester) of nitrous acid, which contains the nitrite ion. Unlike the nitrate used in fertilizers, nitrite is toxic to plants in large concentrations. Soil bacteria form nitrites as a stage in converting nitrogen (ammonia) to usable nitrates. Nitrites such as sodium nitrite are added to foods as a preservative, especially to retain color in meats and to prevent botulism. Harmful nitrosamines are formed during the cooking of foods such as bacon, which contains sodium nitrite, although cigarette smoke, high-nitrate drinking water, and many vegetables are greater sources of nitrosamines for most people. See NITRATE, NITROGEN FIXATION, NITROSAMINES.

nitrite-forming bacteria Any of several bacteria (genus *Nitrosomonas*) that oxidize ammonia to nitrite in the soil. The process is one stage of nitrification, the conversion of nitrogen present in organic matter into inorganic forms that can be utilized by plants. Nitrite-forming bacteria use the ammonia from animal wastes and decaying organic matter, converting the nitrogen into a form that can then be oxidized by nitrate-forming bacteria. Compare DENITRIFYING BACTERIA, NITRATE-FORMING BACTERIA, NITROGEN-FIXING BACTERIA.

Nitrobacter Nitrate-forming bacteria; the genus of nitrifying bacteria that carries out the stage in which soil nitrites are converted to nitrates, the form of soil nitrogen that plants can use. Compare AZOTO-BACTER, NITROSOMONAS, RHIZOBIA.

nitrogen (N) Gaseous element with atomic weight 14.01 and atomic number 7. In its elemental form it exists as a diatomic molecule, N_2. It makes up about 80 percent of the volume of the earth's atmosphere and is found in all plant and animal tissues. It is used commercially as a coolant, an aerosol propellant, and to create an inert atmosphere in food packaging.

nitrogen cycle The process by which nitrogen moves within and between ecosystems and throughout the biosphere. The nitrogen cycle is not a simple circuit; it involves many interconnected processes that are often mediated by bacterial activity. Nitrogen generally enters ecosystems from the atmosphere in the form of nitrate or ammonia through electrical discharge (lightning), biological fixation, or precipitation. These nitrogen compounds are taken up by plants, which in turn are consumed by animals, which release nitrogen back into the environment, largely in the form of urea. The bodies of all organisms release nitrogen from proteins and amino acids into the environment as they decompose; these nitrogen compounds are converted to

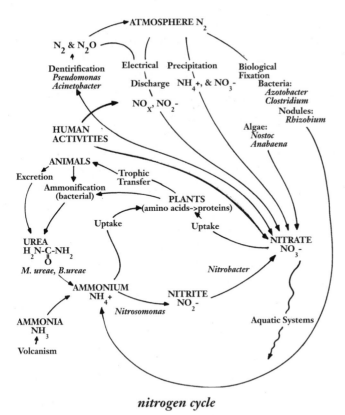

nitrogen cycle

nitrogen fixation The process of converting inorganic, atmospheric nitrogen into an organic form of nitrogen, ammonia. The conversion is carried out by various microorganisms, especially certain bacteria and cyanobacteria found in the soil and bacteria present in nodules on the roots of leguminous plants (such as alfalfa, clover, lupine, peas, and beans). The ammonia formed can be used directly by plants or may undergo nitrification. Nitrogen fixation is also used to describe the chemical processes used in the manufacture of fertilizers. See NITROGEN-FIXING BACTERIA.

Nitrogen-Fixing Aerobic Bacteria A phylum of bacteria that transforms atmospheric nitrogen into organic nitrogen. The genus *Rhizobium*, although incapable of nitrogen fixation in soil, produces organic nitrogen in symbiotic relationship with many plants, particularly legumes. Bacteria induce plants to produce nodules in which the bacteria are transformed into nitrogen-fixing bacteroids. In the absence of nitrogen-fixing bacteria the biosphere would be nitrogen deficient. See MONERA, RHIZOBIA, ROOT NODULES.

ammonia through ammonification. Ammonia is then converted to nitrite and then nitrate through nitrification. Nitrate may be reduced to molecular N_2 through denitrification. See NITROGEN FIXATION, NITRIFICATION, MINERALIZATION, DENITRIFICATION.

nitrogen dioxide NO_2, a rust-colored gas that is a major air pollutant. It forms when nitric oxide from combustion engines mixes with oxygen in the atmosphere; it is also a direct product of combustion. It is a component of photochemical smog and, in the presence of sunlight, reacts with oxygen to form ozone. It also reacts with itself to form another pollutant, nitrogen tetraoxide (N_2O_4). See PHOTOCHEMICAL SMOG, OZONE.

nitrogen-fixing bacteria Bacteria that assist in converting atmospheric nitrogen into ammonia, a form that plants can use. Nitrogen-fixing bacteria include the free-living *Azotobacter* and *Clostridium butyricum* as well as the symbiotic, nodule-forming *Rhizobium*. Some cyanobacteria (formerly called blue-green algae) also fix nitrogen. See AZOTOBACTER, RHIZOBIA, NITROGEN FIXATION, SYMBIOTIC NITROGEN-FIXING BACTERIA, ROOT NODULES.

nitrogen monoxide Another name for nitric oxide. See NITRIC OXIDE.

nitrogenous wastes General term including animal urine and manures as well as refuse with a high nitrogen content, such as residues from processing operations (blood meal, meat tankage). Although nitrogenous wastes have potential as fertilizer, they must be handled carefully in order to prevent formation of offensive odors. In composting, they must be diluted with large quantities of high-carbon wastes (such as leaves or wood chips) in order for composting to proceed and to avoid odor.

nitrogen oxides NO_x, compounds formed by the oxidation of nitrogen, including nitrogen dioxide and nitric oxide. These compounds are air pollutants because they undergo photochemical reactions to form smog and ozone; they are also believed to contribute to the depletion of the earth's protective ozone layer. See PHOTOCHEMICAL SMOG, OZONE, SO_x.

nitrogen-phosphorus detector (NPD) A detector used in gas chromatography to detect substances after they have passed through the column. Nitrogen-phosphorus detectors work by burning the compounds and observing the intensity of the emitted light. The light intensity is translated into an electrical current, which is measured to determine the amount of substance present. Nitrogen-phosphorus detectors are used for the analysis of trace levels of nitrogen and phosphorus-containing pesticides.

nitrogen tetroxide N_2O_4, an air pollutant. Nitrogen dioxide converts to this compound at temperatures below 17°C. Also called dinitrogen tetroxide, this compound is produced along with other nitrogen oxides from internal combustion engines. See NITROGEN OXIDES.

nitrosamines Any of several organic compounds derived from amines and containing the molecular group -NNO. These compounds are considered potent carcinogens. They are formed when nitrites react with certain proteins in the human stomach, when protein foods containing nitrites are cooked, and when tobacco is burned. Although cigarette smoke provides the largest source of nitrosamines, other sources are vegetables with naturally high concentrations of nitrates (which convert to nitrites during storage and after being consumed), meats cured with nitrates or nitrites, and drinking water that contains nitrates. Ascorbic acid (vitamin C) inhibits the converstion of nitrites into nitrosamines in the stomach, so some meats are now cured with ascorbic acid as well as sodium nitrite. See NITRATE, NITRITE.

Nitrosomonas Nitrite-forming bacteria; the genus of nitrifying bacteria that oxidizes ammonia to convert it into nitrites in the soil, the first stage of nitrification.

nitrous oxide N_2O, also called laughing gas and dinitrogen oxide. It is a gaseous compound used for anesthesia and as an aerosol propellant.

nm See NANOMETER.

NNI Short for noise and number index. See NOISE AND NUMBER INDEX.

No Chemical symbol for the element nobelium. See NOBELIUM.

No$_x$ Abbreviation for nitrogen oxides, common air pollutants. See NITROGEN OXIDES.

nobelium (No) An artificial, transuranic, radioactive element with atomic number 102; the atomic weight of its most stable isotope is 255.0. Nobelium is a heavy element belonging to the actinide group. See ACTINIDE.

noble gases   A group of elements whose electronic structure makes them inert to almost all chemical reactions; as a result they are sometimes called the inert gases. They include helium, neon, argon, krypton, xenon, and radon 222.

noble metal   A term for metals that show great resistance to corrosion (oxidation). Gold, silver, and platinum are all noble metals; they do not readily combine with nonmetallic elements.

nocturnal   Becoming active only after dark. Bats and owls are mostly nocturnal. Contrast CREPUSCULAR, DIURNAL.

node   Any joint where leaves attach to a plant stem. See APPENDIX p. 628–629.

nodule   A small, rounded growth on a plant, especially one of a number on the roots of plants that contain symbiotic, nitrogen-fixing bacteria. See NITROGEN-FIXING BACTERIA, RHIZOBIA.

nodum   A unit of vegetation of any size, such as a community or an association.

noise   Undesired sound or sounds of many different frequencies not harmonically related.

noise and number index   (NNI) A rating for evaluating the degree of disturbance to humans from aircraft noise. The index was developed through a survey of persons living in the vicinity of Heathrow Airport outside of London, England. Heathrow is a site of heavy air traffic, including a couple of daily flights of the supersonic transport Concorde, which is distinctly louder than other aircraft.

noise criteria   Sound levels that are acceptable to the people hearing them, and thus

not considered noise pollution.

noise exposure forecast   (NEF) Technique for predicting aggravation caused by fluctuating noises, such as from factories or from airplanes flying overhead. The forecast is based on weighting a number of contributing factors, such as the time of day the noises occur, how often and for how long the noises occur, and flight paths of aircraft.

noise reduction coefficient   (NRC) A number expressing the average of the ratio of sound energy absorbed to the total sound energy striking an area of a given surface or material at different sound levels (250 Hz, 500 Hz, 1 kHz, and 2 kHz). It indicates the ability of the surface or material to absorb sound.

nonbiodegradable   Describing a substance that is not broken down by natural processes and so remains in its original form for long periods of time. Many plastics and some pesticides are non biodegradable.

nonconformity   An erosional surface between two types of rock that could not have formed in contact with one another. A stratified sedimentary rock overlying a granitic igneous rock forms a nonconformity. See UNCONFORMITY, DISCONFORMITY.

nonequilibrium thermodynamics   The study of systems not in thermal equilibrium, such as systems with temperature differences. Nonequilibrium thermodynamics can be contrasted with classical thermodynamics, which describes the behavior of systems at equilibrium.

nonevolutionary responses   Changes, such as different behaviors, made by an organism in order to adapt to changes in the environment.

nongovernmental organization (NGO) Voluntary organizations, usually international nonprofit membership organizations, not affiliated with any government organization. The United Nations (U.N.) gives NGOs official observer status to attend some meetings, or to testify to U.N. committees.

nonmetallic Describing elements that are generally poor conductors of electricity and heat, and readily form negative ions. These qualities are the opposite of those assocated with elements that are metals. See METALLIC, METALLOID.

nonmetallic mineral 1) A description of the physical mineral property of luster as applied to minerals that reflect light but do not have a characteristic sheen like that of a metal. Minerals having lusters that are glassy, vitreous, silky, and resinous may describe nonmetallic minerals. 2) Minerals lacking properties of a metal.

nonparametric statistics Statistical analyses performed on data from populations that are not known to have a normal distribution.

nonpasserine Describing birds that do not belong to the order of Passeriformes (perching birds, including common songbirds). Rollers, kingfishers, and hornbills are nonpasserine species of birds. The term is often used to refer to species of the order Coraciiformes. See CORACIIFORMES.

nonpersistent Potentially polluting chemicals that are broken down or reduced to acceptable levels by natural processes.

nonpoint sources Pollution-producing entities that are not tied to a specific origin such as an individual smokestack. Nonpoint sources of water pollution include runoff washing pollutants from roads into storm sewers and water bodies; runoff carrying agricultural chemicals from lawns, fields, and golf courses; and automobiles, which constitute multiple, small, mobile sources of air pollution. Compare POINT SOURCES.

nonporous Not permeable to water, air, or other fluids.

nonrenewable resources Resources that exist in fixed amounts in various places on the earth's crust and have the potential for renewal only by geological, physical, and chemical processes taking place over hundreds of millions of years. Coal and other fossil fuels are nonrenewable. Contrast RENEWABLE RESOURCES.

nonselective Describing pesticides with a broad range of action, able to kill many different species. Nonselective insecticides such as the chlorinated hydrocarbons are often toxic to nontarget species as well. Also called broad-spectrum.

nonsilicate A mineral with a crystalline structure that is not based on silica tetrahedra.

nontarget species An organism that is killed unintentionally during pesticide application. See NONSELECTIVE.

nontransmissible disease A disease not caused by a living organism that can be spread from one person (or organism) to another person. Cancers, ulcers, and genetic diseases such as hemophilia are nontransmissible.

nori An edible seaweed (usually species of *Porphyra)*, used in Japanese cooking, especially to wrap rice in the form of sushi known as nori rolls.

normal distribution Statistical term for a distribution in which most values fall near (or at) a central, mean point. The decrease in other values is greater with increasing distance from this central point, and the other values fall roughly symmetrically on either side of the mean point. The resulting curve (a normal curve) is shaped like a bell. Also called Gaussian distribution. Compare LOG-NORMAL DISTRIBUTION, RANDOM DISTRIBUTION.

normal fault   A fault in which the apparent sense of displacement of the hanging wall is downward relative to the footwall. Compare REVERSE FAULT, THRUST FAULT.

normally magnetized   A time interval in geologic time in which the geomagnetic field has the same polarity as in the present.

North Atlantic Current   An ocean surface current flowing from the Grand Banks of Newfoundland eastward to northwestern Europe. The North Atlantic Current is a northerly extension of the Gulf Stream.

North Equatorial Current   A westward-moving ocean current that flows from the west coast of central America toward the Philippines. See EQUATORIAL CURRENT.

northern coniferous forest   See BOREAL FOREST.

northern lights   Common term for the aurora borealis, or Northern Hemisphere aurora. See AURORA.

notifiable disease   Diseases that by regulation must be reported to local health authorities, such as the Board of Health or similar state agencies. Notifiable diseases vary in different states and countries, but usually include most serious communicable or contagious diseases such as smallpox, scarlet fever, typhoid, diphtheria, cholera, typhus, encephalitis, tuberculosis, typhus,

and some forms of food poisoning.

no-till agriculture   Cultivation of crops without plowing or turning over the soil. It is the most extreme form of minimum-tillage (reduced-tillage) agriculture. Existing vegetation is usually killed with a broad-spectrum herbicide, and after a waiting period, the new crop is seeded (direct-drilled) into the soil right through the dead vegetation. In some cases, herbicides are not needed and the new crop is seeded over the stubble of the previous crop. Because the soil is never fully exposed, no-till agriculture greatly reduces soil erosion by rain or wind; it also often lowers labor costs and saves energy. See MINIMUM-TILLAGE FARMING.

notochord   A flexible, skeletal rod that runs lengthwise down the back of some invertebrates (Chordata, which includes lancelets, lampreys, and sea squirts). In vertebrates, the notochord is reduced to an embryonic precursor to the backbone and forms the main support structure of the body.

Notogea   See AUSTRALASIAN REGION.

nova   A star that undergoes a sudden and very large increase in brightness. Most novas are believed to belong to a pair of close, orbiting binary stars. Compare SUPERNOVA.

noy   Unit for perceived noisiness as opposed to loudness. The noy is expressed in perceived noise decibels (PNdB).

NPK   Abbreviation for nitrogen, phosphorous, and potassium, the three major plant nutrients used in fertilizers. Bags of fertilizer usually have three numbers listed on them, such as 5-10-5. These indicate the NPK levels: 5 percent nitrogen, 10 percent phosphorous, and 5 percent potassium.

NRC   See NUCLEAR REGULATORY COMMISSION.

NTA   Abbreviation for nitrilotriacetic acid, one of the early replacements for phosphates in detergents.

nuclear aftermath   General term for phenomena occurring as a result of a nuclear destruction.

nuclear autumn effect   A slight reduction in atmospheric temperatures that might theoretically occur after a nuclear war. Because large amounts of soot, dust, and smoke could be suspended in the air following a nuclear exchange, some theorists have suggested that the reduced insolation caused by these particles would reduce global temperatures. It is a modification of the nuclear winter theory. Compare NUCLEAR WINTER.

nuclear breeder reactor   See BREEDER REACTOR.

nuclear change   An alteration in the number of protons, neutrons, or both, in the nucleus of an atom, changing the isotope into a different isotope or a different element. Nuclear change occurs in any nuclear reaction.

nuclear energy   Energy contained within the nuclei of atoms or that can be released by fusing some nuclei. Nuclear energy now most commonly refers to the release of energy from changing the structure of an element's nucleus. Commercial nuclear power plants use one form of nuclear energy, fission, in which large amounts of energy are released when the nuclei of heavy elements such as uranium split. Fusion is another form of nuclear energy, in which light nuclei are fused or joined; it has yet to be harnessed for commercial power generation. Nuclear energy is released by fission or fusion of atomic nuclei. Also called atomic energy, but the preferred term is now nuclear energy.

nuclear fission   The splitting of a heavy isotope into two or more lighter elements, a process which releases large amounts of energy. Fission can be spontaneous or can be induced through bombardment by neutrons. It is the process used in nuclear power plants to produce electricity and in nuclear weapons such as the atomic bomb. See CHAIN REACTION.

nuclear fuel cycle   The entire process of nuclear power generation, from extraction at the mine to storage of radioactive waste.

nuclear fusion   See FUSION.

nuclear holocaust   The heavy destruction that would accompany a full-scale nuclear war.

nuclear magnetic resonance   (NMR) The application of a known radio frequency to an object (such as the human body) or a substance (such as an unknown chemical) within a strong magnetic field. The influence of the radio frequency on the nuclei of the atoms in the object or substance can be analyzed to provide much information about the molecules and their environment. It is used in the analysis of substances to determine the molecular structure of the molecules, a technique called NMR spectroscopy. See MAGNETIC RESONANCE IMAGING.

nuclear power   A general term for electricity produced from nuclear energy.

nuclear power plant   A facility producing electricity from nuclear energy (fission). Although a number of different types of nuclear power plants are in use, most harness the heat from fission to produce steam, which drives the turbines that run electrical generators. See CANDU, STEAM-

GENERATING HEAVY-WATER REACTOR, PRESSURIZED WATER REACTOR.

nuclear reactor   A facility for obtaining a controlled, nuclear fission chain reaction for research, military purposes, medical treatment, or power generation. Nuclear reactors usually consist of a core with fissionable fuel (fuel rods), control rods to control the chain reaction of that fuel, a moderator to slow down the neutrons, plus a system for harnessing the huge quantities of heat produced and for dissipating waste heat. See NUCLEAR ENERGY.

Nuclear Regulatory Commission (NRC) Five-member federal commission responsible for licensing and regulating all commercial uses of nuclear energy to protect the health and safety of the public and the environment. The NRC was formed in 1974, when the responsibilities of its predecessor, the Atomic Energy Commision, were divided between it and the Energy Research and Development Administration (ERDA).

nuclear waste   See RADIOACTIVE WASTE.

nuclear winter   The theory that a nuclear war would cause a significant reduction in global temperatures by sending so much soot, smoke, and dust into the atmosphere that a substantial amount of incoming solar radiation would be blocked out. See NUCLEAR AUTUMN EFFECT.

nucleating agent   A substance such as dry ice or silver iodide used in cloud seeding. A nucleating agent promotes the formation of condensation nuclei, which in turn promote cloud formation. See CONDENSATION NUCLEI, CLOUD SEEDING.

nucleic acid   General term for any of a group of compounds that control cellular heredity and function. Nucleic acids are natural polymers in which purine or pyrimidine bases are attached to a long phosphate-sugar chain. RNA and DNA are the most important nucleic acids.

nucleolus   A small round structure within the nucleus of a cell. The nucleolus contains protein, DNA, and ribosomal RNA; it is important in the synthesis of ribosomes. See RIBOSOMES.

nucleoprotein   Any of a group of compounds consisting of a protein combined with a nucleic acid. They are found in cell nuclei and viruses. Both chromosomes and ribosomes are made of nucleoproteins.

nucleoside   Any of a group of glycoside compounds in which purine or pyrimidine (nitrogen bases) are combined with pentose (a sugar). When joined with a phosphate group, they become nucleotides. See GLYCOSIDE, NUCLEOTIDE, PURINE BASE, PYRIMIDINE BASE.

nucleotide   The basic building block of a nucleic acid, such as DNA or RNA. Nucleotides contain a purine or pyrimidine base, a pentose sugar, and a phosphate group (a nucleoside plus a phosphate group). See NUCLEIC ACID, DNA, RNA.

nucleus   The central region within the atom, containing the protons and neutrons and around which the electrons orbit. Most of an atom's mass is located in the nucleus. Because it contains particles of positive charge (protons) plus particles of neutral charge (neutrons), the nucleus carries a net positive electrical charge. See PROTON, NEUTRON.

nuclide   Any isotope of an element that exists for a measurable duration of time; a type of atom described by its atomic number and atomic weight. Contrast ISOTOPE.

Nudibranchia   An order of carnivorous

marine, gastropod molluscs (often called sea-slugs) that lack shells and have bilateral symmetry in adults. In most species, the body is covered with brilliantly colored respiratory structures such as gill tufts and cerata. They crawl, burrow, and swim; many species have very specialized diets. Approximately 750 species exist. See GASTROPODA.

nuées ardentes   Ash flows from a volcanic eruption consisting of very hot pyroclastic debris, frothy magma, and volcanic gases. The name is French for fiery clouds.

nuisance   A legal term, used in tort law. A broad concept characterizing the defendant's interference of the plaintiff's interest.

null hypothesis   The assumption in statistics that, unless proven otherwise, no relationship exists between two (or more) groups and that any observed differences are due solely to chance.

nullipore   See CORALLINE ALGAE.

nullisomy   A form of aneuploidy in which one chromosome pair is entirely missing from the normal complement in an organism or a cell. See ANEUPLOIDY.

numerical abundance   The number of individuals populating a given area. See ABUNDANCE, RELATIVE ABUNDANCE.

numerical forecast   A weather forecast generated on computer from a model consisting of a number of interrelated equations. Such a model is called a numerical forecast model.

numerical response   Changes in the size (or birthrate) of a population of predators as a result of changes in the density or availability of its prey.

numerical taxonomy   A classification system based on phenetic relationships among groups of organisms. A large number of characteristics are measured for different taxa, and then the taxa are arranged in a phenogram based on the degree of character similarity. See PHENETICS, PHENOGRAM.

nummulite   A large kind of Foraminiferan marine protozoan that forms a coin-shaped test (shell). Nummulite fossils are significant components of Paleocene, Eocene, and Oligocene age rocks formed in warm and shallow marine environments.

nunatak   An isolated mountain summit standing above the surrounding ice sheet of a glacier.

nut   The dry, one-seeded fruit of various trees or shrubs, consisting of a hard, unsplitting shell with a soft kernel, as in acorns and walnuts. Also used just for the kernel within a nut, especially when edible.

nutation   Small deviations in the precession of a spinning object. The wobble in the precession of a spinning top or of the earth are each a nutation. See PRECESSION OF THE EQUINOXES.

nutrient   Anything providing nourishment, especially a mineral element or food compound required for normal functioning of plants or animals.

nutrient budget   A budget drawn up for an ecosystem in terms of nutrient pools (compartments) and transfers (fluxes). The circulation may be determined either for inputs and outputs to the system or for nutrients in circulation within the system, or both. See BIOGEOCHEMICAL CYCLES.

nutrient capital   Total amount of nutrients available for cycling within an ecosystem. See BIOGEOCHEMICAL CYCLES, INTRASYSTEM NUTRIENT CYCLE.

nutrient cycles   See BIOGEOCHEMICAL CYCLES.

nutrient-holding capacity   The ability of soils to retain mineral nutrients and to prevent them from being washed out of the soils. The presence of organic matter (humus) in the soil increases its nutrient-holding capacity. Compare CATION EXCHANGE CAPACITY.

nutrient loading   See CULTURAL EUTROPHICATION.

nutrient stripping   Processing sewage to remove nutrients in order to prevent eutrophication of bodies of water. See CULTURAL EUTROPHICATION.

nutrient-use efficiency   The ratio of nutrients assimilated by a crop or by individual livestock to the amount of nutrients supplied in fertilizers or animal feed.

nyct-   Prefix meaning at night, as in nyctanthous plants, those that flower at night.

nyctinasty   Repetitive raising and lowering of plant parts. Typically, it is a raising of leaves during the day and a lowering or folding of leaves at night; the pattern often continues even after the plants are placed in total darkness. The prayer plants *(Maranata)*, common houseplants, demonstrate nyctinasty. Also called sleep movements. See NASTIC RESPONSE.

Nymphaceae   The waterlily family in the Angiospermophyta. These herbaceous aquatic dicots spread by stout creeping rhizomes and bear conspicuous, floating, solitary flowers on erect stems. Water lilies *(Nymphaea)*, pond lilies *(Nuphar)*, and American lotus *(Nelumbo)* are members of this family, but there are many cultivated members of this family as well. See ANGIOSPERMOPHYTA.

O

O Chemical symbol for the element oxygen. See OXYGEN.

O₂ The chemical symbol for diatomic oxygen—two atoms of oxygen bound together—the form in which oxygen naturally occurs as a gas.

oasis A green vegetated area of isolated desert habitat that is supplied with ground-water throughout the year.

oblate A description applied to a disk-shaped clastic particle that is defined by the relative lengths of axial diameters. An oblate particle has an intermediate axis to long axis ratio greater than 2:3, and a short axis to intermediate axis ratio less than 2:3.

oblateness See OBLATE.

obligate anaerobes Organisms that can only survive in anaerobic conditions. Compare FACULTATIVE ANAEROBES.

obligate halophyte A plant that not only tolerates but requires salty soils in order to thrive, such as salt grass, *Distichlis*.

obligate mutualism A relationship between two coexisting species in which each is dependent on the other in order to survive. See MUTUALISM, PROTOCOOPERATION.

obligate parasite A parasite capable of living naturally only as a parasite, that cannot exist outside of its single host. Contrast FACULTATIVE. See SYMBIOSIS, COEVOLUTION.

obligate predator A predator lives off of only one species of prey. Contrast FACULTATIVE.

obligate relationship A relationship between organisms in which neither can exist without the other, such as obligate mutualism.

oblique slip A description of a fault movement that has a sense of displacement in both the dip-slip and the strike-slip directions along the fault plane.

oblique slip fault A fault characterized by oblique-slip motion.

obliterative shading A form of cryptic coloration in animals in which color gradations provide protection and give a flat appearance.

obsidian A naturally occuring volcanic glass that forms when rhyolitic lava cools too quickly to permit crystal formation. Obsidian is usually black or reddish brown and may be used as an ornamental jewelery stone.

Occam's razor The principle that in scientific evaluations the simplest theory, among alternative theories, that is consistent with the facts should be used. Developed by the 14th-century English philospher, William of Occam, it is also known as the principle or law of parsimony.

occluded front The front resulting from a cold front overtaking a warm front and forcing the warm air upward. Occluded fronts move slowly and are associated with complex weather systems, but usually bring precipitation and sometimes prolonged bad weather. See WARM FRONT, COLD FRONT.

Occupational Safety and Health Act (OSHA) A federal statute (29 U.S.C.A. §651 et seq.) establishing laws and associated regulations for protection of workers in the workplace.

ocean The volume of salt water that covers approximately 71 percent of the earth's surface. The ocean is divided into the geographic areas of Antarctic, Arctic, Atlantic, Indian, and Pacific.

oceanadromous fish Species of fish that both spawn and feed as adults in the open waters of the ocean. Contrast ANADROMOUS, CATADROMOUS.

ocean floor spreading See SEA-FLOOR SPREADING.

oceanic crust The variety of basaltic crust of the earth that is produced at a mid-oceanic ridge. The relatively dense crust, or sima, underlies most of the ocean area of the earth.

oceanic realm A term applied to features or inhabitants of the ocean in areas beyond the continental shelf.

oceanic ridge See MID-OCEANIC RIDGE.

oceanic rise Any elevated broad area of the ocean floor, not part of mid-ocean ridge systems. Compare CONTINENTAL RISE.

oceanic trench See DEEP-SEA TRENCH.

oceanography The science and study of all aspects of the ocean.

ocean shorelines The boundary area between a continental landmass and the ocean.

Ocean Thermal Energy Conversion (OTWC) A system that uses the temperature gradient of the ocean to produce electricity. So far, it has only been used in pilot programs (Hawaii's Mini-OTEC). It must be located in tropical oceans where the surface temperature is very warm. OTEC uses a liquid with a very low boiling point such as ammonia. The warm surface water causes the ammonia to boil, powering a turbine. The vapor is then carried to lower, cooler ocean levels where it condenses. As it is pumped to the surface, it vaporizes to continue the cycle.

ocellus A small, primitive eye or eyespot in insects and other invertebrates. Also, an eye-shaped marking on an animal.

ochre A natural yellow, brown, or red pigment derived from iron oxide that is typically a weathering product of limonite or hematite.

octane A hydrocarbon containing eight carbon atoms (C_8H_{18}). Octane is a liquid and is found in petroleum; it exists as 18 different isomers. The term octane number is used informally for rating gasoline. See OCTANE NUMBER.

octane number A method for rating gasoline that indicates its ability to resist knocking (i.e., to undergo efficient combustion and to burn smoothly). Octane number is the percentage by volume of isooctane (2,2,4-trimethylpentane, C_8H_{18}) in a blend of gasoline containing normal heptane and providing the same amount of knocking as the sample of fuel being rated.

octet rule Atoms usually form bonds by sharing or exchanging electrons with other atoms until their outer shell is filled with eight electrons. The resulting configuration is like that of the noble gases argon, neon, or krypton.

Octopoda An order type of cephalopod molluscs, commonly known as octopuses, and related to cuttlefish and squids. There is virtually no shell to support the rotund, sac-like body that is provided with eight, sucker-bearing arms connected by a web at their base. Most forms are benthic, relatively cryptic and, despite their reputation, essentially harmless to humans. Their advanced nervous systems and learning abilities make them good subjects for laboratory studies. See CEPHALOPODA.

odd-toed ungulates See PERISSODACTYLA.

Odonata A predacious order of insects with powerful mandibles in which the anterior thorax segment is articulated with the head to form a "neck." The paired, membranous wings, which cannot be folded to the sides, are held horizontally (dragonflies) or vertically (damselflies) when at rest. Juveniles (nymphs) of most species are aquatic. The elongate abdomen has 10 segments. See INSECTA, DRAGONFLIES, DAMSELFLIES.

odp See OZONE-DEPLETING POTENTIAL.

OECD See ORGANIZATION FOR ECONOMIC COOPERATION AND DEVELOPMENT.

oestrogens See ESTROGENS.

oestrous cycle See ESTROUS CYCLE.

offlap A distinctive sequence of sedimentary facies in which finer particle layers are found below coarser particle layers. An offlap represents a sedimentary history of marine regression.

offpeak Not occurring during the peak. Offpeak is used in reference to electrical power. Higher rates are sometimes charged during the day when office and home use result in the greatest consumption of electricity; these hours of greatest consumption are called peak hours. The remaining hours when the demand for electricity is less, evening and nighttime, are the hours of offpeak load. Lower rates during these hours encourage consumers to shift optional energy use to these times to reduce the peak load that power plants must supply.

off shore A general reference to the relatively flat ocean floor areas extending from the low water shoreline to the edge of the continental shelf.

ohm (Σ) The unit for electrical resistance in the Système International d'Unités (SI) and meter-kilometer-second (MKS) system. One ohm equals the resistance in a wire in which one volt produces an electrical current of one ampere.

Ohm's law The principle that electric current (amperes) is directly proportional to the supplied voltage and inversely proportional to the resistance (ohms). Ohm's law applies to metallic conductors at constant temperature and zero magnetic field in direct-current electric circuits; also used for alternating current with little resistance, but impedance is substituted for resistance.

oil A naturally occurring fossil hydrocarbon in liquid form. Oil is formed by the anaerobic decay of organic material trapped in sedimentary rocks.

oil pool An accumulation of oil that is naturally contained within subsurface bedrock structures.

oil seep A natural migration of oil to the earth's surface.

oil shale A variety of finely laminated shale that is rich in kerogen.

oil slick The film of oil that floats on the surface of a body of water; oil slicks are the result of oil spills. See OIL SPILL.

oil spill An unintentional release of oil into the environment, especially into a waterway. In waterways, oil spills kill large numbers of plants and animals and can disrupt aquatic ecosystems for long periods of time. Oil spills are typically associated with accidents involving large tankers, but spills associated with the Alaska pipeline have severely damaged tundra.

okta Meteorological term for one-eighth of the visible sky area. The okta is used to describe the total area of cloud cover for weather reports, especially for airports. Also spelled octa.

old-growth forest A mature stand of trees that has not been harvested (at least, not for generations), containing massive trees that are often hundreds of years old. Compare SECOND-GROWTH FOREST, TREE FARM.

old soils See PALEOSOL.

Oligocene The third of five epochs in the Tertiary subera of geologic time. The Oligocene lasted from approximately 38 to 24 million years ago. See APPENDIX p. 610.

Oligochaeta A class of primarily freshwater and terrestrial annelids, characterized by a reduction in size and in number of setae as compared to polychaetes, a simple head, and hermaphroditic reproductive organs. A special organ, the clitellum, secretes a cocoon to hold the eggs. The most well-known members of this group are the earthworms. See ANNELIDA, EARTH-WORMS.

oligoclase A member of plagioclase feldspar mineral series commonly found in igneous rocks and that has a high percentage of silica. Oligoclase has an abundance of sodium cations in the crystal lattice and contains between 10 percent and 30 percent of the mineral albite. See PLAGIO-CLASE.

oligohalabous plankton Describing plankton, especially phytoplankton, that live in water having a salinity of less than 5 percent. See EUHALABOUS, MESOHALABOUS.

oligohaline Describing water (or an aquatic environment) that contains small amounts of salt, less than salt water but more than fresh water (0.5 to 5.0 percent salinity). Also called mesohaline. See FRESHWATER.

oligophagous Subsisting on a diet of relatively few foods.

oligosaprobic Describing water with a high oxygen content in which organic matter is only slowly decomposing.

oligotrophic Not providing much nutrition, especially water containing few if any nutrients. Oligotrophic waters are at the opposite end of the nutrient-organic matter spectrum from eutrophic waters. Compare DYSTROPHIC, EUTROPHICA-TION.

oligotrophic lake A lake, usually with rocky shores and bottom, little vegetation in shallow regions, abundant oxygen but very few nutrients, and thus low primary productivity. It is usually the first stage of the evolution of a lake. Contrast EUTRO-PHICATION, MESOTROPHIC LAKE.

oligotrophic plankton Species of plankton that can survive in nutrient-poor (oligotrophic) regions of oceans or lakes; because such water is relatively barren, light can penetrate to great depths in such areas.

olivine A ferromagnesian silicate mineral formed at high temperatures and composed of individual silica tetrahedra bonded by a mixture of iron and magnesium atoms; is black to greenish with conchoidal fracture and glassy luster; hardness = 6.5 to 7; specific gravity = 3.5 to 4.5. See APPENDIX p. 625.

ombrogenous Obtaining its only water and nourishment from rain, as in some wetlands such as raised bogs. Compare SOLIGENOUS.

ommatidia   The structures resembling simple eyes that make up the compound eyes of insects and crustaceans.

Omnibacteria   A diverse and important phylum of Monera. One class, the Enterobacteria, contain many pathogenic genera, including *Salmonella* and *Shigella*, and also benign forms such as the intestinal bacterium *Escherichia coli*. See MONERA.

omnivore   An animal that feeds in more than one trophic level, for example, potentially eating animal, plant, and fungal material. Compare CARNIVORE, FRUGIVORE, HERBIVORE.

Onagraceae   The evening-primrose family of the Angiospermophyta. These herbaceous dicots have four petals, four sepals, and usually eight stamens in two series. The inferior ovary is frequently elongated and consists of four united carpels. Evening-primroses *(Oenothera)*, fireweed and willow-herbs *(Epilobium)*, fuchsia *(Fuchsia)*, and godetias *(Clarkia)* are members of this family. See ANGIOSPERMOPHYTA.

onchocerciasis   Skin or eye disease caused by infestation with nematodes (small, wormlike organisms of the genus *Onchocerca)* that are found in West Africa. One species causes skin swellings around cysts containing the parasite; another species infects the eye and can cause blindness.

oncosphere   The hook-bearing embryonic stage of tapeworms.

onlap   A distinctive sequence of sedimentary facies in which the coarser particle layers are found below finer particle layers. An onlap represents a sedimentary history of marine transgression.

onshore   A reference to the direction toward the shoreline of a marine environment.

ontogenesis   See ONTOGENY.

ontogeny   The full life cycle, or history of the development of an individual member of a species. Also called ontogenesis. Compare PHYLOGENY.

Onycophora   An animal phylum often generally referred to as Peripatus. The long, cylindrical body bears a varied number of paired fleshy legs whose structure is unique. They are hollow outgrowths from the body, unsegmented and are provided with terminal pads and pairs of claws. The fluid-filled coelom acts as a hydrostaic skeleton supporting the body wall with its sets of circular and longitudinal muscles. The head has paired tentacles and mandibular-like jaws. Onycophorans, which may reach a length of 15 cm, feed on small invertebrates and are found in moist, tropical environments . Once thought to be a missing link between annelids and arthropods, it is now clear that any resemblance they may have to a hypothetical, arthropod ancestor must be superficial. There are 70 species. See ANNELIDA, ARTHROPODA.

onyx   1) A variety of quartz mineral having whitish bands of opal alternating with reddish bands of chalcedony. Onyx is valued as an ornamental jewelry stone. 2) A form of calcite as "Mexican onyx."

oocyte   A female gametophyte, an ovum or egg cell that has not fully developed. Compare SPERMATOCYTE.

oogamy   The union of a large nonmotile egg and a small motile sperm, a form of heterogamy. Compare ISOGAMY.

oogenesis   The formation and development of a mature female reproductive cell, or ovum. Also called ovogenesis. Compare SPERMATOGENESIS.

oogonium 1) The precursor cell in an ovary that splits to create the oocyte. 2) In many algae and some fungal-like organisms (such as powdery mildew), a one-celled structure that produces and contains one or more female gametes (oospheres or eggs). Compare SPERMATOGONIUM.

oolite A variety of limestone composed of ooliths.

ooliths The spherical, sand-sized particles of calcium carbonate that form in shallow marine environments. Oooliths form as concentric rings of calcium carbonate precipitated around a small solid particle.

Oomycota The phylum of the kingdom Protista that includes downy mildews, white rusts, blights, and water molds, organisms with cell walls of cellulose. They feed by extending funguslike hyphae into their hosts. The zoospores produced by oomycotes have a pair of flagella and develop into a new thallus. Sexual reporduction involves gametes produced in specialized structures at the tips of hyphae. The late blight of the potato is caused by *Phytophthora*, damping off is caused by *Pythium*, and the downy mildew of grapes is caused by *Plasmopara*, all members of this phylum. See PROTISTA.

oosphere The nonmotile female reproductive cell (gamete) in plants, which sometimes contains stored food compounds. In lower plants, it may be contained within an oogonium and, when fertilized, becomes an oospore. In higher plants it is contained in an archegonium or embryo sac, and also called an ovum or egg cell.

oospore The zygote, or fertilized female cell, within an oogonium in some algae and fungi. It goes through a resting phase before developing into a sporophyte.

oostegite Slender endites that form the brood pouch in female members of Malocostraca such as shrimps, crabs, and lobsters, krill, and pill bugs. See MALACOSTRACA.

ootid A haploid cell resulting from meiosis of a secondary oocyte.

ooze See PELAGIC SEDIMENT, GLOBIGERINA OOZE, RADIOLARIAN OOZE.

opacity The opposite of transparency; the degree to which a substance blocks the passage of light rays or other radiant energy. Thick smoke has more opacity than light smoke.

opaque Not allowing the passage of electromagnetic radiation. Opaque usually refers to material that blocks light, but can also be used to refer to infrared radiation, x-rays, etc. For example, the earth's atmosphere lets in visible light but is opaque to infrared radiation.

OPEC See ORGANIZATION OF PETROLEUM EXPORTING COUNTRIES.

open burning Burning of garbage or yard waste out in the open, as in a dump or a backyard, instead of in the controlled setting of an incinerator designed to reduce air pollution.

open canopy Describing forest in which the crowns of trees do not often touch each other and in which the crowns shade less than 20 percent of the ground surface. Woodland in the open-canopy stage is not considered sufficiently mature for commercial harvesting. Compare CLOSED CANOPY.

open community A plant community in which the vegetation does not entirely cover the ground, so that in some areas the soil is visible.

open fold Any bedrock fold in which the angle formed between the fold limbs is between 90° and 170°.

open-pit surface mining A technique of mineral extraction in which access is provided by removal of the entire bedrock or overburden above the ore body.

open system Any system that is not contained within boundaries and therefore may have some form of flow (which could be energy or matter) moving into or out of the system. The earth's atmosphere is an open system because it is both a source and a sink for substances such as water and carbon dioxide and because it receives energy from the sun and radiates energy to space.

open woodlands See OPEN CANOPY.

operon A genetic unit for coordinated transcription of DNA code. Operons contain one or more structural genes plus protein coding; these are controlled by adjacent positions (loci) on the gene, an operator site and a promoter site. See REGULATOR GENES.

Ophidia The reptilian suborder that includes snakes. Ranging in length from the Anaconda at 10 m to smaller individuals no more than 10 cm, snakes of the reptilian suborder Ophidia are essentially very specialized, legless squamates in which locomotion is achieved by undulations of the body. Because of their elongated shape, many of the body organs are displaced and the left lung is small or missing. The forked, distensible tongue functions in part as an olfactory organ. With more than 2000 species, snakes have varied habits that include fossorial, arboreal, and aquatic lifestyles. All are carnivorous. Some subdue prey with poisons from venom glands located in the mouth, whereas others crush or suffocate prey by constriction. The prevalence of serpents in folklore and religion may stem from abhorrence of their unusual shape and fear that they may be poisonous. Examples are boas and pythons, typical snakes (colubrids) like garter snakes, coral snakes, cobras and vipers, including the rattle snake and copperhead, etc. See SQUAMATA.

ophiolite A sequence of rock materials consisting from top to bottom of pelagic sediments, basalt pillows and dikes, gabbro, and ultramafic peridotites.

ophiolite suite A series of rocks formed of a regionally metamorphosed ophiolite.

Ophiuroidea See BRITTLE STARS.

opportunistic species Organisms that are adapted for exploiting temporary habitats or conditions. They have a broad niche width, enabling them to quickly colonize openings within an established community. Populations of such species are subject to wide fluctuations. Compare R-STRATEGISTS.

opportunity cost The price of an activity expressed in terms of the cost incurred by sacrificing an alternative activity. A company that continues to produce a particular commodity rather than switching to a commodity that brings in a higher profit may be avoiding risk, but it is also incurring an opportunity cost.

optical activity The degree to which a substance rotates plane-polarized light passing through it. Optical activity is caused by an asymmetrical molecular structure. Substances exhibiting this phenomenon occur in two mirror-image structural forms similar to right and left hands. These molecules are called optical isomers, or enantiomers, in which each enantiomer rotates polarized light in opposite directions. Only one isometric form occurs in most compounds found throughout nature. For example, most naturally occurring amino acids have only *one* enantiomeric structure. Amino acids of the opposite symmetry cannot be used to build proteins and may even be toxic to living systems.

optimal diet model   Mathematical formation of prey selection that would, in theory, provide the best energy return to the predator.

optimal foraging theory   A theory for predicting the food-gathering behavior of predators that will achieve maximum efficiency of intake of food by concentrating foraging in areas where prey is in high density.

optimum population   The population level that produces the maximum sustainable yield, and thus provides the best balance between the health of the ecosystem and the economic reward. See MAXIMUM SUSTAINABLE YIELD.

optimum yield   The greatest yield of a renewable resource which can be achieved over a long time period without jeopardizing the ability of the population or its environment to continue to replace the harvested biomass and to sustain this level of yield. Also called optimum sustainable yield and maximum sustainable yield. See SUSTAINED YIELD.

Opuntiaceae   The cactus family of the Angiospermophyta, sometimes referred to as the Cactaceae. These New World dicots have succulent, green, jointed stems and usually ephemeral, rudimentary leaves. Most members of the family have spines arranged in clusters known as areoles, and minute, barbed bristles called glochids. The flowers have numerous sepals and

petals that intergrade with each other, many stamens arranged in a cyclic fashion. The ovaries in this family are inferior. Prickly pear and cholla (*Opuntia*), barrel cactus (*Echinocactus*) and (*Ferrocactus*), cereus and saguaro (*Cereus*), and fishhook cactus (*Mammillaria*) are all members of this family. See ANGIOSPERMOPHYTA.

OR   Abbreviation for orientation response. See ORIENTATION RESPONSE.

oral contraceptive   Birth-control pill (also called the pill); a combination of synthetic estrogen and progesterone designed to simulate the hormone levels produced during pregnancy thus preventing ovulation and fertilization.

orbit   A circular or elliptical path of one body around another. Orbit has been used to refer to the path of electrons around the nucleus of an atom (in the Bohr model, now usually replaced by the concept of orbitals), as well as the path of a planet around the sun or a satellite around the earth.

orbital   The region within which an electron traveling around an atomic nucleus has a high probability of being located. The orbitals of electrons belonging to the same atom differ in size, in shape, and in spatial orientation. The different atomic orbitals or sublevels, in order of increasing energy, are the s-orbital, p-orbital, d-orbital, and f-orbital. Other than the s-orbital (which contains a single orbital), each of these sublevels contains several orbitals.

Orchidaceae   The orchid family of monocot angiosperms. One of the largest plant families with more than 10,000 species, these perennial herbs are most abundantly represented in the tropics. The flowers have three sepals and three petals, with one of the petals highly differentiated from the others. The stigma and style are fused in a column. The anthers produce sticky pollen bodies called pollinia that are removed en masse by pollinating insects. Orchids produce dust-sized seeds that frequently require symbiotic relationships with fungi early in their life cycles. Lady's-tresses (*Spiranthes*), lady's-slippers (*Cypripedium* and *Paphiopedilum*), and coralroot (*Corallorhiza*) as well as many others are members of this family. See ANGIO-SPERMOPHYTA.

order Name given a group of related families, for example all families of carnivores, Ursidae (bears), Felidae (cats), Canidae (dogs), etc. belong to the order Carnivora; and the Poaceae (grasses) and the Cyperaceae (sedges) belong to the order Poales. See FAMILY, APPENDIX p. 613.

order of magnitude An estimation to the closest order of 10; for example, the numbers between 1 and 10 are a different order of magnitude from the numbers between 100 and 110, or 100 and 1000.

ordination Mathematical method of graphing distribution of communities according to similarities in species composition or by relative position along environmental gradients. Statistical information is gathered from a number of locations with different variables on different axes, so that the most similar communities appear closest together. (From the German *ordenung*, an ordering.)

Ordovician The second geologic time period of the Paleozoic era. The Ordovician period lasted from approximately 505 to 438 million years ago. See APPENDIX p. 611.

ore body A concentration of ore minerals in a rock mass that is distinctly different from the country rock.

ore mineral 1) A mineral that has a metallic luster after polishing. 2) A metalliferous mineral found in an ore body.

organ Any part of an animal or plant with tissues that are adapted and specialized to perform a specific function, such as the stomach or the stamens.

organelle A defined structure within a cell that is specialized to carry out particular functions. Nuclei and mitochondria are organelles.

organic agriculture Growing food crops without the use of synthetic chemicals such as pesticides, herbicides, and fertilizers. Pests are controlled by a variety of techniques including cultivation techniques and timing of planting, biological controls, or use of insecticides derived from natural sources (pyrethrin, rotenone, etc.). Fertilizers are derived from natural sources such as manures, rock phosphate, and composts. Also called organic farming and biological husbandry. Compare BIODYNAMIC FARMING, BIOLOGICAL CONTROL, INTEGRATED PEST MANAGEMENT.

organically bound Held by organic matter. As organic matter decomposes, oxygen atoms are incorporated into the molecular structure of the decomposition products. These oxygen atoms are capable of binding to metal ions such as aluminum or calcium. Organically bound metals are less reactive than free metals.

organic compounds or molecules 1) Compounds or molecules containing carbon bound to hydrogen; these are far more prevalent than inorganic compounds and make up all living matter. 2) Sometimes used to refer to compounds extracted from living organisms rather than being synthesized in the laboratory; as such compounds are synthesized from organisms, they therefore fit within the first definition. See INORGANIC, SYNTHETIC ORGANIC COMPOUNDS.

organic detritus   Relatively small particles of dead and partially decayed plants, animals, and excrement, broken down by the action of microorganisms. Organic detritus usually makes up the greater part of most detritus; weathered minerals make up the rest. See DETRITUS.

organic evolution   See EVOLUTION.

organic fertilizer   Plant fertilizers directly derived from natural sources (usually from living matter), as opposed to inorganic fertilizers made of synthesized chemicals. Organic fertilizers include rock phosphate, greensand, bat guano, kelp, wood ashes, bone meal, blood meal, cottonseed meal, and animal manures.

organic matter   Material derived from decaying organic molecules of natural organisms (the remains of plants and of animals). Organic matter is essential for healthy soil. It may be partially recognizable, as in rough compost or leaves on their way to becoming leaf mold; when fully broken down, organic matter in soils is called humus. See HUMUS.

organic nutrients   Amino acids, carbohydrates, proteins, and other carbon-containing substances that are produced by plants or animals and required by some organisms.

organic phosphate   Another name for an organophosphorous insecticide. See ORGANOPHOSPHOROUS PESTICIDES.

organism   Any unicellular or multicellular living body whose different components work together as a whole to carry out life processes. Animals, plants, fungi, and microbes are all organisms.

Organization for Economic Cooperation and Development   (OECD) An institution whose members are countries that concern themselves with development and economic issues with worldwide implications.

Organization of Petroleum Exporting Countries   (OPEC) An international cartel established in 1960 to regulate the export volume and the price of crude oil on the international market. Its members possess much of the world's known and projected oil supplies.

organochlorine compounds   Another term for chlorinated hydrocarbons. See CHLORINATED HYDROCARBONS.

organogeny   The development of organs from embryonic tissue. Also called organogenesis.

organometallic complex   Compounds in which there is a metal-carbon bond.

organophosphorous pesticides   Any of a class of synthetic insecticides that are very toxic but less harmful to the environment than chlorinated hydrocarbon insecticides (when administered correctly) because they break down relatively quickly into compounds that are not harmful. Organic phosphate insecticides contain sulfur as well as phosphorous and act as nerve poisons; they include malathion and parathion. Also called organophosphates.

organotherapy   The use of animal organs, or extracts of animal organs, to treat disease. Usually, endocrine glands are used; thyroid supplements used to treat hypothyroid conditions are prepared from cleaned and dried animal thyroid glands.

organo-tin paint   Paint containing tin, in the form of organo-tin compounds such as tributyl tin, to make it toxic. It is used on ships' hulls to protect them by discouraging barnacles. Organo-tin compounds are extremely toxic to marine life and are an increasingly serious problem of marines and dry docks where the paint is stripped and often ends up in the water. See ANTIFOULING PAINT.

Oriental region   One of the biogeographical realms into which the earth is divided according to the distribution of animal life (fauna). The Oriental region is made up of southern Asia (east of the Persian Gulf), Indonesia, the Indian subcontinent south of the Himalayan Mountains, southern China, and Malaysia. See ZOOGEOGRAPHY, WALLACE'S LINE.

orientation response   (OR) A change in position in an organism, or a part of an organism, in reaction to a stimulus. See HELIOTHERMISM, NASTIC RESPONSE, TAXIS.

original horizontality, principle of   See LAW OF ORIGINAL HORIZONTALITY.

ornithology   The branch of zoology that studies birds.

ornithophily   Pollination of plants by birds (such as hummingbirds). Compare ANEMOPHILY, ENTOMOPHILY.

orogenesis   Any process of mountain building.

orogenic belt   An area of mountain building characterized by regional forces of lateral compression, as in a tectonic convergence.

orogeny   A mountain-building event. Orogeny frequently refers to regional mountain building as a result of tectonic convergence.

orographic   1) Of or about mountains, as in orographic lifting or orographic clouds. 2) Concerning orography, the branch of geology devoted to the study of mountains.

orographic effect   The increased rainfall on windward mountain slopes and the lack of rainfall on the leeward side of mountain ranges caused when orographic lifting causes an air mass to cool and thus to lose most of its moisture as it moves across mountain ranges. This effect causes the deserts to the east of the Sierra Nevada range. See OROGRAPHIC LIFTING.

orographic lifting   The rising of an air mass caused by mountains or high elevations blocking the flow of air and displacing it upward. Orographic lifting cools the air mass and thus may cause its moisture to condense into clouds or precipitation.

ortho-   A prefix used in describing organic compounds containing a benzene ring. It indicates the presence of two substituents in the benzene ring that are on adjacent carbons of the ring. It is usually italicized, as in *ortho*-dichlorbenzene. The same information can also be conveyed using numbers to indicate the position of substituents, as in 1,2-dichlorbenzene. See PARA-.

orthoclase   A variety of the mineral potassium feldspar that forms pink or white monoclinic crystals in granitic igneous rocks. See APPENDIX p. 626.

orthogenesis   An early theory of evolution proposing that evolution of new species is directed by inherent (possibly genetic) tendencies, not external, environmental influences such as natural selection.

orthograde   Walking with an upright or vertically oriented body.

orthohombic   A description of a crystal shape in which there are three mutually perpendicular axes of unequal lengths.

Orthoptera   Insect order which includes

grasshoppers and locusts, katydids, crickets, roaches, mantids, etc. There is considerable disagreement over the group's classification, some authorities preferring to place distinct types, such as roaches, into separate orders. Basically, the order as recognized here is relatively unspecialized in general body form. The chewing mouth parts are well developed and antennae tend to be long and slender. Wings may be present, vestigial, or missing. In winged forms, the anterior pair of wings is often leathery and can be folded down over the membranous hindwings. In some forms, the forewings are modified to resemble surrounding vegetation such as leaves, and the hind limbs may be specialized for jumping. Stridulating sounds are produced in some groups by rubbing the forewings or the forewings and parts of the hindlimbs together. Many roaches and some large grasshopper species (locusts) are both ecologically and economically of great importance. More than 25,000 species exist. See INSECTA, LOCUSTS.

orthoquartzite   A sedimentary sandstone formed by quartz cementation of sand particles. Compare QUARTZITE.

oscillation   A regular, repeating vibration in which a quantity varies from one limit of its range to another. Alternating (electrical) current is a classic example of oscillation.

oscillation ripples   A type of symmetric ripple mark formed in sand by the oscillating action of waves in a shallow marine or lake environment.

OSHA   See OCCUPATIONAL SAFETY AND HEALTH ACT.

osmoconformer   An aquatic organism whose body fluids contain the same concentration of dissolved ions (salts) as that of the surrounding water; a change in the saltiness of its aquatic environment causes a direct change in the concentration of salts in its body fluids because the organism is not capable of osmoregulation. Compare OSMOREGULATOR.

osmoreceptor   Cells that are specialized to detect changes in osmotic concentrations and to initiate corrective changes to maintain homeostasis. Osmoreceptors in the human hypothalamus regulate the amount of water in the blood to balance overall osmotic pressure. See OSMOREGULATION.

osmoregulation   The balancing of overall salt concentration in living tissues (osmotic levels), often by adjusting water levels for dilution, in order to maintain the best conditions for cellular activity. See OSMORECEPTOR, OSMOREGULATOR.

osmoregulator   An aquatic organism that is able to maintain a specific concentration of dissolved ions in its body fluids, even when the osmotic concentration of the water in which it lives fluctuates; an organism capable of osmoregulation. Compare OSMOCONFORMER.

osmosis   The process in which a solvent diffuses through a semipermeable membrane (a membrane that allows the solvent to pass through, but not the larger solute molecules), moving toward a solution having a greater concentration. Osmosis progresses in the direction of equalizing the concentration on both sides of the membrane. Plants contain relatively high concentrations of dissolved salts, and for that reason the roots take up water from the outside. During osmosis, fresh water moves through the root cell walls to mix with the higher concentrated water on the inside. Compare REVERSE OSMOSIS.

osmotic pressure   Pressure that counteracts the direction of flow of osmosis. When sufficient pressure is applied to one side of a semipermeable membrane, it prevents osmosis even if the concentration is much higher on that side of the membrane.

Osmunda   A genus of ferns characterized by having naked sori (clusters of sporangia).

Osteichthyes   This class of vertebrates contains the bony fish, whose skeletons are formed partially or entirely of bone. It is one of the most widespread and successful of all vertebrate groups, whose habitats range from hot springs to polar waters and from swampy shores to ocean depths. They are divided into two important subclasses: fleshy-finned types such as the lungfishes, and the ray-finned group, which includes the vast majority of bony fishes. Over 18,000 species exist. See VERTEBRATA, DIPNOI, TELEOSTS.

osteology   The study of the structure and the function of bones.

osteophony   The conduction of sound waves to the inner ear via the facial bones and the skull.

osteoporosis   A decrease in the density of bones causing brittleness and increased breakage. It is commonly associated with long-term calcium deficiency and lack of weight-bearing exercise, most often appearing in old age.

Ostracoda   Class of small crustacea with bivalved carapaces that superficially resemble those of bivalved molluscs. They have only seven pairs of appendages and the body is unsegmented. Distributed worldwide in fresh and marine waters, they are principally burrowing or planktonic forms ranging in size from near microscopic to about 30 mm. Ecologically, they are an important food source for fish and for invertebrates. See CRUSTACEA.

otolith   A calcareous or siliceous concretion in the otocyst of aquatic invertebrates or within the inner ear of vertebrates. The otolith is part of the sensory apparatus for equilibrium and for hearing.

outbreak   A sudden appearance of numerous cases of a specific disease, or a sudden population explosion of a pest species. See SECONDARY PEST OUTBREAK.

outbreeding   The opposite of inbreeding; sexual reproduction (or intentional crossing) between organisms that are not closely related. Outbreeding increases the incidence of heterozygous pairs of genes. See CROSS-BREEDING, INBREEDING.

outcrop   Any part of the bedrock that is exposed at the earth's surface.

outcrossing   The breeding of individuals usually from the same breed but from strains having a different genotype.

outer core   See CORE OF THE EARTH.

outfall   A point at which a natural stream or artificial drainage discharges into a body of water.

outgassing   The release of gases from molten rock. In the early stages of the Earth's formation, outgassing helped form the atmosphere.

outlet   A point at which a body of water, such as a lake, discharges into a stream.

outlier   1) Erosional: outcrop of newer rocks surrounded by older, the result of its separation by erosion from the main mass of which it forms a detached portion. 2) Structural: outcrop of newer rocks let down between faults or in a syncline which has survived when adjacent portions of the same rocks have been removed by denudation. 3) Data that lies outside major tendencies of a distribution.

output   The amount or quantity of matter, energy, or information produced by a system. Examples of output include: the force or work done by a mechanical system, such as a machine; the current or power produced by an electronic system, such as an amplifier or an electric generation plant; and the computer-generated information sent to a printer, file, or screen. Compare INPUT.

outwash   The water-transported material carried away from the ablation zone of a melting glacier.

outwash deposit   An accumulation of well-sorted and layered glacial alluvium deposited by debris-laden meltwater streams. Contrast GLACIAL TILL.

outwash plain   A landform composed of glacial outwash, generally forming a broad and flat apron of well-sorted gravel, sand, and silt.

outwelling   The nutrient enrichment of coastal waters from an estuarine source.

ova   Plural of ovum. See OVUM.

ovarian follicle   Another term for Graafian follicle. See FOLLICLE.

overburden   1) Any unconsolidated material above the bedrock. 2) The mass of rock and soil above a specified point.

overcompensating density dependence   See DENSITY OVERCOMPENSATION.

overexploitation   Utilizing renewable resources at a level greater than the system can sustain, causing damage to the ecosystems involved and reducing yields either in the short term or over the long term. Overexploitation of fishing stocks can greatly reduce available fish because the populations are reduced to the point that they can no longer reproduce fast enough to maintain current populations. See OVERHARVESTING, OVERHUNTING.

overgraze   To allow too many animals to graze in one area, causing a gradual change in the vegetation to less desirable species, or causing increased erosion so that portions of the pasture or the rangeland become bare. Although overgrazing may provide short-term economic gain, in the long term it produces land that supports fewer animals or even destroys the land for grazing, greatly reducing the potential for economic gain.

overgrowth competition   Competition among plants or sessile animals involving growth of one individual into or over another, reducing available light (as in taller plants shading lower plants), suspended food, or some other resource.

overharvesting   Exceeding the maximum sustainable yield of renewable natural resources; removing more of a crop than the system can continue to produce. Overharvesting of timber can destroy the land (especially steep slopes) so that it no longer produces economically profitable quantities of lumber. See MAXIMUM SUSTAINABLE YIELD, OVER-EXPLOITATION.

overhunting   A form of overexploitation in which hunters are allowed to take so many individuals from a given population of a game species, such as wild turkeys or deer, that reproduction is hampered, causing reduced populations that may die out if overhunting is allowed to continue.

overland flow   The component of meteoric water that runs off the land surface when the infiltration capacity is exceeded. Compare INFILTRATION.

oversaturated island   An island having

more species than would be predicted by the equilibrium theory of island biogeography. See EQUILIBRIUM THEORY OF ISLAND BIOGEOGRAPHY. Contrast UNDERSATURATED ISLAND.

overshoot   The increase of a population beyond the carrying capacity of its environment.

overshot wheel   The classic waterwheel design of old mills. The water comes in at the top and flows over (overshoots) the wheel, landing in a bucket to pull the wheel. The wheel is thus driven by gravity rather than the force of the current. It depends on having a drop in elevation between the incoming and outgoing water. When such a drop does not exist (low-head hydropower), a less-efficient undershot wheel driven by water current must be used.

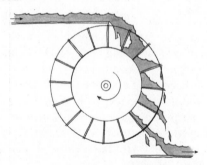

overshot waterwheel

overstory   Uppermost layer of vegetation in a forest, formed by the leaves and the branches of the highest trees. The overstory contributes to the total canopy. Compare UNDERSTORY.

overthrust fault   A type of thrust fault in which a relatively large horizontal displacement is due to the active movement of the hanging wall block over a passive footwall block.

overturn   See VERNAL OVERTURN.

overturn (aquatic)   The mixing of stratified layers in an aquatic environment. Overturn most often refers to the spring- and fall-density current circulation in stratified lakes.

overturned bed   A term applied to rock strata that have been tilted beyond perpendicular so that the top and bottom layers are reversed.

overturned fold   A term applied to a fold whose axial plane has been tilted beyond perpendicular so that one limb is in an upside-down orientation.

overwood   See OVERSTORY.

oviparous   Egg-laying; producing eggs that develop after leaving the body. Birds and amphibians are oviparous, as are most reptiles, fish, and insects. Compare OVOVIVIPAROUS, VIVIPAROUS.

ovisac   A structure containing an egg (ovum) or a brood-pouch.

ovogenesis   See OOGENESIS.

ovotestis   A reproductive organ that produces both male and female gametes. The gonad (genital gland) of slugs, snails, and some other gastropods is an ovotestis.

ovoviviparous   Producing eggs that have distinct shells but hatch inside the body of the mother. Some reptiles, fish, and invertebrates are ovoviviparous. Compare OVIPAROUS, VIVIPAROUS.

ovulate   To produce eggs (ova); in mammals, to discharge eggs from the ovary.

ovulation   The process of ovulating, producing eggs or discharging them from the ovary.

ovule   The structure in seed plants that is contained inside the ovary and, after fertilization, grows into a seed. The ovule is made up of embryo sac, nucellus, and integuments.

ovum An egg or an egg cell; the female reproductive cell (gamete) produced in the ovary of animals, also the oosphere of plants. Plural is ova. See OOSPHERE.

oxbow lake A lake that forms in the abandoned channel of a cutoff meander on a river floodplain.

oxbows A series of well-developed meanders on a river floodplain.

oxic horizon The characteristic B horizon layer of an oxisol. The oxic horizon is rich in kaolinite clay and almost completely lacking in weatherable minerals.

oxidant An oxidizing agent; substance that causes oxidation. Oxidants undergo reduction in the process of causing oxidation of another substance. See OXIDATION.

oxidase Any of a group of plant and animal enzymes that catalyze biological oxidation reactions. Cytochrome oxidase is a particular form of a cytochrome plant pigment that catalyzes part of a chain of reactions that transfer electrons to produce energy for individual cells. See OXIDATION.

oxidation Any chemical reaction that involves atoms or molecules losing electrons. The oxidation of one compound is always coupled with the reduction of another compound, and so the process is sometimes called oxidation-reduction. The term originally described a reaction in which oxygen combined chemically with another substance, but now includes many reactions that do not involve oxygen, such as dehydrogenation. See REDUCTION.

oxidation pond A structure for purifying sewage, consisting of a pond in which

bacterial action is allowed to slowly oxidize (break down) the effluent.

oxidative phosphorylation The chemical reaction producing ATP (adenosine triphosphate) to store energy for cellular metabolism, carried out by the mitochondria of cells during the final stages of respiration. In the process, coenzymes donate electrons (are oxidized) and ADP (adenosine diphosphate) is phosphorylated to produce ATP. See ADENOSINE DIPHOSPHATE, ADENOSINE TRIPHOSPHATE.

oxide 1) A chemical compound that contains oxygen plus one other element. Common rust is iron oxide. 2) The oxide ion, O^{-2}.

oxidize To bring about oxidation. See OXIDATION.

oxidizing agent Any substance that is capable of removing electrons from another substance. An oxidizing agent carries out the process of oxidation and is itself reduced (i.e., it *gains* electrons). See OXIDATION, REDUCTION.

oxisol Any member of a soil order characterized by highly weathered, thick, poorly expressed horizons with few remaining weatherable mineral constituents, and containing a high proportion of iron and aluminum oxide clays.

oxygen (O) Gaseous element essential for all aerobic respiration; it has atomic weight 16.00 and atomic number 8. In its elemental form, it exists as the diatomic molecule O_2. On earth, oxygen is the most abundant of the elements, comprising a fifth of the atmosphere as O_2, almost 90 percent of the oceans as H_2O, and almost 50 percent of the earth's crust as metal oxides and silicates.

oxygen cycle The circulation of oxygen

through organisms and ecosystems and its continuing reutilization throughout the biosphere. Oxygen is taken up from the atmosphere and hydrosphere by the oxidation of iron and other elements and by organisms through metabolic processes where it combines with carbon and is then released in carbon dioxide. Since oxygen is a constituent of water, the oxygen cycle also includes the hydrological cycle (water cycle). Other important aspects of the oxygen cycle are the photolysis (dissociation caused by light) of water in the upper atmosphere and in photosynthesis, the production of carbon monoxide and carbon dioxide by volcanic activity, and the formation-destruction subcycle of ozone (O_3) in the atmosphere. See BIO-GEOCHEMICAL CYCLES.

oxygen-demanding wastes Effluent that requires oxygen for its decomposition. Raw sewage requires oxygen to support the bacteria that decomposes it. Effluent containing inorganic compounds that are readily oxidized can also be termed oxygen-demanding wastes because these too deplete oxygen from the water bodies into which they are discharged. See BIOLOGICAL OXYGEN DEMAND, CHEMICAL OXYGEN DEMAND.

oxygen dissociation curve A graph showing the extent to which hemoglobin molecules in the blood are saturated with oxygen, based upon the partial pressure of oxygen. It thus shows the relative proportions of oxyhemoglobin and reduced hemoglobin. See HEMOGLOBIN, OXYHE-MOGLOBIN.

oxygen minimum layer A layer of water in which the dissolved oxygen content is lower than in the adjacent layers of water above and below.

oxygen sag A decline in oxygen content

that develops downstream from a pollution source, releasing effluent with a high biological oxygen demand (such as raw sewage). See BIOLOGICAL OXYGEN DEMAND.

oxygen sag curve A graph plotting the concentration of dissolved oxygen in a body of water against the distance from a polluting source. It shows a sharp decline in levels of dissolved oxygen at the point where the pollution discharge enters the water. See OXYGEN SAG, OXYGEN-DEMANDING WASTES.

oxygen sink An element or compound that chemically combines easily with oxygen and so removes it from the atmosphere. See OXYGEN SAG.

oxygen technique Method for measuring primary productivity based on the equivalence between oxygen and food produced by autotrophs. One common form of oxygen technique is the light-and-dark-bottle method, in which samples of a body of water are placed in bottles, with one darkened to prevent photosynthesis, for 24 hours. The increase in oxygen in the light bottle (net production) plus the amount of decrease in oxygen in the dark bottle (respiration) equals the gross primary productivity.

oxyhemoglobin Hemoglobin that contains oxygen, the compound that gives arterial blood its bright red color and transports oxygen to all the tissues in the body. Oxyhemoglobin easily releases its oxygen when it is surrounded by low concentrations of oxygen. See HEMOGLOBIN.

oxyphobe A plant that cannot live in acid soil. Calcicoles (calciphiles), calcium-requiring plants, are one form of oxyphobe because calcium soils are alkaline rather than acid. Contrast OXYPHYTE.

oxyphyte   A plant that can live on acid soil. See CALCIFUGE.

oxytocin   A pituitary hormone that stimulates contraction of the uterus during childbirth; it also stimulates the production of milk. Also called oxytocic principle. It is sometimes administered to women during childbirth if they are not producing enough on their own.

Oyashio Current   A western boundary surface Current flowing southwest from the Bering Sea in the northern Pacific Ocean. The Oyashio Current converges with the Kuroshio Current east of northern Japan.

ozone O_3, a compound that is formed when oxygen gas is exposed to ultraviolet radiation. In the outer atmosphere (stratosphere) ozone acts to shield the earth from excessive radiation. In the lower atmosphere (troposphere), however, it forms from combustion gases and is a major air pollutant contributing to photochemical smog. Ozone has commercial uses as a bleaching agent and water purifier. See OZONE SHIELD, PHOTOCHEMICAL SMOG.

ozone-depleting potential   (ODP) An evaluation of how much of a negative effect a substance such as a chlorofluorocarbon has on the stratospheric ozone shield. See OZONE SHIELD.

ozone layer   The common name for the earth's ozonosphere. See OZONOSPHERE.

ozone shield   Another name for the earth's ozonosphere. See OZONOSPHERE.

ozonosphere   A layer of the earth's atmosphere containing gaseous ozone (O_3). It overlaps the lower levels of the mesosphere and the upper stratosphere. Because this layer absorbs the sun's harmful ultraviolet radiation and thus shields the earth, it is also called the ozone shield or ozone layer.

p

P Chemical symbol for the element phosphorous. See PHOSPHORUS.

P₁ Symbol for parental generation. See PARENTAL GENERATION.

Pa Chemical symbol for the element protactinium. See PROTACTINIUM.

Pacific North American floral region One of the phytogeographic regions into which the Holarctic realm is divided according to similarities between plants; it runs from Alaska to Mexico, west of the Rocky Mountains. See HOLARCTIC REALM, PHYTOGEOGRAPHY.

Pacific South American floral region One of the phytogeographic regions into which the Neotropical realm is divided according to similarities between plants; it consists of parts of Colombia, Ecuador, Peru, and northern Chile. See NEO-TROPICAL REALM, PHYTOGEOGRAPHY.

package sewage treatment plant Small-scale, self-contained facility for treating wastewater, used for sites not connected to municipal sewers. Such facilities are used where septic systems are not feasible (as on poorly draining soils) and may be installed for small housing subdivisions or shopping centers. Package sewage treatment plants are problematic if not properly maintained.

packed column A vessel filled with material (often broken or randomly sized pieces of an inert substance) and used either for absorption or for distillation.

packed tower Air pollution control device consisting of a column filled with crushed rock, wood chips, or a similar substance to slow the flow of air. As air is forced upward through the column, liquid is sprayed downward on the solid material. The air pollutants are removed by chemically reacting with the liquid (or dissolving in it). The slower speed of the air flow allows more time for the chemical reaction to proceed, so that more of the pollutants are removed.

PAH See POLYCYCLIC AROMATIC HYDRO-CARBONS.

pahoehoe A subaerial flow structure of fluid basalt forming a complexly folded and extruded, ropy lava with a lustrous and finely lineated upper surface. Contrast AA.

Palaearctic region One of the biogeographical realms into which the earth is divided according to the distribution of animal life (fauna). The Palaearctic region encompasses Europe, northern Asia, and northern Africa. The Palaearctic region is sometimes included, along with the Nearctic region, in a larger Holarctic region. See ZOOGEOGRAPHY.

paleo-, palaeo- Prefix meaning ancient, from the Greek *palaios*.

Palaeognathae The superorder of large, flightless birds with reduced wings and strong legs. The ostrich of Africa and the rheas of South America have soft, plume-like feathers, whereas those of the cassowaries and the emus of Australia and neighboring islands are coarse and hair-like. See AVES, RATITES.

Palaeotropic realm See PALEOTROPICAL REALM.

paleobiology A study of the past environments and biological occurrences by examination of the fossil record, sedimentary rocks, and biological facts and principles. See PALEOBOTANY, PALEONTOLOGY.

paleobotany A branch of botany that studies fossil plants. See PALEONTOLOGY.

Paleocene The first of five epochs in the Tertiary subera of geologic time. The Paleocene lasted from approximately 65 to 54.9 million years ago. See APPENDIX p. 610.

paleoclimatology The science devoted to studying the climates of prehistoric eras such as the ice ages, using geological evidence.

paleoecology Prehistoric ecology; the study of the relationships between prehistoric living things and between fossilized organisms and their environments through analysis of the fossil record.

paleoendemic Having a distribution restricted to a particular area (endemic) in ancient times, as shown by the fossil record. See ENDEMIC.

paleoflora The fossilized plant life (flora) found in a particular geological formation or from a given geological period.

paleogenesis See PALINGENESIS.

paleogeography The study of continent positions and locations of major land and water bodies in prehistoric times, and their associated biota.

paleontology The study of fossilized flora and fauna as represented in the rock record.

paleosol An older soil that is buried beneath a younger material, often serving as a useful indicator of past conditions of erosion, weathering, and climate.

Paleotropical realm One of the floristic (phytogeographical) realms into which the globe has been divided by plant geographers. The Paleotropical realm includes the ancient tropical regions, and now includes most of Africa and South-East Asia; it is divided into 14 floral regions. See FLORAL REALM, PHYTOGEOGRAPHY.

Paleozoic era The first of three eras of the Phanerozoic eon in geologic time. The Paleozoic lasted from approximately 590 to 248 million years ago. The term Paleozoic means early life. See APPENDIX p. 611.

palingenesis 1) The emergence of ancestral characteristics during ontogeny, the developmental stages of an organism, or as abnormalities in an individual. Also called atavism or palingenesis. 2) The melting of existing rocks deep within the earth's crust to form new magma. See ONTOGENY.

palisade mesophyll A layer of vertically elongated parenchyma cells just below the upper epidermis of leaves. Palisade mesophyll is specialized for conducting photosynthesis and so contains a rich supply of chloroplasts. Also called palisade layer. See PARENCHYMA, SPONGY MESOPHYLL.

pallial line A line on the shell of bivalve molluscs (mussels, clams, oysters, scallops) showing the attachment of mantle muscles to the shell. See BIVALVIA.

Palmaceae The palm family. See ARECACEAE.

palp One of the jointed feelers, organs of touch or taste, attached to the mouths of insects and crustaceans. Also called palpus.

paludification The rise in the water table and increase in peat accumulation that results in the expansion of a bog.

paludism See MALARIA.

palynology The study of fossil plant spores and the pollen of bogs and lakes. Pollen and spores found in sediments are often the only fossils sufficiently intact to use for correlating fossil layers with geological time.

pampas A form of grassland vegetation found in Argentina and to a lesser extent elsewhere in southern areas of South America, consisting of a vast, grassy, treeless plain.

pan- Prefix meaning all or throughout. Pan-African distribution occurs in all of Africa.

PAN See PEROXYACETYL NITRATES.

panclimax Two (or more) related climax communities with similar life forms, species, or dominants that exist under similar climatic conditions. See CLIMAX.

pandemic Describing a disease that expands beyond local epidemics and that spreads over a huge area such as an entire continent, or occurs as epidemics in many different areas. The bubonic plague (black death) of medieval Europe was pandemic. Compare ENDEMIC, EPIDEMIC.

panemone A windmill composed of concave or flat surfaces that radiate from and revolve around a vertical axis. A Savonius rotor is a form of panemone. See SAVONIUS ROTOR.

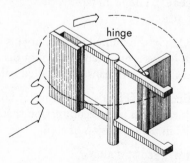

A simple panemone-type windmill.

Pangaea A postulated supercontinent that persisted from the late Permian period through the early Triassic period. Pangaea consisted of all of the earth's major continental crust in one amalgamation.

panicle A loose, branching flower cluster in which several flowers are borne on small stems leading out from the main stem; technically, a branched raceme in which each branch produces a raceme of flowers, as in oats.

panmixis Random crossing within a specific population.

Panthalassa The one universal ocean that hypothetically surrounded the ancient supercontinent Pangaea.

pantomictic plankton Plankton in which no one species predominates. Compare MONOTON PLANKTON.

pantothenic acid $HO-CH_2-C(CH_3)_2-CH(OH)-CONH(CH_2)_2-COOH$, a water-soluble, B-complex vitamin. It is found in all living cells and is a component of coenzyme A, essential for the oxidation of carbohydrates and fats. Pantothenic acid is readily available in many foods.

pantropical Found throughout the tropical regions of the world, as in describing the distribution of a taxonomic group found in South America, Africa, Asia, and Australia.

Papaveraceae The poppy family in the angiosperms. These dicots have four or six petals and numerous stamens. The buds are often nodding, and the sepals are usually shed as the flower bud expands. The fruits of members of this family are usually capsules filled with small seeds. Poppies (*Papaver*), prickly poppies (*Argemone*), and celandine (*Chelidonium*) are all members of this family. See ANGIOSPERMOPHYTA.

para- A prefix used in describing organic compounds containing a benzene ring. It indicates a group bonded to, or substituting for, two carbon atoms located opposite each other in the ring. It is usually italicized, as in *para*-dichlorobenzene. The same information can also be written using the prefix 1,4-. See ORTHO-.

parabolic dune A crescent-shaped sand dune in which the horns of the arc are pointed upwind. Contrast BARCHAN DUNE.

parabolic mirror A concave mirror with the shape of a parabola rotated around its axis. Parabolic mirrors focus incoming parallel light into a point. They are used in reflecting telescopes because they focus light from outer space into clear images. They are also used to concentrate the rays of the sun in solar cookers or collectors.

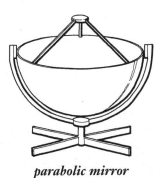

parabolic mirror

paraconformity A type of disconformity in which little or no erosion is evident in marking the surface of unconformity. A paraconformity may closely resemble a bedding plane. See DISCONFORMITY, UNCONFORMITY.

paradox of the plankton The paradox (proposed by G. E. Hutchinson) that aquatic systems appearing be homogeneous habitats support a large number of plankton species in apparent violation of the competitive exclusion principle. The explanation for the paradox is that aquatic systems actually vary sufficiently in time and space that no single plankton species has the competitive advantage long enough for competitive exclusion to take place. See COMPETITIVE EXCLUSION PRINCIPLE.

paraffin oil Another name for kerosene, especially in British usage. See KEROSENE.

paralic A term applied to sedimentary deposits on the continental side of the water edge at a coastline.

parallel drainage pattern A term describing appearance in map view of stream drainage systems in which the streams flow parallel to one another over a large portion of the drainage area. Parallel drainage commonly develops over folded sedimentary rocks.

parallel evolution The independent appearance of similar traits or similar patterns of evolution in groups of related organisms that had become separated geographically or genetically at an earlier stage in their evolution. Also called parallelism or parallel descent. Compare CONVERGENT EVOLUTION, DIVERGENT EVOLUTION. Contrast ADAPTIVE RADIATION.

parallel of latitude One of the imaginary lines on the earth's surface parallel to (coplanar with) the equator. Contrast MERIDIAN.

parallel plate capacitor A device for storing electrical charge. It consists of two thin conducting electrodes separated by an insulator. Its storage capacity is measured in coulombs per volt.

parallel strata Any two or more rock layers in which the thickness of layers remains constant so that the bounding planes of each layer remain parallel. Parallel strata are usually composed of undeformed sedimentary rocks.

parameter   An attribute, a variable, or a physical property in a set of such variables or properties that together characterize or determine the behavior of a system.

paramo   A high, bleak or barren plain in the Andes Mountains in South America and its associated vegetation.

paramorph   1) A taxonomic variation within a species, used as a general term when the exact nature of the subgroup has yet to be determined. 2) A mineral that has the same chemical formula as another, but whose atomic or molecular structure has changed without altering its chemical composition. Aragonite and calcite are paramorphs.

Parano-Amazonian floral region   One of the phytogeographic regions into which the Neotropical realm is divided according to similarities between plants; it consists of Brazil and Bolivia. See NEOTROPICAL REALM, PHYTOGEOGRAPHY.

parapatric   Describing adjacent species, those whose ranges overlap slightly but not by much more than the dispersal range of one individual within its lifetime. Compare SYMPATRIC, ALLOPATRIC.

parapatric speciation   The separation of a group into different, spatially adjacent species, while limited exchange of genetic material still occurs.

parapyle   The two lateral pore fields in the central capsule of Radiolarians (marine plankton).

paraquat   A very poisonous, broad-spectrum, contact herbicide. It is used in some forms of no-till agriculture to kill the cover crop, with the next crop seeded directly into the residue of the first crop. Paraquat only works on foliage, so even when it is still present in the soil, it does not hamper the germination of the second crop. See NO-TILL AGRICULTURE.

parasematic coloration   Markings designed to distract or to mislead predators, as in flush coloration and the eye-spots on the wingtips of butterflies. See APOSEMATIC COLORATION.

parasite   An organism that lives on or in another organism (host) and derives its food from or at the expense of the host. The relationship is usually beneficial to the parasite and harmful for the host. See AUTOECIOUS PARASITE, ECTOPARASITE, ENDOPARASITE, HEMIPARASITE, HETERO-ECIOUS PARASITE, HOLOPARASITE, HYPERPARASITE, NECROPARASITE, SYMBIOSIS.

parasitism   The condition in which one organism lives as a parasite on another organism (the host), deriving benefits, while the second organism is harmed. Compare COMMENSALISM, SYMBIOSIS.

parasitoid   Animal (usually an insect) that is a parasite at only one stage during its development, usually a larval stage, and thus usually a free-living adult. Ichneumon wasps, used for biological pest control, are parasitoids; they lay eggs in moth and butterfly larvae and eventually destroy the host organisms as the eggs develop.

parasympathetic nervous system   A subdivision of the autonomic nervous system in vertebrates that stimulates involuntary responses by releasing acetylcholine. Its effects are specific and counteract those of the sympathetic nervous systems. Actions controlled by stimulation of the parasympathetic nervous system include increasing gastrointestinal activity, contracting the pupils of the eyes, slowing the heartbeat, and a general lowering of blood pressure. Also called the craniosacral system. See AUTONOMIC NERVOUS SYSTEM, SYMPATHETIC NERVOUS SYSTEM.

parathion $NO_2C_6H_6OP(S)(OC_2H_5)$, a highly toxic organophosphorous insecticide. It is no longer available for use in the United States. It is also known as methyl viologen and by the names Gamoxane, Weedol, and Dextrone X. See ORGANO-PHOSPHOROUS PESTICIDES.

paratonic Describing plant movements in response to external stimuli. Paratonic movement includes tropic responses and taxes. See TAXIS.

paratrophical Parasitic; feeding by parasitism. See PARASITE.

parcel A unit of timber for commercial purposes, either standing timber, cut logs, or milled lumber.

parenchyma 1) The basic tissue type of higher plants, composing most of the inner tissue of leaves, stem pith, and fruit pulp; mature, unspecialized cells that may, with proper stimuli, become somewhat specialized and develop into other types of cells. 2) In animals, the tissue making up the functional parts of an organ, rather than the supporting or surrounding tissues (stroma).

parental generation Organisms selected for crossbreeding. Crossing the parental (P_1) generation yields the F_1 generation. See F_1 generation.

parental investment The resources individual organisms devote to their offspring, such as the amount of energy provided in the cotyledons of individual seeds of plants or the amount of prenatal and neonatal care provided by animals.

parent material The underlying bedrock or unconsolidated deposits from which soil is derived.

parhelia Another term for mock suns or sun dogs. Singular is parhelion. See SUN DOGS.

parmal pores The major pores that pierce the shell, surrounding each radial spine in some species of radiolaria (marine plankton).

parr 1) A stage in the young salmon's development before it enters salt water. 2) The young of other species of fish.

parthenocarpy The formation and growth of fruit without fertilization and thus without viable seeds. Parthenocarpy may occur spontaneously but it can also be induced artificially with plant growth substances.

parthenogenesis Reproduction by an unfertilized gamete, usually an egg, forming a normal member of the species. Aphids can reproduce this way, and dandelions sometimes spontaneously undergo parthenogenesis. Compare APOMIXIS.

partial pressure The pressure exerted by one of the component gases in a mixture of two or more gases; for example, the partial pressure of oxygen in the atmosphere.

particle Extremely small bit of matter with finite mass but no appreciable volume. Used under different circumstances for the particles consitituting particulate matter, molecules and atoms, and elementary or subatomic particles.

particulate Having the qualities of a particle. When plural, particulates is short for particulate matter. See PARTICULATE MATTER.

particulate matter   A category of air pollutants that refers to small, solid particles or liquid droplets suspended in air. Such particulates include soot, fumes, dust, pollen and spores, smoke, spray, and even fog.

particulate organic matter   Ocean particulates, small pieces of material, that result from the partial decomposition of dead organisms.

parts per billion (ppb)   Amount of a substance, often a pollutant, in every billion (10^9) parts of another substance such as air, water, or soil.

parts per million (ppm)   Amount of a substance, often a pollutant, in every million (10^6) parts of another substance such as air, water, or soil.

parts per trillion (ppt)   Amount of a substance, often a pollutant, in every trillion (10^{12}) parts of another substance such as air, water, or soil.

pascal   (Pa) The unit for pressure or for stress in the Système International d'Unités (SI) or meter-kilogram-second (MKS) system. One pascal equals one newton per square meter.

passenger pigeon   Although they are frequently hunted for food, many doves and pigeons still exist in reasonable numbers. The passenger pigeon *Ectopistes migratorious* of North America, however, whose vast numbers during migration filled the sky as late as the 19th century, was largely exterminated by hunting coupled with natural disasters and is now extinct. See COLUMBIFORMES.

Passeriformes   With more than 5000 species, the passerine order contains more than half of all living species of birds and is one of the most successful and rapidly evolving orders. The feet are specialized for perching, with three toes forward and a strong, hind toe. None of its members is very large, the raven being the largest, and all are altricial (having the young hatched in an immature condition). Among the most common or well-known groups are the crows or corvids, thrushes, finches, manakins, tyrant-flycatchers, ovenbirds, larks, swallows, orioles, drongos, wrens, shrikes, old world warblers, starlings, sunbirds, vireos, honycreepers, wood warblers, tanagers, weaverbirds, and mockingbirds. See AVES, PASSERINES, CORVIDAE.

passerines   A common name for the numerous and widespread perching birds belonging to the order Passeriformes. See PASSERIFORMES.

passive dispersal   Movement of seeds, spores, or dispersing stages of animals caused by external agents such as winds or water.

passive heat absorption   Using nonmoving structures or materials such as stone floors and walls to gather and to hold heat. This is the original, oldest form of solar heating.

passive solar heating system   A system for providing heat directly from the sun's rays. It uses large expanses of windows as collectors, natural (not forced) convection for circulation, and thermal mass for storage. It does not use pumps, blowers, or pipes filled with liquids to transfer heat. Contrast ACTIVE SOLAR HEATING SYSTEM.

pasteurization   The process used to kill harmful bacteria in milk by heating it to a specified temperature for a specified amount of time. Sometimes used for other liquids; in this country, beer may be pasteurized to kill unwanted yeasts or to stop fermentation.

pasture   Land covered with grasses, mixed with other small plants, that is managed for grazing.

patacole   An animal that is found temporarily in the litter of the forest floor. Compare PATOXENE.

patch dynamics   The idea that communities are a mosaic of different areas (patches) within which nonbiological disturbances (climate, etc.) and biological interactions proceed.

patchiness   Localized diversity of conditions within an ecosystem (terrestrial or aquatic); the existence of smaller areas that differ in one or more physical conditions. For example, an ecosystem might have different microclimates, and therefore support somewhat different species or relative abundances of species, causing aggregations rather than an even distribution.

patch reef   Any isolated mound that grows in a lagoon behind a barrier reef or near an atoll; they may range from small pillars to several-kilometer expanses. Small patch reefs are also called coral heads or coral knolls.

patchwork clearcutting   A method of timber harvesting in which smaller areas of trees are cleared away within larger areas of timber that are left standing. Contrast CLEARCUTTING, SELECTIVE HARVEST.

patchy habitat   An environment within which there are significant variations in size or in quality of habitats suitable for a particular species.

paternoster lakes   A linear sequence of rock basin lakes and streams formed within a glacial valley. The name alludes to the map view resemblance of paternoster lakes to a string of rosary beads.

pathogen   Any microorganism or other agent that can cause disease, including pathogenic bacteria, viruses, worms, or protozoans.

patina   A thin layer of weathering by-products on the surface of a rock. A patina may have a distinctive light color which aids in identification.

patoxene   An animal that occurs only accidentally in the litter of the forest floor. Compare PATACOLE.

patterned ground   The surface appearance of soils showing irregular bands and polygons of rock debris. The pattern formation is associated with ice wedging within the permafrost followed by the infilling of the cracks with sediment. See ICE-WEDGE POLYGON.

pattern intensity   The extent to which a community shows patterning in the distribution of its component species. See PATCHINESS.

Pb   Chemical symbol for the element lead. See LEAD.

PBB   See POLYBROMINATED BIPHENYLS.

P/B ratio   The ratio of annual net primary productivity to total biomass in a biotic community. It is the inverse of the biomass accumulation ratio. See PRIMARY PRODUCTIVITY.

PCBs   See POLYCHLORINATED BIPHENYLS.

peak sound pressure level   Maximum recorded sound pressure expressed in decibels. This value is used in noise pollution abatement ordinances.

pearly   A particular description of the physical mineral property of luster when the surface resembles the sheen of a pearl.

peat   A dense layer (or layers) of water-saturated, dead, and partially decomposed organic matter, usually largely sphagnum moss, that forms in acidic bogs. Cut and dried for fuel in some parts of the world, peat can convert to coal over extremely long periods of time under the right geologic conditions. Peat moss is also harvested for soil mixes at a rate much faster than the ecosystem can replace it.

peaty   A description of a soil that is largely composed of organic material in the form of fibrous plant remains.

pebble   Any clastic rock fragment between 2 and 64 mm in average diameter.

pecking order   See DOMINANCE HIERARCHY.

pectin   A group of water-soluble, colloidal polysaccharides found in cell walls of vascular plants. Acid solutions of pectins form gels in the presence of sufficient quantities of sugar; pectins found in fruits or added to fruit syrups are responsible for the jelling of jams and jellies.

ped   The smallest natural unit of soil classification. Each ped consists of an aggregate of soil particles.

pedalfer   A broad category of soils formed in humid, middle-latitude regions where leaching is a dominant process; characterized by accumulation of silicate clays and iron or aluminum sequioxides in the B horizon. Term derives from Latin words for soil (*pedon*), aluminum, and iron (*ferrum*). Contrast PEDOCAL.

pediment   An erosional surface of low relief, often covered with a veneer of gravel, forming at the foot of a mountain range.

pediplane   An erosion surface of low relief formed by pediment erosion processes in a piedmont area. A coalescence of pediments.

pedocal   A broad category of soils formed in drier, middle-latitude regions where leaching is not a dominant process; characterized by an accumulation of calcium carbonate in the upper horizons. Term derives from Latin word for soil (*pedon*) and the first three letters of calcite.

pedogenesis   The formation of soil, usually involving interaction of parent material, time, biologic activity, climate, and landscape.

pedogenic regimes   The diagnostic categories of pedogenesis characterized by combinations of parent material, climate, topography, biologic activity, and time.

pedology   The general study of characteristics, taxonomy, and origin of soils.

pedon   The smallest unit of soil classification that represents all of the typical horizons and structures of a soil profile.

peduncle   1) The main flower-bearing stem of a plant, usually supporting more than one flower. 2) A stalklike structure in animals.

pegmatite   A light-colored (felsic), coarsely crystalline igneous rock formed in the late stages of magma crystallization. A pegmatite is an extreme example of a granitelike rock with phaneritic texture.

pelagic   Living in or relating to the open sea, especially surface waters to the middle depths. Krill and the whales that feed upon them are examples of pelagic animals.

pelagic ecosystem   The ecosystem of the open ocean.

pelagic sediment   The material that accumulates on the sea floor beneath a pelagic, or open water, environment. The predominant constituents of pelagic sediment are microscopic skeletal remains of foraminifera, diatoms, and radiolaria.

pelean volcano   An extremely energetic eruption of a composite volcano, beginning with the emergence of a viscous lava dome near the vent and culminating with the explosive production of nuées ardentes. See NUÉES ARDENTES.

Pelecanidae   See PELICANS.

Pelecaniiformes   An order of relatively long-billed, fish-eating birds that have the distinctive characteristic of having all four toes joined by webs (e.g., tropic and frigate birds, pelicans, boobies, gannets). Approximately 55 species exist. See AVES, PELICANS.

pelicans   Large lake and shorebirds belonging to a family of the order Pelecaniformes. With wingspans of from six to nine feet, they are strong and graceful flyers, although they may have difficulty becoming airborne. The pelican is perhaps best known for its distensible throat pouch, which is used as a scoop for catching fish and as a cooling surface. See PELICANIIFORMES.

pelitic   A description applied to sedimentary rocks composed of clay.

pellagra   Deficiency disease caused by a lack of niacin or an inability to absorb it (as from some gastrointestinal problems or alcoholism). It often occurs with diets lacking tryptophan, an essential amino acid not present in all proteins. Pellagra is characterized by lethargy, dermatitis, diarrhea, dementia, and in extreme cases, death. See NIACIN, DEFICIENCY DISEASE.

pellets   1) Fuel for nuclear reactions. Pellets are used to describe both the small pieces of uranium oxide filling long fuel rods in nuclear fission, and the fuel used in inertial-confinement nuclear fusion. The fusion pellets are less than a millimeter in diameter and consist of deuterium and tritium contained within spheres of glass, plastic, or a similar material. Also called fuel pellets. 2) Wood processed into small, uniform pieces for use in a pellet stove, a very clean-burning form of woodstove used to heat homes.

Peltier effect   The heating or cooling that occurs when electrical current passes through a junction of two unlike metals or semiconductors. The Peltier effect has been used for small-scale refrigeration. The heat produced or absorbed is sometimes called Peltier heat. Contrast SEEBECK EFFECT.

Pelton wheel   A simple impulse turbine that uses a nozzle to increase the pressure of the incoming water. The water shoots out of the nozzle, hits a divided bucket, and is deflected into a double curve. The Pelton wheel is driven by the velocity of the water rather than its height. With very little water, but with high speed and pressure, the Pelton wheel is able to run efficiently.

A Pelton wheel has buckets on the center shaft.

peneplain A hypothetical erosion surface of low relief and highly developed drainage representing the final stages in landscape erosion. See CYCLE OF EROSION.

penguins See SPHENISCIFORMES.

penicillin Any of a large group of natural and synthetic antibiotic compounds used to treat diseases caused by some strains of bacteria, some molds, spirochetes, and rickettsias. Penicillin is produced by several species of molds belonging to the genus *Penicillium*.

peninsula A projecting landmass almost surrounded by water.

peninsular 1) Of or relating to land that is almost completely surrounded by water, such as the Florida peninsula. 2) Describing any entity (such as a population) that is similarly isolated on most sides and only connected by a narrow passageway.

pennate Winged; having feathers. Compare PINNATE.

Pennsylvanian The sixth geologic time period of the Paleozoic era. The Pennsylvanian period lasted from approximately 320 to 286 million years ago. See CARBONIFEROUS, also APPENDIX p. 611.

pentad Five days, a period used in meteorological record-keeping and forecasting because, unlike a week, it is an even subdivision of the 365-day year.

pentadactyl limb The five-digit appendage characteristic of mammals, birds, reptiles, and amphibians. The human arm ending in a hand with five fingers is the classic pentadactyl limb. In other organisms it is often modified so that it can only be seen in the skeleton, or with some of the digits fused so that only two or three remain.

pentlandite The major ore of nickel in the form of iron nickel sulfide; usually forms cubic crystals in mafic igneous rocks.

pentoses A group of simple sugars (monosaccharides) containing five carbon atoms. Pentoses (deoxyribose and ribose) are constituents of DNA and RNA.

peptidase See PROTEASE.

peptides Biological compounds containing two or more linked amino acids. Proteins are made up of peptides; peptides are also formed by the digestion of proteins. See POLYPETIDE.

perceived noise level The amount of random noise that humans consider equal in noisiness to a desired sound (as opposed to a sound level measured on a sound pressure level meter). It is measured in units of perceived noise decibels (PNdB). See DECIBEL, NOY.

percentage area method A technique for determining the density and the distribution of a species by measuring the percentage of area covered (and the area of bare ground) in a series of quadrats (marked plots). See COVER, QUADRAT.

perched water body An unconfined aquifer separated from the underlying water table by an impermeable layer or unsaturated zone.

perching birds Common name for the order Passeriformes, often called passerine birds. See PASSERIFORMES.

perchloric acid $HClO_4$, a strong acid and strong oxidizing agent. It is a colorless liquid that is unstable and highly explosive in concentrated form. One of its numerous uses includes in analysis of animal or plant tissue for the presence of heavy metals, as it effectively decomposes the organic compounds to free up the heavy metals.

percolate See PERCOLATION.

percolating filter   A device that separates solid material from a liquid by percolation through a semipermeable medium, such as a clay layer.

percolation   The gravity flow of water or liquid down through a medium such as soil, rock, or filter material, under conditions of near saturation.

percolation test   A means of determining the capacity of a soil to accept percolating water. The test is used to determine the suitability of soils for engineering needs, such as domestic sewage waste disposal systems.

perdominant   See DOMINANCE.

perennating   Persisting from year to year. Perennating organs are buds and dormant resting organs (such as roots and tubers) that enable a plant to survive a dormant period. See RAUNKIAER'S LIFE FORMS.

perennial   A plant whose life cycle lasts longer than two years. Although the tops of herbaceous perennials become senescent (die down) at the end of the growing season, buds, roots, and underground portions persist. Woody perennials have above-ground tissues that persist over a dormant season. Compare ANNUAL, BIENNIAL.

perfect flower   An individual flower that contains functioning stamens (male organs) and pistils (female organs). Also called a hermaphroditic, monoclinous, or bisexual flower. See FLOWER.

perfoliate   Describing a stalkless leaf whose base completely encircles the stem, giving the appearance that the stem is growing through the leaf. See APPENDIX p. 628–629.

perianth   The structures surrounding the sexual organs in a flower. The perianth is made up of the calyx (usually protective green, leaflike sepals) and the corolla (inner whorl of colored petals); the term is often used to describe flowers such as the tulip whose calyx and corolla are difficult to distinguish because they all look like petals. See FLOWER.

pericarp   The tissue (fruit wall) surrounding the ripened ovary or fruit of a flowering plant, which may be differentiated into three layers, the exocarp, the mesocarp, and the endocarp. See ENDOCARP, EPICARP, MESOCARP.

pericline   A geologic structure in which the collective dips of rock strata are arranged in a pattern radiating from a central point. A dome or a basin is an example of a pericline. See DIP.

peridotite   A very dark (mafic) igneous rock having a phaneritic texture and composed primarily of the mineral olivine. Peridotite usually forms at the bottom of magma chambers and is uncommon near the surface of the earth.

perigee   The point on the orbit of the moon or a satellite where it reaches its closest distance to earth; the opposite of apogee.

periglacial   A term referring to an environment in close proximity to a glacier, or more generally applied to any environment in which freeze-and-thaw cycles are a dominant surface process.

perihelion   The point on the orbit of a planet, comet, or other sun-orbiting body where it reaches its closest distance to the sun. Compare PERIGEE.

period   1) The second-order unit of the geologic time scale; a period is shorter than an era and longer than an epoch. See APPENDIX p. 610. 2) The time required for one complete cycle of a vibration, which is the time it takes for the wave to return to a similar phase or point. Usually called simply period, this quantity is the reciprocal of the frequency of a wave. Compare FREQUENCY, AMPLITUDE, WAVELENGTH.

periodic Recurring at regular intervals; repeating in a cycle.

periodic plankton Organisms that are regularly found in the plankton only at certain times, such as at dusk. See MIGRATION.

periodic table of elements An arrangement of all chemical elements in order by increasing atomic number. The elements are placed in rows and in columns so that elements with similar chemical and physical properties are located in the same column. See APPENDIX p. 608.

peripheral nervous system (PNS) All of the nerves outside the central nervous system; in vertebrates, all nerves (including the autonomic nervous system) outside of the brain and spinal cord. Compare CENTRAL NERVOUS SYSTEM.

periphyton A complex community of plants and animals that cling to objects at the bottom of bodies of fresh water, such as stems of rooted plants and rocks. Periphyton includes microscopic algae, insect larvae, and small crustaceans. Also called aufwuchs. See AUFWUCHS.

Perissodactyla A heterogeneous order of odd-toed ungulates which includes tapirs, rhinoceroses, and horses. Their feeding mechanisms are less specialized than those of the even-toed ungulates. A highly specialized foot structure, in which the third digit is the principal weight-bearing member, is common to members of the order although all do not have an odd number of digits as the ordinal name

implies. All 16 species are endangered by humans, especially the rhinos which, although heavily protected, have been poached to near extinction for their horns. See EUTHERIA.

permaculture See MIXED PERENNIAL POLYCULTURE.

permafrost An area of perennially frozen subsoil. Permafrost is generally associated with arctic regions.

permafrost table The upper limit of the permafrost in a soil. Compare WATER TABLE.

permanence theory A popular hypothesis of the 19th and early 20th century that suggested that the continents have been fixed in position over geologic time. The permanence theory has been replaced by the plate tectonic theory.

permanent plot Standard technique for studying communities and ecological succession using a marked area of land (plot) that is set aside exclusively for study, and repeatedly sampled over various periods of time.

permanent retrievable storage A method for dealing with radioactive waste and other hazardous materials in which they are stored so that they can be periodically inspected, removed for repackaging, or retrieved for an alternative storage method. Salt mines and stable caverns, as well as secured buildings, have been considered as sites for permanent retrievable storage.

permanent wilting point The condition at which water content in a soil becomes so low that plants are unable to recover from the damaging effects of wilting.

permeability The degree to which a porous substance allows the passage of a liquid or a gas. The permeability of a membrane is its ability to allow the passage of compounds in solution.

permeable   Allowing fluids such as water to pass through. Contrast IMPERMEABLE.

Permian   The seventh and youngest geologic time period of the Paleozoic era. The Permian period lasted from approximately 286 to 248 million years ago. In the European system this is the sixth period. See CARBONIFEROUS, also APPENDIX p. 611.

permineralization   A type of fossilization in which mineral materials are deposited within the pore spaces of the preserved tissue.

perovskite   A yellowish-to-black mineral of calcium and titanium oxide.

peroxisomes   Organelles found in the cytoplasm of most nucleated cells (common in leaf cells) and containing oxidizing enzymes. Peroxisomes protect the cell from perioxides, converting them to oxygen (O_2) and water.

peroxyacetyl nitrates   A group of air-polluting chemical compounds found in photochemical smog; they are produced by the action of sunlight on oxides of nitrogen, hydrocarbons, and ozone. Peroxyacetyl nitrates are usually known by their acronym, PAN. PAN irritate the eyes, nose, and throat; they are toxic to plants. Sometimes called peroxyacyl nitrates.

persistent pesticides   Chemical compounds used for pest control that do not readily break down once released into the environment. They become more or less permanent features of the ecosystem, working their way up the food chain to reach high concentrations in the tissues of higher predators. Chlorinated hydrocarbons such as DDT are classic examples of persistent pesticides.

Peru Current   A cold ocean current flowing northward along the western coast of South America. The nutrient-rich upwelling water of the Peru Current supports an abundance of marine life.

pest   An animal or plant that is directly or indirectly detrimental to human interests, causing harm or reducing the quality and value of a harvestable crop or other resource. Weeds, termites, and rats are examples of pests.

pesticide   A substance used to kill or to control harmful or destructive organisms. Insecticides, herbicides, germicides, fungicides, and rodenticides are all pesticides.

pesticide tolerance   A legal limit established by federal agencies for food and animal feed specifying permissible levels for residues of pesticides.

PET   See POLYETHYLENE TEREPHTHALATE, POTENTIAL EVAPOTRANSPIRATION.

petal   One of the colored parts of a flower corolla, often used loosely for any colored flower structure, including the calyx. See COROLLA.

petiole   1) A leaf stalk, the slender and usually short stem that attaches a leaf blade to a larger stem, a twig, or a branch. See APPENDIX p. 629. 2) Another term for an animal peduncle, especially a slender portion of the abdomen connecting the thorax to the abdomen in some insects.

petrifaction   A type of fossilization in which the organic matter is replaced by minerals deposited from an aqueous solution. Involves a conversion to stone.

petrified rock   Any rock that is formed by

petrifaction. So-called petrified wood is a petrified rock.

petrocalcic horizon   A soil hardpan layer cemented by calcium carbonate or magnesium carbonate.

petrochemical   Any compound derived from natural gas or crude oil. Also used as an adjective to refer to industries or processes related to the production of such compounds.

petrogenesis   The process of rock formation, especially as applied to igneous rocks.

petrographic microscope   A microscope that has specialized features, such as a rotating stage with polarized light, to aid in the study of thin sections of rocks.

petrol   British term for gasoline. See GASOLINE.

petroleum   Crude mixtures of hydrocarbon oils, occurring naturally in the earth's crust as a result of long-term geological action on decaying organic matter from eons ago. Unrefined petroleum often contains minerals such as sulfur and vanadium in addition to hydrocarbons. When coexisting natural gas and water are removed from petroleum, it becomes crude oil and is used for the production of many fuels and chemicals. See CRACKING, FRACTION.

petrological microscope   See PETRO-GRAPHIC MICROSCOPE.

petrology   The comprehensive study of rocks, including rock origins, field relations, mineralogic composition, and texture.

Petromyzoniformes   See LAMPREYS.

pfbc   See PRESSURIZED FLUIDIZED-BED COMBUSTION.

pH   A measure of the relative concentration of hydrogen ions in a solution; this value indicates the acidity or alkalinity of the solution. It is calculated as the negative logarithm to the base ten of the hydrogen ion concentration in moles per liter. A pH value of 7 indicates a neutral solution; pH values greater than 7 are basic, and those below 7 are acidic. Vinegar has a pH of 3; ocean water has a pH value of approximately 8. See PH PAPER, ACID-BASE INDICATOR.

phacolith   A type of lens-shaped concordant igneous rock intrusion that follows the curving structure of folded strata.

Phaeophyta   The brown algae phylum of the Protista kingdom. These are mostly brown seaweeds of coastal and marine environments. Brown algae produce flagellated, motile sperm that fertilize egg cells, thereby giving rise to a sporophyte, diploid thallus. The sporophyte may also produce zoospores that develop directly into multicellular individuals. Typically brown algae have a rootlike hold fast, a stemlike stipe, and a leaflike blade, although they lack the differentiated vascular system of members of the kingdom Plantae. Members of the Phaeophyta include the giant kelp *Macrocyctis*, kelps *Laminaria*, kottted wrack *Ascophyllum*, Sargasso gulfweed *Sargassum*, and rockweed *Fucus*. See PROTISTA.

phaeoplankton   Plankton that reside in the upper regions of bodies of water, within the top 30 meters.

phages   See BACTERIOPHAGE.

phagocyte   A cell that can engulf and destroy waste or other harmful material, such as bacteria, protozoa, dead cells, and cell debris. Phagocytes are found in the blood and body tissues. See MACROPHAGE.

phagocytosis   The process by which a

phagocyte ingests and digests waste or harmful bacteria. Phagocytosis is similar to the way in which amoebae ingest food: A protrusion of the cell enables the particle to be surrounded. Compare PINOCYTOSIS.

phagotrophic Describing the form of heterotrophic feeding in which cells ingest solid particles of food. See HETEROTROPH.

phalanges Bones in the fingers or the toes of vertebrates. Singular is phalanx.

phanerites A general term for crystalline igneous rocks with a crystal texture coarse enough to be viewed without magnification.

phanerophyte A woody plant whose overwintering or dormant (perennating) buds are located more than 25 cm above the soil level. The term comes from the Raunkiaer life form classification system. See RAUNKIAER'S LIFE FORMS.

Phanerozoic eon The geologic time division spanning from approximately 590 million years ago to the present. The term Phanerozoic means visible life. The Phanerozoic is subdivided into the Paleozoic, Mesozoic, and Cenozoic eras. See APPENDIX p. 610.

phase 1) One of the states in which matter can exist depending on temperature and pressure: solid, liquid, gas, or plasma. 2) The varying shape of the brightly illuminated portion of a planet or moon. 3) A means of comparing different electromagnetic waves, usually of the same frequency (as in out of phase or in phase with each other).

phase transition A change in physical state of a substance, as in the melting of a solid to become a liquid.

phellem See CORK.

phellogen See CORK CAMBIUM.

phenetics A system of classification of organisms based on comparing and weighting resemblances in as many observable traits as possible among the organisms to be classified. If phenetic groupings resemble evolutionary groupings, it is through coincidence. Compare PHYLOGENETIC SYSTEMATICS, CLADISTICS.

phenocopy An organism that superficially resembles one having a different genotype, especially one having an abnormality produced by its environment but not passed on genetically.

phenocryst The term for the larger crystals in an igneous rock with porphyritic texture.

phenogram A tree diagram that illustrates the degree of similarity in character states of groups of taxa. The branches of the diagram are arranged by the degree of similarity and characteristics, regardless of the date at which the characteristics were derived. Phenograms are used in phenetic representations of evolution. See PHENETICS, NUMERICAL TAXONOMY.

phenology The study of the relationship between climate and periodic natural phenomena such as migration of birds, bud bursting, or flowering of plants.

phenolphthalein alkalinity The acid neutralizing capacity of a water sample as measured by the amount of acid to lower the pH of the sample to the equivalence point of penphthalein, about pH 8.3. See ACID-BASE INDICATOR.

phenols A group of aromatic alcohol hydrocarbons that are derived from benzene and have a hydroxyl group (OH^-) attached directly to the benzene ring.

Unlike ordinary alcohols, phenols are acidic; the simplest member of the group, carbolic acid (C_6H_5OH), is sometimes known as phenol. Phenols are used as disinfectants. They are also by-products of many industries and produce unpleasant odors and taste when present in drinking water (large quantities poison aquatic organisms).

phenotype   The appearance or observable characteristics of an organism, produced by the interaction of its genetic heritage (genotype) with its environment. Contrast PHENOTYPE.

phenotypic plasticity   The variation in the phenotypic expression of a given genotype; a measure of the amount of variation in the observable aspects of a quantitative genetic character among individuals. See PHENO-TYPE.

phenylalanine $C_6H_5CH_2CH(NH_2)COOH$, an amino acid with an aromatic (carbon-ring) structure that makes up many plant and animal proteins. It is considered an essential amino acid because it must be supplied in human diets.

pheromone   A substance secreted externally by some animal species, especially insects, that influences the behavior of other members of the same species. Pheromones are so concentrated that small quantities can be detected by the organisms at great distances. Functions include attracting the opposite sex, marking trails, and communicating within colonies of social insects. They also include direct physiological effects on the recipients, such as the "queen substance" secreted by queen bees that inhibits the development of other potential queens in the same hive. Humans use pheromones in integrated pest management to lure insects into traps, or to cause disorientation. See SEX ATTRAC-

TANT, CONFUSION TECHNIQUE. Compare ALLOMONE, HORMONE.

phi (grade) scale   A logarithmic scale used to classify the diameter of rock particles. The ϕ value is expressed as the negative logarithm to the base 2 of the particle size measured in millimeters.

philoprogenitive   Prolific; producing large numbers of offspring.

phloem   Conducting tissue in vascular plants that carries sugars and other manu-factured food from their source of produc-tion (usually leaves) to stems and roots that do not photosynthesize but require sugars for metabolic processes. See SIEVE TUBE, XYLEM.

phon   A unit for loudness of sounds. The phon measures increments of loudness that sound equal to the human ear; these intervals are logarithmic, which distinguish them from the simple-interval unit sone. To determine the phons for a tone, its intensity is compared to a reference tone of one-kilohertz frequency and the decibel level is matched to the original tone. Compare SONE, DECIBEL.

phonolite   A porphyritic extrusive igneous rock composed primarily of alkali feldspars with smaller amounts of nephaline, pyroxene, and amphibole. Phonolite is formed in continental and oceanic rift zones.

phoresy   The transportation of one species by another, without involving any form of parasitism. Some mites accomplish dis-persal by attaching themselves to the backs of certain insect species.

phosgene $COCl_2$, a very poisonous gas, also known as carbonyl chloride. It is used as a war gas, causing severe lung irritation. Phosgene is also used as a chlorinating agent in chemical processes such as the manufacturing of dyes.

phosphagen   A substance capable of phosphorylating (adding a phosphate group to) ADP, acting as a reserve of ATP and thus an energy reserve in animal cells.

phosphatase   Any of a group of enzymes that catalyze the splitting (hydrolysis) of the high-energy phosphate bonds in some compounds. Phosphatases are essential in carbohydrate and phospholipid metabolism, and in the formation of nucleotides. See CREATINE PHOSPHOKINASE.

phosphate   1) The phosphate ion, PO_4^{-3}. 2) A salt (or ester) of phosphoric acid (H_3PO_4). Phosphates are essential nutrients for plant and animal metabolism; some forms of rock and decaying organic matter provide sources. Because phosphorous is often the limiting nutrient in an ecosystem, when large quantities of phosphates are released into water bodies, they cause a boom in algae populations that deplete the available oxygen supply and contribute to eutrophication. See SUPERPHOSPHATE, EUTROPHICATION.

phosphate rock   A rock composed primarily of inorganic calcium phosphate minerals.

phosphatidic acids   A group of compounds important in the synthesis of phospholipids and triglycerides. They have the general formula $R_2C(O)$-O-CH-(CH_2OPO_3H) $CH_2OC(O)R_1$. See PHOSPHOLIPID.

phosphocreatine   A compound of creatine and phosphoric acid that occurs in muscle tissue. Also called creatine phosphate, it is one of the primary energy compounds of the phosphagen group. Its high-energy phosphate bonds (ester bonds) are split (by the enzyme creatine phosphokinase) when

energy is needed; during resting periods the compounds are reformed. See CREATINE PHOSPHOKINASE.

phospholipid   Any of a group of lipids containing one or more fatty acids plus a phosphate group. Phospholipids are important structural components of cellular membranes. Lecithin is a common phospholipid. Also called phosphatide.

phosphorescence   1) Giving off light without producing noticeable heat, as in the light given off by organisms such as fireflies. 2) Delayed luminescence, after excitation by visible, ultraviolet, or x-ray radiation. See FLUORESCENCE.

phosphoric acid H_3PO_4, an acid produced from mined phosphate rock. It has many commercial applications: It is used in fertilizers and in detergents and by the food industry. Also called orthophosphoric acid.

phosphorite   A sedimentary rock composed of carbonate hydroxyl fluorapatite that is precipitated in nodules within a shallow marine environment.

phosphorus   (P) $R_2C(O)$-O-$CH(CH_2OP)$-$(CH_2OC(=O)R_1)$, A nonmetallic element with atomic weight 30.97 and atomic number 15. It exists in a number of different forms; one form of pure phosphorus is poisonous and burns skin on contact. Phosphorus is an essential nutrient for plants and animals; it is part of DNA molecules. It is used in matches to make them flare up quickly with the heat of friction.

phosphorus cycle   The process by which phosphorus moves within and between ecosystems and throughout the biosphere. Phosphorus is released in the soil by the weathering of minerals. It is taken up from the soil by plants and passed along to organisms feeding upon the plants. It is an essential component of nucleic acids, bones, teeth, and cellular energy compounds such as ATP and ADP. It is released back to the soil through wastes excreted by animals as well as through decomposition of plants and animals. See PHOSPHORUS, BIOGEOCHEMICAL CYCLES.

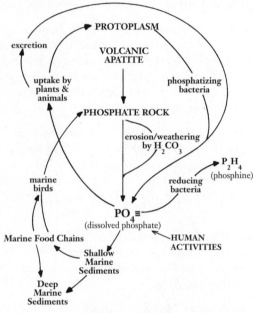

phosphorous cycle

photic Concerning or relating to light, especially as a stimulus to organisms.

photic zone The surface layer of the ocean that receives sufficient light for photosynthesis to occur. The depth of this layer varies depending on the turbidity of the water. Compare APHOTIC ZONE, EUPHOTIC ZONE.

photoautotroph See PHOTOTROPHIC.

photobiology The branch of biology that studies light (or radiant energy) as it affects organisms and biological processes. See PHOTOCHEMISTRY.

photochemical oxidants A major class of toxic secondary air pollutants that are especially damaging to plants. Photochemical oxidants are chemical compounds resulting from the action of sunlight on chemicals such as oxides of nitrogen and hydrocarbons; they are the main components of photochemical smog. Ozone is a photochemical oxidant. See PHOTOCHEMICAL SMOG, OZONE.

photochemical reactions Chemical reactions in which the necessary energy comes from light, usually sunlight. Photooxidation and photosynthesis are examples of reactions that use light as their energy source.

photochemical smog Air pollution in the form of a brown haze often seen over Los Angeles, CA, and other cities. It occurs on sunny days in areas with large volumes of automobile traffic. Photochemical smog is produced when sunlight acts on nitrogen oxides (NO_x), ozone, and hydrocarbons in a series of complex reactions. Photochemical smog is a respiratory irritant and can kill or alter plant tissues.

photochemistry The study of chemical reactions caused by visible and ultraviolet light (sunlight).

photoconverter A device, such as a photoelectric cell, that changes the energy from light into electricity.

photodegradable plastics Special plastics engineered to break down in the presence of sunlight (ordinary plastics are very resistant to weathering and decomposition). Although photodegradable plastics have promise for reducing litter, they will not help reduce the volume of solid waste disposed of in landfills because they must be buried (and thus shielded from light) when disposed of in sanitary landfills.

photodissociation   A chemical process in which light breaks a chemical bond. The breakdown of ozone (O_3) into oxygen (O_2) results from photodissociation.

photoelectric cell   A device that alters (or produces) a flow of electrical current in response to the amount of light falling on it. Photoelectric cells can control automatic door-opening mechanisms; photovoltaic cells used in solar energy are a form of photoelectric cell.

photokinesis   Movement or activity stimulated by light.

photon   A quantum (discrete unit) of electromagnetic energy. A photon is a unit of light behaving more like a particle than a wave; it is considered an elementary particle. See QUANTUM MECHANICS.

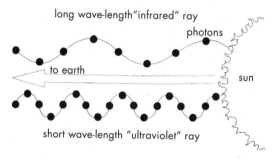

The sun gives off rays of different wavelenths. The shorter the wavelength, the more frequently solar energy units called photons strike the earth.

photonasty   A reversible or permanent movement of a plant in response to changes in light. For example, flowers such as bloodroot and hepatica that open with the sun. See NASTIC RESPONSE.

photooxidant   See PHOTOCHEMICAL OXIDANTS.

photooxidation   Short for photochemical oxidation, chemical reactions that change (oxidize) compounds using energy from sunlight and some oxidant, such as oxygen (O_2) or ozone (O_3). Photooxidation produces photochemical smog. See PHOTOCHEMICAL SMOG.

photoperiod   The length of time that a plant or animal is exposed to light (and, more important for plants, the corresponding length of the dark period) each day, especially in terms of how it affects the growth or physiological responses of a plant or animal. See PHOTOPERIODISM, PHENOLOGY.

photoperiodism   An organism's response to the changing day length associated with the seasons, especially in terms of how it affects vital functions. The sleeping and waking cycles of many animals are governed directly or indirectly by photoperiodism, as are flowering cycles of many temperate-zone plants. See LONG-DAY ORGANISM, SHORT-DAY ORGANISM, DAY-NEUTRAL PLANT, PHYTOCHROME, PHENOLOGY.

photophore   A luminous, cup-shaped organ on the undersides of some ocean species of crustaceans and fish.

photophosphorylation   The chemical reaction that uses the energy of light to add a phosphate group (phosphorylates) ADP to form the energy compound ATP in photosynthesis. See ADENOSINE DIPHOSPHATE (ADP), ADENOSINE TRIPHOSPHATE (ATP).

photoprotection   Protection from effects of short-wave ultraviolet light resulting from a cell's prior exposure to longer-wave ultraviolet light.

photoreactivation   Any chemical reactivation that is caused by the presence of light.

photoreceptor   A structure or molecule

that is responsive to light, such as the rods and cones in vertebrate eyes. See EYE-SPOT, PHYTOCHROME.

photorespiration A stimulation of respiration rates outside the mitochondria of C_3 plants in the light in comparison to their respiration rates in the dark. Although the respiration of C_3 plants is three to five times higher in the light, photorespiration is not found at appreciable levels in C_4 plants. See C_3 PLANT, C_4 PLANT.

photosensitive Showing a response to light or other forms of radiation such as infrared. Lenses for glasses that darken to become sunglasses outdoors are photosensitive. Photographic paper and film are also photosensitive. Adverse reactions to sunlight, caused by medications or disease, are further examples of photosensitivity.

photosphere The intensely bright visible surface of the sun and other bright stars, a layer of extremely hot, glowing gases several hundred miles thick.

photosynthate Any of the energy-rich organic molecules manufactured during photosynthesis, especially one of the sugars.

photosynthesis The series of chemical reactions by which plant cells transform light energy into chemical energy through producing simple sugars (or other energy compounds) and oxygen from carbon dioxide and water. It is measured with a variety of units, including milligrams (mg) of carbon dioxide (CO_2) per gram of leaf tissue per hour, mg CO_2/mg chlorophyll/hour, mg CO_2/square decimeter of leaf surface/hour, and mg CO_2/ square meter aquatic or marine surface/day or hour. See DARK REACTIONS, LIGHT REACTIONS.

photosynthetically active radiation (PAR) Wavelengths of 400 to 700 nm, the portion of the spectrum that is effective in photosynthesis.

photosynthetic bacteria Types of bacteria that are capable of fixing their own energy using a somewhat simpler form of photosynthesis than that found in green plants. The green and purple sulfur bacteria use hydrogen sulfide in place of water during the reaction; purple and brown nonsulfur bacteria use organic compounds instead. See CYANOBACTERIA.

photosynthetic efficiency Percentage of light energy assimilated by plants during photosynthesis. It is based on either net production or gross production.

photosynthetic quotient The ratio of the amount of carbon dioxide used to the amount of oxygen produced by photosynthesis in an organism or community.

photosynthetic rate The production of matter per unit of light absorbed, or per unit of carbon dioxide consumed, in photosynthesis. The photosynthetic rate usually increases with increasing intensity of sunlight, although it often shows a

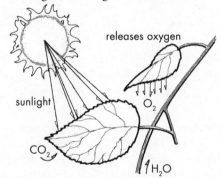

photosynthesis

Chlorophyll-containing plants convert carbon dioxide and water into simple sugars and starches, giving off oxygen as a by-product. The solar energy thereby fixed in the plant can be stored for millions of years, as in the case of fossil fuels.

significant dip during the brightest parts of the early afternoon. It is expressed as oxygen (O_2) liberated or carbon dioxide (CO_2) taken up per area of leaf tissue per unit time.

phototactic Demonstrating, or relating to, phototaxis.

phototaxis The movement of an organism in response to light. Compare PHOTONASTY, PHOTOTROPISM.

phototrophic Describing organisms that derive their energy from the sun through photosynthesis. All green plants are phototrophic. Contrast CHEMO-AUTOTROPH, CHEMOORGANOTROPHIC.

phototropism A change of orientation in plants in response to light. The movement in phototropism differs from that of photonasty because it is a directional

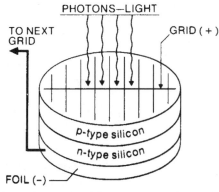

A photovoltaic cell is fabricated in the form of a thin wafer or chip from a semi-conductor material that sustains an electric field when exposed to sunlight, such as silicon or cadmium-copper sulfide. When photons from sunlight are absorbed by the outer surface of the solar cell, free electrons are created, which then cross to the bottom surface that is coated with foil to collect current for the next cell. The cells are attached in a series or parallel in a grid to create an electric current.

movement by a plant either toward or away from the light source. Compare PHOTONASTY, PHOTOTAXIS.

photovoltaic cell A device that converts sunlight or other radiant energy directly into electromotive force (voltage). It is a specialized photoelectric cell that generates voltage and therefore the potential for current, rather than altering an existing current, under light. It is also called a solar cell.

pH paper Paper that has been coated with acid-base indicators to provide an easy-to-use means of testing the pH (acidity or alkalinity) of lakes, ponds, soils, or other substances. The range of pH over which a given type of pH paper is accurate depends upon which indicators are used in the coating. See ACID-BASE INDICATOR.

phreatic eruption A volcanic eruption occurring when lake water, sea water, or groundwater is vaporized by contact with magma.

phreatic explosion A violent phreatic eruption that occurs when confining rock is ruptured by steam pressure.

phreaticolous Describing organisms living in fresh groundwater.

phreatic water The water that occupies the voids within a rock or soil at a level below the water table.

phreatophyte Any plant capable of sending roots deep into the ground to obtain water; a plant whose roots penetrate the water table.

phycobilins A group of accessory photosynthetic pigments, compounds that absorb light energy and transfer that energy to the primary form of chlorophyll during photosynthesis. Phycobilins include phycocyanin, phycoerythrin, and fucoxanthin; in brown, red, and blue-green algae they take over the role performed by chlorophyll b in green plants. Sometimes called biliproteins. See FUCOXANTHIN, CAROTENOID.

phycology   The study of algae. Also called algology.

phyletic evolution   The evolution of taxonomic groups such as species.

phyletic gradualism   Gradual, steady evolutionary change along single, sequential lines of descent. Contrast PUNCTUATED EQUILIBRIUM. Compare ANAGENESIS.

phyletic lines   Theoretical lines between present and past groups of organisms indicating their evolutionary relationships and derivation.

phyllite   A metamorphic rock characterized by a fine-grained schistose texture and a silky sheen. The dominant minerals in phyllite are chlorite, muscovite, and sericite. Phyllite forms between slate and schist in metamorphic grade.

phyllode   An enlarged, usually flattened petiole, serving as the photosynthetic organ in some plants in which the ordinary leaf is absent or much reduced in size, as in some species of *Acacia*.

phylogenetic systematics   A system of classification of organisms based on evolutionary history and relationships. Compare PHENETICS. See SYSTEMATICS.

phylogeny   The evolutionary history or development of a species or higher grouping of organisms. Compare ONTOGENY.

phylosphere, phyllosphere   The microenvironment immediately surrounding a leaf.

phylum   The highest taxonomic grouping within the animal kingdom, corresponding to a division within the plant kingdom. A phylum is made up of several closely related classes (or, rarely, a single class). Names of phyla usually end in -a, as in Porifera, Cnidaria, Arthropoda, Mollusca, and Chordata (Platyhelminthes, another phylum, is an exception). Fungi phyla end in -mycota, as in Ascomycota and Basidiomycota. Plant phyla end in -ophyta, as in Cycadophyta, Coniferophyta, and Angiospermophyta. Monera phyla have various, often non-Latinized endings. See APPENDIX p. 613.

physical barrier   Something that structurally protects a plant from pests, either a natural feature such as fine bristles, or a human device such as a band of sticky material surrounding the trunk of a tree to prevent larvae from crawling up the trunk.

physical factors   Environmental conditions that do not involve living species, such as precipitation, soil type, temperature, and sunlight. See ABIOTIC.

physics   The science of the fundamental properties of matter and energy, including light, electricity, mechanics (classical and quantum), magnetism, thermodynamics, and radioactivity, and excluding chemical changes and biological systems.

physiognomy   The apprearance of vegetation, or the way a biotic community looks.

physiographic   Relating to geographical features of land forms.

physiographic province   A geographic region that has a distinctive pattern of geomorphic features. See BASIN-AND-RANGE TOPOGRAPHY for an example.

physiological ecology   A branch of ecology that studies the changing relationships of individuals to the physical conditions, the nonbiological aspects, of their environments as mediated by the organisms' physiologies.

physiological longevity   The maximum expected life span of an individual that dies of "old age."

physiological race   A grouping of organisms into a race that is defined by genetically determined physiological characteristics. See RACE.

physiological specialization   The development of biochemical differences in a population that superficially seems to be identical, a form of adaptive radiation leading to the production of physiological races. See PHYSIOLOGICAL RACE.

physiological time   A measurement that incorporates temperature in order to evaluate periods of time required for development in ectothermic organisms. Physiological time acknowledges that growth and development proceeds at different rates in different temperatures.

phytoalexins   Substance that a plant produces in response to invasion by pathogenic fungi or other microorganisms. Phytoalexins are usually complex compounds, often terpenoids or phenols.

phytochemistry   The study of chemical constituents of plants. Phytochemistry covers growth and metabolism, but especially secondary products such as alkaloids that are of potential medical and economic use to humans.

phytochrome   A protein plant pigment that serves as a light receptor, sensitive to infrared wavelengths (730 nm) in darkness and red wavelengths (660 nm) in sunlight. It regulates photoperiodic responses and thus many plant activities such as flowering, nocturnal leaf movements, stem elongation, and seed germination.

phytocoenosis   A plant community or the vegetative components of a biotic community.

phytogeocoenosis   The vegetative components of an ecosystem.

phytogeography   The study of plant distribution throughout the world, with a concentration on how environmental factors have influenced evolution and distribution. Also called plant geography. Compare ZOOGEOGRAPHY. See BIOGEOGRAPHY.

phytome   A plant community.

phytophagous   Herbivorous; feeding on plants.

phytoplankton   Small (usually microscopic) aquatic plants; those plankton capable of photosynthesis, such as unicellular algae. Compare ZOOPLANKTON.

phytoplankton bloom   See ALGAL BLOOM.

phytosociology   The branch of plant ecology that studies plant associations and the interrelations among the populations of various plant species. Also called plant sociology.

phytotoxic   Poisonous or harmful to plants. Herbicides are made up of phytotoxic compounds. Compare PHYTOTOXIN.

phytotoxin   Substances produced by plants that are harmful or poisonous to other organisms, usually animals. Ricin, the substance that makes castor beans poisonous in minute quantities to humans, is a phytotoxin.

phytotron   A large plant growth chamber designed to create a controlled environment for the study of plant life.

PIC   Acronym for product of incomplete

combustion, substances formed when combustion is inefficient and does not proceed to completion. Carbon monoxide is a PIC.

pica Craving for nonfood substances such as clay, plaster, or starch. Pica sometimes occurs during pregnancy or in people suffering anemia from a lack of iron in the diet.

Picidae See WOODPECKERS.

Piciformes An order of specialized climbing, largely insectivorous, sometimes fruit-eating birds, most of which have hard, powerful bills. The feet are short and the toes usually zygodactylous with two pointed forward and two behind (e.g., woodpeckers, toucans, honeyguides, puffbirds and barbets). Approximately 330 species exist. See AVES, ALCEDINIDAE.

pico- Prefix in the Système International d'Unités (SI) indicating one trillionth, 1 x 10^{-12}. One picofarad equals one trillionth of a farad, or 10^{-12} F.

picojoule (pj) A very small unit for measuring energy, work, or heat, equal to one trillionth (10^{-12}) of a joule.

piedmont glacier A part of a valley glacier that extends beyond a confining valley and spreads out over a lowland region.

piezoelectric effect The induction of electric polarization and therefore voltage through pressure (mechanical strain). The effect also works in reverse: The application of an electric field to piezoelectric materials causes a mechanical deformation. This effect occurs in some asymmetrical crystals such as quartz.

piezometer A device (type of well) for measuring the elevation of the water table or the hydraulic pressure of the groundwater.

piezometric surface The imaginary surface defined by the level to which water in a confined aquifer would rise if measured with piezometers.

pig iron Crude iron that comes out of a blast furnace; it contains carbon, silicon, manganese, sulfur, or phosphorus as impurities. When further refined, it is used in making steel and different forms of iron.

pigment Any substance that naturally occurs in or colors the tissues of an organism. See MELANIN, CHLOROPHYLL, CAROTENE, XANTHOPHYLLS, PHYCOBILINS.

pigmentation An organism's coloration, or the presence of deposits of pigment within the organism.

pile Short for atomic pile. Sometimes used colloquially for a nuclear reactor. See ATOMIC PILE, NUCLEAR REACTOR.

piliferous Having hair. The piliferous layer refers to the epidermis of young roots, the layer that bears the root hairs.

pill bugs See ISOPODA.

pillow basalt A pod of basalt that has a glassy outer skin and an inner flow structure associated with the formative environment of a pillow lava.

pillow lava A lava with a distinctive shape acquired by basalt that erupts into a submarine environment. The rapid chilling of lava results in a series of convex basalt pods that have the appearance of a stack of stone pillows.

pilot balloons Small rubber balloons used for meteorological observations. The balloons are filled with hydrogen and released into the atmosphere, where they track the direction and speed of high-level air currents.

Pinaceae   The pine family in the phylum Coniferophyta of the plant kingdom. The members of this family are resinous, woody trees and shrubs with seeds attached to a woody cone and leaves that are either needlelike, flattened, or scalelike. Often the family is subdivided into three groups by leaf, bud, and cone characteristics: 1) pines *(Pinus)*, larches *(Larix)*, Douglas fir *(Pseudotsuga)*, hemlocks *(Tsuga)*, spruces *(Picea)*, and firs *(Abies)*;   2) redwoods *(Sequoia, Sequoidendron,* and *Metasequoia)* and bald cypress *(Taxodium)*; and 3) Cypress *(Cupressus)* and Junipers and cedars *(Jupinerus)*. See CONIFEROPHYTA.

pingo   A dome-shaped hill formed by a core of expanding ice in a permafrost soil. Pingos range from 2 to 50 m in height.

pinna   1) The outer, visible portion of a mammal ear, also called an auricle. 2) A bird's feather or wing; a winglike part on other animals, such as fish fins. 3) A primary leaflet in a pinnately compound (featherlike) leaf such as a fern frond. See PINNATE.

pinnate   Featherlike in appearance. A pinnate (or pinnately compound) leaf is one with many leaflets (pinnae) arranged on either side of a central stem, as the leaves of ash trees and roses. See APPENDIX p. 629.

pinnatifid   Divided in a pinnate or featherlike manner; describing a leaf blade or other structure having divisions extending at least halfway down to a central axis, and pointed ends on the divided lobes. A pinnatifid leaf is a single, lobed leaf, rather than many leaflets within a compound (pinnate) leaf. See PINNATE.

pinocytosis   The process used by a cell to take in fluid and nutrients. As in phagocytosis, the cell membrane folds back into itself to engulf the fluid and then closes up to form a vacuole. Compare PHAGOCYTOSIS.

pioneer community   The collection of first organisms to colonize a bare site, such as that formed by severe forest fire, agricultural abandonment, or avalanche; the first stage in ecological succession on land. Lichens, microbes, and particularly hardy species of plants usually make up a significant portion of a pioneer community in primary succession, whereas weedy r-selected species dominate the pioneer communities in secondary succession. See CLIMAX COMMUNITY, PRIMARY SUCCESSION, SECONDARY SUCCESSION.

pioneer species   See COLONIZER.

pipet   A small tube, usually widening into a bulb at the center, used to transfer a calibrated volume of liquid.

pisciculture   Fish farming; raising fish on a large scale.

piscivorous   Fish-eating; describing organisms that subsist exclusively or primarily on fish.

pistil   The female, seed-producing part of a flower, made up of the sigma, style, and ovary. The pistil consists of one or more carpels (depending on whether it is a simple or a complex pistil). Also called gynoecium. Compare STAMEN. See FLOWER.

pistillate   Describing female flowers, those that have one or more pistils but no stamens (male organs). See FLOWER. Compare STAMINATE.

pitch   1) A sticky, resinous substance derived from various conifer tree species; a natural gum or resin, such as pine pitch. Some forms of pitch have been used medicinally. 2) A description of the angle formed between a lineation feature on the surface of a rock and the strike of the rock face. Compare STRIKE, DIP. 3) The property of a sound that allows a human ear to determine its relative position on a frequency scale. Middle C is a specific pitch. 4) Vertical inclination, as of a slope.

pitchblende   A common name for the mineral uranite (uranium oxide). Pitch-blende is a major ore mineral for uranium. It usually occurs in brownish black botryoidal masses associated with pegmatites.

pitchstone   A volcanic glass with a resinous luster imparted by the gradual absorption of water on the specimen surface. Pitch-stone is also known as fluolite.

pituitary gland   The master endocrine gland in vertebrates. A pea-sized structure located at the base of the brain, the pituitary gland secretes hormones that control many bodily functions, including growth, reproduction, and regulation of other glands. Also called the hypophysis. See FOLLICLE-STIMULATING HORMONE, LUTENIZING HORMONE, LUTEOTROPHIC HORMONE, SOMATOTROPIN.

pj   See PICOJOULE.

placenta   1) The spongy tissue in mammals (except marsupial species) that connects the fetus to the wall of the uterus. The placenta delivers oxygen and nourishment from the mother to the fetus, and carries wastes from the fetus back to the mother's body for excretion. 2) Plant tissue to which spores, sporangia, or ovules are attached.

placental mammals   See EUTHERIA.

placer   An unconsolidated mass containing ore minerals such as gold in stream deposits.

placer deposit   A loose deposit of economic minerals that have been concentrated by natural mechanical forces such as weathering and transport.

plaggen   A soil horizon that is artificially produced by a long-term practice of adding manure and fertilizer.

plagioclase   An abundant group of rock that forms minerals which contain sodium- or calcium-rich silicates in a solid solution series. The following individual plagioclase varieties are listed with their associated proportions of calcium plagioclase: albite, less than 10 percent; oligoclase, 10 to 30 percent; andesine, 30 to 50 percent; labradorite, 50 to 70 percent; bytownite, 70 to 90 percent; anorthite, more than 90 percent. See APPENDIX p. 626.

plagioclimax   See DISCLIMAX.

plagiosere   A stage leading to a disclimax or plagioclimax. See DISCLIMAX.

plagiotropism   The tendency for the stems and branches of some plant species to always grow at the same angle from the vertical. Plagiotropism is a form of geotropism, a movement directed by gravity. See GEOTROPISM.

plague   1) General term for any epidemic of an infectious disease causing many fatalities. Often used in a more limited sense to refer to bubonic plague. 2) An unusually severe infestation by a pest, as in a plague of locusts. See BUBONIC PLAGUE.

Planck's constant   The constant of proportionality in Planck's law, a fundamental principle of quantum mechanics that describes the quantization of electromagnetic radiation. Planck's law states that the magnitude of the energy contained in a quantum of radiation is equal to the frequency of the radiation times Planck's constant. The value of this constant is

6.626 x 10^{-34} joule seconds. See QUANTUM MECHANICS.

plane of the ecliptic   The plane of the earth's orbit around the sun. See EQUINOX.

planet   A large celestial object (at least 1000 km in diameter) that does not radiate energy or light and is in orbit around a star. The only planets currently known are those of our own solar system: Mercury, Venus, Earth, Mars, Jupiter, Saturn, Uranus, Neptune, and Pluto.

planetary nebula   An immense, glowing cloud of dust and gas in space surrounding a very hot star. Planetary nebula are formed when stars under specific conditions explode, and the gas cloud is constantly ionized by ultraviolet radiation given off by the remaining core of the star. Compare NEBULA.

planetesimal   A theoretical planet-precursor formed as the initial solar nebula condensed during the formation of our solar system. See SOLAR NEBULA THEORY.

plane wave   A three-dimensional wave whose wave front or surfaces of equal phase is planar.

plankter   A single plankton organism. See PLANKTON.

planktivorous   Feeding upon plankton. Also called planktotrophic.

planktobiont   See HOLOPLANKTON.

plankton   Small, often microscopic, organisms that float in still or moving fresh water, or in salt water. They are an important source of food for many larger animals. Sometimes divided into photosynthetic species (phytoplankton) and heterotrophic species (zooplankton). Contrast BENTHOS, NEKTON, NEUSTON, PERIPHYTON, AUFWUCHS, PROTISTA.

plankton spectrum   The full range of plankton organisms, the species and their relative abundance, at a given time and location.

planktotrophic   Another term for planktivorous. See PLANKTIVOROUS.

Planned Parenthood Federation   (PPF) A nonprofit health and advocacy agency operating clinics that provide family planning information and services throughout the United States. Its international division, the International Planned Parenthood Federation (IPPF) provides similar services throughout the developing world.

planosol   A term applied to soils having a leached surface layer above a hardpan. Planosols are typical of humid and subhumid climates.

plant   Any organism that usually manufactures its own food by photosynthesis and (with the exception of some algae) is not capable of locomotion. Sometimes restricted to multicellular members of the Kingdom Plantae, thereby excluding Monera, Protista, and Fungi. See PLANTAE.

Plantae   The plant kingdom, consisting of multicellular, eukaryotic organisms that produce embryos and, for the most part, photosynthesize. It excludes fungi and algae (kingdom Fungi, kingdom Protista). Plantae includes nonvascular bryophytes (mosses and liverworts), vascular pteridophytes that do not bear seeds (ferns, clubmosses, and horsetails), as well as seed-bearing gymnosperms and angiosperms. See APPENDIX p. 616–617.

plantation   Area of land with trees or shrubs planted for commercial purposes, such as coffee plantations, banana plantations, tea plantations, or pine plantations.

plant-available   Describing amounts of soil nutrients or elements in the soil that can be readily absorbed and assimilated by growing plants.

plant community   All of the plant species found growing together at one time in a given habitat or region. Compare BIOTIC COMMUNITY.

plantigrade   Using the entire sole of the foot in walking, as humans and rabbits do. Compare DIGITIGRADE, UNGULIGRADE.

planula   The free-swimming, cilia-covered, and usually flattened larva of many cnidaria such as jellyfish, coral, sea anemone, and hydra.

plasma   Gas that is mostly ionized and therefore conducts electricity, but whose net charge is zero as positive and negative charges balance. Plasma is often considered a fourth state of matter, after solid, liquid, and gas; most of the universe is plasma. The sun is made up of plasma.

plasmagene   Genetic material outside of the cell nucleus. An example is the DNA occurring in mitochondria and chloroplasts rather than that incorporated into the chromosomes. Such DNA is still able to influence heredity in a manner similar to the chromosomes of the nucleus. Compare PLASMID, PLASTOGENE.

plasmalemma   Another term for plasma membrane, used especially in botany to distinguish the living cell membrane from the inert cell wall. See PLASMA MEMBRANE.

plasma membrane   The cell membrane; the thin, semipermeable boundary that encloses the cytoplasm of the cell and controls which substances enter or leave the cell. Often called a plasmalemma when referring to cells of higher plants, where it is the outermost part of the living tissue of the cell. See FLUID-MOSAIC MODEL.

plasmasphere   A region within a planet's magnetosphere consisting of dense, highly ionized plasma. The earth's plasmasphere overlaps the outer regions of the ionosphere, and both follow the earth's rotation. See MAGNETOSPHERE.

plasmid   Extrachromosomal genetic material in bacteria that is replicated and transferred independently of the chromosomes. Plasmids are used in recombinant DNA research. See EPISOME, PLASMAGENE, PLASMON, GENETIC ENGINEERING.

plasmodesmata   Narrow strands or fine tubes of cytoplasm in plant cells that fill gaps in the membranes and cell walls to connect the protoplasts of adjacent cells.

plasmodial slime molds   See MYXOMYCOTA.

Plasmodiophoromycota   A phylum of the kingdom Protista that includes obligate plant parasites that form plasmodia, multinucleate masses of protoplasm. The life cycle of these organisms is also characterized by the production of zoospores (which fuse to form new plasmodia) and by the production of resting cysts. The clubroot disease of cabbage is caused by *Plasmodiophora*, and the powdery scab of potato tubers is caused by *Spongospora*, both members of this phylum. See PROTISTA.

plasmodium   A mass of protoplasm containing numerous nuclei and able to move in an amoebalike fashion. It is formed by the fusion of a number of cells in which no fusion of the nuclei occurs. In the slime molds (Myxomycota), the zygote matures into a plasmodium thallus. See SYNCYTIUM, PLASTOGAMY, MYXOMYCOTA.

plasmon   All of the genetic material outside of the nucleus of the cell, such as plasmagenes, plastogenes, and plasmids.

plastic 1) Any of a group of compounds synthesized from petroleum by polymerization. They are easily shaped by melting, and are sturdy and resistant to most forms of decay, making them very useful as containers but also very long lasting in the environment. 2) Indicating ability to deform under stress, but unlike elastic materials, the deformation is permanent and the material does not return to its original shape once the stress is removed. See THERMOPLASTICS.

plastic deformation A permanent change in the shape of a rock or a mineral that occurs when the material is stressed beyond the elastic limit without attendant brittle failure.

plastic flow (glaciers) A type of glacial movement characterized by gradual creep related to deformation of ice under the force of gravity. The glacier movement is often faster near the glacier's base where pressure is highest. Contrast BASAL SLIP (GLACIERS).

plastids Small, often pigmented organelles that occur in large numbers in the cytoplasm of most cells. Pigmented plastids such as chloroplasts are called chromatophores (or chromoplasts); colorless ones are leukoplasts. See CHLOROPLAST, CHROMATOPHORE, ELAIOPLAST, LEUKOPLASTS.

plastogamy The fusion of cell tissue (cytoplasm), without fusion of the associated nuclei, in the formation of a plasmodium. See PLASMODIUM.

plastogene Genetic material located within plastids, rather than in the chromosomes of the cell nucleus, that helps control plastid activity. See EPISOME, PLASMAGENE.

plate A semirigid body of the lithosphere that moves independently on the plastic layer of the earth's upper mantle. See PLATE TECTONICS.

plateau A relatively flat and broad upland region that is considerably higher than adjacent regions and is bounded on at least one side by a scarp.

plateau basalt A flood basalt forming a topographic plateau. See FLOOD BASALT.

plate tectonics A unified scientific description combining theories of continental motion, sea-floor spreading, and the rock cycle. Plate tectonic theory holds that the brittle outer surface of the earth's crust is broken into independent plates that are free to move about on the plastic layer of the mantle.

platform 1) A small plateau. 2) The relatively flat sedimentary rocks overlying a craton. 3) A continental shelf.

platinum (Pt) A heavy metallic element with atomic weight 195.1 and atomic number 78. Platinum is a precious metal, grouped with other noble metals because of its resistance to chemical change and corrosion. It is used in jewelry, dentistry, and industrial equipment, and as a catalyst as in catalytic converters. See NOBLE METAL.

Platyhelminthes Phylum of dorsoventrally flattened, wormlike animals with three tissue layers, bilateral symmetry and, usually, a distinctive head. There is no circulatory system or coelom, and the gut, when present, is incomplete bearing only a mouthlike structure and lacking an anus. The phylum is usually divided into three classes, Turbellaria, Trematoda, and Cestoda (tapeworms) of which the latter two are thought to be exclusively parasitic. Approximately 2500 species exist. See

TURBELLARIA, TREMATODA, CESTODA.

playa lake   A temporary lake in a desert environment. Playa lakes are flooded intermittently by runoff from nearby mountains and dry out later through evaporation. Playa lakes are frequently the site of evaporite mineral formation.

Playfair's law   The geologic principle stating that each stream cuts its own valley, which is proportional to the size of the stream, and that two streams have the same grade at the point of confluence.

pleiotropic   Relating to the condition in which one allele controls more than one unrelated genetic trait.

Pleistocene   The first epoch of the Quaternary subera of geologic time. The Pleistocene lasted from approximately 2 to 0.1 million years ago. See APPENDIX p. 610.

pleomorphism   See POLYMORPHISM.

pleon, pleion   The abdominal section of a crustacean of the group Malacostraca.

pleopod   See SWIMMERET.

pleurilignosa   The woody vegetation (trees and shrubs) of the rainforest. See HIEMILIGNOSA, LAURILIGNOSA, LIGNOSA, HERBOSA.

pleurodont   Possessing teeth that grow from the inside of the jawbone without separate sockets.

pleuston   Multicellular plants forming floating mats of vegetation in fresh water.

plinian volcano   A type of composite volcano that erupts with a massive plume of airborne pyroclastic material, especially pumice, which is dispersed over a large land area.

Pliocene   The last of five epochs in the Tertiary subera of geologic time. The Pliocene lasted from approximately 5.1 to 2 million years ago. See APPENDIX p. 610.

ploidy   The number of chromosome complements or sets found in a cell. See HAPLOID, DIPLOID, POLYPLOID.

plucking   A glacial process that pulls away blocks of bedrock in the downstream direction of ice movement. An irregular surface of ledges is often formed by plucking. See ROCHES MOUTONNÉES.

plug   See VOLCANIC NECK.

plug dome   A volcanic feature that forms by dome-shaped accumulation of viscous lava around a volcanic neck. See VOLCANIC DOME.

plumbism   Another term for lead poisoning, from the Latin name for lead, *plumbum*. See LEAD POISONING.

plume   A column of visible exhaust, as from a smokestack. 2) A large upward flow of molten or partially molten rock from the earth's mantle.

plumule   1) The embryonic terminal bud and stem of a sprouting seed, located just above the cotyledon or cotyledons and sometimes containing immature leaves. 2) Down feather on an immature bird, sometimes persisting into adulthood.

plunge (of a fold)   The angular measure between an inclined fold axis and a horizontal plane.

plunging breaker   A type of breaking ocean wave that collapses suddenly into the preceding trough as the crest of the wave curls over and spills.

pluteus larva   The free-swimming larval stage of marine Echinoidea and Ophiurida.

pluton   A body of intrusive igneous rock

that has solidified from magma within the crust of the earth.

plutonic Of or relating to conditions of rock formation from magma within the crust of the earth.

plutonic rock An intrusive igneous rock with a coarse texture indicating that the crystals formed slowly under conditions of burial.

plutonium (Pu) A radioactive, very poisonous, metallic element with atomic number 94; the atomic weight of its most stable isotope is 244.0. Found in very small quantities in different uranium ores, plutonium is usually produced artificially from uranium. Plutonium 239 (Pu-239) is a fissionable, carcinogenic isotope of plutonium used in nuclear reactors and weapons.

pluvial A lake forming during a period of more abundant rainfall. Large-scale pluvial lakes are often associated with glacial climates.

pluvial lake A lake that formed during a pluvial climate.

PNdB Short for perceived noise decibels, the unit for perceived noise level. See PERCEIVED NOISE LEVEL.

pneumatolysis The mineral formation associated with hot gaseous solutions derived from a magma in the late stages of crystallization. The mineral tourmaline is produced by pneumatolysis.

pneumatophore A specialized root in some aquatic plants that grows upward from the rest of the root system, reaching above the surface of the water to increase the gas-exchange capacity. The "knees" of black mangrove trees are pneumatophores.

pneumoconiosis Lung disease caused by repeated inhalation of dust particles, usually seen in people with occupational exposure to coal dust, asbestos, silica, etc. It causes the formation of fibrous tissue forms in the lungs, which impairs breathing. Black lung disease (anthracosis) is a form of pneumoconiosis. See ASBESTOSIS, BERYLLIOSIS, BYSSINOSIS, RESPIRATORY FIBROTIC AGENTS, SILICOSIS.

PNS See PERIPHERAL NERVOUS SYSTEM.

PO Chemical symbol for the element polonium. See POLONIUM.

Poaceae The grass family of the monocot angiosperms. Grasses have hollow stems, except at nodes where leaves are attached. Stems are usually round or flattened and two-ranked leaves have a specialized, linear blade with a sheath that surrounds the stem. The highly specialized flowers have three stamens and a single ovary with two styles. Each individual floret is sub-tended by a pair of bracts and the entire inflorescence by another pair of bracts. Timothy *(Phleum)*, bamboo *(Bambusa)*, rice *(Oryza)*, barley *(Hordeum)*, sugar cane *(Saccharum)*, maize *(Zea)*, fescues *(Festuca)*, oats *(Avena)*, brome *(Bromus)*, rye *(Secale* and *Poa)*, wheat *(Triticum)* and many other grasses are members of this family. See ANGIOSPERMOPHYTA.

pod 1) The exterior shell or casing in which legume species of plants (the bean and pea family) grow their seeds. Also called a legume. 2) A number of animals traveling or clustered together, as in a pod of whales.

Podicepiiformes Grebes. An order of aquatic birds with soft, thick plumage and near functionless tails. Unlike most water birds, the toes are not connected by webbing but rather, each toe is laterally expanded and lobate. This, however, has no special phylogenetic significance as this adaptation is also found in some other groups of birds. See AVES.

podzol   Any of a group of zonal, ash gray pedalfer soil types formed in subarctic and northern humid forested envrironments. See PEDALFER.

podzolic soil   A soil formed by the process of podzolization.

podzolization   A soil-forming process in subarctic and northern humid forested environments; characterized by combined processes of leaching, eluviation, and illuviation. See PODZOL.

poikilohydric   Describing plants that are not able to regulate their rate of water loss and therefore contain roughly the same amounts of moisture as their immediate environment. Nonvascular plants such as algae, lichens, and bryophytes, as well as vascular plants that grow underwater, are poikilohydric. Compare HOMOIOHYDRIC.

poikiloosmotic   Describing aquatic organisms whose body fluids change in concentration if the water in which they live becomes more or less concentrated (they are therefore in osmotic equilibrium with their aquatic environment). Many marine invertebrates are poikiloosmotic. Compare HOMOIOSMOTIC.

poikilotherm   See ECTOTHERM.

point-bar   A depositional bar that forms on the inside of a meander bend. Contrast CUT BANK.

point sources   Stationary, identifiable pollution-generating facilities, such as power plants, industrial plants, and sewage treatment facilities. Point sources usually produce high concentrations of either air or water pollutants. They are much easier to identify, and in theory to control, than nonpoint sources. Compare NONPOINT SOURCES.

poison   Any substance that can kill or seriously injure if consumed, inhaled, or touched.

Poisson distribution   A mathematical method in statistics used to test for (or describe) the degree of randomness in the distribution of a variable whose values are not continuous. It is used especially for relatively rare occurrences.

polar   Referring to an entity having poles, such as a magnet (with north and south poles), an electrical battery (with positive and negative electrical poles), or a chemical dipole (with positive and negative electrical poles). Also used to refer to the regions near the North Pole and the South Pole.

polar body   One of two small cells, each of which develops during the initial two meiotic divisions of a plant egg cell (oocyte).

polar cap absorption   (PCA) A phenomenon in the ionosphere above the earth's poles that occurs while solar flares are active. The ionosphere begins to absorb high-frequency and very high-frequency radio waves (from 3 to 300 megahertz), and reflects low- (LF) and very low-frequency (VLF) radio waves at lower altitudes than normal. The disturbance may last for days or weeks, and is detected by riometers. See HIGH FREQUENCY, LOW FREQUENCY, VERY HIGH FREQUENCY, VERY LOW FREQUENCY.

polar (cold) glacier   A slow-moving glacier in which the base is always colder than the pressure-melting point of the ice. They are frozen to the rocks and contain no meltwater. Polar glaciers move primarily by internal deformation.

polar covalent bond   A chemical bond that has both ionic and covalent characteristics. A polar covalent bond occurs between two elements of significantly different electronegativities, such as between hydrogen and chlorine in the HCl molecule. The resulting bond is polarized, with more of the electron density in the region of the more electronegative atom. Most bonds have a certain amount of both ionic and covalent character. See COVALENT BOND, IONIC BOND, ELECTRONEGATIVITY.

polar desert soil   A type of mineral soil that develops in polar environments where there is an annual precipitation of less than 13 cm. Polar desert soils develop little or no identifiable soil horizons.

polar emergence   The occurrence of deep-water tropical marine species in the shallow water of polar regions. Compare EQUATORIAL SUBMERGENCE.

polarity reversals   See MAGNETIC REVERSAL.

polarizability   The ability to deform the electron cloud of a molecule by subjecting it to an external electric field. Molecules containing heavy atoms or multiple bonds have greater polarizability than molecules lacking either.

polar wandering   The change in position of the geomagnetic poles in relation to the geographic poles of the earth as measured over geologic time. See APPARENT POLAR WANDERING.

polar-wandering curve   A map line connecting geomagnetic pole positions relative to a fixed point on a continent for various times in the geologic past.

polder   An area of dry land that has been maintained by building dikes and pumping out sea water. Polders are commonly created for agricultural land in the Netherlands.

Polemoniaceae   The phlox family of the Angiospermophyta. These dicots usually have 5-petaled, trumpet-shaped, radially symmetrical flowers borne in terminal clusters. The superior ovary has three united carpels and a three-lobed stigma. The leaves are usually in an opposite arrangement. Phlox (*Phlox*), gilias (*Ipomopsis*) and (*Gilia*), and Jacob's-ladder and sky pilot (*Polemonium*) are members of this family. See ANGIOSPERMOPHYTA.

policy   The course or general plan of action adopted by a government, a person, or another party.

polje   A closed interior valley formed in regions of karst topography.

pollard   An extreme form of tree pruning more common in Europe than in the United States. An individual tree is topped, and all branches are cut back to the trunk every year or so. The result is a thick, stubby trunk (usually under 8 feet) with many straight shoots growing out of the top. Pollarding was originally developed to produce large quantities of very small wood (such as willow osiers) at a height that was relatively easy to cut, yet was above the reach of browsing animals. Now it is sometimes done purely for ornament. Compare COPPICE.

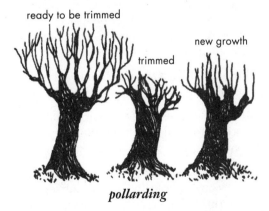

ready to be trimmed

trimmed

new growth

pollarding

pollen   Grains or microspores containing male gametophytes of seed plants, produced by the anthers. Each pollen grain is tiny, although in large quantities it is visible as a fine, usually yellow powder. Fertilization in seed plants is the transfer (by insect, wind, bird, etc.) of pollen from the anthers of flowers to the pistils of either the same or other flowers. Wind-borne pollens, such as those of ragweed (*Ambrosia*), are common human allergens.

pollen analysis   The scientific study of pollen, especially to determine the past existence and abundance of plant species by analyzing peat or sedimentary deposits. Pollen is very resistant to decay and can provide important information about climate and dominant flora for times before any historical record exists. See PALYNOLOGY.

pollen rain   The fallout of pollen from the atmosphere into a bog or a lake. See POLLEN ANALYSIS.

pollination   The act or process of fertilizing a pistil with pollen, usually brought about by wind, water, insects, bats, or birds.

pollinator   An insect or other agent that transfers pollen from anthers to stigma.

pollinia   In orchids and milkweeds, masses of pollen grains held together by a sticky fluid and carried to other flowers en masse by insects or by other pollinators. Singular is pollinium.

pollinosis   Hay fever; nasal congestion or other allergic response to pollen.

pollutant   A substance that is irritating, harmful, or toxic to plant or animal life; something that causes pollution.

pollution   An undesirable change in the environment, usually the introduction of abnormally high concentrations of hazardous or detrimental substances, heat, or noise. Pollution usually refers to the results of human activity, but volcano eruptions and contamination of a water body by dead animals or by animal excrement are also pollution.

polonium   (Po) A radioactive element with atomic number 84; the atomic weight of its most stable isotope is 209. It is found in pitchblende but can also be produced artificially. Polonium has more isotopes than any other element and is used as an emitter of alpha radiation.

polyandrous   1) Describing male flowers with several to many stamens. 2) Characterized by polyandry. Compare ADELPHOUS.

polyandry   The characteristic mating of a single female with more than one male in a breeding season. This mating system is relatively rare, found in several species of birds including jacanas. Compare POLYGYNY.

polybrominated biphenyls   (PBBs) Organic compounds containing more than one bromine atom, produced by the bromination of biphenyl (C_6H_5-C_6H_5, a compound consisting of two benzene rings, also called diphenyl). Polybrominated biphenyls are used as flame retardants; some are used in organic synthesis. In toxicity and persistence in the environment, they are similar to PCBs. See POLYCHLORINATED BIPHENYLS.

polycentric distribution   The presence of a population, a species, or another taxonomic group in several widely detached places.

Polychaeta   A mainly marine class of annelid worms, characterized by the presence of paired, lateral extensions of the body called parapodia, which bear clusters of bristles (chaetae). Tentacles, palps, and eyes may be present on the prostomium and anterior segments of errant or crawling forms. In burrowing species, parapodia are often greatly reduced and, in some, prostomial tentacles are elaborately expanded into filter-feeding organs. Approximately 8000 species exist. See ANNELIDA.

polychlorinated biphenyls (PCBs) A group of toxic, carcinogenic organic compounds containing more than one chlorine atom, formerly used in the manufacture of plastics and as insulating fluids in electrical transformers and capacitors. PCBs behave much like DDT in the environment: They are very stable compounds and therefore persistent, and fat-soluble, so they accumulate in ever-higher concentrations as they move up the food chain. Although not acutely toxic, long-term exposure has been linked with cancer, so the use of PCBs was banned in the United States in 1979.

polyclimax theory The theory (popularized by G. E. Nichols, 1917) that succession in a given climatic region leads to one of several climaxes, depending on local environmental conditions such as soil type. It is an outgrowth of the monoclimax theory. See MONOCLIMAX THEORY.

polycondensed plastic A type of plastic produced by a form of polymerization involving a condensation reaction. Nylon and Bakelite (phenol-formaldehyde resin, the first synthetic plastic) are examples. Most are capable of being recycled.

Polycoplacophora An exclusively marine class of Mollusca, the chitons have ovate, flattened bodies and shells of eight, serially arranged, imbricated plates. The mantles form thick girdles which overlap the plates around their margins holding them together and, in some cases, nearly covering them. The large, muscular foot, together with the girdle, help the animals adhere tenaciously to surfaces while they graze algae. Over 500 species exist. See MOLLUSCA.

polyculture Raising more than one species of plant within a relatively small area. Polyculture can refer to the mixture of plants typical of a home garden, or a mixture of numerous small plots of single species. It can also refer to a more complex method of intercropping (on a small or large scale), in which many different plants that mature at different times are planted together. Polyculture is less susceptible to diseases and insect pests than monoculture crops. See MONOCULTURE, MIXED PERENNIAL POLYCULTURE.

polycyclic aromatic hydrocarbons (PAHs) A group of organic compounds containing more than one aromatic ring structure. PAHs include phenylbenzene and naphthalene; many are carcinogenic and are metabolized in the body to form toxic compounds. Some sources are barbecued meats, tar, and asphalt.

polyelectrolyte Synthetic compounds used in wastewater treatment. Polyelectrolytes such as activated silica are high-molecular-weight polymers that form connections between colloidal particles, increasing the speed of flocculation and therefore speeding the removal of solids.

polyembryony The development of more than one embryo from a single egg, ovule, or seed.

polyenergid Describing a cell nucleus with a polyploid set of chromosomes. See POLYPLOID.

polyethene European term for polyethylene. See POLYETHYLENE.

polyethylene A strong and flexible plastic with a relatively low melting point, used for bottles and other containers, toys, electrical insulation, plastic wrap, and in synthetic fabrics. It is made by polymerizing ethylene ($CH_2=CH_2$). See POLYETHYLENE TEREPHTHALATE, HIGH-DENSITY POLYETHYLENE.

polyethylene terephthalate (PET) A form of plastic used in synthetic fabrics and plastic bottles. It is capable of being recycled. See HIGH-DENSITY POLYETHYLENE.

polygamous 1) A plant that has both unisexual and hermaphrodite flowers on the same plant or on different plants of the same species. 2) Characterized by polygamy.

polygamy The characteristic mating with more than one of the opposite sex in a single breeding season. See POLYANDRY, POLYGYNY.

polygenes Genes whose effects are too small to notice, but which work together to influence complex quantitative hereditary traits such as height. See QUANTITATIVE GENETICS, MULTIPLE GENE INHERITANCE.

polygenetic Relating to or having a polyphyletic origin, from two or more independent ancestral lines of descent.

Polygonaceae The buckwheat or knotweeed family of the Angiospermophyta. This family of dicots typically has alternate, simple leaves with stipules that form a sheath around the stem at their points of attachment. The petals and sepals are generally in whorls of three. The fruits are three-angled achenes. Knotweeds and smartweeds (*Polygonum*), buckwheat (*Fagopyrum*), rhubarb (*Rheum*), docks (*Rumex*), and California buckwheats (*Eriogonum*) are members of the Polygonaceae. See ANGIOSPERMOPHYTA.

polygoneutic See MULTIVOLTINE.

polygynous 1) Describing flowers with many pistils or styles. 2) Characterized by polygyny. See POLYANDROUS, POLYGYNY.

polygyny The characteristic mating of a single male with more than one female in a breeding season. Compare POLYANDRY. See HAREM.

polyhaline A term describing water that has a chlorinity value between 1.0 percent and 1.65 percent.

polymer A compound formed by linking many small molecules into a large, complex molecule with different properties. Biological polymers include carbohydrates (linked sugars) and proteins (linked amino acids). Polyethylene glycol and plastics such as polyethylene are synthetic polymers. See MONOMER.

polymictic plankton Plankton containing several different species, each present in large numbers. Compare MONOTON PLANKTON.

polymorphism 1) The existence of more than one form or type in a species, beyond simple gender differences. Social insects such as honeybees with queens, drones, and workers demonstrate polymorphism. 2) Another term for pleomorphism, the occurrence of distinct forms during the life cycle of a plant or animal, such as the caterpillar and pupa that precede the adult.

polymorphonuclear leukocytes A white blood cell whose nucleus contains two or more parts (lobes) connected by fine strands. Also called a granulocyte.

polyna An expanse of water surrounded by ice. A polyna may be formed where currents distribute sea ice around the mouth of a river.

Polynesian floral region One of the

phytogeographic regions into which the Paleotropical realm is divided according to similarities between plants; it consists of the Pacific islands east of Indonesia, excluding Fiji and Hawaii. See PALEO-TROPICAL REALM, PHYTOGEOGRAPHY.

polynucleotides   A nucleic acid polymer, one made up of a number of individual nucleotides.

polyp   The stationary form of some cnidarians (coelenterates) such as sea anemones, hydras, and corals. It is a cylindrical body attached at the base, with a mouth at the opposite end surrounded by fingerlike, stinging tentacles.

polypeptide   A compound consisting of 10 or more linked amino acids. Proteins are large polypeptides with 100 to 300 amino acids. Smaller polypeptides include some hormones.

polyphagous   Eating many different kinds of food.

polyphyletic   1) Describing organisms that are presumed to have descended from more than one independent ancestral group. 2) Describing a grouping in cladistics that does not include the most recent ancestral form of the organisms in the group. See MONOPHYLETIC, CLADISTICS.

polyphyodont   Describing animals with more than two successive sets of teeth. Compare DIPHYODONT, MONO-PHYODONT.

polyploid   Having three or more sets of chromosomes. Compare DIPLOID, HAP-LOID.

polyprotic acid   An acid in which each molecule is capable of giving up more than one hydrogen ion. Sulfuric acid (H_2SO_4) and carbonic acid (H_2CO_3) are both polyprotic acids.

polyribosome   A grouping of ribosomes that are linked by a molecule of messenger RNA and work together as a unit in synthesizing proteins. Also called polysome.

polysaccharide   A biological polymer made up of many sugar molecules. Polysaccharides are complex carbohydrates including starch and cellulose. They can be broken down into their simple sugar components and used by organisms as a source of energy.

polysaprobic   Describing an organism that can survive in highly polluted water that usually has a low oxygen concentration. See OLIGOSAPROBIC.

polysomes   See POLYRIBOSOME.

polystyrene   ($CH_2{=}CHC_6H_5)_n$, a polymer commonly known as Styrofoam. It is used extensively in packaging and as insulation in refrigerators. See HIGH DENSITY POLYETHYLENE.

polythalamous   Describing shells that have many chambers, such as those of some species of Foraminifera. See FORAMIN-IFERA.

polythetic   A method of defining taxonomic classifications based on many characteristics rather than just one, as in a polythetic key. Compare MONOTHETIC.

polytocous   Producing many eggs, or bearing many young, at one time. Compare MONOTOCOUS, DITOCOUS.

polytopic   Occurring in more than one area.

polytrophic   See POLYPHAGOUS.

polytypic   Possessing more than one variant form, especially species having subspecies or varieties.

polyunsaturated fat   A fatty compound having double bonds replacing many of the single bonds linking its carbon atoms, and thus having fewer hydrogen atoms attached to the carbon chain. Polyunsaturated fats are liquid at room temperature and include many vegetable oils. See SATURATED FAT, UNSATURATED FAT.

polyurethane   Any of a group of polymer resins made by combining diisocyanate compounds with phenol, amines, hydroxylic compounds, or carboxylic compounds. Uses of polyurethane include in elastic fabrics (Spandex), adhesives, rubbers, paints, varnishes, protective coatings, and molded plastics such as automobile fenders. See POLYMER.

polyvinyl chloride   (PVC) The best known and most used of a group of synthetic compounds called vinyl plastics. PVCs are particularly sturdy plastics and are not biodegradable. All contain many repeating units of the same formula linked together, with the general structure $(CH_2CHCl)_n$. Rigid forms of PVC are used for plumbing; flexible forms are used for vinyl fabrics.

polyxenous   Describing parasites that have more than one species as their hosts. Compare MONOXENOUS.

pome   The fruit type characteristic of apples and pears. The flower carpels form the seed-containing core, and the flower receptacle expands to form the large volume of flesh surrounding the core.

Pongidae   A primitive family of the great apes which belong to the primate order. Their large size, barrel-shaped chest, shortened back, elongated forearms, and relatively large brain differentiate them from old world monkeys. Chimpanzees and gorillas spend much time on the ground walking on all fours in contrast to orangutangs which are almost completely arboreal. The bonobos, sometimes called pigmy chimpanzees, which are generally

assigned species rank, are the most bipedal of the great apes. All pongids are endangered by hunting and loss of habitat. See PRIMATES.

pool   A relatively deep, well-scoured zone along a streambed. Pools are often found alternating with riffles. Contrast RIFFLE.

population   1) Organisms of the same species that inhabit a specific area. 2) The total count of individuals within such a group, such as the population of the United States.

population cycle   A pattern of regularly repeating changes in population levels; repeating increases followed by decreases.

population density   Number of individuals in a given population within a specific area (such as per square meter, acre, or square mile) or within a volume of water (cubic centimeter, cubic meter).

population distribution   The variation of population density over a specific geographical area.

population dynamics   The study of fluctuations in population levels and the major biotic and environmental factors that cause these fluctuations.

population ecology   The study of population levels, fluctuations, and distribution.

population momentum   The potential for increased population growth as a large group of young individuals mature and become able to reproduce.

population profile   An analysis of the different sizes of age groups within a given population, often expressed as a bar graph representing number of individuals of each age, or (for long-lived organisms) number of individuals in five-year age groups. See AGE DISTRIBUTION.

population pyramid A visual representation of the age distribution within a population. The youngest, usually largest, age group is graphed as the base, and successive age classes (typically increasingly small in size) are stacked above it. See AGE DISTRIBUTION, POPULATION PROFILE.

population regulation Factors leading to adjustment for population levels, creating a tendency for some density-dependent factor to cause increased growth when population levels are low, and to decrease growth when high population levels are reached. See DENSITY DEPENDENT.

population trajectory Movement of a point on a population graph that illustrates simultaneous changes in two populations, usually in a predator-and-prey system or two populations in competition with each other.

population vulnerability analysis (PVA) An evaluation of a population's likelihood of extinction, used to evaluate populations or entire species that are at risk.

Porifera Phylum of aquatic, multicellular animals, the sponges have few cell types and a unique method of filter feeding in which water is moved in through external pores into internal canal systems by action of flagella located on specialized cells called choanocytes. The sponges lack true symmetry and discrete organs, and their body shape may vary with environmental circumstances. Simple to complex skeletal systems are formed from calacareous or siliceous spicules and protein fibers. Their worldwide distribution includes approximately 10,000 species.

porosity Degree of permeability to water or to other fluids caused by the presence of holes or spaces. In geology, porosity refers specifically to the percentage volume of empty space within a rock or soil sample.

porphyritic texture A description applied to igneous rocks having larger, early-formed crystals surrounded by a ground-mass of smaller, later-formed crystals.

porphyry A general term for an igneous rock with a porphyritic texture.

positive feedback 1) Describing any system in which a response to a stimulus or initial cause exerts an enhancing effect on that stimulus or cause (an increase causes an additional increase in the system, or a decrease causes an additional decrease). Economic inflation is an example of positive feedback. 2) In population dynamics, positive feedback is a factor that works to stimulate population growth. A newly encountered energy source can cause positive feedback and thus a rapid population increase, as when sheep were introduced on Tasmania. Contrast NEGATIVE FEEDBACK. See DENSITY DEPENDENT.

positron A positive electron; an elementary particle with the same mass as an electron, but with a positive instead of a negative electrical charge. The positron is the antiparticle to the electron.

post- Prefix meaning after, following later, or behind.

postclimax A relict biotic community of a formerly cooler, moister climate, as in patches of grasslands found surrounded by desert. Contrast PRECLIMAX.

postlarva A crustacean with the morphology of an adult that has not yet reached sexual maturity. See SUBADULT.

potable water A term that describes water that is safe and palatable for human consumption.

potamobenthos   The bottom-dwelling organisms of rivers and of streams.

potamon   Organism that lives in moving, freshwater habitats (streams and rivers).

potash   A common name for the chemical compound potassium carbonate.

potassium   (K) A soft metallic element with atomic weight 39.10 and atomic number 19. One of the most abundant elements in the earth's crust, it occurs naturally only in compounds. It is an essential nutrient for living organisms and is an ingredient in most fertilizers. See APPENDIX p. 626.

potassium-argon dating   A radiometric dating technique based on comparison of the proportion of ^{40}K to ^{40}Ar isotopes in a mineral or rock sample. Potassium-argon dating is only effective on rocks that are at least 250,000 years old.

potassium permanganate $KMnO_4$, a purple salt of potassium. It is a strong oxidant; it is used in analytical chemistry and as a disinfectant.

potential energy   Energy stored in a substance or a body because of its position (or the position of its component parts) or state, rather than its current motion. Water pumped up into a storage tank has potential energy because when it is allowed to fall, gravity enables the water to accomplish work. Gravitational energy, electrical energy, nuclear energy, and chemical energy are all forms of potential energy. Contrast KINETIC ENERGY.

potential evapotranspiration   The theoretical maximum amount of water lost through combined evaporation and transpiration that could occur from a surface with an unlimited supply of water. This value helps predict water consumption needs in different climates.

potentiation   1) The amount of damage predicted from the interaction of two or more toxic substances. 2) The degree of synergistic increase in the combining of two drugs, two hormones, etc.

potentiometric surface   The hypothetical surface of underground water as determined by piezometric measurements. The potentiometric surface is also known as the piezometric surface.

pothole   A smooth cylindrical depression excavated by the grinding action of rock debris within the eddies of a stream.

power   The rate at which work is accomplished; the number of joules of work accomplished per second of time, measured in watts (1 watt = 1 joule/sec). (Formerly measured in horsepower.)

power plant   A facility for generating electricity, especially a large-scale facility. See COAL-FIRED POWER PLANT, NUCLEAR POWER PLANT.

power tower   Another term for a solar furnace. See SOLAR FURNACE.

pozzolana   Any siliceous volcanic tuff or burnt clay used as a component of cements that harden under water. Pozzolana is named for a source in Puzzuoli, Italy.

ppb   See PARTS PER BILLION.

ppm   See PARTS PER MILLION.

ppt   See PARTS PER TRILLION.

prairie   A form of grassland vegetation typical of semiarid temperate latitudes in continental interiors, as in the United States. Prairies are typically large, flat areas covered with a mixture of grass species and a few herbs. See SAVANNA, SHORTGRASS PRAIRIE, TALLGRASS PRAIRIE.

prairie soil   An older soil classification term

describing thick soils that develop under tall grass prairies in temperate climates. See MOLLISOL.

pre-adaptation   Having traits that permit a shift into a new niche or habitat, or a change in the structure of an organism that occurs before a corresponding change in behaviors.

Precambrian   The informal term for all geologic time before the beginning of the Cambrian period (approximately 590 million years ago). The Precambrian is subdivided into Priscoan, Archaean, and Proterozoic eons.

Precambrian shield   A common reference in North America to the Canadian shield.

precession of the equinoxes   The slow westward drift in the position of the vernal and the autumnal equinoxes caused by the gyrating motion of the axis around which the earth rotates, over a period of 26,000 years.

precipitant   Compound added to a solution to cause a dissolved substance in the solution to precipitate out. See PRECIPI-TATE.

precipitate   1) To cause a dissolved substance to chemically solidify out of solution, sometimes settling to the bottom, so it can be collected or removed. 2) A substance that has been chemically removed from a solution by precipitation.

precipitation   The deposition on the ground of any form of condensed water vapor from the atmosphere. Precipitation includes rain, snow, hail, and mist.

precipitation titration   A technique used to analyze the concentration of molecules, ions, or atoms in solution or in a solid mixture. A standard solution of known concentration that will precipitate out the

desired molecule or ion is added to the sample in the presence of an indicator compound that will change color when all of the desired molecule or ion has been used up. The unknown quantity of the desired molecule or ion can be calculated from the volume of standard solution added.

precipitator   Any device that is used in air pollution control to remove particulates from exhausts. Precipitators can be mechanical or electrical. See ELECTRO-STATIC PRECIPITATOR.

preclimax   A relict biotic community of a formerly warmer, drier climate, as in patches of desert vegetation found in areas that generally support grassland, or patches of grassland surrounded by deciduous forest. Contrast POSTCLIMAX.

precocial   1) Referring to those species of birds born covered with down or feathers and with open eyes. Unlike altricial species, precocial species such as chickens and wading ducks require little care after hatching. 2) Describing mammals, such as horses and giraffes, whose young are capable of walking immediately after birth. Contrast ALTRICIAL.

precursor   A compound or molecule from which another compound or molecule develops. In plants, protochlorophyllide is a precursor of chlorophyll.

predation   An interaction between two organisms of different species in which one (usually larger) individual or species hunts, kills, and eats another individual or species. See PREDATOR, PREY.

predator   An animal that hunts and kills other animals for its food.

predator-prey relationship   See PREDA-TION.

preferential species A plant species found in a variety of communities, but showing greater abundance in one particular type of community. See PHYTOSOCIOLOGY.

preheating One of the simplest applications of solar energy: Solar panels are used to heat water (especially cold well water) to lukewarm, reducing the supplemental energy required to bring the water up to the temperatures required for domestic hot-water use.

prescribed burning A method of forest management in which relatively small, controlled fires are set under favorable conditions to prevent build-up of large quantities of brush or dead wood, and thereby prevent more destructive crown fires during dry seasons that may rage out of control. Also called controlled burning.

preservation 1) Conservation. 2) Often used in a more restricted sense for the maintenance or conservation of the natural environment as it is, without change or extraction of resources, as opposed to a more utilitarian, multiple-use approach to land management. Such hands-off preservation, sometimes called the preservation ethic, is almost exclusively an American concept; supporters of this view are called preservationists.

preservative A substance added to another substance (especially food) to prolong its useful life or to maintain its appearance. Sodium nitrate, alone or in combination with ascorbic acid, is a food preservative used in processed meats such as bacon to maintain pink color.

pressure 1) The force acting over a given area divided by that area, measured in pascals; this is also called stress. 2) Atmospheric pressure, the force per unit area exerted by the atmosphere on the surface of the earth, caused by the weight of the air. At sea level, atmospheric pressure is approximately 10^5 pascals, or 1013 millibars, and decreases with increasing altitude. Also called barometric pressure, this is the most important variable in weather forecasting.

pressure-volume work The work done on or by a system due to gas expansion or contraction. For small volume changes, the pressure-volume work equals the pressure times the change in volume. Pressure-volume work is used in evaluating gas exchange accomplished by animal organs (lungs).

pressurize To put under pressure. Pressurizing fluids such as gases causes them to become liquids at higher temperatures than they would at normal atmospheric pressure.

pressurized water reactor (PWR) One of two forms of the light-water reactor. A nuclear reactor is one in which the water in the reactor is kept under pressure to keep it from boiling despite its high temperature (600°C). The superheated water moves through a heat exchanger that produces steam to power turbines and generate electricity. The Yankee Atomic nuclear power plant in Rowe, MA (1960–1992) was a pressurized water reactor. Compare BOILING WATER REACTOR.

pretreatment (waste water) A preliminary treatment of a wastewater stream that involves the use of skimming, flocculation, or preaeration to remove specific contaminants before it enters a primary treatment process.

prevailing wind Any wind that flows consistently from one direction. Although often used to describe local winds, the prevailing westerlies are global west-to-east wind currents found between 30° and 60° above and below the equator. See TRADE WINDS.

prevention of significant deterioration Provision of the Clean Air Act Amendmendts of 1977, 42 U.S.C.A. §7470-7491. Limitations on the incremental amounts of pollution allowed in clean air areas, with smaller increments allowed in areas where there are special national or state interests served by limiting increases in pollution. The process begins by classifying all attainment air quality control regions (AQCRs) or parts into three categories (Class I, II, or III) and setting air standards for each class.

prey Any animal that is killed and eaten by a predator. See PREDATION.

primary Describing a structure that was formed first or has greatest importance, as in a primary meristem of plants or primary feathers.

primary ambient air quality standards Legal allowable levels for pollutants in outdoor (ambient) air quality to avoid harm to human health. Levels are established by the Environmental Protection Agency as part of the regulations of the Clean Air Act. (As opposed to secondary standards designed to protect public welfare.) See NATIONAL AMBIENT AIR QUALITY STANDARDS.

primary consumer An herbivore; an organism belonging to the second trophic level feeding directly upon photosynthesizing plants (primary producers). Butterflies and deer are primary consumers. Compare SECONDARY CONSUMER.

primary division freshwater fish Fish with little or no tolerance for salt water. Compare SECONDARY DIVISION FRESHWATER FISH.

primary forest A forest that has not been subjected to clearing or harvesting by human activity, a remnant of the prehistoric forest. Compare SECOND-GROWTH FOREST.

primary minerals The minerals that crystallize directly from a magma.

primary oil recovery A term applied to the first extraction of oil from a well or an oil field. Compare SECONDARY OIL RECOVERY.

primary pollutants Substances that are harmful when directly released into the environment. Carbon monoxide is a primary air pollutant; oil from an oil spill is a primary water pollutant. See SECONDARY POLLUTANTS.

primary producer Another term for autotroph; an organism at the bottom of a food chain or a food web. See AUTOTROPH.

primary production The creation of organic matter (biomass) through photosynthesis or chemosynthesis. Compare SECONDARY PRODUCTION. See NET PRIMARY PRODUCTION.

primary productivity The rate at which energy, in the form of organic compounds produced by photosynthetic and chemosynthetic organisms, is stored in an ecological community or group of communities. See GROSS PRIMARY PRODUCTION, NET PRIMARY PRODUCTION.

primary rock An older rock classification term for the postulated "primitive" igneous and metamorphic rocks of the Neptunism concept, as proposed by Abraham Gottlob Werner (1749–1817).

primary sewage treatment The first stage in the purification of wastewater. In this stage, most floating and larger suspended solids are mechanically removed, either by

settling out or by screening. See SECOND-
ARY SEWAGE TREATMENT, ADVANCED
SEWAGE TREATMENT.

primary standards   An allowable concentration for a pollutant, established at a level
that is believed to allow a margin of safety
to protect human health. Primary standards were established in the 1970 Clean
Air Act. See SECONDARY STANDARD.

primary succession   The sequential
development of communities on an area
that was not previously occupied by
organisms, such as the surface of a recent
volcano or deglaciated land. See PIONEER
COMMUNITY, SECONDARY SUCCESSION.

primates   An order of mammals that retain
a number of primitive, structural characteristics such as the five digits on forelimb
and hind limb, but that also show many
advanced features like forward-facing eyes
with binocular vision, enlarged cerebrums,
prehensile hands and feet, and forelimbs of
great flexibility, which evolved for climbing and swinging in trees. The order is
divided into several groups including tree
shrews, lemurs, and the new and old world
monkeys. It is from the latter group that
the pongids (great apes) and hominids
(humans) have evolved. See MAMMALIA,
LEMURS, PONGIDAE, HOMINIDAE.

Primulaceae   The primrose family of
herbaceous dicot angiosperms. Flowers
have five sepals and five petals that are
united at least at the bases. The stamens
are attached to the petals opposite the
corolla lobes. The hypogenous ovary have
free-central placentation and produce a
capsule fruit. Members frequently exhibit
heterostyly. Shooting stars (*Dodecatheon*);
primroses, oxslips, and cowslips (*Primula*);
loosestrifes (*Lysimachia*); and cyclamens
(*Cyclamen*) are all members of this family.
See ANGIOSPERMOPHYTA.

principal components analysis   (pca) A
form of statistical analysis that organizes
information along multidimensional axes.
The first axis is used for the quantity
demonstrating the maximum amount of
variance; at right angles to this, the next
axis demonstrates the maximum of the
remaining variance. The dimensions of the
resulting space will equal the number of
principal components in the analysis.

principle   1) A basic truth or fundamental
assumption, as in a scientific principle
explaining a phenomenon. 2) A code or
individual rule governing conduct.

principle of original horizontality   See
LAW OF ORIGINAL HORIZONTALITY.

principle of stratigraphic superposition
See LAW OF SUPERPOSITION.

principle of superposition   See LAW OF
SUPERPOSITION.

principle of uniformity   A principle
advanced by the 18th-century natural
philosopher James Hutton that states that
the processes occurring in the present are
uniformly equivalent to the processes
operating in geologic time. The principle
is often paraphrased as "the present is the
key to the past."

principle of parsimony   See OCCAM'S
RAZOR.

prior appropriation   Legal principle that
the first user of water from a stream or
river establishes the legal right to continue
to use the same amount that was initially
withdrawn. Called "first use" for short,
this principle is the basis of water rights in
the western United States but contravenes
the basis of water rights in the eastern
United States. Prior appropriation governs
water use in Arizona, Nevada, Utah,
Idaho, Montana, Wyoming, Colorado, and
New Mexico; in adjacent states (California,

Oregon, Washington, North Dakota, South Dakota, Nebraska, Kansas, Oklahoma, and Texas) a combination of riparian rights and prior appropriation governs water rights. Compare RIPARIAN RIGHTS.

Priscoan eon The earliest eon of all geologic time, which lasted from approximately 4500 to 4000 million years ago. There is no rock record from the Priscoan eon.

prisere Natural, undisturbed primary succession.

probabilistic models See STOCHASTIC MODELS.

probability 1) Likelihood of occurring. 2) In statistics, a ratio expressing the chances that a certain event will occur, based upon previous measurements of the frequency of occurrence of the same kind of event.

Proboscidia Elephants. This mammalian order includes the largest of living, terrestrial animals. There are only two living species, the African elephant, *Loxodonta africana*, and the Indian elephant, *Elephas maximus*. In addition to such well-known features as their trunks and large ears, the elephants have a very specialized dentition. In the adult the only teeth remaining are a pair of upper incisors which form the tusks and the enormous, grinding molars. All elephants are threatened by habitat restriction and ivory poaching. See EUTHERIA.

procaryote See PROKARYOTE.

Procellariiformes Oceanic order of birds with tubular nostrils which extend over the beak. Well adapted to flying and soaring, they only come on land to mate and to nest. A single, white egg is laid. Examples are albatrosses, shearwaters, petrels; 92

species exist. See AVES, ALBATROSSES.

proclimax A stable (climax) plant community that appears to have become established under climatic conditions or soil conditions that differ from those found in a region today. See PRECLIMAX, POSTCLIMAX.

prodelta The offshore region of a delta. The prodelta is an area that experiences active deposition of clastic particles.

producer Another term for autotroph (also called primary producer). See AUTOTROPH.

production General term for biomass created by a trophic level in a given area or ecosystem. The rate of production per specified period of time is technically called productivity, although the terms are often used interchangeably. See GROSS PRIMARY PRODUCTION, NET PRIMARY PRODUCTION, PRIMARY PRODUCTION, SECONDARY PRODUCTION.

production efficiency The amount of biomass stored compared with the total amount synthesized (primary production) or ingested (secondary production).

productivity The rate of production, biomass created in a given area or ecosystem, over a specified period of time. See PRIMARY PRODUCTIVITY.

profile 1) A representation of a vertical cross section of a geologic material. 2) A transverse line of data acquisition in seismology. 3) Vertical section of soil showing connected horizons. 4) Graph showing values of elevations and distance along a section of the earth's surface.

profundal zone The region extending from the lower limit of vegetation to the deepest part of a lake.

progesterone   A female sex hormone produced by the adrenal glands, corpus luteum, and placenta. Progesterone initiates changes in the lining of the uterus during the second half of the menstrual cycle, preparing it to receive a fertilized ovum. In pregnant women, progesterone stimulates development of the placenta and the milk-producing glands.

prograde   1) The direction of advancing deposition in a developing sedimentary structure such as a delta or an alluvial fan. 2) A change toward increasing temperature or pressure during metamorphism.

prograde metamorphic effects   The changes in mineral structure that occur during increasing metamorphic intensity.

prokaryote   A cell or organism lacking a distinct nucleus. Prokaryotes include bacteria, blue-green algae, and actinomycetes. Contrast EUKARYOTE.

prokaryotic   Referring to or concerning prokaryotes. See PROKARYOTE.

prolactin   A hormone produced by the pituitary gland. When other hormones (estrogen and progesterone) are present in the right balance, prolactin stimulates the mammary glands to give milk. Prolactin also stimulates the production of pigeon's milk in doves and in pigeons.

proline C_4H_8N-COOH, an amino acid found in plant and animal proteins. Proline is a constituent of the fibrous proteins of tendons and bones known as collagen.

prominence   A luminous eruption of hydrogen gas above the chromosphere of the sun.

promiscuous   A form of breeding in which organisms mate with several individuals over a season. See POLYGAMY.

promontory   An elevated topography with an abrupt termination above a lowland or a body of water.

promoters   1) Substances that increase the activity or the effectiveness of a biological catalyst. 2) Sites on strands of DNA that, in conjunction with the operators, control the action of genes and therefore the synthesis of particular proteins.

propagation   Reproduction of a plant or an animal, especially intentional multiplication or increase of stocks carried out by humans. Propagation includes sexual reproduction as well as asexual methods of plant reproduction, such as sending out runners, or horticultural techniques of layering, taking cuttings, and grafting.

propagule   Another term (chiefly British) for disseminule. See DISSEMINULE.

propane H_3C-CH_2-CH_3, a simple hydrocarbon gas found in petroleum and in natural gas. When pressurized, propane becomes liquid; in this form it is used for gas stoves and similar gas appliances, especially in rural areas where propane tanks may be seen adjacent to homes. Propane is also used to fuel buses and automobiles, and in the chemical industry. The British term for pressurized propane is LPG (liquified petroleum gas).

propellant   1) Substance used in aerosol cans to force out the contents; CFCs have been used extensively as aerosol propellants. 2) Substances used to power rockets, consisting of fuel plus oxidant.

prophase   The first stage of mitosis (ordinary cell division). During prophase, the genetic material condenses into recognizable chromosomes, which divide lengthwise. See ANAPHASE, METAPHASE, MITOSIS, TELOPHASE.

prophylactic   A substance used as a preventative, such as a medicine used to prevent disease. Condoms are sometimes called prophylactics, because they are used to prevent pregnancy and transmission of disease.

prophylaxis   Prevention of disease; practices that promote health and prevent the spread of disease organisms.

propolis   A reddish brown resin gathered by social bees from tree buds and used to strengthen the hive. It is darker, stronger, and more brittle than the wax they produce to contain honey.

proprioceptor   A sensory nerve ending that receives stimuli from inside the body, especially one involved in communicating the relative position of body structures.

protactinium   (Pa) An uncommon radioactive element with atomic number 91; the atomic weight of its most stable isotope is 231. It is one of the actinide series and is found in all uranium ores.

protandry   The condition in which male organs of a flower or hermaphrodite animal mature before female organs, or before the female organs are ready for fertilization. Protandry in flowers (a form of dichogamy) is a mechanism for preventing self-fertilization. Contrast DICHOGAMY, PROTOGYNY.

protease   Any of a group of enzymes that break down proteins into peptides or peptides into their component amino acids. Also called peptidases or proteolytic enzymes.

protein   Any of a large group of organic compounds containing nitrogen (and often sulfur), found in and synthesized by all living organisms. Proteins are large molecules made up of one or more linked long chains (polypeptides), each made up of amino acids. Protein is an essential part of animal diets.

protein-calorie malnutrition   A form of malnutrition usually seen in young children or infants and interfering with growth. It may simply be a lack of sufficient food, or a lack of sufficient calories and the necessary balance of amino acids in protein. It can also be caused by intestinal parasites or disease. See MARASMUS.

proteoclastic enzymes   See PROTEOLYTIC.

proteolytic   Describing enzymes that break down proteins into simpler peptides and amino acids by breaking the peptide bonds (proteolysis). See PROTEASE.

Proterozoic   The last of three eons in Precambrian geologic time. The Proterozoic eon lasted from approximately 2500 to 570 million years ago. The term Proterozoic, meaning before life, should not be taken literally since microscopic life forms did exist during this eon.

prothallus   The small, flattened body of the gametophyte generation of ferns, clubmosses, and horsetails (pteridophytes). The mature prothallus bears archegonia, antheridia, or both.

Protista   The kingdom of living organisms that includes eukaryotes that are not plants, animals, or fungi. Many members of this kingdom are unicellular like the protozoa, although many red, brown, green, and other algae are multicellular. Most protists live in aquatic environments, or in the body fluids of other organisms, and have flagella at some time during their life cycles. This kingdom includes protozoa, algae, and slime molds. Sometimes referred to as Protoctista. See APPENDIX p. 617–618.

protium The most common isotope of hydrogen. It has one proton, one electron, and no neutrons, giving it an atomic weight of one. Protium is the lightest of hydrogen's three isotopes. See DEUTERIUM, TRITIUM.

proto-, prot- Prefix meaning first; also, occuring first or first-formed.

protocooperation A form of mutualism, a relationship which is beneficial to both species but not obligatory. Contrast OBLIGATE MUTUALISM.

Protoctista See PROTISTA.

protogyny The condition in which female organs of a flower or hermaphrodite animal mature before male organs, or before the male organs produce gametes. Protogyny in flowers (a form of dichogamy) is a mechanism for preventing self-fertilization. See DICHOGAMY, PROTANDRY.

protomyxa Slime molds other than Myxomycota. See ACRASIOMYCOTA.

proton 1) A subatomic particle that resides in the nucleus of the atom and carries a positive charge of the same magnitude as the negative charge of an electron. Protons have an atomic mass of one, approximately the same as the mass of a neutron. 2) The entity H^+ is often referred to as a proton because it is a hydrogen atom that has lost its only electron and therefore consists entirely of one proton (hydrogen atoms have no neutrons). An acid is thus often defined as a proton-donor and a base as a proton acceptor. See ELECTRON, NEUTRON, ATOMIC NUMBER.

protoplasm The living portion of a cell, a complex colloidal mixture of proteins, carbohydrates, fats, inorganic and organic compounds in water. Much of the proto-plasm is organized into discrete organelles; it is the medium in which all metabolic activity occurs.

protoplast The living material (protoplasm) within the cell wall of plant cells and bacteria. The protoplast is made up of nucleus, cytoplasm, organelles, and plasma membrane.

protore A rock that contains a concentration of economic minerals so low that profitable extraction would not be possible without a technologic advance or shift in market values.

protosoil A partly weathered regolith material that precedes soil formation. See SOIL.

Prototheria The monotremes are primitive mammals which share many skeletal characteristics with birds and reptiles and lay eggs. However, they have hair, mammary glands, and other distinct mammalian features. Examples are duck-billed platypuses, spiny anteaters. See MAMMALIA.

Protozoa Originally referred to a phylum of single-celled, eukaryotic "animals," now placed in the kingdom Protista. Protozoa do not form a natural group, but are merely a group formed by convenience. See PROTISTA.

provenance The original region in which a plant or animal species was found; the native range for an exotic species.

proven resources A known deposit of ore, gas, coal, or oil that may be legally and profitably extracted under current economic conditions.

proventriculus Technical name for the gizzard, the second stomach of a bird, an insect, or an earthworm, and the stomach (gastric mill) of a crustacean.

proximal   Towards the center. Contrast DISTAL.

psammitic   1) A term describing a metamorphic rock derived from psammite (arenaceous rock). 2) A rock that is arenaceous. See ARENACEOUS ROCK.

psammon   Organisms living in sand, or the water between grains of sand.

psammophyte   A plant that grows well on sand or in sandy soil.

psammosere   The sequence of plant communities that develops on sand, as on coastal dunes.

pseudoaposematic coloration   Markings or colors on harmless species that resemble markings on dangerous or distasteful species sufficiently to protect the harmless species from predation. See BATESIAN MIMICRY, APOSEMATIC COLORATION.

pseudochitinous   Made of a substance that resembles chitin, but that is chemically different.

pseudoextinction   The evolution of a species causing it to change so significantly that it is considered a new species, and the old species essentially becomes extinct by evolution rather than by dying out.

pseudofeces   Substances that have been regurgitated into the water by suspension feeders or deposit feeders, after being rejected as potential food material and before entering the gut.

pseudoforce   An illusion of a force acting; a displacement caused by a lack of force that appears to be an actual force. Centrifugal "force" is really the action of inertia tending to keep an object moving in a straight line. As a car turns, the passengers inside continue briefly to move in a straight line; although this feels like a force pulling them to the outside of the curve, it is really inertia. The coriolis "force" is another pseudoforce. See CORIOLIS EFFECT, CENTRIFUGAL FORCE.

pseudofossil   A natural inorganic rock feature that appears similar to and may be mistaken for a fossil.

pseudogamy   A form of apomixis (asexual reproduction) in plants in which the deposition of pollen is required to stimulate the pistil and the ovary to produce a seed, even though the pollen never fertilizes the ovule. See APOMIXIS.

pseudointerference   A pattern in which rates of predation decrease with increasing population density of the predator, as if mutual interference were operating, but instead caused by predators congregating in areas where more prey are located.

pseudokarst   A term applied to landscapes, such as a lava field, that have karstlike features that are not related to the carbonation of limestone.

pseudoplankton   Organisms that are not normally found floating among plankton, but have been swept up by currents or turbulence.

pseudopodium   A temporary extension of the protoplasm of a one-celled organism or amoeboid cell, either to provide locomotion or to surround and absorb food. Pseudopodia are characteristic of amoebae.

Psittaciformes   The sturdy head, heavy and sharply hooked bill, and brilliant plumage distinguish this bird order, which contains the parrots, cockatoos, and macaws, etc. of the Southern Hemisphere. Their well-developed feet have two toes forward and two toes behind (zygodactyous), which give them great agility in climbing. The

Carolina parakeet, the only species endemic to the United States, became extinct in the 1920s. See AVES, MACAWS.

psychosomatic Describing illnesses, or genuine physical disorders, that are partially caused by emotional stress or other mental states rather than by disease organisms. Also, an individual experiencing such symptoms. Also called psychophysiological disorders.

psychrometer Device for measuring relative humidity in the atmosphere, also called a wet and dry bulb hygrometer and swing psychrometer. It consists of two thermometers mounted together, one of whose bulb is kept wet. As the thermometers are swung through the air or as a fan blows air over them, evaporative cooling lowers the temperature of the wet bulb. Comparing the two thermometer readings to a table yields values for relative humidity and dew point. See RELATIVE HUMIDITY.

psychrometry The study or measurement of the thermodynamic properties of moist air.

psychrophilic Describing organisms that thrive in low-temperature environments, generally below 20°C. Compare MESOPHILIC, THERMOPHILIC.

Pt Chemical symbol for the element platinum. See PLATINUM.

Pteridophyta See FILICINOPHYTA, LYCOPODOPHYTA.

pteroylglutamic acid (PGA) A synonym for folic acid (tetrahydrofolic acid). See FOLIC ACID.

Pu Chemical symbol for the element plutonium. See PLUTONIUM.

pubescent Describing parts of plants or animals that are covered with fine hairs.

public trust doctrine A doctrine under which the state is said to own lands lying under navigable waters, and to hold such lands in trust for the benefit of the people of the state.

puddling 1) The process by which fine, especially high-clay, soils become compacted when wet, destroying natural structure and making the soil almost impervious to water or to gas exchange. Puddling can result if clay soils are plowed while they are too wet; it is then difficult to restore the soil to its original porosity. 2) Behavior displayed by some Lepidoptera (moths and butterflies) of congregating around puddles to consume brackish water.

pull-apart zone A structure of sedimentary rock in which beds have stretched and parted into slabs before compaction of the grains occurs.

pull factors (in urbanization) Attractive conditions that lead people to move from the country into the city. They are sometimes contrasted with push factors, features that drive people from rural areas to more urban areas out of necessity.

pulp 1) Any soft, fleshy mass of tissue in plants or animals, such as the mesocarp of many fruits or the interior of some animal organs. 2) Wood processed for paper manufacturing.

pulsation 1) An incidence of a recurring change in an ecosystem, such as a fire in chaparral and other ecosystems adapted to periodic burning, or a seasonal rise or drop in water levels. 2) An alternation of increases and decreases in a quantity such as voltage or the earth's magnetic field.

pumice A volcanic glass with a foamy or frothy vesicular structure that is imparted by trapped gas during cooling. Pumice may be light enough to float on water.

pumped storage   A mechanism for accommodating the extra demand for electrical power during peak periods. In offpeak hours when demand is relatively low, electrical energy is used to pump water into an elevated storage tank. This water is available to generate electricity, using gravity to drive turbines, and can therefore cover surges of demand that occur during peak hours.

punctuated equilibrium   A variation of Darwinism proposing that natural selection acts on a species to keep it stable for long periods, interrupted by relatively short bursts of rapid change during which new species develop.

pupa   A stage in the development of many insects when the organism appears to be inactive but is, within its protective case, undergoing metamorphosis from the larval stage into its adult form (imago). The pupa stage is found in those insects with a complete metamorphosis, including beetles, butterflies and moths, flies, bees, and ants.

pupate   To become a pupa or to go through a pupal stage. See PUPA.

pure line   A group of individuals within a species that breed true to type when crossed with others of the same type, especially a group descended from a common ancestor. Also called an inbred line, such individuals have a high percentage of homozygous alleles. In plants, this is more often called a pure strain, and is maintained by self-fertilization within the desired strain.

purge and trap   A method used to analyze for the presence of volatile organic compounds in water samples. The sample is purged by bubbling air through it, a process which expels the volatile organic compounds from the liquid. The air is then passed through an adsorbant material that traps the organic compounds. After the purge cycle is complete, the adsorbent is heated to release the organic molecules and the vapor is passed into a gas chromatograph for separation and quantification.

purine base   A structural component of nucleic acids (RNA and DNA) and coenzymes. Purine bases are derived from the organic base purine ($C_5H_4N_4$) and characterized by containing a purine fused double ring (a six-carbon ring fused to a five-carbon ring, with two nitrogen atoms replacing two of the carbon atoms in each ring). Adenine and guanine are both purine bases. See PYRIMIDINE BASE, BASE-PAIRING.

pustule   A small, blisterlike spot. In animals, they usually contain pus or lymph; the pox of smallpox is a pustule. On plants, such spots may occur on fruit, stems, or leaves, and usually contain developing spores of a fungus.

putrefaction   Decomposition of organic matter, especially when accompanied by foul-smelling gases caused by anaerobic conditions (and the decay organisms active under such conditions). See ANAEROBIC DECOMPOSITION.

PVC   See POLYVINYL CHLORIDE.

P wave (primary wave)   A faster-moving type of seismic body wave propagated by motion parallel to the direction of travel, like the stretching and recoil of a spring. This faster-moving wave is known as a primary wave, because it is the first to arrive at any given point after an earthquake. Compare S WAVE, BODY WAVES.

PWR   See PRESSURIZED WATER REACTOR.

pycnocline   A region of a lake or an ocean where the density changes rapidly with increasing depth.

pyramid   1) Any tapering structure of animals, such as the bundle of nerve fibers in the medulla oblongata of vertebrates. 2) A hierarchical system in which quantities or classes become smaller and smaller as they approach the top of the hierarchy, as in the biomass pyramid and the pyramid of numbers in ecology. See BIOMASS PYRAMID, ENERGY PYRAMID, PYRAMID OF NUMBERS.

pyramid of biomass   See BIOMASS PYRAMID.

pyramid of energy flow   See ENERGY PYRAMID.

pyramid of numbers   A diagram showing the decreasing populations at each trophic level of an ecosystem. Although the numbers pyramid illustrates numbers of individuals, it corresponds roughly with the biomass pyramid because of decreasing tissue growth efficiency with successive trophic levels. Also called Elton pyramid (for Charles S. Elton, who recognized this trend in 1927). Compare ENERGY PYRAMID, BIOMASS PYRAMID.

pyrenoid   A small, spherical protein structure found in the chloroplast of most eukaryotic algae. Pyrenoids are involved in the storage of starch and are often surrounded by a layer or layers of starch.

pyrethrins   See PYRETHRUM.

pyrethrum   One of a number of species of chrysanthemum, or the organic insecticide produced from these species. The active agents extracted from the flowers are known as pyrethrins. Pyrethrins are used on garden insects and fleas; they are relatively nontoxic to warm-blooded organisms and break down quickly in sunlight.

pyridoxine CH_3-$CH_5HN(OH)(CH_2OH)_2$, the full name for the water-soluble vitamin B_6. Pyridoxine is an essential nutrient required for protein metabolism and the production of red blood cells.

pyrimidine base   A structural component of nucleic acids (RNA and DNA) and coenzymes. Purimidine bases are derived from the organic base pyrimidine ($C_4H_4N_2$), and characterized by containing the pyrimidine ring (a six-carbon ring in which two nitrogen atoms replace two of the carbon atoms). Cytosine, thymine, and uracil are all pyrimidine bases. See PURINE BASE, BASE-PAIRING.

pyrite   A brass-yellow mineral of iron sulfide that forms cubic crystals in a wide variety of rock forming environments. Also known as fool's gold. See APPENDIX p. 626.

pyroclastic   The collective term applied to rock fragments produced by volcanic explosions.

pyroclastic cones   The accumulations of loose debris ejected from volcanic eruptions. The cinders and tephra settle into a relatively steep-sided hill surrounding a central vent. Compare COMPOSITE VOLCANO, SHIELD VOLCANO.

pyroclastic flow   See ASH FLOW.

pyroclastic rock   A rock derived from materials ejected in a volcanic eruption. Pyroclastic rocks often exhibit a combination of brecciated igneous and laminar sedimentary textures.

pyroclasts   An individual particle or fragment explosively erupted from a volcano.

pyroclimax   A disclimax vegetation type

caused by periodic fires. Also called a fire climax. See DISCLIMAX.

pyrolusite   A black mineral of manganese oxide that often forms masses in the bedding or cleavage planes of rocks. Pyrolusite may be formed as a secondary mineral in manganese oxide nodules.

pyrolysis   Chemical decomposition or conversion of a compound through exposure to high temperatures, usually in the absence of oxygen.

pyrometer   An instrument for measuring high temperatures. Pyrometers are based on any of a number of different physical properties, including the radiation given off by hot materials.

pyroxene   Any of a group of minerals containing silicon and often found in igneous rocks. Pyroxenes include silicates of magnesium, calcium, and iron; they may also contain manganese, titanium, lithium, or sodium. See APPENDIX p. 626.

Pyrrhophyta   See DINOFLAGELLATA.

q

Q₁₀ Symbol for temperature coefficient. See TEMPERATURE COEFFICIENT.

quad Abbreviation for quadrillion (10^{25}) Btus.

quadrat A small plot of known area, used for sampling and studying larger ecological communities, especially plant populations. Compare TRANSECT.

quadrat method A technique for comparing plant communities based on sampling the vegetation in different marked plots (quadrats). See QUADRAT, TRANSECT.

quake See EARTHQUAKE.

quaking bog A type of bog characterized by a floating mat of vegetation (primarily sphagnum moss—peat—in various stages of decay) that "quakes" when walked upon. Quaking bogs often support a distinctive plant community. See BOG. Compare RAISED BOG.

qualitative 1) Descriptive and not expressible in numbers. A qualitative evaluation of a site might include a description of its scenic beauty, or the extent to which its vegetation appears to form distinct layers, rather than an estimate of the quantity of timber it could produce. 2) Referring to or involving a description of types rather than quantities. Qualitative analysis in chemistry determines what compounds are present, rather than how much of a particular compound is present.

quantitative Numerical, or capable of being expressed in numbers; derived from measurements or other numerical values. Contrast QUALITATIVE.

quantitative genetics The branch of genetics devoted to the study of quantitative characters, those traits such as height or foot size that show continuous variation because they are controlled by multiple factors (as opposed to traits such as eye color). See MULTIPLE FACTORS, MULTIPLE GENE INHERITANCE.

quantum mechanics The theory that unifies the wave and particle aspects of light, electrons, atoms, and molecules. Quantum mechanics replaces classical Newtonian mechanics for explaining circumstances involving speeds approaching that of light, particles the size of individual small molecules, and regions in which gravity is especially strong. See NEWTONIAN MECHANICS, PLANCK'S CONSTANT.

quantum yield The probability that a photon (quantum of energy) will produce a specific action. It is the ratio between the number of photons that induce the desired reactions to total number of photons absorbed.

quarantine A period of isolation to prevent the spread of disease (originally, such isolation lasted for 40 days). Quarantine is still imposed on patients with scarlet fever, to avoid spreading the disease to others. Quarantine of varying lengths is imposed on animals and on certain cargo arriving from other countries to see if they have transported a disease or a pest organism; it may also be imposed on people arriving from a country in which an infectious disease is epidemic.

quarry An open bedrock area where ore or building stone is extracted.

quarrying A synonym for glacial plucking.

Plucking is the preferred term. See PLUCK-ING.

quartz   A nonferromagnesian framework silicate mineral consisting entirely of silica tetrahedra; has clear or variable color related to minor inclusions; has chonchoidal fracture, no cleavage, and vitreous to glassy luster; hardness = 7; specific gravity = 2.65. See APPENDIX p. 627.

quartzarenite   A variety of quartz sandstone composed of at least 95 percent sand grains and less than 5 percent cement or matrix.

quartzite   Any hard and dense metamorphic rock derived from quartz sandstones. Compare ORTHOQUARTZITE.

quartzose sandstone   Any sandstone composed of at least 95 percent clear quartz grains and less than 5 percent feldspar grains and cement matrix.

quartz porphyry   A variety of porphyritic rock in which quartz is the predominant mineral. See PORPHYRITIC TEXTURE.

quassia   A tree, *Quassia amara*, found in the West Indies and tropical regions of the Americas. The wood of this tree has been used as a bitter tonic, for fevers and for intestinal parasites.

Quaternary   The second of two suberas of the Cenozoic era in geologic time. The Quaternary spans approximately 2 million years from the end of the Tertiary to the present. The Quaternary is a period, rather than a subera, in some systems of measurement. See APPENDIX p. 610.

quaternary ammonium salts   A class of organic compounds that are ionic, in which the cation consists of a nitrogen atom bound to four organic (hydrocarbon-containing) groups and the anion is generally a halogen (Cl^-,Br^-) or hydroxide. The general formula of a quaternary ammonium salt is $R_4N^+X^-$, where R is an organic group and X is the counterion. The nerve transmitter acetylcholine is a quaternary ammonium salt, and many neuroloically active drugs are quaternary ammonium salts that mimic the action of acetylcholine.

quench layer   A build-up of unburnt hydrocarbons on the walls of cylinders in an internal combustion engine such as an automobile. The quench layer develops from incompletely combusted fuel.

quick clay   Any unconsolidated sediment composed of clay-sized particles that has a low shear strength when saturated with water. Quick clay can be easily liquefied during a seismic disturbance.

quicklime   The common name for calcium oxide (lime). See CALCIUM OXIDE.

quicksilver   An old name for mercury. See MERCURY.

quinine   A bitter, alkaloid drug extracted from the bark of the South American tree *Cinchona officinalis*, used to treat malaria and other fevers.

quininism   Poisoning from cinchona or its quinine extract. Also called cinchonism.

r

R Abbreviation for the unit roentgen. See ROENTGEN.

r_{max} Symbol for biotic potential (innate capacity for increase). See BIOTIC POTENTIAL, INNATE CAPACITY FOR INCREASE.

Ra Chemical symbol for the element radium. See RADIUM.

rabbits Previously classified with the rodents, the 63 species of rabbits and their allies are now recognized as a distinct order, the Lagomorpha, whose resemblance to the rodents is only superficial. Each pair of the large incisors is accompanied by a second, much smaller pair and, in most forms, the rear legs are strongly modified for springing and jumping. Examples are rabbits, hares, and conies. See EUTHERIA, RODENTIA.

race A nonspecific term for populations of organisms within a species that are not geographically isolated and that have characteristics (such as physiology) that are distinct enough to be recognizable from generation to generation. Different breeds of domestic animals are races. The term is roughly equivalent to subspecies, but is often used in the same manner as ecotype and called a physiological race. Compare SUBSPECIES, VARIETY, FORM.

raceme A type of flower cluster with separate flowers on unbranched stalks of nearly equal length along a stem, with the lower flowers blooming first. Lily of the valley bears flowers in racemes.

rachion The part of a shoreline where substrate disturbance related to wave action and undertow is the greatest. The rachion often forms a step-shaped ridge parallel to the shoreline.

rachis 1) The main stem of a flower cluster or pinnately compound leaf. 2) A stalklike animal structure, such as the shaft of a feather, or the vertebrate spine.

rad A unit of absorbed dose of ionizing radiation, equal to 0.01 joule per kilogram or 10^{-2} gray. The rad has been replaced by the unit gray in the Système International d'Unités (SI). Absorbed dose is affected both by the kind and strength of the radiation and by the ability of a substance to absorb radiation.

radar Acronym for radio detection and ranging. It is a technique for measuring distances over a long range by bouncing radio waves off the target and analyzing the reflections of those waves. Radar can also be used to pinpoint locations or speeds of objects. Compare SONAR.

radial canals Tubes running from the stomach (located in the center of the organism) to the outer edge of the disc, part of the gastrovascular system in the medusa stage of cnidarians. See CNIDARIA.

radial drainage A term describing the appearance in map view of stream drainage systems that radiate out from a central point. Radial drainage often forms above a regional uplift such as a volcano or a structural dome.

radial growth The type of secondary growth produced by the cambium in some dicotyledon plants, consisting of concentric rows of new cells. See CAMBIUM.

radial symmetry Describing structures or organisms whose parts radiate out from a central axis. Such structures can be divided into two similar halves by any imaginary plane running through the center axis in any direction. Starfish and buttercups are radially symmetrical. See ACTINOMORPHIC, BILATERAL SYMMETRY.

radiant energy   Energy moving in the form of electromagnetic energy, such as x-rays, radio waves, infrared radiation, or visible light.

radiate   1) To spread out from a central point, or to give off waves of energy. 2) Characterized by radial symmetry, especially flowers that are like daisies in structure, having petals (ray florets) radiating out from a center of densely packed central flowers (disk florets).

radiation   1) Energy given off or traveling in the form of electromagnetic waves, photons, acoustic waves, or subatomic particles. 2) Ionizing radiation, such as the subatomic particles and electromagnetic radiation given off by substances undergoing radioactive decay. 3) A form of heat transfer. Contrast CONVECTION, THERMAL CONDUCTION.

radiation dose equivalent   A measurement of the degree to which a specific quantity of ionizing radiation can cause harm to tissues with different degrees of susceptibility to radiation. Also called dose equivalent, this value is calculated by multiplying absorbed dose by different values for different types of tissue. The resulting units are called sieverts (Sv). See SIEVERT.

radiation fog   Fog created when the ground and ground-level air cool by radiating the heat they have absorbed during the day, causing moist air to reach its dew point and condense. Such fog occurs most often on nights with clear skies or in early morning hours when air temperatures reach their lowest point. Radiation fog

collects in valleys and hollows. Compare STEAM FOG.

radiation sickness   Illness caused by exposure to radiation, either accidental or subsequent to medical treatment. Short-term effects include loss of appetite, nausea, vomiting, and diarrhea. Long-term exposure can produce sterility, anemia and impaired blood-cell formation (suppression of bone-marrow function), lowered resistance to infection, cataracts, and cancers or leukemia.

radiation window   The ability of a substance to allow the passage of electromagnetic radiation for only selected wavelength regions. The earth's atmosphere has radiation windows for visible light and radio waves because it lets radiation at some of these wavelengths pass through while absorbing radiation of other wavelengths.

radiative evolution   See ADAPTIVE RADIATION.

radical   A group of atoms or a single atom with an unpaired electron and no charge. In general, radicals are quite reactive and as a consequence have short lifetimes. Radical species result from homolytic bond cleavage reactions such as the decomposition of CFCs when exposed to ultraviolet light: $F_2ClC-Cl$ + UV light ——>$F_2ClC\cdot$ + $Cl\cdot$. Radicals have been implicated in the ozone destruction process.

radical chain reaction   A self-perpetuating chemical reaction promoted and sustained by the presence of radicals. An example is the reaction by which ozone is destroyed in the stratosphere. See RADICAL.

radicle   1) The first root produced by a germinating seed, the first structure to emerge from most seeds. 2) A rootlike structure, as in the fibril of a nerve cell.

radioactive decay   The process causing radioactivity; the spontaneous or induced disintegration of a substance in which ionizing radiation such as alpha particles, beta particles, or gamma rays is given off. Radioactive substances have characteristic rates of radioactive decay called half-lives.

radioactive disintegration   Another term for radioactive decay. See RADIOACTIVE DECAY.

radioactive emissions   Particles or rays given off by unstable isotopes in the process of radioactive decay. See ALPHA RADIATION, BETA RADIATION.

radioactive half-life   The period of time required for half of a quantity of radioactive nuclides to undergo decay. Half-life is independent of conditions and specific to each nuclide, so the property is used to characterize radioactive substances. Often simply called half-life. See RADIOACTIVE DECAY, NUCLIDE.

radioactive isotope   Isotope of an element that is unstable because its nuclei spontaneously emit high-energy particles, rays (alpha, beta, or gamma radiation), or both, in the process of decaying either into another radioactive isotope of lower atomic mass, or eventually into a stable, nonradioactive substance such as some isotopes of lead. Also often called radioisotope or radionuclide. See X-RAY.

radioactive substance   Material exhibiting radioactivity, either natural or artificially induced.

radioactive waste   Waste material sufficiently radioactive to be of concern.

radioactivity   The property of emitting subatomic particles and electromagnetic radiation shown by substances undergoing radioactive decay. The primary forms of radioactive emissions are alpha particles, beta particles, and gamma rays.

radiocarbon   Any radioactive isotope (^{14}C, ^{11}C, ^{10}C) of the element carbon. Radiocarbon ^{14}C undergoes radioactive decay, which may be measured and applied as an absolute dating technique.

radiocarbon technique   Using radioactive carbon as a tracer to follow and to measure uptake of labelled carbon dioxide in order to estimate primary productivity for aquatic ecosystems.

radiodermatitis   Skin inflammation (or, at extreme levels, burns) caused by exposure to ionizing radiation.

radioecology   A branch of ecology that studies the effects of radioactivity on living things, especially the plants and the animals of a particular community or population.

radio frequency   Electromagnetic radiation between 10^4 hertz and 3×10^{12} hertz that can be used for radio transmissions (very high frequency and ultra high frequency are also used for transmitting television signals). For example, AM radio stations use frequencies from approximately 500 to 1600 kilohertz (kHz); FM radio stations use 88 to 108 megahertz (MHz). Radio frequencies are divided into bands known as extremely low frequency, very low frequency, low frequency, medium frequency, high frequency, very high frequency, and ultra high frequency. Also called the radio spectrum.

radiogenic   Created by the process of radioactive decay, as in stable radiogenic isotopes.

radiogenic heat   Heat generated by radioactive decay.

radiograph   An x-ray; a photograph of an opaque object in which the image is formed on film (or a fluorescent screen) by ionizing radiation rather than visible light.

radioisotope   See RADIOACTIVE ISOTOPE.

radiolarian ooze   A pelagic sediment in which at least 30 percent of the composition is made up of the siliceous tests of radiolarian protists. Radiolarian ooze commonly forms below the carbonate compensation depth.

radiolarians   See ACTINOPODA.

radiometric   A term that relates to the measured radioactive decay of elements.

radiometric age   An estimated absolute age of a rock or a mineral that is based on radiometric dating methods.

radiometric dating   The determination of the estimated absolute age of a rock or a mineral by a comparison of the proportions of radioactive parent isotopes and daughter nuclides to a known half-life of a radioactive element.

radionucleotide   See RADIOACTIVE ISOTOPE.

radiosensitive   Easily or rapidly injured or changed by exposure to ionizing radiation. Cancer cells that are radiosensitive respond best to radiation therapy. Human organs such as the cornea and blood-forming bone marrow are most easily damaged by radiation because they are the most radiosensitive.

radiosonde   A device for measuring temperature, pressure, humidity, and upper-atmosphere winds. It consists of a balloon carrying measuring instruments and a radio transmitter to convey the readings to earth-bound observers. Radiosondes designed to study upper-atmosphere winds use radar to track the balloon's location. Wind speeds and direction can be calculated from the balloon's changes in position. See DROPSONDE.

radiotherapy   The use of radiation (usually x-rays, but can technically include ultraviolet light) to treat disease, usually forms of cancer. Sometimes called therapeutic radiology.

radium   (Ra) A naturally occurring, radioactive, fluorescent, metallic element with atomic weight 226 and atomic number 88. Radium is found in small amounts in uranium ores such as pitchblende. It is very unstable. It is used in research and some medical treatments.

radon   (Rn) An inert, gaseous, radioactive element with atomic number 86; the atomic weight of its most stable isotope is 222.0. Radon is formed by the decay of ^{226}radium. Radon exists in minute quantities in the atmosphere. It also exists in groundwater, making it unfit for human use, and in some soils and construction materials containing granite and similar rocks. In houses built of such materials, or built over granite deposits, radon (which is a carcinogen) can accumulate to hazardous levels, especially in basements, if proper ventilation is not installed.

radon daughters   The four decay products of radon: polonium 218, lead 214, bismuth 214, and polonium 214. All are short-lived radioactive metals.

radula   A tonguelike, stiff strip in the mouth of a mollusc, set with very small teeth and used for rasping food.

rafting   Dispersal of organisms by floating on objects.

rainbow   The arch composed of bands of color caused by water droplets causing sunlight to disperse into its component colors. Rainbows can be seen sometimes in a fine spray of water, as well as in the sky when the sun is behind the observer. They display the entire visible spectrum, with red at the outer edge of the curve and orange, yellow, green, blue, indigo, and violet toward the inside.

rainforest A large, dense forest found in regions receiving very heavy rainfall (usually more than 80 inches a year). Rainforests are characterized by very high numbers of different species; huge, broad-leaved evergreen trees; many vines (lianas); and abundant epiphyte plants. See TROPICAL RAINFOREST, TEMPERATE RAINFOREST.

rainmaking Seeding clouds with the intention of producing precipitation rather than of dispersing clouds. See CLOUD SEEDING.

rainout A form of wet deposition in which atmospheric particulates serve as condensation nuclei, eventually producing rain that carries the particulates out of the atmosphere and onto the earth's surface.

rain shadow Lack of rainfall on the leeward side of mountain ranges. It is caused when orographic lifting causes air to lose most of its moisture as it moves upward before crossing mountain ranges. This effect, which increases with the altitude of the mountains, causes the deserts to the east of the Sierra Nevada range and Mongolia's Gobi Desert. See OROGRAPHIC LIFTING.

rainwash A transport process that moves particles down a hill by a combination of raindrop splash energy and of downslope flow of rainwater.

raised beach See RAISED BEACH PLATFORM.

raised beach platform A landform composed of a beach or marine terrace that has been raised by regional uplift or abandoned by marine regression.

raised bog A bog in which dead *Sphagnum* moss has accumulated but not decomposed, building up layers that raise the bog vegetation growing through the moss above the surrounding land. Compare BLANKET BOG, QUAKING BOG.

ramet An individual shoot of a modular plant emerging from the ground, especially a clone produced by vegetative reproduction. Ramets may be visible, above-ground portions of the same genet. Compare GENET.

ramus A branch, especially of a forked structure; a primary division in a blood vessel or nerve; one half of the lower jawbone; a barb on a feather.

random distribution 1) A statistical distribution that follows no apparent pattern and is the result of pure chance rather than observable relationships. 2) Random spatial distribution. Compare NORMAL DISTRIBUTION. See RANDOM SPATIAL DISTRIBUTION.

random spatial distribution A form of dispersion of organisms within an ecosystem in which individuals are found at random distances from other individuals, showing no tendency to clump together in a consistent manner. Random spatial distribution is relatively rare and tends to be found in environments showing little local variation. Contrast CLUMPED DISTRIBUTION, UNIFORM SPATIAL DISTRIBUTION.

range 1) The limits of geographic distribution of a species. 2) A sequence of measured values between a highest and a lowest measured value, as in a range of temperatures. 3) A large area of grassland, scrub, or open woodland used for grazing livestock (or supporting foraging animals). A range denotes a much larger area than a field, one that may have poorly vegetated areas, and is often used to describe grazing lands of the western United States. Also called rangeland. See HOME RANGE.

range condition   A method for estimating how close a particular area of rangeland is to reaching its potential for producing (and sustaining) vegetation that can be used by grazing or browsing animals. See RANGE.

range management   A form of intentional land use that attempts to maximize the number of grazing animals that can be kept on a given area of rangeland. Good range management avoids overgrazing, which can result in the replacement of good forage species with less nutritious weed species and can eventually damage grassland so extensively that it converts to desert.

range of tolerance   Scope of chemical and physical conditions that provides optimum conditions for the survival of an individual or a population; the limits of temperature, moisture, nutrients, and so forth that cannot be exceeded in order for the individual, the population, or the species to continue to exist. Also called tolerance. See LAW OF LIMITING FACTORS, LAW OF TOLERANCE.

rank-abundance diagram   See DOMINANCE-DIVERSITY CURVE.

ranked preference   An order of preference shown by consumer animals between different food species.

Ranunculaceae   An angiosperm family often called the buttercup family. These herbaceous dicots have a diversity of flower patterns and of fruit types, but often have divided leaves, numerous stamens spirally arranged, and free carpels. See ANGIOSPERMOPHYTA.

Raoult's law   A chemical principle stating that the vapor pressure for ideal solutions at any pressure equals the sum of the vapor pressures of each component multiplied by the mole fraction of that component. Raoult's law describes the effect of a solute on the vapor pressure of the liquid in which it is dissolved. See VAPOR PRESSURE.

raphe   A crease or ridge marking the area where two structures join, such as the junction of the two hemispheres in vertebrate brains, or a central ridge on the valve of a diatom. In plants, raphe refers to a ridge on seeds or ovule where the stalk joins the ovule.

raptor   A bird of prey, such as an eagle or a hawk.

rare   Not commonly found; unusual and occurring in only a few cases. A rare species is one that is not found anywhere in large numbers, occurring only as small, local populations, but not currently believed to be in danger of extinction. See ENDANGERED SPECIES, THREATENED SPECIES.

rassenkreis   See POLYTYPIC.

rate   A quantity, amount, or degree of something measured in relation or proportion of another quantity or amount (or a standard value). Examples include miles per hour or carbon dioxide (CO_2) taken up per area of leaf tissue per unit time (photosynthetic rate).

rate of natural change   The difference between the crude birth rate and the death rate, showing the speed at which the size of a population is increasing (or decreasing).

rate of population growth   The increase in the population in a specific area (death rate subtracted from birth rate, plus migration into the area) divided by the size of the population.

ratites   Common name for large, flightless birds of the superorder Palaeognathae, such as the ostrich. See AVES, PALAEOGNATHAE.

Raunkiaer's life forms   A widely used system for classifying vegetation based on the position of a plant's dormant (perennating) buds during its resting season (winter or dry season). The degree to which a plant's resting buds are hidden or protected shows its adaptation to extreme conditions. The different life forms, in increasing order of protection of bud, are epiphyte, phanerophyte, chamaephyte, hemicryptophyte, cryptophyte (often divided into geophyte, hydrophyte, and halophyte), and therophyte.

raw sewage   A general term for untreated waste materials, especially as applied to human metabolic waste.

raw sludge   Semisolid sewage before it undergoes a primary treatment process.

raw wastewater   Civil or industrial aqueous waste that has not been purified or treated.

ray   A thin line or beam of radiant energy, usually one that is radiating out from its source rather than focused into a beam. Also, the path followed by a wave of electromagnetic energy.

Rayleigh scattering   One form of deflection of electromagnetic radiation as it travels through matter. Also called elastic scattering, this does not affect the energy of the molecules and atoms. The photons simply bounce off some of the molecules and atoms as they travel through the substance. Rayleigh scattering occurs when light passes through soot or fine dust suspended in air. Because blue light scatters more effectively than other colors (such as red), Rayleigh scattering is also responsible for the blue of the sky.

Rayleigh-Taylor instability   A wave or ripple of instability that forms on a surface and moves along it. The Rayleigh-Taylor instability is used in theories explaining phenomena in the earth's magnetosphere and ionosphere.

Rayleigh wave   A type of surface seismic wave characterized by an ellipsoidal particle motion in a vertical plane that is oriented parallel to the direction of wave propagation. See SURFACE WAVE.

RCRA   See RESOURCE CONSERVATION AND RECOVERY ACT.

RDF   Abbreviation for refuse-derived fuel. See REFUSE-DERIVED FUEL.

reactivity   The tendency for a chemical to combine with other elements and compounds.

reactor   See NUCLEAR REACTOR.

reactor vessel   A reinforced container within fission nuclear reactors surrounding core, control rods, moderator, and coolant. It is usually made of thick steel. See NUCLEAR REACTOR.

reagent   A compound involved in a chemical or biochemical reaction, especially one used in chemical analysis to produce a characteristic reaction in order to determine the presence of another compound.

realized niche   The actual niche (ecological role) filled by an organism or a species within an ecosystem. The realized niche is smaller than the theoretical (fundamental) niche because of the natural constraints imposed by other organisms, including competitors, predators, and disease organisms. Contrast FUNDAMENTAL NICHE.

reauthorization   An act of the U.S. Congress (or a U.S. state legislative body) to extend the effective life of a law, usually involving amendments.

receiving water   Any stream, lake, or ocean into which treated or untreated wastewater is ultimately discharged.

recent   See HOLOCENE.

receptacle   The structure that supports the reproductive organs of a plant, especially the end of a stem supporting a flower cluster or the various parts of a single flower.

receptor   1) A chemical group within a large molecule or a cell that can combine selectively with other such molecules or cells to activate or to deactivate an enzyme, a hormone, an antigen, a neurotransmitter, or another chemical substance. 2) Short for receptor cell. See RECEPTOR CELL.

receptor cell   Sensory nerve endings, sense organs, or other cells within the nervous system that are specialized for detecting stimuli (internal or external changes), translating them into signals that can be transmitted to the brain. See CHEMORECEPTOR, EXTEROCEPTOR, PHOTORECEPTOR, PROPRIOCEPTOR.

recessive   Describing an allele in a pair of contrasting alleles whose expression is suppressed by the different allele (the dominant one). The inherited trait governed by that allele only appears if both alleles in the pair are identical for the recessive trait (homozygous). Contrast DOMINANCE.

recharge   The volume of meteoric water that is added to the water table, usually by the process of infiltration.

recharge area   The portion of a catchment basin or a watershed area that contributes to groundwater recharge.

recharge well   A well that serves as a water source in the forced or artificial recharge of an aquifer.

recharge zone   The subsurface portion of the soil that contributes to groundwater recharge at the top of the zone of saturation. See ZONE OF SATURATION.

reciprocal predation   A competitive interaction between two species in which each preys on the other, as when adult insects of different species eat the eggs, the pupae, or both of the other species.

reclamation   A general term for the filling, grading, and reseeding or replanting of land that has been disturbed by a natural disaster such as fire or flood.

recombination   The phenomenon in which offspring can display new combinations of traits not seen in either parent, resulting from new genetic combinations caused either by crossing-over or by independent assortment. Recombination can be produced artificially in the laboratory by disrupting and recombining DNA from similar or different organisms, producing recombinant DNA. See GENETIC ENGINEERING.

record (geologic)   See GEOLOGIC RECORD.

recruitment   The addition of individuals to a population through reproduction and immigration.

rectangular drainage pattern   The map-view appearance of a stream drainage system that develops over zones of weakness in regularly jointed or faulted bedrock. The streams and tributaries in rectangular drainage exhibit many equiangular bends that separate reaches of approximately equal length.

recumbent fold   A fold in bedrock in that the limbs and the axial plane of the fold are oriented within approximately horizontal planes.

recurrence interval   Time duration between repeat occurrences of a phenomenon.

recycle   To collect and to reprocess a resource so that it can be made into new products, such as salvaging bottles or aluminum cans for processing into new bottles or cans. Recycling differs from reuse by involving reprocessing; reuse refers to using a resource again in its original form, as in washing and refilling a container.

red algae   See RHODOPHYTA.

red bed   A general reference to a sedimentary rock which is strongly colored by reddish hematite.

red blood cells   See ERYTHROCYTES.

red clay   A very fine-grained pelagic sediment composed of brown or red clay from terrigenous sources. The clay material is distributed to deeper parts of the ocean by wind and currents.

red data book   A catalog published by the International Union for the Conservation of Nature and Natural Resources (IUCN) that lists rare species and those in danger of extinction.

Red Queen hypothesis   The hypothesis (proposed by L. Van Valen) that the evolutionary advance of one species represents the deterioration of the envi-

ronment for the other species with which it interacts. Therefore, like the Red Queen in L. Carroll's *Through the Looking Glass* who had to run as fast as she could just to stay in place, organisms must evolve as fast as they can merely to survive.

red tide   A reddish bloom that sometimes occurs in surface waters of the ocean, caused by rapid population increases of dinoflagellate algae (phytoplankton) species. Such algae can produce toxins (similar to the botulism toxin) that poison fish and persist in shellfish that feed upon them, making the shellfish toxic to humans. Although red tide occurs naturally, it may be stimulated by increases in nutrients such as phosphates that result from human activity. See ALGAL BLOOM, DINOFLAGELLATA.

reduce   1) To lessen or to make smaller. For example, a reduced tillage system requires less plowing and turning of the soil than do traditional systems of cultivation. 2) To cause a substance to undergo chemical reduction, either by removing oxygen from a compound or by adding electrons.

reducer   See DECOMPOSER.

reducing agent   A chemical substace that is capable of donating electrons to another substance. A reducing agent carries out the process of reduction and is itself oxidized (i.e., it loses electrons). See OXIDATION, REDUCTION.

reducing environment   An environment that is low in free oxygen, such as that found in the lower levels of marine sediments or in waterlogged soils. Aerobic organisms cannot live in reducing environments, although chemosynthetic bacteria and anaerobic organisms can live there.

reduction   The opposite of oxidation; a chemical reaction in which a compound undergoes a decrease in oxidation state by gaining electrons. Reduction of one compound is always accompanied by oxidation of another compound. Many reduction reactions involve removing oxygen atoms from a compound or liberating metals from their compounds. Contrast OXIDATION.

reduction division Another term for meiosis, so called because it is the form of cell division that produces cells with a reduced number of chromosomes, a haploid rather than a duplicate diploid set. See MEIOSIS.

reductionism Process of breaking down a system into its components and examining the components independently to resolve complex scientific questions. Knowledge about separate components is used to make generalizations about the system as a whole. It is the opposite of a holistic approach. Contrast HOLISTIC.

reef A ridge or mass of rocks or of coral (or an extended sand bar) near the surface of the ocean. See BARRIER REEF, CORAL REEF ECOSYSTEM.

reef flat In a coral reef, a flat expanse of dead coral forming a rocky platform, which is partly or completely dry at low tide and often contains small pools of various depths.

refine The process of separating impurities from oil, coal, or ore material.

reflection The turning back of a ray of electromagnetic radiation or high-energy particles, as when bouncing off an impermeable surface. Light is reflected off of mirrors. Contrast REFRACTION.

reflux Simmering; a laboratory technique in which liquids are boiled in a container attached to a condenser that returns condensed vapor to the container, preventing it from boiling dry. Refluxing is used for organic synthesis reactions that need a long period of time to reach completion.

refoliation The leafing out of a plant whose leaves have been stripped, as of a tree that has been attacked by large numbers of gypsy moths or other defoliators, or one that has been exposed to a late spring frost.

reforestation Planting trees (or tree seeds) in an area that was once forested but where all the trees have been cut.

refraction The bending of direction of a wave that occurs as the wave crosses the boundary between two different media, or travels through a substance whose density is not uniform. Contrast REFLECTION.

refraction (seismic) A change in direction or propagation that occurs as a seismic wave crosses a boundary between Earth materials of different densities. A strong refraction occurs at the Gutenberg and Mohorovicic discontinuities.

refractory period A temporary period of lessened or zero responsiveness that immediately follows a previous response in an organism or a tissue such as a muscle.

refrigerant Fluids used in refrigerator systems. Freon and ammonia are common refrigerants because of their low boiling points. However, research is being directed at discovering a non-chlorofluorocarbon replacement for Freon to reduce atmospheric ozone depletion.

refuge 1) An area designated for the preservation of a number of species (such as a bird refuge) where they can escape destruction of their habitats and continue to live safely. Also called a wildlife refuge or preserve. 2) A device used by an organism to avoid predation.

refugium   A localized area containing species that have survived environmental changes that have occurred around them, usually because of a favorable microclimate. In some U.S. midwestern states where the prairie has vanished beneath the plow, remnants of prairie ecosystems are preserved in rural graveyard plots. See RELICT DISTRIBUTION.

refuse-derived fuel   (RDF) Energy derived from materials that would otherwise be disposed of as waste. RDF is a broad category that includes biomass converted into methane, ethanol, and biocrude as well as municipal garbage burned in cogeneration incinerators (resource recovery plants such as the Vicon facility in Pittsfield, MA). See BIOMASS FUEL.

reg   A region characterized by stony or gravelly surface. A reg typically develops in a desert region where fine particles are removed by wind erosion.

regional metamorphism   A type of metamorphism that occurs on a regional scale (for example, in an orogenic belt) and that is most often a result of tectonic plate convergence. Regional metamorphism produces rock fabrics that decrease in metamorphic grade with increasing distance from the source of tectonic activity.

regolith   The loose and unconsolidated, but otherwise unaltered, layer of rock overlying the bedrock. See WEATHERING, SOIL PROFILE.

regression analysis   Statistical term for the analysis of the relationship between two variables when one variable is dependent upon another independent variable, as expressed by a mathematical equation of a line.

regular distribution   See UNIFORM SPATIAL DISTRIBUTION.

regulator genes   Any genes that control the rate of synthesis of molecules by working with an operator site or repressor molecule to control the function of an operon. See OPERON.

regulatory response   A rapid behavioral or physiological reaction by an organism to a change in its environment, capable of being reversed if the environment changes back to its original state.

rejuvenation   The latest stage in a cycle of erosion during which a regional uplift increases stream gradients and a new phase of erosional dissection of the landscape begins. See CYCLE OF EROSION.

relative abundance   A rough estimate of population density of a given species. It is calculated from the number of individuals of a certain species sighted over a certain period of time (number of birds counted per hour) or in a particular place, divided by the total of all species in a community. Relative abundance is usually expressed as a percentage.

relative age   A determination of relative chronology based upon physical position, stratigraphy, fossil content, degree of weathering, or soil development. Contrast ABSOLUTE AGE.

relative basal area   The basal area of a species divided by the total basal area of all species in a community; usually expressed as a percentage. See BASAL AREA.

relative date   An age measured in relation to a rock unit of known relative age and without reference to fixed time units, such as years.

relative density   The density of a species divided by the total density of all species in a community; usually expressed as a percentage. See DENSITY.

relative frequency 1) The number of samples in which a species occurs divided by the total number of samples; usually expressed as a percentage. If a species occurs in half of the samples, it has a relative frequency of 50 percent. 2) The number of occurrences of a particular event divided by the total number of trials.

relative humidity The ratio comparing the amount of moisture actually present in the air with the maximum amount of moisture that could be present in the air at that temperature. It is expressed as a percentage; this value is often used to express levels of discomfort during hot weather in humid locations such as Washington, DC. See HUMIDITY, ABSOLUTE HUMIDITY, SPECIFIC HUMIDITY.

relative resource scarcity Condition in which supplies of a resource exist, but are prevented from becoming available in a particular area, or are not available in sufficient quantities to meet the demand and thus cause a localized shortage. See ABSOLUTE RESOURCE SCARCITY.

relative transpiration The rate at which a plant loses water through transpiration per unit of surface area divided by the rate of evaporation from the surface of open water under exactly the same conditions. See TRANSPIRATION.

relativity The theories advanced by Albert Einstein concerning space and time. Einstein assumed that laws governing matter, space, motion, and time remain the same in all frames of reference. As a result, it is not possible to draw conclusions about the absolute motion of a frame of reference. Relativity also assumes that the speed of light is constant throughout the universe and independent of the velocity of any observer. From these assumptions can be derived space contraction, time dilation, relativity of simultaneity, and equivalence of mass and energy—the familiar $E = mc^2$.

relay floristics A theory of succession proposing that groups of species arrive on a site, establish dominance, and then are replaced by other groups of species until a climax community is reached. The changes in species composition through succession are viewed as relays of different species groups. Contrast INHIBITION.

release 1) Short for ecological release. 2) A forestry practice of cutting down large, poor-quality overstory trees to allow high-quality understory or suppressed trees to grow more rapidly; often called release cutting.

releaser A behavioral signal that provokes a particular, predictable reaction in an animal, such as the cry of a young animal that stimulates its parent to feed it. Also called sign stimulus.

relevé 1) Vegetation sampling technique in which sample stands are chosen to reflect homogenous, ideal examples typifing a community (in contrast to the random sampling of a community or a group of communities). Popular in Europe, relevé is French for abstract. 2) List of plant species found in a specific area (with a visual estimate of the amount of canopy cover in each layer) and a description of soils and other habitat characteristics.

relict A community fragment that has survived environmental change.

relict distribution The localized occurrence throughout a region of a species or a genus that now occupies only a fraction of the geographic area that it previously occupied. See NUNATAK, REFUGIUM.

relict sediment   A continental shelf deposit that was emplaced by a process that is no longer active. A shallow marine deposit that remains intact after a marine transgression is an example of a relict sediment.

relief   1) A general reference to the degree of variation in elevation between parts of a landscape. 2) The variations in the refractive indices of minerals and the mounting medium in thin-section microscopy; the mineral features appear to stand out prominently if the difference is high.

relief map   A map that illustrates the topographic relief of the map area, either by shading or by raised three-dimensional forms.

rem   Acronym for roentgen equivalent man, a unit for biological damage of an absorbed dose of radiation. It takes into account the different amounts of biological damage caused by the same amount of absorbed energy from different types of ionizing radiation. It uses as a standard of comparison the biological damage caused by 1 roentgen of x-rays. The rem is gradually being replaced by the sievert (1 sievert = 100 rems). See ROENTGEN, REP, RAD, SIEVERT.

remanent magnetization   The natural magnetism acquired in rocks during igneous, thermal, depositional, or chemical processes. Remanent magnetization remains fixed after an externally applied field is removed. See CHEMICAL REMANENT MAGNETISM, DEPOSITIONAL REMANENT MAGNETISM.

remobilization   The plastic or fluid movement of country rock materials that are subjected to the effects of a body of magma.

remote sensing   Collecting information without having direct contact of the entity being studied. Remote sensing from aircraft or from satellites provides essential information about the earth. It includes such techniques as radar, aerial photography, and infrared photography.

rendzina soils   An older soil classification term applied to brown grassland soils that develop over calcareous parent materials. See INCEPTISOLS, MOLLISOL.

renewable energy   Sources of energy that are not depleted by consumption, such as solar and wind energy.

renewable resources   A resource that potentially can last indefinitely without reducing the available supply because it is replaced through natural processes. Timber, shellfish, and grasslands are examples of renewable resources. Nonrenewable resources such as coal and oil may eventually be replaced by natural processes, but these occur over long periods of geologic time rather than within the time frame of current civilization. Contrast NONRENEWABLE RESOURCES.

Rensch's laws   1) In cold climates, birds have larger clutches of eggs and mammals have larger litters than the same species in warmer climates. 2) Mammals have less insulating fur and birds have shorter wings in warmer environments. 3) Land snails of colder climates have darker shells than those of warmer climates. 4) Shell thickness in land snails is related to aridity and sunlight. See ECOGEOGRAPHIC RULES.

rep   A unit of measurement of radiation absorbed equal to the amount of ionizing radiation that transfers 93 ergs of energy to each gram of living tissue. Rep, an acronym for roentgen equivalent physical, is often used to measure beta radiation. See RAD, RADIATION.

repellent   A substance that scares off or

repels unwanted organisms, such as an insect repellent.

replacement capacity   1) The ability for a biological system to regenerate its original state after harvest (or other form of use). 2) Net reproductive rate.

replacement-level fertility   The fertility rate required to maintain the current size of the reproducing population. The average rate worldwide is just over two children per family (2.1 to 2.5 children, because sometimes children die before their parents).

replication   1) Repetition or duplication. 2) The duplication of genetic material, as in the formation of new DNA, or the duplication of chromosomes prior to cell division. 3) Replacing destroyed habitat, especially wetlands, by artificially creating new ones (by earth moving and by revegetation with wetland species, or with other appropriate species).

reprocessing   Mechanical and chemical treatment of used nuclear fuel to recover fissionable fuel and separate out waste.

reproduction   The creation of offspring, new generations of individuals, by living organisms, either through sexual or asexual mechanisms. Reproduction perpetuates the species.

reproduction curve   A graph showing the number of individuals at a given stage in a particular generation, relative to the previous generation.

reproductive age   The age (or age range) at which an organism is sexually mature and capable of bearing young. For humans, the range is approximately ages 15 to 44 (the prime reproductive age covers a smaller range).

reproductive allocation   The proportion of

an organism's available resources that is devoted to reproduction over a specific period of time. Sometimes used for the percentage of an organism's body (mass or volume), or a plant's net primary production, that is specialized as reproductive tissue.

reproductive cost   A lowering of rates of survivorship or of growth caused by an organism increasing its reproductive allocation, and the subsequent decrease in the potential for future reproduction. See REPRODUCTIVE ALLOCATION.

reproductive effort   The total number of seeds and vegetative disseminules produced by a modular organism. Compare REPRODUCTIVE ALLOCATION.

reproductive failure   The inability for a population or individual to successfully bear young. Exposure to DDT caused reproductive failure in raptor species at the top of the food chain because it resulted in eggshells that were too weak to protect the developing young. Compare REPRODUCTIVE SUCCESS.

reproductive isolation   The inability for a given population to interbreed with other populations of the same species, which may be caused by a geographical barrier or by intrinsic factors.

reproductive output   The offspring of a given individual or population.

reproductive potential   1) The absolute number of offspring that a given population of a species could produce if provided with unlimited resources and ideal environmental conditions. 2) The number of offspring produced by a single female in a population.

reproductive rate   The number of offspring produced over a specified period of time, such as a year.

reproductive success   The number of offspring of an individual surviving at a particular time.

reproductive value   The expected contribution to the future population made by an individual female of a certain age from that age to the end of her reproductive life. See RESIDUAL REPRODUCTIVE VALUE.

Reptilia   Because of its many divergent lines of evolution, including those that led to mammals and birds, this class of vertebrates is difficult to define with precision. The skin, as compared to that of amphibians, is dry, lacking in glands, and covered with scales. Bony plates may develop in the dermis, particularly in the head region. Claws are present and skull, limb bones, vertebrae, muscles, and so forth are stronger and more advanced than those of reptiles. Fertilization is internal, there is no larval stage, and eggs are cleidoic. See CLEIDOIC EGG, AMNIOTA, VERTEBRATA.

reserves   Resources that have been identified and from which a usable mineral can be extracted profitably at present prices with current mining technology. Reserves are subdivided into proven, probable, possible, and undiscovered, depending on the extent to which they have been identified and mapped and how profitable they are expected to be.

reservoir   1) An artificial body of surface water that is retained by a dam. 2) A natural underground rock formation that retains water, oil, or natural gas.

reservoir effect   The rapid surge in biological production that follows the flooding of land (as in the construction of a reservoir) because of the accompanying deposition or release of nutrients. The effect is temporary, unless the land is repeatedly flooded,

as some riverbanks.

reservoir rock   Porous rock that serves as an underground reservoir for water, oil, or natural gas.

residence time   The average length of time a molecule of a substance such as a nutrient takes to pass through a specific compartment of a model (such as a watershed) or ecosystem. For example, the residence time of water in a lake is calculated by dividing the volume of the lake by the mean annual water flux into the lake.

residual oil   The viscous and combustible bottom product of a crude oil distillation process. Residual oil is also known as liquid asphalt.

residual reproductive value   (RRV) The probable contribution of organisms of specific age groups within the reproductive ages to produce offspring in the future, excluding the present progeny. RRV refers only to expected future progeny. Compare REPRODUCTIVE VALUE.

residue   Material remaining after some process has occurred, such as pesticide residues that stay in the soil after the pest(s) have been killed.

resilience   The ability of an ecosystem or other natural system to return to original or steady-state conditions after a disturbance.

resin   1) A synthetic polymer that has not yet undergone a final curing process. Also the general term used for any plastic; sometimes called synthetic resin. 2) The sticky sap of some tree species, such as pines. Such resins as amber and rosin are sometimes called natural resins to distinguish them from synthetic resins. 3) An adsorbant used in ion chromatography or an ion-exchange process. See POLYMER.

resinous   Containing or resembling resin, a sticky vegetable substance exuded from several species of plants or trees (rosin, amber, copal).

resistance   The opposition to the flow of electricity through objects, equal to the potential difference across the object divided by the resulting current. Resistance is expressed in units of volts per ampere (ohms); it is the reciprocal of conductance.

resistivity   The opposite of conductivity; the degree of resistance characteristic of a substance. Resistivity is measured in ohm meters.

resonance   1) The relatively high amplitude oscillation that occurs when a system (atoms, molecules, electric circuits, acoustic and mechanical devices) is stimulated by an oscillating driving force at or very close to its natural frequency. This vibration is noticeably greater in amplitude than vibrations induced at other frequencies. Resonance is the minimum point of mechanical or acoustical impedance. 2) A particle that exists for very brief periods when a more stable particle is excited.

resonant frequency   The frequency at which resonance occurs in a system. See REASONANCE.

resorption   1) The act or process of reabsorbing, as in removing pus or other matter by absorption or absorbing digested food into the circulatory system. 2) Retranslocation of nitrogen and of potassium from leaves to woody tissues in perennial plants at the onset of dormancy.

resource   1) A component of the environment (often related to energy) that is utilized by an organism. 2) Anything obtained from the living and nonliving environment to meet human needs and wants.

resource allocation   The distribution of limited resources throughout an ecosystem; also, the way in which an individual organism uses limited resources, such as the proportion of (food) energy devoted to various tissues or to various activities such as hunting, reproduction, etc.

resource competition   Density-dependent population responses within or between species that result from the competition for resources in short supply. Also called scramble competition. Compare INTERFERENCE COMPETITION.

Resource Conservation and Recovery Act   (RCRA, pronounced "rickra") Federal law (42 U.S.C.A. § 6901 et seq.) mandating appropriate identification, tracking, and disposal of hazardous waste. It established a system of Environmental Protection Agency permits and fines for enforcement. Passed in 1976, the law was amended in 1984.

resource depletion zone   The area surrounding a consumer organism in which the availability of its food resource is reduced.

resource development   Generating or promoting techniques or systems for the utilization of natural resources. See EXPLOITATION.

resource exploitation   See EXPLOITATION.

resource partitioning   Specialization within an ecosystem allowing different populations and species to reduce direct competition for limited resources, as by eating slightly different types of food. See NICHE.

resource recovery   1) The somewhat euphemistic industry term for incineration of municipal and industrial solid waste in which the heat produced by burning waste is "recovered," harnessed for energy generation (driving turbines) or space heating. Resource recovery plants are an increasingly popular form of waste disposal, especially as landfills reach capacity or are closed for environmental reasons, because they reduce the volume of the waste stream to a very small volume of ash. 2) Salvaging of usable metals, paper, and glass from solid waste and selling them to manufacturing industries for recycling or reuse.

resource switching   The shift of an organism from exploiting one resource (a food species) to another that has become more abundant. Compare APOSTATIC SELECTION.

respiration   The process in which a living organism or cell utilizes oxygen to convert food compounds into carbon dioxide and water, releasing energy. See AEROBIC RESPIRATION, ANAEROBIC RESPIRATION, GLYCOLYSIS, TRICARBOXYLIC ACID CYCLE.

respiration efficiency   A ratio of the energy in respiration to energy ingested (or absorbed by a plant) for an organism or trophic level. Compare ASSIMILATION EFFICIENCY, ECOLOGICAL GROWTH EFFICIENCY, TISSUE GROWTH EFFICIENCY.

respiratory fibrotic agents   Any substance that damages the lungs and causes inflammation, hardening of tissues, or growth of fibrous tissues, reducing breathing capacity. This class of irritants includes corrosive chemicals and especially particulate materials such as coal dust, asbestos, or cotton and paper fibers. See ASBESTOSIS, BERYLLIOSIS, BYSSINOSIS, PNEUMOCONIOSIS, SILICOSIS.

respiratory pigment   A protein that carries oxygen throughout an organism for cellular respiration, such as hemoglobin and hemocyanin (an analogue of hemoglobin found in crustaceans and molluscs).

respiratory quotient   (RQ) The ratio of the volume of carbon dioxide released to that of oxygen consumed in respiration over a specified time period (CO_2 evolved: O_2 absorbed). For humans, the normal respiratory quotient is 0.9.

resting potential   The difference in electrical potential across a cell membrane, between the outside and the inside, of a cell at rest.

rest mass   The mass of a particle at rest, that is, when the observer measuring the particle is at rest relative to the particle. Unless particles move at speeds approaching the speed of light, the value for their relativistic mass will be nearly the same as their rest mass.

restoration ecology   The study of primary succession and revegetation of derelict land, and the application of such knowledge to reestablish cover on land that has been strip-mined or otherwise stripped of any vegetation. Wetlands replication is a new branch of restoration ecology.

resurgence in pests   A rapid increase in pest populations after the immediate impact of a control measure, usually chemical pesticides, has worn off. It is caused by destruction of the natural predators and parasites that normally slow the growth of pest populations.

resurgent caldera   A volcanic caldera that becomes reactivated with flowing magma. A resurgent caldera may form a new volcanic dome within the existing crater.

rete   A network or a netlike structure in animals, usually a particular network of blood vessels or nerves.

retention pond A small body of water created to allow solids to settle out of water. Retention ponds are required on large construction sites to allow silt washed from the site to settle out rather than to be carried into neighboring waterways and to clog them.

retention time The amount of time a chemical is held on the surface of the adsorbing medium during the process of chromatography. Different chemicals are held for different lengths of time so that retention time can be used to identify different chemicals. See CHROMATOGRAPHY.

reticulate evolution Evolutionary change caused by recombination and splitting among several interbreeding populations. If the generations in reticulate evolution are plotted, they show as interweaving lines of descent among the different populations.

reticulum 1) Any netlike (reticulated) system or structure, such as the endoplasmic reticulum of cells, a network of membranelike tubules. 2) The second stomach division in ruminating animals (cud-chewing animals such as cows and sheep). See RUMEN, RUMINANT.

retinol Another term for vitamin A derived from animal sources, as opposed to carotene, the vitamin A precursor that comes from plant sources. See VITAMIN A.

retro-arc basin A type of back-arc basin formed in continental crust and that has an abundance of terrigenous clastic sediments. See BACK-ARC BASIN. Contrast INTER-ARC BASIN.

retrofit To renovate a building or a piece of equipment using improved technology, especially to meet new environmental standards.

retrograde metamorphic effects A type of metamorphism that occurs while rocks are cooling in the presence of fluid phase materials. Retrograde metamorphic effects occur most commonly where regional metamorphism does not reach a grade high enough to drive off all water within the rock.

retrovirus Any of several viruses that contain RNA and which are able to manufacture DNA copies of their RNA and to insert these into the chromosomes of their human or animal hosts. AIDS is caused by a retrovirus, and these microorganisms often produce cancerous tumors. See AIDS.

retting A step in the processing of flax into linen fiber. The harvested flax stems are soaked in water so that bacteria can help soften the tough parts of the stems. After retting, the fibers can be separated from the stem by beating.

reverberation time The amount of time it takes for a sound of average intensity to fall 60 decibels to the point where it is barely audible in a closed room. Reverberation time, measured in seconds, is important in designing auditoriums with good acoustics.

reverberatory furnace A device used in refining metals, in which heat is applied to the roof of the chamber and carried to the ore by radiation. This keeps the ore from being contaminated by the fuel or by combustion fumes.

reverse fault A fault in which the sense of displacement of the hanging wall is upward along a fault plane that has a dip measuring between 45° and 90°. Contrast NORMAL FAULT.

reverse magnetized   A term applied to a rock in which the natural remanent magnetism indicates a polarity which is opposite that of the earth's present magnetic field.

reverse osmosis   A process in which water is forced under pressure through a selectively permeable membrane. It is called reverse osmosis because the water is driven in the opposite direction—toward a less concentrated solution—than the direction in which osmosis naturally occurs. Reverse osmosis is used to purify water and to speed the process of converting maple sap to maple syrup. Also called microfiltration and ultrafiltration. See OSMOSIS.

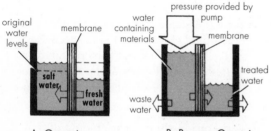

A. Osmosis   B. Reverse Osmosis

A) Fresh water will migrate through a membrane toward a saltier, more concentrated solution. B) But if pressure is applied to the more concentrated side, with a pump for example, osmosis can be reversed, and minerals will be strained by the membrane. Many water purification systems operate on the principle of reverse osmosis.

revolution   The motion of one object around another, such as the orbit of a planet around the sun. Also, one complete cycle of such motion is called a revolution. Contrast ROTATION.

reworked   A term applied to a rock or a mineral that has been displaced from a point of origin and incorporated into a newer rock or deposit.

rheophyte   A plant adapted for growing in flowing water, such as watercress.

rheotaxis   The movement of a cell or an animal in response to a stimulus of flowing water or air; the resulting movement is against the direction of flow of the current. See TAXIS.

rheotrophic   Describing or relating to organisms that extract resources (nutrients and energy) from flowing water.

rhizobia   Members of the genus *Rhizobium*, rod-shaped, nitrogen-fixing bacteria that exist symbiotically in root nodules of leguminous plants. See NITROGEN-FIXING BACTERIA.

rhizoid   One of the filaments that functions as a root in mosses, ferns, liverworts, and fungi, anchoring them to the ground and possibly absorbing some nutrition. Rhizoids lack the complex vascular system found in true roots or rhizomes. Compare RHIZOME, ROOT.

rhizome   A horizontal, often-enlarged underground stem of some plants, from which both roots and leafy shoots sprout. It stores food to be used by the new plant the following year and has nodes, buds, and small-scale leaves. Irises and a number of other herbaceous perennials form rhi-

rhizome

zomes. Sometimes called rootstock. Compare STOLON, CORM, BULB, TUBER.

rhizomorph　A dense, rootlike strand of hyphae by which fungi (such as the fungus responsible for causing dry rot) spread.

rhizoplane　The outer root surface of plants, one component of the rhizosphere. See RHIZOSPHERE.

Rhizopoda　A phylum of the Protista that includes the amoeba. Although lacking flagella, amoebae have motility through cytoplasmic extension of flowing, footlike pseudopods. Reproduction of these organisms is by mitotic fission, lacking meiosis and sexuality. Members of this phylum include the carnivorous aquatic amoeba *Amoeba proteus*, the organism causing amoebic dysenteries *Entamoeba*, and the herbivorous *Saccamoeba*. See PROTISTA.

rhizosphere　The plant-root zone of soil, that immediately surrounding the roots of a plant. Uptake of plant nutrients occurs in the rhizosphere, and the population of microorganisms in this zone is markedly different from that of the rest of the soil.

rhodochrosite　A translucent, rose-colored mineral of manganese carbonate that occurs in hydrothermal veins and in metasomatically altered sedimentary rocks. Rhodochrosite is a minor ore of manganese.

Rhodophyta　A phylum of the kingdom Protista, the red algae are commonly found in marine coastal environments. All members of the phylum reproduce sexually, but not through the production of flagellated sperm cells. The sperm is merely released into the water near the large egg cells, followed by fertilization and then meiosis to produce haploid cells. Some members of the phylum produce diploid carpospores that have the capacity to produce a small thallus with haploid–spore-producing tetrasporangia. In addition to chlorphylla a, red algae have rhodoplasts which contain phycocyanin, allophycocyanin, and phycoerythrin. Familiar members of the phylum include Irish moss *Chondrus*, coraline algae *Coralina*, laver *Porphyra*, and the many-branched tubed weeds *Polysiphonia*. See PROTISTA.

R horizon　A layer composed of underlying consolidated bedrock, or unaltered sediment beneath the soil profile; is not necessarily the parent material of overlying horizons. See REGOLITH, SOIL PROFILE. See APPENDIX p. 619.

rhyolite　A light-colored extrusive igneous rock with an aphanitic-to-porphyritic texture. Rhyolite has essentially the same mineralogy and composition as granite.

rhythm method　A form of birth control involving close monitoring of a woman's monthly reproductive cycle to determine the days of the month during which fertilization is most likely to occur. Intercourse is avoided on those days, reducing the likelihood of pregnancy. The rhythm method is not a reliable means of preventing pregnancy, even when monitoring basal body temperature is added to counting the days of the cycle.

ria　A flooded river valley in a region of well-developed topographic relief. A ria may form during a relatively rapid rise in sea level, such as occurs at the end of an ice age.

ria coast　A geological term referring to the submergence of any land margin dissected more or less transversely to the coastline.

riboflavin　The common name for the water-soluble vitamin B_2. Riboflavin is an essential nutrient used for energy metabolism, protein synthesis, and growth and repair of tissues.

ribonuclease   An enzyme that catalyzes the cleavage (hydrolysis) of RNA, ribonucleic acid. It is sometimes abbreviated RNAase.

ribonucleic acid   The full name for RNA. See RNA.

ribose $HOCH_2(CHOH)_3CHO$, a simple sugar (monosaccharide). Although it rarely occurs alone, it is common in biological tissue because it is a constituent of RNA. Deoxyribose, equally important as a constituent of DNA, is derived from ribose. See RIBONUCLEIC ACID.

ribosomes   Numerous, beadlike structures of RNA and protein in the cytoplasm of living cells that synthesize proteins under the direction of messenger RNA.

Richardson number   A number used in meteorological studies to evaluate shearing flow. It gives an indication of the likelihood of turbulence developing.

Richter magnitude scale   A measure of the energy of seismic waves generated by an earthquake. The Richter scale is based on wave amplitude and location of the receiving station in relation to the earthquake focus. In practical use, Richter values range from 1 to 9 in logarithmic units, but the scale is mathematically open at both ends. Compare MERCALLI SCALE.

rickets   A deficiency disease that interferes with bone formation in children, causing abnormal bone shape and structure. It occurs in children whose diet lacks vitamin D and who are exposed to very little sunlight, preventing the body from manufacturing its own vitamin D. It can also be caused secondarily from extreme loss of, or inability to absorb, calcium. Also called rachitis.

Rickettsia   A genus of gram-negative, parasitic microorganisms that causes numerous diseases, including the different forms of typhus and Rocky Mountain spotted fever. The parasites are intermediate between viruses and bacteria; they are transmitted to humans by the bites of ticks, mites, lice, or fleas. See SCRUB TYPHUS.

ridge   1) A long, elevated structure or landform most often found separating two stream drainages. 2) An extension of an atmospheric high-pressure zone into an adjacent low pressure zone. See MID-OCEANIC RIDGE.

ridge-ridge transform fault   A transform fault that develops as a displacement between sections along the axis of a mid-oceanic ridge. See TRANSFORM FAULT.

riffle   A relatively shallow zone in a stream produced by deposition where the water flows with gentle turbulence over a gravelly substrate. Contrast POOL.

rift   An extensional joint forming between solid rocks that originally composed a single unit.

rift valley   A regional scale trough or graben bounded by a network of extensional faults. A rift valley may represent the early stages in the formation of a tectonic spreading center.

right ascension   One of two values used to pinpoint locations of celestial objects such as planets and stars; it corresponds to longitude on earth. It is measured along the celestial equator in hours of time east of the vernal equinox (the point where the sun crosses the celestial equator as it moves north). See CELESTIAL EQUATOR, DECLINATION.

right-lateral fault   See DEXTRAL FAULT.

right-of-way   A legal easement for passage

or access upon or across the lands of another.

rill A small depression through which surface water drains off the soil surface.

rill erosion The development of small and preferred drainage channels at the soil surface as soil particles are carried away by surface water runoff. Contrast SHEET WASH.

rime Frost, especially a heavy build-up of frost on the windward side of objects. The feathery ice crystal formations grow into the wind, and at high elevations can often become quite long.

ring barking See GIRDLE.

ring current A current of electricity flowing westward through the earth's magnetosphere, near the geomagnetic equator. During magnetic storms, this current causes a global depression of the earth's magnetic field.

Ringelmann chart A series of charts used to make visual measurement of smoke density for enforcing emissions standards. Each chart consists of a progressively darker series of shades of gray. These are compared with the color of emissions coming out of a smokestack to find the shade that most closely matches the emissions. Ringelmann No. 0 is the lowest, least dense chart; smoke at 20 percent density corresponds to Ringelmann No. 1, and a 5 percent increase in density corresponds to Ringelmann No. 5.

ring of fire See CIRCUM-PACIFIC BELT.

ring-porous wood structure Wood characterized by larger sizes and greater numbers of vessels in the early wood of the annual growth ring, as compared to the wood laid down later in the season. Cross sections of such trees show ring-shaped arrays of clearly visible holes alternating with rings of solid wood. Ring-porous wood structure is only found in northern tree species such as ash, locust, hickory, and some oaks. Compare DIFFUSE-POROUS WOOD STRUCTURE.

riparian A term pertaining to features or land use along the banks of a stream or a river.

riparian habitat The environment found on banks of streams and rivers; sometimes also used to refer to lake shores. See RIVERINE ENVIRONMENT.

riparian rights Legal principle that anyone whose land adjoins a flowing stream or river has the right to use water from the stream as long as enough is left for downstream users and the quality of the water is maintained. The riparian doctrine is the common-law principle governing water use in the eastern United States and in the United Kingdom.

rip current An often strong, usually temporary, current that flows seaward from a shoreline. Rip currents may develop as a consequence of waves and wind driving water against a shoreline.

ripple See RIPPLE MARK.

ripple mark A small bedform of silt or sand formed by currents of air or water, or by the oscillation of waves in shallow water. Ripple marks may be symmetrical or asymmetrical in cross section.

riprap A general term for large blocky stones that are artificially placed to stabilize and to prevent erosion along a riverbank or shoreline.

rise A natural surface of land or of ocean floor that increases gradually in elevation over a broad area. See CONTINENTAL RISE.

risk The probability that undesirable events (injury, property damage, environmental damage) will happen from a potential hazard or from deliberate or accidental exposure to an existing hazard.

risk analysis Identifying hazards and evaluating the nature and severity (especially, economic costs) of their associated risks. This information is used to determine options and to make decisions about reducing or eliminating risks and communicating information about risks to the public. Also called risk assesment.

risk-benefit analysis Evaluation of short- and long-term risks in relation to overall benefits associated with using a particular product or technology. The estimated societal benefits are divided by the estimated societal risks to produce a value called a desirability quotient. For example, x-rays have a large desirability quotient because the medical benefits far outweigh the costs.

river A natural surface drainage channel that has a comparatively large annual discharge. A river usually terminates at the ocean.

river capture See STREAM CAPTURE.

riverine environment 1) Land adjacent to a river. 2) A lotic environment. See RIPARIAN HABITAT.

river order See STREAM ORDER.

river terrace Any terrace formed by a river. See TERRACE.

river zones Zone 1: Headwater region source of water and sediment. Zone 2: Middle reaches of a river system, where storage and transfer of sediment occurs. Zone 3: Where deposition occurs on a delta or an alluvial fan.

rivulet erosion The erosion of a land surface by the action of small-scale streams.

RMBK reactor A nuclear reactor design common in Russia and the Federated Republics that uses unenriched uranium as fuel, graphite as moderator, and water as coolant. The reactor at Chernobyl in the Ukraine was an RMBK reactor. Russia has now halted further construction of RMBK plants and joined other countries in rejecting the RMBK design as unsafe.

rms value See ROOT-MEAN-SQUARE-VALUE.

Rn Chemical symbol for the element radon. See RADON.

RNA Short for ribonucleic acid, a complex organic molecule found in all living cells and many viruses, where it controls protein synthesis and the transmission of genetic information. RNA is made up of a long chain of nucleotides containing the sugar ribose and the bases adenine, cytosine, guanine, and uracil. Its structure is similar to that of DNA, but each sugar (ribose) in RNA contains a hydroxyl group in place of one of the hydrogens in the DNA sugar (deoxyribose). See DNA.

rock salt A common name for halite, naturally occurring sodium chloride ($NaCl$). See HALITE.

roche moutonnée A bedrock landform smoothed on the upstream side and quarried on the downstream side by glacial abrasion and plucking. The term is French for fleecy rock.

rock A naturally occurring, solid cohesive aggregate formed as a mixture of one or more minerals or mineral materials.

rock avalanche A type of rapid mass wasting that consists almost entirely of loose rock debris tumbling or free falling down a steep slope or cliff. See MASS WASTING.

rock basin lake   A lake forming in a depression within a valley formerly occupied by an alpine glacier.

rock cleavage   A description of the character of a broken rock surface. Slaty or schistose rock cleavage occurs along the foliation planes of metamorphic rocks. Fracture rock cleavage may occur along closely spaced fractures or crenulations in the rock fabric.

rock cycle   A theoretical cycle of rock formation through sequential stages of uplift, weathering, erosion, transportation, deposition, lithification, metamorphism, and magmatism of rock materials.

rockfall   See ROCK AVALANCHE.

rock flour   The finely powdered silt and clay-sized particles forming from rocks that are pulverized as they are carried by glacial ice.

rock glacier   An accumulation of rock rubble, snow, and ice that flows downslope in a glacierlike manner.

rock salt   A common name for halite, naturally occurring sodium chloride (NaCl). See HALITE.

rockslide   A type of mass wasting characterized by large blocks of rock that slide as a relatively coherent mass along a steeply sloping surface or slip plane.

rock-stratigraphic unit   Any distinct body of rock that is identifiable by lithologic characteristics alone. Rock stratigraphic units are ranked into supergroups, groups, formations, members, and beds.

rocky intertidal   A form of littoral habitat, that between the high-water mark and the low-water mark, found on rocky beaches rather than on sand or mud. The rocky intertidal zone supports a unique group of organisms. See INTERTIDAL, LITTORAL ZONE.

rod   1) In most vertebrates, one of two types of cells of the retina of the eye that respond to light. Rods are thin stacks of disk-shaped structures that pick up dim light and so are used for night vision. 2) A unit of fissionable material used to fuel nuclear reactors; also called fuel rod or fuel element. See FUEL ELEMENT, FUEL ROD.

Rodentia   With over 1500 species, the order of rodents are among the most successful and highly evolved of all mammals despite the fact that they have many primitive, mammalian characteristics. The one pair of incisor teeth in each jaw is curved, grow continually, and have a hard exterior coat and a softer interior core so that the sharpness of the teeth is maintained by gnawing. Their worldwide distribution includes mice, voles, lemmings, rats, beavers, squirrels, gophers, porcupines, guinea pigs, and capybaras. See EUTHERIA.

rodenticide   Pesticide that kills rodents. See WARFARIN.

roentgen   (R) Unit for absorbed dose of x-rays or gamma rays. One roentgen equals the absorption of 0.00878 joules of energy per kilogram of dry air at standard conditions. The unit was named for the discoverer of x-rays (Wilhelm Roentgen, 1845–1923). The roentgen has largely been replaced by the rad and the gray.

room-and-pillar mining   A mining technique in which pillars of country rock are left in place to support the weight of the overburden within a large room or cavern created by excavation.

root   The part of a plant that grows downward into the ground and serves as a conduit for the uptake of water and dissolved minerals from the soil. Roots generally have a well-developed vascular system organized into a solid central stele surrounded by cortex tissue. They are distinguished from shoots by their general lack of chlorophyll and of buds, although adventitious buds occur in some species. See TAPROOT SYSTEM, FIBROUS ROOT SYSTEM.

root cap   A mass of cells at the tip of growing roots that protects the actively dividing apical meristem directly behind it. The root cap is constantly worn away as the root tip pushes through soil particles, and is continually replaced with new cells produced by the apical meristem. Also called calyptra.

root hairs   The fine hairlike growths from the outer layer of plant roots. Root hairs greatly increase the surface area for absorption of water and dissolved minerals from the soil.

root-mean-square value   (rms value) For any fluctuating quantity such as an alternating waveform, a value calculated by taking the square root of the mean (average) of the squares of the continuous series of values for voltage or current (or other ordinate) over one full cycle. Also called the effective value, it is used in calculating sound pressure level and in statistical analysis (root-mean-square deviation, more commonly known as standard deviation). See STANDARD DEVIATION.

root-mean-square velocity   The average speed of molecules in an ideal gas or solution. The formula for calculating this velocity incorporates taking the square root of an average (mean) of the squares of the values.

root, mountain   The body of crustal rock that extends beneath a mountain into the dense material of the mantle. The mountain root provides buoyancy to keep the mountain in isostatic equilibrium.

root nodules   Bulbous growths on the roots of legumes that contain symbiotic nitrogen-fixing bacteria. See NITROGEN-FIXING BACTERIA, RHIZOBIA.

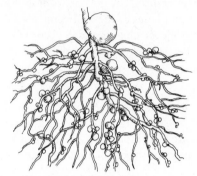

root nodules

rootstock   1) Another term for rhizome, or sometimes any underground plant part. 2) A plant base used for grafting, as in the creation of dwarf fruit trees. See RHIZOME, GRAFT.

Rosaceae   The rose family, a large and important group of angiosperm dicots. The alternate leaves usually have stipules and the flowers have flowers and sepals in multiples of fives. The numerous stamens are arranged in a hypanthium ring surrounding the superior and free, or inferior and united, carpels. This family includes spireas (*Spirea*), cinquefoils (*potentilla*), strawberries (*Fragaria*), raspberries (*Rubus*), antelope-brush (*Purshia*), apples and pears (*Pyrus*), cherries and peaches (*Prunus*), firethorn (*Pyracantha*), and many others. See ANGIOSPERMOPHYTA.

Rossby waves   Huge, slow-moving air waves within the westerlies, the west-to-east air currents in the upper half of the troposphere at middle latitudes. These waves contribute to the jet stream. The wavelengths range from 4000 to 6000 km.

rostrum A bird beak, or beak-shaped structure on animals.

rotation The process of spinning around an axis; also, one complete movement. The earth completes one rotation each day around the imaginary axis running through its poles. Contrast REVOLUTION.

rotation time The number of years between the cutting down of a tree or a forest until that tree or forest grows again to an equivalent size.

rotational landslide A type of mass wasting in which the entire mass of debris moves around a central point of rotation above a concave upward slide surface. In a rotational landslide, the displaced mass acquires a backward tilt relative to the initial position.

rotational slump A small-scale rotational landslide.

rotenone $C_{23}H_{22}O_6$, a natural contact insecticide extracted from the root of the *Derris* tree and occurring in a few other plants. Although toxic to fish, it is relatively safe for animals and breaks down rapidly into harmless components, making it a common pesticide for home gardens. It is also used in flea powders and for mothproofing.

Rotifera An animal phylum sometimes referred to as wheel-animals because the ring of cilia that surround their heads appears to rotate. Rotifers are generally small (less than 2.0 mm), urn shaped, and covered with a thin chitin layer. Mostly living in freshwater environments, rotifers eat a variety of bacteria and planktonic species. See ANIMALIA.

rotor 1) The central shaft that turns between two fixed magnets in an electrical motor or a generator; also, the turning portion of a windmill. 2) Short for rotor flow. Compare STATOR.

rotor flow A large, horizontal eddy of air that moves in a closed loop, often surrounded by clouds. Such air-flow patterns can form alongside mountains, underneath large lee waves. They may be associated with extreme turbulence and can be strong enough to pull the roof off of a house. Also called simply a rotor. See LEE WAVES.

rottenstone A highly weathered and decomposed limestone. Powdered rottenstone is used as an abrasive material.

roughage Foods containing fibrous, indigestible material that stimulate the movement of food and waste products through the intestines. When referring to human foods such as bran and fruit skins, it is also called dietary fiber.

round worms See NEMATODA.

RQ See RESPIRATORY QUOTIENT.

r selection Ecosystem pressures that favor rapid population growth at low population density, encouraging species that devote most of their energy to reproduction, producing many seeds or offspring. Such species are not necessarily able to survive increased competition or higher population densities. Contrast K SELECTION.

r-strategists An r-selected (ruderal) species, one that is able to colonize early in succession. Many are opportunistic weed species, although other r-strategists have tolerances that are too narrow to be opportunistic. Compare K-STRATEGISTS.

ruby See CORUNDUM.

ruderal Describing habitats of rubble,

rubbish, or disturbed areas such as roadsides, or plants that grow in those habitats (such plants are not necessarily weeds). Plantain, a lawn weed *(Plantago)*, is a classic ruderal species that Native Americans called "the white man's footprint"; it grows on the worst waste ground, even in the compacted dirt of a parking lot.

rudite   A general reference to sedimentary rocks composed of grains larger than 2 mm in average diameter. Rudite is coarser than sandstone.

rumen   The first stomach division in animals that chew cud (ruminant), such as cows and goats. Rumen is also the term for the first stomach division in whales and in dolphins. See RETICULUM, RUMINANT.

ruminant   An animal that chews its cud (food that has been chewed once, swallowed, and regurgitated). Ruminants have stomachs with three or four separate cavities and include cows, sheep, goats, camels, and deer. See RUMEN, RETICULUM.

runner   A thin stem that grows along the ground surface and takes root to produce new plants, a form of vegetative (asexual) reproduction. Also, a small plant formed by such a process. Strawberry plants produce

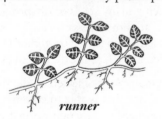

runner

numerous runners. Sometimes called stolon, but runners are usually distinguished by being thinner and not arising from a branch. Compare STOLON.

runoff   Precipitation that flows freely away from soil into streams.

run-of-the-river flow   An engineering hydrologic term applied to water power generation that varies with normal water level fluctuation rather than with artificial control of impoundments.

rust   1) Iron oxide (Fe_2O_3); the rough reddish brown layer that forms on exposed iron or steel as it oxidizes. 2) Common name for a group of fungal diseases of plants characterized by raised red or orange spots on leaves or stems. Rusts attack a number of cereal crops and conifers as well as garden ornamentals. They are caused by fungi of the class Heterobasidiomycetae. See BASIDIO-MYCOTA.

rutile   A reddish brown mineral of titanium oxide that occurs as an accessory mineral in a wide variety of rocks.

R value   The thermal resistance (thermal insulating ability) of a material or a structure. It is the inverse of the thermal conductance. Fiberglass batts of 3-inch thickness have an R value of 10 (Btu per hour per square foot per degree Fahrenheit temperature difference); batts of 6-inch thickness have twice the insulating power, or an R value of 20.

R wave   See RAYLEIGH WAVE.

S

S Chemical symbol for the element sulfur. See SULFUR.

S₁ Symbol for selfing, the first generation of a self-cross breeding. Compare FIRST FILIAL GENERATION, SECOND FILIAL GENERATION.

sabkha A lagoonal feature of the north African coastal plain where carbonate and sulfate minerals are formed by intense evaporation.

saccharin $C_7H_5NO_3S$, an artificial sweetener that is many times sweeter than ordinary sugar. Its use in the United States has been discouraged since 1977 because it is carcinogenic in animals.

saccharoidal Having a granular texture, used to describe some forms of limestone and marble.

Saccharomyces A genus of single-celled yeasts that do not have a real mycelium and reproduce by budding. One species, *Saccharomyces cerevisiae*, is the yeast used in baking bread and in fermenting alcohol; it is also used in the laboratory as an eukaryote host for producing DNA. *Saccharomyces ellisoides* is used in winemaking.

Safe Drinking Water Act, 1974 Federal law providing mandatory standards for drinking water, including safe acceptable levels of many possible contaminants (42 U.S.C.A. §3007 et seq.).

safe yield The maximum amount of water in groundwater management that may be pumped from a well or a series of wells in a groundwater basin without depletion of the aquifer. Economic yield is a preferred synonym because of the uncertainty in estimating yield in relation to the variable boundaries of an aquifer.

Sagebrush Rebellion A term for a popular movement in the 1960s and 1970s among ranchers, advocating privatization of public lands in the western United States. It was a response to increasing regulation by the Bureau of Land Management.

sagittal Describing a vertical plane through an organism that parallels the median plane (the one dividing it into left and right halves); one that cuts through from the back (dorsal) surface to the front (ventral) surface. The sagittal crest or suture along the top of the skull follows such a plane; a sagittal section is equivalent to a cross section but is cut along a sagittal plane.

Sahara A geographic name for the desert region of northern Africa that extends from the Atlantic coast to the Red Sea.

Saharan green belt A pair of proposed regions for revegetation along the southern edge of the Saharan desert, designed to stop the spread of the desert while providing limited grazing.

Saharo-Arabian floral region One of the phytogeographic regions into which the Paleotropical realm is divided according to similarities between plants; it consists of the Arabian Peninsula and the eastern side of the Persian Gulf. See PALEOTROPICAL REALM, PHYTOGEOGRAPHY.

Sahel A semiarid region to the south of the Sahara in West Africa in which food shortages and even famines occur frequently as seasonal rains are often insufficient for food production. The area includes parts of the following countries:

Senegal, Mauritania, Mali, Burkina Faso, Ghana, Niger, Nigeria, Chad, Cameroon, and Central African Republic.

Salicaceae   The willow family of the Angiospermophyta. These dioecious woody dicots are most abundant in the mid-to-high latitudes of the Northern Hemisphere. Small flowers lacking petals are borne in dense catkins. The female catkins mature into clusters of capsules that rupture at maturity, releasing small, wind-borne seeds covered by fine hairs. Willows (*Salix*) and aspens, cottonwoods, and poplars (*Populus*) are members of this family. See ANGIOSPERMOPHYTA.

salination   The process of salt accumulation in the soil that is usually caused by evaporation of saline groundwater moving upward through capillary action, or by continued evaporation of water from irrigation of crops.

saline   Resembling or containing ordinary salt (NaCl) or similar salts. A saline solution used for medicinal purposes is one with a concentration of salt approximating that found in the blood.

saline seep   A natural spring where briny groundwater flows to the surface.

salinity   Degree of saltiness. Natural salinity of soils is often greater in arid regions. In addition, it becomes a problem in areas irrigated for agriculture because salts brought in with irrigation water remain behind when the water evaporates. In combination with salts applied as fertilizers, the salts brought to the surface by irrigation can result in such high salinity levels in the soil that it can no longer support plant growth.

salinity meter   A device used to measure the salinity of a solution. There are two different types of salinity meters. One measures the density of the solution, a quantity that is proportional to the salinity. The other type measures the ability of the solution to conduct electricity, a property that is related to the amount of dissolved salts in solution. See SALINITY.

salinization   See SALINATION.

salt   1) Sodium chloride (NaCl), common table salt. 2) Any ionic compound. Salts such as sodium bicarbonate are formed when strong acids and strong bases react and neutralize each other.

saltation   A particle transportation process that occurs in the turbulent flow of water or of air. Particles transported by saltation appear to be bouncing and skipping along rather than to be flowing smoothly with the current.

salt dome   A circular mass of salt in an anticlinal fold or bedrock. A salt dome often migrates upward by arching and breaking through brittle overlying rock.

salt field   A diapir of salt that migrates upward and forms a structural dome within the overlying bedrock. An oil reservoir may form at the top or on the flanks of a salt dome.

salting out   A chemical procedure in which salt is added to an aqueous solution to cause an organic compound to precipitate out so that it can be physically removed. Salting out is also used in a two-phase extraction process. Addition of salt to an organic/aqueous two-phase mixture will increase the polarity of the solution and drive the less-polar organic substances into the organic layer.

salt marsh   A marsh having soils with periodically high concentrations of salt, usually found in estuaries or other areas subject to flooding by ocean water.

salt pan   A flat basin containing concentrations of chemically precipitated minerals formed by evaporation in arid lands.

salt playa   A salt pan that forms on the edge of a playa lake. See PLAYA LAKE.

salt pump   A mechanism in aquatic organisms for maintaining a higher concentration of ions (dissolved salts) than is present in the surrounding water, called a pump because it uses energy to oppose the natural flow of the osmotic gradient.

saltwater   Of, relating to, or living in water that contains significant concentrations of salt (greater than 3.0 percent), such as that found in oceans. Compare BRACKISH, FRESHWATER.

saltwater intrusion   A displacement of a lens of fresh water because of the encroachment of saltwater from natural causes or from groundwater withdrawal. See GHYBEN-HERZBERG PRINCIPLE.

samara   A dry fruit that is wing shaped and does not split open when ripe. A maple seed is the classic example of a samara. Sometimes called a key fruit.

sand   Any clastic rock fragment between 1/16 and 2 mm in average diameter.

sand bar   An unconsolidated accumulation of sand built up by wave action and currents near a shoreline. A sand bar may also form by accumulation of sand in a river channel. See BAR.

sand dollar   A very flat, round type of sea urchin (class Echinoidea) that lives on the ocean floor. The shell-like skeleton of dead sand dollars is collected by beach walkers and shell enthusiasts for its beautiful form. See ECHINODERMATA.

sand dune   Any dune composed of sand. See DUNE.

sandfly fever   Viral infection caused by the bite of a sandfly *Phlebotomus papatasii*; it resembles a mild influenza (with no respiratory infection) and can be caused by any of three arboviruses. Sandfly fever occurs in warm regions with long dry periods, such as the Middle East.

sand sea   A broad expanse of gently undulating sand dunes in a desert region that has a large supply of sand. A sand sea lacks the typical directional indicators commonly found in other dune forms.

sandstone   Any clastic sedimentary rock composed of sand-sized particles cemented together during burial diagenesis.

sandstorm   A strong wind carrying large quantities of dust and of sand. Sandstorms occur most often in dry areas, because dry soil is more easily lifted by wind.

sandy soil   Any soil that contains at least 70 percent sand and less than 15 percent clay particles.

sanitary landfill   A municipal waste disposal area where layers of trash are spread out and covered with layers of compacted soil. Modern sanitary landfills may be underlain and overlain by impermeable materials to prevent groundwater contamination.

sanitation   1) Public hygiene; practicing hygienic methods and maintaining sanitary environments to prevent disease. 2) A technique for plant disease control in which dead tissues are removed to prevent sites of entry for pathogenic organisms. For example, dead branches of elms are often removed to reduce sites for bark beetle entry and thereby lessen the spread of Dutch Elm disease.

sap   The liquid that travels through the vascular system of plants. It carries water and dissolved minerals upward to the leaves and water and compounds produced by photosynthesis downward to the roots. Maple syrup is produced by boiling down large volumes of maple sap into a much smaller volume of concentrated sugar syrup.

saponins　A group of organic compounds found in plants; they are forms of glycosides. They emulsify oils and in water tend to lather like soap, giving rise to the common names soapwort, soapbark, and soapberry of plants containing high concentrations of saponins. A number of saponins, such as digitonin (from *Digitalis)* act as steroids.

sapphire　See CORUNDUM.

sapro-　Prefix meaning decay, rotten, or relating to decomposition.

saprobe　A microorganism that derives its nutrition from decaying matter. Many bacteria and fungi are saprobes; they differ from saprophytes in that they can only absorb dissolved nutrients from organic matter. Also called saprotrophs or osmotrophs.

saprobic　Referring to or living in an environment with abundant organic matter (especially sewage or animal wastes) and relatively low levels of oxygen because of the high biochemical oxygen demand of the organic matter. See SAPROBIC CLASSIFICATION.

saprobic classification　A system for classifying organisms according to the amount of pollution they can tolerate. See OLIGOSAPROBIC, POLYSAPROBIC.

saprobiont　See SAPROBE.

saprogenous　Causing, produced by, or relating to decay or putrefaction. Also called saprogenic.

saprolite　A red or brown earthy, clay-rich

material formed by the chemical decomposition of igneous or metamorphic rocks.

sapropel　An accumulation of organic material deposited in submarine anaerobic conditions.

saprophage　Organism that feeds on decaying matter. See SAPROBE.

saprophagous　Feeding on dead organic matter.

saprophyte　A plant such as Indian pipes *(Monotropa uniflora)* that lives on decaying matter. Compare SAPROZOIC.

saprozoic　Describing animals such as tapeworms that derive nourishment from decaying organic material and dissolved salts. Compare SAPROPHYTE.

sapwood　The outer, new layer of wood found between the bark and the heartwood on shrubs and trees. Sapwood differs from heartwood in being softer, usually lighter in color, and in still conducting fluid (sap) through the tree. Also called alburnum. Compare HEARTWOOD.

sarcolemma　The thin membrane that encloses each contracting muscle fiber (myofibril) within skeletal muscle tissue. See MYOFIBRIL.

sarcoma　Any of a group of cancers or growths that originates in nonepithelial or underlying tissues such as muscle, bone, etc., and is capable of spreading to other tissues. Kaposi's sarcoma is one of the diseases frequently associated with AIDS. See CANCER, CARCINOMA, LEUKEMIA, LYMPHOMA.

Sarcomastiophora　In some classification schemes a phylum in the kingdom Protista that includes dinoflagellates, euglenoids, cryptomonads, zoomastigotes, amoebae, caryoblasteas, slime molds, slime nets, forams, actinopods, and coccolithophorids. See DINOFLAGELLATA, EUGLENOPHYTA, CRYPTOPHYTA, ZOOMASTIGINA, RHIZOPODA, CARYOBLASTEA, LABYRINTH-ULAMYCOTA, MYXOMYCOTA, PLASMO-DIOPHOROMYCOTA, FORAMINIFERA, ACTINOPODA, HAPTOPHYTA.

sarcomere One of the sections into which a myofibril of skeletal (voluntary) muscle is divided by thin dark bands. The alternation or striping of light and dark bands within voluntary muscle tissue gives rise to its alternate name, striated (finely striped) muscle tissue. See MYOFIBRIL.

sarcoplasm The semifluid, translucent material (cytoplasm) between individual myofibrils in the cells of skeletal muscle tissue. See SARCOPLASMIC RETICULUM.

sarcoplasmic reticulum A network of minute tubelike structures within the cells of skeletal muscle fibers that conducts the calcium ions, the stimuli for muscle contraction, from the outside surface of the cell to the interior.

Sargasso Sea The calm, central region enclosed by the currents of the large gyre in the north Atlantic Ocean. The Sargasso Sea is characterized by vast floating masses of *Sargassum* brown algae.

satellite A body in orbit around a celestial body. This includes natural satellites, such as the moon of earth or the moons of Jupiter, plus artificial satellites such as the geosynchronous weather satellites above earth.

satin spar A crystal of selenite gypsum with a fibrous texture and silky luster. See SELENITE.

saturated adiabatic lapse rate Another term for wet adiabatic lapse rate. See WET ADIABATIC LAPSE RATE.

saturated fat Fats containing no double or triple bonds connecting the carbon atoms into a chain, and therefore containing the maximum possible number of hydrogen atoms attached to the carbon atoms. Saturated fats, which include many animal fats such as lard, are usually solid at room temperature. Consumption of a high proportion of saturated fats has been linked with a number of health problems, from heart disease to cancers.

saturated flow Movement of water through the saturated pore spaces of a soil or other material when all the pore spaces are filled with water (as below the water table).

saturated zone See ZONE OF SATURA-TION.

saturation point The point at which one substance has incorporated as much as possible of another substance; the point of maximum possible concentration. Often simply called saturation. Water vapor in air must reach a saturation point before precipitation can occur. At 100 percent relative humidity, the level of water in the atmosphere has reached saturation. See SUPERSATURATION.

saturnism See LEAD POISONING.

savanna A grassland/woodland mosaic vegetation type found in tropical or subtropical regions with long dry periods, and receiving more rainfall than desert areas but not enough to support complete forest cover. Savanna is characterized by scattered trees, or scattered clumps of trees; fire often plays an important role in maintaining the vegetation. Also spelled savannah.

Savonius rotor A windmill design based upon a vertical axis (a panemone); it is not as efficient as some other windmills, but can deliver a more steady supply of energy. Compare DARRIEUS GENERATOR.

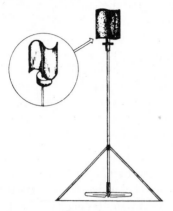

Savonius rotor

sawlog A log that is large enough to saw into lumber, as measured by a variety of scales. For eastern hardwoods, a sawlog is a minimum of 8 to 10 feet long with a minimum diameter (inside the bark) of 8 to 10 inches; for eastern softwoods, a minimum diameter (inside the bark) of 6 to 8 inches; for western softwoods, minimum lengths of 20 to 40 feet are common.

saxicolous Describing plants that grow among rocks.

Saxifragaceae The saxifrage family of the Angiospermophyta. Mostly composed of perennial herbaceous dicots with alternate leaves that form moundlike rosettes, but some shrubs and trees as well. The flowers are bisexual with five or four sepals and an equal number of petals. Usually there are five to ten stamens arranged in two series. Saxifrage (*Saxifraga*), gooseberries and currants (*Ribies*), hydrangea (*Hydrangea*), grass-of-parnassus (*Parnassia*), and coral-bells (*Heuchera*) are all members of this family. See ANGIOSPERMOPHYTA.

Sb Chemical symbol for the element antimony. See ANTIMONY.

Sc Chemical symbol for the element scandium. See SCANDIUM.

scale 1) A small, semitransparent, leaflike structure on a plant, especially one of those on most deciduous plants that cover a bud during the winter (bud scale). 2) A platelike, tough, or bony structure on animals, such as the scales covering fish or butterfly wings. 3) A graduated series of measurement units, as in a temperature scale. 4) To measure timber volume.

scandium A rare metallic element with atomic weight 44.96 and atomic number 21. Although it has no major industrial uses, it is sometimes used in semiconductors and as a tracer.

scansorial Characterized by, or suited for, climbing. Nuthatches are scansorial birds, noted for scrambling up and down tree trunks.

Scaphopoda A class of Mollusca that includes the tooth and the tusk shells. These benthic marine animals have bilaterally symmetrical, tubular shells that are open on either end. The head and the tentacles project from the larger end of the shell. See MOLLUSCA.

scarification The use of a mechanical device to break up the surface layer of a soil. Scarification may be used to even out the surface irregularities in a soil that has become deeply rutted.

scarp A sloping cliff or embankment that forms a barrier at the edge of an area of flat land. A scarp may be formed by faulting or erosional processes.

scattering Irregular deflection of waves or of high-energy particles as they pass through a substance. See RAYLEIGH SCATTERING.

scavenger An animal that feeds on the dead

bodies of other animals it has not killed, or on refuse. Compare PREDATOR, OMNIVORE.

scheelite   A whitish or yellowish mineral of calcium and tungsten oxide that forms tetragonal crystals in high-temperature mineral veins. Scheelite is an ore mineral of tungsten.

schist   A type of metamorphic rock characterized by megascopically visible platy or elongated minerals in approximately parallel orientation. The chemical composition of schist is quite variable depending on the parent rock type and metamorphic history.

schistocity   A general term describing the wavy texture and appearance of any schist-like rock.

Schistosoma   Genus of blood fluke which causes schistosomiasis (formerly called bilharziasis). The parasites develop in water snails, their intermediate host; they enter humans or animals through skin or mucous membranes where they travel to the blood vessels of organs (different species settle in different organs). The genus was formerly called *Bilharzia*. See SCHISTOSOMIASIS, TREMATODA.

schistosomiasis   A disease of humans caused by trematodes, blood flukes of the genus Schistosoma. Worms mature in the blood of the host where eggs are produced that finally reach the intestine or the bladder. When eggs from urine or from feces are deposited in water they mature into free-swimming larvae that penetrate the tissues of snails, the intermediate hosts. After several generations within snails, small forked-tailed larvae (cercariae) emerge into surrounding waters, where they penetrate the skin of humans who may be bathing, wading, or washing clothes. Various types of schistosomiasis, sometimes called bilharziasis, are widespread throughout Africa, the Middle East, parts of India and South America, China, and the Philippines. Liver cirrhosis, diarrhea, fever, toxemia, and lesions of the nervous system may result from infection. See TREMATODA.

schizocarp   A fruit type that is dry when ripe and splits into two or more seed vessels; each vessel contains one seed and does not usually split open when ripe. Carrot seeds are schizocarps.

schizogenesis   See SCHIZOGONY.

schizogony   Vegetative (asexual) reproduction by cell division (fission). Schizogony occurs in some protozoa and bacteria. Also called schizogenesis.

schwingmoor   Vegetation that grows on land but forms a mat that grows out to float on the surface of the water next to the shore.

scientific method   Investigating a system by formulating hypotheses (educated guesses based on initial observations) about the behavior of the system, then making predictions based upon these hypotheses, and finally designing experiments (or making observations) to test these predictions. After several tests validate different predictions, a hypothesis becomes a scientific theory or law. This process is the basis of Westen science.

scientific name   The formal Latin name of a taxon. See TAXON.

scintillation counter   A device for measuring the intensity and the energy spectrum of radiation. It uses a crystalline or liquid substance that emits flashes of light (scintillations) when radiation falls on it. A photomultiplier amplifies these flashes and translates them into pulses that can be analyzed to calculate the levels of the radiation that contacted the crystal. Scintillation counters are used extensively in radioisotope tracer studies in biochemistry.

scion A part removed from a plant, usually a green shoot or bud used for grafting. See GRAFT.

sciophyll A shade leaf; a leaf modified to perform in shady conditions. One plant develops different leaf structures when transplanted from bright sun into shade: The leaves become thicker, with longer cells in the palisade mesophyll providing more chloroplasts to scavenge available light, and less well-developed air spaces. Compare HELIOPHYLL.

sclereid A stone cell (or similar cell). Sclereids are short, thick-walled (sclerenchyma) cells that may occur together forming hard layers, such as the shell of a nut, or be distributed throughout other tissue, as in the gritty stone cells found in pear fruits. See SCLERENCHYMA, STONE CELLS.

sclerenchyma Nonliving plant tissue made up of thickened and hardened cells from which the protoplasm has disappeared and has usually been replaced by lignin (and sometimes minerals). Its main purpose is to support plants, protecting tissue and providing strength to withstand wind and other forces. Sclerenchyma is divided into sclerids (stone cells) and fibers (the usable part of flax, cotton, and other fiber plants). See SCLEREID.

sclerophyll A tough or leathery, usually evergreen, leaf that is adapted to resist water loss, and thus most often found on species growing in warm, relatively dry climates. The leaves of eucalyptus and

many chaparral trees and shrubs are sclerophylls.

sclerophyllous Describing leaves possessing sclerophylls, or vegetation characterized by plants with sclerophylls. The Mediterranean forests, somewhere between savannas and ordinary forests, are sometimes called sclerophyllous forests because they consist primarily of scattered clumps of shrubs and trees with sclerophyllous leaves. See SCLEROPHYLL.

scleroproteins A group of fibrous proteins, including keratin and elastin, that make up hardened tissues such as ligaments, cartilage, nails, hair, and hooves.

sclerotium A hardened body of compact hyphae or protoplasmic material formed in some fungi and slime molds as a resting stage, becoming dormant but at some later time developing an active mycelium or spores.

Scolopacidae Birds belonging to the sandpiper family with long, downturning bills and long legs. They all breed in the tundra and migrate long distances, often to equatorial regions. The extinct Eskimo curlew was slaughtered by hunters and all curlews suffer from overshooting. See CHARADRIIFORMES.

scorbutus See SCURVY.

scoria A variety of basalt characterized by abundant vesicles. The vesicles are formed by gas bubbles that are trapped as a lava cools.

scoriaceous Having the properties of scoria.

scorpions Members of the chelicerate class Arachnida, scorpions have pedipalps provided with pinching claws. The cephalothorax is compact, whereas the abdomen is distinctly segmented. The last five abdominal segments are relatively thin and terminate in a poisonous stinger. Many species are relatively harmless but others can inflict fatal stings. They are all terrestrial, mostly nocturnal in habit, and are widely distributed in the tropics and the warmer temperate areas. Approximately 1200 species exist. See ARACHNIDA.

scotoplankton   Plankton that live in poorly illuminated environments.

scramble competition   See RESOURCE COMPETITION.

scraper   An organism that obtains food by scraping environmental surfaces, as in aquatic insect larvae that feed on algae and on bacterial films growing on rock surfaces.

scree   A sloping accumulation of coarse cobble-sized rock debris that forms against the base of a cliff.

screening   The process of separating soil materials into particle size classes by passing samples through a grate, a sieve, or a screen. Also called sieving.

Scrophulariaceae   The snapdragon or figwort family of the dicot angiosperms. A highly variable family in terms of flower shapes. The flowers have five sepals and five petals, the latter often zygomorphically arranged into an upper lip of two lobes and a lower lip of three lobes. The fruit is a locular capsule that contains many seeds, a characteristic, along with its usually round stems, that separates this family from the Lamiaceae. Monkey flower *(Mimulus)*, foxglove *(Digitalis)*, Indian paintbrush *(Castilleia)*, snapdragon *(Antirrhinum)*, figwort *(Scrophularia)*, speedwell *(Veronica)*, and mullein *(Verbascum)* are all members of this family. See ANGIOSPERMOPHYTA.

scrub   Small (or stunted) trees and shrubs, also land covered with such vegetation (scrubland or scrub forest).

scrubber   A device for reducing flue gas emissions, used in air pollution control. Liquid is sprayed into the flue gases; the liquid removes pollutants either by reacting with them to produce less harmful chemicals, or by simple adsorption. Scrubbers also reduce the temperature of stack gases. Also called flue gas scrubber or wet scrubber. See FLUE GAS DESULFURIZATION.

scrub forest   Land covered with shrubs and short stunted trees, characteristic of very poor soils or areas with too little rainfall to support larger trees. See MAQUIS.

scrub typhus   A less severe form of typhus caused by *Rickettsia tsutsugamushi* and transmitted to humans by two species of mites. It is common in Southeast Asia. Also called tsutsugamushi disease. See RICKETTSIA.

scurvy   Disease caused by deficiency of vitamin C (ascorbic acid), usually because of a lack of fresh fruits or vegetables in the diet. Scurvy is characterized by spongy, bleeding gums; pain in joints, muscles, and limbs; and hemorrhages under the skin or of mucus membranes. British sailors earned the name Limeys when they began carrying barrels of limes on long sailing voyages to prevent scurvy.

scutum   1) A hard, shieldlike outer shell, as that of the turtle or the armadillo. 2) An individual hard scale or plate on an animal covered with numerous, overlapping scales or plates, such as a snake.

scyphistoma   The rarely present polyp stage of a jellyfish (scyphozoan). See SCYPHOZOA.

Scyphozoa   An exclusively marine class of the phylum Cnidaria, in which the medusoid or typical jellyfish form is predominate in the life cycle. Medusae have gastrovascular cavities divided into four compartments by septae and may range in size from 20 cm to over 2 m in diameter, depending on the species. Approximately 200 species exist. See CNIDARIA, MEDUSAE.

Se   Chemical symbol for the element selenium. See SELENIUM.

sea   A general reference to an area of ocean that is relatively well enclosed by landmasses.

seaboard   The area of a landscape that borders on an ocean. The term seaboard is frequently applied to a coastal plain.

sea breeze   A local wind occurring at the coast in good, otherwise calm weather. During the afternoon as the sun warms the land, cool air over the water begins to blow toward the warmer, less dense air above the land. As the land cools at night, the process reverses to create a land breeze.

sea cliff   A coastal cliff that forms by wave action at the landward limit of beach erosion.

sea cucumbers   See HOLOTHUROIDEA.

sea-floor spreading   The plate tectonic process by which new oceanic crustal plates diverge from one another at a midoceanic ridge.

sea fog   A persistent fog that forms when the air above the ocean is very stable. It is essentially a huge cloud at the ocean surface and is a form of steam fog. In the winter, dense sea fogs form at high latitudes when cold air flows over ice and frozen ground and reaches the ocean, which may be 30° warmer; such fogs are sometimes called arctic sea smoke because of their appearance. See STEAM FOG.

sea level   The mean elevation of the ocean water level between high and low tides. Sea level serves as the standard topographic elevation reference point for surveying and oceanography.

sea loch   The Scottish term for fjord. See FJORD.

seam   A bed or layer of rock that separates two adjacent layers of differing composition.

seamount   A submarine volcanic mountain peak with a summit that rises at least 1000 meters above the sea floor.

search image   A term used in behavioral ecology for the preformed image presumed to be in the minds of the predators concerning the appearance and the most likely location of their prey.

searching efficiency   The probability that a predator will kill a particular organism at a specific point in time. Also called attack rate.

seasonal estuary   An estuary characterized by a pattern of river discharge that is highly variable on a seasonal basis.

sea urchins   See ECHINOIDEA.

seawall   1) An engineered structural wall of stone or concrete designed to reduce wave erosion at a shoreline. 2) A steep accumulation of debris deposited high on a beach by the action of storm waves.

sea water   The saline water of the ocean. The dissolved components of sea water amount to an average of 34 parts per 1000 measured by weight.

secchi depth   A measure of water transparency obtained by measuring the depth at which a lowered secchi disk disappears from view. The secchi disk is a 20-cm disk marked with alternating quarter sections of black and white.

second 1) (s) The standard fundamental unit for measuring time in metric and non-metric systems. It is now defined as the duration of 9,192,631,770 periods of the radiation corresponding to the transition between two energy levels of the cesium-133 atom. 2) (") A unit for measuring angles, as of longitude or right ascension, equal to $1/60$ of a minute and $1/3600$ of a degree.

secondary ambient air quality standards Legal allowable levels for pollutants in outdoor (ambient) air quality to avoid harm to the public welfare. Levels are established by the Environmental Protection Agency as part of the regulations of the Clean Air Act. See NATIONAL AMBIENT AIR QUALITY STANDARDS.

secondary consumer A carnivore; an animal that feeds off of smaller, plant-eating animals. Compare PRIMARY CONSUMER.

secondary division freshwater fish Fish that, although usually restricted to fresh water, are sufficiently tolerant to be able to cross narrow saltwater barriers. Compare PRIMARY DIVISION FRESHWATER FISH.

secondary energy Energy that undergoes a conversion before use (as opposed to the direct heat from a wood stove). Electricity is a form of secondary energy because the turbines that produce it are driven by another (primary) energy source, such as coal burning or hydropower.

secondary enrichment The redeposition of secondary sulfide and oxide minerals that have been leached by groundwater from an overlying deposit.

secondary forest See SECOND-GROWTH FOREST.

secondary minerals A mineral formed by weathering or by metamorphic solution activity, which alters the primary minerals within an igneous rock. Serpentine is a secondary mineral produced by the alteration of olivine.

secondary oil recovery Any technique of increasing the flow of oil from depleted wells by gas or chemical injection, hydrofracturing, or other recovery methods.

secondary pest outbreak A population explosion elevating to pest status a species of plant-eating insect that normally exists in such small quantities that it isn't considered a pest. Secondary pest outbreaks can be produced by use of chemical pesticides that inadvertently kill off the natural predators that formerly held these (relatively) harmless species in check. See RESURGENCE IN PESTS.

secondary poisoning Contamination or death of nontarget species of organisms that feed on poisoned animals. Compound 1080, a now-banned chemical once used for poisoning coyotes on rangeland, was notorious for causing secondary poisoning of many species.

secondary pollutants Chemicals that are not released directly into the water or the air, but are formed from chemical reactions between pollutants that are directly released (primary pollutants). Ozone and peroxyacetylnitrates are examples of secondary air pollutants.

secondary producers Herbivores; animals that produce biomass and grow by consuming the biomass produced by plants (primary producers). Compare PRIMARY PRODUCER.

secondary production   The creation of organic matter (biomass) by consumption of other organic matter (plants or herbivores). Compare PRIMARY PRODUCTION.

secondary productivity   The rate at which energy is stored as biomass by consumers, those organisms that feed on producers (plants). Compare PRIMARY PRODUCTIVITY.

secondary rock   A general term for a clastic sedimentary rock composed of rock particles derived from a preexisting source rock.

secondary sewage treatment   The second stage in the purification of wastewater. In this stage, bacteria decompose the solids remaining after primary treatment (screening and settling). Oxygen is bubbled into the effluent to supply oxygen to the decomposing bacteria so that as many of the solids as possible can be broken down. See PRIMARY SEWAGE TREATMENT, ADVANCED SEWAGE TREATMENT.

secondary sexual characteristics   Features that distinguish the two genders of any species, excluding the reproductive organs. These usually develop at puberty or during the breeding season as a result of hormonal secretions of reproductive glands.

secondary standards   Environmental standards established to protect public welfare rather than to protect human health. These may cover improved visibility, avoiding structural damage to buildings, or crop protection. Contrast PRIMARY STANDARDS.

secondary substances   Chemical substances produced by plants that have no known biochemical function providing an adaptive advantage; they may be metabolic "waste" products. Often found in high concentrations in leaves and in shoots, many secondary substances provide an indirect advantage to plants containing them by deterring herbivores.

secondary succession   Ecological succession in an area where a climax terrestrial community (or other vegetation) previously existed and was removed, without removing or completely destroying the underlying soil. Secondary succession occurs after timber harvesting and abandonment of agricultural lands. Compare PRIMARY SUCCESSION.

secondary thickening   The increase in diameter of a plant stem, resulting from new cells produced by a cambium layer; also called secondary growth. Wood is produced by secondary growth forming secondary xylem and secondary phloem. See CAMBIUM.

secondary wave   See S WAVE.

secondary xylem   The xylem formed by a cambium layer as part of secondary thickening; formed as the plant stems increase in diameter, as opposed to the xylem that forms during the initial vertical growth of a shoot. Most wood is made up of secondary xylem and associated cells. See CAMBIUM, XYLEM.

second filial generation   (F_2) The second generation in an experimental cross or breeding experiment; the results of crossing two individuals from the F_1, or first filial, generation. See F_1 GENERATION, S_1.

second-generation pesticides   Synthetic organic compounds developed around the time of World War II to replace older plant-derived pesticides and those based on toxic heavy metals (arsenic, copper, etc.). DDT was the first of this new generation, and numerous other chlorinated hydrocarbon insecticides and herbicides followed.

second-growth forest   Stands of trees

undergoing secondary succession on land that was previously logged, cleared of vegetation, or otherwise disturbed by human activity. The species composition in a second-growth forest generally differs from that of the primary (virgin) or climax forests and usually has less biomass as well. Also called secondary forest. See SECONDARY SUCCESSION, OLD-GROWTH FOREST.

second law of thermodynamics   This law of thermodynamics states that natural processes tend to work in one direction only. It has many formulations, but can be generalized as systems naturally progress from order toward chaos (or increase in entropy), and heat flows only from a hot body to a cooler body. See FIRST LAW OF THERMODYNAMICS.

secretion   A substance that is produced by cells or by glands to perform certain functions, such as saliva and other digestive juices. Also, the manufacture and discharge of such substances, as in the secretion of milk.

section   A 1-square mile (640-acre) unit of the U.S. public land survey system. Each township, which is established by meridians and parallels, is subdivided into 36 consecutively numbered sections.

secured landfill   A landfill designed for hazardous waste sealed off and covered by impermeable materials. A secured landfill is designed to remain isolated from the environment.

sediment   A general term for any unconsolidated particulate material that has been deposited by an agent of transport, such as water, ice, or wind.

sediment trap   Any natural or artificial basin that slows the movement of water and promotes the settling of suspended sediment.

sedimentary   A term relating to any feature or process involving sediment.

sedimentary cycle   A continous series of processes involving weathering, erosion, transportation, deposition, and burial of rock particles; lithification and uplift; and repetition of the sequence. Less-resistant minerals are destroyed in the cycle, thus the rocks formed in a sedimentary cycle are increasingly dominated by resistant minerals.

sedimentary facies   A group of sedimentary structures, particle compositions, or bedding characteristics that serves to define an environment of deposition for a sedimentary rock by contrast with other features of the same stratigraphic unit.

sedimentary mineral deposit   A general term for any mineral deposit that forms by a sedimentary process.

sedimentary nutrient cycles   The biogeochemical cycle of elements, such as potassium, calcium, sodium, and magnesium, that lack a gaseous phase and tend to be in circulation either as dissolved or particulate material. The weathering of geological sediments, the atmospheric transfer of dust or salt particles, the washout by precipitation, the transport of dissolved nutrients in flowing water, and the formation of sedimentary deposits are all important events in the transfer of sedimentary nutrients. See BIOGEOCHEMICAL CYCLES.

sedimentary rock   A rock that forms by sedimentary processes at or near the earth's surface. Sedimentary rocks are classified as being of clastic, chemical, or biogenic origin.

sedimentary stratification   The appearance of horizontal and subhorizontal layers that is produced by variation in the composition and texture of a sedimentary rock.

sedimentary structure   An external or internal form of sedimentary rock that provides an indication of depositional environmental conditions. External sedimentary structures include ripple marks, scour marks, and raindrop imprints. Internal sedimentary structures include cross-stratification, lamination, and bioturbation.

sedimentation   1) The deposition of sedimentary particles, which includes gravity settling, chemical precipitation, and biogenic accumulation. 2) A general term applied to the infilling of a basin with clastic particles.

sedimentation potential   An estimate of the amount of solid sediment that may be yielded by a fluid through gravity settling or flocculation.

sedimentation tanks   A holding tank designed to separate suspended particulate matter from a fluid medium.

Seebeck effect   An electromotive force (EMF) created in a circuit consisting of two metals (or semiconductors) with two junctions that are at different temperatures. The metals or semiconductors and the temperatures of the two junctions control the strength of the resulting electromotive effect. The Seebeck effect is the basis for the thermocouples used as thermometers and in nuclear "batteries" used in some space probes. See PELTIER EFFECT.

seed   The mature embryo of a plant that develops from the fertilized ovule and from which a new individual grows. It is surrounded by a seed coat that protects it through dormancy until it is ready for germination. Often, it contains endosperm or cotyledon tissue that provides an energy reserve for the developing seedling.

seed bank   1) The reservoir of viable seeds contained on top of or within the soil, or in underwater sediments. 2) Collection of seeds that are used in research or as a genetic resource to maintain rare or endangered species of plants. Compare GENE BANK.

seed tree   A tree that is left standing when all the others around it are harvested, to serve as a source of seeds to revegetate the cleared area.

seed-tree cutting   A practice of forestry in which most trees on a site are harvested, but several mature parent trees of economic value are left standing as sources of seeds to revegetate the cleared area (seed trees). It is a form of selective harvest. Compare WHOLE-TREE HARVESTING, CLEARCUTTING, SHELTERWOOD CUTTING.

seep   A place where an aquifer or an oil reservoir intersects the ground surface and produces a gradual discharge of fluid. A seep does not produce runoff at a visible rate. Compare SPRING.

seepage   1) The gradual release of water from an aquifer or of oil from a reservoir at a seep. 2) The gradual movement of a fluid, such as water or oil, through a porous rock.

segmentation   1) The early stage of development in a fertilized ovum, a series of cell divisions form a multicellular, undifferentiated structure called the blastula. Also called cleavage. 2) Short for metameric segmentation. See METAMERIC SEGMENTATION.

segregation   The process during meiosis in which opposing pairs of genes separate, enabling the different alleles on the two genes to be inherited independently of each other. See MENDEL'S LAWS.

seiche A slow-moving, oscillating wave in a lake or an enclosed bay area. A seiche may be caused by persistent wind or by earthquake motions.

seif See LONGITUDINAL DUNE.

seismic A term that relates to the release of energy and consequent rock motion produced by an earthquake.

seismic belts A geographic region of high earthquake frequency. Seismic belts are common along the margins of tectonic plates. See CIRCUM-PACIFIC BELT.

seismic gap 1) A subregion within a defined earthquake belt where no recent earthquakes have been recorded. 2) An interruption in the data collection of a seismic record.

seismic-reflection profiling The technique of gathering seismic information along a transect line in order to construct a seismic profile.

seismic shadow zone A region of the earth's surface that does not lie along the propagation path of seismic waves. For example, a seismic shadow zone may be created where a seismic wave is attenuated by the core of the earth.

seismic stratigraphy The collection of stratigraphic information through the study of seismic wave reflections from subsurface layers of rock.

seismic tomography A technique of mapping subsurface geologic formations by studying a three-dimensional array of seismic records.

seismic wave 1) A wave of energy that propagates from an earthquake through the crust by the elastic strain of rock material. 2) A Tsunami or seismic sea wave. See TSUNAMI.

seismogram A graphic seismic activity record produced by seismic recording equipment.

seismograph A mechanical or electric device designed to sense and to record seismic waves.

seismonasty A curling or bending nastic movement of a plant tissue in response to a mechanical shock.

selection Any processes, either natural or artificial (human-directed), in which some animals or plants survive, reproduce, and contribute more individuals to the next generation than other organisms, which die, fail to reproduce, or are prevented from producing as many offspring as the other individuals. See NATURAL SELECTION, ARTIFICIAL BREEDING.

selection pressure Any force (or combination of forces) that affects populations by causing some individuals to produce more healthy offspring relative to other individuals, resulting in a greater proportion of the genes of those successful offspring in the total gene pool and directing the evolutionary process. Also called selective pressure.

selective breeding See ARTIFICIAL SELECTION.

selective death A death that would have been avoided had the individual possessed the ideal genotype rather than a deleterious one.

selective harvest A forestry practice in which certain trees, or groups of trees, are selected for harvest based on size, shape, growth potential, or competition with neighboring trees. Other trees are left standing for later harvest, producing a stand with trees of varying ages and in which the most valuable species predominate. Also called selection cut, selection-tree cutting, and selective cut. See SEED-TREE CUTTING, SHELTERWOOD CUTTING, STRIP CUT. Contrast CLEAR-CUTTING, HIGH-GRADING.

selective herbicide A chemical that kills only certain types of plants without harming desirable species in the area. Many lawn chemicals are selective herbicides; when used correctly, they kill only broadleaved species of weeds without harming grasses.

selective species A species of low fidelity, one found most commonly in a particular community, but also infrequently in other communities. See FIDELITY.

seleniferous 1) A term applied to an ore that contains selenium. 2) A term applied to a plant that absorbs selenium from the soil and concentrates the selenium within its tissues.

selenite A colorless and transparent form of the mineral gypsum. A satin-spar variety of selenite mineral with a fibrous texture. See GYPSUM.

selenium (se) A nonmetallic element with atomic weight 78.96 and atomic number 34. It occurs in different forms and in small quantities in a number of different ores. Selenium is used in photoelectric cells because of its property of increased conductivity with increased intensity of light falling on it. It is also used in glass and in enamels. A large portion of the U.S. Great Plains has soils with high selenium contents that are toxic to nonnative plants and to range animals.

selenodont Describing organisms that have cheek teeth whose chewing surfaces show characteristic ridges. Compare ACRODONT, BUNODONT, LOPHODONT.

self-compatible Describing a plant that is capable of self-pollination. See SELF-FERTILIZATION, SELF-INCOMPATIBLE.

self-fertilization 1) The successful formation of a zygote from male and female gametes derived from the same organism (or clone). 2) The fertilization of a flower by pollen from the same flower, also called self-pollination. Compare CROSS-FERTILIZATION.

self-incompatible Describing a plant that is not capable of self-fertilization and that must be pollinated by another plant. Also called self-sterile.

selfing Self-fertilization; artificial or natural fertilization of a flower with its own pollen, a common technique in plant breeding. See SELF-COMPATIBLE.

self-limitation See SELF-REGULATION.

self-pollination Fertilizing by transferring pollen from the anther to the stigma of the same flower. Also called SELF-FERTILIZATION.

self-regulation 1) A process of automatic internal control in a system, as when a population maintains a relatively constant size through changes in birth and death rates rather than through external forces. 2) Referring to human actions involving voluntary changes in behavior. See NEGATIVE FEEDBACK, POSITIVE FEEDBACK.

self-sterile The condition in hermaphroditic species in which male and female gametes from the same organism cannot successfully fertilize to form a viable zygote. In plants, this is usually called self-incompatible.

self-thinning A gradual decrease in population density as organisms in a population of growing individuals increase in size. See THINNING LINE.

selvedge 1) A distinctive edge zone of a rock mass, such as a chill zone at the edge of a pluton. 2) A fault gouge composed of clay-sized particles. See FAULT GOUGE.

sematic coloration Conspicuous markings on organisms that serve to signal warning or danger. The yellow and black markings on wasps are sematic coloration. Compare APOSEMATIC COLORATION.

semelparity The production of all of an individual's offspring in one event, which may occur over a short period of time. Compare ITEROPARITY, DISCRETE GENERATIONS.

semi- Prefix meaning half, partial, or incomplete as in semiannually or semi-darkness.

semiaquatic Growing or living near water and also capable of living in it.

semiarid Describing climates that are not as dry as desert, having light rainfall (usually 10 to 20 inches) capable of sustaining some grasses and shrubs but not enough for woodland.

semiconductor A material having a greater ability to conduct than an insulator, but not as great an ability as a conductor such as metal. Silicon is the best-known material for semiconductors, but germanium, lead sulfide, and selenium are also used. Semiconductors are essential for modern electronics; they are used in all integrated circuits as well as in diodes and in transistors. Specific uses include converting alternating current into direct current and generating electricity from sunlight.

semidesert Vegetation found in semiarid climates and having some of the same

species as a desert, often located between grassland and desert vegetation.

semipermeable membrane A membrane that allows some substances to pass through but not others. Semipermeable membranes used in chemical processes such as reverse osmosis allow solvent to pass through, but not the compounds (ions) dissolved in the solvent. Biological membranes such as those surrounding cells are semipermeable membranes. See OSMOSIS, REVERSE OSMOSIS.

senescence Aging process in mature individuals; the period near the end of an organism's life-cycle. Also, in deciduous plants, the process that occurs before the shedding of leaves.

sensitive clay A type of clay that experiences a great reduction in shear strength when it is deformed by remolding.

sensitize To make particularly sensitive to an antigen, such as a drug, or pollen.

sensor Any mechanical device or biological mechanism that registers and transmits a physical stimulus such as light or sound. See SENSORY RECEPTOR.

sensory impulses Any stimuli capable of being detected by the sense organs or sensory receptors.

sensory receptor A sense organ; any biological structure that registers stimuli.

sepal The green, leaflike structures that make up the calyx, which covers the unopened bud of a flower. Sepals are sometimes modified to be almost indistinguishable from the flower petals. See FLOWER.

Sepiidae See CUTTLEFISH.

septic tank A form of sewage disposal common in rural areas where municipal sewers, and their associated sewage treatment plants, are not available. It is a tank buried in the ground away from the house and any water supplies; it receives wastewater and sewage from the household and allows sewage and other solids to settle to the bottom, where they are digested by bacteria (and should be pumped out every couple of years). The liquid passes out to the perforated pipes comprising the septic (or leaching) field, draining into the soil.

sequential hermaphrodite An organism that first produces one form of gamete, and next produces the other form, switching from production of male gametes to female gametes or vice versa.

sequestering agent A chemical compound added to a solution to form complexes (or chelates) with the metallic ions in solution, preventing them from precipitating out. Gardeners in areas with alkaline soil use fertilizers containing sequestered (or chelated) iron to enable the iron to remain in a form usable to plants. See CHELATE.

seral Of or relating to the series of sequential stages in ecological succession. See SERE.

seration A series of communities within a biome, such as the group of communities in a valley and extending up the slopes of adjacent hills. See ZONATION.

serclimax A stable successional community representing an interruption of a succession by repeated natural disturbance such as fire or ocean spray. See SERE.

sere The complete successional sequence of ecological communities that have occupied a specific area, from the first species to colonize the area to the final climax vegetation. See SUCCESSION, MICROSERE, XEROSERE, LITHOSERE, CLISERE, SUBSERE.

series A group of related compounds, especially a group of organic compounds sharing the same basic structure and recognized by having the same suffix. The alkanes are a series of straight-chain, saturated hydrocarbons; the series includes methane, ethane, propane, butane, pentane, hexane, etc.

serine $HO-CH_2-CH(NH_2)COOH$, an amino acid found in many proteins and important in bacterial metabolism.

serotinous cones Pine cones, produced by some species such as jack pine *(Pinus banksiana)*, that remain on the tree for many years and are tightly closed until stimulated by the heat of a forest fire to open and release seeds.

serpentine Any of a group of greenish, secondary iron magnesium silicates that have a silky luster and form monoclinic crystals; produced by the alteration of olivine and pyroxene. Antigonite, chrysotile, and lizardite are serpentine minerals. See APPENDIX p. 627.

serpentinite Any rock that is composed mostly of serpentine minerals.

serule A very small-scale succession within a microhabitat.

servosystem A system that operates by negative feedback. See NEGATIVE FEEDBACK, POSITIVE FEEDBACK.

sesquioxide Any compound containing oxygen and another element in the proportion of three atoms of oxygen for every two atoms of the other element. Aluminum oxide (Al_2O_3) is a sesquioxide.

sessile 1) Describing organisms such as barnacles, corals, or rooted plants that are not mobile because they are permanently fixed to a base. 2) Describing stalkless structures, those attached directly to the main part of an organism rather than being located at the end of a stem.

seston Particulates that are suspended in water, including very fine particles of silt as well as organic detritus.

setae Any slender, bristlelike structures on an organism, such as the bristles on appendages of some crustaceans and insects. Also called chaetae.

set-aside 1) A portion saved or reserved. 2) An agricultural support and conservation program in which farmers are paid not to grow certain crops on their land.

settlable solids A measure of the proportion of sludge that forms solids in the suspended solid component of sewage.

settling chamber A container in which solids or heavy liquids are allowed to separate by gravity settling.

settling pond A basin that is designed to allow solid material to be separated from effluent by gravity settling. Settling ponds are often a component of sewage waste treatment systems.

Sevin 1-napthyl-N-methyl carbamate, one of the insecticides developed as a replacement for DDT. A carbamate insecticide, it breaks down very quickly in the environment and is not toxic to animals. It is, however, very toxic to bees. Also known as carbaryl.

sewage Aqueous discharge from municipal or industrial sanitary collection systems, especially as pertains to human fecal waste.

sewage fungus A slimy bacterial colony that forms on water containing sewage or similar organic pollutants.

sewage lagoon An open holding basin in a sewage waste treatment system.

sewage sludge A sludge that is a product of a sewage waste treatment system. See SLUDGE.

sewer An engineered collection system that carries sewage to a treatment plant.

sewerage A collective term for the sewage collection system and treatment apparatus in a particular district or area.

sex Gender; the totality of characteristics that distinguish male organisms from females of the same species.

sex attractant A natural compound secreted by females of a number of different insect species to attract males for mating. Sex attractants, a form of pheromone, are used in biological pest control to lure species into traps or to cause disorientation. See PHEROMONE, CONFUSION TECHNIQUE.

sex chromosomes The pair of chromosomes that in combination determine the sex of an organism. In humans, the sex chromosomes are called the X chromosome and THE Y chromosome. See X CHROMOSOME, Y CHROMOSOME.

sex-linkage The phenomenon in which some inherited traits are associated with a particular gender because they are carried on a sex chromosome (the X chromosome, or its analogue). Sex-linked traits in humans characteristically appear only in males; although the gene for the trait may be transmitted by females, they contain an additional X chromosome that usually masks the particular trait. Examples include ordinary baldness, red-green color blindness, and hemophilia.

sex ratio The proportion of males and females within a specific population; it is usually expressed as the number of males per 100 females.

sexual selection Natural selection promoted by factors affecting the choice of mates, often specific characteristics related to courtship behavior.

shade-intolerant species Plant species that cannot grow in the shade of other species. Many pioneer species that thrive under harsh conditions and bright sunlight are shade-intolerant. Sometimes simply called intolerant species.

shade-tolerant species Plant species that grow well in the shade of another species. The species found in climax forest communities are usually shade-tolerant, even though they may eventually outgrow the species shading them and reach direct sunlight. Sometimes simply called tolerant species.

shadow 1) The dark region formed (in the direction away from the incoming light) when an object interrupts light. 2) An analogous interruption of other forms of radiation, such as the poor radio or television reception directly behind a mountain that blocks the incoming signal. See UMBRA, SOUND SHADOW.

shadow zone See SEISMIC SHADOW ZONE.

shale Any fine-grained clastic sedimentary rock composed of clay and silt-sized particles. Shale tends to break along the aligned mineral grains of the compact bedding planes.

shale oil See OIL SHALE.

shallow-focus earthquake An earthquake that has a focus, or point of brittle failure of rock, located less than 70-km deep within the crust of the earth.

Shannon-Weiner diversity A measure of equitability (H') calculated to incorporate the sum of the proportional contributions of an individual species to the total popula-
tion, biomass, dominance, productivity, etc. of a community. Minimal values occur when one species has a disproportionate dominance whereas maximum values occur when all species share equally in the dominance of the community. Compare EQUITABILITY, EVENNESS, SIMPSON DIVERSITY INDEX. Contrast SIMPSON DOMINANCE INDEX.

shear 1) The equal and opposite tangential forces acting on opposite sides that cause layers of a material to tend to slide past each other (also called shearing force). Also, the reaction to such an applied force, the deformation in which components of the body under stress are moved parallel to each other. Shear is a type of deformation (response to stress) that results in no overall change in volume. 2) Short for wind shear. See WIND SHEAR.

shear strength The ability of a material (such as rock, or geological layers) to withstand sheer stress. Also called shearing strength.

shear stress The stress that results when a shear force is applied to a body or substance. The forces causing shear response can be sudden or gradual.

shear waves Transverse waves that cause movement of an elastic medium, but no compression. An S wave within the earth's crust is a type of shear wave, as are some acoustical waves. See S-WAVE (SECONDARY WAVE).

sheep-month The amount of forage consumed by an average sheep, or a ewe with a nursing lamb, over one month. About five sheep-months are equivalent to one cow-month. See ANIMAL UNIT, COW-MONTH.

sheeted flows See FLOOD BASALT, OVERLAND FLOW.

sheet erosion   The removal of a uniform thickness of particles from the soil surface by rainsplash and runoff processes.

sheet flood   The accumulation and runoff of storm water above the soil surface as water quantity exceeds depression storage and infiltration capacity.

sheeting   The development of tension cracks (sheet joints forming roughly parallel to surface topography) in crystalline bedrock as outer layers expand more than inner layers because of unloading. See EXFOLIATION (SHEETING).

sheet piling   A structural foundation element, such as a steel bulkhead, that is designed to resist lateral forces or lateral seepage within a soil.

sheet wash   The movement of debris and soil particles washed downhill by overland flow. See OVERLAND FLOW.

shelf   An area of sedimentation on a stable craton. See CONTINENTAL SHELF.

shelf-slope break   The line of transition between the edge of a continental shelf and the top of a continental slope.

shell   The hard exterior of a fruit or a nut, or of an animal such as snail.

shelter belts   Rows of trees planted as a soil conservation measure to protect smaller crops from the wind, to reduce soil erosion, or to slow the formation of large snow drifts. Shelter belts were planted extensively in the U.S. Great Plains to curb and prevent a recurrence of the Dust Bowl of the middle and late 1930s, which swept topsoil from the plains states as far east as New York City. Outside of the Great Plains, these are usually called windbreaks. See DUST BOWL.

shelterwood cutting   The practice of leaving a number of large trees standing while others are cut; the standing trees serve to shade and shelter young seedlings. It is a form of selective harvest, typically spreading out the cutting of mature trees over two to three different harvests over a ten-year period. Compare SEED-TREE CUTTING, WHOLE-TREE HARVESTING, CLEARCUTTING. See SHELTER BELTS.

shield volcano   A large volcano composed of gently sloping, solidified lava flows. The islands of Hawaii are typical examples of shield volcanos. Compare COMPOSITE VOLCANO, PYROCLASTIC CONES.

shifting cultivation   A form of crop growing usually practiced in conjunction with slash-and-burn agriculture in tropical woodlands. Plots of land are abandoned after two to five years because the tropical soils, ideally suited to growing large forests, cannot support extended cultivation (they may also be invaded by forest vegetation). The farmer must therefore shift to another plot to grow food crops. See SLASH-AND-BURN AGRICULTURE.

shingle   A term applied to the loose water-worn rock particles that accumulate on the upper part of a beach. Shingle may consist of flattened beach pebbles, cobbles, and boulder-sized materials.

shoal   A shallowly submerged bank or bar of unconsolidated gravel, sand, and mud. A shoal forms where coarse particles accumulate at the mouth of a river.

shock   1) Any sudden, violent disturbance of a system in equilibrium. 2) A medical condition that accompanies severe injuries. It is characterized by pallor, shallow breathing, low blood pressure, and weak pulse as the body reduces blood flow to the extremities. 3) A group of cornstalks tied together and standing on their ends, or sheaves of grain (as a shock of wheat). 4) An abrupt change in pressure, density, and particle velocity (localized temperature), traveling as a wave through a fluid or plasma. Lightning, nuclear explosions, and supersonic aircraft generate shock waves. See SONIC BOOM.

shock wave A compressional wave that travels through rock at a velocity that exceeds the elastic limit of the material. A shock wave, such as the wave generated by a meteorite impact, may cause distinctive deformation of crystalline rock materials.

shorebirds A very general term often applied to members of the order Charadriiformes such as gulls and sandpipers. However, it may also apply to species of the orders Ciconiiformes and Gruiformes. See CHARADRIIFORMES, CICONIIFORMES, GRUIFORMES.

shoreline A narrow strip of substrate that separates a land area from the water. The shoreline is between the upper limit of wave action and the lower limit of the waterline.

shore zonation The series of distinct ecological communities usually present at the edge of the ocean. They are distinguished by whether the surface is usually covered with water, periodically covered as high tide approaches, or only rarely covered by the highest tides or storms. See INTERTIDAL.

short-day organism Organism that needs short periods of light, or long periods of dark, in order to mature and to undergo sexual reproduction, such as short-day plants. Contrast LONG-DAY ORGANISM.

short-day plant A plant that flowers or undergoes some morphological change only when stimulated by relatively short day length, requiring dark periods that are longer than some given time to trigger the response. Chrysanthemums, asters, and traditional varieties of strawberries are short-day plants. Contrast LONG-DAY PLANT, DAY-NEUTRAL PLANT.

shortgrass prairie North American prairie vegetation characterized by shorter species of grasses (buffalo grass, blue gramma, side oats gramma), once the characteristic vegetation of the drier, western regions of the Great Basin in the United States. Also called xeric prairie. Compare TALLGRASS PRAIRIE.

shredders Aquatic animals that consume coarse particles of organic matter, which they shred into smaller pieces in the process.

shrub A short, woody, perennial plant, usually having many stems.

shrubland Vegetation in which shrub forms predominate, as in hot and dry areas in the Southwest where creosote bush *(Larrea)* dominates the desert, or in chaparral where *Ceanothus* dominates. See MAQUIS.

shrub layer A horizontal layer in a plant community containing primarily shrubs. Compare HERB LAYER, UNDERSTORY.

shutdown A temporary idling of a nuclear reactor by inserting control rods to reduce the fission reaction to its lowest possible level.

shuttle box A construction for animal behavior experiments, consisting of a narrow box in which the animal "shuttles" from one end to the other in response to stimuli.

SI See SYSTÈME INTERNATIONAL D'UNITÉS, also APPENDIX p. 605.

si Chemical symbol for the element silicon. See SILICON.

sial   A collective term describing the granitic nature of rocks found on continental tectonic plates. The name combines the first two letters of silica and aluminum. Contrast SIMA.

sibling species   Different species of animals that look very similar or identical, but are not able to interbreed with one another (are reproductively isolated). Also called twin species and species pair.

siccicolous   Describing species that live in dry environments.

sidereal day   An earth day as measured with respect to the stars. Because the revolution of the earth causes it to travel while it is rotating, a stellar or sidereal day is not the same length as the solar day with which we are familiar. A sidereal or stellar day is almost four minutes shorter than a solar day. Compare SOLAR DAY.

siderite   A gray-to-brown mineral of iron carbonate forming trigonal crystals or botryoidal masses in a wide variety of sedimentary rocks. Siderite is an ore mineral of iron.

siderophile   1) Describing a species that thrives in an iron-rich environment. 2) Describing elements that readily combine with metallic iron, and which tend to be located in conjunction with deposits of metallic iron.

siemen   Unit for electrical conductance in the Système International d'Unités (SI). One siemen is the reciprocal of one ohm (1/ohm); it is equal to the conductivity of a circuit or an element having a resistance of one ohm (that is, through which one ampere of current flows when the potential difference is one volt). The unit was formerly called mho. See OHM, CONDUCTANCE.

Sierra Club   A citizen environmental organization founded in 1892 by John Muir. Its programs include public education, advocacy, litigation, and publishing on environmental issues and wild areas, as well as wilderness outings and conferences.

sievert   (Sv) A unit for the effective dose equivalent of radiation. It is proportional to the expected biological harm of the dose. One sievert equals 1 joule of energy per kilogram of absorbing tissue, or 100 rem. The sievert is gradually replacing the rem.

sieve tube   An individual component of plant phloem. Sieve tubes are a series of elongated, hollow cells connected through small openings in their end walls and running the length of the stem. Each sieve tube carries food compounds synthesized in the leaves downward to shoots, stems, and roots. See PHLOEM.

sigma-t   The oceanographic symbol used to denote density and usually stated as gm/m^3.

sigmoid growth   Population increase that begins at a slow rate and gradually increases until a sufficient population exists for it to increase exponentially. As the population reaches carrying capacity, growth levels off. Sigmoid growth is named for the characteristic "s" shape of the curve that results when it is graphed; it is characterized mathematically by the logistic equation. See LOGISTIC EQUATION, S-SHAPED CURVE.

signal-to-noise ratio   The ratio of the decibels of a desired signal to decibels of unwanted, residual noise in a system; it provides a measurement of the efficiency of a communication.

silage   Cut grass, cornstalks, or other plants that are harvested while green and stored in silos or concrete bunkers to undergo fermentation. Silage can be stored for long periods of time and provides important winter feed for cattle, containing more of the nutrients of fresh grass than are present in hay.

silica SiO_2, silicon dioxide. Silica is a mineral that makes up about 12 percent of the earth's crust. Sand, opals, flint, and quartz are all forms of silica.

silicate   Any member of a mineral group in which silica tetrahedra ($Si^{4+}O^{2-}$) form the fundamental structural unit within the crystal lattice.

silicate mineral   A mineral based on the silicate structure; examples are olivine, pyroxene, amphibole, hornblende, mica, quartz, and feldspar.

siliceous   A term applied to material containing an abundance of silica.

siliceous sinter   A white and porous opaline silicate deposited by precipitation from the hot mineral water of a geyser or a hot spring. Also called geyserite.

silicic   A general term applied to a magma or igneous rock that has a composition of approximately two-thirds silica.

siliciclastic sediment   A term applied to any accumulation of unconsolidated clastic particles primarily composed of silicate minerals.

silicification   1) A type of fossilization in which the original organic material is replaced by silica. 2) The process of diagenetic rock formation in which minerals are replaced or pores are filled by silica material.

silicon   (Si) Abundant, nonmetallic element with atomic weight 28.09 and atomic number 14. Silicon is the main constituent of many rocks and the soils formed from such rocks. It is used in glass, alloys, and semiconductors.

silicon-controlled switch   (SCS) A switching device using semiconductors that is used to control the voltage of power circuits.

silicon tetrahedron   The fundamental SiO_4 structural unit making up all silica-containing minerals (silicates). It consists of a silicon atom surrounded by four oxygen atoms; the oxygen atoms outline a tetrahedron. Silicates are classified by how the tetrahedra are linked together to form crystals.

silicosis   A chronic disease of the lungs caused by the inhalation of silica (quartz) dust. It is a form of pneumoconiosis found in people who mine and refine silica, or in stonemasons working with rock containing much quartz; it can lead to emphysema. See PNEUMOCONIOSIS.

silicula   A type of dry seed pod that spontaneously releases its seed when ripe, formed from a flower with two fused carpels in each ovary. A silicula is as long as it is wide. Also spelled silicle. Compare SILIQUA.

siliqua   A dry seed pod that spontaneously releases its seed when ripe; it is similar to a silicula, but is significantly longer than it is wide. Also spelled silique. Compare SILICULA.

sill   A horizontal or tilted tabular pluton that has intruded parallel to strata of the country rock or flowed in a sheetlike manner across the surface. A sill is one type of concordant intrusion.

sillimanite   An aluminum silicate mineral forming light-colored orthorhombic crystals in relatively high-pressure, high-temperature, metamorphic environments. Chemically equivalent to andalusite and kyanite. See APPENDIX p. 627.

silt Any clastic rock fragment between $1/256$ and $1/16$ mm in average diameter.

siltation A term applied to the gradual sedimentation of silt-sized particles in a body of water. Siltation often occurs behind the dam in a reservoir.

silt-clay fraction The portion of a soil or sedimentary material that includes only particles of silt and clay size. The silt-clay fraction includes all particles less than $1/16$ mm in diameter.

siltstone A sedimentary rock composed of at least two-thirds silt-sized particles. Siltstone, which lacks the fissile texture of shale, is a mudstone containing more silt than clay.

Silurian period The third geologic time period of the Paleozoic era. The Silurian period lasted from approximately 438 to 408 million years ago. See APPENDIX p. 611.

silva Tree flora; the tree species found in a forest in a certain area.

silver (Ag) Metallic element with atomic weight 107.9 and atomic number 47. Like other noble metals, it does not corrode readily (although it tarnishes). It occurs in compounds and in pure form; the pure form is easily shaped and is the best known conductor of heat and electricity. It is used in photographic emulsions, jewelry, and for industrial uses requiring a metal that does not corrode easily. See APPENDIX p. 627.

silviculture Forest management; the cultivation and the harvest of trees to provide a renewable supply of wood.

sima A collective term describing the basaltic nature of rocks found on oceanic tectonic plates. The name combines the first two letters of silica and of magnesium. Contrast SIAL.

simazine $ClC_3N_3(NHC_2H_5)_2$, a potent, broadleaf herbicide used to prevent weeds from sprouting around ornamental plantings and orchards (it does not control existing weeds). It is a member of a group of herbicides called triazines.

similarity index See COEFFICIENT OF COMMUNITY.

Simpson diversity index (D) A measure of the diversity in a community, based on the probability of randomly picking two individuals of different species from a community. It is calculated from the Simpson Dominance index. Compare EVENNESS, EQUITABILITY, SHANNON-WEINER DIVERSITY. Contrast SIMPSON DOMINANCE INDEX.

Simpson dominance index (C) A measure of the dominance concentration within a community; it expresses the probability of randomly picking two individuals of the same species in a sampling of a community. Contrast EVENNESS, EQUITABILITY, SHANNON-WEINER DIVERSITY, SIMPSON DIVERSITY INDEX.

Simpson's index of floristic resemblance An index of similarity (coefficient of community) calculated by dividing the number of species common to two communities by the total number of species in the smaller of the two communities. Compare JACCARD INDEX, SØRENSEN SIMILARITY INDEX, GLEASON'S INDEX, MORISITA'S SIMILARITY INDEX, KULEZINSKI INDEX.

sine wave The wave-shaped curve of a sine function; also the curve that results when simple harmonic motion is graphed. In the equation for a sine wave, one variable is proportional to the sine of the other variable. Also called sinusoidal wave.

single bond A covalent chemical bond

between two atoms that consists of one pair of shared electrons. Single bonds are characteristic of saturated carbon compounds such as saturated fats. See COVALENT BOND, DOUBLE BOND, TRIPLE BOND.

single-cell protein   A food supplement providing an excellent source of protein. It is made from culturing and harvesting one-celled organisms including yeasts, algae, or bacteria. Single-cell protein is added as a supplement for animal feeds and, as nutritional yeast, is a human dietary supplement providing many B vitamins as well as amino acids.

sinistral fault   The term applied to a lateral fault in which the sense of displacement is a strike-slip motion toward the left of the block from which the observation is made.

sink   A part of a system that absorbs a substance (or heat), removing it naturally or artificially from the system. In plants, a sink is an area with a demand for metabolic substances; growing meristems are sinks for energy compounds from photosynthesis and mitochondria are oxygen sinks. Undisturbed tropical rainforests and deep oceans may act as carbon sinks, absorbing carbon dioxide from the atmosphere. Compare SOURCE.

sinkhole   A rounded topographic depression formed by solutional processes in rocks such as limestone or rock salt. Some form by the collapse of the roof over a cavern.

Sino-Japanese floral region   One of the phytogeographic regions into which the Holarctic realm is divided according to similarities between plants; it consists of Japan and most of China, including the northern Himalayan mountains. See HOLARCTIC REALM, PHYTOGEOGRAPHY.

sinter   A hard crust or a deposit of siliceous (silica-containing) or calcareous (containing calcium carbonate) minerals formed when mineral waters evaporate. Sinter forms near geysers and so is also called geyserite; it can also form stalactites from cold mineral waters in caves.

sinuous channel   A description of a meandering stream channel. See MEANDER.

sinusoidal wave   See SINE WAVE.

Siphonoptera   Order of insects, the fleas have laterally compressed, hairy bodies. Mouthparts are modified to form a tube-like structure for piercing and sucking. There are no wings. Most species are ectoparasitic on mammals but some infest birds. Fleas are important health hazards because they not only cause intense itching and discomfort but are also vectors of murine typhus and plague. Because they are not as host-specific as many parasites, fleas may sequester human disease organisms in other mammals, particularly domestic and wild rodent populations. Distribution is worldwide. Approximately 1700 species exist. See INSECTA.

sirocco   A hot wind originating in northern Africa and blowing across the Mediterranean to southern Europe (Italy, Malta, and Sicily). The sirocco is usually very dry and carries large quantities of dust picked up in the Libyan deserts. Sometimes the sirocco picks up enough moisture as it crosses the Mediterranean to become a hot, moist wind.

sister taxa   A pair of taxa that have arisen from a single evolutionary event (speciation) and therefore are closely related descendents of a common ancestor taxon.

site   1) The combination of environmental conditions that affect the type of vegetation at a particular place. Site includes soil type, drainage, slope, and climate, but not the existing plant cover. 2) The location or the position at which something occurs or is built, such as the site of the chemical spill or the site for a new power plant.

site index   A quantitative measure of a site's ability to produce timber that is derived from a graphical analysis of tree age at breast height (on the x-axis) and height of tree in feet (on the y-axis).

site quality   The potential of a site with particular species to produce timber as measured by the site index. See SITE INDEX.

site type   A group of biotic communities sharing a particular set of site characteristics.

SI units   See SYSTÈME INTERNATIONAL D'UNITÉS, see also APPENDIX p. 605.

skarn   A rock formed by the contact metamorphism of limestone or of dolomite. Skarn may be an ore source for magnetite or copper sulphides.

skid trail   A small, temporary trail used to move logs to a staging area during a logging operation.

skimmming (of oil slick)   A technique of capturing and removing floating oil by erecting a mechanical device to drag across the water surface.

slag   1) A general term applied to a fragment of scoriaceous pyroclastic rock.
2) A furnace cinder that has a scoriaceous appearance.

slaked lime   A white crystalline industrial form of calcium hydroxide. Slaked lime is formed by treating natural lime with water.

slash   The residue of treetops and branches produced during a logging operation.

slash-and-burn agriculture   A method of farming common in tropical forests, in which the forest is cut down to create open space for growing crops, and after being allowed to dry, the wood is burned to provide short-term nutrients for crop plants. The plot must be abandoned after several crops have been grown because the process destroys the nutrient reserves normally tied up in the standing trees and it usually destroys soil humus; the same area cannot be reused for many years, often up to 30 years. A new plot is prepared in the same way in a continuing cycle. See SHIFTING CULTIVATION.

slate   A fine-grained rock formed by the low-grade metamorphism of shale or of volcanic ash. Slate has distinctive cleavage which is directed by platy minerals that are aligned at right angles to the direction of metamorphic compression.

slaty cleavage   A distinctive type of foliation that develops in slate. Slaty cleavage breaks along lines parallel to the orientation of chlorite and of sericite mineral grains.

sleet   Frozen precipitation in the form of minute pellets. Sleet forms when rain falls through a layer of air whose temperature is below the freezing point. When falling rain is supercooled to temperatures just below freezing, it does not form sleet but may become freezing rain if it lands on (and freezes to) objects that are colder than the freezing point. Compare HAIL.

slick   A calm area of open water that is maintained by a thin film of oil, which alters the normal surface tension of the water.

slime nets   See LABYRINTHULAMYCOTA.

sling psychrometer   Another name for a psychrometer (wet and dry bulb hygrometer). See PSYCHROMETER.

slip 1) The measure of relative displacement between fault blocks. 2) The amount of cleavage movement, distortion, or rotation between the grains of a deformed rock.

slip face The steeply sloping surface, standing at the angle of repose, on the downwind side of a dune. Sand grains in a typical dune are blown up to the dune crest and then fall down the slip face in the leeward wind shadow of the dune.

slope A measure of the steepness of an inclined surface. Slope is stated as a measure of the difference in elevation between two points divided by the horizontal distance between them. See CONTINENTAL SLOPE.

slough 1) Dead tissue that is separated from living tissue. Also, (of a living organism) to cast off such tissue, as when a snake sheds its old skin. 2) A swamp, marsh, or muddy backwater.

sludge 1) A general term for solid or semisolid waste materials produced in an industrial process. 2) A muddy stream deposit or sea-floor ooze. 3) A thickened accumulation of sea ice particles that have not yet frozen into a solid mass.

sludge cake A mass of sea ice sludge that has frozen into a solid body, or floe, and that is large enough to support the weight of a person.

sludge digesters A device designed to heat sewage sludge and to promote decomposition of organic material by anaerobic bacteria.

sluice An artificial channel that controls the discharge of water from an impoundment.

slump (rotational slide) See ROTATIONAL LANDSLIDE.

slurry A mobile suspension of particulate material in a fluid medium. The material in hydraulic mining is transported to the surface in the form of a slurry.

smithsonite A whitish secondary mineral of zinc carbonate formed in association with sphalerite and often found in limestone. Smithsonite is an ore mineral of zinc.

smog Photochemical smog: The air pollution resulting when hydrocarbon and nitrogen oxides (NO_x) are exposed to sunlight, causing them to undergo a photochemical reaction to produce much more harmful chemicals. Often used in a more general sense to refer to any visible air pollution, especially at levels high enough to reduce visibility and to cause irritation to eyes, lungs, etc. The word was coined in 1905 as a combination of smoke and fog. See PHOTOCHEMICAL SMOG.

smoke Very fine particles of ash and/or partially burned or unburned fuel, suspended in the air as visible clouds. Smoke is often the result of the incomplete combustion of compounds or substances containing carbon.

smudge pot An intentional, very smoky fire used as an attempt at frost prevention, especially in orchards. Smudge pots attempt to create a cloud of particulates above orchards to reduce radiational cooling when frosts are predicted (sometimes by burning piles of old tires). Smudge pots are not very effective, plus the cloud may actually reduce solar warming on the following day. Heaters, wind-generating machines, fine sprinklers, and cloth or paper coverings are all more effective than smudge pots.

smut Any of a group of parasitic fungi (class Heterobasidiomycetae) that causes plant diseases characterized by the production of masses of sootlike black spores. Many are economically important, such as corn smut and the stinking smut disease of wheat, which differs from other smut fungi in creating a foul odor in addition to the sooty spores. Smuts are not controlled in the field with fungicides; the best controls are use of clean or disinfected seed and practicing crop rotation. Compare RUST.

Sn   Chemical symbol for the element tin. See TIN.

snag   1) An upright stump or trunk of a tree that provides habitat for a broad range of wildlife, from beetle larvae (and the birds such as woodpeckers that feed upon them) to dens for racoons. Sometimes called a stub. 2) A tree or a branch embedded in a river or lake. 3) A small branch of an antler.

snakes   See OPHIDIA.

snipes   See SCOLOPACIDAE.

snowline   1) The topographic boundary line marking the lower limit of perennial snow in a glacial snow field. The snowline also represents the outer limit of the zone of accumulation of a glacier. 2) The elevation above which rain turns to snow in a particular storm.

snowpack   A general reference to the snow covering a landscape. The snowpack in mountainous areas is often monitored to provide data for estimating spring runoff.

SO_x   A symbol for the group of sulfur oxide air pollutants, such as sulfur dioxide and sulfur trioxide. The full list is SO_2, SO_3, $(S_2O_7)^{-4}$, and $(SO_4)^{-2}$. They produce acid rain when they interact with water vapor in the atmosphere. Compare NO_x.

social facilitation   An increase of any behavior by association with other individuals engaged in similar behavior.

Yawning in primates is a classic example of social facilitation.

social group   A group of individuals of the same species that live and travel together and to some degree depend on each other for well-being.

social mimicry   See CHARACTER CONVERGENCE.

social parasite   1) A parasite that exploits the normal behavior, such as nesting or hunting, of a host species. Birds such as the cowbird and the cuckoo, which lay their eggs in another bird's nest to be hatched by the other bird (brood parasitism), are social parasites. 2) A parasite of social insects, such as ants or bees. A different insect that invades the colony and consumes the food reserves (heptoparasitism) or the young, or both.

social pathology   Physiological and behavioral disturbances in animals caused by crowding that result in reduced fecundity and increased mortality.

social selection   Natural selection favoring genetic traits promoted by social behavior; nonreproductive interactions between members of the same species that are not closely related. See SELECTION.

society   In the monoclimax thoery, a subdivision of a faciation in which subdominant species exert local control in localized communities that are different from the regional norms.

sociobiology   The study of the biological basis of social behavior (nonreproductive interactions between members of the same species that are not closely related) in a variety of organisms, and the evolution of such behavior. Sociobiology includes the study of human behaviors. See KIN SELECTION.

soda A general term for substances containing sodium (Na). Soda can refer to sodium bicarbonate ($NaHCO_3$, baking soda), sodium carbonate ($NaCO_3$, washing soda), sodium hydroxide (NaOH, caustic soda), or any other sodium-containing salt. See CAUSTIC SODA.

soda lake An alkaline lake in an arid environment where evaporation promotes the accumulation of dissolved sodium chloride and sodium sulfate.

sodium (Na) A soft, metallic element with atomic weight 22.99 and atomic number 11. It is very reactive and so is only found in nature in compounds with other elements. Liquid sodium is used as a coolant in some forms of nuclear reactors and as a heat sink in solar collectors. See SODIUM CHLORIDE, CAUSTIC SODA.

sodium chlorate $NaClO_3$, a compound containing sodium, oxygen, and chlorine. It is a strong oxidizing agent. Uses of sodium chlorate include in matches, in explosives, as a systemic herbicide, and in dye manufacturing and paper-pulp processing.

sodium chloride NaCl, ordinary table salt, the primary salt found in the oceans. Sodium chloride is essential for maintaining the electrolyte balance in animals. It is probably the oldest food preservative and a universal seasoning agent. Naturally occurring sodium chloride is also called halite (rock salt).

sodium silicofluoride $NaSiF_6$, one of the compounds used in the fluoridation of municipal water supplies. See FLUORIDATION.

soft-energy paths Energy policies emphasizing conservation and renewable, low-polluting sources of energy such as the sun and the wind. Soft-energy paths shift society away from depending on large-scale use of fossil fuels and nuclear power. These are replaced with solar power, wind power, tidal power, and biomass fuels, with a greater dependence on less-centralized and often relatively simple technology (although computers may control the systems). Amory Lovins coined the term; he published a book by the same title in 1977.

softwood A coniferous, needle-leaved, and usually evergreen tree (a gymnosperm) such as pine or spruce. Also, the wood from such a tree, which is characteristically less dense and more resinous than that from hardwood (broad-leaved deciduous) species. Contrast HARDWOOD.

soil 1) A combination of mineral and organic matter with water and air above the bedrock surface. 2) The upper part of the regolith that supports life. 3) Earth material modified by physical, chemical, and biological processes such that it supports rooted plant life. See REGOLITH.

soil-acting herbicides Plant-killing compounds that work either by killing seeds before they emerge or as systemics (taken up by the roots). Soil-acting herbicides differ from the contact herbicides used in most lawn weed-killers (and in herbicides used for military purposes, such as Agent Orange), which work by direct contact with foliage.

soil association Two or more similar soils that are mapped, for convenience, as a single unit on a soil map. A soil association may be mapped where adjoining soils are intricately distributed.

soil chronosequence An array of similar soils that differ in the degree of horizon development because of differences in age. For example, a soil chronosequence may form on a sequence of stream terraces or of glacial moraines.

soil classification A method of characterizing and naming soils. Examples include the Unified Classification used by engineers and the Seventh Approximation of Soil Taxonomy used by soil scientists. See APPENDIX P. 619.

soil compaction The reduction in pore space and the consequent increase in density of a soil. Mechanical soil compaction can be used to increase the engineering suitability of a soil.

soil conditioner Any substance, such as fertilizer, mulch, or manure, that is added to improved the agricultural quality of soil.

soil conservation The management of land use practices to prevent chemical deterioration or physical loss of a soil.

soil creep See CREEP.

soil drainage A characterization of the percolation behavior of a soil. See PERCOLATION.

soil erosion The loss or removal of soil material by a transport process, such as rill erosion, gully erosion, sheet erosion, and wind erosion.

soil flow See SOLIFLUCTION.

soil horizon A zone or fairly distinct layer of the soil parallel to the surface that has a definitive identity related to composition and to physical characteristics. See APPENDIX p. 619.

soil map A two-dimensional representation of the areal extent of soils in relation to topographic or cultural features.

soil moisture The amount of water held in voids between soil particles as determined by weight or volume.

soil moisture belt The zone of a soil where plant roots are found. The soil moisture belt is the uppermost part of the zone of aeration, or vadose zone, within a soil.

soil phase A subdivision of a soil series based upon characteristics affecting use and management (slope, wetness, thickness, or stoniness).

soil porosity A measure of the percentage of soil volume occupied by voids between particles.

soil profile A description of composition, texture, and structure of soil horizons taken in vertical sequence. See APPENDIX p. 619.

soil series A grouping of soils that are similar in composition, thickness, and arrangement, but not in texture, of soil profiles.

soil sterilant A substance used to kill all the organisms in soil, especially soil-borne diseases, mites, and insects. Soil sterilants are often toxic compounds that must be handled carefully (methyl bromide, formaldehyde). Heat is a very effective soil sterilant (as is steam), although cumbersome on a large scale because temperatures of 180°F must be maintained for 30 minutes. Many soil sterilants are fumigants, although some are liquid drenches. See FUMIGANT.

soil structure A description of soil particle aggregates or compound particles. See FERET TRIANGLE.

soil survey A compilation of map-based field data pertaining to soil locations, descriptions, and classifications for land-use planning.

soil texture A description of the relative proportions of sand, silt, and clay in a mass of soil. See FERET TRIANGLE.

soil type A subdivision of a soil series. Soil types are distinguished by differences in the textures of surface layers. See APPENDIX p. 619.

Solanaceae The nightshade family of the angiosperm dicots. The typically alkaloid-rich leaves are often glandular and sticky. The flowers have five united petals that are often rotate or tubular and have five equal stamens that are usually appressed to the style. The fruits are fleshy berries or dry capsules containing many seeds. Tomato (*Lycopersicon*), potato, eggplant, and nightshade (*Solanum*), tobaccos (*Nicotiana*), peppers (*Capsicum*), petunia (*Petunia*), and Jimson weeds (*Datura*) are members of this family. See ANGIOSPERMOPHYTA.

solanine A complex, poisonous plant alkaloid found in potato sprouts and in tomatoes (both members of the plant family *Solanaceae*). Solanine is a steroid glycoside and a narcotic (cholinergic) drug.

solar Derived from or relating to the sun, as in a solar eclipse or solar energy.

solar cell Common name for a photovoltaic cell, a device that converts sunlight directly into electrical energy. See PHOTOVOLTAIC CELLS.

solar collector The part of an active solar energy system that absorbs radiant heat from the sun and transfers it to where it is needed or to a storage system. Solar collectors are usually large black areas (plates mounted on rooftops, or entire south-facing walls) with water, antifreeze, or air circulating through them to remove the heat.

solar constant The intensity of radiant energy the earth receives from the sun at the outside edge of the earth's atmosphere, a quantity that only varies by 0.1 percent. The solar constant includes all wavelengths of radiation. It is calculated for the earth's average distance from the sun, and is 430 BTU/ft^2/hr, or 1.37 kw per square meter, or about 2 calories per square centimeter per minute.

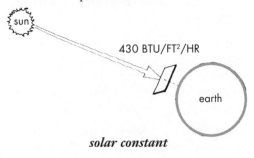

solar constant

solar day The 24-hour interval between the time the sun crosses a meridian on earth, and the time the sun returns to that meridian. Because this is calculated for a "mean sun," an average of the varying speed at which the sun moves, it is technically called a mean solar day. Compare SIDEREAL DAY.

solar design Architecture that incorporates some form of solar energy, such as passive solar heat or active solar hot-water systems.

solar energy Any form of energy derived from the energy coming from the sun, excluding naturally modified plant and animal matter, such as coal, natural gas, or oil. Solar energy can be direct or indirect, and includes production of heat from solar collectors or passive solar design, production of electricity from photovoltaic cells, and indirect forms of solar energy such as wind power, burning of methanol, and heating with wood. See DIRECT SOLAR ENERGY, INDIRECT SOLAR ENERGY.

solar farm A large-scale system for generating large amounts of electric power. It could utilize banks of photovoltaic cells (as in California's Solar One), or possibly a solar furnace.

solar furnace   A device that collects and often concentrates solar radiation for producing high-temperature heat, as for cooking. Solar furnaces usually incorporate mirrors to focus the sun's rays. One application is called a power tower or solar power tower. See SOLAR POWER TOWER.

solarimeter   An instrument for measuring the intensity of solar radiation (insolation) received per unit area of the ground. It works by measuring the temperature difference created when the sun heats two strips of metal, each a different thickness but both colored black. Also called a pyranometer.

solar maximum   The point in the 11-year sunspot cycle where the number of visible sunspots is greatest. It occurs five and a half years before or after solar minimum. See SUNSPOT CYCLE.

solar minimum   The point in the 11-year sunspot cycle when the fewest sunspots are seen. It occurs five and a half years before or after solar maximum. See SUNSPOT CYCLE.

solar nebula   The primeval, spinning cloud of gas and dust from which the sun and our solar system formed, according to the solar nebula theory.

solar nebula theory   The theory that our solar system began as a cloud of gas and dust. Some event, such as the shock wave from a neighboring supernova, may have caused the cloud to begin to condense and to spin. This spinning cloud collapsed into a protosun and planetesimals, which in turn evolved into our sun and planets. It is the currently accepted theory of the formation of our solar system.

solar plexus   A major nerve center in mammals located just behind the stomach. It consists of two large nerve ganglia that interconnect with sympathetic nerves leading to all of the visceral organs. See SYMPATHETIC NERVOUS SYSTEM.

solar pond   A small body of water that develops a strong temperature gradient when heated by the sun. This temperature difference between the warmed surface and cooler bottom layers can be used to extract energy (an indirect form of solar energy) in a manner similar to ocean thermal energy conversion. See OCEAN THERMAL ENERGY CONVERSION.

solar power   Another term for solar energy, although solar power is sometimes used specifically for electrical power derived from the sun, as through photovoltaic cells.

solar power tower   Another term for a solar furnace. See SOLAR FURNACE.

solar radiation   An inclusive term for all electromagnetic energy given off by the sun. See illustration p. 506.

solar spectrum   The electromagnetic spectrum emitted by the sun. See illustration p. 506.

solar wind   The flow of ionized particles generated by the sun and traveling throughout space. The solar wind is an extension of the sun's corona; it affects the earth's magnetic field.

sol brun acide   An older classification for zonal soils with poorly developed horizons. See INCEPTISOLS.

solfatara   A volcanic fumarole that vents primarily sulfur-rich gases.

solid   Not fluid; the state of matter characterized by having molecules densely

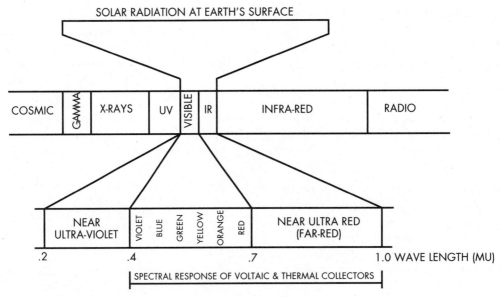

solar spectrum

packed, having a definite shape, and having much greater resistance to deforming forces than either a gas or a liquid. Solids are often crystalline in structure, and the movement of the molecules is very restricted in comparison with liquids or gases. Compare LIQUID, GAS.

solid load   The total material in stream transport processes that are carried as solid particles. Solid load includes suspended load and bed load. Compare DISSOLVED LOAD (STREAM).

solid solution   A form of metallic alloy in which the atoms of a different element are incorporated into the crystals of a metal. The atoms of the element may either replace some of the atoms in the metal molecules, or in some cases may lodge in between the metal molecules. Also called a primary solid solution.

solid waste   Municipal or industrial waste product that is transported and disposed of in solid form.

solifluction   The gradual downslope movement of viscous regolith materials under saturated conditions. Solifluction is especially characteristic in periglacial environments, but the term is also applied to the movement of thick tropical soils.

soligenous   Describing a wetland fed by water that has passed through mineral soil and therefore contains dissolved nutrients; the kinds and the amounts of nutrients depend on the composition of the soil. See FEN. Compare OMBROGENOUS.

solstice   Either of two points on the ecliptic (the sun's apparent yearly path) when the sun reaches its greatest distance north or south of the equator; these points fall halfway between the two equinoxes and mark the beginning of summer and of winter. Solstice is also used to refer to the time in late June and in late December when the sun reaches those points. Compare EQUINOX.

solubility   The readiness with which a substance dissolves in another substance, also stated as how much of a substance will dissolve at a given temperature. Solubility increases with temperature for most solids in liquids and decreases with temperature for gases.

solubility product constant, K_{sp}   The equilibrium constant for the chemical reaction involving the dissolving of a slightly soluble salt. The value of K_{sp} gives a measure of how soluble a substance is in water, with larger K_{sp} values indicative of greater solubilities.

solubilization   The process of making a substance soluble.

soluble   Capable of being dissolved. Ionic compounds are water-soluble. Fat-soluble compounds such as vitamins A and D and pesticides dissolve in fats. As a result, the former can be removed from foods by cooking in fats, and the latter can build up to harmful levels in fatty tissues.

solum   The upper part of a soil, typically the A and B horizons, where soil-forming processes occur.

solute   Any substance that dissolves in another substance (the solvent) to form a solution. Contrast SOLVENT.

solution   A mixture of two or more compounds that forms a homogenous liquid, gas, or even solid (amalgams are solutions of metals in mercury). Solutions usually refer to liquid mixtures, often ones containing water as the solvent (technically, an aqueous solution).

solvation   The process by which an ion or molecule of a solute is surrounded by a solvent, causing it to dissolve. Solvation results from the interactions between solute and solvent molecules, including hydrogen bonding, ion-dipole interactions, and Van der Waals interactions. See HYDRATES.

solvent   Any substance used to dissolve another substance (the solute) to form a solution. Often used for substances that are capable of dissolving a wide variety of other compounds; acetone is a good solvent for a variety of compounds, but water and kerosene are also good solvents for many compounds.

soma   Those portions of an organism excluding the reproductive tissues; the body except for the germ cells. See SOMATIC GROWTH.

somaclonal variation   Differences in plants that have been propagated asexually by tissue culture. See TISSUE CULTURE.

somatic growth   Growth of the body of an organism rather than of the reproductive tissues.

somatic mutation   Changes in the genetic material within nonreproductive (somatic) cells of an organism; such changes are not inherited. Somatic mutations are usually rare, except in the genes that control the synthesis of some proteins such as antibodies. Somatic mutations can be caused by exposure to radiation.

somatotrophic hormone   See SOMATOTROPIN.

somatotropin   A hormone secreted by the anterior lobe of the pituitary gland to stimulate the growth of bones and of muscle. Also called somatotrophic hormone and human growth hormone.

sonar   Acronym for sound navigation and ranging. Sonar is the technique using pulses of sound to detect the distance and the direction of objects under water, based on analyzing the echo (reflected sound pulse). It is the underwater equivalent of radar but uses audible sound rather than radio waves. Compare RADAR.

sonde   Any small system for remote measurement (of temperature, of air pressure, of radiation intensity, etc.) transported by balloon, parachute, rocket, or satellite. Radiosondes, the most common form, are often used in weather studies. See DROP-SONDE, RADIOSONDE.

sone   A unit of measurement for loudness. The sone is designed to indicate equal increments of loudness. One sone equals a simple tone at a frequency of one kilohertz and a level of 40 decibels above the threshold of the listener; a tone twice as loud would equal two sones. One sone is equivalent to 40 phon. Compare PHON, DECIBEL.

songbirds   Common name for those birds of the order Passeriformes which have have musical and often complex calls. See PASSERIFORMES, PASSERINES.

sonic   Relating to sound (frequencies audible to the human ear), sound waves, or speed of sound. The speed of sound is approximately 343 meters per second or 767 miles per hour at 20°C.

sonic boom   The noise associated with the strong shock wave produced in the atmosphere by an aircraft or rocket when it travels at speeds faster than the speed of sound (approximately 343 meters per second or 767 miles per hour at 20°C). When this shock wave reaches the ground, it sounds like the boom of an explosion and an instantaneous increase in air pressure can often be felt.

soot   Carbon particles produced by incomplete burning of carbon-containing substances such as oil or wood. It may be released to the air as particulate matter, or may form a black deposit on surfaces such as the lining of chimneys.

Sørensen similarity index   A coefficient of community, a measure of the degree to which two plant communities resemble each other based on their species composition. It is calculated by doubling the number of species common to two communities, and by dividing this value by the sum of the total number of species in each of the two communities. Compare JACCARD INDEX, SIMPSON'S INDEX OF FLORISTIC RESEMBLANCE, GLEASON'S INDEX, MORISITA'S SIMILARITY INDEX, KULEZINSKI INDEX.

sorption   1) A general term including processes such as absorption and adsorption; many are used in pollution control. 2) Absorption of a gas by a solid. See ABSORPTION, ADSORPTION.

sorting (of a sediment)   The mechanical separation, whether natural or artificial, of unconsolidated rock particles into distinct fractions that are based on clast size. A well-sorted sediment has a relatively narrow range of particle sizes, and a poorly sorted sediment combines a wide variety of particle sizes.

sound   Mechanical vibrations occurring in matter that fall within the limited frequency range that activates the human eardrum, 20 hertz to 20,000 hertz. Vibrations above the 20,000-hertz limit of human hearing are called ultrasound.

sounding balloons   Small balloons carrying instruments for measuring temperature, pressure, humidity, or other upper-atmosphere phenomena. Sounding balloons have their own recording equipment. Such devices are equipped with instruments for relaying the information back to earth. See SONDE, RADIOSONDE.

sound navigation and ranging   The full name for the acronym sonar. The technique is now almost always referred to by its acronym. See SONAR.

sound pressure level   The level of sound pressure (compression of air caused by a sound wave) compared to a reference pressure of 20 micropascals. This comparison or ratio, which can be measured with a sound pressure level meter, gives the intensity of a sound and is usually expressed in decibels. See DB, DBA, DECIBEL.

sound propagation   The process of transmitting the mechanical vibrations constituting acoustic (sound) waves. Sound travels as longitudinal waves through gases and liquids, but may travel as longitudinal or transverse waves through a solid medium.

sound shadow   An interruption of sound or radio reception caused by an object blocking the path of the sound waves or the radio waves, analagous to the light shadow produced when an object blocks the path of light. An example is the dead spot, where radio reception is very poor or nonexistent, directly behind a mountain that blocks the incoming signal. Also called an acoustic shadow.

sound waves   The mechanical vibrations that produce a response in the human eardrum. Sound waves through air or a similarly elastic medium are longitudinal, but through a solid medium, sound may travel as either longitudinal or transverse waves.

source   1) The opposite of a sink; a point of origin or part of a system that generates an excess of a substance or (heat), supplying it naturally or artificially to a system. 2) The point of origin of a river or a stream. Compare SINK.

source category   A classification of primary pollutants based on their origin.

source reduction   Reducing consumption of nonrenewable minerals, plastics, hazardous substances, or paper by reducing uses at the level of production, especially of products that cannot be recycled or reused. Source reduction, such as discontinuing the use of unneccessary packaging, is the most effective means of reducing the overall waste stream.

source rock   1) The ancestral rock from which a sediment is derived. 2) A sediment or rock in which petroleum originates.

source separation   The sorting of various components of municipal solid waste at the site (for recycling and composting, and to reduce the volume of the waste stream) at which it is produced, as opposed to attempting to sort it after it has all been thrown together. Recycling paper in the office or bottles at home and composting yard waste are all forms of source separation.

sour gas   Natural gas containing impurities that give it an acrid odor. Poisonous hydrogen sulfide is the most prevalent of such impurities.

South African floral region   One of the phytogeographic regions into which the Austral realm is divided according to similarities between plants; it consists of South Africa, Botswana, and Namibia. See AUSTRAL REALM, PHYTOGEOGRAPHY.

South-East Asian floral region   One of the phytogeographic regions into which the Paleotropical realm is divided according to similarities between plants; it consists of southern Burma (Myanma), northern and central Malaysia, and Indochina. See PALEOTROPICAL REALM, PHYTOGEOGRAPHY.

southerly Used to describe winds blowing from the south toward the north.

South Oceanic floral region One of the phytogeographic regions into which the Austral realm is divided according to similarities between plants; it consists of the islands in the southern Atlantic Ocean and Indian Ocean south of 50° latitude. See AUSTRAL REALM, PHYTOGEOGRAPHY.

soybean Leguminous plant *(Glycine max)* grown for its edible beans, which are high in protein and in oil but have very little starch. Soybeans are used as a source of vegetable oil, as a green manure crop, in animal feeds, and are also an excellent protein source in human diets. Also called soya. See GREEN MANURE.

spadix A spike made up of tiny flowers on a fleshy stem, usually surrounded by a large, petal-like structure called a spathe. The spadix is prominent in the flowers of calla lilies *(Zantedeschia)*, anthurium (peace lily), and Jack-in-the-pulpit *(Arisaema)*. See SPATHE.

spathe A large, often showy, modified leaf (bract) that surrounds or subtends a spadix (flower spike). The striped outer portion of the Jack-in-the-pulpit *(Arisaema)* is a spathe. See SPADIX.

spatial distribution The physical arrangement of organisms in three-dimensional space. See DISPERSAL, RANDOM DISTRIBUTION, CLUMPED DISTRIBUTION, UNIFORM SPATIAL DISTRIBUTION.

spawn 1) Eggs laid directly in the water by aquatic or amphibian animals, such as fish or frogs. Less often used for the young organisms immediately after hatching from such eggs. 2) Mycelia or spores prepared for cultivating mushrooms.

specialist An organism that has strict requirements for survival, depending on a limited source of food or living in only restricted areas or habitats. Specialists have often evolved to have specific behaviors or structural adaptations tailored to their limited requirements.

specialization The adaptation of an organ or an organism by structure of function to specific life requirements, increasing its adaptation for exploiting a particular niche but reducing its ability to change to other niches.

speciation The evolutionary development of new species, usually as one population separates into two different populations no longer capable of interbreeding.

specient An individual organism of a given species.

species A naturally occuring population or a group of potentially interbreeding populations that is reproductively isolated (i.e., cannot exchange genetic material) from other such populations or groups. This definition does not apply to asexually reproducing forms such as many types of Monera or Protista, etc.

species-abundance curve A graph illustrating the relationship between the number of species (y-axis) and the number of individuals per species (x-axis). The curve provides information about the species diversity and the species richness of an area. See DOMINANCE-DIVERSITY CURVE.

species-area curve A graph illustrating the number of species (y-axis) found in a given area (x-axis). The shape of the resulting species-area curve provides information about the species diversity and the species richness of an area. If linear scales are used, the curve usually shows saturation with increasing area. If logarithmic scales are used, the different values usually sketch a straight line. The species-area curve can be used to determine the most efficient size of plot for sampling a community. See SPECIES DIVERSITY, ALPHA DIVERSITY.

species-area relationship A relationship often found on islands in which smaller islands show a smaller number of species. See EQUILIBRIUM THEORY OF ISLAND BIOGEOGRAPHY.

species composition The particular species found in a given community or area.

species-deletion stability In a model community, the tendency for the remaining species to remain stable at a local level after the extinction of another species.

species diversity The number and relative abundance of all the species within a given area. See ALPHA DIVERSITY, BETA DIVERSITY, EVENNESS, EQUITABILITY.

species packing The number of niches that can be filled by different species within a given community or segment of hypervolume. See NICHE, N-DIMENSIONAL HYPERVOLUME.

species pair See SIBLING SPECIES.

species recovery plan An action plan for aiding endangered and threatened species.

species richness See ALPHA DIVERSITY.

species selection A type of group selection in which populations of species with different characteristics increase or decrease in number at different rates because of the differences in their characters. See NATURAL SELECTION.

specific gravity The density of an object or a substance compared to water. Specific gravity is the ratio between the mass of an object or a substance to the mass of an

equal volume of water at 4°C. The specific gravity of solids can be calculated by measuring the volume of water an object of known weight displaces and using the known density of water (1 gram per cubic centimeter) to calculate the water's mass. Specific gravities of liquids can be measured with hydrometers. Also called relative density. For the specific gravity of minerals, see APPENDIX p. 620-628.

specific heat The quantity of heat in calories required to raise the temperature of one gram of a substance by one degree Celsius.

specific humidity The ratio of the mass of water vapor to the total mass of a given sample of moist air. Specific humidity is usually expressed as grams of water vapor per kilogram of air. Compare RELATIVE HUMIDITY, ABSOLUTE HUMIDITY.

spectator ions An ion that takes no part in a chemical reaction, even though it is present in the system. Spectator ions are thus not often included in the chemical equation for the reaction.

spectrography 1) The study of electromagnetic spectra recorded photographically (as opposed to using spectrometers, similar instruments that show the component wavelengths of radiation without recording them on film). 2) Short for mass spectrography, a precursor to the use of the mass spectrometer. Mass spectrographs record charge-to-mass ratios of isotopes on photographic plates, rather than electronically, as in a mass spectrometer. Mass spectrographs were used in the first identification of the isotopes of uranium. Compare MASS SPECTROMETER.

spectroscope A laboratory device for producing and analyzing spectra. Spectroscopes use either a slit or a diffraction grating to break a ray of light into its component wavelengths. Spectrographs and spectrometers are both forms of spectroscopes. See DIFFRACTION GRATING.

spectrum 1) Electromagnetic radiation separated into its component wavelengths or frequencies, and displayed. Often used to refer specifically to the component colors of white light, as shown when light is passed through a prism. 2) The display of mass-to-charge ratios of ionized atoms or molecules shown by a mass spectrometer. This display indicates the relative proportions of different isotopes of an element present. Plural is spectra. See ELECTROMAGNETIC RADIATION.

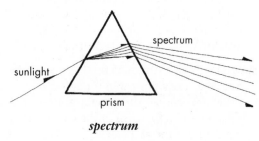

spectrum

specularite See HEMATITE.

speed of sound The speed at which sound waves travel, approximately 343 meters per second or 767 miles per hour at 20°C. Unlike light, sound waves do not travel at a constant velocity throughout the universe; the speed is affected by such factors as temperature and humidity.

spent fuel Fuel that has been used in a nuclear reactor and typically has been "poisoned" by fission products to the point where it no longer sustains a chain reaction. Spent fuel is one form of high-level radioactive waste. See NUCLEAR REACTOR.

sperm A mature male gamete or reproductive cell of an animal, capable of moving

because of its flagellalike tail. Also called spermatozoon.

spermaceti A clear, high-grade oil found in the head of sperm whales and of porpoises; it is processed into a white wax used in the cosmetics industry as well as in ointments and candle wax.

spermagonium One of the flask-shaped cavities where the nonmotile male gametes (spermatia) of some lichens, rust fungi, and some ascomycetes are produced. Also spelled spermogonium.

spermatid A haploid cell formed from the meiotic division of a secondary spermatocyte cell. Spermatids develop into spermatozoa without further cell division. See MEIOSIS.

spermatocyte A male germ cell, especially at the stage just before development into a spermatozoid or spermatozoon.

spermatogenesis Sperm formation; the meiotic cell divisions of spermatogonium and the subsequent differentiation that produce a spermatozoon or spermatozoid.

spermatogonium A primordial germ cell that undergoes division to produce a spermatocyte, and eventually develops into a male gamete.

spermatophore A complex case or mass containing numerous spermatozoa, found in males of many species of insects, squids, copepods, and some vertebrate animals.

spermatozoid One of the microscopic, flagellated, and motile male gametes produced by most algae, some fungi, and green plants other than seed plants (mosses, liverworts, ferns, horsetails, clubmosses, cycads, and *Ginkgo* trees). Also called antherozoids, these usually develop within a structure called the antheridium. See ANTHERIDIUM.

spermatozoon   See SPERM.

spermicide   Substance that kills sperm, especially one used as a contraceptive.

sphagnicolous   Describing organisms that live in peat moss, and therefore tend to be found in peat bogs.

sphagniherbosa   Describing a plant community dominated by peat moss. See HERBOSA.

Sphagnum   The genus containing peat mosses that are commonly found in bogs in the Northern Hemisphere. See MUSCI.

sphalerite   A yellowish brown ore mineral of zinc, iron, and sulfur that forms cubic crystals in low-temperature vein deposits. See APPENDIX p. 677.

sphene   A yellowish brown mineral of titanium and calcium silicate that forms monoclinic crystals in granitic or intermediate coarse-grained igneous rocks.

Sphenisciformes   An order of marine birds found in the Southern Hemisphere and commonly known as penguins. They lack flight feathers and have stiff, flipperlike wings used in swimming. The group is an ancient one and its relationship to other birds is unclear. There are 15 species. See AVES.

Sphenophyta   A phylum of relatively primitive, spore-bearing, vascular plants characterized by jointed stems and surfaces that are roughened due to the presence of the mineral silica in their epidermal cells. This phylum, formerly included as part of the phylum Pteridophyta and sometimes called the Equisetophyta, contains horsetails (*Equisetum*), which are plants of damp terrestrial habitats. Horsetails bear spores in sporangia clustered in a terminal strobilus.

spheroidal weathering   The formation of concentric spherical shells as angular fragments become progressively rounded through physical and chemical weathering.

spherulite   A coarsely crystalline body of rock composed of radiating accicular crystals. Spherulites are commonly formed of feldspar, chalcedony, or calcite.

sphingosine   A long-chain base (an amino alcohol) found in cerebroside and those phosopholipids known as sphingomyelins, which are present in blood lipids and in nerve tissues. See CEREBROSIDE.

spicule   Any calcareous or siliceous internal test of a marine organism such as a sponge, a radiolarian, or a diatom. Spicules may accumulate to significant proportions in a pelagic sediment.

spiders   See ARACHNIDA.

spike   A flower cluster with a long major axis to which many flowers are attached directly, without stalks, and frequently are crowded together. Plantain flowers are spikes. See INFLORESCENCE.

spikelet   A small spike or flower cluster; also a secondary spike within a compound spike, the tiny cluster of flowers forming the basic unit of the flowers of wheat and other grasses. See INFLORESCENCE.

spilite   A vesicular or an amygdaloidal basalt that has been altered by low-grade metamorphism to contain albite, chlorite, actinolite, sphene, and calcite minerals. Spilite is formed in midoceanic ridge basalts and in greenstones.

spilling breaker   An ocean wave in which water breaks at the crest and cascades down the steep advancing surface of the wave. Compare PLUNGING BREAKER.

spillway   1) A glacial drainage channel that

forms at the outlet of a glacial lake or at the bedrock surface of a stream confined by glacial ice. 2) The channel designed to direct overflowing water at the top of a dam.

spindle The group of fibers named for their spindle shape, each made up of microtubules, that form a framework along which the chromosomes align at the end of the prophase of mitosis and of meiosis.

spindle bomb A type of volcanic bomb that is shaped like an elongated cylinder with tapered ends.

spinel A hard mineral of magnesium aluminum oxide that forms variably colored octahedral crystals in mafic igneous rocks and in contact metamorphosed limestones. Spinel crystals are valued as a gemstone known as spinel ruby.

spiracle A small opening for breathing in insects and in some arachnid species (external part of the trachea), amphibian larvae, cartilaginous fish, and aquatic mammals (the blowhole of a whale).

spiraling In riverine (lotic) ecosystems, the concept that nutrients are carried downstream as they are in circulation, creating a spiral-shaped trajectory. A spiraling index can be expressed as the distance downstream an atom of a nutrient travels to complete a cycle from the water through organisms and back to the water.

spirillum Any bacterium having a spiral, or nearly spiral, shape. Spirilla are commonly motile, aerobic organisms found in soil and in water and feeding on organic matter. One such species, *Spirillum minus*, causes a form of ratbite fever.

Spirochaetae A phylum of the Monera, characterized by spiral shape and unique arrangement of internal flagella (axial filaments). Many are free living in waters

and in mud deposites, whereas others, such as the genus *Treponema*, which causes syphilis and yaws, are endoparasitic. See MONERA.

spit A curving projection of accumulated gravel, sand, or silt deposited by longshore currents at a shoreline.

splash erosion A type of soil erosion caused by splashing raindrops.

splenic fever See ANTHRAX.

spodic A term applied to a soil horizon that is characterized by illuvial accumulations of amorphous iron and aluminum with organic matter. A spodic horizon is the diagnostic feature of spodosols.

spodosol Any member of a soil order characterized by subsurface horizons showing cementation with iron oxides or other amorphous accumulations of organic matter and iron or aluminum clays; typically occurs in leached sandy soils in cool coniferous forests.

spoil banks A submerged deposit of mining or dredging waste.

spoil deposit A heap of waste material accumulated from an excavation or mining operation.

sponges See PORIFERA.

spongin Skeletal material, either fibrous or hard, produced by various groups of animal sponges (*Porifera*).

spongy mesophyll Spherical parenchyma cells containing relatively few chloroplasts and interspersed with air spaces that fill most of the leaf tissue within the epidermis that is not taken up by palisade mesophyll. See PALISADE MESOPHYLL, PARENCHYMA.

spontaneous generation The obsolete theory, disproved by Pasteur, that living organisms could be produced from nonliving matter. Before the discovery of microorganisms, made possible by the invention of the microscope, it was believed that rotting organic matter spontaneously produced microbe colonies and the maggots that were found in it. Also called abiogenesis.

sporadic E (E_s) Short for sporadic-E ionization; irregular patches of increased ionization in the E-layer of the earth's ionosphere that improve high-frequency radio wave propagation. Sporadic E occurs at any time of day or any location, although not predictably; it may be caused by a number of factors including meteor showers, thunderstorms, or solar activity. Also called sporadic-E layer refraction. See HIGH FREQUENCY.

sporangium The structure in bacteria, algae, and plants where spores (asexual reproductive cells) are produced.

spore A reproductive cell or body capable of developing into a new organism asexually, without fusing with another reproductive cell. Bacteria, fungi, some protozoa, and plants produce spores; in those plants demonstrating alternation of generations, the spores produced by the sporophyte generation develop into gametophytes. Spores are often adapted to serve as disseminules. Compare GAMETE.

sporicide A substance that kills spores.

sporogonium The sporophyte generation of mosses and of liverworts, supporting the spore-producing capsule.

sporophyll A leaf or an organ that supports the spore-producing sporangia. In ferns, the entire frond is a sporophyll; in higher plants, the sporophyll is an organ derived from a modified leaf. See MEGASPORO-PHYLL, MICROSPOROPHYLL.

sporophyte In plants showing alternation of generations, the usually diploid generation (or individual plant) that undergoes meiosis to produce haploid spores. See ALTERNATION OF GENERATIONS. Contrast GAMETOPHYTE.

Sporozoa A class of parasitic protozoa that are nonmotile or amoeboid and which reproduce both sexually and asexually in alternate generations. It includes the disease-causing genera *Isospora*, *Toxoplasma* (which causes toxoplasmosis), and *Plasmodium* (which causes malaria). See APICOMPLEXA.

SPOT Acronym for Système Probatoire d'Observation de la Terre, a French observation satellite launched in 1986. It provides earth images at a resolution of 10 meters for use in agriculture, geology, and land-use planning. See LANDSAT.

spotted slate A slate that exhibits graphite-rich rounded spots on the cleavage surfaces. The spots are formed from the aggregation of organic material trapped within the parent shale.

spreading axis A central line marking the position of divergence in a mid-oceanic ridge.

spreading edge The edge of a tectonic plate that is nearest a spreading center. See MID-OCEANIC RIDGE.

spreading pole The point of intersection between the earth's surface and an imaginary axis line through the center of the earth. The spreading pole serves as a fixed reference point in describing the relative motion between two diverging tectonic plates.

spring A natural point of visible groundwater discharge formed where an aquifer intersects the ground surface.

spring meltwater flush   A seasonal increase in stream runoff caused by the melting of a snowpack.

spring overturn   A phenomenon occurring in temperate-climate lakes that form thermoclines. With cooler weather, the epilimnion (warm surface water) cools to 4°C—the temperature at which it reaches greatest density—and sinks to the bottom, aided by wind stirring the surface. This mixes the intermediate layers, evening up the distribution of oxygen and nutrients at different depths. A similar overturn occurs in the spring. See EPILIMNION, HYPOLIM-NION, STRATIFICATION, THERMOCLINE, VERNAL OVERTURN.

spring tide   The time when periodically high tidal ranges coincide with the new and the full phases of the moon. Spring tides are higher than the high tides of other phases because the gravitational pull during spring tides reaches a maximum as sun, moon, and earth align in a common direction. Compare NEAP TIDE.

Squamata   The order of reptiles that includes lizards, snakes, and *Sphenodon* (tuatara). Like the archosaurs, the squamates have a skull with two temporal openings on each side. See LACERTILLA, OPHIDIA.

squids   Cephalopods with a tubular body and a head provided with two prey-catching tentacles and eight arms which bear sucking disks. Both the eyes and the nervous system are well developed. These strong, jet-propulsion swimmers have only the vestige of a shell. They are widely distributed in coastal waters and in deep seas and vary in size from minute forms to the giant squid, *Architeuthis*, which may attain a length of 20 m, including the tentacles, and is, by far, the largest inverte-brate. Squids have giant axons that are used by neurophysiologists to study nerve action. See CEPHALOPODA.

Sr   Chemical symbol for the element strontium. See STRONTIUM.

S-shaped curve   A curve resulting from the graphing of population growth rates in which population growth is initially slow, then exponential (as in a J-shaped curve), and finally shows a levelling off caused when a rapidly growing population exceeds the carrying capacity of its environment and ceases to grow in numbers. See SIGMOID GROWTH. Compare J-SHAPED CURVE.

St. Anthony's fire   Historic name for ergotism. See ERGOTISM.

St. Elmo's fire   Electrical discharge (brush discharge) usually occurring on or around metal objects such as the mast of a ship or an airplane wing. Named for the patron saint of sailors. See BRUSH DISCHARGE.

stability   The inherent ability of an ecosystem (or any system) to resist change, or to maintain steady-state conditions when confronted by a disturbance. Compare RESILIENCE.

stability-time hypothesis   The theory that the highest species diversities are found in stable environments, in communities that have existed for a very long time. The hypothesis predicts that the number of species decreases along a gradient from low to high environmental stress.

stabilization lagoon   A holding pond that retains wastewater for sedimentation, anaerobic or aerobic decay, or reduction of the level of nuisance odors.

stabilize   To make stable, or to reach a condition in which opposing forces maintain a continuing balance so that little change occurs.

stabilizing factors   Characteristics of an ecosystem (or of an organism or any system) that increase its ability to resist change. In particular, stabilizing factors are those density-dependent influences that compensate for population changes to return the population to its original size. See DENSITY DEPENDENT.

stabilizing selection   Natural selection favoring individuals that are closer to the population average for a specific trait, rather than those individuals showing extreme forms of the trait. Also known as normalizing selection, centripetal selection, and survival of the mediocre. See NATURAL SELECTION.

stable age distribution   The establishment and maintenance of relatively constant proportions of each age class within a population over time, especially during increases or decreases in the total size of the population. A stable age distribution develops in populations that exist for a long time in a stable environment.

stable air   Air whose temperature and pressure, and sometimes moisture content, keep it from rising or falling—it is in a state of convective equilibrium. If it is pushed upward, it will cool and become colder than the surrounding air, causing it to return to its original altitude. Because stable air resists movement, it can contribute to a build-up of air pollutants.

stable limit cycles   See LIMIT CYCLES.

stable population   A population that remains at the same level, where the number of births plus immigration equals the number of deaths plus emigration.

stable runoff   An average amount of runoff that is available throughout the year.

stack effect   The phenomenon that causes fumes to rise within smokestacks. The hot gases are much higher in temperature and therefore less dense than the air surrounding the smokestack, and so they naturally move upward.

stage   1) A reference in chronostratigraphy to a particular body of rock that has distinguishing physical properties and has accumulated within a known absolute age unit. 2) Elevation of water level in a river measured at a gaging station.

stain   1) A colored mark on the surface of something porous, or a substance used to impart a color to a porous material such as wood. 2) A dye that differentially colors transparent tissues that will be examined under a microscope; also, to treat tissue with such a dye in order to highlight certain features such as chromosomes. See GRAM'S REACTION.

stalactite   A dripstone deposit of calcium carbonate which forms a conical structure that tapers down from the roof of a cave. A stalactite may form a columnar union with a stalagmite.

stalagmite   A dripstone deposit of calcium carbonate that forms a conical structure that tapers up from the floor of a cave. A stalagmite may form a columnar union with a stalactite.

stamen   The male reproductive structure of a flower, consisting of a slender filament supporting the pollen-bearing anther. Also called androecium. Compare PISTIL.

staminate   Describing male flowers, those possessing stamens but no pistils. Compare PISTILLATE.

stand   1) A group of land plants, especially trees, growing in the same locality. 2) A plant community growing in a specific, defined area.

standard   A legal level established at the state or federal level for pollutants, such as an emissions standard or a water quality standard.

standard alkalinity   The acid-neutralizing capacity of a water sample as measured by the amount of acid required to lower the pH of the sample to pH 4.5. Also known as total alkalinity.

standard deviation   A statistical measure of variation in a given set of data. In a normal frequency curve, a distance of one standard deviation on each side of the mean (average value) will include 68 percent of all cases. It is calculated as the root-mean-square value of the deviations from the mean (average value), which equals the square root of variance. See ROOT-MEAN-SQUARE VALUE, VARIANCE.

standard error   Statistical term for the standard deviation of the sample (or a given parameter) in a frequency distribution. Sometimes called the standard error of the mean, it is calculated by dividing the standard error by the square root of the number of observations in the sample minus 1 (n-1), or if the sample is large, n is used.

standard metropolitan statistical area   (SMSA) A U.S. designation for metropolitan areas with populations of 100,000 or greater.

standard project flood   An estimate of the most extreme flood conditions that are likely to develop within the project area of an engineered structure such as a dam or a bridge.

standing crop   Another term for biomass, used for aquatic as well as terrestrial communities. See BIOMASS.

standing wave   A fluid waveform that either remains stationary or oscillates between two points. A standing wave may develop in association with boulders obstructing water flow within a stream channel or by the interference between two waves which are traveling in opposite directions.

starch   A polysaccharide (complex carbohydrate) found in all green plants, where it serves as an energy storage compound. It is therefore an essential energy source in animal diets; it breaks down into glucose. Starch is also used in adhesives, fabric and paper treatments, and medicines.

star dune   A free-standing dune that has several ridges that diverge from a central crest. Star dunes tend to form in areas where the wind direction is highly variable.

stare decisis   A legal term from Latin, meaning to stand by that which was decided. It is the rule making common law courts slow to interfere with principles announced in former decisions, often upholding them even though they would decide otherwise if the question were new. As a result, a judicial decision will be overturned only upon showing of good cause. The doctrine is limited in the field of constitutional law.

starfish   See ASTEROIDEA.

stasipatric speciation   The formation of new species that results from a chromosomal rearrangement producing superior homozygous individuals that are reproductively isolated from the ancestral species.

stasis   1) A stoppage, especially cessation of growth in plants or stagnation of normal flow of any body fluids such as the blood in the blood vessels. 2) A condition of stability, or lack of change, in an ecosystem or other system. 3) A lack of change in a species over a great many years (often millions). See HOMEOSTASIS.

static electricity   Electrical charges that are stationary, as opposed to the moving charges of lightning or electrical current. Something producing a stationary electrical charge is described as electrostatic. See ELECTRICITY.

static life table   A life table showing the age structure of a given population at a specific point of time. See LIFE TABLE, COHORT LIFE TABLE.

statics   A branch of physics (mechanics) devoted to the study of bodies at rest, or of forces that counterbalance each other to produce equilibrium. Statics is in contrast to kinetics, which concentrates on the study of bodies in motion.

stationary age distribution   See STABLE AGE DISTRIBUTION.

stationary growth phase   A period when a population stops increasing in size, which shows as a levelling off on a population growth curve.

stationary source   A pollution emitter whose location is fixed, such as a power plant or an industrial complex, as opposed to moving cars and trucks (mobile sources). Compare MOBILE SOURCE.

statocyst   1) An organ in some crustaceans, flatworms, and other invertebrates that promotes balancing by providing information on the organization's location in space. Also called an otocyst. 2) A plant cell containing statoliths, which move in response to gravity to help the plant maintain a constant orientation with respect to gravity. See STATOLITH.

statolith   A solid grain of calcareous material that is secreted by a living organism.

stator   1) The circle of fixed blades that rotates around a fixed axis within a turbine. 2) The nonmoving casing surrounding the rotor in a generator.

statute   An act of the legislature, a law. Enacted to prescribe conduct, define crimes, create lower governmental bodies, promote public good, or raise funds.

statute of limitations   Any law that fixes the time within which parties must take judicial action to enforce rights or else be thereafter barred from enforcing them.

statutory law   An act of the legislative branch of government, often simply called a statute or a law.

staurolite   An iron and magnesium aluminum silicate mineral that forms brown monoclinic crystals in metamorphic rocks. Staurolite is often found with kyanite and with garnet. See APPENDIX p. 627.

steady state   1) A condition of equilibrium in which opposing forces balance so that they cancel each other out, as when inputs to a system balance outputs, or death rate balances birth rate to maintain a stable population size. 2) A form of economic homeostasis, popularly called a no-growth or sustainable economy.

steam fog   Fog created when water evaporating from warm water is added to a mass of cooler air above the water. Such fog occurs when a cold air mass moves across relatively warm water (sea fog); it resembles steam coming off the water. Compare RADIATION FOG.

steam-generating heavy-water reactor (SGHWR) A nuclear reactor that uses water containing deuterium (heavy water) as a moderator, and ordinary water as a coolant, and includes a steam generator. Hot coolant (deuterium) from the reactor core heats water in the steam generator to the boiling point to drive a turbine. The Canadian deuterium-uranium reactor (CANDU) is a commercial steam-generating heavy-water reactor.

steel   Any of several strong alloys of carbon and iron, usually with small amounts of other minerals, such as manganese. Steel has a wide variety of physical properties depending on its exact composition and the different heat treatments it receives. Stainless steels contain significant quantities of chromium.

Stefan-Boltzmann law   Another name for the Stefan-Boltzman law, which states that the total energy radiated by a perfectly radiating source (an idealized black body) per square centimeter per second increases proportionally with the fourth power of the source's absolute temperature. A gas at 10,000 K thus gives off 16 times more energy than the same quantity of gas at 5000 K.

stele   The vascular tissues (xylem and phloem), and any ground tissues immediately surrounding them, in the stems and roots of plants. The structure of the stele is one of the major differences between mature monocotyledon and dicotyledon species. Also called the vascular cylinder.

stellar   Of or referring to stars, as in stellar spectra. A synonym for sidereal.

stem   1) The main above-ground structure of a vascular plant, supporting leaves, buds, and reproductive tissues, and containing the vascular tissue.

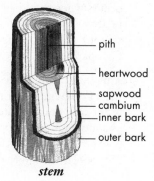

pith
heartwood
sapwood
cambium
inner bark
outer bark

stem

2) A stalklike structure in animals, such as the brain stem.

stem flow   The movement of precipitation intercepted by a plant canopy down the stems of plants and into the soil. Stem flow is a component of leaching transfers in terrestrial nutrient cycles. Compare THROUGHFALL. See BIOGEOCHEMICAL CYCLES.

steno-   Prefix meaning narrow. Contrast EURY-.

stenobathic   Describing organisms only capable of living within a restricted range of depth in water.

stenohaline   Able to live in water with a narrow range of salinity; not adaptable to great differences in the salt content of water. Contrast EURYHALINE.

stenoplastic   An organism with little or limited variation in the phenotypic (observable) expression of a given genotype. See PHENOTYPIC PLASTICITY.

stenothermal   Existing within a narrow temperature range; unable to survive large temperature variations. Contrast EURYTHERMAL.

stenotopic   Restricted to living in one habitat; not able to tolerate a wide variation in environmental conditions.

step cline   An abrupt change in the characteristics of a species over a short distance of a geographic gradient (cline), rather than a gradual change in the characteristics.

steppe   A temperate grassland vegetation consisting of extensive dry, almost treeless plains in southeastern Europe, southwestern Asia, and Siberia. Steppes are equivalent to North American prairies.

stereoscopic vision   Depth perception; the ability to discern three dimensions that is possible with binocular vision. See BIN-OCULAR VISION.

stereotaxis   A traveling movement of an organism because of contact with a solid body. Also called thigmotaxis. See TAXIS.

stereotropism   See THIGMOTROPISM.

sterile male technique   A form of biological pest control in which males of a pest species are reared in large numbers and sterilized by exposure to radiation. When these males are released in large numbers, they outnumber normal males available for mating. As a result, many normal females mate unsuccessfully, causing a tremendous reduction in the size of the subsequent generation of pests, especially if the species is one that mates only once in its lifetime.

sterilization   1) Causing an organism to lose its ability to produce offspring, such as exposing insects to radiation, or vasectomies in men and tubal ligations in women. 2) Destroying all microorganisms on an object by exposing it to steam, chemicals, radiation, special filtration to remove microorganisms, or other agents. See STERILE MALE TECHNIQUE, VASECTOMY.

sterilize   1) To disinfect; to kill all microorganisms on a substance such as a surgical instrument or in a medium such as soil, by exposure to high temperatures, radiation, or strong chemicals. Some herbicides and fungicides kill a wide range of organisms to effectively sterilize soil. 2) To render incapable of reproduction. Exposing males of certain insect species to controlled doses of radiation sterilizes them; when released in large quantities, these insects mate unsuccessfully with females, reducing populations without interfering with natural predators.

steroid   Any of a large group of organic compounds containing four linked six-carbon rings and related to fats. Steroids include sterols (such as cholesterol), vitamin D, bile acids, some alkaloids, and steroid hormones (the hormones, including sex hormones, produced by the adrenal cortex). Often used to refer exclusively to steroid hormones, used in medicine (for osteoporosis treatment), in sports (for body building), and in agriculture (for fattening cattle). See CHOLESTEROL, ESTROGENS.

sterol   Any of a group of organic compounds that includes cholesterol. They are steroid-based alcohols, usually unsaturated, found in animal and plant tissues. All contain a steroid nucleus (17 carbon atoms linked in a multiple-ring structure) plus an 8-, 9-, or 10-carbon side chain containing the hydroxyl (OH^-) group characteristic of all alcohols. See CHOLESTEROL.

Stevenson screen   A small shed for protecting meteorological instruments at a weather station. It is designed to protect instruments from sunlight and wind while allowing good ventilation. It consists of a solid roof and louvered walls, and is located about 3 feet above the ground.

stigma   1) The uppermost part of the pistil (female reproductive structure) of a flower, adapted for receiving pollen for fertilization. Located on top of a necklike structure (style) above the ovary, the stigma is often sticky to encourage pollen to adhere for germination. 2) Any spot, mark, or opening on an animal, especially with distinctive coloring such as the blotches on some butterfly wings; also, the eyespot of algae or the spiracle of arthropods. See EYESPOT, FLOWER, SPIRACLE.

still   A device used to distill liquids, consisting of a vessel in which liquids are boiled connected to a condensation chamber for cooling and collecting the vapor. Stills often contain a reflux apparatus as well. See REFLUX.

stimulus   Any internal or external agent that causes an organism to respond. Light, heat, and sound are examples of stimuli that provoke a behavioral or a physiological response.

stipule   One of the pair of tiny leaflike parts found at the base of a leaf stem (petiole) in dicotyledon plants. Stipules are sometimes modified into spines, protective structures for winter buds, or photosynthesizing small leaves. See APPENDIX p. 628-629.

stochastic   Referring to a random element or a process in which there is some element of chance.

stochastic models   Models of systems that predict probabilities of various outcomes given certain initial conditions. Also called probabilistic or Markovian models. Compare DETERMINISTIC MODELS.

stock   1) A group or strain of plants or animals (or bacteria) maintained for breeding. 2) The main stem of a plant, especially a tree trunk, that receives a graft. 3) Short for rootstock; also, the species used as rootstock. See ROOT-STOCK. 4) Short for livestock. 5) A discordant intrusion of igneous rock having an outcrop area less than 100 km². Compare BATHOLITH.

stocking   1) Supplying a stream, a lake, or a river with fish reared in a hatchery. Trout and other sport fish are often stocked for recreational fishing. 2) The density of individuals per unit area.

stocking rate   The numbers of a particular species of animal allowed to graze on an area of pasture or rangeland. The stocking rate is different for different species and

different mixtures of species because of different food preferences (as in cattle and in sheep).

stock recruitment models   Mathematical models used to predict how many recruit fish will be produced at various levels of stocking.

stockwork   A complex network of ore veins compact enough to be mined as a single vein.

stoichiometric   Referring to chemical reactions whose components can be described using whole numbers; these reactions can be used to specify masses of reactants and their products (masses can be measured in grams or in moles). Stoichiometry describes precise quantitative relationships between atoms and molecules, giving rise to exact formulas for compounds and balanced equations for chemical reactions. See BALANCED CHEMICAL EQUATION.

stolon   1) Another term for runner. 2) A slender branch that resembles a runner but grows underneath the ground, rooting at the tip to grow into a new plant. Sometimes also used for the stem of plants like blackberries, which arches above ground and forms a new plant where it touches the ground. 3) A stemlike structure in animals, especially the tubelike outgrowths of sessile, hydroid coelenterates that may develop into new individuals or colonies Compare RUNNER.

stolon

stomata   The gas-exchange ("breathing") pores in the outer (epidermis) layer of leaves and of stems. Each stoma is surrounded by two guard cells that expand to open the stoma to allow oxygen, water vapor, or carbon dioxide to be circulated between the plant tissues and the atmosphere. As the guard cells dehydrate and become flaccid, the stomata close, preventing gas exchange and further loss of water from the leaf. Stomata often refers to the pores and their associated guard cells.

stone   1) A general term for a hand specimen of rock. 2) Any rock that has been quarried. 3) A reference applied to an ore before processing.

stone cells   An irregularly spherical form of sclerid (technically, a brachysclerid); hardened cells found distributed throughout other tissues, especially seeds and fruit. The gritty cells surrounding the core of a pear are stone cells. See SCLEREID, SCLERENCHYMA.

stony meteorite   A variety of meteorite composed primarily of mafic and ultramafic silicate minerals such as pyroxene or olivine with plagioclase. Stony meteorites are the most abundant of all meteorites.

stoping   A process by which blocks of country rock are wedged loose by the upwelling of magma. The blocks may then drop into the magma chamber as the magma migrates upward.

stoplogs   A plank or a beam that is placed across a spillway or sluice to prevent the flow of water from an impoundment.

storks   See CICONIIDAE.

storm   Technically, a very strong wind (above 10 on the Beaufort scale) accompanied by precipitation. Commonly used for any significant atmospheric disturbance, especially one producing unpleasant or destructive weather. See APPENDIX P. 630.

storm flow   The volume of runoff, groundwater flow, or stream flow attributed to a storm event. Storm flow is a quantity of discharge in excess of base flow conditions.

storm sewer   A runoff collecting system designed to drain storm runoff away from a roadway or other impervious surface.

storm surge   An increase in sea level above the expected tidal range during a storm event. Storm surge occurs as wind drives water against a coastline, often in combination with reduced atmospheric pressure.

storm water   See STORM FLOW.

storm water retention reservoirs   A type of artificial basin that is designed to collect and detain storm water in order to reduce peak discharge during a flood.

storm water runoff   The component of direct runoff which occurs as a consequence of a storm event. See STORM FLOW, RUNOFF, OVERLAND FLOW.

stoss-and-lee topography   A landscape dominated by the smooth abraded hills and rugged plucked surfaces of roches moutonnées. See ROCHES MOUTONNÉES.

strain   Changes in shape, structure, or volume of a body that result from applied forces (stress). The distortion that occurs when a board sags under a heavy weight is strain. Also, to cause such a change. See ELASTIC DEFORMATION, PLASTIC DEFORMATION.

strain rate   The ratio of change in volume or in length of a body compared to the original volume or length. For shear strain, it is the ratio between the amount of deflection in the direction of the shear force and the distance between sheer forces.

strait   A relatively narrow channel of water that separates nearby landmasses and links adjacent bodies of water. A classic example is the Strait of Gibraltar, which connects the Atlantic Ocean and the Mediterranean Sea.

strata   An identifiable layer of bedrock or sediment. The term strata does not imply a particular thickness of rock.

strategic minerals   Nonrenewable minerals with important military uses (as opposed to critical minerals, those essential for a country's civilian economy).

strategy   The particular set of adaptations (genetic traits and behaviors) associated with a particular species that make it better suited to some environments than to others. Number of offspring produced, growth rate, and longevity are different aspects of strategy. See K-STRATEGISTS, R-STRATEGISTS.

stratification   1) Separation into distinct vertical layers within an environment, characterized by distinct groups of species. Stratification includes the formation of temperature layers in lakes (thermal stratification), the distinct vertical regions within soils, and the vegetation layers (herbaceous layer, shrub layer understory, canopy) in a forest. 2) The separation of materials, such as rock layers or zones of water, into an array of layers. 3) An artificial technique for breaking dormancy in those seeds that require cold temperatures prior to their germination. It consists of sowing them in moist soil and storing them at near-freezing temperatures for a given period of time. See SYNUSIA, THERMAL STRATIFICATION. Compare VERNALIZATION.

stratified drift   A deposit of layered and sorted glacial sediment that has been transported and deposited by the action of running meltwater. Compare GLACIAL TILL.

stratigraphy   The scientific study of composition, correlational position, and age relations of rock strata.

stratocumulus   Low, gray or whitish clouds occurring in rounded puffs, sometimes joined together into heavy rolls or a broad layer. Darker gray areas of shading are usually present. Stratocumulus clouds are found below 2000 meters of elevation and may be accompanied by some rain. Contrast ALTOCUMULUS, CIRROCUMULUS, STRATUS.

stratopause   The boundary between the stratosphere and mesosphere layers of the earth's atmosphere. It lies approximately 55 km above the earth's surface.

stratosphere   The region of the earth's atmosphere between the troposphere and the mesosphere. It is characterized by horizontal winds and little temperature change as altitude increases. It extends from roughly 20 to 50 km above the earth's surface. See TROPOSPHERE, MESOSPHERE, IONOSPHERE.

stratovolcano   See COMPOSITE VOLCANO.

stratum   A layer, especially a layer of life forms (often plants) within a biotic community or a layer in sedimentary rocks.

stratus   A continuous layer of low-lying, gray clouds resembling a sheet of elevated fog. Stratus clouds cover much of the sky and may be accompanied by drizzle. Compare NIMBOSTRATUS, CUMULUS, CIRRUS.

straw   Dried stems of grasses or of cereal crops. Straw differs from hay in that it is mostly stem, because traditionally it is what remains after grain has been threshed, whereas hay also has many seeds.

streak   The color displayed by a mineral in powdered form; is conventionally tested by scraping the mineral across a piece of unglazed porcelain, called a streak plate.

stream   A natural body of water that flows in a natural channel from higher elevation to lower elevation by the force of gravity. The term stream usually applies to natural channels that are smaller than rivers. See RIVER.

stream capture   The diversion of the headwater channels of a stream system into a channel of another stream system. Stream capture occurs where an adjacent stream system has a higher degree of erosional activity and a lower stream channel elevation than those of the captured stream.

stream flow   The volume of stream channel flow that is derived from direct runoff processes.

stream order   A dimensionless integer that describes the hierarchical position of stream reaches in relation to other confluent reaches of a drainage network. A stream with no tributaries is a first-order stream, the confluence of two first-order streams produces a second-order stream, and so on.

stream tin   A term applied to the alluvial occurrence, or placer deposit, of the mineral cassiterite. Stream tin is recognized by the edge-worn appearance of water-transported pebbles.

stream-water yield   A calculation of the volume of stream flow under stated conditions of input, runoff, evapotranspiration, and withdrawal.

strength   Nontechnical term for ability to withstand applied forces without developing strain (deformation).

stress   1) The force per unit area applied on a substance. The application of stress causes strain in a substance. See STRAIN. 2) Strong, negative environmental factors or constraining forces that provoke physiological responses in an organism or alter the equilibrium of an ecosystem. Pollution, lack of water, lack of nutrients, and disease are all examples of stress. Stress is used both for the agent and the resulting state of tension; stressor is sometimes used for the agent provoking the condition.

stress-shock   General term for any of a number of physical, psychological, and/or behavioral changes believed to be caused by the stress of intense competition and extreme closeness to other individuals of the same species.

striations, glacial   See GLACIAL STRIATIONS.

Strigiformes   This order contains the owls, whose large, forward-directed eyes, acute sense of hearing, and soft wing beat are well adapted for hunting at night. Their hooked beaks and heavily taloned feet resemble those of the raptors. The bones, fur, and feathers of their prey are regurgitated as pellets. There are 11 species of barn owls and some 120 species of typical owls. See AVES, RAPTORS, CIRRIPEDIA, BARN OWLS.

strike   The azimuth direction of a line formed by the intersection of a structural rock feature with the surface of an imaginary horizontal plane. Compare DIP.

strike-slip fault   A fault in which the sense of displacement between the fault blocks is parallel to the strike of the fault plane.

Dextral and sinistral faults are examples of strike-slip faults.

strip-cropping An agricultural technique used to reduce soil erosion on sloping land or on land subjected to high winds. Two (or more) crops are planted in alternating bands paralleling the contours of the slope or perpendicular to the prevailing wind; plows also cultivate the soil in the same direction. The crops are chosen for harvesting at different times, so that strips of the ground are always covered; typically, corn (a crop that contributes greatly to erosion) is alternated with hay.

strip cut A forestry practice in which bands of trees are left standing in between bands that are clearcut, to reduce erosion on slopes and to speed reforestation. Generally, no more than a third of a forest stand is strip cut at a time, and several years may elapse between strip harvests. This is usually sufficient time for reforestation to progress.

strip mining A mining technique in which large areas of land are excavated to expose an underlying ore. Strip mining, unless done carefully, can be an environmentally damaging extraction technique.

strobilation Asexual reproduction in which segments (strobilae) break off to form new individuals. Strobilation occurs in tapeworms, in marine bristleworms, and (in slightly modified form) in some jellyfish. Also spelled strobilization. A similar process in plants is called gemmation.

strobilus 1) A cone, the reproductive structure of pines and other gymnosperms. 2) The cone-shaped reproductive structure of club-mosses and horsetails, a group of scalelike, spore-producing leaves (sporophylls) clustered around a central stem. 3) A compound fruit of some angiosperm

species consisting of achenes surrounded in bracts, and shaped somewhat like a cone when mature. Hop plants form strobili. See SPORANGIUM.

stroma 1) The supporting tissues surrounding an organ; in mammals, stroma includes nerves and blood vessels. 2) The colorless, spongy framework of protoplasm around a red blood cell or surrounding the grana in a chloroplast. 3) Another term for the thallus, or main body, of a mushroom. Contrast PARENCHYMA.

stromatolite A dome- or mushroom-shaped sedimentary structure with a laminated internal structure. Stromatolites form by an association of shallow marine microorganisms, which trap silt and precipitate calcium carbonate in alternating depositional layers.

strombolian volcano A type of composite volcano that erupts in a continuous sequence of small explosions generated by the gases trapped within a viscous basaltic lava.

strong acid An acid that undergoes essentially complete dissociation in water into its component ions. Hydrochloric acid (HCl), sulfuric acid (H_2SO_4), nitric acid (HNO_3), and perchloric acid ($HClO_4$) are all strong acids. See ACID, PH.

strong base 1) A base that in water undergoes essentially complete dissociation to form high concentrations of hydroxide ions (e.g., $NaOH \longrightarrow Na^+ + {}^-OH$). 2) Strong bases can also be compounds that react with water to form high concentrations of hydroxide ions (e.g., ${}^-NH_2 + H_2O \longrightarrow NH_3 + H_2O$). Aqueous solutions of strong bases have high pH values. See BASE, PH.

strong electrolytes Ionic substances that fully dissociate into their component ions when dissolved in water. Solutions of these substances conduct electricity very well. Compare WEAK ELECTROLYTES, ELECTROLYTE.

strontium (Sr) Metallic element with atomic weight 87.62 and atomic number 38. It is very reactive and so always found in compounds with other elements. One radioactive isotope, strontium 90, is produced by nuclear fission of uranium. Present in radioactive fallout from atomic explosions, it is easily absorbed by animals because it resembles calcium; it becomes concentrated in cow's milk and human bone tissue.

structural diversity Variations in the physical characteristics of an environment that create a variety of habitats within a community, increasing the diversity of species that can live there. See FOLIAGE HEIGHT DIVERSITY.

structural genes Sequences of nucleotides within a strand of DNA that direct the manufacture of messenger RNA to control the sequence of amino acids in the synthesis of particular proteins.

structural geology The scientific study of the three-dimensional forms, arrangements, and relations of rocks.

structural terrace A type of terrace that has a shape controlled by the structure of the underlying bedrock. A structural terrace may be formed by the differential erosion of a tilted stratum or fold.

structureless soils Soils which are without peds, or recognizable aggregates of soil particles. Structureless soils may be single-grained or massive in character.

strychnine $C_{21}H_{22}N_2O_2$, a very toxic plant alkaloid. It is found in a couple of species of *Strychnos* (nightshade). Like Compound 1080, it has been used to poison large predators on rangelands.

stubble The stems and stumps left in the field after harvesting a crop, especially corn or cereal grains. When these plant residues are left on the soil surface when a new crop is planted, they are often called a stubble mulch.

stump sprouting The phenomenon in which new shoots grow from the bases of trees that have been harvested or broken by the wind. Not all trees are capable of producing such shoots; red maple is one species able to produce a prolific crop of sprouts from cut stumps. See COPPICE.

style The usually elongated, stemlike neck of the reproductive structure of a flower (pistil) that connects the stigma to the ovary.

stylet 1) A long, thin mouthpart on an insect or other animal, adapted for stinging, piercing prey, or sucking sap. 2) Any bristle-shaped projection on an animal. 3) A thin probe used in surgical examinations, or a thin, hollow tube for clearing or stiffening a catheter.

styliform Shaped like a bristle; slender and pointed.

sub- Prefix meaning below or subordinate to, as in subspecies.

subadult An organism that is past the juvenile stage but that has not yet developed adult characteristics, or is breeding because of behavioral or sexual immaturity. Also used to describe an age class of organisms at this stage of development. See POSTLARVA.

subaerial erosion An erosion that proceeds directly on the land surface exposed to the atmosphere.

subalpine Describing the region, the climate, the vegetation, or all three found just below alpine regions, usually on mountainsides at 1300 to 1800 meters in elevation. Subalpine vegetation is that just below treeline, often dominated by pine or spruce trees.

subatomic particles Particles smaller than, and usually contained within, atoms; these include electrons, protons, and neutrons as well as the smaller fundamental particles. See FUNDAMENTAL PARTICLES.

sub-bituminous coal A type of black coal that is intermediate between lignite and bituminous coal. Sub-bituminous coal has a higher energy value and a lower moisture content than lignite.

sub-bottom profile A seismic profile of the sea floor that is based on reflections from a surface deeper than the bottom of the ocean.

subclimax A stage late in the development of a plant community in which development stops before the formation of the climatic climax community because of some kind of interference, such as fire or excessive browsing by deer. Compare DISCLIMAX.

subcontinent 1) Any large landmass, such as Greenland, that is smaller than a formally named continent. 2) A region of a continent that is significantly distinct from other parts of the continent. For example, India forms a subcontinent that is separated from the rest of Asia by the Himalayan Mountains.

subdelta A small delta formed by a distributary channel of a larger delta complex.

subdominant An important species within a community, but one that is less prevalent, smaller, or of less importance than the dominant species. See DOMINANCE.

subduction The process by which a tectonic plate descends below another converging plate. The subducting plate is ultimately consumed by the mantle.

subduction zone A general term for a region of tectonic plate convergence. A subduction zone slopes down from an oceanic trench to the lower limit of the lithosphere. See BENIOFF ZONE.

suberin A complex, fatty compound found in some of the walls of some plant cells, especially in cork. See CUTIN.

suberization The conversion of cell walls into cork by impregnation with suberin. See SUBERIN.

subharmonic A repeating vibration or wave whose frequency is a fraction of the fundamental frequency. Compare HARMONIC.

sublimation The phase transition of a substance directly from solid to vapor, without first becoming a liquid. Compare EVAPORATION, MELT.

sublime To pass directly from a solid phase to a vapor phase, without first becoming liquid.

sublittoral The ocean-floor region extending from the lower limit of the littoral zone to the edge of the continental shelf. In fresh water, the sublittoral zone is equivalent to the limnetic zone. See LITTORAL ZONE.

submarine canyon A V-shaped valley cut into a continental shelf or continental slope. Submarine canyons may be scoured by turbidity currents.

submarine fan A sloping apron of terrigenous or shallow marine sediment that accumulates at the mouth of a submarine canyon.

submarine plateau   A relatively flat-topped elevation of the sea floor, usually rising 200 m or more above the surrounding abyssal plain.

submerged aquatic vegetation   (SAV) Plants that grow in the sediments at the bottom of a body of water.

subpolar glacier   A glacier in which some seasonal melting occurs within the zone of accumulation. Subpolar glaciers have temperate interiors that are maintained near the pressure-melting point of ice, and colder margins that remain frozen.

subsequent   A term applied to a geologic feature that develops after a related consequent feature whose position has been determined by a natural gradient. The erosion of a canyon in a plateau occurs subsequent to the erosion of the plateau surface.

subsequent stream   Any stream in which the direction of the water course is controlled primarily by bedrock features such as faults, folds, or joints. Compare CONSEQUENT STREAM.

subsere   A secondary series of successional stages in plant communities, especially one following human disturbances such as an abandoned field being invaded by goldenrod and eventually reverting back to forest. See SERE.

subsidence   A term applied to the progressive depression of a land surface or a crustal elevation because of natural or artificial causes. Phenomena such as groundwater withdrawal, mining, or karst activity may cause subsidence.

subsidence inversion   A form of thermal or temperature inversion caused by adiabatic warming of a layer of sinking (subsiding) air. Subsidence inversions are among the most intense and long-lasting. See THER-MAL INVERSION, HIGH-LEVEL INVERSION.

subsoil   A general reference to the zone of accumulation in a soil profile. See B HORIZON.

subsong   A bird song that appears to serve no purpose, unlike functional songs that advertise, territorize, or attract mates.

subsonic   1) Moving at speeds below the speed of sound, as in subsonic aircraft. 2) Concerning sound waves with frequencies below the range of human hearing (below 10 hertz); also called infrasonic. See SUPERSONIC TRANSPORT.

subspecies   A taxonomic category that subdivides species into morphologically distinct groups of individuals representing a step toward the production of a new species, although they are still fully capable of interbreeding. Subspecies are usually geographically isolated. Compare RACE, VARIETY, FORM.

substrate   1) Surface or medium that serves as a base for something. Substrate refers to the nutrient medium for an organism (as in a laboratory Petri dish), or to a physical structure on which it grows (as in barnacles growing on ship hulls). It is also used for crystals in a semiconductor providing a surface on which to build an integrated circuit or semiconductor. 2) The substance that is acted upon by an enzyme in a biochemical reaction.

substratum   A term applied to the soil layers situated beneath the solum layers. The substratum generally includes all layers beneath the B horizon. See SOLUM.

subsurface mining   A general term for mining operations carried out below the earth's surface that are accessed by a shaft or an adit.

subsurface water   A term that applies to all water below the earth's surface, whether solid, liquid, or gaseous in form. Subsurface water includes that of soil, bedrock, and lithosphere. Contrast SURFACE WATER.

subtidal   A term applied to the shallow marine or tidal flat environment that is below the mean low water level of spring tides.

subtropical   Describing regions, climate, or vegetation found in the regions bordering on the tropics, sometimes described as the portion of the temperate zones closest to the equator.

subtropical high   One of the semipermanent high-pressure (anticyclonic) regions lying within a broad belt from 25° to 35° latitude. Subtropical highs occur over oceans and are more pronounced during the summer. A belt of generally low pressure near the equator (equatorial low) separates the two belts where subtropical highs occur. Also called oceanic anticyclone.

subtropical jet   An almost permanent, westerly, fast stream of air that occurs at about 25° latitude (above and below the equator), at altitudes of approximately 13 km. The subtropical jet streams have not been studied as extensively as the polar jet stream (which is often called simply the jet stream). See JET STREAM.

subumbrella   The usually concave underside of a jellyfish, where the mouth (manubrium) is located. See EXUMBRELLA.

subzone (biostratigraphic)   A subdivision of a biostratigraphic zone. A subzone is usually based on continent-wide correlations. See BIOSTRATIGRAPHIC UNIT.

succession   The gradual change between types of transient communities or ecosystems that involves changes in the plant and animal species composition. Given a sufficient period of time, and a lack of repeated interruptions, succession eventually leads to a steady-state community, a climax community. See CLIMAX, DISCLIMAX, PRIMARY SUCCESSION, SECONDARY SUCCESSION, SERE.

successional trajectory   In stochastic (Markovian) models of succession, the direction that transitions in species composition and dominance are taking at a specific point in time.

successive percentage mortality   Mortality for successive age groups of a population expressed as a percentage of the number of living individuals at the beginning of each development stage.

succulent plants   Plants that have thick, fleshy leaves or stems, adapted for storing water. Most succulents are xerophytes, plants preferring dry climates, such as cactus and aloe; some are halophytes such as glasswort (*Salicornia*), adapted for living in salt soils where water retention is a problem. Many succulent plants have a specialized form of photosynthesis, the crassulacean acid metabolism that allows them to reduce water loss on hot days. See CRASSULACEAN ACID METABOLISM.

sucrose $C_{12}H_{22}O_{11}$, cane sugar (common table sugar). It is a disaccharide sugar that can be extracted from sugar cane and sugar beets. It consists of a fructose molecule linked to a glucose molecule. Also called saccharose.

suction dredge   A device that incorporates a centrifugal suction pump that drives underwater excavation or sediment removal.

Sudanian-Sindian floral region   One of

the phytogeographic regions into which the Paleotropical realm is divided according to similarities between plants; it consists of the African Sahel, Sudan, and northwestern India. See PALEOTROPICAL REALM, PHYTOGEOGRAPHY.

sudden ionospheric disturbance (SID) Disruptions or enhancements of high-frequency radio communications caused by a variety of phenomena, including solar flares, lightning, and geomagnetic storms. The disruptions may include fading of short-wave communications, selective fading, and decrease in daytime absorption causing an enhancement of daytime signal strength. See HIGH FREQUENCY.

suffruticose Describing small plants that are woody only at the base and herbaceous at the top, as in alpine species of willow and thyme. Also called suffructescent.

suite A group of associated characteristics or species.

sulfate 1) The $(SO_4)^{-2}$ anion. 2) A salt or ester containing the sulfate $(SO_4)^{-2}$ anion, such as gypsum $(CaSO_4 \cdot 2H_2O)$.

sulfate mineral A nonsilicate mineral composed of one or more metallic ions combined with a radical group containing proportions of one sulfur to four oxygen atoms; is generally light in color, translucent, and fragile; an example is gypsum.

sulfate-reducing bacteria Several species of bacteria belonging to two genera (*Desulfovibo* and *Desulfotomaculum*) that reduce sulfates present in the soil to form sulfides. They accomplish this by using the oxygen in the sulfate to in turn oxidize organic compounds; some forms convert organic alcohols to acids.

sulfide 1) The S^{-2} anion. 2) A salt containing the S^{-2} anion. Many metals combine with sulfur to form sulfides, including galena (PbS), iron pyrite (FeS_2), and cinnabar (HgS).

sulfite 1) The $(SO_3)^{-2}$ anion. 2) A salt or ester containing the sulfite $(SO_3)^{-2}$ anion. Sulfites are often good reducing agents.

sulfur (S) A flammable, nonmetallic element with atomic weight 32.06 and atomic number 16. It is an essential nutrient required for the synthesis of proteins in animals and plants. Sulfur has many commercial uses, including in gunpowder, matches, bleaches, plant fungicides and insecticides, and medicines (sulfa drugs).

sulfur dioxide SO_2, a major air pollutant; a gas formed when sulfur burns in the presence of oxygen. It is corrosive and harmful to plants and animals, and is especially damaging to trees, causing chlorosis and dwarfing. It reacts readily with oxygen to form SO_3; it also dissolves in water (as when it comes in contact with water vapor in the atmosphere) to form a mixture of sulfurous acid and sulfuric acid, causing acid rain. Power plants burning sulfur-containing coal and smelters used for refining metal ores are the major sources of this pollutant. Sulfur dioxide is also used as a disinfectant, bleach, and food preservative. See SULFUR TRIOXIDE.

sulfuric acid H_2SO_4, a strong and very corrosive acid. It forms when sulfur oxides (SO_x) dissolve in water, as when emissions from sulfur-containing fuels come in contact with water vapor to produce acid rain. Large quantities of sulfuric acid are used in the production of phosphate fertilizers. It is also used to refine petroleum and to produce explosives. Also called oil of vitriol.

sulfur trioxide (SO_3) A secondary air

pollutant produced from sulfur dioxide (from burning sulfur-containing fuels) by the action of sunlight. Sulfur trioxide forms sulfuric acid when it dissolves in water or water vapor, producing acid rain.

sullage water   A wastewater stream that drains from a farmyard, a refuse heap, or a street.

sump   A natural lake or an excavation that collects water to be pumped into an irrigation system or a drainage system.

sun dogs   One of two bright spots appearing slightly below and on either side of the sun at an angle of 22°. They are caused when the ice crystals within cirrus clouds refract sunlight. Also called mock suns or parhelia.

sun pillar   A vertical column of light that appears to extend from the sun at sunrise or sunset. The visual effect is cased by reflection of the sun's rays through horizontal ice crystals. If the sunlight is reddish at that time of day, the sun pillar will be reddish, also.

sunspot   A disturbed region of the solar surface that appears darker than the surrounding regions. Sunspots are intense localized magnetic fields; they appear darker because they are lower in temperature than the surrounding photosphere. They usually consist of a dark central umbra surrounded by a less dark region, the penumbra.

sunspot cycle   The 11-year periodic variation in the number of sunspots visible on the sun. Many related signs of solar activity such as plages and flares also increase with increasing sunspots. See SOLAR MAXIMUM, SOLAR MINIMUM.

sunspot number   A measure of solar activity based upon the number of sunspots visible on the sun, calculated daily. It is not a simple count, but incorporates individual spots plus sunspot groups, multiplied by a factor calculated for each observing site. The sunspot numbers follow an 11-year periodic cycle; many forms of solar activity such as plages and flares correlate with this cycle. See SOLAR MAXIMUM, SOLAR MINIMUM.

supercooled fog   A fog in which the water droplets have been cooled to a temperature at or below freezing, but still remain in liquid form. Although not common, such fog is more easily dissipated by artificial cloud seeding than a warm fog.

supercooling   Lowering the temperature of a liquid (especially water) below its freezing point, without causing it to change from liquid to solid.

superfamily   A taxonomic group of organisms ranking above family but below order.

Superfund   Common name for the complex federal law, Comprehensive Environmental Response, Compensation and Liability Act (CERCLA), or the funds associated with that statute. Superfund sites are those on the National Priorities List established by the Environmental Protection Agency. See CERCLA.

supergene   A deposit of minerals that has been formed by secondary enrichment. See SECONDARY ENRICHMENT.

superinsulated house   A well-insulated and very airtight house. Such houses may have managed air exchange systems using air-to-air heat exchangers to avoid a build-up of excess moisture or indoor air pollutants while conserving heat. Superinsulated houses have much more insulation in their walls and roofs than ordinary houses, and all pipes, electrical wires, windows, and other structures piercing the walls are carefully caulked to prevent air leaks.

superior image   A mirage in which the false image appears above the location of the object. It occurs when the air close to the ground is colder than the air higher up, bending light rays and allowing observers to see objects that would normally lie below the horizon. A Fata Morgana mirage is a superior image. Also called a superior mirage. See FATA MORGANA.

supernova   A violent explosion of a star; supernovae occur when massive stars undergo gravitational collapse. The explosion releases tremendous amounts of energy (radiation). Supernovae are much brighter than unrelated ordinary novae.

superorganism concept   The theory that communities develop as if they were different parts of one organism; their constituent species are so tightly bound together that they will continue to evolve together and will share a common evolutionary history. See GAIA HYPOTHESIS, MONOCLIMAX THEORY.

superphosphate   A principal form of phosphorous used in plant fertilizers. It consists of rock phosphate that has been treated with sulfuric acid to make it more soluble and thus faster acting. Superphosphate contains a higher percentage of P_2O_5, the form of phosphorous most easily assimilated by plants. Although superphosphates give fast, impressive results, they are more easily leached away than the slower-acting, untreated rock phosphate. They can thus more easily contribute to eutrophication.

superposed stream   A term applied to a stream drainage system that has eroded down through an existing landscape pattern and crosscuts the underlying rock structure.

supersaturation   A solution containing more of a dissolved substance than is usually possible at equilibrium (saturation point). In the atmosphere, supersaturation is a condition in which localized relative humidity exceeds 100 percent. It can occur when no particles are present to serve as condensation nuclei. See SATURATION POINT.

supersonic transport   (SST) Commercial airplanes designed to travel at speeds exceeding the speed of sound (about 767 miles per hour or 343 meters per second). The Concorde is an example of supersonic transport.

superspecies   A group of species that is related, or a group of incipient species that is regarded as too distinct to be considered a single species. See SUBSPECIES.

supertramp species   Species having a wide geographical range but lacking the ability to persist under competition. See R-STRATEGISTS, RUDERAL.

supply/demand curve   An economic term for a plot of prices and quantities indicating the relationship between consumer demand and manufacterer's supply of a commodity at given prices in a free market economy.

suppression   A condition in which understory or small trees growing in deep shade exhibit minimal growth without dying for a prolonged period of time. Under extreme suppression, trees may stop diameter growth entirely.

supragenus   Describing taxa above the level of genus.

supratidal zone   A term describing the region of shoreline immediately above the high-tide level of the littoral zone.

supreme court, state   Highest court of

appeals in most states, as established by state laws. States vary in numbers of justices and judges.

Supreme Court, U.S   The highest court in the U.S. federal court system, and thus the highest court of appeal. Established by the Constitution, it is made up of one chief justice and eight associate justices, all appointed by the president.

surface-active agent   Another term for surfactant. See SURFACTANT.

surface deposits   See SURFICIAL DEPOSITS.

surface fire   A forest fire that burns undergrowth but rarely becomes hot enough to damage higher levels of vegetation. Plant communities recover relatively quickly from most ground fires. Compare GROUND FIRE.

surface flow   See OVERLAND FLOW.

surface layer   1) In meteorology, the thin zone of airflow near the earth's surface where shear forces are relatively constant and the wind pattern is determined by temperature gradient and topography. 2) The uppermost horizon of a soil profile.

surface mining   The extraction of ore minerals directly at the surface of the land. The term surface mining is often applied to the collection of placer deposits.

surface ocean currents   A general term applied to ocean currents that flow at an average depth of 1 to 3 m.

surface pressure   1) The atmospheric pressure measurement obtained inside a Stevenson screen; essentially, the atmospheric pressure at ground level. 2) A measurement of the tendency for a surfactant (a substance such as soap or detergent) to spread over the surface of another liquid such as water. Surface pressure is calcu-

lated as the difference between the surface tensions of a pure liquid and that of a surfactant (surface-active solution). See SURFACE TENSION.

surface processes   Mechanical, chemical, or physical processes that affect geologic materials at or near the surface of the earth. Weathering and erosion are primarily surface processes.

surface tension   The tendency for liquids to keep their surface at a minimum. It is caused by surface molecules only experiencing molecular attraction from below, whereas those in the interior of a liquid experience molecular attraction from all directions. The strong surface tension of water is what allows water bugs to skate across the top of a pond; it also allows water to be drawn up into capillary tubes.

surface water   An occurrence of water on the earth's surface. Contrast SUBSURFACE WATER.

surface wave   1) A seismic wave that propagates parallel to or along the surface of the earth. Rayleigh waves and L waves are types of surface waves. 2) A deep-water ocean wave in which the rapid particle movement is confined to the ocean surface.

surface wind   Wind measured at 33 feet (10 meters) above a smooth stretch of earth. Unlike the gradient and the geostrophic winds, surface winds are greatly affected by friction over land, and are thus also called friction layer winds. Compare GEOSTROPHIC WIND, GRADIENT WIND.

surfactant   Any compound that reduces the surface tension of liquids, increasing its wetting ability and spreadability. Soaps and detergents act as surfactants in water; however, additional surfactant compounds are added to detergents to increase lathering. Such compounds often contain phosphates, causing sudsing if released directly to streams and contributing to eutrophication. Also called surface-active agent.

surficial deposits   An unconsolidated deposit of rock particles overlying the bedrock. Alluvium and glacial till are examples of surficial deposits.

surge   A turbulent flow of expanded volcanic gas with ash clouds, phreatic eruptions, or pyroclastic material.

surplus yield model   An approach to managing harvested plant or animal populations that attempts to balance harvesting and recruitment rates by harvesting no more than the maximum sustainable yield. See MAXIMUM SUSTAIN-ABLE YIELD. Compare DYNAMIC POOL MODEL.

survival of the fittest   The principle behind natural selection, that those organisms best suited to their environments are the most likely to survive and produce offspring. See NATURAL SELECTION.

survival rate   See SURVIVORSHIP.

survivorship   1) The proportion of a given population that reaches a particular age. 2) (l_x) The number of individuals in a particular age group counted at the beginning of a given period that are still alive at the end of the period.

survivorship curve   A plot illustrating the survivorship of age classes in a given population. Survivorship curves are usually constructed for a specific group of individuals, such as a cohort (age group); they show how the size of the group declines with age, and where in its life span mortality occurs. See MORTALITY RATE.

susceptibility   The degree to which an organism is prone to infection by a particular disease, or is sensitive to a particular drug or poison.

suspect terrane   A regional tectonic structure or orogenic belt that is suspected to represent a terrane but whose fault boundaries have not been identified. See TERRANE.

suspended load   The portion of stream load that is carried along in suspension by flowing water. Compare BED LOAD.

suspended sediment   See SUSPENDED LOAD.

suspended solids   (ss) A mixture of dispersed solids carried within a fluid medium. Fine detrital particles may be transported as suspended solids within a stream.

suspension feeder   An aquatic animal that feeds by straining phytoplankton or suspended plant debris from water.

sustainable agriculture   An economically viable method of agriculture that emphasizes stewardship (long-term rather than solely short-term returns), soil conservation, and integrated pest management to ensure that there is no degradation of the environmental quality or the capacity of the system to continue to produce. See ORGANIC AGRICULTURE.

sustainable capacity   See SUSTAINED YIELD.

sustainable development   Economic growth and activities that do not deplete or degrade the environmental resources upon which present and future economic growth depend.

sustained yield   A level of harvest of a renewable resource per year (or other time period) that can be continued without jeopardizing the ability of the ecosystem to be fully renewed, and thus to continue to provide an undiminished level of harvest each year long into the future. Usually used for timber management, but also used for wildlife and other natural resources. Also called sustainable yield. See MAXIMUM SUSTAINABLE YIELD.

suture 1) A line formed by the junction of two structures, such as fused carpels or other plant parts, or a nonmoving junction between bones (as in the skull). 2) In plants, especially fruits that split open when ripe, a line of weakness in a tissue or structure across which a split occurs. 3) A distinct geographic zone of hybridization between two species or two subspecies.

Sv Abbreviation for the unit of radiation dose, sievert. See SIEVERT.

swallow hole A place where a stream drains down into a sinkhole.

swamp A wetland dominated by woody plants such as trees or shrubs, and in which peat does not accumulate. See MANGROVE SWAMP ECOSYSTEM. Compare MARSH, BOG.

swans The largest waterfowl of the order Anseriformes, the swans have five species distributed in the Northern Hemisphere and two in the Southern Hemisphere. The trumpeter swan of North America, which came close to extinction but is now breeding in reasonable numbers, is a good example of how active government and private programs can serve in species preservation. See ANSERIFORMES.

S wave (secondary wave) A type of seismic body wave that is propagated by a shearing motion, like that of a rope that is shaken at one end. This slower-moving wave is known as a secondary wave and arrives at any given point after the P wave. Compare P WAVE (PRIMARY WAVE), BODY WAVES.

sweepstakes dispersal Chance dispersal of organisms across water by rafting. See RAFTING.

swidden agriculture Another term for slash-and-burn agriculture, practiced widely in Neolithic times throughout what is now called Europe. Also called milpa agriculture. See SLASH-AND-BURN AGRICULTURE.

swimmeret An abdominal limb or appendage in crustaceans usually used for swimming; it may be used to help females in carrying eggs. Also called a pleopod.

switching The ability of a predator to change its prey depending upon which species is most abundant within a region.

syenite An intermediate intrusive igneous rock composed of more than 65 percent alkaline feldspar, with the remaining portion consisting of mafic minerals such as hornblende and biotite.

sylvite A whitish mineral of potassium chloride forming cubic crystals in evaporite deposits. Sylvite is used as a fertilizer.

symba process A method for producing protein from starch wastes remaining after food processing. The waste is first innoculated with the fungus *Endomycopsis fibuliger*, which hydrolizes the starch. The product is then used to grow *Torula* yeast (nutritional yeast), an excellent source of amino acids and vitamins.

symbiont An organism living in symbiosis.

symbiosis 1) Broadly, any long-term association of two or more organisms from different species, especially those that are obligatory or involve coevolution. 2) Often restricted to associations that are mutually beneficial (mutualism), but sometimes used to include commensalism and parasitism (which are not beneficial—or are harmful—to one of the organisms). See MUTUALISM, OBLIGATE MUTUALISM, PROTOCOOPERATION, COMMENSALISM.

symbiotic nitrogen-fixing bacteria Those bacteria that form nodules in the roots of plants, where they fix elemental nitrogen from air in the soil into forms that plants can use. They provide nitrogen to the host and benefit from the energy compounds produced by the host. Most symbiotic nitrogen-fixing bacteria are associated with legumes (peas, beans, alfalfa, etc.) and belong to the genus *Rhizobium*. Several species of angiosperm trees and shrubs that grow in nitrogen-poor soils also have nodule-forming, nitrogen-fixing bacteria; these include alder, ceanothus, bayberry, sweet gale *(Myrica)*, and members of the Oleaster family (buffalo-berry, Russian-olive). Also called nodule organisms. See NITROGEN FIXATION, RHIZOBIA, ROOT NODULES.

symmetrical fold A bend in layered rock in which an identical angle is formed between each limb and the axis of the fold. Compare ASYMMETRICAL FOLD.

sympathetic nervous system The part of the autonomic nervous system that uses adrenaline (or noradrenaline) to produce involuntary responses, including increased heartbeat and circulation, increased activity, and slowing of the digestive system. Its action tends to oppose that of the parasympathetic nervous system. It is centered in two groups of ganglia connected by nerve cords, running down either side of the spine. See AUTONOMIC NERVOUS SYSTEM, PARASYMPATHETIC NERVOUS SYSTEM.

sympatric Describing two populations or species that live in the same region without merging into one population through interbreeding. Contrast ALLOPATRIC.

sympatric speciation The evolution of distinct species, populations incapable of interbreeding, from organisms living in the same area. Contrast ALLOPATRIC SPECIATION.

symplesiomorphy The common possession by two taxa of an ancestral trait or a character.

synanthrope An organism often found associated with human settlements, such as plantain, stinging nettle, and the Norway rat.

syncarp A plant reproductive structure (gynoecium) containing two or more carpels fused to form a compound ovary. See CARPEL. Compare ADELPHOUS.

synchorology The study of the occurrences and the distribution ranges of plant communities or larger phytosociological regions, including current patterns of plant migration.

synchronous satellite An artificial satellite traveling in a geosynchronous orbit around earth, or traveling around another planet or star at the same speed as the speed at which the planet or star is rotating. Synchronous satellites remain over a fixed point above the object they are orbiting. See GEOSYNCHRONOUS.

syncline A sag in the rock strata in which the layers bend upward away from the fold axis. The oldest layers in an eroded syncline are exposed farthest from the fold axis. Contrast ANTICLINE.

synclinorium A composite group of synclines and anticlines forming a large synformal structure on a regional scale. Contrast ANTICLINORIUM.

syncytium Plant tissue containing several nuclei and not divided up by cell membranes into separate cells. It is formed either by fusion of several cells without fusion of their nuclei (forming a plasmodium), or by division of the nucleus without division its cytoplasm. See PLASMODIUM.

syndynamics The study of plant succession, especially the changes and their causes.

synecology The ecological study of groupings of organisms, such as associations or communities, in relation to each other and to their environment. Synecology includes population, community, and ecosystem ecology. Compare AUTECOLOGY.

synergy A condition in which two or more factors interact, with the net effect being greater than the sum of the independent effects of the factors.

synform A synclinelike sagging rock structure for which the relative ages of the layers have not been determined. Therefore, a synform may be a syncline or an inverted anticline. Contrast ANTIFORM.

synfuels Short for synthetic fuels, the industry name for hydrocarbon fuels processed from coal, oil shale, or tar sand so that they resemble liquid petroleum fuels derived from crude oil and natural gas. Synfuels make high-grade fuels out of very dilute, low-grade forms but consume great quantities of energy and water during the processing, resulting in no net energy gain (and perhaps a net loss of energy). See OIL SHALE, TAR SAND.

syngameon A grouping of species within which hybridization is occurring, or can occur.

synodic Concerning celestial motions as viewed from the vantage point of earth (technically, the line joining the earth and the sun), rather than compared to the fixed background of the stars (sidereal). The phases of the moon and of Venus follow a synodic period.

synrock A refined petroleum product used as a sealant that is injected into bedrock to fill pores and fractures.

syntectonic A term applied to processes occurring at the same time as tectonic deformation of rocks. For example, regional metamorphism due to tectonic convergence is said to be syntectonic.

synthesis 1) The combining of separate parts or ideas to form a single entity. 2) The formation of a chemical compound or complex substance from its component elements, or from simpler compounds, through one or more chemical reactions.

synthetic chemicals Compounds produced in the laboratory or in large-scale chemical plants as opposed to being extracted from living organisms. For example, the heart medicines digoxin and digitoxin were originally extracted from the leaves of foxglove *(Digitalis)* and were thus organically derived chemicals. These medications are now produced in the laboratory, and thus are now considered synthetic chemicals. Fertilizers can similarly be synthetic chemicals as well as be derived from natural sources. Note that confusion arises when carbon-containing compounds are synthesized; these are called organic compounds, yet they are synthetic chemicals. See ORGANIC COMPOUNDS OR MOLECULES, SYNTHETIC ORGANIC COMPOUNDS.

synthetic natural gas Methane manufactured from coal, oil shale, or tar sands as opposed to the usual natural gas extracted from liquid petroleum deposits. See SYNFUELS.

synthetic organic compounds Carbon-containing chemicals that are produced

SYNFUELS.

synthetic organic compounds   Carbon-containing chemicals that are produced rather than extracted from living organisms.

syntype   A specimen used for defining a species when the holotype has not yet been specified. Also called cotype. See TYPE SPECIMEN.

synusia   A horizontal layer at a specific height in a stand of vegetation, including different species of plants with the same general life form. See STRATIFICATION, CANOPY, UNDERSTORY, SHRUB LAYER, GROUND LAYER, HERB LAYER.

Système International d'Unités   (SI) The standard international scientific system of measurement. It differs from the metric system, upon which it is based, because three additional units—candela, kelvin, and mole—have been added to meter, kilogram, second, and ampere. From these seven basic units, all the other units (joule, etc.) are derived. This system is a decimal system and uses prefixes such as deci-(deca-), centi-, kilo-, and so forth to indicate the sizes of units in mutiples or divisions of ten. See APPENDIX p. 605.

systematics   The branch of biology concerned with the relationships between organisms, including classification and the naming of organisms and taxonomic groupings. Often considered a synonym for taxonomy. See TAXONOMY.

systemic pesticide   A compound absorbed by leaves or roots and transported throughout the plant. Systemic pesticides, usually organic phosphate compounds, are very effective in controlling sucking insects and mites on ornamental species (most cannot be used on food crops). They are usually applied to the soil at the base of the plant, causing very little damage to nontarget species. Often simply called systemics. See ORGANIC PHOSPHATE.

t

Ta Chemical symbol for the element tantalum. See TANTALUM.

tableland A large elevated region or landscape of low relief, such as a plateau or a mesa.

taconite A general term applied to unleached iron ore containing hematite, magnetite, siderite, and hydrous iron silicate minerals.

taiga See BOREAL FOREST.

tailings The fine-grained waste materials from an ore processing operation.

taking A complex legal concept in U.S. law, drawn primarily from the language of the Fifth Amendment to the Constitution as incorporated in the Fourteenth Amendment for state actions, and substantially replicated in corresponding provisions of most state constitutions. It provides that when private property is taken for public use, there must be just compensation.

tallgrass prairie Prairie vegetation characterized by taller species of grasses (such as big bluestem and Indian grass), once the characteristic vegetation of the eastern, moister regions of the U.S. Midwest. Also called mesic prairie. Compare SHORT-GRASS PRAIRIE.

talus An accumulation of coarse, angular rock debris at the base of a cliff or steep rocky slope.

tannins A group of acidic, organic compounds extracted from plants such as oak bark or leaves. Ordinary black tea is high in tannins, which give it its astringent quality (coffee contains smaller quantities). The name comes from the long-standing use of tannins in tanning leather; they are also used in dyes, inks, and medicines.

tantalum (Ta) A hard, metallic element with atomic weight 180.9 and atomic number 73. It resists corrosion and is used in alloys for surgical equipment (including pins to join bones), electrical components, and nuclear reactors. It is sometimes used as a substitute for platinum when a corrosion-resistant metal is required, as in the chemical industry or in some laboratory instruments.

tapetic Felt-forming; describing the growth habit of some species of blue-green algae, diatoms, or similar organisms.

tapeworms See CESTODA.

taproot system A common form of plant root structure in which one large, primary root grows vertically downward, and from it numerous lesser (often very fine) roots grow laterally. Dicotyledon species and conifers have taproot systems. Compare FIBROUS ROOT SYSTEM.

taproot

tar Dark, viscous, carcinogenic liquid usually derived from coal (coal tar) but also derived from other carbon-containing material such as wood or peat. The thick liquid may contain hydrocarbon oils, phenols, bases (pyridine, pyrrole, etc.), plus derivatives of these compounds. Dehydrated forms are used as a binder in road surfacing. Coal tars are a source for the synthesis of many organic compounds. See COAL TAR.

tarantulas A term loosely applied to a number of large, hairy spiders of threatening appearance belonging to several different families. They are often distributed in cargoes of fruit and, popular sentiment to the contrary, produce bites that may be discomforting but are not usually life threatening. See ARACHNIDA, CHELICERATA.

tar balls   A descriptive term for the viscous spheres of liquid petroleum compounds that aggregate in water after an oil spill.

target organism   The organism(s) a pesticide is intended to kill.

target theory   The theory that ionizing radiation causes cellular or chromosomal damage when it reaches a small, sensitive area of the chromosome, resulting in alteration or in breakage of the chromosome. Associated with this theory is the trend for species with larger-sized cell nuclei to be more sensitive to the effects of ionizing radiation than those whose nuclei or chromosomes have relatively small volumes.

tarn   Any small rock basin lake formed on a glaciated landscape. Tarns are often formed at the base of glacial cirques. See PATERNOSTER LAKES.

tar sand   A deposit of sand or sandstone that contains a residual oil filling the interstitial spaces.

tarsus   1) The bones of the ankle on humans and terrestrial vertebrates. 2) The lowest portion of the leg in insects, centi-pedes, millipedes, and mites. 3) Fibrous tissue that supports the eyelid in vertebrates.

Taxaceae   The yew family of the phylum Coniferophyta in the plant kingdom. The members of this family are shrubs and small trees with linear, spirally arranged, evergreen, flat leaves, and seeds enclosed in a fleshy aril. Members include yews (*Taxus* spp.) and California nutmeg (*Torreya*). See CONIFEROPHYTA.

taxis   An involuntary or reflex movement in a particular direction by a free organism or cell in reaction to an external stimulus such as light. Taxis differs from tropic responses in that it involves an actual change in location rather than mere bending. See CHEMOTAXIS, GEOTAXIS, HELIOTAXIS, HYDROTAXIS, PHOTOTAXIS, RHEOTAXIS, STEREOTAXIS, THERMO-TAXIS, KINESIS.

taxon   A grouping of organisms given a formal taxonomic name at any rank: species, genus, family, order, class, division, phylum, or kingdom. Plural is taxa. See SYSTEMATICS.

taxon cycle   A theory that generalist species colonizing islands can evolve into specialists that in turn are replaced by other colonizing species.

taxonomic key   A guide for identifying plants or animals, using a systematic arrangement of the defining characteristics so that an unknown species can be pinpointed by gradually eliminating other families, genera, and species. The defining characteristics are usually arranged in pairs of questions called couplets.

taxonomy   The science of classification as applied to organisms (living or extinct). Classification of individual organisms or higher groupings is based on anatomy, morphology, characteristics of genetic material (chromosomes, genes, and nucleic acids), biochemical relationships (such as protein structure, metabolic pathways), and statistical analysis to interpret combinations of the above characteristics. See SYSTEMATICS.

Tb   Chemical symbol for the element terbium. See TERBIUM.

Tc   Chemical symbol for the element technetium. See TECHNETIUM.

TCDD   Abbreviation for tetrachloro-dibenzoparadioxin, one of the persistent and very poisonous dioxin by-products formed during the manufacturing process of the herbicide 2,4,5-T. It is also formed as a by-product as 2,4,5-T breaks down. See 2,4,5-T.

TDI   See TOLUENE DIISOCYANATE.

Te   Chemical symbol for the element tellurium. See TELLURIUM.

tear fault   A strike-slip fault that has a small component of reverse fault type overthrusting. See STRIKE-SLIP FAULT.

technetium   (Tc) An artificial, radioactive, metallic element with atomic number 43. The atomic weight of its most common and most stable isotope is 99. It is a fission product of uranium and has been identified in the spectra of some stars. It is used in medicine as an imaging agent.

technological assessment   The process of studying the probable environmental and social impacts of the introduction of new or continuation of existing technology options. Technological assessment attempts to anticipate and plan for the consequences of technological changes by analyzing various scenarios from interdisciplianry perspectives.

technology   Practical application of science to the creation of products and processes intended to improve life.

tectonic   A term relating to large-scale movements of earth's crust, such as the motions of plate tectonics activity. See PLATE TECTONICS.

tectonic creep   The description applied to the slow and continuous movement of a fault.

tectonic cycle   A cycle of large-scale crustal motions involving orogenic and tectonic genesis, deformation, and destruction of crustal rocks. See PLATE TECTONICS.

tectonic dam   A naturally formed dam that results from tectonic motions of the earth's crust. For example, a lake may be formed in a graben when stream drainage is blocked by the uplift of a tectonic dam.

tectonic plates   See PLATE, PLATE TECTONICS.

tektite   A small, spherical, or teardrop-shaped fragment of siliceous glass. Tektites may be formed by the impact of meteorites or comets on silicate rocks.

TEL   Abbreviation for the gasoline additive tetraethyllead. See TETRAETHYLLEAD.

teleology   1) The doctrine that natural phenomena are caused or shaped by utility or purpose. 2) Explaining animal and plant structures in terms of purpose and design.

teleosts   General term for a group of bony fish which, with well over 17,000 species, is the most widespread and numerous of all vertebrate subgroups. It includes eels and tarpons, herring and anchovies, minnows and carp, pikes and trout, deep-sea lantern fish, cods and toadfish, flying fish and the perchlike fish. This latter group, which has more than 10,000 species alone, includes basses, perches, sunfish, smelts, bluefish, snappers, cichlids, blennies, gobies, tunas and mackerels, swordfish, gouramis, puffers and so forth. See OSTEICHTHYES.

teleplanic larvae   Aquatic larvae that can become dispersed over tremendous distances, even across entire oceans.

tellurium   (Te) An element with atomic weight 127.6 and atomic number 52. It is called semimetallic or metalloid because it has only some of the properties of a metal.

It mostly occurs in ores (tellurides) with other metals such as gold and silver. It is used in semiconductors and some steel and lead alloys.

telolecithal Describing eggs that are large, with the yolk taking up most of the volume, and the remaining tissue concentrated at one end of the egg. Birds, reptiles, and sharks have telolecithal eggs.

telophase The final stage of cell division, either meiosis or mitosis. In mitosis, the now-separated chromosomes return to their usual less-visible state and the nuclear membrane reappears. Meiosis has a first and second telophase, during which the dividing nucleus reorganizes into two distinct daughter nuclei. See PROPHASE, ANAPHASE, METAPHASE. Compare MEIOSIS.

telson The portion of the body behind the last true section of the abdomen in certain crustaceans and arachnids. It contains the anal opening and forms, with the uropods, the tail fan of scorpions and lobsters.

temperate 1) Of or relating to moderate climates, intermediate between tropical and polar and with distinct warm to hot summer seasons and cool to cold winter seasons. 2) Of or relating to the region between 23°27' and 66°33' north or south latitude in which temperate climates are found. 3) The vegetation in such regions. Compare TROPICAL.

temperate deciduous forest The major biome found in regions with moderate temperatures, distinct seasons, and abundant rainfall (30 to 60 inches). It generally corresponds to some of the most densely populated regions of the globe and thus is often greatly modified. The forest has well-developed shrub and herb layers. In North America, this biome contains several important, distinct types of vegetation, including beech-maple and maple-basswoods of north-central states, oak-hickory forests in western and southern regions, oak (originally oak-chestnut) and mixed hardwood forests of the Appalachians, and the southern pine disclimax forest. Compare MIXED-HARDWOOD FOREST, TEMPERATE EVERGREEN FOREST.

temperate evergreen forest The vegetation type found in the warm-temperate marine climate of Japan and along the Gulf and South Atlantic coastal plains (in areas protected from fire), warm regions with high levels of precipitation. It is characterized by broad-leaved evergreen species such as rhododendron, live oak, magnolia, and hollies, with abundant vines and epiphytes. Compare TEMPERATE DECIDUOUS FOREST.

temperate rainforest The conifer-dominated vegetation found along the coast from Alaska to central California, areas with high humidity (frequent fog), abundant rainfall (30 to 150 inches), and moderate temperatures restricted to a fairly small range. It has more understory than the northern boreal forests and includes the redwood forest of California. Also called moist coniferous forest. Compare TROPICAL RAINFOREST.

temperate (warm) glacier A type of glacier in which most of the ice mass is near the melting point throughout most of the year. A temperate glacier moves mainly by basal slip.

temperature The radiation of a body or a substance that determines the transfer of heat to or from other bodies or substances. Temperature is determined by the average kinetic energy in the molecules of the substance being measured; it is measured using a thermometer calibrated to a scale of degrees such as the Celsius, Fahrenheit or Kelvin scales. See CELSIUS, KELVIN, FAHRENHEIT.

temperature coefficient   The change in any physical quantity (such as electrical resistance) per unit increase in temperature. In biological systems, the temperature coefficient or Q_{10} is the increase in a physical process caused by a 10-degree increase in temperature. Q_{10} is often associated with a doubling of the rate of physiological processes (within liveable limits).

temperature inversion   Another term for thermal inversion. See THERMAL INVERSION, INVERSION LAYER.

temporal variation   Changes that occur over a period of time, such as on a daily, hourly, or seasonal basis. Temperature shows temporal variation, changing between day and night and from season to season. See DIURNAL.

tendril   A slender, spiralling shoot on a climbing plant that attaches itself to an object or surface to support the plant. Tendrils are initially soft and pliable, but after they find an object to secure themselves to, they eventually harden.

tensile stress   The opposite of compressive stress; equal and opposite forces pulling on opposite ends of a body that tend to cause it to strain and change shape by stretching it along the line of the force. Compare SHEAR STRESS.

tensiometer   An instrument for measuring tension. Some tensiometers measure the surface tension of liquids (such as the soil water available to plant roots in the root zone); others measure the degree of tautness (tension) in an object, such as a wire or fabric. See SURFACE TENSION.

tension   The condition of tautness caused when a body or substance is subjected to tensile stress; the opposite of compression. Tension is caused by a pair of forces tending to pull a body or substance apart.

tentacle-tube foot suspension feeder   An aquatic organism, a form of suspension feeder, that obtains the food particles on which it feeds by trapping them on the end of long tentacles or, in echinoderms, on tube feet. See SUSPENSION FEEDER.

tentaculocyst   See LITHOSTYLE.

tepal   A part of a flower perianth that cannot be visually classified as either a calyx or corolla. See CALYX, COROLLA.

tephra   A type of pyroclastic rock fragment ejected in a volcanic explosion. Tephra usually refers to particles settling through the air rather than pyroclasic flow over the ground. See PYROCLASTIC.

tera-   (T) Prefix in the Système International d'Unités (SI) indicating one trillion, 1×10^{12}. One terawatt-hour equals 10^{12} watt-hours.

teratogen   A substance, such as thalidomide, that causes birth defects. See THALIDOMIDE.

terbium   (Tb) An element with atomic weight 158.9 and atomic number 65. It is used in semiconductors.

terbutryne $C_{13}H_{19}N_5S$, a member of the triazine class of herbicides. It is used to control grasses and broadleaved weeds, for submerged plants and algae in slow-moving or still water, and for weed control in fields of cereal grains such as wheat and barley.

tergum   The plate covering the back of each segment (somite) of an arthropod or other segmented animal.

terminal anchor   Any structure used to anchor the leading portion of burrowing aquatic organisms, permitting them to pull the rest of the body into the sediment using muscular contraction.

terminal curvature   The distortion of strata or rock cleavage near the surface of a fault plane. Terminal curvature occurs by frictional forces at the fault plane surface.

terminal moraine   A type of end moraine marking the farthest point of glacial advance. See END MORAINE.

termitarium   A termite nest. See ISOPTERA.

termiticole   Organisms such as some fungi and insects that live alongside termites in termite nests.

terpenoids   A group of hydrocarbon compounds found in plants that have a chemical structure resembling terpene. They are made up of one or more five-carbon chains. Many important plant compounds are terpenoids, including essential oils, gibberellins (growth substances used as weed-killers), carotenoids (plant pigments), and rubber.

terrace   A landform that is nearly flat on top and has an abrupt slope at the edges. Usually formed by stream or marine processes.

terracing   An agricultural practice of planting crops on flat areas built up on or carved out of slopes, to retain water and reduce soil erosion or, on very steep slopes, to make it possible to grow any cultivated crops at all.

terrane   A fault-bounded region having a distinctive stratigraphy or geologic structure. A terrane has a geologic history that is distinctly different from that of adjacent areas.

terra rossa   A soil classification term applied in the European system to reddish, oxidized soils developing over limestones. These soils would be classified as mollisols in the U.S. soil taxonomy.

terrestrial   Of or relating to the land rather than the water; the opposite of aquatic. Terrestrial organisms live or grow on the land.

terrestrial deposit   Any accumulation of unconsolidated material on the land surface and not in the water.

terrestrial planets   Those planets resembling the Earth in size and rocky structure: Mercury, Venus, Earth, and Mars. Contrast JOVIAN PLANETS.

terrigenous deposit   Any accumulation of clastic material derived from a nearby land-based source. The term is usually applied to sediment in a shallow marine environment.

terrigenous sediment   A shallow marine deposit of clastic particles that were eroded and transported from the land surface.

terriherbosa   Nonwoody vegetation that grows on dry land. See HERBOSA.

territoriality   A type of behavior in some animals in which individuals, mates, or social groups defend the area in which they live against encroachment from others, usually of the same species. Territoriality may involve direct aggression, but often involves more indirect mechanisms such as marking the boundaries of the defended area. Territoriality is common in birds and in carnivorous mammals.

territory   An area with defined boundaries where an animal lives, and which it defends to keep out other individuals of the same species. Compare HOME RANGE.

tertiary 1) Third in rank or importance. 2) A division (period) of geological time during which mountains such as the Rockies, Andes, and European Alps were formed. It is the earlier part of the Cenozoic era. In Britain it is regarded as one of the four eras (rather than a period), with its corresponding groups of rocks, and with its four divisions (periods) of Eocene, Oligocene, Miocene, and Pliocene. 3) Rocks or rock formations from the Tertiary period. See APPENDIX p. 610.

tertiary consumers An animal that feeds on the secondary consumers (other carnivores) in a food chain. They are predators at or near the top of food chains, such as raptors, lions, and sharks; they are most susceptible to pollutants such as DDT that undergo bioaccumulation. See BIOACCUMULATION, TOP CARNIVORE.

tertiary production The biomass, or living material, created by organisms that consume herbivores, usually expressed per unit of area (or volume, in aquatic ecosystems). In some schemes, tertiary production is included as part of secondary production.

tertiary sewage treatment The final stage in sewage waste water treatment. May include filtration, chemical or biochemical treatment. See ADVANCED SEWAGE TREATMENT, PRIMARY SEWAGE TREATMENT, SECONDARY SEWAGE TREATMENT.

tesla (T) Unit for magnetic field in the Système International d'Unités (SI). One tesla equals one weber per square meter.

test reactor A noncommercial nuclear reactor, one used for research and development. See NUCLEAR REACTOR.

Testudinata An order of reptiles characterized by the presence of a dorsal shell or plastron and a lower one, the carapace.

The skull has no temporal openings and the beaked jaws are toothless. All are oviparous. Over 200 species exist including turtles and tortoises. See TORTOISES.

Tethys Sea A sea that separated the supercontinents of Gondwana and Laurasia during the Mesozoic era.

tetrachloroethylene $Cl_2=CCl_2$, a carcinogen that has been used as a dry-cleaning solvent, a degreaser, and as fumigant.

tetraethyllead (TEL) $Pb(CH_2CH_3)_4$, a compound formerly added to gasoline to increase its octane rating and to reduce knocking (to make it burn more smoothly). Gasoline containing this antiknock additive is called leaded or ethyl gasoline. Its use has been discontinued because it resulted in high concentrations of lead (a toxic heavy metal) in areas with high vehicle traffic.

tetrahedrite A sulfide mineral forming tetrahedral crystals in hydrothermal deposits. Tetrahedrite, also known as gray copper, may be found with deposits of copper, antimony, and silver.

tetraploid Possessing four times the haploid number of chromosomes characteristic of a particular species. Some ornamental plants such as daylilies have been chemically induced to be tetraploid. See POLYPLOID, HAPLOID, DIPLOID.

tetrapod 1) A quadruped; an organism that has four feet. 2) A member of the Tetrapoda superclass of vertebrates, those having four limbs. Tetrapods include amphibians, reptiles, mammals, and birds (but not fish).

tetrapoda A general name having no taxonomic status, for four legged vertebrates (e.g., amphibians, reptiles, birds, and mammals). In some forms such as snakes, limbs may be missing or degenerate vertebrata.

Teuthoidea   See SQUIDS.

textured vegetable protein   (TVP) Soybeans that have been processed to change their texture to resemble various kinds of meat (usually ground beef). It is used to produce meatless processed foods that resemble dishes usually prepared with meat, and is added to meat as a high-protein extender to lower the fat or the cholesterol content of foods.

texture, rock   A description of size, shape, and arrangement of minerals in a rock. See APHANITIC, PHANERITIES, PORPHYRITIC TEXTURE.

Th   Chemical symbol for the element thorium. See THORIUM.

thalidomide   (N-phthalimido) glutarimide, a compound once used extensively as a sedative and a sleeping pill. Its use was discontinued in the early 1960s because it is teratogenic, causing extreme malformation of limbs in fetuses exposed to it early in their development.

thallium   (Tl) A soft, metallic element with atomic weight 204.4 and atomic number 81. Thallium compounds are very toxic and used in insecticides; thallium sulfate is a rodenticide.

thallus   A plant body that is not divided into leaves, stem, and roots and that is lacking a true vascular system, as in lichens. It may consist of a single cell or a colony of cells, as in algae; a mycelium or collection of packed mycelia, as in mushrooms; or a complex and branched multicellular structure.

thalweg   A line connecting the deepest points in a series of cross-sections through a valley or river channel.

thanatocoensis   A fossil assemblage representing a group of organisms that were deposited together by the same currents but that did not necessarily live together in the same environment.

theca   A sac, spore case, or covering in an organism, often used as another term for an animal capsule or a lorica. See CAPSULE, LORICA.

thecodont   Describing animals whose teeth are embedded in sockets within the bone. Compare ACRODONT, BUNODONT, LOPHODONT, SELENODONT.

theodolite   An instrument used to measure horizontal and vertical angles in surveying land.

theory   A scientific hypothesis that has withstood several experimental tests and so has a good probability of being a true explanation for a given phenomenon. A theory is an explanation that appears to be true but has not conclusively been proven. Compare HYPOTHESIS, LAW.

therm   Any of several different units for heat or energy. Therm is used for calorie, kilogram calorie, and even 1000 kilogram calories. In British usage, therm is also used for a unit of energy equal to 1.06×10^8 joules or 10^5 Btu, used in marketing natural gas.

thermal   1) Of or relating to heat. 2) A localized, ascending current of air. Thermals are caused when ground heated by the sun heats a region of air above it; as this air warms, it rises. Thermals are often found near mountain ridges, where glider planes, hang gliders, and hawks may be seen floating on them.

thermal conduction   Transmission of heat energy through molecular contact. In gases and liquids, the transfer occurs as colliding molecules and atoms encounter those with less kinetic energy. In metals, fast-moving electrons accomplish the heat transfer. Compare CONVECTION, ELECTRICAL CONDUCTION, RADIATION.

thermal convection See CONVECTION.

thermal energy 1) Internal energy, the sum of kinetic and potential energy of the molecules in a system. 2) Energy derived from a naturally occurring, large difference in temperature, as between the top, sun-warmed layers and the cold bottom layers of a body of water. OTEC harnesses thermal energy for electricity production. See OCEAN THERMAL ENERGY CONVERSION, SOLAR POND.

thermal enrichment An enhancement in an underwater ecosystem caused by a small increase in water temperature. Excessive temperature increases become destructive. See THERMAL POLLUTION.

thermal inversion A reversal of the normal atmospheric temperature gradient, in which a layer of cooler air is trapped near the ground by a layer of warmer air sitting at a higher elevation. Thermal inversions interrupt the normal air circulation patterns and can lead to unusually high levels of air pollution. Also called a temperature inversion. See INVERSION LAYER, HIGH-LEVEL INVERSION, SUBSIDENCE INVERSION.

thermal metamorphism See CONTACT METAMORPHISM.

thermal neutral zone See THERMO–NEUTRAL ZONE.

thermal oxide reprocessing plant (THORP) A facility that recovers usable fissionable material (U-235 and plutonium) from spent reactor fuel. A THORP facility is located in Sellafield in the United Kingdom.

thermal plume A current of hot water discharged into a natural body of water from a power plant or other facility. Thermal plumes result when water from an outside source is used as a coolant in a facility and is not cooled to its original temperature before it is released back into the original body of water. Cooling towers reduce the temperature of such water and thus lessen the damaging effects of thermal plumes, or eliminate entirely the need for an external water source.

thermal pollution An increase in temperature of a body of water, as from a thermal plume, that damages the aquatic ecosystem. Thermal pollution of a body of water can be reduced or eliminated by using a cooling tower. Severe thermal pollution caused by sudden, intense changes in water temperature is also called thermal shock. Thermal pollution can also refer to changes in localized weather patterns caused by the emission of hot stack gases. Compare THERMAL ENRICHMENT.

thermal shield The inner shield of a nuclear reactor. It is located on the inside of a reactor vessel to protect the vessel walls (and the biological shield surrounding the reactor vessel) from excess heat generated by the reactor core.

thermal shock The negative ecological effects produced by a sudden, dramatic change of temperature in an aquatic ecosystem, as when large quantities of hot water produced by an industrial process are released into a small river. Large die-offs of fish and other aquatic organisms can result.

thermal soaring See SOAR.

thermal stratification The development of distinct temperature layers in temperate lakes, which results from the density-temperature relations of water. Often a thermocline, a zone of rapidly decreasing temperature and increasing density, separates the warm, less-dense surface layer (epilimnion) from the cold, denser water below (hypolimnion). See THERMOCLINE, EPILIMNION, HYPOLIMNION.

thermionic valve An electronic valve in which a cathode is heated to produce thermions, electrons, or other particles with electrical charge given off by a heated body. Also called a thermionic tube. Thermionic diodes have been replaced by semiconductor diodes; triodes (thermionic valves with three electrodes, the tubes of old radios) have been replaced with transistors.

thermocline A zone of rapidly decreasing temperature and increasing density in a lake or pond. The thermocline develops during the summer and separates the warm surface layer (epilimnion) from the cold bottom water (hypolimnion). See EPILIMNION, HYPOLIMNION, VERNAL OVERTURN.

thermocline circulation The vertical currents induced by thermal convection within a body of water.

thermodynamics The branch of physics devoted to the study of heat and its conversion into or from other forms of energy. Thermodynamics includes studies of the maximum efficiency with which heat can be converted to other, more useful forms of energy. See FIRST LAW OF THERMODYNAMICS, SECOND LAW OF THERMODYNAMICS, THIRD LAW OF THERMODYNAMICS.

thermogenesis Generating body heat through increased metabolic rate in response to a drop in temperature.

thermograph A recording thermometer; a device that makes a continuous recording of temperature. Also, the chart produced by such an instrument.

thermokarst A surface landscape of shallow pits and depressions caused by thawing of permafrost in periglacial environments.

thermonasty A reversible or permanent plant movement in response to temperature changes. Some tulips and crocuses open or close their flowers when stimulated by very small changes in temperature; the change from horizontal to near-vertical leaf orientation seen in rhododendrons at extreme cold temperatures is also thermonasty. See NASTIC RESPONSE.

thermoneutral Describing a chemical reaction in which no net energy (such as heat) is either given off or absorbed from the surroundings. Contrast ENDOTHERMIC, EXOTHERMIC.

thermoneutral zone The temperature range over which an endothermic organism is best adapted for living; those temperatures that require the least metabolic exertion to maintain normal body temperature.

thermonuclear reaction A nuclear fusion reaction induced by very high temperatures, as in fusion power research or in a hydrogen (thermonuclear) bomb. Thermonuclear reactions also occur inside stars such as the sun. See FUSION, NUCLEAR ENERGY.

thermophilic Describing organisms that require or tolerate very high temperature environments. Compare PSYCHROPHILIC, MESOPHILIC.

thermoplastics A large group of soft plastics capable of being melted, reshaped, and cooled over and over again without changing the properties of the material. Thermoplastics are thus good candidates for recycling. They include vinyl polymers, polyamides (Nylon), polyester, acrylic resins, and polystyrenes. Compare THERMOSET POLYMERS.

thermoregulation   The control of metabolic rate in order to maintain a constant body temperature. See ENDOTHERM, THERMOGENESIS.

thermoremanent magnetism   (TRM) Magnetism resulting from a substance that was formerly molten but has cooled past its Curie point. Thermoremanent magnetism is responsible for the magnetic properties of many rocks. See CURIE TEMPERATURE.

thermoset polymers   A group of hard plastics not capable of being remelted; their cross-linked molecular structure is destroyed by being subjected to additional heat. Chemical reactions are initiated by their original heating during molding when they are produced; these reactions determine their character but are destroyed upon further heating. Thermoset polymers and plastics include urea formaldehyde and phenol formaldehyde (Bakelite, Formica). Compare THERMOPLASTICS.

thermosphere   Another name for the earth's ionosphere. See IONOSPHERE.

thermotaxis   An organism's response to change in temperature by moving. See TAXIS.

therophyte   An annual; a plant that grows for only one season. Therophytes can be said to overwinter or avoid drought as seeds. The term comes from the Raunkiaer life form classification system. See RAUNKIAER'S LIFE FORMS.

thiamine $C_{12}H_{17}ClN_4OS$, a water-soluble vitamin in the B complex, formerly called vitamin B_1. It is essential for metabolizing carbohydrates and proteins to release energy, and indirectly involved in the metabolism of fats. A severe thiamine deficiency causes beriberi; this disease has become prevalent in the Orient since Western traders in the 19th century introduced polished rice, replacing brown rice as the dietary staple. Also spelled thiamin. See BERIBERI.

thigmotaxis   See STEREOTAXIS.

thigmotropism   The tendency for a plant (or an attached animal, one not capable of traveling) to bend or to turn in response to being stimulated by touching a solid body or hard surface. The curling and growth of a tendril around a support is thigmotropism. Also called haptotropism and stereotropism.

thin   A forestry management technique of removing some trees from a stand, usually the smaller or misshapen individuals and those species that are not marketable. This reduces the competition for the remaining trees, allowing them to grow larger for harvesting. See CULL.

thinning line   The line resulting from graphing the log of the mean individual weight (y-axis) against the log of the density (x-axis). Self-thinning populations of a sufficiently high biomass tend to progress along this line, which for many species has a slope of - 3/2. See 3/2 THINNING LAW.

thin section   An extremely narrow cross-section cut from a sample of tissue or from rock for mounting on a glass slide and examining under the microscope. Rocks and minerals for sectioning are ground to a thickness of about 0.03 millimeters, and used for studying their optical properties.

thiosulfate   1) The thiosulfate ion, $S_2O_3^{-2}$.

2) A salt (or ester) of thiosulfuric acid, $H_2S_2O_3$. Thiosulfate is frquently used as a reducing agent.

third law of thermodynamics   For systems consisting of perfect crystals at absolute zero, the total entropy of the system is zero. This law creates a reference point for a scale of entropy values; as the temperature of a substance approaches absolute zero, its entropy will also approach zero. Compare FIRST AND SECOND LAWS OF THERMODYNAMICS.

thixotropic   The physical property describing materials that change from a gel to a liquid state while under shear stress. Some clay-rich soils are thixotropic.

tholeiite   An abundant variety of basalt in which calcium plagioclase is the dominant mineral. Tholeiite forms from the lava of midocean spreading centers or from flood basalt flows.

Thomson effect   Heat is produced or absorbed when electrical current is run through a conductor whose two ends are kept at different temperatures. The larger the difference in temperature, the greater the heat produced or absorbed. Also called Kelvin effect.

thorium   (Th) A radioactive, metallic element with atomic weight 232.0 and atomic number 90. It is added to some alloys to strengthen them, to produce uranium fuel for nuclear reactors, or in breeder reactors.

THORP   See THERMAL OXIDE REPROCESSING PLANT.

thread cell   See NEMATOBLAST.

threatened species   Species with populations that are declining sharply in parts of their range, and which may be in danger of becoming extinct in specific areas, as a result of direct or indirect actions by humans. Threatened is a classification between rare and endangered. Sometimes called vulnerable species.

3/2 thinning law   The observed general rule that the average weight of a mature plant varies inversely with the thickness at which the plant is seeded, raised to the 3/2 power. Also known as Yoda's law.

threonine CH_3-$CH(OH)$-$CH(NH_2)$-$COOH$, an amino acid found in plant and animal proteins. It is one of the essential amino acids required in the human diet.

threshold   1) The point or level at which a given stimulus or variable begins to have an effect. 2)The minimum energy required to cause a physical process to take place. See THRESHOLD LEVEL.

threshold dose   The minimum level of radiation or of a drug needed to produce a desired physiological effect. Sometimes also used for the maximum level of radiation, of a drug, or of a pollutant that produces no noticeable negative effects. Also called the tolerance dose.

threshold effect   A detrimental or even fatal effect resulting when a small change in environmental conditions pushes an organism or species past its limit of tolerance. For example, a stream-dwelling organism may easily adjust to a 5° increase in water temperature from 55° to 60°, but if the organism rarely survives at temperatures exceeding 80°, then the same small 5° increase of water at 77° can be enough to kill the organism. See TOLERANCE LIMITS, LAW OF TOLERANCE, THRESHOLD LEVEL.

threshold level   The minimum level of a value required to cause a particular response in a system. See THRESHOLD DOSE, THRESHOLD OF AUDIBILITY, THRESHOLD OF PAIN, THRESHOLD

EFFECT, TOXIC THRESHOLD, TRANSMISSION THRESHOLD.

threshold of audibility   The lowest level (of sound intensity or sound pressure) at which a sound becomes audible to a human ear at a given frequency.

threshold of pain   1) Minimum sound pressure level or intensity at which a sensation of discomfort or pain is registered in the ears of the average human listener. It is generally considered to fall between 130 and 140 dB. 2) The point at which any stimulus starts to cause a sensation of pain.

threshold shift   A temporary or permanent change in a person's threshold of audibility, as from sustained exposure to loud noises. See THRESHOLD OF AUDIBILITY.

thrombocyte   Former term for blood platelet.

throughfall   The movement of precipitation intercepted by a plant canopy to the ground, carrying substances washed off and out of plant surfaces. Throughfall is a component of leaching transfers in terrestrial nutrient cycles. Compare STEM FLOW. See NUTRIENT CYCLES.

throw   1) A measure of the vertical displacement between fault blocks. 2) The total distance traveled by a crank or a similar structure, also called the stroke.

thrust fault   A type of low-angle reverse fault in which the hanging wall overrides the footwall along a dip-slip face that is angled at less than 45°. Thrust faults are typical in regions of tectonic convergence.

thrust sheet   A body of rock of which the hanging wall block or a thrust fault is composed. See NAPPE.

thulium   (Tm) A metallic element with atomic weight 168.9 and atomic number 69. Thulium is one of the rarest elements. One of its isotopes (thulium 170) is used as a source of x-rays in portable radiography units.

thymine   5-methyl-2,6-dioxytetrahydro-pyrimidine, a pyramidal base that is one of the major components of DNA. Thymine pairs with adenine; both bases are found only in DNA. See DNA.

thymine dimer   A DNA molecule in which two adjacent thymine bases on the same strand of DNA are bound together. Formation of thymine dimers is a mutation of the DNA structure and commonly occurs as a result of ultraviolet radiation of DNA. This mutation stops DNA replication.

thyroid gland   Endocrine gland of vertebrates, located in the neck below the larynx in humans. It secretes two hormones (thyroxine and triiodothyronine) that control the metabolic rate of body processes and therefore exert an influence on physical development, especially nervous system development. In amphibians, the thyroid controls metamorphosis. See THYROXINE.

thyroxine   One of two hormones produced by the thyroid gland. Thyroxine controls metabolic rate, conversion of food into heat, and the growth rate of children; it is administered to treat insufficient thyroid activity (hypothyroidism).

Ti   Chemical symbol for the element titanium. See TITANIUM.

tidal bore   A rapidly rising wave forming where tidewater is funneled into a shallow bay or an estuary.

tidal current   The periodic ebb and flow of ocean water in reversing horizontal motions because of gravitational interactions of the earth, sun, and moon.

tidal delta A deltaic deposit of sediment reshaped by the action of intertidal waters. See DELTA.

tidal energy Energy derived from the movement of tides. For example, water can be trapped at high tide, and then this water can power gravity-driven turbines at low tide. See TIDAL POWER.

tidal flat A flat-topped deposit of unconsolidated sand, silt, or mud that is exposed to the air at low tide.

tidal power Electrical energy generated using tidal energy. Tidal power plants are only feasible where local topography creates a large difference between high and low tides; the Nova Scotia Power Corporation has tidal power facilities, and another exists on the River Rance in Brittany, France.

tidal wave See TSUNAMI.

tide A natural fluctuation of ocean-water level caused by the rotation of the earth in combination with the gravitational forces of the earth, the moon, and the sun.

tied ridges A network of connected artificial ridges enclosing an area so that runoff is contained and allowed to soak into the ground.

tight fold A fold of bedrock that is nearly isoclinal, having the axial plane inclined at less than 10°. See ISOCLINE.

till See GLACIAL TILL.

tillage The cultivation of land, especially with a plow or disc harrow. See MINIMUM-TILLAGE FARMING, NO-TILL AGRICULTURE.

tiller 1) A lateral shoot that develops at the base of some plants, or from the axils of lower leaves; common in grasses. 2) A small machine for cultivating the ground, guided by a person rather than pulled behind a tractor.

tillite A sedimentary rock formed of lithified glacial till.

tiltmeter A surveying instrument used to measure the amount and rate of surface displacement measured with respect to a horizontal position. See CLINOMETER.

timberline The transition from subalpine wooded vegetation to treeless vegetation, especially alpine tundra. Timberline (also called treeline) occurs at higher elevations at low latitudes.

timbre Characteristic quality of a sound, caused by its distinctive combination of overtones (harmonics) plus the fundamental frequency. Timbre allows human ears to distinguish different musical instruments when they are playing the same note. See HARMONIC.

time-delay population model A model for population growth that describes the lags in predator population oscillations behind those of prey populations, often showing a lag of ¼ the length of the cycle.

time-rock unit In chronostratography, a sequence of identifiable rock layers for which an absolute age has been determined. Time-rock units are classified in hierarchical groups from erathems to the smaller stages, series, and systems.

time-stratigraphic unit A sequence of rocks formed within a specific interval of the geologic time scale.

tin (Sn) Metallic element with atomic weight 118.7 and atomic number 50. Tin is soft and easily shaped. Because it resists corrosion, tin is used in plating metals such as steel. Its low melting point makes it a useful addition to lead in soldering compounds; it is also used in alloys, including bronze and pewter.

tincture of iodine   A dilute (2 percent) solution of iodine and sodium iodide in dilute alcohol; it is a powerful disinfectant.

tissue   An aggregation of similar cells (with any intercellular substance) that performs one or more particular functions within an organism, such as vascular tissue or connective tissue. Organs are usually made up of several different tissues.

tissue culture   The growth or maintenance of tissues, cells, or even organs taken from living organisms in an artificial nutrient medium. Tissue culture is used for commercial propagation of some plants such as orchids. See SOMACLONAL VARIATION.

tissue growth efficiency   The ratio of energy in production to energy assimilated by an organism or trophic level, sometimes referred to as net growth efficiency. Endothermic organisms, because of their high respiration rates, tend to have lower tissue growth efficiencies than do ectothermic organisms. Compare ASSIMILATION EFFICIENCY, RESPIRATION EFFICIENCY, ECOLOGICAL GROWTH EFFICIENCY.

titanium   (Ti) A metallic element with atomic weight 47.90 and atomic number 22. It is used extensively in alloys because it is strong, lightweight, resists corrosion, and combines easily with almost all other metals. It is used as a paint pigment; it is a much less hazardous replacement for lead in white paint.

titration   A laboratory process in which one solution is slowly added to a known volume of another. The two undergo a reaction; the solution is added just until it can be seen that the reaction has reached comple-tion usually because of a color a change or similarly dramatic end point. If the concentration of one of the solutions is known, the concentration of the other can be calculated with accuracy from the known volume of one solution and the measured volume of the other solution.

Tl   Chemical symbol for the element thallium. See THALLIUM.

Tm   Chemical symbol for the element thulium. See THULIUM.

tocopherol   Any of four related alcohol compounds having some of the biological properties of vitamin E, which is alpha tocopherol. See VITAMIN E.

tokamak   A magnetic "bottle" in the shape of a torus (a doughnut), used to contain the hot plasma in nuclear fusion. The name is a Russian acronym for toroidal magnetic chamber. The Joint European Torus facility and Princeton's Tokamak Fusion Test Reactor use tokamaks. Compare INERTIAL CONFINEMENT.

tolerable erosion   An arbitrary measure in soil science of the calculated maximum allowable rate of soil erosion.

tolerance   See RANGE OF TOLERANCE.

tolerance limits   1) The range of environmental factors such as heat, salinity, or moisture that an organism can tolerate, or the levels at which harmful effects begin. 2) The maximum levels for food additives permitted by the U.S. Food and Drug Administration. Also called tolerance levels. See LAW OF LIMITING FACTORS, LAW OF TOLERANCE, RANGE OF TOLERANCE.

tolerance model   A model of succession that is intermediate between facilitation and inhibition models. It characterizes succession as a series of species with increasing competitive abilities and therefore more tolerant of thriving under conditions of limited resources. Contrast FACILITATION, INHIBITION.

tolerant 1) Relating to an organism's capacity to withstand specific environmental conditions. Calciphile plants are tolerant of highly alkaline soils. 2) Foresters and forest ecologists use the term in reference to plants that are able to live in shady environments. See RANGE OF TOLERANCE, SHADE-INTOLERANT SPECIES, SHADE-TOLERANT SPECIES.

toluene diisocyanate (TDI) CH_3-C_6H_3-$(NCO)_2$, the most common form of isocyanate used to produce polyurethane foam.

tombolo A bar of sand or of gravel that joins an island to a coastline.

tonalite A type of quartz diorite in which hornblende and biotite are the principal mafic minerals.

tongue The movable, fleshy organ in the mouth of most vertebrates; it is the main organ of taste and assists in the chewing and swallowing of food. It is also essential for normal human speech.

tonoplast A thin membrane that surrounds each vacuole in plant cells. Unlike most biological membranes, it is a single rather than a double-layer membrane. See VACUOLE.

tooth shells 1) Members of a class of marine molluscs (Scaphopoda) having tubular shells that are open at each end. 2) The shells produced by such organisms. See SCAPHOPODA.

top carnivore A predator at the top of a food chain or food web, such as eagles and other raptors. See TERTIARY CONSUMERS.

topographic map A two-dimensional representation of the earth's surface used to illustrate topographic relief by the use of contour lines. The shape and spacing of contour lines portrays the size, shape, and elevation of landscape features. See CONTOUR.

topographic profile A representation of a vertical cross-section of a topographic transect.

topographic relief See RELIEF.

topography A description of the size, shape, and elevation of the land surface.

toposequence A description of an array of soils according to their relative positions on a hillside.

topotype 1) A specimen collected from the same locality as the original type specimen used in defining the species. 2) A population of species that has become differentiated from other populations of the same species because of adaptations to local geographical features. See TYPE SPECIMEN.

topset layer A horizontal or subhorizontal layer of fine-grained clastic particles deposited on the top of a prograding delta. See DELTA.

topsoil A general term for the upper part of a soil, especially the A horizon. The topsoil is valued as an essential medium for plant growth.

tornado A tube of air in violent, cyclonic rotation that extends downward from a cumulonimbus cloud. Although never measured, wind speeds within the tornado vortex are estimated to be 100 to 300 mph, making these the most destructive localized weather phenomena. Also called a twister or cyclone. See WATERSPOUT.

torpor   Sluggishness; lack of activity, as in rest or sleep. Also called torpidity. Contrast HIBERNATION.

torque   Informally, a twisting force. Formally, the force times the effective lever arm between the point of application of the force and the axis of rotation (torsion). Also, the measurement of such a force in pound-feet or newton-meters.

torsion   1) The act of twisting, or the state of being twisted (subjected to torque). 2) The restoring or resisting torque from the object being twisted.

tort   A wrong; a private or civil wrong or injury resulting from a breach of a legal duty that exists by virtue of society's expectations regarding interpersonal conduct rather than by contract.

tort law   Body of common law governing situations when a tort has been alleged by one of the parties. See TORT.

tortoises   A family of terrestrial turtles characterized by the presence of moderately to highly domed carapaces and elephantlike feet. They are often found in extremely arid environments. All 39 species are threatened or endangered. The giant tortoises of the Galapagos Islands, which can weigh up to 400 pounds, have been subjects of a very successful captured breeding and resettlement program, in which great care has been taken to maintain the genetic identity of the different, island subspecies. See CHELONIA.

torus   A doughnut-shaped surface, especially a magnetic confinement system with such a shape used for nuclear fusion. Torus designs used in fusion facilities include the Elmo Bumpy torus and the tokamak.

total alkalinity   See STANDARD ALKALINITY.

total dissolved solids   (TDS) A measure of the total quantity of dissolved substances contained in water or effluent, including organic matter, minerals, and other inorganic substances. Advanced sewage treatment removes some dissolved (and suspended) solids, because water containing excessive levels of dissolved solids is unfit for industrial use and is inferior for use as drinking water. See SUSPENDED SOLIDS, DISSOLVED ORGANIC CARBON.

total fertility rate   An estimate of the number of living offspring a woman will produce during her lifetime.

total organic carbon   (TOC) All of the carbon atoms present in wastewater or other water that are bound into organic molecules; carbonate, bicarbonate, and dissolved carbon dioxide are excluded from this category because they are considered inorganic forms of carbon. TOC is a measure of the organic matter present in water, used along with other measurements to determine water quality. Although it can sometimes be used to estimate biochemical oxygen demand or chemical oxygen demand, it does not replace testing for those quantities. TOC includes nondissolved organic carbon (NDOC, all particulates trapped by a filter with pores 0.45 micrometers in diameter) and dissolved organic carbon (DOC, what remains after NDOC is removed). See BIOCHEMICAL OXYGEN DEMAND.

tourmalinization   The pneumatolytic replacement of igneous rock with crystals of tourmaline. See PNEUMATOLYSIS.

tow   1) A collection of plankton captured in a net that is towed behind a boat or cast into the water. 2) The coarse or short (broken) fibers of flax, hemp, or jute (or, less commonly, synthetic fibers). Also, a cloth made from such fibers.

toxic   Poisonous; caused by, producing, or relating to a poison or toxin.

toxicant   A poisonous substance or agent, especially one used to kill rather than repel pests.

toxicity   1) The potency of a poisonous substance, the degree to which it is harmful to organisms. 2) The amount of poison found in a substance or produced by an organism. See LD_{50}.

Toxic Substances Control Act   (TSCA) A 1976 federal law controlling the output of toxic substances, their transportation between locations, and their disposal. It requires industries to notify the Environmental Protection Agency of the storage, transport, and disposal of any hazardous materials. It also requires the screening of new compounds for safety before they are approved for general use (15 U.S.C.A. §2601 et seq.).

toxic threshold   The level at which a substance causes poisonous effects or death. See LD_{50}.

toxic waste   Hazardous waste that is capable of causing severe injury (burns, tissue damage, cancer) or death in humans or other organisms. See HAZARDOUS WASTE.

toxins   Poisons (usually proteins) produced by living organisms, especially those capable of stimulating the production of antibodies. Botulism and tetanus are both caused by bacterially produced toxins.

trace element   An element considered critical to the proper functioning of an organism but only required in small amounts and often toxic in high concentrations. The most important trace elements are iron, manganese, zinc, copper, iodine, cobalt, selenium, molybde-num, chromium, and silicon. Also called micronutrient. Compare ESSENTIAL ELEMENTS. See MICRONUTRIENT.

trace fossil   A structure in sedimentary rock showing evidence of the behavioral activity of living organisms. Tracks, burrows, and nests are examples of trace fossils.

tracer   1) A radioactive isotope introduced into a chemical or biological system under study, whose progress through the system can be monitored. Tracers are isotopes that behave identically to elements routinely metabolized, but whose radioactivity labels them for easy monitoring. Tracers have revealed a great deal about how compounds are translocated throughout plants; they are also used in medical diagnosis. Nonradioactive isotopes can be used if they show up clearly in magnetic resonance imaging (MRI). 2) A mineral in soil, stream gravel, or rock deposits whose presence indicates a good probability for the presence of another mineral of economic significance. See MAGNETIC RESONANCE IMAGING.

trachea   1) Windpipe; the tube that carries air to and from the lungs in terrestrial vertebrates, or one of the air-carrying tubes in the respiratory system of insects and some other arthropods. 2) Formerly used for some primary xylem elements (modified tracheids) in plants, because of their resemblance to insect trachea.

tracheid   One of the fundamental structures in plant xylem. An elongated cell with perforated walls that has lost its living contents, occurring in woody plant tissues. It conducts water and dissolved minerals and contributes to the support of the plant. See XYLEM, VESSEL ELEMENTS.

trachyte   An intermediate extrusive igneous rock composed of more than 65 percent alkaline feldspar and with the remaining portion consisting of mafic minerals such as hornblende and biotite.

track   1) Path followed by a subatomic particle, especially the visible path recorded in a cloud chamber or a similar device. 2) Path on which electronic information is stored, as on an audiotape or a computer disk. See CLOUD CHAMBER.

trade cumulus   Cumulus clouds that often form in the belts where the trade winds blow, and usually associated with these winds.

trade winds   Easterly prevailing winds that blow toward the equator from latitudes 30° north and south of the equator. Also called the northeast trades and southeast trades. The trade winds get their name because they propelled the ships that explored the Americas and carried cargo-bearing ships to the Americas from Africa and Europe.

tragedy of the commons   See COMMONS, TRAGEDY OF THE.

trait   A stable characteristic or physical feature that is particular to an organism, varying from one individual to another. Eye color and flower color are traits.

transad   Closely related species that were separated at some point in their mutual development by a physical barrier, such as an ocean forming from plate tectonics. The North American caribou and European reindeer are transads.

transduction   The conversion of an electrical impulse into another form of signal, or the reverse process. Transduction occurs within a microphone or stereo speaker.

transect   A term applied to a reference line in any field survey procedure. A straight-line sequence of plot surveys is said to be conducted along a transect. Transects are

particularly useful in studying the changes in vegetation across a particular gradient. Compare QUADRAT.

transfer efficiency   The percentage of energy that is passed along through various stages in a food web, or through the different trophic levels of a community. See FOOD CHAIN EFFICIENCY.

transfer RNA   A form of RNA that assembles the amino acid building blocks needed for protein synthesis and delivers them to the ribosomes, where they will be assembled. Each amino acid has its own transfer RNA. Compare MESSENGER RNA.

transformation   A genetic alteration caused by the incorporation into a cell of DNA from another cell or from a virus. Transformation can occur in eukaryotic cells as well as in bacteria.

transformation series   A series of homologous characteristics in sequence of their evolutionary development.

transform fault   A type of strike-slip fault that forms between offset sections along the axis of a midoceanic ridge. The sense of displacement across the transform fault is opposite to the sense of displacement of the offset sections. A transform fault usually crosses the midoceanic ridge at a right angle and the strike of the fault aligns with the direction of sea floor spreading.

transfrontier pollution   Pollution transported across political boundaries by natural forces such as wind, rivers, etc.

transgenic   Describing animals produced by one form of genetic engineering. Isolated DNA for genomes from another species are introduced into embryos at an early stage of development. The foreign genes may become incorporated into the nucleus and chromosomes of the embryo, creating an animal with some of the genetic characteristics of the donor species. See GENETIC ENGINEERING.

transgression See MARINE TRANSGRESSION.

transient polymorphism A temporary occurrence of two or more forms of a gene, or an entire organism, within a population as one is gradually replaced by the other.

transition element Another term for transition metal. See TRANSITION METAL.

transition metal Any of the elements in Groups 3-12 in the periodic table. Transition metals have partially filled d-orbitals, or readily form ions with partially filled d-orbitals. Transition metals typically exhibit a variety of oxidation states (e.g., Fe^{+2}, Fe^{+3} are both common), and thus form many complex ions and compounds. Copper, titanium, molybdenum, platinum, and zinc are all examples of transition metals. Also called transition elements. See APPENDIX p. 608.

transition temperature 1) The temperature at which a material becomes a superconductor. 2) Any temperature at which a major change occurs, such as a change of phase or magnetic properties (Curie point).

translation (slab) landslide A type of mass wasting in which the body of material in transport slides as a single mass down a steeply sloping surface. See MASS WASTING.

translocated herbicides Weed-killing compounds that work by absorption through plant leaves; they are then conducted to other tissues and eventually kill the entire plant. Compare CONTACT HERBICIDE, SOIL-ACTING HERBICIDES.

translocation 1) The movement of nutrients or metabolites from one location to another within a plant. 2) The transfer of part of a chromosome from one position to another in the same or in another chromosome.

transmissible disease An infectious disease that can be spread directly from one person to others.

transmission loss 1) A measure of the effectiveness of sound-insulating materials, calculated from the difference in sound levels on either side of the material. 2) Power lost in a line transmitting electricity or in audio signals.

transmission threshold The population level of a disease organism (or parasite) required in order for it to spread.

transparency 1) The degree to which a substance allows the passage of light rays or other forms of radiant energy; the opposite of opacity. The earth's atmosphere shows transparency to sunlight, but blocks infrared radiation. 2) The ratio of intensity of transmitted light to intensity of incident light, also called transmission ratio.

transpiration 1) Loss of water vapor through the stomata (gas-exchange pores) in plants and, to a lesser extent, by evaporation from cell tissue. 2) The movement of any vaporized liquid or gas across a membrane or similar surface, such as perspiration through animal skin. See EVAPOTRANSPIRATION.

transpiration efficiency The amount of plant tissue produced (biomass, measured in grams) per thousand grams of water transpired. Compare TRANSPIRATION RATIO.

transpiration ratio The amount of water

transpired per pound of plant tissue produced; it is the reverse of transpiration efficiency. Compare TRANSPIRATION EFFICIENCY.

transuranic elements   Artificial radioactive elements with atomic numbers greater than 92 (the atomic number of uranium). They can be produced from uranium by neutron bombardment; all are unstable. Neptunium, plutonium, americium, curium, berkelium, californium, einsteinium, fermium, mendelevium, nobelium, and lawrencium are all transuranic elements.

transverse dune   A type of dune in which the longest axis is oriented at an approximate right angle to the prevailing wind.

trap   1) A device for capturing animals, used either for pest control or as an alternative to hunting. 2) Formerly used for basalt or any other finely grained rock with a sheetlike structure, used especially in road-making. Also called traprock.

trapping technique   Pest control in which target species are lured into traps, often by use of sex attractants. An example is Japanese beetle traps that incorporate sex lures.

traveling wave   Wave in which the medium carrying the wave moves with, and gains energy from, the wave. Unlike stationary waves, traveling waves continuously radiate energy away from their source. See P WAVE, SHEAR WAVES.

travertine   A mineral deposit of calcium carbonate that forms by precipitation at a hot springs or from the calcium-rich water on cave dripstones. A porous form of travertine is known as tufa.

tree   A perennial plant that usually grows taller than four to five meters, with a single woody trunk that lacks lower branches but which supports branches well above the ground. Generally, it is a plant capable of attaining a diameter of 10 cm (4 inches) at breast height.

tree farm   A stand of trees planted and managed for commercial harvest, as for Christmas trees or for pulpwood. Tree farms are usually devoted to even-aged stands of a single species, or only a few species.

tree layer   See CANOPY.

tree line   See TIMBERLINE.

tree ring   See GROWTH RINGS.

tree-ring dating   See DENDROCHRONOLOGY.

trellis drainage   The map view appearance of a drainage system characterized by a regular pattern of elongated tributary streams that flow parallel to the main stream. Each tributary approaches the main stream by a short reach oriented at right angle to the main stream. A trellis drainage pattern often forms above openly folded or tilted strata.

Trematoda   Class of the phylum Platyhelminthes whose species are exclusively parasitic and commonly known as flukes. Adults carry structures such as suckers and hooks for attachment to host organisms and are usually hermaphroditic. The class is divided into two distinct groups, the Monogenea, whose members are mainly fish ectoparasites and have simple life cycles, and the Digenia, all species of which have complex life cycles including two or more hosts and most of which are endoparasites of vertebrates. This group is responsible for many diseases of humans, such as schistosomiasis and bilharzia. See PLATYHELMINTHES, SCHISTOSOMIASIS, BILHARZIASIS.

trespass   1) A form of action in common law, instituted to recover damages for any unlawful injury to the plaintiff's person, property, or rights, involving immediate force or violence. 2) A wrongful disturbance of, or interference with, the possession of property.

triage   A method for categorizing degree of urgency in order to allocate a limited resource. The term originated on the battlefield, where the wounded were categorized as those who would live without treatment, those who would probably die even with treatment, and those who would survive only if given prompt treatment. When medical personnel or supplies are limited, they concentrate on the latter group. Triage has been extended to other situations, such as rescuing animals from oil spills or theorizing how to distribute aid to developing countries.

Triassic period   The first of three geologic time periods of the Mesozoic era. The Triassic period lasted from approximately 248 to 213 million years ago. See APPENDIX p. 610.

tricarboxylic acid cycle   (TCA cycle) A repeating series of metabolic reactions that are essential for energy production and synthesis of important molecules in the body. The cycle changes dicarboxylic and tricarboxylic acids from one form to another and back to oxidize acetyl coenzyme A and produce ATP (adenosine triphosphate). More commonly known as the Krebs cycle. See ACETYL COENZYME A.

trichloroethane H_3C-CCl_3, methyl chloroform. A nonflammable solvent of low toxicity used in industry for cleaning electrical equipment. It is much less hazardous than carbon tetrachloride (tetrachloromethane).

trichloromethane   Another name for chloroform. See CHLOROFORM.

trichogyne   A hairlike outgrowth of the female reproductive organ, used for the reception of the male gamete, in red algae and a few green algae as well as some lichens and fungi.

trichomes   Plant hairs; outgrowths from the epidermis of plants that contain one or more cells but not vascular tissue.

trickling filter   A device used for secondary sewage treatment. The filter consists of a bed of crushed stone containing aerobic bacteria. As wastewater seeps through the stones, the bacteria break down organic wastes into harmless components. Compare ACTIVATED SLUDGE PROCESS.

trigonal   A variety of crystal symmetry characterized by one axis of threefold symmetry intersected at right angles by three axes of twofold symmetry. A crystal within the trigonal system forms a six-sided rhombohedron that may be viewed as three separate pairs of parallel faces.

triiodothyronine $HOC_6H_3IOC_6H_2I_2$-$CH_2CH(NH_2)COOH$, an amino acid produced by the thyroid gland. It is one of two forms of the gland's principal hormone. Natural or synthetic triiodothyronine is used to treat hypothyroidism (insufficient thyroid production).

triple bond   A chemical bond between two atoms that is covalent and consists of three shared electron pairs. Triple bonds are found in some unsaturated carbon compounds such as acetylene. See COVALENT BOND, DOUBLE BOND, SINGLE BOND.

triple junction   The point of intersection between three separate tectonic plates of the lithosphere. Triple junctions occur in a variety of configurations determined by a combination of midoceanic ridges, oceanic trenches, and transform faults.

triple point   The temperature and pressure at which a substance can exist simultaneously in three phases (solid, liquid, gas) at equilibrium. The triple point of water occurs at 273.16 K and 610 N m^{-2} in a sealed vacuum; at this temperature and pressure, ice, water, and water vapor exist in equilibrium.

triple superphosphate   A high-grade, very concentrated form of superphosphate produced by treating phosphate rock with phosphoric acid rather than with the sulfuric acid used to produce ordinary "single" superphosphates. Also called treble superphosphate, this compound is increasingly replacing superphosphate in concentrated fertilizers. See SUPERPHOSPHATE.

triploblastic   Describing animals with three tissue types or germ layers: the ectoderm, the endoderm, and the mesoderm. Vertebrate embryos are triploblastic. Compare DIPLOBLASTIC.

triploid   Possessing three times the haploid number of chromosomes that is characteristic of the species. See HAPLOID, DIPLOID, TETRAPLOID, POLYPLOID.

trisomic   Having three of one chromosome type rather than the usual two; having a normal complement of chromosomes except for an extra chromosome attached to one pair. Down's syndrome is produced by trisomy of chromosome number 21.

tritium   (T, also $^{3}_{1}$H) A very rare, unstable radioactive isotope of hydrogen that is three times as heavy as ordinary hydrogen (its atomic weight is 3.022) because it contains two neutrons and one proton. Tritium is used as a tracer for labelling aqueous compounds in medical research. It

is also used in nuclear fusion research and particle accelerators. See DEUTERIUM, PROTIUM.

trochophore   A cilia-covered, free-swimming pelagic larva characteristic of most molluscs and of some segmented marine worms, bryozoans, and brachiopods. In molluscs, the trochophore develops into a veliger. See VELIGER, MOLLUSCA, POLYCHAETA.

trombe wall   An interior wall that serves as a heat storage device, such as a wall of water-filled columns.

trophallaxis   Reciprocal feeding; the mutual exchange of food between organisms such as social insects, especially between adults and larvae.

trophic   Relating to nutrition or food.

trophic level   A stage in a food web occupied by organisms that feed on the same general type of food, used in diagramming the energy flow within an ecosystem. Carnivores, herbivores, and plants constitute different trophic levels.

trophic level efficiency   The ratio of production by a trophic level to the production by the previous trophic level, as in the ratio of production of herbivores to that of primary producers. Compare ECOLOGICAL EFFICIENCY, LINDEMAN EFFICIENCY, CONSUMPTION EFFICIENCY, UTILIZATION EFFICIENCY.

trophic structure   The energy flow of an ecosystem as illustrated by the feeding relationships in food chains and food webs.

trophobiont   An organism participating in trophobiosis. See TROPHOBIOSIS.

trophobiosis   A form of symbiosis in which different species feed each other, such as ants that feed plant roots to aphids living in their nests, and in turn feed off of the honeydew that the aphids secrete.

trophogenic zone   The region within a body of water in which photosynthesis is occurring at a rate sufficient to create organic material. See EUPHOTIC ZONE.

tropical   1) Of or relating to warm, equatorial climates. 2) Referring to the region between 23°27' north latitude and 23°27' south latitude in which such climates are found. 3) Referring to the vegetation of such regions. Compare TEMPERATE.

tropical climate   A general term for any climate with year-round high temperatures and high annual precipitation. Such conditions are typical of equatorial and tropical regions.

tropical cyclone   An air mass with cyclonic circulation, originating over a tropical ocean and developing high winds and torrential rains. Wind speeds of tropical cyclones may exceed 300 km/hour (190 mph); hurricanes are well-known examples.

tropical rainforest   A forest biome found in the tropics near the equator, in climates with continually warm to hot weather and very heavy rainfall. Tropical rainforests are green all year long, characterized by high species diversity, soils that are usually old and nutrient-poor, stratification into distinct layers of trees and shrubs, and many epiphytes and vines. The largest tropical rainforest is in the Amazon region, which covers a large portion of South America. Compare TEMPERATE RAINFOREST, TROPICAL SEASONAL FOREST.

tropical seasonal forest   A forest vegetation found in the tropics near the equator, in hot climates with distinct rainy and dry seasons. The drought-resistant trees are semi-evergreen (partly deciduous), with many shedding leaves during the dry season; patches of grassland interrupt the trees, which often border on savannas. Compare TROPICAL RAINFOREST.

tropism   An involuntary or reflex turning by a cell, a plant, or a fixed animal (one not capable of travelling) in response to a stimulus. Unlike a nastic response, tropic responses are oriented with respect to the direction of the stimulus; phototropism is movement toward or away from light. Also called tropic response. See CHEMOTROPISM, GEOTROPISM, HELIOTROPISM, HYDROTROPISM, PHOTOTROPISM, PLAGIOTROPISM, THIGMOTROPISM.

tropoparasite   An organism that is an obligate parasite during some period of its life, but a free-living organism for the remainder of its life.

tropopause   The upper region of the earth's troposphere in which the decrease in temperature associated with increased altitude becomes very small. The tropopause is a transition zone between the troposphere and the stratosphere.

tropophyte   1) A plant that can grow in climates that alternate between dry and moist or cold and hot, such as deciduous species. 2) A tropical plant.

troposphere   The lowest layer of the earth's atmosphere, extending from the earth's surface up to about 7 to 17 km (5 to 10 mi). Within the troposphere, temperature usually decreases at a regular rate with increasing altitude. Weather is confined to this turbulent layer of the atmosphere.

trough   Elongated region of a low-pressure air mass. See COLD TROUGH.

trough cross-stratification   A type of cross-stratification formed by migrating dunes or by the progressive movement of linguoid ripple marks. Trough cross-stratification is characterized by a concave lower erosional surface that forms a base for later deposition.

true fat   See NEUTRAL FAT.

true predator   An organism that kills its prey soon after attacking, and that will go on to attack several other individuals during its lifetime, as opposed to more parasitic organisms that feed off of and slowly kill only one or two hosts.

true wood   See HEARTWOOD.

truncated spur   The eroded triangular rock face of an arete that has been cut off by a valley glacier flowing at an angle to the ridge. See ARETE.

tryptophan $C_{11}H_{12}N_2O_2$, an amino acid found in animal and plant proteins. It must be supplied in the diet for normal growth and development; tryptophan is converted to niacin by the body. It is a limiting amino acid meaning that if it is not present in sufficient quantities in the diet, other amino acids cannot be utilized by the body; a complete protein or protein combination must contain tryptophan.

TSCA   See TOXIC SUBSTANCES CONTROL ACT.

tsunami   A large tidal wave caused by a submarine earthquake or volcanic eruption, or by a tectonic upheaval of the ocean floor.

tubal ligation   A form of permanent birth control (sterilization) in which the fallopian tubes of a female are cut and tied off to prevent eggs from being released. Compare VASECTOMY.

tuber   A solid swelling or outgrowth of an underground plant stem that bears axillary buds from which new plants may grow. Tubers such as the potato act as storage and overwintering organs. Compare CORM, BULB, RHIZOME.

tuber

tufa   A porous, chemical sedimentary rock formed near saline springs. Tufa is composed of calcium carbonate precipitated by the evaporation of water. See TRAVERTINE.

tuff   A general term for extrusive igneous rock consolidated from volcanic ash or pyroclastic ejecta.

tuff breccia   A volcanic rock consolidated from the coarse, angular particles of volcanic debris.

tumor   An abnormal growth on some part of the body that serves no physiological function. Tumors are classified as either benign or malignant (cancerous). See CANCER.

tundra   A biome found above treeline in arctic regions, on usually waterlogged soil sitting on permafrost, ground that is frozen all year. A variant, alpine tundra, is found at high elevations above treeline. Tundra is characterized by grasses and grasslike plants (such as sedges), lichens, and dwarf forms of woody plants. Large parts of Siberia and northern North America are tundra. See ALPINE TUNDRA.

tundra soils   An older soil classification term applied to cold-climate soils that have a dark, organically rich upper horizon lying over a gray-colored lower horizon. See INCEPTISOLS.

tungsten   (W) Heavy metallic element with atomic weight 183.9 and atomic number 74. It is used in alloys for high-speed cutting tools and surgical instruments, and in light-bulb filaments. It was formerly called wolfram, after an ore in which it is found.

tunneling 1) A mode of failure within an earthen dam or embankment. Tunneling occurs when desiccation cracks formed under dry conditions or other pathways in a soil are filled with water and collapse in a tunnel-shaped failure zone within the dam or embankment. 2) The process of excavating a tunnel within an earth material.

Turbellaria Class of the phylum Platyhelminthes whose members are generally free living, although some are ectocommensal. They function as predators and scavengers in both aquatic and terrestrial habitats. Most are relatively small but some land forms may be 45 cm or longer. The ciliated epidermis has a locomotive function in small forms but all turbellarians can move about by undulations of the body. Because of their remarkable regenerative capacity, some types such as planaria are used in biology courses for experiments on regeneration. See PLATYHELMINTHES.

turbidimeter An optical device designed to measure the turbidity within a liquid.

turbidite Sediment or rock that forms from material deposited by turbidity currents.

turbidity A lack of clarity in a fluid, usually caused by turbulent flow picking up large quantities of particulates. Turbidity can refer either to air hazy with pollutants, or to water carrying large quantities of suspended silt or organic matter.

turbidity current A submarine density current caused by a suspension of clastic particles. A turbidity current may form at the forefront of a delta or at the edge of a continental shelf.

turbine A machine driven by the pressure of a current of fluid such as water or steam and producing rotary motion directly. The fluid pushes against a rotor or stator, a series of curved vanes arranged around a rotating shaft. See IMPULSE TRUBINE, PELTON WHEEL, TURBOGENERATOR.

The Francis turbine is a reaction turbine intended for use submerged in water. Reaction turbines are designed to be operated with a specific amount of head and rate of flow. Any variation reduces their efficiency substantially.

turbogenerator A turbine (usually steam-driven) connected to an electrical generator. Almost all power plants (nuclear, coal, hydropower) employ turbogenerators. See TURBINE.

turbulence Fluid flow characterized by constant, apparently random, small-scale changes in speed and direction. Also called turbulent flow. Contrast LAMINAR FLOW.

turgor The normal rigidity (caused by turgor pressure within the cell) of living cells. See TURGOR PRESSURE.

turgor pressure The pressure against a cell membrane from within the cell contents. Turgor pressure is usually positive in plant cells, although negative in some of the xylem cells in plants undergoing transpiration. It holds plants erect.

turnover rate 1) The fraction of an element or biochemical compound that is released from organisms to the environmental medium per unit time. 2) The ratio of nutrient releases from living organisms per unit time to the amount of those nutrients contained in their biomass.

twister Colloquial term for any funnel cloud phenomenon, including tornados, dust devils, and waterspouts. See TOR-NADO, DUST DEVIL, WATERSPOUT.

2,4-D Short for 2,4-dichlorophenoxyacetic acid, an herbicide widely used to control broadleaved weeds in lawns. It is a synthetic compound whose effects resemble those of auxin (a natural plant growth hormone). 2,4-D was used as a defoliant during the Vietnam War and is still mixed with fertilizers for use in lawn conditioning products. See HORMONE WEEDKILLERS.

2,4,5-T Short for 2,4,5-trichlorophenoxyacetic acid, an herbicide formerly widely used to control unwanted woody plants and other weeds. It is a synthetic compound whose effects resemble those of auxin (a natural plant growth substance). 2,5,5-T is one of the ingredients in Agent Orange, a defoliant used by the U.S. military in Vietnam; toxic dioxins are produced as a by-product of the manufacture and the decomposition of this substance. The use of 2,4,5-T is severely restricted now because of public outcry in the early 1970s. See AUXIN, HORMONE WEEDKILLERS, DIOXIN.

tychoplankton Organisms that are not usually found in with plankton, but have been carried upward from the bottom by unusual turbulence.

type locality 1) The particular location from which a type specimen was collected. 2) The original location of a rock formation or a fossil used as the basis for naming and describing the type of rock formation or fossil. See TYPE SPECIMEN.

type specimen A single specimen used as the basis for the official description of species, subspecies, variety, or lesser taxonomic group. Also called a holotype, although that term is technically used only if the type specimen is the initial sample collected by the author who named the species. A replacement for an initial type specimen (holotype), chosen from the same original material, is called a lectotype; a replacement from new material is a neotype.

Typhaceae The cattail family of the monocot angiosperms. All members of this family, the single genus *(Typha)*, are perennial wetland herbs. They are monoecious, with many minute male flowers in dense clusters above the equally dense clusters of female flowers. The wind-borne fruits are minute achenes attached to a tuft of hairs. See ANGIOSPERMOPHYTA.

typhoon A violent, tropical cyclonic storm occurring over the Indian Ocean or western Pacific. It is essentially another name for a hurricane. See CYCLONE.

typhus An epidemic disease caused by the bacterium *Rickettsia prowazeki*, transmitted to humans by the bite of lice. It is characterized by high fever, weakness, headache, and sometimes a rash; it is often fatal. See RICKETTSIA, SCRUB TYPHUS.

tyrosinase A plant and animal enzyme that converts (oxidizes) tyrosine into pigments such as melanin. See TYROSINE.

tyrosine $OHC_6H_4CH_2CH(NH_2)COOH$, an amino acid found in many proteins. It is a precursor of adrenaline, thyroxine (a thyroid hormone), and melanin.

Tytonidae See BARN OWLS.

u

U Chemical symbol for the element uranium. See URANIUM.

UHF Short for ultra high frequency radiation. See ULTRA HIGH FREQUENCY.

ultisol Any member of a soil characterized by acidic highly weathered horizons with accumulation of silicate clays in subsurface layers; usually forms in tropical and subtropical climates.

ultrabasic See ULTRAMAFIC ROCK.

ultra high frequency (UHF) Electromagnetic radiation of frequencies from 300 to 3000 megahertz. UHF frequencies fall between those used for broadcasting AM radio signals and those of microwave frequencies. UHF is used for broadcasting some television signals. See EXTREMELY LOW FREQUENCY, HIGH FREQUENCY, LOW FREQUENCY, MEDIUM FREQUENCY, ULTRA HIGH FREQUENCY, VERY HIGH FREQUENCY.

ultramafic rock Any dark igneous rock predominantly composed of ferromagnesian minerals. Dunite and peridotite are examples of ultramafic rocks.

ultraviolet (UV) Utilizing or concerning electromagnetic radiation of ultraviolet wavelengths. See ULTRAVIOLET RADIATION.

ultraviolet light Ultraviolet radiation can initiate many chemical reactions and is one of the causes of skin cancer. Fortunately, much of the sun's UV rays are blocked from reaching the earth's surface by a protective layer of ozone in the atmosphere, one of the important reasons for protecting the earth's ozone layer.

ultraviolet radiation Electromagnetic energy with wavelengths from 10 to 400 nanometers. UV radiation wavelengths fall between those of visible (violet) light and x-rays. Compare INFRARED RADIATION.

umbel A broad, flat-topped cluster of flowers; it differs from a corymb in that the small stalks of the flower cluster arise from a single point at the top of a main stalk. Queen Anne's lace is an umbel. Compare CORYMB, CYME.

Umbelliferae The parsley family. See APIACEAE.

umbra The part of the eclipse shadow in which the solar disk is completely blocked.

UNCED See UNITED NATIONS CONFERENCE ON ENVIRONMENT AND DEVELOPMENT.

UNCLOS See UNITED NATIONS CONFERENCE ON THE LAW OF THE SEA.

unconfined aquifer An aquifer in which the upper boundary is determined by the height of the water table and not by a confining rock layer.

unconformity A gap in the depositional history of a rock layer. An unconformity is recognized by the presence of an erosional surface or a surface representing a period of no deposition between layers of rock. Unconformities are classified as angular unconformities, disconformities, or nonconformities according to the characteristics and relationships of the adjacent rock layers. See ANGULAR UNCONFORMITY, DISCONFORMITY, NONCONFORMITY.

unconformity-bounded sequence A sequence of rock strata that terminates

567

above and below by the presence of unconformities.

undercompensating density dependence Density-dependent reactions in which the change in death rate (increase), birth, or growth rate (decrease) is less than the original increases in density that triggered the reaction. The result is an increase in the original density, but a smaller increase than would be expected in the absence of density-dependent correcting factors. See DENSITY OVERCOMPENSATION.

underdominance Natural selection that favors individuals with homozygotic rather than heterozygotic gene pairs.

underground injection wells See DEEP-WELL DISPOSAL.

underground storage tanks (USTs) Any buried vessel used for storage, particularly for petroleum products. Formerly, home heating oil and the fuel sold at gas stations were usually stored in metal USTs. Corrosion poses a great health risk, however, as very small quantities of gasoline and of oil can travel through the soil to poison home wells and entire municipal water supplies. New federal regulations have outlawed USTs for many uses, as above-ground tanks are much easier to monitor. For uses requiring USTs (such as gas stations), double-walled plastic tanks are now required.

undergrowth Shrubs, tree seedlings, and other plants that grow underneath the canopies of large trees. Undergrowth is a general term, less technical than under-story. Also called underbrush. See UNDER-STORY, SHRUB LAYER, HERB LAYER.

undernourished Describing humans (or animals) receiving less than 90 percent of their minimum dietary requirements over a sustained period of time, reducing their

resistance to infections and their mental and physical functioning.

undersaturated island An island having fewer species than would be predicted by the equilibrium theory of island biogeography. See EQUILIBRIUM THEORY OF ISLAND BIOGEOGRAPHY. Contrast OVER-SATURATED ISLAND.

understory The lowest layer of trees in a forest; the layer between the overstory tree layer and the shrub layer. Compare CANOPY.

undetectable concentration Pollution levels that are too low to be measurable by current methods.

UNDP See UNITED NATIONS DEVELOP-MENT PROGRAMME.

UNEP See UNITED NATIONS ENVIRON-MENT PROGRAMME.

uneven-age stand A group of trees of all different ages, usually also containing a variety of tree species. Compare EVEN-AGE STAND.

ungulate Describing hoofed mammals that graze, such as horses, deer, cows, tapirs, bison, and rhinoceroses. Ungulates do not necessarily belong to closely related genera.

Ungulates A general term for hoofed, herbivorous mammals. See ARTIODACTYLA, PERISSODACTYLA.

unguligrade Using only the tips of en-larged toes (hoofs) in walking, as in horses, cattle, and deer. Compare DIGITIGRADE, PLANTIGRADE.

unified soil classification system A soil classification system that is based on the relative proportions of particle sizes in the soil. The unified soil classification system is used primarily for engineering purposes.

uniformity   See PRINCIPLE OF UNIFOR-MITY.

uniform spatial distribution   A form of dispersion of organisms within an ecosystem in which individuals are found at fairly regular intervals throughout the community, with roughly equal spacing in all directions from other individuals compared to a purely random distribution. Uniform spatial distribution tends to be found where competition between individuals is intense (as when plants compete for limited water), in territorial animals, or when some form of antagonism occurs between individuals (such as plants releasing allelopathic chemicals into the soil.) Also called even distribution. Contrast CLUMPED DISTRIBUTION, RANDOM SPATIAL DISTRIBUTION. See EQUITABILITY.

uniovular twins   See MONOZYGOTIC TWINS.

uniparous   Producing a single egg or offspring at birth.

Uniramia   The largest and most important, subphylum of Arthropoda. The group has unbranched (uniramic) limbs and stout mandibles. The head, which is compact and encapsulated, bears one pair of antennae in all forms. Because the uniramia are so diverse, authorities often disagree on details of their classification. Broadly, the subphylum can be broken into two groups: the myriapod forms with many leg-bearing segments and the hexapod forms in which a pair of legs occurs on each of three thoracic segments. Five or six classes are usually recognized, including centipedes (Chilopoda), millipedes (Diplopoda), and insects (Insecta). See DIPLOPODA, INSECTA.

unisexual   Describing organisms that are distinctly male or female, possessing the essential reproductive organs of just one sex.

United Nations Conference on Environment and Development   (UNCED) An international conference colloquially referred to as the "Earth Summit" held in Rio de Janeiro, Brazil, in June, 1992. Attending were delegates from 150 countries, approximately 800 representatives of indigenous peoples, and thousands of representatives of local, national, and international non-governmental organizations. It was the successor to the 1972 United Nations Conference on the Human Environment held in Stockholm. UNCED produced Agenda 21, a nonbinding blueprint for pursuing sustainable development in the next century. While a climate change convention was passed, the United States insisted on removing the convention's concrete targets for cutting greenhouse gases and also refused to sign the biodiversity convention.

United Nations Conference on the Law of the Sea   (UNCLOS) A series of meetings between 1974 and 1982 sponsored by the United Nations to develop new international laws regulating the exploitation of marine resources while maintaining the right of ships to move without interference. The agreements, called the Law of the Sea Convention, establish exclusive economic zones extending 320 kilometers (200 miles) from a country's shoreline. See EXCLUSIVE ECONOMIC ZONE.

United Nations Development Programme   (UNDP) The agency of the United Nations responsible for international technical funding for aid to Third World countries. It works in cooperation with more specialized UN agencies. Compare CONSERVATION DISTRICT.

United Nations Environment Programme (UNEP) The UN organization responsible for advocating environmental management among the various UN agencies as well as national, regional, and nongovernmental organizations. UNEP activities include monitoring, training, technical cooperation, information exchange, and legal initiatives for oceans and coastal areas, loss of biological diversity, hazardous chemicals, fresh-water systems, and the atmosphere. Formed in 1972, UNEP operates the Earthwatch Programme.

United States Department of Agriculture (USDA) The federal agency responsible for regulating agriculture, promoting domestic and foreign marketing, scientific research, and good nutrition. Through the Forest Service, it is also responsible for managing national grasslands and natural forests.

unit leaf rate Over a specific period of time, the increase in dry weight produced by a plant, compared to the plant's leaf area. Also called net assimilation rate.

unit membranes A model for the structure of biological membranes consisting of a phospholipid bilayer with embedded proteins. The bilayer has a hydrophobic core and hydrophilic surfaces. Most membranes (those around the nucleus, mitochondria, and chloroplasts) are double, but the membrane surrounding a vacuole (the tonoplast) is a single unit membrane. Compare FLUID-MOSAIC MODEL.

universal oil product (UOP) A substance used in scrubbers for flue gas desulfurization. The percentage of sulfur recovery by the oil is high enough that sulfur can be extracted, allowing the oil to be used over again in the scrubber. See FLUE GAS DESULFURIZATION, SCRUBBER.

universal time (UT) Any of several international time standards, but usually used to refer to Coordinated Universal Time. See COORDINATED UNIVERSAL TIME.

univoltine Describing an organism that produces only one brood per year. Compare BIVOLTINE, MULTIVOLTINE.

unsaturated fat A fatty compound containing some double bonds connecting the carbon atoms into a chain, and therefore containing less than the maximum possible number of hydrogen atoms attached to the carbon atoms. Unsaturated fats are usually liquid at room temperature and include many vegetable oils. Unsaturated and polyunsaturated fats are healthier to consume than saturated fats, because consumption of a high proportion of saturated fats has been linked with a number of health problems, from heart disease to cancers. See POLYUNSATURATED FAT, SATURATED FAT.

unsaturated flow The flow of water within the soil zone of aeration (vadose zone). Unsaturated flow is strongly affected by capillary forces within the pore spaces of the soil.

unsaturated zone See VADOSE ZONE.

unsorted A term applied to any sediment that consists of a wide variety of particle sizes.

unstable equilibrium An equilibrium that is subject to change or easily disrupted. See EQUILIBRIUM. Compare DYNAMIC EQUILIBRIUM.

UOP See UNIVERSAL OIL PRODUCT.

updraft A current of air that is rising; usually warm air. Cumulus clouds often contain strong updrafts.

upland game Animal species, especially game animals such as bighorn sheep, living in mountainous areas.

upper air contours A drawing of the elevation contours for a surface of given atmospheric pressure (often 500 milibars), equivalent to isobars for that pressure. Upper air contours show the location of high- and low-pressure regions. Also called constant-pressure charts or pressure-contour charts, upper air contours appear on the national weather maps on television or in newspapers.

upper shore A general term for the area of a beach that extends from the lower limit of wave activity to the base of a cliff, a dune, or a vegetated beach ridge.

upright fold A fold in rock strata in which the axial plane orientation is approximately vertical.

upset A narrow slice excavated within an inclined seam of coal.

upstream flood A flood that is localized in a relatively small region of a watershed, usually at higher elevations.

upwarp A general term for a regional geologic uplift. An upwarp may be the result of tectonic forces or of regional unloading and isostatic adjustment.

upwelling 1) The term applied to a persistent and rising cold-water current in an oceanic circulation system. 2) The gentle eruption of a volcanic lava.

uracil 2,6-dioxypyrimidine, one of the four pyrimidine bases that make up the structure of RNA; it pairs with adenine. Uracil is the only base not found in DNA; it is replaced there with thymine. See RNA.

uraninite Another name for the uranium-containing ore pitchblende. See PITCH-BLENDE, also APPENDIX p. 628.

uranium (U) A radioactive element with atomic weight 238.0 and atomic number 92. Its three naturally occurring isotopes are uranium 234, uranium 235, and uranium 238; the last is most common, making up over 99 percent of the uranium in ores. It is the major fissionable material used in nuclear reactors and nuclear weapons and has a half-life of 4.5×10^9 years. See URANIUM 235.

uranium 235 (u-235) A fissionable isotope that makes up about 0.7 percent of naturally occurring uranium. It is used in nuclear reactors and weapons because of its ability to sustain a rapid chain reaction. See CHAIN REACTION.

uranium enrichment The process of increasing the concentration of desirable uranium 235 above the low levels found in nature, to make it suitable for use in nuclear reactors and weapons. Gaseous diffusion, centrifuging, and laser isotope separation are some of the techniques used. See GASEOUS DIFFUSION.

urban heat island Another term for the heat island effect. See HEAT ISLAND.

urbanization 1) Process of change characterized by movement of populations from rural to urban areas of a country. Urbanization is associated with the conversion of farmland or undeveloped land into housing, businesses, and other forms of development. 2) The percentage of a country's population that lives in urban (as opposed to rural) areas.

urban runoff A general term for the rain water runoff of impervious surfaces, such as buildings and pavement, in developed areas. Urban runoff may be contaminated with salt, soot, oil residue, and other pollutants.

urea $H_2N\text{-}CO\text{-}NH_2$, a compound produced in the liver of mammals as a by-product of the breakdown of amino acids during protein metabolism. It is present in mammalian blood, lymph, and urine. Urea is synthesized for fertilizer and animal feed supplements, and for use in plastics such as urea formaldehyde. See UREA FORMALDE-HYDE.

urea formaldehyde An organic compound manufactured by heating urea with formaldehyde. Urea formaldehyde is resistant to weathering, alcohols, greases, and oils. It is a thermoset plastic and is also used in adhesives. Urea formaldehyde was formerly used in foam insulation for buildings (UFFI, short for urea formaldehyde foam insulation), but the use of UFFI has been banned in some areas because of associated health problems. See THERMOSET POLY-MERS.

ureotelic Describing animals that produce urea as the main constituent of their nitrogenous waste. Mammals are ureotelic. Compare URICOTELIC.

uric acid $C_5H_4N_4O_3$, 2,6,8-trihydroxy-purine. A purine acid produced by the breakdown of purine metabolism (purines are components of DNA and RNA). In insects, reptiles, and birds, a solution of uric acid replaces urea as the primary fluid used to excrete excess nitrogen. Uric acid is also produced by mammals.

uricotelic Describing animals such as birds and reptiles that excrete nitrogen as uric acid. Compare UREOTELIC.

Urodela Salamanders. Although many species are highly specialized, most urodeles have a very generalized, tetrapod body form with two pairs of limbs, elongate body, and tails. Parts of the cranial skeleton may be reduced in number and formed of cartilage. The aquatic larvae have gills that may or may not be retained in the adult. See AMPHIBIA.

U.S. AID See AGENCY FOR INTERNATIONAL DEVELOPMENT.

USDA See UNITED STATES DEPARTMENT OF AGRICULTURE.

U-shaped valley An erosional landform carved by a valley glacier formed by the plucking and abrasion of a preexisting valley. The cross-section profile, shaped like the letter U, is a characteristic of glacial erosion. Compare V-SHAPED VALLEY.

UT See UNIVERSAL TIME.

utilization efficiency The ratio of energy assimilated by a given trophic level to the energy produced by the previous trophic level. Utilization efficiency generally increases with trophic level. Compare ECOLOGICAL EFFICIENCY, LINDEMAN EFFICIENCY, TROPHIC LEVEL EFFICIENCY, CONSUMPTION EFFICIENCY.

UV See ULTRAVIOLET.

UV-A The low-energy part of the ultraviolet radiation spectrum extending from 320 nanometers to 400 nanometers in wavelength. This is the least harmful form of UV light. See ULTRAVIOLET LIGHT.

UV-B The high-energy part of the ultraviolet radiation spectrum, extending from 290 nanometers to 320 nanometers in wavelength. This is the most harmful form of UV light and is usually filtered out by the protective ozone layer in the stratosphere. Thinning of the ozone layer will result in greater exposure of terrestrial inhabitants to dangerous UV-B radiation. See ULTRAVIOLET LIGHT.

UV detector A detector used for high pressure liquid chromatography (HPLC) to quantify the amount of a substance present in a mixture after separation from other components. The UV detector works by shining ultraviolet light on a sample and detecting how much of the light was absorbed. The amount of light absorbed is proportional to the amount of substance present.

UV light Short for ultraviolet light, another term for ultraviolet radiation. See ULTRA-VIOLET RADIATION.

V

V Chemical symbol for the element vanadium. See VANADIUM.

vacuole A sac within the cytoplasm of a cell, bound by a single membrane and containing air or fluid (water or sap), food, and waste products of cell metabolism. Plant cells usually have one large vacuole; animal cells typically have several smaller vacuoles. Protozoa often have temporary vacuoles used for ingesting or digesting food. See TONOPLAST.

vadose water See PHREATIC WATER.

vadose zone A soil zone between the land surface and the water table. The vadose zone is subdivided into the belt of soil water, the intermediate belt, and the capillary fringe. Also called the zone of aeration. See WATER TABLE.

valence electrons The outer electrons of an atom (those located in the outermost shell or orbital), those used for forming chemical bonds with other atoms. See COVALENT BOND, IONIC BOND, OCTET RULE, ORBITAL.

valine $(CH_3)_2CH(NH_2)$-COOH, an amino acid found in animal and plant proteins. It is required for the healthy growth of infants and for nitrogen metabolism in adults.

valley fog Radiation fog occurring in a valley. See RADIATION FOG.

valley glacier An alpine glacier that flows down through topographic valleys adjacent to mountain flanks or summit areas. The shape and the flow of valley glaciers is generally controlled by the underlying topography. Compare MOUNTAIN GLACIER.

valley train A glacial outwash deposit originating from the zone of wastage of an alpine glacier. Valley trains are confined to the lower part of a U-shaped valley. Contrast OUTWASH PLAIN.

vanadium (V) A hard, metallic element with atomic weight 50.94 and atomic number 23. It is used in steel alloys.

Van Allen belts Regions of strong magnetic fields above the earth's equator. They trap incoming high-energy, charged particles from the solar wind. Also called Van Allen radiation belts.

Van der Waals forces A general term for intermolecular forces, weak forces of attraction between molecules or crystals that are not bound to each other. These forces are much weaker than the forces of covalent or ionic bonding. These forces include dipole-dipole and nonpolar interactions. See DIPOLE-DIPOLE INTERACTIONS.

vaporization The conversion of a liquid to a vapor. Vaporization can be accelerated by heating, by reducing the pressure, or by increasing the surface area by breaking the liquid up into minute droplets, as in an ultrasonic humidifier.

vapor plume Smokestack emissions that are made visible by the condensation of water droplets or other vapor.

vapor pressure Pressure exerted by the molecules of a specified vapor. Vapor pressure in a closed container containing only one compound is the force exerted against the walls of the vessel; in an open system (necessarily involving more than one material) it is the partial pressure that the substance contributes to the overall pressure of the system. Used in meteorology to refer to the partial pressure of water vapor in the atmosphere.

variable   A quantity that can assume any value in a set of values and therefore is not constant.

variance   In statistics, the square of the standard deviation, which is equal to the mean (average) of the sum of the squares of the values of deviations of a series of observations from the corresponding central value or mean. See STANDARD DEVIATION.

variance-to-mean ratio   A measurement of the dispersion of a population in which numbers of individuals are defined using quadrats, and the variance is then divided by the mean. If the dispersion is random as described by the Poisson distribution, the variance divided by the mean is 1.0. Values less than 1.0 indicate a uniform dispersion, whereas values greater than 1.0 indicate a clumped distribution.

variation   1) Deviation from expected occurrence or motion, as in an animal or a plant that is different from the normal or recognized form. 2) In statistics, the distribution of values about the mean or central value.

variety   1) A group of similar organisms within a species that clearly differs from other members of the species; a subspecies, a race, or a breed. Organisms of one variety transmit their characteristics to their offspring, but are also capable of interbreeding with other varieties within the same species. 2) Loosely used for any variation within an animal or a plant species.

varve   A pair of light- and dark-colored sediment layers representing 1 year of depositional activity within a glacial lake. The light-colored silt is deposited by summer meltwater and the dark-colored clay settles out during the winter.

vascular bundle   A strand of conducting tissue in plants, containing xylem and phloem and extending from the roots through the stem and into the leaves. In monocotyledon stems, the vascular bundles are spread out through the stem. In dicotyledons, they are arranged in a circle around the pith, with a ring of cambium tissue separating the xylem from the phloem. Also called fascicle, all of the vascular bundles together make up the stele. See CAMBIUM, STELE.

vascular cryptogam   A nonflowering vascular plant. In early systems of classification, cryptograms were those plants whose reproduction was hidden (i.e., not readily seen as flowers, cones, or seeds). Vascular cryptogram was used to distinguish those plants with primitive conducting tissue (ferns, club-mosses, and horsetails) from other cryptograms (algae, fungi, liverworts, and mosses).

vascular cylinder   See STELE.

vascular plant   Any plant having an organized system for transporting water and nutrients. Vascular plants include many organisms that do not produce seeds, such as the Filicinophyta (ferns), Lycopodophyta (club-mosses), and Sphenophyta (horsetails). They also include the seed-bearing gymnosperms (Coniferophyta, Cycadophyta, Ginkgophyta, and Gnetophyta) and the flowering plants (Angiospermophyta).

vasectomy   A form of permanent birth control (sterilization) in which all or part of the vas deferens is cut and removed to prevent sperm from being released. Compare TUBAL LIGATION.

vector 1) A disease carrier, an animal that transmits a disease organism to other organisms. The mosquito that transmits the malaria protozoan to humans and the tsetse fly that transmits African sleeping sickness are vectors. 2) A pollinator; a bee, a bird, a bat, or other organism that transfers pollen from anther to stigma, or from one flower to another. 3) An agent such as a bacteriophage or a plasmid used in recombinant genetic engineering to insert a foreign piece of DNA into a gene in the host cell. See GENETIC ENGINEERING. 4) A quantity having direction as well as magnitude, as opposed to a scalar quantity that has only magnitude. Velocity is a vector quantity because it incorporates direction as well as speed. Speed is a scalar, not a vector, quantity.

veering wind Wind that is shifting to a clockwise direction. See BACKING WIND.

vegetarian 1) Describing a diet that includes no meat. 2) A person adhering to such a diet. Vegetarians may consume eggs and milk products, although some (vegans) refrain from consuming any such animal products. 2) An herbivore. See HERBIVORE, MACROBIOTICS.

vegetation All of the plants growing in and characterizing a specific area or region; the combination of different plant communities found there.

vegetative propagation Any asexual method of reproducing plants, as by cuttings, tissue culture, or grafting rather than by seeds. Vegetative propagation produces a clone, an organism genetically identical to the parent plant.

vegetative reproduction Asexual reproduction; the creation of offspring without the union of male and female gametes. Vegetative reproduction includes budding by yeasts and coelenterates (sponges, etc.) and plants sending out runners (stolons) that root to form new plants.

vein 1) A vessel that conducts blood back to the heart. Compare ARTERY. 2) A vascular bundle, with its associated tissues, in a plant leaf. 3) A wirelike strut that strengthens and supports an insect wing; also called a nervure. 4) In geology, a sheetlike or branchlike intrusion of an igneous material into fractures or weak zones of a country rock.

veld An open, grassy plain of southern Africa. Velds often contain some bushes but very few trees. Veld (also spelled veldt) means field in Afrikaans (and Dutch, from which South African Afrikaans derives). Compare SAVANNA.

veliger The secondary larval stage of molluscs, characterized by a cilia-covered swimming membrane (velum). See TROCHOPHORE.

velocity The direction and rate at which an object moves. Velocity resembles speed, but includes the direction of the motion; it is a vector quantity.

vent agglomerate See VOLCANIC AGGLOMERATE.

vent bacteria Bacteria that live in the harsh environment and the extremely hot, sulfurous water surrounding hydrothermal vents ("black smokers") on the ocean floor. Vent bacteria such as hydrogen sulfide bacteria derive nutrition from chemosynthesis, as they live far below the depths that light can penetrate the ocean.

ventifact A wind-sculpted pebble that features distinctive facets on the upper abraded surfaces.

ventral Located on or referring to the belly or underside of a plant structure, an animal body, or an organ. The ventral side is usually that facing the ground, but in humans it is the front. Compare DORSAL.

ventriculus   1) A digestive cavity, such as the stomach of some insects, fish, and reptiles; the gizzard in birds is called the proventriculus. Also called midgut. 2) The body cavity of a sponge. 3) Another term for ventricle, such as the ventricles (lower chambers) of the heart.

vents   Fissures and openings that allow volcanic or geothermal materials to pass to the earth's surface.

Venturi effect   The increase in speed of a fluid (liquid or gas) that occurs when it flows through a constriction in its channel. The Venturi effect applies to stream flow speeding up as the stream channel narrows, and to water speed increasing as the nozzle of a hose is adjusted to a smaller opening.

vermiculite   A term applied to a variable group of platy minerals derived from the alteration of biotite and phlogopite micas. Vermiculite has water-absorbent properties and is used in industrial processes.

vernal equinox   Technically, the point on the sun's apparent path where it crosses the celestial equator when moving from south to north. Also used colloquially for the date the sun reaches this point, on or near March 21 each year, which marks the beginning of Northern Hemisphere spring. See AUTUMNAL EQUINOX.

vernal overturn   A phenomenon occurring in lakes of temperate climates. As warming spring temperatures melt the lake ice and the surface water warms to reach 4°C (the temperature at which it reaches greatest density), the dense water (aided by wind stirring the surface) sinks to the bottom, displacing the cold, nutrient-rich water

there. This mixes the intermediate layers, creating a uniform distribution of oxygen and nutrients at different depths. Lakes undergoing temperature stratification (thermocline formation) during the summer show a similar autumnal overturn. Also called spring overturn. See EPILIMNION, HYPOLIMNION, STRATIFICATION, THERMOCLINE.

vernalization   The act of forcing a plant to grow or to flower by exposing the plant or bulb to very cold temperatures. (Treating seed in a similar fashion to break dormancy is usually called stratification.) Gardeners in southern portions of the United States have to subject tulip bulbs to vernalization, mimicking climates with cold winters, in order for them to bloom. Winter wheat requires vernalization in order to flower in the spring. Compare STRATIFICATION.

Vertebrata   A subphylum of Chordata distinguished by possession of a cartilagenous or bony, axial endoskeleton that forms a braincase and a vertebral column supporting the nerve cord. Except for the agnatha, vertebrates have jaws and most have two pairs of appendages such as fins and limbs With more than 40,000 species, the vertebrates are one of the most diverse and successful life forms. See CHORDATA, AGNATHA.

vertically homogeneous estuary   A term applied to an estuarine system that does not typically develop a stratified water circulation pattern. A vertically homogeneous estuary is typically formed in shallow water where water is mixed by wind and tides.

vertical stratification   1) The separation of water masses into distinct horizontal layers. Vertical stratification is caused by differences in density related to variations in temperature and water chemistry. 2) The arrangement of rock layers into a vertical array of horizontal layers. See THERMOCLINE, STRATIFICATION.

vertisols Any member of a soil order characterized by evidence of shrinking and cracking when dry, usually very old, weathered soils with moderate-to-high concentrations of clays expand and contract with changes in moisture content.

very high frequency (VHF) Electromagnetic radiation with wavelengths between 1 and 10 meters and frequencies from 30 to 300 megahertz. Like UHF, these frequencies fall between those used for broadcasting AM radio signals and those of microwave frequencies. They are used for broadcasting FM radio and some television signals. See EXTREMELY LOW FREQUENCY, HIGH FREQUENCY, LOW FREQUENCY, MEDIUM FREQUENCY, ULTRA HIGH FREQUENCY, VERY LOW FREQUENCY.

very low frequency (VLF) Electromagnetic radiation falling in the radio spectrum, with wavelengths from 10 to 100 km and frequencies from 3 to 30 kilohertz. VLF radio waves are used in communicating with underwater submarines. Compare VERY HIGH FREQUENCY.

vesicular Bladderlike; containing, resembling, or relating to vesicles (small cavities or sacs).

vesicular-arbuscular mycorrhiza (VAM) A symbiotic (probably mutualistic) association between a fungus and a plant root in which the fungus grows between and into the host cells, forming vesicles and arbuscules (branched structures), and spreading into the surrounding soil. Unlike ectomycorrhiza, these form no thick layer surrounding the root. VAMs form with many species of herbaceous plants, including crop plants; they improve the efficiency of nutrient absorption and use in the host plant and may improve resistance to environmental stresses. Compare ECTOTROPHIC MYCORRHIZA, NITROGEN-FIXING BACTERIA.

vessel elements A perforated form of tracheary element making up the xylem of a plant. Also called vessel members or vessel segments.

vestigial Degenerate or rudimentary; reduced to the point of no longer being useful. For example, the human appendix, or the wings of an ostrich.

vesuvian volcano Any composite volcano which erupts explosively, as typified by Mt. Vesuvius.

VHF See VERY HIGH FREQUENCY.

vibration Oscillating motion. Vibration refers both to the repeated shaking of an elastic substance in response to a disturbance (the movement of a plucked guitar string) and to the movement of a waveform signal (the frequency of a radio wave is its number of vibrations per second).

vicariad See VICARIANTS.

vicariants Plant or animal taxonomic groups that are very closely related and so are presumed to have evolved either from a common ancestral group that was geographically isolated or from each other, but which have now spread out to inhabit nonoverlapping geographic areas.

vigor Vitality, strength, or robustness; the degree of heath and healthy growth exhibited by an organism. See HYBRID VIGOR.

vinyl chloride monomer (VCM) $CH_2=CHCl$, a molecule of vinyl chloride forming a single unit that, when subjected to polymerization, forms polyvinyl chloride. See POLYVINYL CHLORIDE.

Violaceae The violet family of the Angiospermophyta, containing dicotyledonous herbs and tropical shrubs with zygomorphic flowers that have five unequal petals, the lower of which has a pronounced spur. Violet flowers have five stamens clustered around the short, beaked pistil. See ANGIOSPERMOPHYTA.

virga Precipitation that evaporates as it falls and never reaches the ground. Virga can refer to rain, but is usually ice or snow. The name is Latin for streak, because the precipitation often falls in visible streaks.

virion A mature and complete virus, made up of a strand of RNA or DNA encased in a protein shell (capsid) and constituting the form in which it spreads from one cell to another.

viruses The simplest living organisms, tiny particles lacking cell structure and consisting only of genetic material encased in protein. Viruses are smaller than bacteria, too small to be seen under ordinary microscopes. They can only reproduce inside of a host cell and are capable of manipulating the host cell to promote their own reproduction. Viruses cause many important infectious diseases, including polio, rabies, smallpox, and AIDS, as well as plant diseases; retroviruses can cause cancer. See BACTERIOPHAGE, RETROVIRUS.

viscosity The property of fluids that creates resistance to flowing as a result of their resistance to shear forces. All fluids possess some degree of viscosity. The SI unit for viscosity is pascal seconds. Viscosity is determined by measuring the rate at which a fluid flows through a capillary (very narrow) tube.

visible light Electromagnetic radiation of wavelengths that are visible to the human eye. Such wavelengths range from approximately 380 to 710 nanometers (4000 to 7000 angstroms), from violet light through red light. Visible light falls between ultraviolet and infrared radiation in the electromagnetic spectrum. See ELECTROMAGNETIC SPECTRUM, SPECTRUM. Compare WHITE LIGHT.

visible radiation Another term for visible light. See VISIBLE LIGHT.

Vitaceae The grape family of the Angiospermophyta. These dicots are largely woody vines that climb by tendrils. The small unisexual or bisexual flowers are borne in clusters of cymes. The fruits are berries. Grapes (*Vitis*), possum grape (*Cissus*), and Virginia creeper and Boston ivy (*Parthenocissus*) are members of the Vitaceae. See ANGIOSPERMOPHYTA.

vitamin Any of a group of organic compounds required in relatively small amounts for normal growth and proper functioning of the body. Many are produced by the body, but not in quantities sufficient to meet metabolic requirements. Severe deficiencies of specific vitamins cause diseases such as scurvy, beriberi, and rickets. Vitamins fall into two groups, water-soluble compounds that must be replenished daily because they are easily washed out of the body, and fat-soluble compounds that can be stored temporarily but must be periodically replenished. Fat-soluble vitamins can build up to toxic levels in fatty tissues if mega-doses of these vitamins are taken over a prolonged period.

vitamin A A fat-soluble vitamin essential to the growth of young organisms and for healthy skin and mucous membranes. One of the first symptoms of deficiency is temporary night blindness. Vitamin A can be synthesized by the body from the carotene pigment found in plants, especially in yellow vegetables such as carrots and winter squash. Also called retinol.

vitamin B complex A group of water-soluble micronutrients with different functions; all are found in high concentrations in yeast and liver. Most are better known by their other names: thiamine (vitamin B_1), riboflavin (vitamin B_2), niacin (vitamin B_{12}), pyridoxine (vitamin B_6), pantothenic acid, folic acid, choline, and cobalamin (vitamin B_{12}). See THIAMINE, RIBOFLAVIN, NIACIN, PYRIDOXINE, PANTOTHENIC ACID, FOLIC ACID, CHOLINE, and COBALAMIN.

vitamin C Ascorbic acid; a water-soluble micronutrient required for a wide variety of body functions. A deficiency of vitamin C causes scurvy, a fact discovered in the 18th century when sailors learned to bring fresh fruit such as limes along on long voyages. Vitamin C helps many metabolic reactions, especially the formation of connective tissue. It is an antioxidant.

vitamin D A fat-soluble micronutrient required for regulating calcium and phosphorous in bone formation and throughout the body. The most effective source of vitamin D is not food but exposure to sunlight. Extreme lack of sunlight and food sources causes rickets, so milk is now fortified with this vitamin as a preventative.

vitamin E Alpha tocopherol; a fat-soluble micronutrient required for animal reproduction and healthy cell membranes. Vitamin E is found in unsaturated vegetable oils and unrefined grains; it is an antioxidant.

vitamin H See BIOTIN.

vitamin K A fat-soluble micronutrient required for blood-clotting. Deficiencies of vitamin K are rare in healthy individuals because much of the body's requirement is synthesized by intestinal bacteria and it is readily available in leafy green vegetables.

vitamin P See BIOFLAVANOIDS.

vitelline membrane 1) A membrane that forms around a fertilized egg to prevent additional sperm from penetrating. 2) The transparent membrane that encases the yolk of eggs such as those produced by chickens.

vitreous Resembling glass. Vitreous substances are solids with a noncrystalline structure that resembles the structure of a liquid—and flows—even though it is hard. Many vitreous substances such as certain kinds of rock are formed in the same manner as glass (the classic vitreous substance), by supercooling the substance from a molten state.

vitreous luster A characterization of a mineral luster that gives a glasslike appearance to the material.

viviparous 1) Giving birth to live offspring that are already well developed, rather than eggs. Mammals are viviparous. 2) Describing plants that can produce (either vegetatively or sexually) new plantlets that begin development while still attached to the parent plant. Compare OVIPAROUS, OVOVIVIPAROUS.

VLF See VERY LOW FREQUENCY.

VOCS See VOLATILE ORGANIC COMPOUNDS.

volatile Easily changing into a vapor at normal temperatures and pressures. Volatile liquids are those that evaporate—exist as gases—at room temperature, such as rubbing alcohol.

volatile organic compounds Hydrocarbon compounds that have low boiling points, usually less than 100°C, and therefore evaporate readily. Some are gases at room temperature. Propane, benzene, and other components of gasoline are all volatile organic compounds. Often abbreviated VOCs when used to describe air and water pollutants.

volatiles Short for volatile organic compounds. See VOLATILE ORGANIC COMPOUNDS.

volatilization Conversion of a liquid or a solid into a vapor. Volatilization of hydrocarbons from gasoline is a source of air pollution. See VAPORIZATION.

volcanic Associated with movement, flow, or eruption of magma at the surface of the earth.

volcanic agglomerate Any coarse-grained volcanic rock composed of rounded to subangular particles larger than 2 cm in diameter. Agglomerate is a term usually applied to pyroclastic rock but may also refer to lahar deposits.

volcanic ash The sharply angular glass particles ejected from a volcano as pyroclastic debris. Volcanic ash is a tephra which is less than 2 mm in size.

volcanic bomb A lump of ejected volcanic lava that becomes streamlined during flight and congeals before falling to the ground.

volcanic breccia Any coarse-grained volcanic rock composed of sharply angular fragments larger than 2 cm in diameter. Volcanic breccia forms during pyroclastic activity.

volcanic cone Any type of accumulation with a conic shape. Specific examples are pyroclastic cones, composite volcanoes, and spatter cones.

volcanic dome A convex structure composed of viscous lava that has solidified in or above a volcanic vent. A dome may sometimes form in the caldera of a composite volcano.

volcanic dust Pyroclastic debris that is small enough to remain airborne for long periods of time (days to weeks) after a volcanic eruption.

volcanic glass A pyroclastic material that has cooled too quickly to permit mineral crystallization. Obsidian is a typical example of this natural glass.

volcanic-hazards assessment In the practice of environmental geology, a calculation of the risks associated with human habitation or activity in close proximity to a volcanic region.

volcanic neck The lava-bearing conduit that feeds a volcano from a magma chamber. Solidified volcanic necks may be resistant to erosion and develop into spectacular landforms, such as the one in Shiprock, NM.

volcanic rock Any extrusive igneous rock that has formed in association with volcanic activity.

volcanic sediment A recently formed deposit of pyroclastic material that is washed into place by a lahar or by flood activity associated with a volcanic eruption.

volcanism The physical and chemical processes associated with volcanoes and volcanic activity.

volcano A naturally occuring fissure or vent associated with the transfer of magma and volatiles to the surface of the earth.

volt Unit for potential difference and

electromotive force (EMF) in the Système International d'Unités (SI). One volt equals the difference in potential required to cause a one-ampere electrical current to flow through a resistance of one ohm (the one ampere of electrical current will dissipate one watt of power).

volume growth   Production of an increase in girth or volume rather than mass.

voluntary muscle   Any of the muscle tissues that are can be controlled consciously, generally those attached to the skeleton. Voluntary muscles are made up of striated muscle tissue.

Volz photometer   A device for measuring luminous intensity of direct sunlight. It is used to give general (rather than precise) readings for air-pollution monitoring.

vortex   A spiral flow of a fluid; a whirlwind or whirlpool.

V-shaped valley   A stream valley that has a characteristic V-shaped profile when viewed in a cross section that is oriented at a right angle to stream flow. V-shaped valleys develop where downcutting and sidecutting erosion operates simultaneously.

vug   Any small void or crystal-filled cavity in a vein or a rock.

vulcanian volcano   A type of volcano which erupts with episodic explosions that occur when trapped gases blow off a thin crust of hardened lava.

vultures   A common and confusing name because it includes members of two distinct groups of raptors, the family of New World vultures (Cathartidae) and the family (Acciptridae) that contains the Old World vultures, as well as the eagles, hawks, etc. Although both groups of vultures have superficial similarities such as bare heads or faces, the New World vultures such as condors and turkey vulture are voiceless and lack the strong beak and talons of their Old World counterparts. See FALCONIFORMES.

W

W   Chemical symbol for the element tungsten. See TUNGSTEN.

wad   A secondary mineral of barium and manganese oxides with copper and cobalt. Wad precipitates from groundwater in boggy environments.

waders   Species of shore birds and inland water birds that wade into shallow water to feed on aquatic animals or plants. Sandpipers, snipes, cranes, and herons are wading birds.

wadi   An Arabic name for an ephemeral stream that terminates in a closed basin of a desert environment.

wake   The disturbed region or track left behind an object traveling through a fluid. Aircraft leave a wake in the air behind them; ships leave a wake on the surface of the water.

waldsterben   Rapid decline and death of extensive tree populations in Central Europe, especially the vast pine and spruce forests of Germany (where the term originated; it is German for "the dying of trees") that began in the 1970s. No known diseases caused the dieback; heavy doses of air pollutants from Eastern Europe are believed to be partly or wholly responsible. See DIEBACK, FOREST DECLINE.

Wallace's line   The imaginary line that divides the Australasian zoogeographical region from the Oriental region, running between Bali and Lombok in Indonesia. Named for naturalist A. R. Wallace, who first noted the marked differences in the mammals living in these two regions. See ZOOGEOGRAPHY.

Walther's law   A stratigraphic principle that relates sediments in vertical sequence to identical sediments that are laterally continuous and where there is no gap in depositional history. Walther's law predicts that sedimentary facies that are laterally adjacent may also be found vertically superimposed.

warfarin   3-(alpha-acatonyl-benzyl)-4-hydroxycoumarin), a common rat poison. Warfarin works as a stomach poison, causing severe hemorrhaging. Vitamin K is used as an antidote for accidental ingestion. It is also used as a blood anticoagulant.

warm-bloodedness   See ECTOTHERM.

warm front   The forward boundary of a warm air mass that is rising over and slowly displacing cooler air as its advances. Warm fronts are usually preceded by increasing cloudiness followed by precipitation. See FRONT, COLD FRONT.

warm ridge   A northward bend in the jet stream causing an elongated high-pressure region. The high-pressure ridge draws warm air toward the North Pole. Contrast COLD TROUGH.

warm-season plant   A plant that grows most during the warmest seasons of the year. Compare COOL-SEASON PLANT.

washout   A recently formed channel or gully that is eroded on a land surface after a period of heavy rainfall.

wash zone   The area of a shoreline, just above the intertidal zone, that is regularly overwashed by waves.

wasps   Common name for an extremely diverse group of Hymenoptera, including parasitic and predatory forms as well as plant and nectar feeders. Most well-known species belong to the family Vespoidea (mud daubers, velvet "ants," hornets, yellow-jackets, etc.) in which larvae are reared in cells constructed by females of mud or paper (chewed plant fibers) located in or above ground. Their stings and bites can be extremely painful and can provoke extremely serious, allergic reactions in some people. There are some 4000 species of Vespoidea distributed in temperate and tropical areas. See HYMENOPTERA.

waste stream   The continuous flow of an immense volume of all components of solid waste, especially the municipal solid waste produced by homes and businesses.

wastewater   1) A general term for the seepage of water from a reservoir. 2) A general term for the effluent from a residential or municipal sewage collection system.

wastewater lagoon   A general term applied to an open holding area, such as an excavated pond, in a wastewater treatment plant.

wastewater pond   See WASTEWATER LAGOON.

water   A liquid form of the chemical compound H_2O. Water is essential for all life.

water balance   1) The requirement for organisms to control the relative amounts of water inside of and outside of their cells, replacing water lost through excretion by drinking water. 2) The extent to which the water lost to a region through evaporation, transpiration, or runoff is replenished by precipitation. See HYDROLOGIC CYCLE.

water body   A general term for any mass of water on the earth's surface, such as a lake, a pond, a stream, or an ocean.

water budget   An accounting of the input, storage, and output of water within a hydrologic system, such as a watershed.

water consumption   1) The amount of water artificially withdrawn from a natural hydrologic system. 2) Water that on use does not return to stream or groundwater.

water cycle   See HYDROLOGIC CYCLE.

water diversion   The practice of transferring the flow of water from one stream system to an irrigation system or to another stream system.

water droplet coalescence   The merging of two small drops of water into one larger drop. Most rain is believed to be formed by a collision-coalescence process. In this process, droplets just large enough to begin to fall collide with much smaller droplets along the way, coalescing into larger drops until they reach the size of raindrops and can fall to the ground without evaporating.

water gap   A narrow valley or ravine cut through a ridge or a mountain by the action of an antecedent stream. A water gap may be formed during a gradual regional uplift.

water-holding capacity   A measure of the smallest water content of the soil after a period of gravity drainage.

water injection   (wi) 1) A technique of secondary oil recovery in which water is pumped into a well to drive up the remaining oil. 2) Injection of water down wells into aquifers as recharge.

waterlogging   A term applied to a soil in which the water table stands at or near the land surface. Only hydrophytic plants can survive in a waterlogged soil.

water mass   1) A volume of ocean water characterized by a chemical composition that indicates the mixing of two or more distinct water sources. 2) A volume of ocean water that has a characteristic temperature and salinity.

water potential   The tendency for water to cross a semipermeable membrane as a result of osmosis or diffusion; water moves from solutions that are less concentrated toward those that are more concentrated. Formerly called osmotic pressure. See OSMOREGULATION.

watershed   The total area of land surface from which an aquifer or a river system collects its water.

waterspout   The column of water caused by a tornado occurring over the ocean. When ocean waterspouts move over land, they often release saltwater rain (and sometimes even fish!).

water table   The upper limit of groundwater within an unconfined aquifer of soil or bedrock. The water table forms the boundary between the zone of saturation and the zone of aeration, or vadose zone.

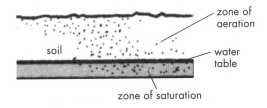

water-use efficiency   The ratio of photosynthesis to transpiration, the dry matter (net primary production) produced per amount of water consumed by a plant.

water vapor   Water in a gaseous rather than a liquid state. The moisture in the atmosphere is water vapor. In addition to becoming precipitation, water vapor contributes to the formation of storms as it absorbs heat from the sun and the earth's surface.

water withdrawal   The pumping or drainage of water from an aquifer.

water yield   A measure of the runoff from a drainage basin. The water yield is theoretically equal to the precipitation minus the evapotranspiration.

watt   (W) Unit for power in the Système International d'Unités (SI) and MKS system. One watt equals one joule per second, or .0013 horsepower (1 kilowatt = 1.34 horsepower). Electrical power is measured in watts, equal to voltage (electrical potential) multiplied by current (in amperes). Compare VOLT.

wattles   1) The fleshly, often brightly colored tissues that dangle from the throats on males of birds such as roosters and turkeys. 2) A similar structure on other animals such as reptiles, or a barbel on a fish.

wave   A disturbance, a change in pressure (as in a sound wave) or a signal (as in a radio or other electromagnetic wave), that follows a periodic rise and fall as the wave travels through an elastic medium. Electromagnetic waves take the form of changes in electric and magnetic fields rather than changes in the pressure of a medium. Waves can transfer energy through a medium without causing any permanent displacement of that medium.

waveband   Range of wavelengths occupied by a particular kind of radio-wave transmission. Very low frequency (VLF) radio waves and ultra high frequency (UHF) radio waves are examples of wavebands.

wave base   The lower limit of wave disturbance in a water body. The wave base generally occurs at a depth equal to ½ the wave length.

wave cloud   A cloud associated with lee waves in mountainous areas. Such clouds form at the crest of these standing waves and, like the waves, do not travel much relative to the ground. Wave clouds are smooth and sometimes lens-shaped, giving rise to the alternate name lenticelular clouds. See LEE WAVES.

wave-cut bench   The gently sloping erosional surface of the shore zone between a wave-cut cliff and the waterline.

wave-cut cliff   An erosional precipice formed by the undercutting action of waves at a shoreline.

wave-cut platform   A flat-topped or gently sloping coastal erosional surface produced by wave action. A wave-cut platform is composed of a wave-cut bench and the adjacent submarine abrasion platform.

wave cyclone   The primary weather-producing feature in the continental United States that travels from west to east. Wave cyclones form along a front and move with it; they may last from a few days to over a week. The name comes from the large-scale wave-shaped circulation they produce. Also called frontal lows or, because they occur at middle latitudes such as the United States, middle-latitude cyclones. See CYCLONE.

wave front   The line or surface connecting points in the same phase of a wave. When a pebble is dropped into a pond, the circles radiating out along the pond surface are each wavefronts.

wavelength   The distance covered during the period, one complete cycle, of a vibration: the distance required for the wave to return to a similar phase or point. Radio wavelengths range from about a meter to about a kilometer; visible light and x-ray wavelengths are much smaller, expressed in nanometers or angstroms. Compare PERIOD, AMPLITUDE.

wave (ocean surface)   A deep-water ocean wave in which the rapid particle movement is confined to the ocean surface.

wave refraction   The changes in the direction of propagation that occur as a wave approaches shallow water or a submarine obstruction. For example, a submerged reef or bar may cause refraction as waves approach a lagoon.

Wb   Abbreviation for weber. See WEBER.

weak acid   An acid that undergoes only partial dissociation into its component ions in water. Most acids are weak or moderately weak. Acetic acid, better know as vinegar, is a weak acid. Compare STRONG ACID.

weak base   A base that undergoes only partial dissociation into its component ions in water. Ammonia and most other bases are weak or moderately weak. Compare STRONG BASE.

weak electrolytes   Substances that are poor conductors of electricity when dissolved in water, because they undergo only partial ionization even in moderately concentrated solutions. Compare STRONG ELECTRO-LYTES.

weather   Short-term variations in atmospheric conditions (from minutes to months) in a certain locality. Weather is made up of temperature, precipitation, wind speed, cloud cover, and similar physical characteristics. Compare CLIMATE.

weathering The disintegration or decomposition of rock and surface material in response to contact with air, water, chemicals, and biologic agents.

weathering series An array of silicate minerals presented in their relative order of susceptibility to weathering. The weathering series is typically the reverse of the crystallization order in an igneous rock.

weather map A graphic representation of weather conditions for a specified region at a specified time. Weather maps, which show temperature, atmospheric pressure, wind velocity, cloud cover, and similar features, are essential for forecasting.

weather modification Deliberate attempts to steer or to alter natural weather phenomena. Constructing windbreaks and cloud seeding for fog dispersal are forms of weather modification. See CLOUD SEEDING.

weather radio Continuous weather reports broadcast by the NOAA (National Oceanographic and Atmospheric Administration) on restricted wavelengths that can be picked up by receivers designed to tune in to those bands.

weather ship A ship outfitted for measuring weather conditions. It may be used for research or for relaying this information to a weather station.

weber (Wb) The unit for magnetic flux in the Système International d'Unités (SI) and MKS system. One weber is equal to one volt multiplied by one second, or one joule per ampere.

Weberian apparatus A structure in carp, catfish, and similar fish species that conveys sound vibrations to the ear; it corresponds roughly to the ossicles in the middle ear of vertebrates. It consists of several small bones that connect the air bladder to the ear. Also called Weberian ossicles. See OSSICLES, AUDITORY.

weed Any plant growing where it is not wanted, usually a wild plant that grows without much care or cultivation and may be invasive in cultivated areas. See R-STRAGETISTS, RUDERAL.

Wegenerism A term applied to the theories of continental drift as presented by Alfred Wegener (1880–1930).

weight A measurement of the gravitational force acting on a body or a substance. Weight is equal to mass multiplied by acceleration due to gravity, and has units of force (newtons or dynes). Compare MASS.

weir A small dam or channel designed to regulate or measure the flow of water. See the chart on the next page to determine the flow or discharge.

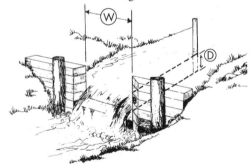

Measuring the flow of water using the weir method.

Weismannism A theory of heredity sometimes called "the continuity of the germ plasm," stressing that genetic material (germ plasm) is passed along unchanged from one generation to the next. This precludes the passing along of

Measuring the Flow Using the Weir Method

Depth in Inches on Stake D	Discharge cubic feet/minute/inch of weir width (W)
1 inch	.40
2	1.14
3	2.09
4	3.22
5	4.50
6	5.90
7	7.44
8	9.10
9	10.86
10	12.71
11	14.67
12	16.73
13	18.87
14	21.09
15	23.38
16	25.76
17	28.20
18	30.70
19	33.29
20	35.94
21	38.65
22	41.43

See the diagram on the previous page to determine where "D" and "W" are measured.

characteristics acquired during one generation. Weismannism is a cornerstone of neo-Darwinism. See NEO-DARWINISM.

welded ash-flow tuff   See WELDED TUFF.

welded tuff   A hard, dense volcanic rock accumulated from ash flows that were partially melted during formation.

Wellman-Lord process   Air pollution control technique, a form of flue gas desulfurization that can be used in coal-burning power plants. It uses a combinination of an electrostatic precipitator (to remove fly ash) and a scrubber. The process is 90 percent efficient; the sulfur dioxide (SO_2) removed is so concentrated that it can be used as a source of sulfur, or for the manufacture of sulfuric acid. See FLUE GAS DESULFURIZATION, SCRUBBER.

well sorted   A term applied to any sediment that consists of clastic particles that are of approximately equal size.

West African floral region   One of the phytogeographic regions into which the Paleotropical realm is divided according to similarities between plants; it extends along the coast from West Africa south to central Angola and east to Lake Tanganyika. See PALEOTROPICAL REALM, PHYTOGEOGRAPHY.

westerly   1) Describing a wind that is blowing from west to east. 2) The west-to-east air currents that dominate the temperate latitudes, between approximately 35° and 65° both north and south of the equator. They are often called prevailing westerlies. Sailing ships crossing the Atlantic Ocean by following the trade winds used the less-reliable westerlies to make their return trip. See TRADE WINDS.

wet adiabatic lapse rate   A variation of normal lapse rate, in which the average drop in temperature with increasing altitude in the lower atmosphere is calculated for saturated air, rather than for dry air. The wet adiabatic lapse rate is always less than the dry rate. It is also more variable; the rate is less for very moist air (roughly 0.5°C for every 100 meters of altitude), and gradually approaches the dry lapse rate as the moisture content of air decreases. Also called saturated adiabatic lapse rate. Contrast DRY ADIABATIC LAPSE RATE.

wet-bulb temperature   The reading from the thermometer whose bulb is covered with water-soaked fabric in a psychrometer (wet-and-dry-bulb hygrometer). Comparing the wet-bulb temperature to the dry-bulb thermometer with tables gives the relative humidity and dew point of the air being sampled.

wet-bulb thermometer   Half of a psychrometer (wet-and-dry-bulb hygrometer): the thermometer whose bulb is covered with water-soaked fabric to give the wet-bulb temperature. Comparing this wet-bulb temperature and the reading from the dry-bulb thermometer with tables gives the relative humidity and dew point of the air being sampled.

wet deposition   Particulate air pollutants from the atmosphere carried to the earth's surface by some form of precipitation. A rainout (or washout) is a form of wet deposition. Compare DRY DEPOSITION.

wetlands   Areas of land that are covered with water for at least part of the year, have characteristic hydric soils, and have one of a number of distinct vegetation types: swamps, marshes, salt marshes (and other coastal wetlands), and bogs. Wetlands have important functions including purifying the water that recharges aquifers, providing food and habitat for many different species, and providing temporary stopover sites for migrating waterfowl.

wet meadow   An area of grass next to a river, subject to periodic flooding. The British call this a water meadow, in which the flooding is often artificial and done to provide a yearly source of fertilizing silt to boost the hay crop.

wet scrubber   Another term for scrubber, used to distinguish ordinary scrubbers from dry scrubbing (dry limestone process). See DRY LIMESTONE PROCESS, SCRUBBER.

whales   See CETACEA.

whirlwind   General term for any small-scale vortex of air. Tornados, dust devils, and waterspouts are all whirlwinds.

white blood cells   See LEUKOCYTES.

white light   Sunlight, or similar artificial light, that has not been divided up into its different component wavelengths (colors). When passed through a prism, white light breaks up into the colors of the rainbow: red, orange, yellow, green, blue, indigo, and violet. Compare VISIBLE LIGHT.

white noise   Sound that includes frequencies from the entire range of audible frequencies, all at once; analogous to white light being a combination of all of the colors of the rainbow/spectrum.

Whittaker plots   See DOMINANCE-DIVERSITY CURVE.

WHO   See WORLD HEALTH ORGANIZATION.

whole-tree clearcut   A logging technique in which the entire above-ground portions of all trees are removed from a site that is clearcut. Compare CLEARCUT, WHOLE-TREE HARVESTING.

whole-tree harvesting   A method of harvesting trees in which machines pull entire trees (above- and below-ground portions) from the ground and reduce them to small chips. Compare SEED-TREE CUTTING, SELECTIVE HARVEST, SHELTERWOOD CUTTING.

Wien effect   An increase in the observed electrical conductivity of an electrolyte when subjected to very high voltage gradients. The Wien effect may be caused by the increased dissociation of the electrolyte ions. Also called the field dissociation effect.

wilderness   General term for a large area

that has not been significantly disturbed by humans, one in which they are only temporary visitors leaving little evidence. See WILDERNESS AREA.

Wilderness Act   Federal legislation passed in 1964 (16 U.S.C.A. $1131 et seq.) designating areas of public lands as wilderness areas. Its stated goal is to preserve areas "where the earth and its community of life are untrammeled by man, where man himself is a visitor who does not remain."

wilderness area   Roadless areas within national parks, national forests, or national wildlife refuges protected from development and set aside for the preservation of its wild characteristics. Recreation such as camping and sport fishing are allowed, but no commercial activities are allowed unless they predate the area's wilderness designation. The areas were authorized by Congress through the Wilderness Act of 1964, creating a national wilderness preservation system.

Wilderness Society   A nonprofit membership organization devoted to the preservation of wild lands and wildlife, and to fostering an American land ethic. Organized in 1935, it lobbies for the designation of wilderness areas (as part of the national wilderness preservation system) and for related conservation issues. See NATIONAL FOREST.

wildlife   Any organisms living independently of humans; undomesticated species of plants and of animals.

Wilson cycle   A theoretical cycle of plate tectonic activities causing the opening and closing of an ocean basin. The cycle begins with continental rifting, production of oceanic crust, and ocean formation, and then proceeds through subduction, conti-nental convergence, and final closing of an ocean basin.

wilting point   The point at which the water content of a soil becomes low enough to induce wilting in plant tissues. See PERMANENT WILTING POINT.

windbreak   A row of trees (sometimes a double row) planted perpendicular to the direction of the prevailing wind. Windbreaks are planted to reduce soil erosion on cultivated land that is exposed to a lot of wind, to reduce the wind speed around houses built in exposed locations (thus making them easier to heat), and to protect houses and gardens along the coast from ocean winds. In the Great Plains area of the western U.S., windbreaks used to reduce soil erosion and snow drifting are usually called shelter belts. See DUST BOWL.

wind-chill factor   An apparent lowering of temperature caused by wind. Although the temperature measured on a thermometer does not change as wind increases, the atmosphere's cooling power (especially on the human body) corresponds to decreasing temperatures at no wind speed. For example, the danger of frostbite increases as higher wind speeds draw more heat away from the body. Also called simply wind chill.

wind-chill index   A chart showing the relationship between increasing wind speed and apparent or perceived temperature due to the increased rate of cooling of the human body. The wind-chill index shows approximate loss of body heat in kilogram calories per hour per square meter of skin surface.

wind erosion   A form of removal, transport, and deposition of soil material caused by the action of wind.

wind farms   Sites with a large number of wind turbines (100 to 1000, large enough to produce energy on a level comparable to conventional power plants), used to convert wind energy into electricity for the power grid rather than to power individual dwellings.

wind measurement   The determination of the speed and the direction of wind (wind velocity). Wind direction is measured with a wind vane, and speed is measured with an anemometer; an aerovane makes both measurements at once. Wind can also be measured by tracking balloons optically or using radar. See BEAUFORT SCALE.

wind power   The energy of the wind harnessed to drive either a machine (a windmill) or to drive an electrical generator (a wind turbine).

wind profile   A description of the change in wind characteristics with altitude. Wind profile includes parameters such as direction, average speed, and turbulence level.

wind rose   Any of several types of charts showing the direction, force, and frequencies of prevailing winds for a given site. A common form uses a star of eight (or sometimes sixteen) lines, for the different compass direction.

windrows   1) Organic material spread out into long rows to facilitate mechanical turning and aeration, used to speed large-scale composting. 2) Debris floating on the surface of a lake or ocean in long rows paralleling the direction of the prevailing wind. They are created by Langmuir circulation within the upper layers of the water.

wind shear   A change in speed or direction of the wind (measured along a line perpendicular to the wind) divided by the distance over which the change occurs. Wind shear can be either horizontal or vertical, but usually implies a change of wind with increasing altitude.

wind turbines   Electrical generators driven by wind power. See DARRIEUS GENERATOR, SAVONIUS ROTOR, WIND FARM.

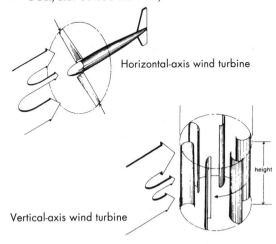

Horizontal-axis wind turbine

Vertical-axis wind turbine

height

windward   Facing the direction from which the wind is blowing; upwind. Compare LEEWARD.

wing   1) A movable organ on animals such as birds, insects, and bats, originally designed for flight even if no longer used (as on the ostrich). 2) A structure on a seed or a fruit that aids in wind dispersal. 3) Any thin, membranous outgrowth on a plant, especially a flange on a stem, as in some species of *Euonymous*.

wintergreen   1) A condition in which a plant, usually an herbaceous species, retains its leaves through the dormant season and then sheds them as new leaves emerge during the following growing season; unlike evergreens, on which some leaves that are more than one year old are always present. Christmas fern (*Polystichum acrostichoides*) are wintergreen. 2) The common name of several different species that have wintergreen or evergreen leaves, such as *Gaultheria procumbens*, and various *Chimaphila* and *Pyrola* species. Contrast EVERGREEN, DECIDUOUS.

wintering area The place where migratory (or sometimes even nonmigratory) animals congregate during the winter season. See DEER YARD.

winze A vertical shaft that connects two different work levels within a mine.

witches' broom An abnormal growth of closely bunched, tuftlike twigs at the ends of branches of a number of different woody plants, including spruce, fir, pine, cherry, blueberry, and hackberry. It is a disease with a number of different causes, usually a virus or fungus but also dwarf mistletoe (a parasitic plant) and, in the hackberry, a combination of a mite and a fungus.

within-habitat comparison Studying and contrasting the species diversity in two different areas having a similar overall type of habitat.

within-habit specialization The development of restricted utilization of habitat factors by different individuals, populations, or species.

wolframite A gray-to-black mineral of iron, manganese, and tungsten oxide that forms monoclinic crystals in pneumatolytic and high-temperature granitic rocks. Wolframite is a major ore mineral of tungsten.

wood The hard, lignified, cellulose tissue that forms the main body of some perennial plants. It is made up of secondary xylem produced by the cambium during secondary thickening. See LIGNIN, XYLEM.

wood alcohol Another name for methanol. See METHANOL.

woodland Vegetation consisting of numerous trees, generally more spread apart than those of a forest, not close enough to form a continuous canopy. See FOREST.

woodpeckers Although not found in Australia or Oceania, the woodpeckers are widely distributed elsewhere. These birds are well adapted for climbing on vertical surfaces such as tree trunks, with their feet gripping the surface and their tails bracing them erect. The woodpecker's slender but strong neck, reinforced skull, and hard straight bill are used for digging holes in trees to obtain boring insects, which are extracted by the long, barbed tongue. Unfortunately, one of the most spectacular woodpeckers, the ivory-billed of North America is probably extinct due to habitat destruction, although the slightly smaller pileated woodpecker may still be found in mature forest areas. See PICIFORMES.

woolsorters' disease The pulmonary form of anthrax, common in workers who inhale spores of the *Bacillus anthracis* bacterium as they handle contaminated fleeces, wool, or other animal hair. See ANTHRAX.

work The work done by a force on an object is the force multiplied by the distance the object moves in the direction of the force. Work describes and quantifies all classical energy transfers.

workability A description of the ease with which a geological material may be processed.

WRI See WORLD RESOURCES INSTITUTE.

World Bank Popular name for the International Bank for Reconstruction and Development, an institution that works with other agencies to promote international trade and development. It grew out of the 1945 United Nations Monetary and Financial Conference at Bretton Woods,

NH, with the original goal of aiding postwar Europe. Compare INTERNATIONAL MONETARY FUND.

World Health Organization (WHO) An agency of the United Nations, founded in 1948 to promote cooperation between nations in disease control and in improving the health of populations throughout the world, especially in developing countries.

World Resources Institute (WRI) A policy research center created in 1982 to assist government and private organizations in addressing issues of natural resource management, economic growth, integrity of local environments, and international security.

Worldwatch Institute (WWI) A nonprofit research organization concerned with identifying and analyzing emerging global problems and trends, and bringing them to the attention of the public and opinion leaders. Founded in 1974, it produces papers on specific global issues as well as an annual report, *State of the World*.

World Wildlife Fund (WWF) The largest private U.S. organization working globally to protect endangered wildlife and wild lands, especially in tropical regions. Founded in 1961, it is now affiliated with the Conservation Foundation, which promotes wise use of natural resources.

wrack zone The portion of the intertidal zone dominated by a marine algae of the genus *Ascophyllum*, such as the common marine algae knotted wrack.

wrench fault See STRIKE-SLIP FAULT.

WRI See WORLD RESOURCES INSTITUTE.

WWF See WORLD WILDLIFE FUND.

X

xanthism Replacement of normal pigmentation with yellow or gold pigments, as in the goldfish. Also called xanthochroia.

xanthophylls A group of yellow (carotenoid) pigments present in plant cells and acting as accessory pigments to chlorophyll in photosynthesis. Different plant and algal groups have different characteristic sets of xanthophylls. Xanthophylls contribute to the yellow and brown colors of autumn leaves. See CAROTENOID.

Xanthophyta The yellow-green algae phylum of the kingdom Protista. These algae have photosynthetic plastids known as xanthoplasts, containing chlorophylls, xanthnins, and carotene. Xanthophytes usually form colonial masses and are frequently associated with scummy mats on turbid fresh waters. Familiar members of the phylum are the filamentous mat-forming *Vaucheria*, the balloonlike *Botrydium* of damp soils, and the brownish pond-scum of alkaline lakes *Botryococcus*. See PROTISTA.

X chromosome One of the two chromosomes in humans that determine sex. If an egg contains two X chromosomes, it will develop female sexual characteristics; if one of these is instead a Y chromosome, it will be a male. See Y CHROMOSOME.

xenia The genetic effects caused by pollen on the endosperm (rather than on the ovary), resulting from the double fertilization that occurs in seed plants and affecting the material tissues of the fruit. An example of xenia is the different colors of corn seed produced depending upon the pollen source. Compare METAXENIA.

xenogamy Cross-fertilization between genetically different flowers, those on different plants of the same species.

xenolith A term applied to a fragment of unrelated country rock that is included as a foreign body within an igneous rock.

xenoparasite Organisms infesting an organism that is not usually a host species, or those that can only infect an organism that is injured.

xerarch succession A sere, or series of successional communities, that begins on a dry site such as a bare sand dune or a rock surface. Compare HALARCH SUCCESSION, HYDRARCH SUCCESSION, XEROSERE.

xeric 1) Lacking available moisture for organisms to utilize. 2) Referring to xerophytes, plants adapted to survive environments where moisture is quite scarce. Contrast MESIC.

xeriscaping Landscape design using native and drought-tolerant species of plants and appropriate water-management design. It is increasingly used in areas recieving very little rainfall, such as the U.S. Southwest and Great Plains. In xeriscaping, much of the lawn is replaced with gravel surrounding cacti and other plants, dramatically reducing the water use. See DROUGHT TOLERANT, XEROPHYTE.

xeromorphy The development of adaptations enabling plants to retain water in order to survive environments lacking fresh water, including deserts but also including saltwater marshes or highly alkaline soil where the concentration of salts tends to draw water out of plant tissues. Xeromorphy can include a thickening of the outer protective layer (epidermis), reduction of leaf size, or alterations of the normal pathway for photosynthesis (crassulacean acid metabolism). See DROUGHT RESISTANCE, CRASSULACEAN ACID METABOLISM.

xerophyte A plant that is very efficient at retaining water and can grow in deserts, on very dry ground, or in environments with high salt concentrations that tend to draw water out of plant tissues by osmosis. Cactus, creosote bush, and sagebush are all xerophytes. Compare HYDROPHYTE, MESOPHYTE.

xerosere The succession of communities growing in dry conditions, such as that starting on a rock surface or a sand dune. The subsequent communities often gradually change the xeric environment to a mesic state. Also called xerarch succession. See SERE, CLISERE, LITHOSERE, SUBSERE.

x-ray 1) Electromagnetic radiation of extremely short wavelength, approximately 10^{-3} to 10 nanometers. X-rays can be generated by bombarding a solid substance, such as metal, with high-speed electrons; they can penetrate many substances including human tissue. X-rays are also produced by high-energy objects in space, such as the sun and other stars, galaxies, and quasars. 2) A photograph made using x-rays, such as those used for medical diagnosis or for analyzing the integrity of structures.

x-ray analysis A technique used to study the physical and chemical characteristics of a crystal lattice by using short-wavelength electromagnetic radiation (x-rays).

xylem The tissue in vascular plants that transports water and dissolved mineral nutrients upward from the roots to the shoots and leaves. It consists of elongated cells linked end to end, with perforated end walls that allow water to pass from one to another. The tracheids and fibers of secondary xylem (that formed by the cambium) also contribute to supporting the plant; thickened with lignin, they form wood. The parenchyma cells associated with xylem can serve as storage tissue. See TRACHEID, VESSEL ELEMENTS, PHLOEM, SECONDARY XYLEM.

xylophagous Feeding on or in wood; sometimes used to describe organisms that bore into wood, destroying it without necessarily consuming it.

y

yard The British standard unit of length, equal to 0.91 meters.

yardang An elongated hill that is sculpted by wind in compact, coherent, but unconsolidated sediment such as clay-rich silt. A yardang aligns with the direction of the prevailing winds.

yarding 1) A winter behavior of deer in which they congregate in sheltered areas known as yards, for protection from extreme weather. 2) The gathering together of slash (residue of tree tops and branches) during or after a logging operation. See DEER YARD.

yazoo stream A small tributary stream that flows parallel to the main stream on a floodplain. A yazoo stream is separated from the bank of the main stream by a levee.

Y chromosome The male sex chromosome (in male heterogamic species), one of the two chromosomes that determine sex and the one that is present only in males. If a fertilized egg contains a Y chromosome along with its initial X chromosome, it will develop male sexual characteristics. See X CHROMOSOME.

year-class effect The common tendency for the surviving offspring from a given year to dominate a given population of a species. This condition is often found in fisheries where the highly variable stock-recruitment relationships lead to some year-class cohorts failing while others succeed.

yeasts Unicellular organisms, especially those belonging to the genus *Saccharomyces*, of the Ascomycota phylum of the kingdom Fungi. Best-known are those species that produce the enzyme zymase, which allows them to break down sugars into carbon dioxide; they are used for baking, winemaking, and brewing. Some are used as a source of protein, amino acids, or B vitamins (*Torula*, nutritional yeast). See FUNGI, ASCOMYCOTA, SACCHAROMYCES.

yellowcake A mixture of uranium oxides extracted from uranium ore. It is used primarily for making fuel for nuclear reactors.

yellow-green algae See XANTHOPHYTA.

yellows A general term for several different plant diseases sharing as a major symptom the yellowing of the plant. Examples include lethal yellowing, a disease of coconut palms that causes yellowing of foliage and death, and cabbage yellows, caused by a fusarium fungus. See YELLOWS.

yield point The stress at which a substance undergoes a significant change (increase in deformation) without an increase in the load (applied stress). Both iron and annealed steel are characterized by an abrupt yielding to stress that occurs at a specific level of stress (the yield point).

Yoda's 3/2 power law See 3/2 THINNING LAW.

yolk sac In some animal species such as squids, octopuses, and lower vertebrates, a sac filled with yolk, which provides the only food for the embryo and is attached to the embryo by the yolk stalk.

young soils See AZONAL SOIL, INCEPTISOLS.

Z

Z (Zulu time) Z is a suffix used internationally by meteorologists to indicate Greenwich Mean Time. Although written as Z, it is pronounced and transmitted as Zulu. The standard hours used by meteorologists are 0000Z, 0600Z, 1200Z, and 1800Z (1800 GMT is 6 PM in England, or 1 AM EST). See GREENWICH MEAN TIME.

zeatin A natural plant growth substance of the cytokinin group that is widely distributed among plants. The name comes from Zea mays (corn), because it was originally isolated from young corn kernels. See CYTOKININS.

zeitgeber Any event, such as a drop in temperature or light intensity in the evening, that synchronizes the biological clock of an organism with external phenomena. The word means time-giver in German.

Zero Population Growth (ZPG) A nonprofit organization founded in 1968 to inform people about the problems accompanying rapid global population growth. It sponsors educational and advocacy programs to reduce the rate of American and global population growth theoretically to zero or replacement level.

zero sum game Originally a term in game theory, but now generalized to any system whereby when one party wins or moves ahead, the other party loses or falls behind. The exploitation of the earth's nonrenewable resources is often described as a zero sum game.

ZETA Acronym for Zero Energy Thermonuclear Apparatus, an experimental device used in nuclear fusion research at the Atomic Energy Research Establishment in Harwell, United Kingdom. Like a tokamak, it is a torus, or doughnut-shaped, apparatus.

zinc (Zn) A hard, metallic element with atomic weight 65.37 and atomic number 30. Its resistance to corrosion makes it useful for galvanizing, providing a rust-resistant coating for iron or steel, as in galvanized roofing nails. Zinc has many commercial uses including in alloys such as brass, in batteries, and as a paint pigment; it is also a trace nutrient for plants and animals.

zircon A mineral of zirconium silicate that forms tetragonal crystals in a wide variety of igneous, metamorphic, and sedimentary rocks. Zircon, which may form gemstone-quality crystals, is the principal ore of zirconium. See APPENDIX p. 628.

Zn Chemical symbol for the element zinc. See ZINC.

zonal flow The large-scale circulation of atmospheric currents within a specific range of latitude.

zonal index A measure of the atmospheric pressure differential between 35° and 55° of latitude. The zonal index provides an indication of the strength of westerly winds in the middle latitudes.

zonal soil A term applied to a soil which has well-developed horizon characteristics. Comparison of zonal soils is the basis of the United States Department of Agriculture soil classification system.

zonation 1) The occurrence of distinct distributions of different species, forming recognizable (usually parallel) bands of characteristic vegetation, along ecological gradients. Zonation occurs along the increasing altitude of a mountain slope as well as across the intertidal zone. Unlike succession, these different communities all occur at the same time but spread out over different areas. Compare STRATIFICATION. 2) The state of separation of earth materials, such as minerals or rocks, into distinct areas or regions. 3) The distribution of organisms or of fossils into biostratigraphic regions.

zone (biostratigraphic)   A particular rock unit that is defined by paleontologic characteristics, such as a fossil assemblage, rather than by lithologic features. The biostratigraphic zone is the fundamental type of biostratigraphic unit. See BIOS-TRATIGRAPHIC UNIT.

zone fossil   An index fossil of a particular biostratigraphic zone.

zone of ablation   An area in the lower part of a glacier where the net loss of ice exceeds the net gain. This zone is usually characterized by a surface of bare ice, old snow, and an abundance of rock particles. This area is also known as the zone of wastage. Contrast ZONE OF ACCUMULA-TION.

zone of accumulation   An area in the upper part of a glacier where the net gain in ice exceeds the net loss. The zone is usually characterized by a build-up of snow from the previous winter's snowfall. Contrast ZONE OF ABLATION.

zone of aeration   See VADOSE ZONE.

zone of convergence   See DESTRUCTIVE PLATE MARGIN.

zone of deposition   An area covered by unconsolidated materials deposited by a receding continental glacier.

zone of divergence   See MIDOCEANIC RIDGE.

zone of leaching   A general reference to the A horizon of a soil.

zone of saturation   The subsurface portion of the soil in which the interstitial spaces are completely filled with water. Found below the water table.

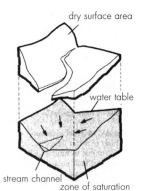

dry surface area

water table

stream channel   zone of saturation

zone of silence   An area in which electromagnetic signals or sounds from a specific source cannot be received strongly enough to be useful. See SOUND SHADOW.

zones of stress   Areas in which a species can survive, but not easily, because the environmental conditions are far enough from the ideal for that species to cause stress. See RANGE OF TOLERANCE.

zoning   A legal mechanism, usually at municipality level, delineating districts for the purpose of regulating or controlling, or in some way limiting the use of private property, and the construction of buildings within the zones. Zoning authority (a form of police power) generally derives from the state legislature which, in turn, by statutes defers or delegetes it to the municipality.

zoobenthos   Aquatic animals living in or on bottom sediments. See BENTHOS.

zoobiotic   Living in association with, or parasitic upon, an animal species.

zoochore   A seed or spore that is dispersed by being carried around by animals, such as burrs that cling to the fur of an animal and are thereby transported away from the parent plant. Compare ANEMOCHORE, ANTHROPOCHORE, ENDOZOOCHORE, HYDROCHORE.

zoocoenose   An animal community. See BIOCOENOSIS.

zoogeography   The study of the geographical distribution of animals, especially the causes and effects of such distribution and the relationships between the areas and the animals that inhabit them. Zoogeography divides the earth into different faunal regions, usually the Holarctic (sometimes divided into the Palaearctic and Nearctic regions), Ethiopian (African), Oriental, Australasian, and Neotropical regions. Compare PHYTOGEOGRAPHY.

zoography   See ZOOGEOGRAPHY.

zooid   1) Animal-like; having a form resembling that of an animal. 2) Any of the distinct individuals making up a colonial or a compound animal organism, as in corals, hydras, sea squirts, and bryozoans. 3) An animal cell capable of locomotion or independent existence, such as a phagocyte or a sperm cell.

Zoomastigina   The zoomastigotes are unicellular (although sometimes aggregating), flagellated organisms that are heterotrophic and compose a phylum of the kingdom Protista. These organisms typically are aquatic, either parasites or symbionts, and represent a considerable diversity of taxonomic classes. Some of these "animal flagellates" have a single undulipodium, whereas others have thousands. Some of the members of this phylum are the urogenital pathogen *Trichomonas*, the *Giaria* organism causing beaver-fever, and African sleeping-sickness *Trypanosoma*, as well as the cellulose-digesting protozoa of termites *Trichomitopsis*, *Devescovina*, and *Gigantomonas*. See PROTISTA.

zoomastigotes   See ZOOMASTIGINA.

zoonosis   Any disease that can be transmitted from animals to humans under natural conditions, such as rabies. Plural is zoonoses.

zoophagous   Carnivorous; feeding on animals.

zoophyte   Any of various invertebrate animals such as corals and sea anemones that look like plants, or have a branching, plantlike growth form (as in sponges).

zooplankton   Animal species of plankton in freshwater or marine environments; plankton that do not produce their own energy but feed on phytoplankton (plant plankton) or smaller animal plankton. They are barely able to swim, if at all, and so are carried by water currents. Rotifers and microcrustaceans are zooplankton. See PLANKTON, PHYTOPLANKTON, KRILL, ACTINOPODA, COPEPODA, DINOFLAGELLATA, FORAMINIFERA, OSTRACODA.

zoosis   Any disease caused by a parasitic animal.

zoospore   An asexual reproductive cell (a spore rather than a gamete) that swims by using cilia or flagella. Zoospores are produced by some algae and fungi.

zooxanthellae   Photosynthetic, symbiotic yellow or green algae living in association with actinopods and coral colonies.

ZPG   See ZERO POPULATION GROWTH.

zygomorphic   Describing flowers that are bilaterally symmetrical; divisible into halves around a single plane, such as monkshood (*Aconitum*) and most orchids.

Zygomycota   The zygomycetes phylum of the kingdom Fungi. The members of this phylum lack cross walls and reproduce by the asexual production of conidia (vegetative spores) or by conjugation of cells of different mating types to produce a zygospore. The phylum consists of the Mucorales (including bread mold *Rhizopus*), the Entomophthorales (a group of fungi that parasitize insects, for example fly fungus *Entomophtora*), and the Zoopagales, a group of fungi such as *Cochlonema* that parasitizes protists. See FUNGI.

zygote The diploid cell formed by the union of a haploid male and haploid female gamete (reproductive cell).

zymase A mixture of enzymes present in yeast that changes sugar into alcohol and carbon dioxide under anaerobic conditions (absence of oxygen). See SACCHAROMYCES.

zymogenous 1) Relating to or producing fermentation. 2) Relating to a proenzyme or precursor to an active protein (zymogen). 3) Pertaining to soil organisms that show an increase in metabolic activity with the addition of new organic matter to the soil surface.

APPENDIX

Ecological Models

EVERY SCIENCE INVOLVES THEORY, and every theory involves a model of how the object of study works. We think and explain science through models of reality. A model is an intentional simplification of reality, simplified so that phenomena of interest can be examined, analyzed, and understood. We use models in two ways, conceptually and formally. In a conceptual model, the model is a metaphor, used to explain how something works. In a formal model, the model is explicit in mathematical or numerical (e.g., a computer simulation) terms.

Some ecologists are uncomfortable with theory, and there has been a tendency in the past to view "doing science" and "creating theroy" as separate activities, a tendency that is decreasing today. In reality, there is a model behind every scientific explanation. In some cases, the model is implicit; in those cases, we may be unaware of important assumptions. We gain when we make models explicit and formal.

While a model is necessary in science, it is also true that any model involves specific assumptions about how the world works (the world is assumed to be like the model). Thus the choice of a model is extremely important in the development of theory and of explanations. It is to our advantage to bring the underlying assumptions to the surface and make them part of active discussion. Ecologists and environmental scientists sometimes believe that mathematical formulations are free of assumptions that might restrict their perspective, but this is not the case. For example, differential and integral calculus,

which was the foundation for most formal ecological theory during the first seven decades of the twentieth century, was derived to explain certain kinds of observations in physics. Calculus applied to real-world ecological processes carries with it many strong assumptions, such as the assumption that events are continuous and deterministic.

In the first half of this century, ecologists tended to believe that their science lacked theory. In reality, ecology had much theory, but the theory that was in use and influential had not been thoroughly checked against or developed from observations. When projections from prevailing theory were so compared, the theory generally contradicted observations. Thus early ecological theory was separated from observation, and ecological theorists and empiricists functioned too independently of each other.

During much of the first half of the twentieth century, most ecological theory concerned population dynamics typically applied to animals, and there were three dominant formal models of population growth: (1) exponential and (2) logistic models of the growth of a single population; and (3) Lotka-Volterra equations for predator-prey interactions. Each was the simplest expression in the mathematics of calculus of an idea about population growth.

Ecological modeling began a period of rapid development during the late 1960s and 1970s. This period included the beginning of development of systems analysis models in ecology; in some cases the methods used were borrowed directly from engineering methods. In addition, development began of biophysical models, concerning energy flow between an individual plant or animal.

In the 1970s, models increased in variety and in realism. Plant ecology, for which few formal models had existed, became the subject of much modeling. One group of models considered a fixed vegetaion structure, such as a crop or stand of trees, through which water and chemical elements flowed. In this case, the vegetation maintained a static structure. Another group of models considered the dynamics of vegetation change over time. These included models now referred to as "gap-phase" models, which began with the JABOWA model of forest growth.

Research on chemical cycling and global biogeochemical cycles began to involve models. The first models were essentially budgetary, keeping track of the storage and flux of a specific chemical element or compound. Concern with the possibility of the build-up of carbon dioxide in the atmosphere due to human activities led to an emphasis on global carbon models. In the earliest of these models, carbon flux was primarily perceived as linear and influenced only by the concentration of carbon in a single storage compartment, and not influenced by other environmental factors.

From the 1980s to the present, models have continued to increase in complexity and in realism. Some ecological aspects are beginning to be added to very complex global models, such as general circulation models (GCMs), the large and complex models of atmospheric circulation. In addition, the process of modeling is now viewed more commonly as an integral part of ecological research.

The Uses of Models and Modeling

Models provide projections of future states of a system, given the definition of the system structure and its dynamics as specified in the model. It is common to believe that such projections are the only utility of a model. In many practical applications in environmental sciences, the projections can be the primary benefit of a model. However, in scientific research, the development of a model often yields insights and new understandings, even if the model fails in its projections; indeed, the failures of a model can be as instructive as the successes to the present state of our understanding. A formal model — i.e., one in which the assumptions are stated explicity as mathematical relationships or as computer code — demonstrates for us the implications of our assumptions. Such demonstrations show us the extent and the limits of our understanding. A model can be subjected to tests of sensitivity, which show how responsive the model is to errors in input and in estimation of parameters. This sensitivity can give us insight into which variables and parameters need to be measured with the greatest accuracy, which are likely to have major effects on projects, and which are not important for a specific question.

A major challenge for ecologists is to develop the means to test models against observations. Almost invariably, observations of ecological phenomena either occur over too short a time period, or lack the measurement of some crucial variable, to be useful in standard tests of a model.

There is no one model for all purposes. Ecology and its applications require a suite of models. Modeling in ecology is as diverse as the subject of this science, and this brief discussion is meant only to point out some of the major concepts and trends, not to provide a definitive discussion or definition of all kinds of models in use; such a discussion would exceed the space available here.

Daniel B. Botkin

SI Units

Quantity	Name of SI Unit	SI Unit Symbol or Abbreviation	Definition of SI Unit
acceleration	meter per second squared		m/s^2
amount of substance	mole	mol	
angular acceleration	radian per second squared		rad/s^2
angular velocity	radian per second		rad/s
area	square meter		m^2
capacity rate	watt per kelvin	W/K	$kg.m^2/s^3.K$
density	kilogram per cubic meter		kg/m^3
electric capacitance	farad	F, A.s/V	$A^3.s^4/kg.m^2$
electric conductivity	ampere per volt meter	A/V.m	$A^2.s^3/kg.m^3$
electric current	ampere	A	
electric field strength	volt per meter	V/m	
electric resistance	ohm	Ω, V/A	$kg.m^2/A^2.s^3$
electromotive force	volt	V, W/A	$kg.m^2/A.s^3$
force	newton	N	$kg.m/s^2$
frequency	hertz	Hz	s^{-1}
heat capacity	joule per kilogram kelvin	J/kg.K	$m^2/s^2.K$
heat flux density	watt per square meter	W/m^2	kg/s^3
heat transfer coefficient	watt per square meter kelvin	$W/m^2.K$	$kg/s^3.K$
illumination	lux, lumen per square meter	lx, lm/m^2	$cd.sr/m^2$
length	meter	m	
luminance	candela per square meter		cd/m^2
luminous flux	lumen	lm	cd.sr
luminous intensity	candela	cd	
magnetic field strength	ampere per meter		A/m
magnetic flux	weber	Wb, V.s	$kg.m^2A.s^2$
magnetomotive force	ampere		A
mass	kilogram	kg	
power, heat flux	watt, joule per second	W, J/s	$kg.m^2/s^3$
pressure	newton per square meter, pascal	N/m^2, Pa	$kg/m.s^2$
quantity of electricity	coulomb	C	A.s
surface tension	newton per meter, joule per square meter	N/m J/m²	kg/s^2
thermal conductivity	watt per meter kelvin	W/m.K, $J.m/s.m^2.K$	$kg.m/s^3.K$
thermodynamic temperature	kelvin	K	
time	second	s	
velocity	meter per second		m/s
viscosity, dynamic	newton-second per square meter, pascal-second	$N.s/m^2$, Pa.s	kg/m.s
viscosity, kinematic	meter squared per second		m^2/s
volume	cubic meter		m^3
volumetric flow rate	cubic meter per second		M^3/s
volumetric heat release rate	watt per cubic meter	W/m^3	$kg/m.s^3$
work, torque, energy, quantity of heat	joule, newton-meter, watt-second	J, N.m, W.s	$kg.m^2/s^2$

continued on next page

SI Units – *continued*

TABLE OF SI – FPS CONVERSION

UNIT/QUANTITY	ABBREVIATION	METRIC	IMPERIAL
LENGTH		*METER*	*FOOT*
meter	m	–	3.280 839
foot	ft	0.3048	–
micron	μm	1×10^{-6}	$3.280\ 839 \times 10^{-6}$
angstrom	Å	1×10^{-10}	$3.280\ 839 \times 10^{-10}$
inch	in	2.54×10^{-2}	8.333×10^{-2}
yard	yd	0.9144	3.0
chain		20.1168	66.0
furlong		201.168	660.0
mile	mi	1609.344	5280.0
UK nautical mile		1853.184	6080.0
International nautical mile		1852.0	6076.114
fathom		1.8288	6.0
AREA		*SQUARE METER*	*SQUARE FOOT*
square meter	m²	–	10.763 904
square foot	ft²	$9.290\ 304 \times 10^{-2}$	–
square inch	in²	6.452×10^{-4}	6.944×10^{-3}
square yard	yd²	0.836 127	9.0
acre		4046.856	43 560
hectare	ha	10 000	107 631
square mile		$2.589\ 988 \times 10^{6}$	$2.787\ 840 \times 10^{7}$
VOLUME		*CUBIC METER*	*CUBIC FOOT*
cubic meter	m³	–	35.31467
cubic foot	ft³	2.832×10^{-2}	
cubic inch	in³	$1.638\ 706 \times 10^{-5}$	$5.787\ 036 \times 10^{-4}$
cubic yard	yd³	0.764 555	27.0
liter	1	$1.0 \times 10{-3}$	$3.531\ 467 \times 10^{-2}$
FLUID VOLUME		*CUBIC METER*	*CUBIC FOOT*
fluid ounce	fl oz	$2.841\ 306 \times 10^{-5}$	$1.003\ 398 \times 10^{-3}$
pint	pt	$5.682\ 613 \times 10^{-4}$	$2.006\ 796 \times 10^{-2}$
bushel	bu	3.6369×10^{-2}	1.285 313
Imperial gallon	gal	4.54609×10^{-3}	$1.605\ 437 \times 10^{-1}$
US gallon		$3.785\ 412 \times 10^{-3}$	$1.336\ 806 \times 10^{-1}$
quart	qt	$1.136\ 523 \times 10^{-3}$	4.01359×10^{-2}
MASS		*KILOGRAM*	*POUND*
kilogram	kg	–	2.204 623
pound	lb	0.453 592	–
ounce	oz	$2.834\ 952 \times 10^{-2}$	6.25×10^{-2}
stone	st	6.350 586	14
quarter	qr	12.700 586	28
hundredweight	cwt	50.802 344	112
ton	t	1016.047	2240
tonne	t	1000	2204.623

SI Units – *continued*

SI Prefixes and Multiplication Factors

Multiplication Factor		Prefix	Symbol
1 000 000 000 000 000 000	10^{18}	exa	E
1 000 000 000 000 000	10^{15}	peta	P
1 000 000 000 000	10^{12}	tera	T
1 000 000 000	10^{9}	giga	G
1 000 000	10^{6}	mega	M
10 000	10^{4}	myria	
1 000	10^{3}	kilo	k
100	10^{2}	hecto	h
10	10^{1}	deca	da
0.1	10^{-1}	deci	d
0.01	10^{-2}	centi	c
0.001	10^{-3}	milli	m
0.000 001	10^{-6}	micro	μ
0.000 000 001	10^{-9}	nano	n
0.000 000 000 001	10^{-12}	pico	p
0.000 000 000 000 001	10^{-15}	femto	f
0.000 000 000 000 000 001	10^{-18}	atto	a

Roman Numerals

I	1
II	2
III	3
IV	4
V	5
VI	6
VII	7
VIII	8
IX	9
X	10
XX	20
XXX	30
XL	40
L	50
LX	60
LXX	70
LXXX	80
XC	90
C	100
CC	200
CCC	300
CD	400
D	500
DC	600
DCC	700
DCCC	800
CM	900
M	1000

Greek Alphabet

Letter	Name	Transliteration	Letter	Name	Transliteration
A α	alpha	a	N ν	nu	n
B β	beta	b	Ξ ξ	xi	x
Γ γ	gamma	g	O o	omicron	o
Δ δ	delta	d	Π π	pi	p
E ε	epsilon	e	P ρ	rho	r
Z ζ	zeta	z	Σ σ, ς[1]	sigma	s
H η	eta	e (or $\bar{e}$)	T τ	tau	t
Θ θ	theta	th	Y υ	upsilon	y
I ι	iota	i	Φ φ	phi	ph
K κ	kappa	k	X χ	chi	ch, kh
Λ λ	lambda	l	Ψ ψ	psi	ps
M μ	mu	m	Ω ω	omega	o (or $\bar{o}$)

[1] At end of word.

Periodic Table of the Elements

Atomic No. / Symbol / Atomic Wt.

Representative Elements s block | **Transition Elements d block** | **Representative Elements p block** | **Noble gases**

Period	IA	IIA	IIIB	IVB	VB	VIB	VIIB	VIII	VIII	VIII	IB	IIB	IIIA	IVA	VA	VIA	VIIA	O
1	+1 1 H 1.0079																	2 He 4.003
2	+1 3 Li 6.941	+2 4 Be 9.012											+3 5 B 10.81	+4 +2 6 C 12.011	+5 +3 -3 7 N 14.007	-2 8 O 15.999	-1 9 F 18.998	10 Ne 20.18
3	+1 11 Na 22.99	+2 12 Mg 24.30											+3 13 Al 26.98	+4 +2 14 Si 28.08	+5 +3 -3 15 P 30.97	+6 +4 -2 16 S 32.06	+7 +5 +3 +1 -1 17 Cl 35.45	18 Ar 39.95
4	+1 19 K 39.10	+2 20 Ca 40.08	+3 21 Sc 44.96	22 Ti 47.90	23 V 50.94	+6 +3 24 Cr 52.00	+5 +4 25 Mn 54.94	+3 +2 26 Fe 55.85	+3 +2 27 Co 58.93	+3 +2 28 Ni 58.71	+2 +1 29 Cu 63.55	+2 30 Zn 65.38	+3 31 Ga 69.72	+4 +2 32 Ge 72.59	+5 +3 33 As 74.92	+6 +4 -2 34 Se 78.96	+5 +3 +1 -1 35 Br 79.90	36 Kr 83.80
5	+1 37 Rb 85.47	+2 38 Sr 87.62	+3 39 Y 88.91	40 Zr 91.22	41 Nb 92.91	42 Mo 95.94	43 Tc 98.91	44 Ru 101.07	45 Rh 102.91	46 Pd 106.4	+1 47 Ag 107.87	+2 48 Cd 112.40	+3 49 In 114.82	+4 +2 50 Sn 118.69	+5 +3 51 Sb 121.75	+6 +4 -2 52 Te 127.60	+7 +5 +3 +1 -1 53 I 126.90	54 Xe 131.30
6	+1 55 Cs 132.91	+2 56 Ba 137.34	+3 57 La 138.91	72 Hf 178.49	73 Ta 180.95	74 W 183.85	75 Re 186.2	76 Os 190.2	77 Ir 192.22	78 Pt 195.09	79 Au 196.97	+2 +1 80 Hg 200.6	+3 +1 81 Tl 204.4	+4 +2 82 Pb 207.2	+5 +3 83 Bi 209.0	+6 +4 84 Po 210	85 At 210	86 Rn 222
7	+1 87 Fr 223	+2 88 Ra 226.0	89 Ac 227	104 Unq	105 Unp	106 Unh												

Inner Transition Elements f block

Lanthanum Series:

58 Ce 140.12	59 Pr 140.1	60 Nd 144.24	61 Pm (147)	62 Sm 150.4	63 Eu 151.96	64 Gd 157.2	65 Tb 158.93	66 Dy 162.50	67 Ho 164.93	68 Er 167.26	69 Tm 168.93	70 Yb 173.04	71 Lu 174.97

Actinium Series:

90 Th 232.0	91 Pa 231.0	92 U 238.0	93 Np 237.0	94 Pu (242)	95 Am (243)	96 Cm (247)	97 Bk (247)	98 Cf (247)	99 Es (254)	100 Fm (253)	101 Md (256)	102 No (254)	103 Lr (257)

Mass numbers of the most stable or most abundant isotopes are shown in parentheses.
The elements to the right of the bold lines are called the nonmetals and the elements to the left of the bold line are called the metals.
Common oxidation numbers are given for the representative elements and some transition elements.
Reprinted with permission from T.R. Dickson, *Introduction to Chemistry*, 2nd ed. Copyright © 1971 by John Wiley & Sons, Inc.

Chemical Elements

Symbol	Name	Atomic No.	Symbol	Name	Atomic No.
Ac	Actinium	89	Mo	Molybdenum	42
Ag	Silver	47	N	Nitrogen	7
Al	Aluminum	13	Na	Sodium	11
Am	Americium	95	Nb	Niobium	41
Ar	Argon	18	Nd	Neodymium	60
As	Arsenic	33	Ne	Neon	10
At	Astatine	85	Ni	Nickel	28
Au	Gold	79	No	Nobelium	102
B	Boron	5	Np	Neptunium	93
Ba	Barium	56	O	Oxygen	8
Be	Beryllium	4	Os	Osmium	76
Bi	Bismuth	83	P	Phosphorus	15
Bk	Berkelium	97	Pa	Protactinium	91
Br	Bromine	35	Pb	Lead	82
C	Carbon	6	Pd	Palladium	46
Ca	Calcium	20	Pm	Promethium	61
Cd	Cadmium	48	Po	Polonium	84
Ce	Cerium	58	Pr	Praseodymium	59
Cf	Californium	98	Pt	Platinum	78
Cl	Chlorine	17	Pu	Plutonium	94
Cm	Curium	96	Ra	Radium	88
Co	Cobalt	27	Rb	Rubidium	37
Cr	Chromium	24	Re	Rhenium	75
Cs	Cesium	55	Rh	Rhodium	45
Cu	Copper	29	Rn	Radon	86
Dy	Dysprosium	66	Ru	Ruthenium	44
Er	Erbium	68	S	Sulphur	16
Es	Einsteinium	99	Sb	Antimony	51
Eu	Europium	63	Sc	Scandium	21
F	Flourine	9	Se	Selenium	34
Fe	Iron	26	Si	Silicon	14
Fm	Fermium	100	Sm	Samarium	62
Fr	Francium	87	Sn	Tin	50
Ga	Gallium	31	Sr	Strontium	38
Gd	Gadolinium	64	Ta	Tantalum	73
Ge	Germanium	32	Tb	Terbium	65
H	Hydrogen	1	Tc	Technetium	43
He	Helium	2	Te	Tellurium	52
Hf	Hafnium	72	Th	Thorium	90
Hg	Mercury	80	Ti	Titanium	22
Ho	Holmium	67	Tl	Thallium	81
I	Iodine	53	Tm	Thulium	69
In	Indium	49	Unh	Unnilhexium	106
Ir	Iridium	77	Unp	Unnilpentium	105
K	Potassium	19	Unq	Unnilquadium	104
Kr	Krypton	36	U	Uranium	92
La	Lanthanum	57	V	Vanadium	23
Li	Lithium	3	W	Tungsten	74
Lu	Lutetium	71	Xe	Xenon	54
Lr	Lawrencium	103	Y	Yttrium	39
Md	Mendelevium	101	Yb	Ytterbium	70
Mg	Magnesium	12	Zn	Zinc	30
Mn	Manganese	25	Zr	Zirconium	40

Geological Table: The Phanerozoic Time Scale

EON						
	Era					
		Period				
			Epoch	Began (yr ago x 10^6)	Duration (yr) 10^6	Evolutionary Events
PHANEROZOIC						
	Cenozoic					
		Quaternary				
			Recent (Holocene)	0.011		Modern humans
			Pleistocene	0.6	2	Early humans
		Tertiary				
			Pliocene	12	5	Large carnivores
			Miocene	20	19	Whales, apes, grazing forms
			Oligocene	35	11	Large browsing animals
			Eocene	55	16	Flowering plants
			Paleocene	65	12	First placental animals
	Mesozioc					Extinction of dinosaurs, flora with modern aspects
		Cretaceous				
			Upper (Base Santonian)	90		
			Middle (Base Albian)	120		
			Lower	140		
		Jurassic			45	Dinosaurs zenith, primitive birds, and first small mammals
			Upper (Base Callovian)	155		
			Middle (Base Bajocian)	170		
			Lower	185		
		Triassic			45	Appearance of dinosaurs
			Upper	200		
			Middle	215		
			Lower	230		

continued on next page

Geological Table: The Phanerozoic Time Scale – *continued*

Paleozoic				
Permian			45	Conifers abundant and reptiles developed
	Upper (Base Ochoan)	245		
	Middle (Base Guadalupian)	260		
	Lower	275		
Pennsylvanian			35	First reptiles, great forests from which coal was later formed
	Upper			
	Middle			
	Lower	Carboniferous		
Mississippian		310	40	Sharks abundant
	Upper			
	Middle			
	Lower	350		
Devonian			50	Fish abundant, amphibians appear
	Upper	365		
	Middle	385		
	Lower	400		
Silurian			20	Earliest land plants and animals
	Upper			
	Middle			
	Lower	420		
Ordovician			70	First primitive fish
	Upper	440		
	Middle	460		
	Lower	490		
Cambrian			50+	Large fauna of marine invertebrates
	Upper (Base Croixian)	510		
	Middle	540		
	Lower			

U.S.D.A. Zone Map

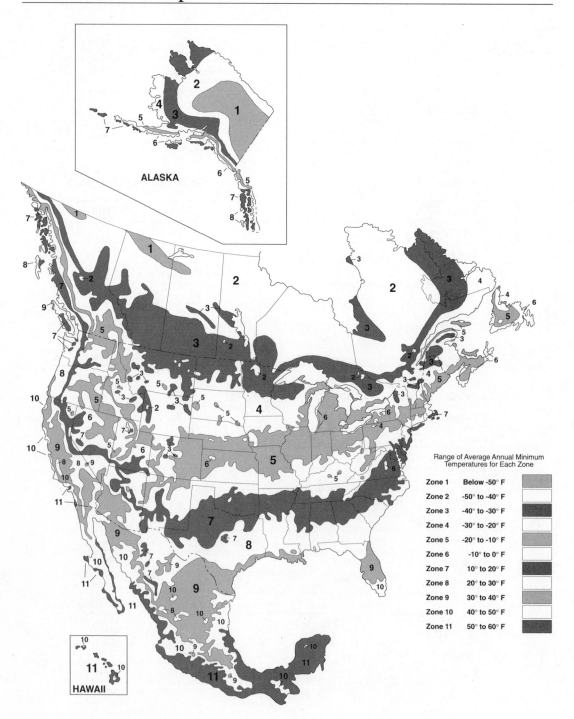

ALASKA

HAWAII

Range of Average Annual Minimum
Temperatures for Each Zone

Zone 1	Below -50° F
Zone 2	-50° to -40° F
Zone 3	-40° to -30° F
Zone 4	-30° to -20° F
Zone 5	-20° to -10° F
Zone 6	-10° to 0° F
Zone 7	10° to 20° F
Zone 8	20° to 30° F
Zone 9	30° to 40° F
Zone 10	40° to 50° F
Zone 11	50° to 60° F

Table of Taxa Included in this Dictionary

KINGDOM ANIMALIA

PHYLUM	SUBPHYLUM	CLASS	SUBCLASS	ORDER	SUBORDER	FAMILY
Annelida		Clitellata	Hirudinoidea Oligochaeta			
		Polychaeta				
Arthropoda	Chelicerata	Arachnida Merostomata		Xiphosura		
	Crustacea	Cirripedia Copepoda Malacostraca		Decapoda	Brachyura	
				Euphausiacea Isopoda		
		Ostracoda				
	Uniramia	Chilopoda Diplopoda Insecta				
				Anoplura Coleoptera Diptera		Culicidae
				Hemiptera Homoptera Hymenoptera Isoptera Lepidoptera Odonata Orthoptera Siphonoptera		
Bryozoa (Ectoprocta)						
Chaetognatha						
Chordata	Cephalochordata Vertebrata	Agnatha		Petromyzoniformes		
		Amphibia		Anura Urodela		

Table of Taxa Included in this Dictionary (continued)

KINGDOM ANIMALIA (continued)

PHYLUM	SUBPHYLUM	CLASS	SUBCLASS	ORDER	SUBORDER	FAMILY
		Aves				
				Anseriformes		
				Apodiformes		
				Caprimulgiformes		
						Caprimulgidae
				Charadriiformes		
						Alcidae
						Scolopacidae
				Ciconiiformes		
						Ardeidae
						Ciconiidae
				Columbiformes		
				Coraciiformes		
						Alcedinidae
				Cuculiformes		
						Culicidae
				Falconiformes		
						Acciptridae
				Galliformes		
				Gaviiformes		
				Gruiformes		
						Gruidae
				Passeriformes		
						Corvidae
				Pelecaniiformes		
						Pelecanidae
				Piciformes		
				Podicepiiformes		
				Procellariiformes		
						Diomedeidae
				Psittaciformes		
				Sphenisciformes		
				Strigiformes		
						Tytonidae
			Palaeognathae			
		Chondrichthyes				
			Elasmobranchii			
		Mammalia				
			Eutheria			
				Artiodactyla		
						Bovidae
				Carnivora		
				Cetacea		
				Chiroptera		
				Edentata		
				Insectivora		
				Perissodactyla		
				Primates		
						Hominidae
						Pongidae

KINGDOM ANIMALIA (continued)

PHYLUM	SUBPHYLUM	CLASS	SUBCLASS	ORDER	SUBORDER	FAMILY
		Mamalia (cont.)		Rodentia	Proboscidia	
			Metatheria Prototheria			
		Osteichthyes				
			Coelocanthini Dipnoi			
		Reptilia				Testudinata
				Chelonia		
				Crocodilia Squamata	Lacertilia Ophidia	
			Archosauria			
Cnidaria						
	Anthozoa	Alcyonaria Zoantheria				
	Medusozoa	Hydrozoa Scyphozoa				
Echinodermata		Asteroidea Echinoidea Holothuroidea				
		Ophiuroidea		Apodida		
Mollusca		Bivalvia Cephalopoda				
				Octopoda Teuthoida		Sepiidae
		Gastropoda		Nudibranchia		
		Polyplacophora Scaphopoda				
Nematoda						
Onycophora						
Platyhelminthes						
		Cestoda Trematoda Turbellaria				
Porifera						
Rotifera						
Ascomycota						
Basidiomycota						

Table of Taxa Included in this Dictionary (continued)

KINGDOM FUNGI

PHYLUM	SUBPHYLUM	CLASS	SUBCLASS	ORDER	SUBORDER	FAMILY
Chytridiomycota						
Deuteromycota						
Mycophycophyta						
Zygomycota						

KINGDOM MONERA

PHYLUM	SUBPHYLUM	CLASS	SUBCLASS	ORDER	SUBORDER	FAMILY
Anaerobic Photosynthesizing Bacteria						
Anaerobic Endospore-forming Bacteria						
Archaebacteria						
Chemoautotrophic Bacteria						
Cyanobacteria						
Nitrogen-fixing aerobic bacteria						
Omnibacteria						
Spirochaetae						

KINGDOM PLANTAE

PHYLUM	SUBPHYLUM	CLASS	SUBCLASS	ORDER	SUBORDER	FAMILY
Bryophyta						
		Anthocerotae				
		Hepaticae				
		Musci				
Angiospermophyta						
		Dicotyledoneae				
						Apiaceae
						Asclepiadaceae
						Asteraceae
						Betulaceae
						Boraginaceae
						Brassicaceae
						Bromeliaceae
						Campanulaceae
						Caprifoliaceae
						Caryophyllaceae
						Chenopodiaceae
						Corylaceae
						Crassulaceae
						Cucurbitaceae
						Ericaceae
						Euphorbiaceae
						Fabaceae
						Fagaceae
						Geraniaceae
						Juglandaceae
						Lamiaceae
						Magnoliaceae
						Malvaceae

KINGDOM PLANTAE (continued)

PHYLUM	SUBPHYLUM	CLASS	SUBCLASS	ORDER	SUBORDER	FAMILY
		Dicotyledoneae (cont.)				Moraceae
						Myrtaceae
						Nymphaceae
						Onagraceae
						Opuntiaceae
						Papaveraceae
						Pinaceae
						Polemoniaceae
						Polygonaceae
						Primulaceae
						Ranunculaceae
						Rosaceae
						Salicaceae
						Saxifragaceae
						Scrophulariaceae
						Solanaceae
						Taxaceae
						Violaceae
						Vitaceae
		Monocotyledoneae				Amaryllidaceae
						Araceae
						Arecaceae
						Cyperaceae
						Iridaceae
						Juncaceae
						Lilaceae
						Orchidaceae
						Poaceae
						Typhaceae
Coniferophyta						
Cycadophyta						
Filicinophyta						
Ginkgophyta						
Gnetophyta						
Lycopodophyta						
Sphenophyta						

KINGDOM PROTISTA

PHYLUM	SUBPHYLUM	CLASS	SUBCLASS	ORDER	SUBORDER	FAMILY
Acrasiomycota						
Actinopoda						
Apicomplexa		Sporozoa				
Bacillariophyta						
Caryoblastea						

Table of Taxa Included in this Dictionary (continued)

KINGDOM PROTISTA (continued)

PHYLUM	SUBPHYLUM	CLASS	SUBCLASS	ORDER	SUBORDER	FAMILY
Chlorophyta						
Chrysophyta						
Ciliophora						
Cnidosporidia						
Cryptophyta						
Dinoflagellata						
Euglenophyta						
Eustigmatophyta						
Foraminifera						
Gamophyta						
Haptophyta						
Hypochytridiomycota						
Labyrinthulamycota						
Myxomycota						
Oomycota						
Phaeophyta						
Plasmodiophoromycota						
Rhizopoda						
Rhodophyta						
Xanthophyta						
Zoomastigina						

Hypothetical Soil Profile and Master Horizon Nomenclature

HORIZON		DESCRIPTION	
O	Oi	Litter layer; loose fragmented organic matter in which original plant structures are still readily discernible	Horizons of maximum biological activity and eluviation
	Oe	Fermentation layer; partially decomposed organic matter	
	Oh	Humified layer; heavily transformed organic matter with no discernible macroscopic plant remains	
A	A	Dark-colored horizon, with high organic content intermixed with minerals	
E	E	Light-colored horizon, with low organic content due to high eluviation	
	EB	Transitional layer, sometimes absent	
B	BE	Transitional layer, sometimes absent	Horizons of illuviation; accumulations of clay, humus or iron oxides leached or translocated from the upper layers
	B	A dark layer of accumulation of transported silicate, clay, minerals, iron and organic matter showing the maximum development of blocky and/or prismatic structure	
	BC	Transitional layer	
C	C	Weathered parent material, comprising mineral substrate with little or no structure; containing a gleyed layer or layers of accumulated calcium carbonate or sulfate in some soils	
R	R	Underlying rock	

Note: Capital letters, singly or in pairs, indicate master horizons
A lower case suffix indicates subordinate features of master horizons.

Table of Minerals and Diagnostic Properties

MINERAL	FORM	CLEAVAGE	HARDNESS	SPECIFIC GRAVITY	OTHER PROPERTIES
Actinolite. Calcium iron silicate, $Ca_2(Mg,Fe)_5Si_8O_{22}(OH)_2$.	Slender crystals, usually fibrous.	Two good cleavages, meeting at angles of 56° and 124°.	5-6	3.0-3.3	Color white to light green. Transparent to translucent. Colorless streak. Vitreous luster.
Albite (sodic plagioclase feldspar). Sodium aluminum silicate, $NaAlSi_3O_8$	Tabular crystals.	Good in two directions at 93° 34'.	6	2.62	Colorless, white, or gray. Transparent to translucent. Streak colorless. Opalescent variety is moonstone.
Amphibole. A group of complex, solid-solution silicates, chiefly of calcium, magnesium, iron, and aluminum. Similar to pyroxene in composition, but containing a little hydroxyl (OH^-) ion. The commonest of the many varieties of amphibole is hornblende.	Long, prismatic, 6-sided crystals; also in fibrous or irregular masses of interlocking crystals and in disseminated grains.	Two good cleavages meeting at angles of 56° and 124°.	5-6	2.9-3.2	Color black to light green; or colorless. Opaque. Highly vitreous luster on cleavage surfaces. Distinguished from pyroxene by the difference in cleavage angle and in crystal form. Amphibole also has much better cleavage and higher luster than pyroxene.
Anhydrite. Anhydrous calcium sulfate, $CaSO_4$.	Crystals rare. Commonly in massive fine aggregates.	Three directions at right angles, forming rectangular blocks.	3-3.5	2.89-2.98	White with various tinges. Transparent to translucent. Streak colorless.
Anorthite (calcic plagioclase feldspar). Calcium aluminum silicate, $Ca(Al_2Si_2O_8)$.	Prismatic crystals, also massive.	Good in two directions at 94° 12'.	6	2.76	Colorless, white, gray, green, red, or yellow. Vitreous to pearly luster. Colorless streak. Striations.
Aragonite. Calcium carbonate, $CaCO_3$	Needle-shaped crystals with rectangular cross section.	One distinct cleavage.	3.5-4	2.93-2.95	Color usually white. Vitreous luster. Colorless streak. Forms compound (twinned) crystals.
Augite. A ferromagnesian silicate, $Ca(Mg, Fe, Al)(Al,Si_2O_6)$.	Short stubby crystals.	Prismatic along two planes nearly at right angles.	5-6	3.2-3.4	Dark green to black. Translucent only on thin edges. Streak greenish-gray. Vitreous luster.
Azurite. Blue copper carbonate. $Cu_3(CO_3)_2(OH)_2$.	Complex crystals, sometimes in radiating groups.	Fibrous.	4	3.77	Intense azure blue. Opaque. Vitreous to dull earthy luster. Streak pale blue. Effervesces with hydrochloric acid.
Bauxite. Hydrous aluminum oxides, indefinite composition.	Rounded grains, earthy masses.	Uneven.	1-3	2-3	Yellow, brown, gray, white. Opaque. Dull to earthy luster. Colorless streak.

Table of Minerals and Diagnostic Properties – *continued*

Mineral	Form	Cleavage	Hardness	Specific Gravity	Other Properties
Beryl (emerald). Beryllium aluminum silicate. $Be_3Al_2(SiO_3)_6$.	Hexagonal prisms.	Conchoidal to uneven fracture.	7.5-8	2.63-2.80	Many colors. Chiefly green. Transparent to subtranslucent. Vitreous or resinous luster. Streak white.
Biotite (black mica). A complex silicate of potassium, iron, aluminum, and magnesium, variable in composition but approximately $K(Mg,Fe)_3AlSi_3O_{10}(OH)_2$.	Thin, scalelike crystals, commonly 6-sided, and in scaly, foliated masses	Perfect in one direction, yielding thin, flexible scales.	2.5-3	2.7-3.2	Black to dark brown. Translucent to opaque. Pearly to vitreous luster. White to greenish streak.
Bornite (peacock ore). Copper iron sulfide, Cu_5FeS_4.	Some cubic crystals. Usually massive.	Uneven.	3	5.06-5.08	Brownish-bronze on fresh fracture. Tarnishes to variegated purple, blue, and black. Metallic luster. Opaque. Streak grayish black.
Calcite. Calcium carbonate, $CaCO_3$.	"Dog-tooth" or flat crystals showing excellent cleavages; granular, showing cleavages; also masses too fine-grained to show cleavages distinctly.	Three highly perfect cleavages at oblique angles, yielding rhomb-shaped fragments.	3	2.72	Commonly colorless, white, or yellow, but may be any color owing to impurities. Transparent to opaque, transparent varieties showing strong, double refraction (e.g., 1 dot seen through calcite appears as 2). Vitreous to dull luster. Effervesces readily in cold dilute hydrochloric acid.
Carnotite. Potassium uranyl vanadate, $K_2(UO_2)_2(VO_4)_2 \cdot 8H_2O$.	Earthy powder.	Not apparent.	Very soft	4.1, approx.	Brilliant canary-yellow color. An ore of vanadium and uranium.
Cassiterite. Tin dioxide, SnO_2.	Well-formed, 4-sided prismatic crystals terminated by pyramids; 2 crystals may be intergrown to form knee-shaped twins; also as rounded pebbles in stream gravels.	None; curved to irregular fracture.	6-7	7	Brown to black. Adamantine luster. White to pale-yellow streak. Chief ore of tin.

Table of Minerals and Diagnostic Properties – *continued*

MINERAL	FORM	CLEAVAGE	HARDNESS	SPECIFIC GRAVITY	OTHER PROPERTIES
Chalcedony (cryptocrystalline quartz). Silicon dioxide, SiO_2.	Crystals too fine to be visible; some are conspicuously banded, or in masses.	None; conchoidal fracture.	6-6.5	2.6	Color commonly white or light gray, but may be any color owing to impurities. Distinguished from opal by dull or clouded luster.
Chalcocite (copper glance). Cuprous sulfide, Cu_2S.	Massive, rarely in crystals of roughly hexagonal shape. May be tarnished and stained to blue and green.	Indistinct, rarely observed.	2.5-3	5.5-5.8	Blackish-gray to steel gray, commonly tarnished to green or blue. Dark gray streak. Very heavy. Metallic luster. An important ore of copper.
Chalcopyrite. Copper iron sulfide, $CuFeS_2$.	Compact or disseminated masses, rarely in wedge-shaped crystals.	None; uneven fracture.	3.5-4	4.1-4.3	Brassy to golden-yellow. Tarnishes to blue, purple, and reddish iridescent films. Greenish-black streak. Distinguished from pyrite by deeper yellow color and softness. A common copper ore.
Chlorite. A complex group of hydrous magnesium aluminum silicates containing iron and other elements in small amounts.	Commonly in foliated or scaly masses; may occur in tabular, 6-sided crystals resembling mica.	One perfect cleavage, yielding thin, flexible, but inelastic scales.	1-2.5	2.6-3.0	Grass-green to blackish-green color. Translucent to opaque. Greenish streak. Vitreous luster. Very easily disintegrated.
Chromite. Iron chromium oxide, $FeCr_2O_4$.	Massive.	Uneven fracture.	5.5	4.6	Black to brownish-black. Subtranslucent. Metallic luster. Dark brown streak.
Copper (native copper). An element, Cu.	Twisted and distorted leaves and wirelike forms; flattened or rounded grains.	None.	2.5-3	8.8-8.9	Characteristic copper color, but commonly stained green. Highly ductile and malleable. Excellent conductor of heat and electricity. Very heavy.
Corundum (ruby, sapphire). Aluminum oxide, Al_2O_3.	Barrel-shaped crystals.	Basal or rhombohedral parting.	9	4.02	Many colors, depending on impurities. Transparent to translucent. Adamantine to vitreous luster. Colorless streak.
Diamond. High-density form of the element carbon, C.	Octahedral crystals.	Octahedral.	10	3.5	Many colors, depending on impurities. Transparent. Adamantine luster. Colorless streak.

Table of Minerals and Diagnostic Properties – *continued*

Mineral	Form	Cleavage	Hardness	Specific Gravity	Other Properties
Dolomite. Calcium magnesium carbonate, $CaMg(CO_3)_2$.	Rhomb-faced crystals showing good cleavage; also in fine-grained masses.	Three perfect cleavages at oblique angles, as in calcite.	3.5-4	2.9	Variable in color, but commonly white. Transparent to translucent. Vitreous to pearly luster. Powder will effervesce slowly in cold dilute hydrochloric acid, but coarse crystals will not.
Epidote. A complex calcium iron-aluminum silicate, $Ca_2(Al,Fe)_3(SiO_4)_3 (OH)$.	Short, 6-sided crystals or radiate crystal groups, and in granular or compact masses.	One good cleavage; in some specimens, a second poorer cleavage at an angle of 115° with the first.	6-7	3.4	Characteristic yellowish-green (pistachio-green) color. Vitreous luster.
Fluorite. Calcium fluoride, CaF_2.	Cubic crystals, also massive.	Four good cleavages parallel to faces of an octahedron.	4	3.18	Many colors. Transparent to translucent. Vitreous luster. Colorless streak.
Galena. Lead sulfide, PbS.	Cubic crystals common, but mostly in coarse to fine granular masses.	Perfect cubic cleavage (three cleavages mutually at right angles).	2.5	7.3-7.6	Silvery-gray color. Metallic luster. Silvery-gray to grayish-black streak. Chief ore of lead.
Garnet. A group of solid solution silicates having variable proportions of different metallic elements. The most common variety contains calcium, iron, and aluminum, but garnets may contain many other elements.	Commonly in well-formed equidimensional crystals, but also massive and granular.	None; conchoidal or uneven fracture.	6.5-7.5	3.4-4.3	Commonly red, brown, or yellow, but may be other colors. Translucent to opaque. Resinous to vitreous luster.
Gold. An element, Au.	Massive or in thin plates; also in flattened grains or scales; distinct crystals very rare.	None.	2.5-3	15.6-19.3	Characteristic gold-yellow color and streak. Rarely in crystals. Extremely heavy. Very malleable and ductile. Variable density reflects impurities.
Graphite. An element, C.	Foliated or scaly masses.	Good in one direction.	1-2	2.3	Black to steel-gray. Opaque. Metallic or earthy luster. Black streak. The "lead" of pencils.

Table of Minerals and Diagnostic Properties – *continued*

Mineral	Form	Cleavage	Hardness	Specific Gravity	Other Properties
Gypsum. Hydrous calcium sulfate, $CaSO_4 \cdot 2H_2O$.	Tabular crystals, and cleavable, granular, fibrous, or earthy masses.	One perfect cleavage, yielding thin, flexible folia; 2 other much less perfect cleavages.	2	2.2-2.4	Colorless or white, but may be other colors when impure. Transparent to opaque. Luster vitreous to pearly or silky. Cleavage flakes flexible but not elastic like those of mica.
Halite (rock salt). Sodium chloride, $NaCl$.	Cubic crystals, granular masses.	Excellent cubic cleavage (3 cleavages mutually at right angles).	2-2.5	2.1	Colorless to white, but of other colors when impure. The color may be unevenly distributed through the crystal. Transparent to translucent. Vitreous luster. Salty taste.
Hematite. Ferric iron oxide, Fe_2O_3.	Highly varied, compact, granular, fibrous, or earthy, micaceous; rarely in well-formed crystals.	None, but fibrous or micaceous specimens may show parting resembling cleavage; splintery to uneven fracture.	5-6.5	4.9-5.3	Steel-gray, reddish-brown, red, or iron-black in color. Metallic to earthy luster. Characteristic brownish-red streak. Hematite is the most important iron ore.
Hornblende. A complex ferromagnesian silicate.	Long, prismatic crystals. Fibrous masses.	Perfect prismatic at 56° and 124°.	5-6	3.2	Dark green to black. Translucent on thin edges. Vitreous luster; fibrous variety, silky. Colorless streak.
Kaolinite. Hydrous aluminum silicate, $H_4Al_2Si_2O_9$. Representative of the 3 or 4 similar minerals common in clays.	Commonly in soft, compact, earthy masses.	Crystals always so small that cleavage is invisible without microscope.	1-2	2.2-2.6	White color, but may be stained by impurities. Greasy feel. Adheres to the tongue, and becomes plastic when moistened. "Claylike" odor when breathed upon.
Kyanite (disthene). Aluminum silicate, Al_2SiO_5	Long, bladelike crystals.	One perfect, and one poor cleavage, both parallel to length of crystals; and a crude parting across the crystals.	4-7	3.5-3.7	Colorless, white, or a distinctive pale blue color. Can be scratched by knife parallel to cleavage, but is harder than steel across cleavage.

Table of Minerals and Diagnostic Properties – *continued*

Mineral	Form	Cleavage	Hardness	Specific Gravity	Other Properties
"Limonite." Microscopic study shows that the material called limonite is not a single mineral. Most "limonite" is a very finely crystalline variety of the mineral **Goethite** containing absorbed water. Hydrous ferric oxide with minor amounts of other elements, roughly $Fe_2O_3 \bullet H_2O$.	Compact or earthy masses; may show radially fibrous structure.	None; conchoidal or earthy fracture.	1-5.5	3.4-4.0	Yellow, brown, or black in color. Dull earthy luster, which distinguishes it from hematite. Characteristic yellow-brown streak. A common iron ore.
Magnetite. A combination of ferric and ferrous oxides, Fe_3O_4.	Well-formed, 8-faced crystals, more commonly in compact aggregates, disseminated grains, or loose grains in sand.	None; conchoidal or uneven fractures; may show a rough parting resembling cleavage.	5.5-6.5	5.0-5.2	Black. Opaque. Metallic to submetallic luster. Black streak. Strongly attracted by a magnet. Magnetite is an important iron ore.
Muscovite. (white mica; isinglass). A complex potassium aluminum silicate, $KAl_3Si_3O_{10}(OH)_2$ approximately, but varying.	Thin, scalelike crystals and scaly, foliated aggregates.	Perfect in one direction, yielding very thin, transparent, flexible scales.	2-3	2.8-3.1	Colorless, but may be gray, green, or light brown in thick pieces. Transparent to translucent. Pearly to vitreous luster.
Olivine. Magnesium iron silicate, $(Fe,Mg)_2SiO_4$.	Commonly in small, glassy grains and granular aggregates.	So poor that it is rarely seen; conchoidal fracture.	6.5-7	3.2-3.6	Various shades of green, also yellowish; opalescent and brownish when slightly altered. Transparent to translucent. Vitreous luster. Resembles quartz in small fragments but has characteristic greenish color, unless altered.
Opal. Hydrous silica, with 3% to 12% water, SiO_2nH_2O. Because it does not have a definite geometric internal structure, it is a mineraloid, not a true mineral.	Amorphous. Commonly in veins or irregular masses showing a banded structure. May be earthy.	None; conchoidal fracture.	5.0-6.5	2.1-2.3	Color highly variable, often in wavy or banded patterns. Translucent or opaque. Somewhat waxy luster.

Table of Minerals and Diagnostic Properties – *continued*

MINERAL	FORM	CLEAVAGE	HARDNESS	SPECIFIC GRAVITY	OTHER PROPERTIES
Orthoclase (potassium feldspar). Potassium aluminum silicate, $KAlSi_3O_8$.	Prismatic crystals and formless grains.	Good in two directions at or near 90°.	6	2.57	White, gray, pink. Translucent to opaque. Vitreous luster. White streak.
Plagioclase feldspar (soda-lime feldspars). A solid solution group of sodium calcium aluminum silicates, $NaAlSi_3O_8$ to $CaAl_2Si_2O_8$.	In well-formed crystals and in cleavable or granular masses.	Two good cleavages nearly at right angles (86°). May be poor in some volcanic rocks.	6-6.5	2.6-2.7	Commonly white or gray, but may be other colors. Some gray varieties show a play of colors called opalescence. Transparent in some volcanic rocks. Vitreous to pearly luster. Distinguished from orthoclase by the presence on the better cleavage surface of fine parallel lines or striations.
Potassium feldspar (orthoclase, microcline, and sanidine). Potassium aluminum silicate, $KAlSi_3O_8$	Boxlike crystals, massive.	One perfect and 1 good cleavage. making and angle of 90°.	6	2.5-2.6	Commonly white, gray. pink, or pale yellow; rarely colorless. Commonly opaque but may be transparent in volcanic rocks. Vitreous. Pearly luster on better cleavage. Distinguished from plagioclase by absence of striations.
Pyrite ("fool's gold"). Iron sulfide, FeS_2.	Well-formed crystals, commonly cubic, with striated faces; also granular masses.	None; uneven fracture.	6-6.5	4.9-5.2	Pale brassy-yellow color; may tarnish brown. Opaque. Metallic luster. Greenish-black or brownish-black streak. Brittle. Not a source of iron, but used in the manufacture of sulfuric acid. Commonly associated with ores of several different metals.
Pyroxene. A solid-solution group of silicates, chiefly silicates of calcium, magnesium and iron, with varying amounts of other elements. The commonest varieties are augite and hypersthene.	Commonly in short, 8-sided, prismatic crystals; the angle between alternate faces nearly 90°. Also as compact masses and disseminated grains.	Two cleavages at nearly 90°. Cleavage not always well developed; in some specimens, conchoidal or uneven fracture.	5-6	3.2-3.6	Commonly greenish to black in color. Vitreous to dull luster. Gray-green streak. Distinguished from amphibole by the right-angle cleavage, 8-sided crystals, and by the fact that most crystals are short and stout, rather than long, thin prisms, as in amphibole.

Table of Minerals and Diagnostic Properties – *continued*

Mineral	Form	Cleavage	Hardness	Specific Gravity	Other Properties
Quartz (rock crystal). Silicon dioxide, SiO_2.	Six-sided prismatic crystals, terminated by 6-sided triangular faces; also massive.	None or very poor; conchoidal fracture.	7	2.65	Commonly colorless or white, but may be yellow pink, amethyst, smoky-translucent brown, or even black. Transparent to opaque. Vitreous to greasy luster.
Serpentine. A complex group of hydrous magnesium silicates, roughly $H_4Mg_3Si_2O_9$.	Foliated or fibrous, usually massive.	Commonly only one cleavage, but may be in prisms. Fracture usually conchoidal or splintery.	2.5-4	2.5-2.65	Feels smooth, or even greasy. Color leek-green to blackish-green but varying to brownish-red, yellow, etc. Luster resinous to greasy. Translucent to opaque. Streak white.
Sillimanite (fibrolite). Aluminum silicate, Al_2SiO_5.	In long slender crystals, or fibrous.	Parallel to length, but rarely noticeable.	6-7	3.2	Gray, white, greenish-gray, or colorless. Slender prismatic crystals or in a felted mass of fibers. Streak white or colorless.
Silver. An element, Ag.	In flattened grains and scales; rarely in wirelike forms, or in irregular needle-like crystals.	None.	2.5-3	10-11	Color and streak are silvery-white, but may be tarnished gray or black. Highly ductile and malleable. Very heavy. Mirrorlike metallic luster on untarnished surfaces.
Sphalerite. Zinc sulfide (nearly always containing a little iron), ZnS.	Crystals common, but chiefly in fine to coarse-granular masses.	Six highly perfect cleavages at 60° to one another.	3.5-4	3.9-4.2	Color ranges from white to black but is commonly yellowish-brown. Translucent to opaque. Resinous to adamantine luster. Streak white, pale yellow or brown. Most important ore of zinc.
Staurolite. Iron aluminum silicate, $Fe(OH)_2(Al_2SiO_5)_2$.	Stubby prismatic crystals, and in cross-shaped twins.	Poor and inconspicuous.	7-7.5	3.7	Red-brown or yellowish-brown to brownish-black. Generally in well-shaped crystals larger than the minerals of the matrix enclosing them.
Talc. Hydrous magnesium silicate,. $Mg_3(OH)_2Si_4O_{10}$.	In tiny foliated scales and soft compact masses.	One perfect cleavage, forming thin scales and shreds.	1	2.8	White or silvery-white to apple-green. Very soft, with a greasy feel. Pearly luster on cleavage surfaces.

Table of Minerals and Diagnostic Properties – *continued*

MINERAL	FORM	CLEAVAGE	HARDNESS	SPECIFIC GRAVITY	OTHER PROPERTIES
Topaz. Aluminum fluorosilicate, $Al_2SiO_4(F,OH)_2$.	Prismatic crystals.	Good in one direction.	8	3.4-3.6	Many colors, yellow common. Transparent to translucent. Vitreous luster. Colorless streak.
Tourmaline. A complex silicate of boron, aluminum, and other elements.	Crystals commonly with curved triangular cross section.	Poor.	7-7.5	3-3.25	Many colors, green common. Translucent. Vitreous to resinous luster. Colorless streak.
Uraninite (pitchblende). Uranium oxide, UO_2 to U_3O_8.	Regular 8-sided or cubic crystals; massive'	None, fracture uneven to conchoidal.	5-6	6.5-10	Color black to brownish-black. Luster submetallic, pitchlike, or dull. Chief mineral source of uranium.
Zircon. Zirconium neosilicate, $Zr(SiO_4)$.	Tetragonal pyramid and dipyramid.	None.	7.5	4.68	Many colors. Translucent, some transparent. Adamantine luster. Colorless streak.

Adapted from *Geology, Resources, and Society* by H.W. Menard. Copyright © 1974 by W.H. Freeman and Company. Reprinted with permission.

Terms Used to Describe Leaves

(See facing page for illustrations.)

Generalized simple leaf: 1, b, blade; c, stipule; p, petiole; a, leaf axil; n, node; i, internode.
Arrangement: 2, alternate (spiral); 3, opposite; 4, perfoliate; 5, whorled.
Composition: 6, simple; 7 pinnately compound; 8, palmately compound; 9, trifopliate; 10, decompound
Venation: 11, parallel; 12 pinnately netted; 13, palmately netted.
Lobing: 14, pinnately lobed; 15 palmately lobed.
Margin: 16, entire; 17, serrate; 18 dentate; 19, crenate; 20, undulate; 21 ciliate; 22, doubly-serrate.
Outline: 23, acicular; 24, linear; 25, lanceolate; 26, oblanceolate; 27, ovate; 28 obovate; 29, oblong; 30 eliptical; 31 deltoid; 32 orbicular; 33, subulate.
Apex: 34, acuminate; 35, acute; 36, obtuse; 37, truncate; 38, mucronate; 39, cuspidate; 40, emarginate; 41, retuse.
Base: 42, acuminate; 43, cuneate; 44, acute; 45, obtuse; 46, oblique; 47, cordate; 48, truncate; 49, auriculate; 50 sagittate; 51, hastate.

Adapted from *Keys to Woody Plants* by W.C. Muenscher. Copyright © 1930 by Comstock Publishing Company, Inc.

Terms Used to Describe Leaves

(See facing page for key to illustrations.)

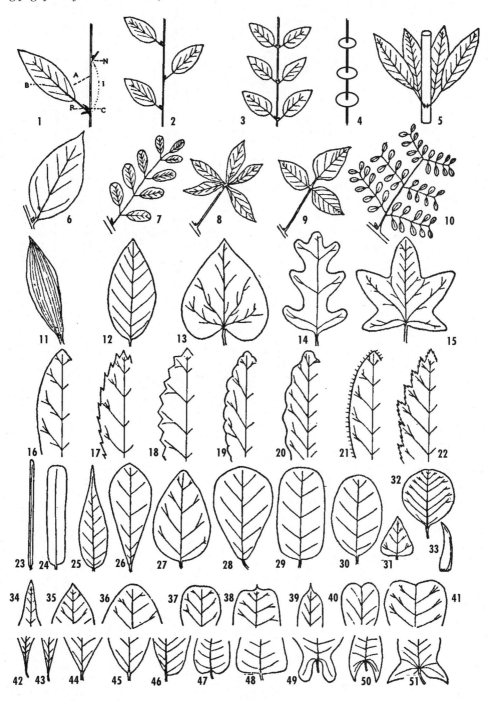

The Beaufort Wind Scale

BEAUFORT NUMBER	WIND	WIND SPEED			EFFECTS CAUSED BY THE WIND	
		KNOTS	MILES/H	M/S	AT SEA	ON LAND
0	Calm	<1	<1	<0.2	water surface smooth, mirror-like	air still; smoke rises vertically
1	Light Air	1-3	1-3	0.3-1.5	only ripples or small wavelets, no foam crests	smoke drifts downwind but vanes remain stationary
2	Light Breeze	4-6	4-7	1.6-3.3	distinct wavelets, small and short but not breaking	wind felt on face; leaves rustle, vanes moved
3	Gentle Breeze	7-10	8-12	3.4-5.3	large wavelets beginning to break, glassy foam, occasional white horses	leaves and twigs move constantly; wind extends small flag
4	Moderate Breeze	11-16	13-18	5.5-7.9	small waves somewhat longer with quite frequent white horses	wind raises dust, loose paper and moves small twigs and branches
5	Fresh Breeze	17-21	19-24	8.0-10.7	pronounced waves, distinctly elongated with many white horses, perhaps isolated spray	small trees in leaf begin to sway; crested wavelets form on inland waters
6	Strong Breeze	22-27	25-31	10.8-13.8	large breaking waves with extensive white foam crests; spray probable	large branches move, telegraph wires whistle, umbrellas difficult to control
7	Strong Wind	28-33	32-38	13.9-17.1	sea heaps up; streaks of white foam begin to be blown downwind	whole trees move; some wind resistance offered to walkers
8	Fresh Gale	34-40	39-46	17.2-20.7	moderately high waves with crests of considerable length, streaks of white foam well marked; spray blown from crests	twigs break off trees; greatly impedes progress of walkers
9	Strong Gale	41-47	47-54	20.8-24.4	high waves, rolling sea and dense streaks of white foam spray begin to reduce visibility	slight structural damage; chimney pots and roof tiles removed
10	Whole Gale	48-55	55-63	24.5-28.4	heavy rolling sea with very high waves; overhanging crests; great patches and dense streaks of foam	much structural damage; trees uprooted; seldom experienced inland
11	Storm	56-65	64-75	28.5-32.6	extraordinarily high waves and deep troughs; sea surface with streaky foam; strong spray impedes visibility	widespread damage; very rarely experienced inland
12	Hurricane	>65	>75	32.7-36.9	sea surface entirely white; air full of foam and spray	–

Average Abundances of Elements in the Earth's Crust, in Three Common Rocks, and in Seawater

Quantities are in parts per million.

ELEMENT	CRUST	GRANITE	BASALT	SHALE	SEAWATER
O	46.4×10^4*	48.5×10^4	44.1×10^4	49.5×10^4	880,000
Si	28.2×10^4	32.3×10^4	23.0×10^4	23.8×10^4	2
Al	8.1×10^4	7.7×10^4	8.4×10^4	9.2×10^4	0.002
Fe	5.4×10^4	2.7×10^4	8.6×10^4	4.7×10^4	0.002
Ca	4.1×10^4	1.6×10^4	7.2×10^4	2.5×10^4	412
Na	2.4×10^4	2.8×10^4	1.9×10^4	0.9×10^4	10,770
Mg	2.3×10^4	0.4×10^4	4.5×10^4	1.4×10^4	1,290
K	2.1×10^4	3.2×10^4	0.8×10^4	2.5×10^4	380
Ti	5,000	2,100	9,000	4,500	0.001
H	1,400				110,000
P	1,100	700	1,400	750	0.06
Mn	1,000	500	1,700	850	2×10^{-4}
F	650	800	400	600	1.3
Ba	500	700	300	600	0.002
Sr	375	300	450	400	8.0
S	300	300	300	2,500	905
C	220	320	120	1,000	28
Zr	165	180	140	180	3×10^{-5}
Cl	130	200	60	170	18,800
V	110	50	250	130	0.0025
Cr	100	20	200	100	3×10^{-4}
Rb	90	150	30	140	0.12
Ni	75	0.8	150	80	0.0017
Zn	70	50	100	90	0.0049
Ce	70	90	30	70	1×10^{-6}
Cu	50	12	100	50	5×10^{-4}
Y	35	40	30	35	1×10^{-6}
La	35	55	10	40	3×10^{-6}
Nd	30	35	20	30	3×10^{-6}
Co	22	3	48	20	5×10^{-5}
Li	20	30	12	60	0.18
N	20	20	20	60	150
Sc	20	8	35	15	6×10^{-7}
Nb	20	20	20	15	1×10^{-5}
Ga	18	18	18	25	3×10^{-5}
Pb	12.5	20	3.5	20	3×10^{-5}
B	10	15	5	100	4.4
Th	8.5	20	1.5	12	1×10^{-5}
Pr	8	10	4	9	6×10^{-7}
Sm	7	9	5	7	5×10^{-8}
Gd	7	8	6	6	7×10^{-7}
Dy	6	6.5	4	5	9×10^{-7}
Er	3.5	4.5	3	3.5	8×10^{-7}

Average Abundances of Elements – *continued*

Quantities are in parts per million.

Element	Crust	Granite	Basalt	Shale	Seawater
Yb	3.5	4	2.5	3.5	8×10^{-7}
Be	3	5	0.5	3	6×10^{-7}
Cs	3	5	1	7	4×10^{-4}
Hf	3	4	1.5	4	7×10^{-6}
U	2.7	5	0.5	3.5	0.0032
Br	2.5	0.5	0.5	5	67
Sn	2.5	3	2	6	1×10^{-5}
Ta	2	3.5	1	2	2×10^{-6}
As	1.8	1.5	2	10	0.0037
Ge	1.5	1.5	1.5	1.5	5×10^{-5}
Mo	1.5	1.5	1	2	0.01
Ho	1.5	2	1	1.5	2×10^{-7}
Eu	1.2	1.0	1.5	1.4	1×10^{-8}
W	1.2	1.5	0.8	1.8	1×10^{-4}
Tb	1	1.5	0.8	1	1×10^{-7}
Tl	0.8	1.2	0.2	1	1×10^{-5}
Lu	0.6	0.7	0.5	0.6	2×10^{-7}
Tm	0.5	0.6	0.5	0.6	2×10^{-7}
Sb	0.2	0.2	0.2	1.5	2.4×10^{-4}
I	0.2	0.2	0.1	2	0.06
Cd	0.15	0.1	0.2	0.3	1×10^{-4}
Bi	0.15	0.2	0.1	0.2	2×10^{-5}
In	0.06	0.05	0.07	0.06	1×10^{-7}
Ag	0.07	0.04	0.1	0.1	4×10^{-5}
Se	0.05	0.05	0.05	0.6	2×10^{-4}
Hg	0.02	0.03	0.01	0.3	3×10^{-5}
Au	0.003	0.002	0.004	0.003	4×10^{-6}

Te, Re, and the platinum metals are less than 0.03 ppm in rocks and less than 10^{-5} ppm in seawater. Concentrations of inert gases in seawater: He, 6.8×10^{-6} ppm; Ne, 1.2×10^{-4} ppm; Ar, 4.3×10^{-3} ppm; Kr, 2×10^{-4} ppm; Xe, 5×10^{-5} ppm.

Notes: The heading "crust" means the continental crust only, and this part of the crust is assumed to be made up of roughly equal parts of basalt and granite. "Granite" includes silica-rich rocks ranging from alkali granite to grano-diorite and their volcanic equivalents; "basalt" includes the more common varieties of basaltic lava, diabase, and gabbro; "shale" includes recent clays as well as shales, but not the fine-grained sediments of the deep sea. "Seawater" is normal surface water with chlorinity of 19 $^0/oo$. No great accuracy can be claimed for any of the values, because they depend on subjective judgments about the kinds of material to be included in each category, because they are subject to change as analytical techniques improve, and because sampling is often inadequate.

Reprinted with permission from Konrad B. Krauskopf, *Introduction to Geochemistry*, Second Edition, Appendix III, pp. 544-546. Copyright © 1979by McGraw-Hill.